LANDSLIDES

in research, theory and practice

VOLUME 1

Edited by E. Bromhead, N. Dixon
and M-L. Ibsen

Proceedings of the 8th International Symposium on Landslides
held in Cardiff on 26–30 June 2000

VIII ISL
CARDIFF 2000

Thomas Telford

Published for the Organizers by Thomas Telford Publishing, Thomas Telford Ltd, 1 Heron Quay, London E14 4JD.
URL: http://www.t-telford.co.uk

Distributors for Thomas Telford books are
USA: ASCE Press, 1801 Alexander Bell Drive, Reston, VA 20191-4400, USA
Japan: Maruzen Co. Ltd, Book Department, 3–10 Nihonbashi 2-chome, Chuo-ku, Tokyo 103
Australia: DA Books and Journals, 648 Whitehorse Road, Mitcham 3132, Victoria

First published 2000

Also available from Thomas Telford Books
Risk and reliability in ground engineering, Skipp, 0 7277 1986 6
Slope stability engineering, Chandler, 0 7277 1660 3
Urban ground engineering, The Institution of Civil Engineers, 0 7277 2786 9

A catalogue record for this book is available from the British Library

ISBN: 0 7277 2872 5.

Printed and bound in Great Britain by MPG Books Limited, Bodmin

Preface

In 1994, the Honorary Technical Secretary of the British Geotechnical Society, Christine Cooling, wrote to me to say that the UK place on the ISSMGE Technical Committeee TC11 had become vacant, and to ask if I was still interested in serving as the UK representative. One of the important missions was to try to attract an ISL event to this country. I started my work on TC11 with this in mind. Preliminary researches showed Cardiff to be the optimum location for the event, and with the blessing of the BGS Committee, I formed a group of three to formulate a serious bid. My colleagues on this bid committee were Mike Davies (then at Cardiff University, and on the BGS Committee) and Howard Siddle (of Halcrow UK, and based in Cardiff).

The bid we assembled was successful against strong competition when presented at the Trondheim 7th ISL, and work started on the event. This was organised through a Symposium Secretariat at Cardiff School of Engineering, led by Cherrie Summers, and Prof Hywel Thomas, who had stepped in as the senior Cardiff academic staff member when Mike Davies took up a Chair at Dundee.

Technical matters were addressed by an organizing committee that included the following members in addition to those on the original bid committee:

Mike Cooper (Southampton University and Gifford Consulting Engineers), Neil Dixon (Nottingham Trent and latterly Loughborough University), Mark Lee (Newcastle University), John Perry (Mott MacDonald Consulting Engineers), Bob Seedhouse (Neath & Port Talbot Borough Council, and BGS Liaison) and Prof Hywel Thomas of Cardiff University.

These Proceedings have been edited by myself and two colleagues: Neil Dixon of Loughborough University and Maia Ibsen, who is a close colleague at Kingston University. They are the written record of a six-year effort, and what is certain to be a successful event, with nearly 300 papers from the worldwide fraternity of landslide engineers and scientists. A complementary book, summarising knowledge on the Welsh Coalfield Landslides has been prepared under the editorship of Howard Siddle (in part as a field trip guide) and is published through the National Museum of Wales.

One rarely gets a chance to put thanks on record, but these proceedings could not have been published, nor the symposium held, without the efforts of the many authors, especially keynote lecturers; the editing work of colleagues and the team at Thomas Telford; the secretariat and organizing committee; support from successive chairmen of BGS and of TC11 (Dave Cruden and Jacques Locat) and supportive friends, colleagues at Kingston and in the profession; and my family. To all of you, for contributions great and small, thank you.

Eddie Bromhead
Professor of Geotechnical Engineering, Kingston University
Chairmain of the Organizing Committee, 8th ISL and Editor in Chief of the Proceedings
May 2000

Contents

Geological Factors Influencing the Naketiya Landslide

A. W. W. ABEYKOON, Research & Development Division, Road Development Authority, Ratmalana, Sri Lanka

ABSTRACT

In Sri Lanka, the landslide at Naketiya, as the local name of the place indicates accumulation of landslides, is reported to be the largest, being about 2.5km in total length and about 300m in width of the displaced mass. It is located in the vicinity of the 198km post on the Colombo-Ratnapura-Wellawaya-Batticaloa road. In November 1997, a serious event of the landslide resulted in destroying about a 110m section of main road following heavy rain in the preceding week.

The landslide continually occurs in a huge mass of colluvium deposit accumulated in a catchment of the Meeriyagaha stream over geological time. The colluvium deposit is made up of a high percentage of soil and a low percentage of rock fragments. The landslide begins as regressive multiple slumps in colluvium in the head region and excessive saturation converting them into an earthflow where a mixture of water, soil and rock fragments move downwards as a slurry mass. The earthflow displays undulating topography with an average slope of 15° and ends in forming its toe at a distance of about 500m from the main road. The developing toe consists of a cohesive fine soil mass. This displaced material spreads out with a uniform slope along the stream up to the lower Lemastota stream.

The landslide and its surroundings are underlain by charnockite and undifferentiated rocks which are highly dissected by two or more joint sets. The landslide lies along a major joint valley. Another major joint following the Monarakanda stream is located at a higher elevation in the northeastern part of the landslide.

It is concluded that the Naketiya landslide is old, repetitive, retrogressive and enlarging at the head of the landslide. It is geotechnically classified as a composite landslide consisting of an earthflow and a spreading failure. Its nature is geologically controlled. There may be subsurface leakage paths from the Monarakanda stream to the slide area.

INTRODUCTION

Naketiya is one of ancient villages in the southeastern part of Sri Lanka (Figure1). The term "Naketiya" indicates the place where landslides accumulate. The Colombo-Ratnapura-Wellawaya-Batticaloa (CRWB) road crosses this place in the vicinity of the 198km post. A series of tension cracks were observed in 1990 at the hillside of the location, just 500m away from the main road. In November 1997, a serious event of landsliding resulted in destroying about a 110m section of the main road following heavy rain in the preceding week. This paper describes how the geological factors influence the nature of the landslide.

Landslides in research, theory and practice, Thomas Telford, London, 2000.

METHODS OF STUDY

A geological study in and around the landslide was carried out by using 1:10,000 scale contour maps, aerial photo interpretation, surface geological mapping and analysis of available drillhole data. The study area is covered by the catchment in which the landslide lies and a further northwestern part, out of the catchment area. A geological map was prepared on a scale of 1:10,000 in and around the landslide (Figure 1).

GEOMORPHOLOGY IN AND AROUND THE LANDSLIDE

The study area is found to be at the intermediate slope of the Monarakanda ridge which is generally in the direction of north-east/south-west and ranges in elevation from 400m (at the lower confluence of the Lemastota and Meeriyagaha streams) to 1200m (at the top of the catchment) above MSL. The catchment exhibits irregular topography in its middle area and is bound by a rock escarpment around a part of its upper periphery. This rock scarp is riddled with old rockfalls. The landslide lies along a south-west strike valley of the upper part of the catchment and extends upto the confluence of the Lemastota and Meeriyagaha streams. The CRWB road crosses the landslide at the elevation of 745m above MSL.

Irregular topography of the landslide is exhibited up to a distance of 500m down from the main road. Slope profiles of this zone vary considerably in detail and their local variations are common. The slope ranges from 5° to 35° with an average of 15°. Beyond this zone, the displaced material lies in a valley bottom forming a uniform slope with a gradient of about 10°. Physical attributes of the landslide such as length, width, internal morphology and drainage are constantly changing. However, the overall picture is still representative. According to the measurements taken on 8th August 1999, the landslide was 2.5km in total length and ranged from about 20 to 250m in width. The width of 250m was in the hillside about 400m away from the main road. Its width was 110m along the main road. The morphology of the displaced material was continuously modified by erosion, transportation and deposition of soils carried by streamlets. Small scale vertical cuttings and temporary bars were commonly associated with the meandering streamlets.

Slumping was one of the prominent features observed in the main scarp of the landslide and within the landslide itself. Larger slumping occurred in a coluvium deposit of the main scarp and was relatively large, about 20m in width. They have radially propagated downward towards the centre of the landslide. The slumping within the landslide was relatively small being a maximum of about 5m in width and was commonly associated with the zone of accumulation and upheavals.

Curvilinear tension cracks (some having an opening of about 1.5m and a length of several metres) occurred above the crown of the landslide with a high density and were found in the zone of accumulation with a low density. Their directions were perpendicular to the axis of the landslide, but some were parallel to it.

HYDROLOGY

The study area is influenced by both south-west and north-east monsoons. Daily rainfall for the period of 1st January 1997 – 30th July 1999 was 150mm (maximum) and 20mm (minimum). The landslide is dendritically drained out by seasonal and perennial streams. A stream starting above the crown of the landslide runs irregularly through the displaced material and is it's course is fed by both seasonal and perennial springs and streamlets. The stream disappears at some places above the crown of the landslide and reappears within the main body of landslide.

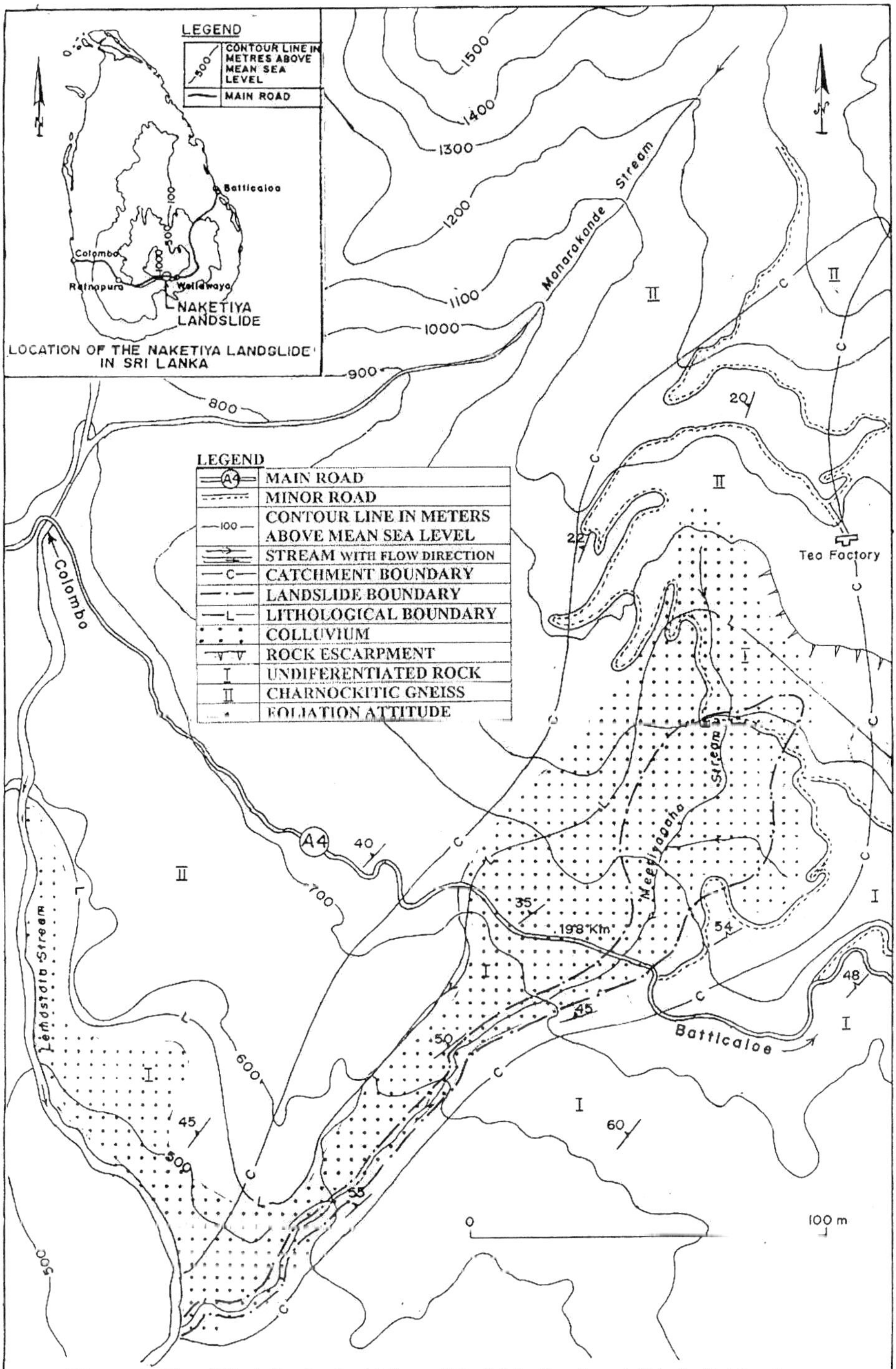

Figure 1. Simplified Geological Map of the Naketiya Landslide in Sri Lanka

HYDROGEOLOGY

In the context of hydrogeology, geological materials occurring in the landslide area can be divided into two groups as far as water permeability is concerned.

i) Overburden consisting of colluvium, soil and fully weathered rocks;
ii) Underlying fissured rocks consisting of moderately, slightly weathered and fresh rocks.

OVERBURDEN GEOLOGY IN AND AROUND THE LANDSLIDE

The overburden of the surrounding area consists of weathered rocks, residual soils and colluvium. The thickness of the overburden varies from a few centimeters to more than 30m as observed at cut slopes. The colluvium is composed of rock fragments (less than 3m in diameter) and soils. Weathered and fresh boulders are randomly distributed over the slopes. Clusters of rock fragments are encountered at some places on the slopes.

Two subdivisions can be recognized from surface and subsurface observations.

a. Superficial Currently Displaced Material;
   As far as type and morphology of displaced material are concerned, two major zones can be recognised.
   1. The displaced material exists upto a distance of 500m down from the main road and consists of yellowish brown coloured sandy, silty and clayey soil (about 85%), fully weathered and fresh rock fragments (about 15%) are observed at the ground surface. Rock fragments are scattered and form clusters at some locations. Boulder sizes vary upto about 3m. Weathered rocks occur as blocks in which the original fabric is still preserved. Kaolin and partially weathered feldspar pockets are seen within these blocks.
   2. The displaced material originates from the above (1) and consists of fine cohesive soil with little or no rock fragments.

b. Complex Mass;
   Complex mass is identified between underlying hard rock and the displaced material and is composed of soil, highly to moderately weathered undifferentiated rocks. Its depth varies from about 5m to a maximum of 25m.

BEDROCK GEOLOGY IN AND AROUND THE LANDSLIDE

Two lithological units can be recognized from geological mapping and drill hole data (Figure 1):

- Undifferentiated rocks
- Charnockitic gneiss.

Undifferentiated rocks

Rock exposures of this unit are generally weathered, but fresh rocks are found in drill holes. Garnet granulitic gneiss is the main rock type in this unit. The landslide is deeply underlain by hard rocks of Garnet granulitic gneiss. Fresh rock of this type is greyish white in colour, medium grained, foliated and quartzo feldspathic. Feldspars are present as layers and lenses concordant to the foliation. Their thickness ranges from a few centimetres to a few metres. Biotite gneiss, Garnet biotite gneiss, Hornblende biotite gneiss and garnet sillimanite graphite gneiss are interlayered with the host rock of garnet granulitic gneiss. Sheared weathered green coloured rocks are found at some places of the landslide base. Granulitic gneiss is exposed at the left flank of the valley in which the landslide takes place and at some places of the valley base. Weathered rocks range from a few meters to a maximum of about 15m.

Charnockitic gneiss

Charnockitic gneiss occurs as an intermediate variety. Hypersthene is the characteristic mineral of this rock. The width of these layers ranges from several centimetres to a maximum of 50m.

GEOLOGICAL STRUCTURE IN AND AROUND THE LANDSLIDE

The strike of foliation varies considerably from north-north-east/east-west with an average of north-east/south-west (Figure 1). The dip of foliation varies from horizontal to vertical. In the north-easten part of the map, foliation attitude is highly variable. In its lower part, the dip of foliation varies from 30° to 60° west.

The whole area is highly dissected by joints which are generally represented by valleys and some of them are followed by streams. The foliation joints are the most prominent in the average direction of north-west/south-east. The landslide follows one of the foliation joints which represents a major lineament in the area, running for about 2.5km. This lineament starts above the landslide and runs further to the south-east. At this narrowing place, no lithological displacements and no sudden relief changes are observed on either side of the lineament. The other joints are north-east/south-west, north-north-east/south-south-west and north-west/south-east in order of prominence. In the northwestern part of the map (Figure 1), a well-defined lineament follows the Monarakanda stream which is located at a higher elevation behind the landslide.

DISCUSSION

The presence of a huge mass of colluvium in the catchment reveals that the landslide and its surroundings have been subjected to landsliding in the past. The disruption to vehicular traffic and consequent temporary realigning of the main road show that the landslide has reactivated from time to time during rainy seasons. The date of first occurrence is not known, but its history may perhaps date back nearly a century according to the informants in the village.

In the first 800m length of the landslide, high percentage of fine soil and low percentage of rock fragments of the displaced material form less fluid consistency. It is not discernible from discrete shear boundaries. Therefore, The first part of the landslide can be classified as an earthflow. The toe of the earthflow develops into a spread due to the narrowing of the gap between bedrock exposures on either flank of the valley. Considering preserved fabric and types of minerals of soil fraction of colluvium, it is mainly derived from undifferentiated rocks. High content of fine soil fraction causes low permeability. Irregular topography, curvilinear tension cracks, depressions and upheavals reveal that the surface of rupture is neither single nor planar in the earthflow and is planar in the spread.

The comparison of the field sketches drawn from time to time for the last two years establishes the retrogressively growing nature of the landslide. As a result, the landslide becomes enlarged in dimensions through retrogressive movement of the rear scar and also due to slumping weathered rocks and old colluvium from each flank above the main road.

Disappearance and appearance of stream water within the main body indicate that ground water flow is taking place through open fractures within the complex mass.

The intensive fracturing of rocks has led to high degree of weathering in the area above the main road. As a result of this, the landslide becomes wider and deeper in this area.

As shown in aerial photographs, the valley in which the landslide lies is found to be a lineament parallel to the regional foliation. However, lack of sudden morphological relief and lithological displacements on either side of the lineament are not sufficient to prove whether this lineament is a fault or not. Green coloured sheared material found at some places below the landslide area is apparently a product of foliation shears.

It appears that the elevation difference of the Monarakanda stream and the landslide plus a high degree of jointing of underlying rocks may form favourable conditions for underground water paths from this stream to the landslide area. This should be examined with the help of tracer techniques, drilling and piezometric observations.

CONCLUSIONS

The Naketiya landslide is geotechnically classified as a composite landslide consisting of an earthflow and a spreading failure. It is old, repetitive and retrogressive for a distance of about 300m and the head area of the earthflow becomes enlarged in both directions normal to the longitudinal axis of the landslide. The surface of rupture of the earthflow is neither single nor planar where as the spreading failure is most probably single. The nature of the landslide, such that can be stated, distributions and styles of activity, types, dimensions and features are geologically controlled.

REFERENCES

International Geotechnical Societies' UNESCO Working Party on World Landslide Inventory, 1993. Multilingual Landslide Glossary.

Monitoring and Modelling of Unsaturated Flow and Effects on Stream-bank Failures

A. M. AFFUSO, N. CASAGLI, S. DAPPORTO, G. GABBANI, A. GARGINI and M. RINALDI, Dipartimento di Scienze della Terra, Università di Firenze, Italia

ABSTRACT

Pore water pressures have been monitored for 3 years using a series of tensiometer-piezometers at increasing depths in a streambank of a river in Tuscany, Central Italy. Saturated/unsaturated flow during the main event recorded was modelled using a finite element seepage analysis and comparing measured with computed values of pore water pressures. A bank failure occurred during the period of monitoring. Based on the reconstructed distribution of pore water pressures within the whole bank and on the geometry of the failed block, an analysis of the failure was conducted applying the limit equilibrium method, in order to establish the variations of the safety factor during the flow event in response to changes in water stage and rainfall.

INTRODUCTION

Streambank failures represent a type of mass movement very common along rivers. In many cases they constitute a threat to human activities or structures crossing the channel or adjacent to the banks. For the design of riverbank stabilization projects, it is important that a stability analysis be performed, in order to assess critical conditions in terms of river stage and related pore water pressures within the streambank. Nevertheless, many aspects concerning the relationship between river stage and pore water pressures during flow events are still not well understood because of difficulties associated with monitoring the parameters involved

A monitoring program was started in February 1996, installing a series of tensiometer-piezometers at increasing depths in a bank of the Sieve River, Tuscany (Italy), with the overall objectives of investigating the role of saturated/unsaturated flow within a streambank in response to rainfall and flow events, and improving methods of riverbank stability analysis.

Results from the first 16 months of monitoring are reported and discussed in Casagli et al. (1999) and Rinaldi & Casagli (1999). Canuti et al. (1998) investigated the aspects concerning pore water pressure, unsaturated flow, and streambank stability at the scale of a single flow event, introducing a preliminary analysis of the saturated/unsaturated flow through a numerical model. In this paper a more complete description of modelling results, obtained by combining finite element seepage analysis with limit equilibrium method, is presented.

The specific objectives of the research were: a) to calibrate the seepage numerical model based on the measured values of pore water pressures at the tensiometers; b) to perform a stability analysis of the bank based on the geometry of the failed block and on the reconstruction of pore water pressure conditions in the entire bank by the previous phase; c) to investigate on the role of the factors determining instability during the flow event.

Landslides in research, theory and practice, Thomas Telford, London, 2000.

STUDY SITE, MONITORING INSTRUMENTATION, AND SOIL PROPERTIES

The monitored streambank is located along an alluvial reach on the lower part of the Sieve valley, about 4 km upstream from the confluence with the Arno River (Fig.1B). The Sieve's catchment is about 840 km^2 and the total length of the river is 58 km.

The streambank is composed of a basal layer of sandy-gravel deposits, 1.7 m thick, and of an upper layer of predominant sandy-silt deposits, 3.1 m thick (Fig.1C). The basal layer can be distinguished into a higher part composed of sandy gravel closely packed, and a lower portion of loose gravel. A series of grain size analyses carried out on cores extracted from hand-drilled boreholes have revealed the presence of a sub-layer at a depth between 1.5 m and 2.0 m of fine gravel included in a sandy-silt matrix, with an upper transitional sub-layer about 0.25 m thick, of intermediate grain size.

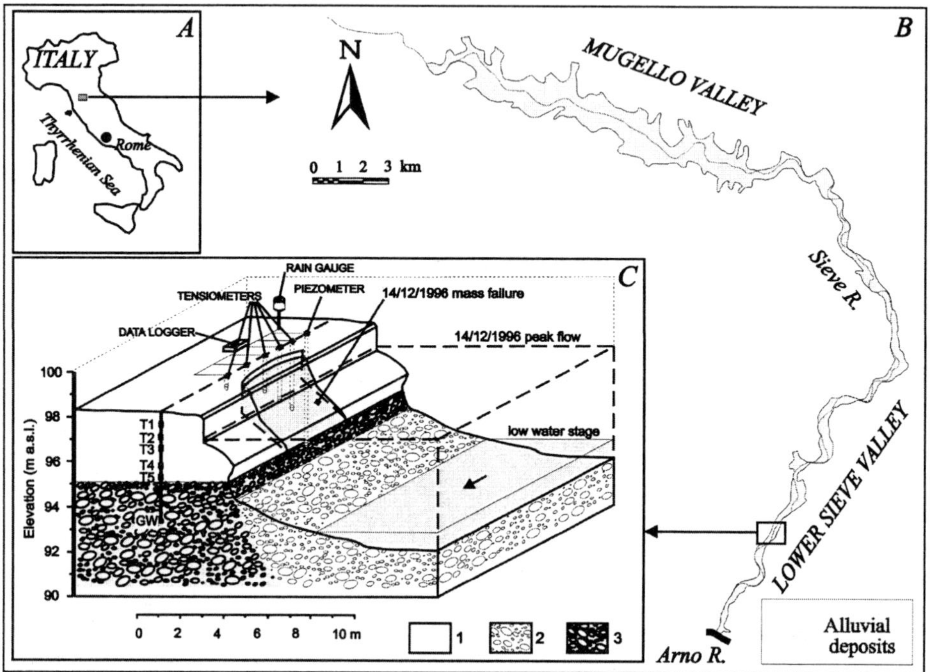

Figure 1. Description of the study site. A: Location of Sieve River. B: Sieve River course. C: Monitoring station. 1: Silt and sand with sub-layers of fine gravel in a sandy-silt matrix; 2: loose gravel; 3: packed gravel with interstitial sand.

The monitoring instrumentation consists of 5 pressure transducer tensiometers placed at different depths, a piezometer, a rain gauge and a hydrometric level gauge (Fig.1C). Bank retreat was measured by repeated cross-profiling and erosion pins. Recently, a new series of 4 tensiometers and a TDR (Time Domain Reflectometry) have been added to the monitoring site, while morphological streambank changes are measured applying terrestrial photogrammetry techniques.

Laboratory and in situ tests were concentrated on the sandy-silt material, which constitutes the dominant component of the upper bank. Laboratory triaxial tests have been performed on saturated samples, while borehole shear tests (BST) have been carried out in the unsaturated portion of the bank (Casagli et al, 1999). Saturated permeability at different depths has been assessed using an Amoozemeter permeameter. Measurements of water content was made at a depth of 0.6 m for a wide range of matric suctions.

UNSATURATED FLOW MODELLING

Unsaturated flow within the streambank was modelled using a finite element code for the analysis of the unsaturated flow in transient conditions (SEEP/W v.4 produced by *Geo-Slope International Ltd.*). The analysis concentrated on the main flow event occurred on December 14th 1996, during which a peak stage of 97.0 m a.s.l. was reached and a mass failure occurred on the upper portion of the bank.

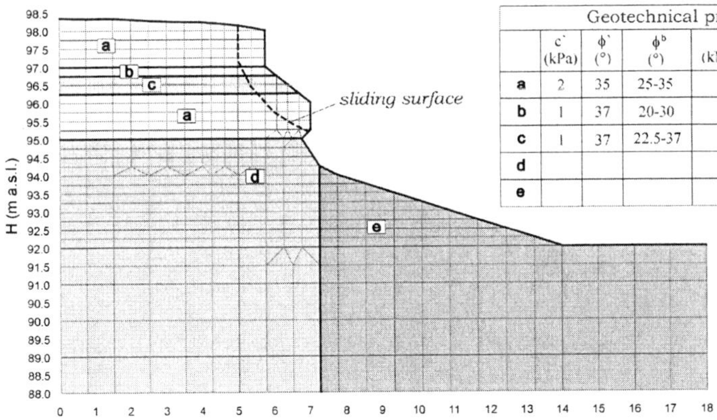

Geotechnical properties						
	c' (kPa)	φ' (°)	ϕ^b (°)	γ (kN/m³)	n (%)	k_{sat} (m/s)
a	2	35	25-35	18	39.5	1.4E-6
b	1	37	20-30	18	39.0	1.0E-5
c	1	37	22.5-37	18	39.0	5.4E-5
d					35.0	1.0E-4
e					35.0	6.0E-4

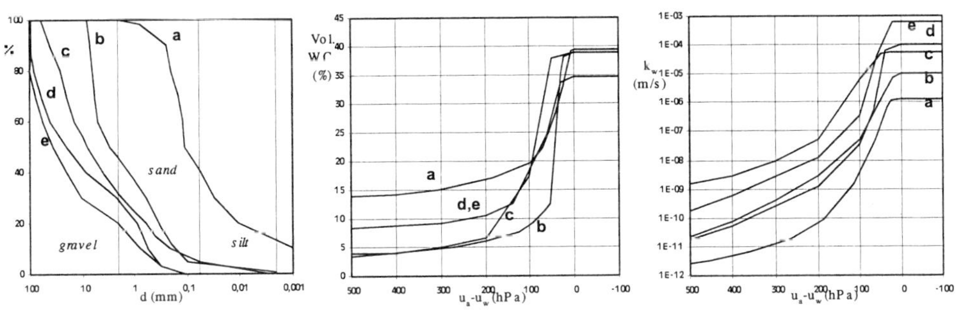

Figure 2. Finite element seepage analysis: geometry of the problem, geotechnical properties of the soils, grain sizes, soil water characteristic curves, and k-curves.

The streambank was discretised into 517 quadrilateral and triangular elements and divided in six regions, composed by five materials with different properties (Fig.2). As regards the boundary conditions, a total head versus time function for the nodes along the bank profile was defined using the hydrograph of the event, while a rainfall intensity versus time function was assigned to the nodes along the top of the streambank. The initial conditions were defined considering the water table at the same level of the river stage before the start of the flow event and the values of matric suction within the sandy-silt layer were defined according to the values measured by the tensiometers. The simulation was performed dividing the event in 24 time steps.

The application of a numerical model for seepage analysis requires that characteristic and k-curves for each of the materials constituting the streambank be known. However, given the difficulties in measuring unsaturated permeabilities for different pore water pressures and

water contents, particularly for packed gravel material, some of the curves had to be estimated or assumed. Only for the sandy-silt material of the upper bank, the characteristic curve was obtained coupling suction readings with water content measurements. For the other materials, different water content curves were tested on the basis of the grain size of the matrix using the software SOILPARA (produced by *Draper Aden Environmental Modelling, Inc.)* and the corresponding permeability functions were determined by the algorithm of Brooks & Corey (1964). For this purpose, the model was also applied to the other three main flow events that occurred during the 3-year period besides the December 14th 1996 event.
Water content and permeability functions giving the best accordance between computed and measured values for the 4 events were selected (Fig.2). In particular, for the December 14th 1996 event, computed values compared to the pore water pressure readings at the nodes corresponding to the tensiometers are shown in Fig.3, while pore water pressure distribution within the bank at the peak stage and in the istant of failure are shown in Fig.4.

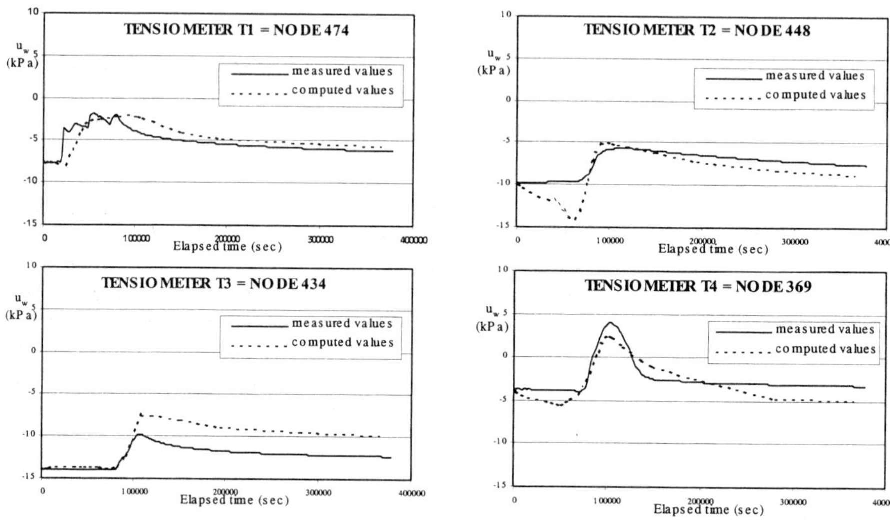

Figure 3. December 14th 1996 event: comparison between measured and computed values in the four tensiometers.

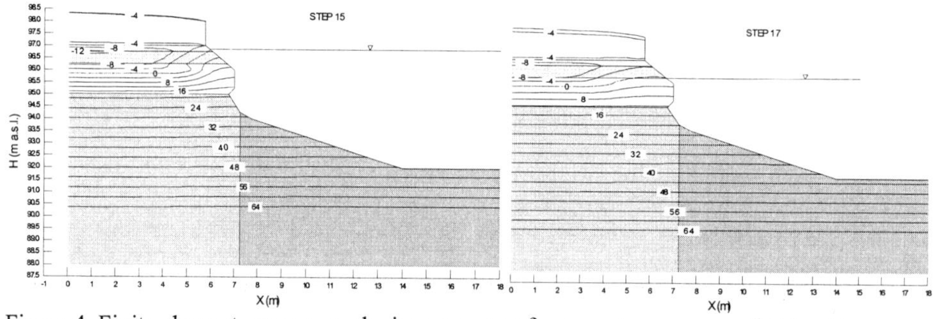

Figure 4. Finite element seepage analysis: contours of pore water pressures (kPa) with vectors of total flux during the peak flow (step 15) and the instant of mass failure (step 17).

BANK STABILITY ANALYSIS

The streambank stability has been analysed for the December 14th 1996 event using a

software (SLOPE/W v.4 produced by *Geo-Slope International Ltd.*) for the application of limit equilibrium methods. The geometry of the failed block was reconstructed through repeated bank profile surveys, before and immediately after the failure. The mechanism of failure was rotational with a slight curvature of the failure surface and with a tension crack. The factor of safety of the upper portion of the streambank was computed with the Morgestern-Price method imposing the observed failure surface for each of the steps before the failure used in the finite element seepage analysis. Pore-water pressures at the base of each slice have been specified importing the values obtained from the SEEP/W finite element analysis for each step of the computation.
The Mohr-Coulomb criterion in terms of effective stress was used in case of positive pore pressures and the Fredlund et al. (1978) criterion in case of negative pore pressures. Shear strength parameters used in the analysis are reported in Fig.2. In order to take into account the variability of ϕ^b during the flow event, a function relating the angle with matric suction was defined based on the soil water characteristic curves, varying the angle from a minimum of 15° to a value equal to ϕ' when the soil approaches saturation. The failure envelope was discretised into a series of linear segments with varying ϕ^b angles and different intercepts, each segment corresponding to a range of matric suction (Fredlund & Rahardjo, 1993).

DISCUSSION

Variations of the factor of safety through time compared with the river stage during the December 14th 1996 event are reported in Fig.5. It can be noted that the safety factor, for both the first minor relative peak and the second greater peak stage, increases slightly as consequence of the stabilising confining pressure of the water in the river, and then decreases before the peak stage, due to the decrease in matric suctions and the development of positive pore pressures along part of the failure surface, despite the increase of confining pressure. After the peak stage, the safety factor continues to decrease because of the loss of confining pressure, although the pore water pressures reverse their trend, until it falls below unity at step 17, 3 hours after the peak.

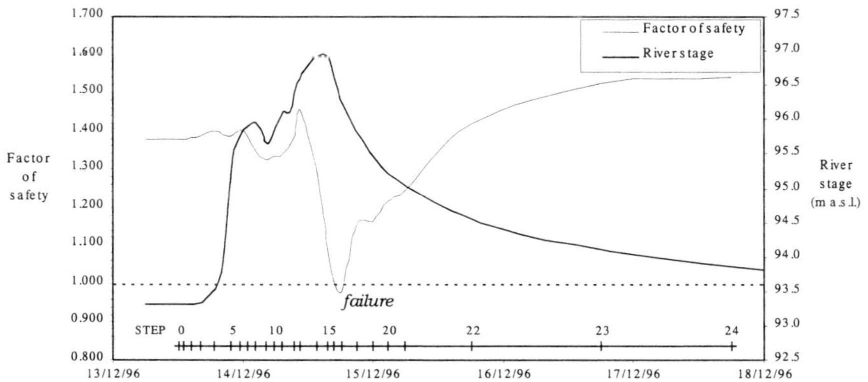

Figure 5. Trend of the safety factor (F) compared with the river stage during the December 14th 1996 event.

A previous bank stability analysis was performed for the same event and reported in detail in Casagli et al. (1999), conducting to partly different results in terms of shear strength parameters, trend of safety factor during the event, and istant of failure. In fact, results of that analysis show a rise in the safety factor up to the peak river stage, due to the prevailing effect

of the increasing confining pressure, and a rapid decrease of the safety factor during the drawdown, due to loss of confining pressure in spite of an increase in matric suction.
The differences in the results reported in Casagli et al. (1999) and in this paper are due to the following differences between the two types of analyses: a) in Casagli et al. mean pore water pressure along the failure surface was obtained from a weighted average of the values measured at the tensiometers; b) geometry of the streambank and of the failure surface are more detailed in this analysis; c) two sub-layers of different permeability included in the sandy-silt material are considered in this analysis; c) the present analysis takes into account the different geometry after failure, the effects of seepage, of rainfall, and the variability of unsaturated shear strength parameters; d) the rigorous Morgestern-Price method was adopted in this analysis compared to the closed-form solution for planar failure used in Casagli et al.

CONCLUSIONS

The method used in this paper, coupling finite element seepage analysis for the reconstruction of pore water pressure conditions within the streambank and a limit equilibrium method for bank stability analysis at different time steps during the flow event, has been successfully applied in modelling the streambank failure. The results permit a better comprehension of the mechanisms of riverbank failure and of the variations of instability conditions at the scale of a flow event. For the type of streambank and for the flow event modelled, the results suggest that the most critical conditions for bank stability occur slightly after the peak stage, when there is a partial loss of confining pressure combined with relatively high pore water pressures. The type of approach used represents an improvement of stability analysis methods previously applied to riverbanks, being capable of taking into account the effects of rainfall and river stage on the variability of matric suction conditions during the different phases of a flow event.

ACKNOWLEDGEMENTS

The research is funded by the National Group for the Hydrogeological Disaster Prevention (CNR-GNDCI). *CNR-GNDCI Publication No. 2047.*

REFERENCES

Brooks, R.H. & Corey, A.T. (1964) - Hydraulic properties of porous media. *Hydrol.Paper* No.3, Colorado State University, Ft.Collins, Co, pp.27.

Canuti, P., Casagli, N. & Rinaldi, M. (1998) - Unsaturated flow and mechanisms of failure of a streambank during flow events. In: Moore, D. & Hungr, O. (Eds), Proceedings VIII Congress of the IAEGE, Vancouver, Canada, 21-25 September 1998, 1865-1870.

Casagli, N., Rinaldi, M., Gargini, A. & Curini, A. (1999) - Monitoring of pore water pressure and stability of streambanks: results from an experimental site on the Sieve River, Italy. *Earth Surface Processes and Landforms*, in Press.

Fredlund, D.G. & Rahardjo, H.(1993) – *Soil Mechanics for Unsaturated Soils*. John Wiley & Sons, New York, 482 pp.

Fredlund, D.G., Morgenstern, N.R. & Widger, R.A. (1978) -The shear strength of unsaturated soils. *Canadian Geotechnical Journal*, 15, 3, 312-321.

Rinaldi, M. & Casagli, N. (1999) - Stability of streambanks formed in partially saturated soils and effects of negative pore water pressures: the Sieve River (Italy). *Geomorphology*, 26, 253-277.

Hydrogeological Risk Assessment of the Po River Basin (Italy)

P. ALEOTTI, P. BALDELLI, G. POLLONI, Aquater S.p.A., San Lorenzo in Campo, Italy, M. GOVI, Po River Basin Authority Technical Commitee, Parma, Italy, and B. VILLANI, Arpa Agenzia, Bologna, Italy

INTRODUCTION

The ever-increasing amount of literature regarding the definition and assessment of hydrogeological hazards derives essentially from the rising demand for protection from natural disasters. Particularly in the present day society, the shortage of building land around urban centers leads inevitably to the increased use of uninhabited or unsuitable areas for development. The assessment of hydrogeological hazards falls within the prevention policy that also includes the definition and implementation of appropriate mitigation measures (Aleotti et al., 1998). Hazard prevention implies both understanding and acceptance of the phenomena; in fact the solution does not always lie in abandoning or avoiding development in hazard-prone areas, both for economic reasons and for the difficulty of precise forecasting geomorphic processes (Leroi, 1996). The methodology adopted to construct hazard forecasting maps of the Po River basin (70.000 km^2) was based on the use of G.I.S. ARC/INFO that enabled the implementation of several intermediate tests and thus the progressive enhancement of various evaluation algorithms. When operating in a vast area, that implies the use of a small work scale, besides the scale factor and the availability of sufficient information (Aleotti and Chowdhury, 1999), the assessment of risk in the proper meaning of the word is particularly difficult due to a series of factors inherent essentially to the hazard component: in the case of landslides, for example, the problematic defining of the relationship between triggering causes and induced effects that is the basic aspect of every probabilistic evaluation (Hutchinson, 1992).

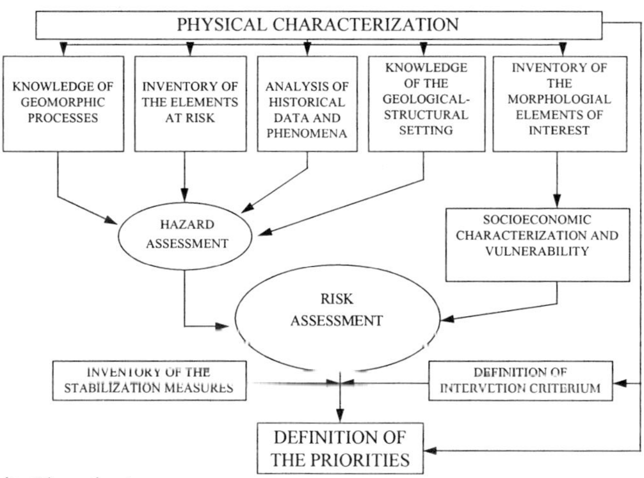

Figure 1. Activity Flow chart

Landslides in research, theory and practice, Thomas Telford, London, 2000.

As the main aim of risk assessment in the Po River basin management plan was to identify the intervention criteria and priorities, for the difficulties stated above a simplified methodology was adopted in this project: hazard was defined simply as susceptibility, i.e. the possibility (not the probability) of occurrence of a given phenomenon within a defined area.

The hydrogeological hazards examined include landslides, floods, erosions, overfloodings, avalanches and mass transport processes on alluvial fans. The input data used for the evaluations are summarized in the flow-chart in Figure 1. For practical advantages in administrative priority classification, the reference land unit used was the municipal area (total of 1878 municipalities).

HAZARD ASSESSMENT

Hazard was assessed for every type of hydrogeological hazard by means of G.I.S. procedures, and significant analytical indexes representative of the actual and/or potential hazards were calculated for each land unit. The values of hazard obtained for each type of hydrogeological process were divided into 4 distinct classes as follows: (a) selection of Land Unit with index $I > 0$, (b) transformation of index I into logarithmic terms: $Y = \ln(I)$; (c) calculation of mean M_Y and of standard deviation δ_Y of every index Y; (d) definition of classes according to formulas [1]-[4] (Table 1). The subdivision into the limits shown in table 1 ensures adequate representativity of classes with frequency typical of the standard gaussian distribution.

Landslide susceptibility I_L, formula [8], was assessed on the basis of two main indexes: observed landsliding index "I_{LO}" and potential landsliding index "I_{LP}". The former [5] represents the objective pronesses to sliding based on the percentage of area affected by landslides and the latter [7] the susceptibility to sliding due to lithological conditions (Table 2). For the summary map, shown in Figure 2, over 40.000 landslides were taken into consideration.

Table 1. Index boundaries among the different classes

Formula	Class	Boundaries
[1]	low (LO)	$I_{LO} < exp\ (M_Y - \delta_Y)$
[2]	medium low (ML)	$exp\ (M_Y - \delta_Y) < I_{ML} < exp\ (M_Y)$
[3]	medium high (MH)	$exp\ (M_Y) < I_{MH} < exp\ (M_Y + \delta_Y)$
[4]	high (HH)	$I_{HH} < exp\ (M_Y + \delta_Y)$

Table 2. Definition of evaluation indexes

Formula	Index		Meaning
[5]	$I_{LOj} = \frac{LS_j}{TS_j} * 100$	observed susceptibility	LS_j = slided surface of unit area TS_j = total unit area surface
[6]	$I_{lith} = \frac{LS_l}{TS_l} * 100$	lithological proneness to sliding	LS_l =slided surface for lith. class "l" TS_l = lith. class "l "total surface
[7]	$I_{LP\,j} = \sum_L I_{lith} * \frac{Ts_{jl}}{TS_{jl}}$	susceptibility due to lithological conditions	Ts_{jl} =surface of lithological class "l" in the unit area j
[8]	$I_{Lj} = I_{LOj} + \alpha\, I_{LPj}$	final susceptibility	$\alpha = 0.5$

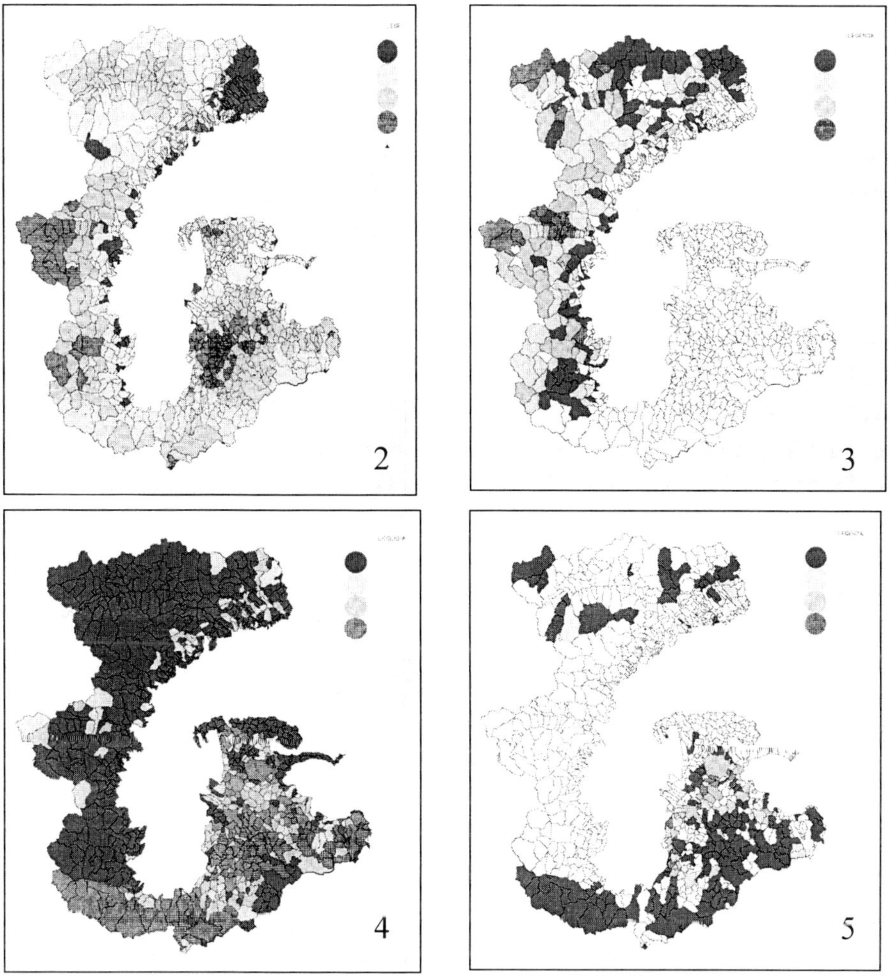

Figures 2, 3, 4, 5. Hazard maps for landslides (2), fans (3), stream processes (4) and floods (5) (Piedmont Region)

The value of hazard for other geomorphic processes was calculated using the indexes summarized in Table 3. Given the adopted work scale, these indexes are necessarily simple, but nonetheless provide an objective (and updatable) representation of the area and of the hazard conditions in each land unit. With reference to the indexes in Table 3 it must point out the following: (i) the areaFS_j , formula [9], was obtained by examining past alluvial processes in the municipalies derived from available maps; (ii) the debris torrent processes [10] comprise both erosion and overflow of river beds; (iii) only active alluvial fans [11] were considered, i.e. those in which flooding and/or mass transport could still occur; (iv) the number of avalanche corridors [12] is considered as directly linked to the probability of occurrence of the phenomenon, independent of the amount of material effectively or potentially involved. Hazard maps concerning the examined hydrogeological hazards are provided in Figures 2, 3, 4 and 5. The results in percentages are summarized in Figure 6a.

Table 3. Hazard indexes for different types of geomorphic processes

Formula	Index	Geomorphic process	Meaning
[9]	$IE_J = \frac{FS_j}{TS_j}100$	floods	FS_j = flooded unit surface TS_j = total unit surface
[10]	$IH_J = \frac{Sl_j}{Tl_j}100$	stream processes	Sl_j = length of affected streams Tl_j = total length of hydr. network
[11]	$IF_J = \frac{AF_j}{TS_j}100$	alluvial fans	AF_j = alluvial fan surface
[12]	$IA_J = \frac{\#A_j}{TS_j}$	avalanches	$\#A_j$ = number of aval. corridors

VALUE, VULNERABILITY AND DAMAGE

The economic value (E_J) of every unit area was calculated by summing the characteristic indexes (I_n) of the elements at risk, appropriately "valued" [13]. These can be distinguished in (i) "widespread value indexes" if they refer to the level of anthropization in the area (density of population, buildings, industries, transport lines) and (ii) "punctual value indexes" if they express the presence of important economic or social infrastructures.

[13] $E_j = I_1 + I_2 + ... + I_n$

Then a value of vulnerability (V_n) is assigned to every index I_n, based on expected loss in case of occurrence of the damaging event (0 = no loss, 1 = total loss), and a weight W_n (between 0 and 1), according to the different importance that is given to the indexes considered in the final algorithm. The value of damage (D_j) within every unit area was assessed according to the equation [14]:

[14] $D_j = I_{J1}*V_1*W_1 + I_{j2}*V_2*W_2 + ... + I_{jN}*V_N*W_N$

A total of 6 damage classes were defined as follows: nil (NN), low (L), medium-low (ML), medium-high (MH), high (H) and very high (HH) (Table 4).

Table 4. Risk matrix

Damage Susceptibility	NN	L	ML	MH	H	HH
LO	RN	RL	RL	RM	RM	RM
ML	RN	RL	RM	RM	RH	RH
MH	RN	RL	RM	RH	RH	RH
HH	RN	RM	RM	RH	RH	RH

RN= risk nil RL= risk low RM = risk medium RH = risk high

SINGLE RISKS AND TOTAL RISK

Risk maps were achieved by overlaying the various hazard maps (one for every type of hydrogeological process) with the damage maps using the key matrix shown in table 4. According to the general equation [15], nil damage ($V*E$) corresponds to nil risk, regardless of the value of susceptibility. At the same time, an area with high susceptibility and high social-economic value must be considered at a high risk.

[15] $R_S = S*V*E$

where R_S = single risk (for every geomorphic process), V = vulnerability, E = economic value.

The distribution of unit areas in the single risk classes for the various geomorphic processes is presented in Figure 6b.

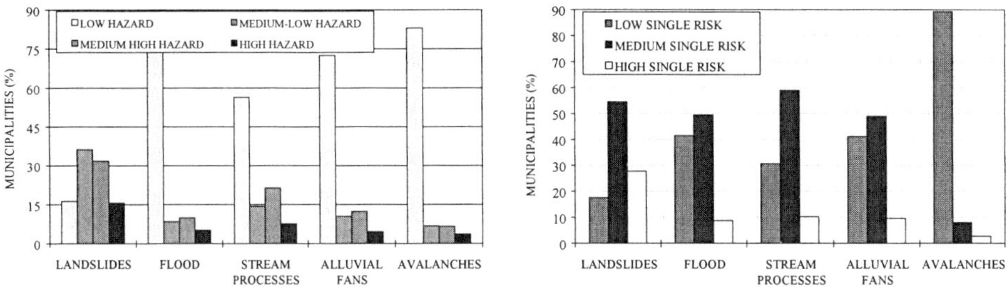

Figure 6a, b - Hydrogeological hazard (a); single-risk classes in the various processes (b)

Total risk R_T was derived by "summing" the single risks and by overlaying the single risk maps. In practice two numeric values K (Table 6) and W (Table 7) were assigned to the single risk classes, the former to express the severity of the single risk, the latter to weigh the different geomorphic processes as a function of their pronesses to induce damages. This subjective factor partly compensates the vulnerability value, fixed for a given element at risk, regardless of the type of geomorphic process. The value of total risk (R_T) for a given area is calculated by means of the following formula:

[17]

$$R_{Tj} = \sum_{i=1}^{5} W_i * K_{i,j} \Big/ \sum_{i=1}^{5} W_i$$

The total risk classes are listed in Table 8 and their areal distribution is shown in Figure 7.

Table 6 - Value for the different risk classes, K

Specific Risk	K
Nil	0
Low	1
Medium	2
High	4

Table 7 - Weight for the different kind of geomorphic processes, W

Geomorphic process	W
Landslides	0.3
Floods	0.4
Stream Processes	0.2
Alluvial Fans	0.4
Avalanches	0.1

Table 8 - Classes of total risk, R_T

Total Risk (R_T)	Definition
Low (L)	$R_T < 1,5$
Medium (M)	$1,5 \leq R_T \leq 2,5$
High (H)	$R_T > 2,5$

SUMMARY AND CONCLUSIONS

The assessment and mapping of hydrogeological hazards constitutes a fundamental tool for all decision-makers and, generally speaking, for all those involved in land use management.

Within this context the Po River basin represents an extremely interesting area for the following reasons: (i) its vast extension; the Po River is the largest basin in Italy; (ii) its highly diversified geomorphic processes; (iii) high population density and concentration of economic activities.

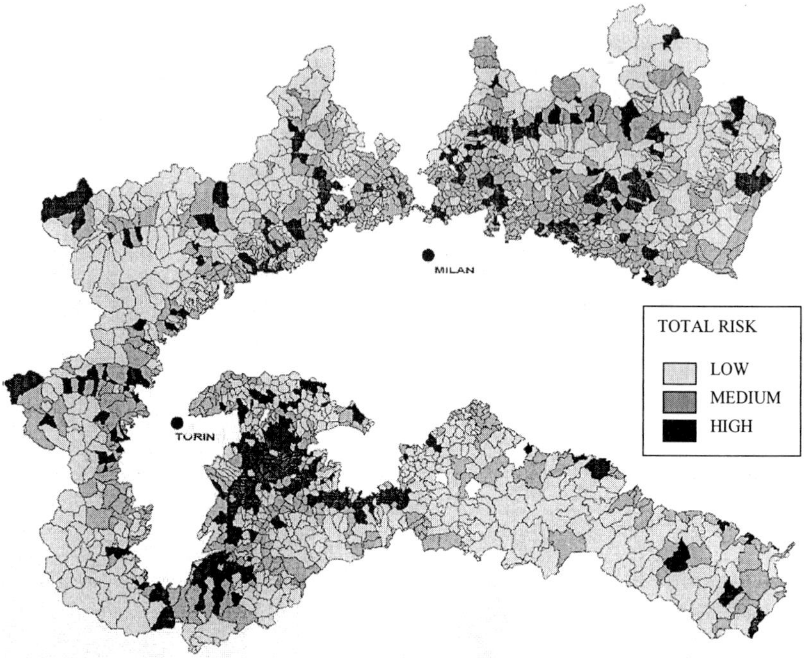

Figure 7. Final risk map

In this project considerable importance was placed on preliminary activities: (i) collecting and updating all available information: due to the extent of the basin an enormous deal of data was examined; b) processing algorithms that can be easily applied to the adopted work scales and addressed to the most objective and updatable assessment, tailored specifically to deal with the examined geomorphic processes; c) produce user-friendly maps that are extremely useful in the selection of strategic options. It is necessary to underline that the hazard and risk classes adopted are of a relative nature as they do not take into consideration any fixed external references. Nevertheless, the adopted methodology enabled the identification of the major risk-prone areas where to address priorities both in terms of additional studies and definition of specific mitigation measures.

BIBLIOGRAPHY

Aleotti P., Baldelli P. Polloni G., Puma F. (1998). Keynote paper: Different approaches to landslide hazard assessment. Proc. 2nd Conf. on Env. Manag. (ICEM-2), Wollongong (Australia), **1**: 3-10.

Aleotti P., Chowdhury R.N. (1999) Landslide hazard assessment: summary review and new perspectives. Bull of Eng. Geol and the Env.

Hutchinson J.N. (1992). Landslide hazard assessment. Proc VI Int. Symp. on Landslides, Christchurch, **1**, 1805-1842.

Leroi E. (1996). Landslide hazard - Risk maps at different scales: objectives, tools and developments. Proc. VII Int. Symp. on Landslides, Trondheim, **1**: 35-51.

Debris Flow Hazard and Risk Assessment Using Airborne Laser Terrain Mapping Techniques (ALTM)

P. ALEOTTI, G. POLLONI, Aquater S.p.A. via Tolstoj 86, San Giuliano Milanese, Italy, P. CANUTI, and A. IOTTI, Earth Science Dept., University of Florence via La Pira 4, Florence

INTRODUCTION

On May 1998 the Sarno area (South Italy) (Fig. 1) was affected by a dramatic event with many debris flows triggered by a severe rainstorm. The result was 161 casualties and heavy, widespread damages in the centres of Sarno, Quindici, Siano and Bracigliano. The research presented in the paper is based upon a detailed topographical map obtained by means of a laser altimeter (Airborne Laser Terrain Model, ALTM). The data, achieved by means of a survey carried out soon after the disaster occurrence, allowed a precise morphological characterisation of the slopes and of the landslides in particular. ALTM is a highly accurate laser scanning system designed to acquire morphological data in the XYZ terrain co-ordinates. The spatial resolution depends upon the airborne flight elevation and velocity, ranging from 40 cm (300m elevation) to 7m (1200m). Aquater has developed a methodology based on an airborne laser system that provides data for the construction of DTM characterised by high density of spot elevation, high geometric accuracy, immediate availability of acquired data and input in GIS. The ALTM System includes (i) airborne instruments: laser, GPS, INP (inertial navigator system) and (ii) ground-based GPS station. The data processing system allows: (i) geometric correction and geocoding; (ii) development of DTM; (iii) development of specific products such as derived maps. The geocoding accuracy of reference point is: X, Y < 1.0m; Z < 15cm.

Figure 1. Location of the studied area. 3D view of the affected area derived from ALTM data

GENERAL SETTING

The bedrock of the studied area is composed of limestones of Cretaceous age, dipping up to 25-30° from the horizontal, with bedding thickness that varies from few centimetres to several metres. Above the limestones the pyroclastic deposits of the Vesuvius eruptions are present (Santacroce, 1987). The stratigraphical evolution of the affected slopes and in particular of the alluvial fans at the exit of the gullies that incise the slopes can be summarised as follows (Fig. 2): (i) development of calcareous alluvial fans at the toe of the slopes (calcareous debris flows); (ii) deposition ofpyroclastic deposits from Somma Vesuvius Volcano (pyroclastic falls); (iii) erosion, transport and deposition of pyroclastic deposits (pyroclastic debris flows) and development of pyroclastic alluvial fans; (iv) lateral and forward evolution of the pyroclastic alluvial fans (Aleotti et al., 1999). The development of deeply incised gullies, first in the calcareous rock and subsequently in the pyroclastic covers, has a major influence on the runout of these phenomena: in fact, the areas involved by the debris flows have migrated towards the plain in consequence of the channel formation and of the "telescopic" evolution of the fans.

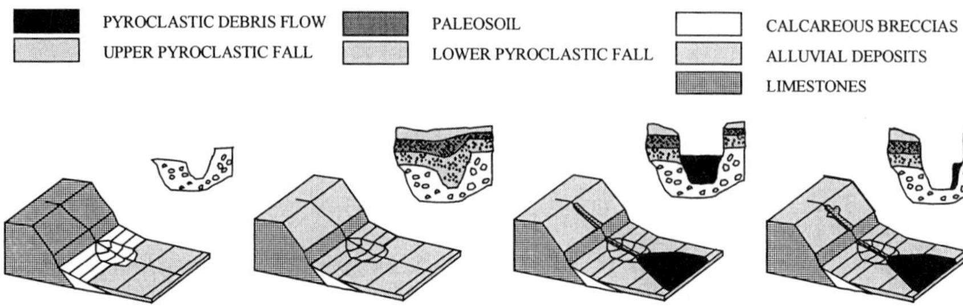

Figure 2. Evolution of the gullies and of the alluvial fans

DESCRIPTION OF THE EVENT

On 4-5 May 1998 an intense and short lasting rainstorm triggered more than 100 soil slips that evolved in debris flows, most of them reaching the high densely built up bottom floor. The recorded rain data showed that the hydrological event was intense but not exceptional: in fact the daily value has 10-15 years return time (Del Prete et al., 1998).

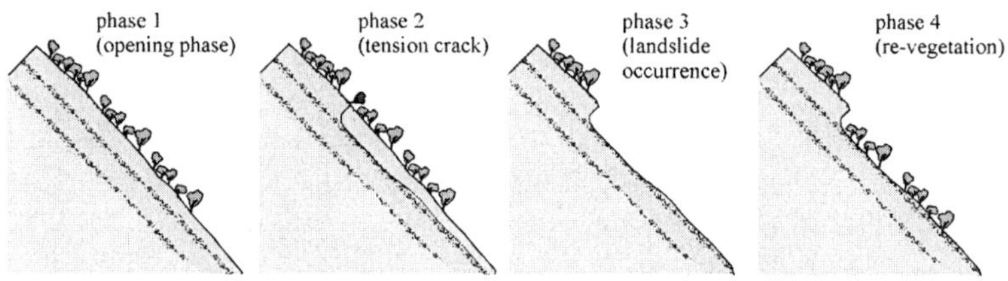

Figure 3. Landslide evolution (dark grey: pyroclastic deposits; light grey: limestones)

However it is likely that the measured rainfall values were underestimated (30-40%) because the rain gauge station is located at the base of the slopes and not in the higher parts of the reliefs where the landslides took place. The hydrological event, besides the great deal of debris flows, caused the formation of many tension cracks that represent the first step of a schematic landslide evolution (Fig. 3). Debris flows have been classified and differentiated according to their location on the slopes (gully failures, F; or planar slope failures, P) and to the main scarp position (Fig. 4).

STABILITY ANALYSES

Volcanic deposits and in particular pyroclastic falls related to the Somma Vesuvius activity now mantle the hill slopes of the studied area. The thickness of such deposits is quite variable being greater at the toe of the slopes and along the axes of the topographic depressions than in the upper part of the reliefs and in the ridge or in the watershed.

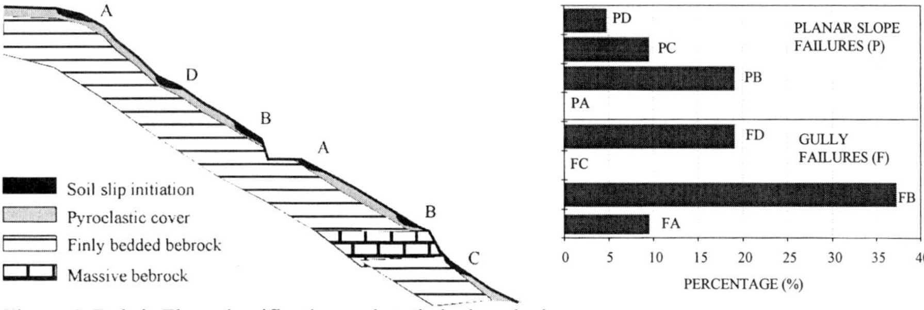

Figure 4. Debris Flow classification and statistical analysis

Figure 4 shows that failures usually initiate just above a limestone bank in correspondence of topographic concavities (hollows) (FB). The availability of a detailed DTM allowed the definition of the geometrical characteristics of the triggered landslides (Fig. 5).

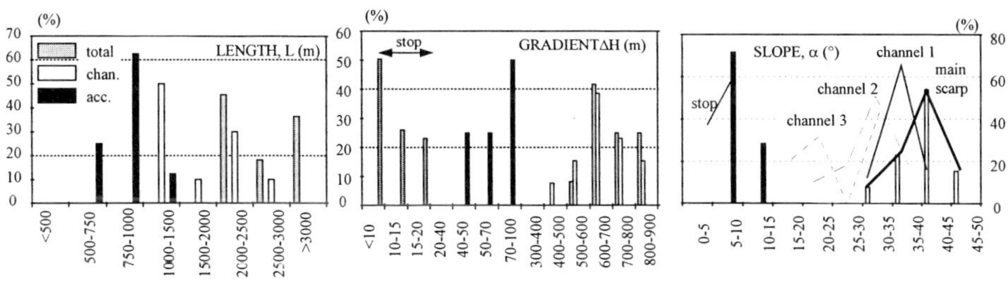

Figure 5. Statistical analyses of the morphometric parameters (channel 1: upper part of the channel, channel 2: middle part; channel 3: final part of the channel before the exit in the valley bottom; stop: last 200 m of the accumulation)

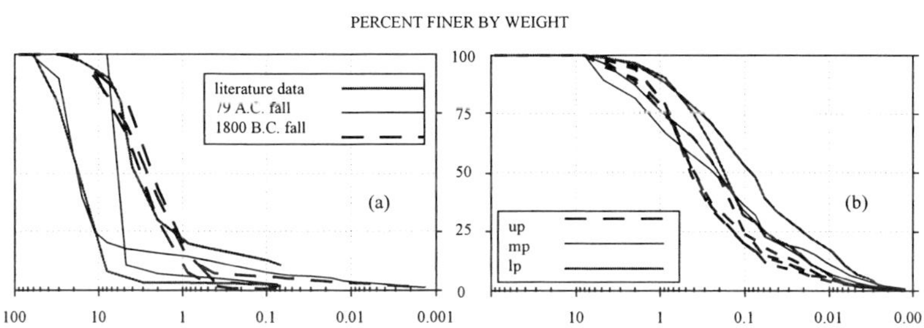

Figure 6. Grain-size analyses in pyroclastic falls (a) and paleosoils (b)

An ideal section-type in the upper part of the Pizzo d'Alvano, where the crowns are located, is the following (from above): (i) modern soil: brown silty sand with rare pumices (thickness, th.= 0.4 m); (ii) 472 A.D. fall (Pollena Eruption): porphyritic greenish pumices (ϕ: mm-cm), vesiculated, with rare lithic fragments (th.= 0.2 m); (iii) upper paleosoil (up): fine sand, dark brown with very rare pumices (th.= 0.7 m); (iv) 79 A.D. fall (Pompei Eruption): white rounded pumices (ϕ: cm) with very good sorting (th.= 0.5 m); (v) middle paleosoil (mp): slightly silty, dark brown sand, with rare white pumices (ϕ: cm) (th.= 0.6 m); (vi) 1800 B.C. fall (Avellino Eruption): white vesicular pumice, with good sorting. Different horizons with direct gradation (various volcanic impulses) (th.= 1.3 m); (vii) lower paleosoil (lp): fine, medium dense brown sand with rare pumices (ϕ: mm) and roots (th.= 0.6

m); (viii) limestones with joints and karst structures (be). Stability analyses were carried out on a test landslide [18], representative of the main group of failures (FB), on the basis of the accurate topographical profiles provided by ALTM. Some laboratory tests were conducted on samples collected from the main scarp: grain-size analyses showed that the involved pyroclastic soils are mainly constituted by gravel while the paleosoils can be classified as sand; the silty fraction is generally low but it increases with depth (from the upper paleosoil to the lower paleosoil) (Fig. 6).

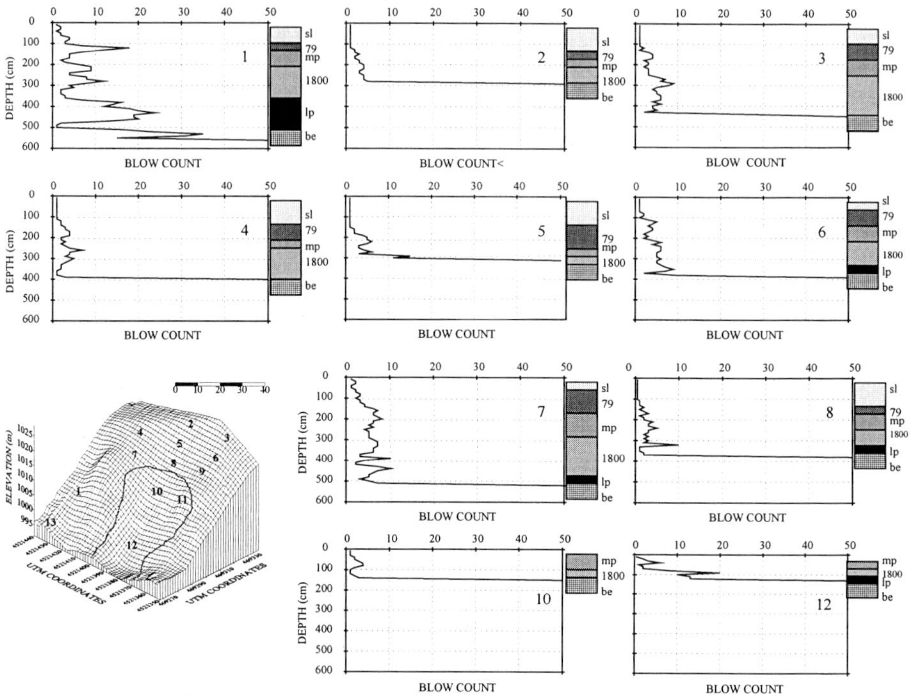

Figure. 7. Penetrometric tests (1800: 1800 B.C. fall; 79: 79 A.C. fall; sl: upper paleosoil, 472 A.C. fall and modern soil); 3D view of landslide [18] derived from ALTM data

The shear strength parameters of the paleosoils have been evaluated by means of direct shear tests under normal loads ranging between 25 and 350 kPa: the peak friction angle ranges from 40° to 44°, the intercept cohesion has very low values, from 0 to 17 kN/m^2. In addition, some in-situ penetrometric tests were carried out near the crown and the main scarp permitting the reconstruction of the stratigraphical setting of this area (Fig. 7). Safety factors were calculated assuming a progressive deepening of a wetting front from the top to the bottom of the stratigraphical section. Figure 8 shows the relationship between water table level and safety factor. The geometrical shape of the slip surface and in particular its depth are consistent with the field observations; in fact the failure surface in the tested landslide is located on the contact between the middle paleosoil and the 1800 B.C. fall deposits.

SUSCEPTIBILITY ASSESSMENT

When dealing with hazard (susceptibility) assessment, it is necessary to identify both where the debris flows will occur and where they will arrive after travelling over long distances (runout prediction). For the first problem the precise DTM provided by means of ALTM system allows the preparation of thematic maps relevant to those parameters that are closely linked to the triggering mechanisms of debris flows: slope, concavity/convexity, distance to the hill crest. More than 6% of the studied area presents the most prone values of steepness (40°-45°) and 22% of the territory has critical slope angle (35°-50°).

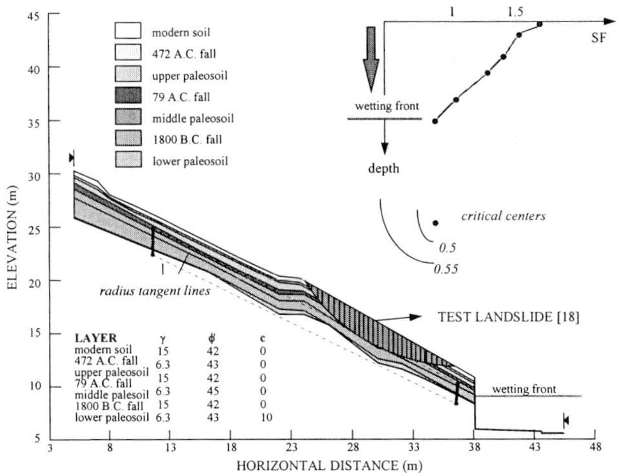

Figure 8. Stratigraphic section of the main scarp and stability analysis of the test landslide [18]

The importance of hollows in debris flows initiation is widely recognised: the topographic convergence, in fact, favours the accumulation of colluvial deposits and causes the concentration of both subsurface and surficial runoff during intense rainfalls (Wieczoreck et al., 1997). In agreement with this general observation, in the Sarno area about 70% of the main scarps are located in correspondence of ground depressions. Such geometrical setting represents 20% of the whole area.

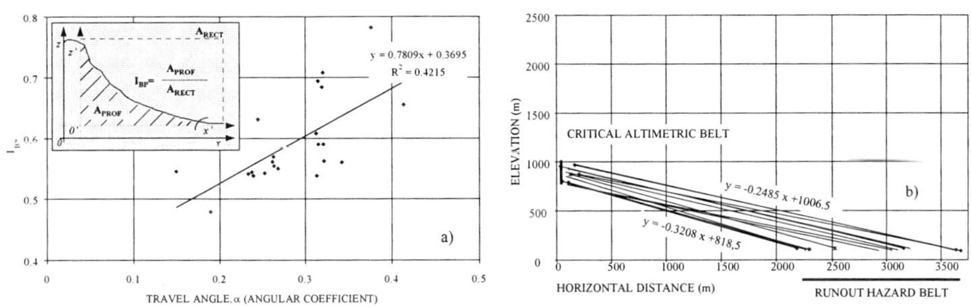

Figure 9a, b. Index of bottom profile (I_{BP}) vs travel angle (a); energy lines (b)

The crowns of the landslides are frequently located near the crest of the hill. It is likely that in such zones some geological and geomorphological predisposing factors coexist: (i) slope steepness; (ii) critical thickness of the surficial deposits (pyroclastic falls and paleosoils) in relationship with their permeability and with the characteristics of triggering rainfalls (intensity and duration); (iii) natural or artificial (pathways) breaks in slope: this particular topographic feature favours development and propagation of a wetting front resulting in faster saturation of soil covers. For these reasons, among the predisposing factor maps, we considered the slope and the concavity maps, as well as the map that indicates the distance of the crowns from the watershed. The final susceptibility map is the result of the superimposition of these parameter maps.

When dealing with the runout prediction it must be noted that the shape and extent of the deposition area are strongly controlled by morphology and topography. A direct relationship between the debris flow travel distance and the drainage basin morphometric characteristics, i.e. drainage surface and density, average slope, relief ratio, etc. can be expected, as in the case of the alluvial fans (Aleotti and Polloni, 1998).

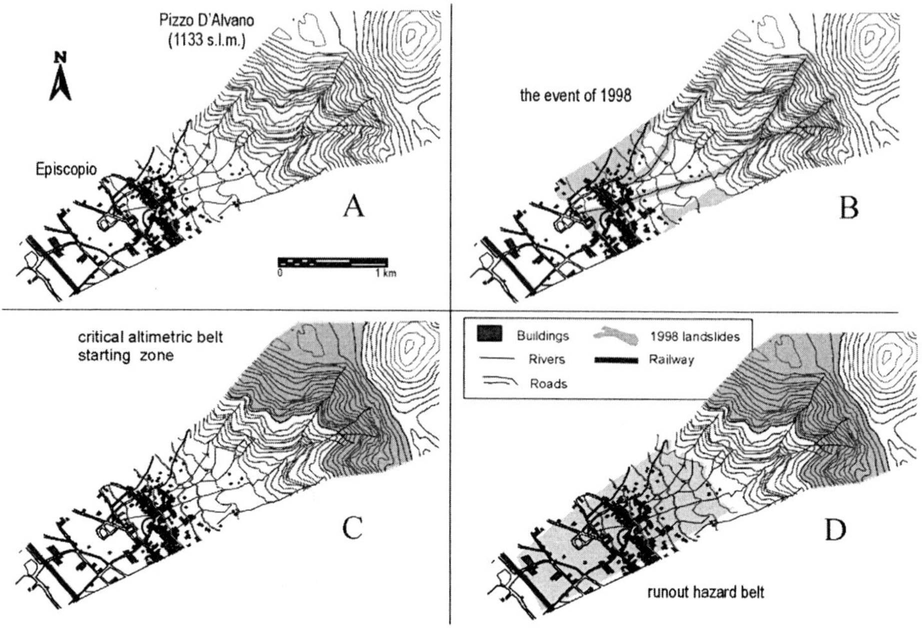

Figure 10. Hazard and risk assessment

Among the above-mentioned parameters we include the so-called "index of bottom profile (I_{BP})" defined as the ratio between the area subtended by the slope (debris flow channel) profile and the area of the circumscribed rectangle. Such a parameter has an approximate inverse relation with the hydrographic basin degree of evolution. Figure 9a shows a direct relationship between I_{BP} and the travel angle, α, pointing out that the debris flow travel distance is related with the basin degree of evolution and, therefore, with its hierarchical structure. The energy line equations of each landslide were calculated allowing the location of both a critical altimetric belt and a critical runout hazard belt (Figs. 9b, 10c, 10d). By overlaying this map with the drainage basin map it is possible to define the hazard areas. Starting from this rough delimitation of hazard areas, the detailed analysis of the topographic map, achieved by data collected with ALTM, allows the subdivision of debris flow hazard zones into direct impact zones, indirect impact zones and flood zones.

RISK ASSESSMENT

Debris flows imply great hazard and risk due to their high spatial density and high destructive power, especially in the direct impact zones. The latter is linked to the high velocity of the debris flows and therefore to the inherent kinetic energy. Many methods have been developed to assess the velocity profile and estimate the impact force of a debris flow for a given set of physical and geometric conditions (Rickenmann and Koch, 1997). The mean velocities of debris flows have been derived from the empirical relationship proposed by Seminara and Tubino (1993) [1]:

[1] $$V = C\sqrt{g\sin(\vartheta)D}$$

where ϑ is the bed slope, g is the gravity acceleration, D is the maximum flow depth and C is an empirical coefficient assumed equal to 8 in the computation. Figure 11 shows the results for a test landslide: maximum velocities (35 m/s) are reached in the upper part of the channel, where the bed slopes are greater, and a general trend, decreasing downstream, is evident although large fluctuations in velocity can be observed. An interesting factor is that the speed is still high when the flow reaches the populated urban centres (15-20 m/s). With reference to this aspect it is important to analyse the housing development in Sarno area during the last century (Fig. 12).

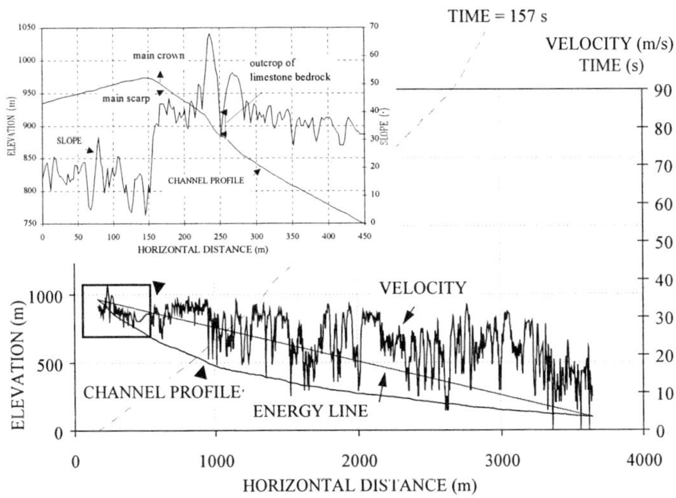

Figure 11. Mean velocity of a test landslide

Even if the dramatic event that struck the Sarno area on May 1998 is unprecedented for number and density of landslides, many significant debris flows have affected the zone in the historic past (Del Prete et al., 1998). Nevertheless urban development increased steadily, especially after the second world war, and many buildings were constructed even in hazardous areas. The majority of the houses destroyed during the May 1998 episode were built in the last few years.

CONCLUSIONS

The catastrophic effects of the studied event are due both to its exceptionality (over 150 landslides), and to uncontrolled urban development in the area. The availability, after a short period of time, of an accurate DTM, through the use of ALTM, allowed the construction of precise topographic and numerical maps, important tools in risk analysis and management.

The field investigations carried out proved that, for the majority of cases, the rupture surface is located entirely within the pyroclastic layer. The stability tests that were carried out are in good agreement with a saturation front that starts at the surface and moves downwards. The improvement in the estimates of the maximum velocity reached by the flows (variable mass models) and the capacity to calculate the distribution of the mass in the deposition areas based on an accurate topographic model (DTM) are fundamental; these consent to organise future urban development and to effectively mitigate existing risk conditions in areas such as Sarno. Future research will concentrate on this topic, which in this paper has only been presented in a preliminary form.

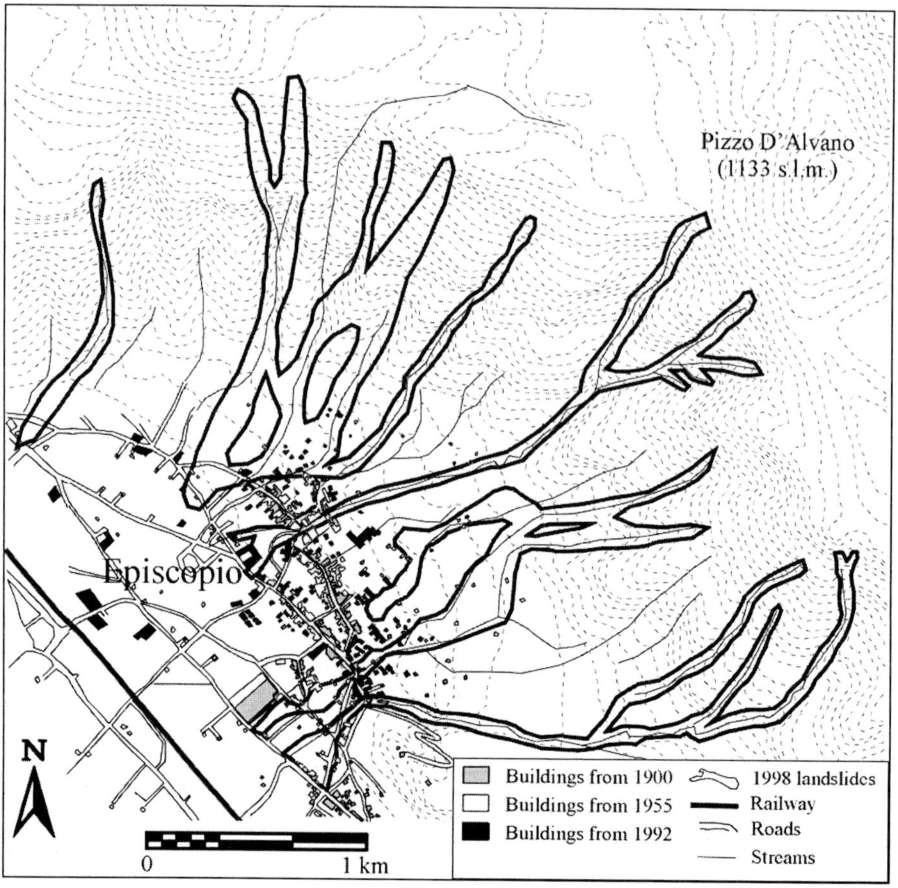

Figure 12. Urban settlement variation in Sarno from 1900 to present day

BIBLIOGRAPHY

Aleotti P., Polloni G. (1998): Risk scenarios in mountain alluvial fans: some examples in Italian Central Alps (Italy). Proc. 8° Congress of the IAEG, Vancouver (Canada), **3**: 2051-2058.

Aleotti P., Bastianelli L., Canuti P., Iotti A., Polloni G., Vannocci P. (1999). The use of new airborne laser terrain mapping techniques in slope instability assessment: an example in theSarno area. In: Geophysical Research Abstract: Non Linear Geophysics and Natural Hazards - 24th General Assembly of the EGS, The Netherlands, 19-23 April 1999, **1** (4), p. 881.

Del Prete M., Guadagno F.M., Hawkins A.B. (1998). Preliminary report on the landslides of 5 May 1998, Campania, southern Italy. Bull. Eng., Geol., Env., **57**: 113-129.

Rickenmann D., Koch T. (1997). Comparison of debris flow modelling approaches. In: Proc. of the First International Conference "Debris flow hazard mitigation: mechanics, prediction and assessment" (Cheng-lung Chen editor), San Francisco California, August 7-9, 1997, 576-585.

Santacroce R. (1987) (editor). Somma Vesuvius. CNR - Quaderni de "La Ricerca Scientifica" - Progetto Finalizzato Geodinamica, **8**, 251.

Seminara G., Tubino M. (1993). Debris flows: meccanica, controllo e previsione. GNDCI-CNR, Italy.

Wieczorek G.F., Mandrone G., DeCola L. (1997). The influence of hillslope shape on debris flow initiation. In: Proc. of the First International Conference on "Debris flow hazard mitigation: mechanics, prediction and assessment" San Francisco (California), August 7-9, 1997, 21-31.

Impact of Mass Movement on Landscape and Nature Transformation in the Polish Carpathians

Z. ALEXANDROWICZ and W. MARGIELEWSKI, Institute of Nature Conservation of Polish Academy of Sciences, Kraków, Poland

INTRODUCTION

The prevailing part of the Polish Carpathians mainly consists of strongly folded and dislocated flysch deposits. Sequences of deposits composed of thick sandstone beds intercalated with shales and thinly-bedded flysch are particularly susceptible to mass movements creating the relief of a considerable part of the mountains. In general, landslide areas cover about 7% of the region in question, but sometimes this proportion increases up to 50% or more in some mountain ranges (Ziętara, 1988).

Recurrent sliding processes are regarded as destructive events for the landscape including plant cover and land use. However, landslide areas have specific values in relation to their surroundings. They reflect the evolution of relief and are substantial components of landscape and biotopic diversity. Numerous Carpathian landslides are connected with particular phases of the Holocene history. They are very interesting both as fields for studies of environmental changes and areas of high tourist attraction due to their very differentiated morphology. Many of them occur in national parks, landscape parks and nature reserves, but they are usually regarded as less significant elements of the relief. Only few landslides are protected due to this phenomenon exclusively, protection of several is planned according to the recently project (Alexandrowicz ed. et al. 1996, Alexandrowicz & Poprawa eds. et al. - in print).

LANDSLIDES AS ELEMENTS OF RELIEF DIVERSITY

Mass movements strongly influence the Carpathians landscape shaping and forming characteristic elements of the relief (Figures 1 and 2). Large landslides are particularly spectacular forms with their concave or even rocky head scarps and convex colluvial mass at the foot of the slope. In relation to geological structures recurrent landslides can be characterized by a complicated profile of the slope, distinctly different from the surrounding area. Large flat areas and depressions without outflow are typical features of sliding slopes. Mass movements developed on both sides of a ridge form sharp and narrow crests limited by shear planes (Baumgart-Kotarba 1994). Double ridges with top trenches, open fissures, rifts, rocky walls or niches, sandstone tors and block fields are other elements typical of landslides (Alexandrowicz 1978, Alexandrowicz & Alexandrowicz 1988, Margielewski 1997a). Some of them are enriched in pseudokarstic caves and picturesque ponds and dam-lakes. Numerous water bodies and marshes are progressively transformed into peat bogs.

In the Polish Carpathians differentiated relief of numerous landslide areas developed during the last 12000 years since the Late Vistulian (Alexandrowicz 1997, Alexandrowicz & Alexandrowicz 1998, 1999, Margielewski 1997b, 1998, Starkel 1997). Organic deposits and fragments of wood found within colluvial masses allow dating of particular mass movements

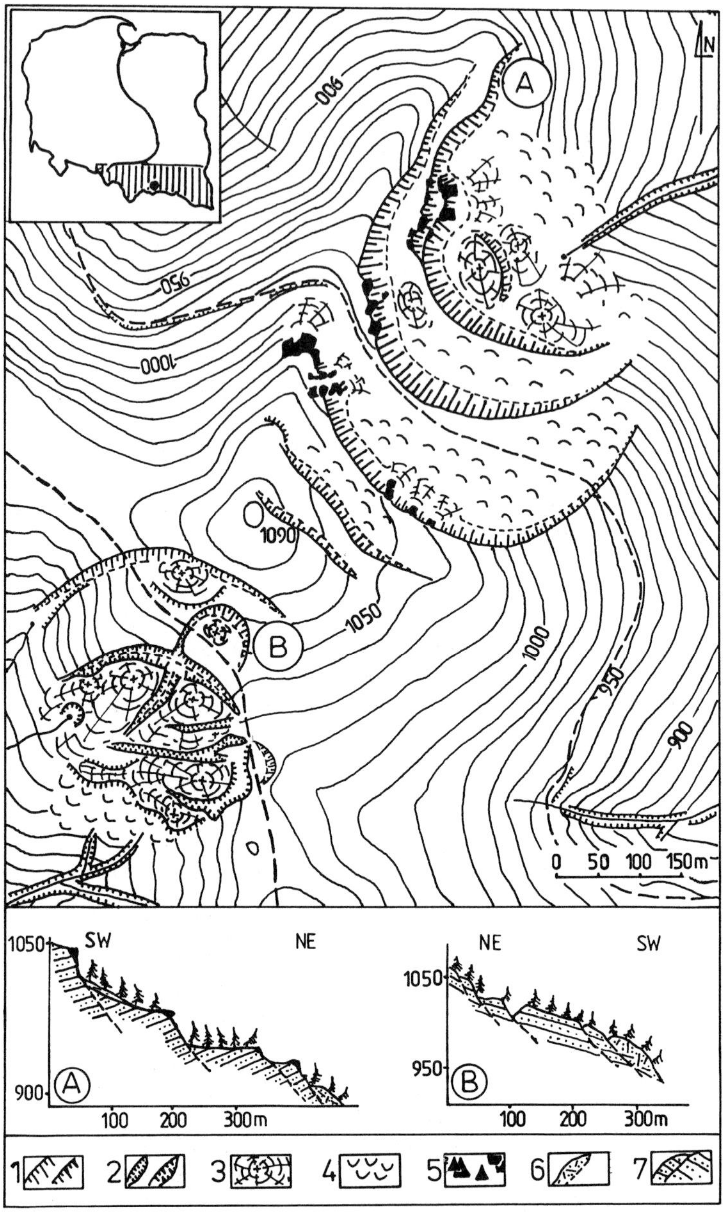

Figure 1. Example of landslide area in Poprad Landscape Park – Jaworzyna Krynicka Range (after Margielewski 1997a): A - northern slope modelled by obsequent landslide with sandstone tors as a remnant of niche; B - southern slope modelled by landslide rifts; 1 - landslide niches, 2 - trenches, 3 - colluvial swell, 4 - colluvial tongue, 6 - sandstone tors, 5 - mixed colluvial material, 7 - sandstones of Magura Formation (on cross-section)

using the radiocarbon method. Biostratigraphic analysis of fossil plant remains (pollen, seeds and fruits, macro-fragments) and mollusca are useful in reconstructing environmental changes, human impact and phases of sliding activity. The age of young landslides is known from historic data and about 65 landslides have been dated by different methods.

LANDSLIDES AS A BIODIVERSITY FACTOR

Landslides have significant influence on biotopes and their transformation. Mosaic habitats are covering surfaces deformed by mass movements. Specific distribution of plant habitats connected with soil and water conditions is characteristic for such places. A lot of different plant communities occur close to each other. Landslide areas with diversified relief are difficult to penetrate for foresters. These areas are very interesting from the natural point of view as they have been overgrown with the rest of the virgin Carpathian forest till now. They abound with places which are particularly suitable for wildlife (refugium for wild animals). These refugia are also connected with landslide ponds and swamps.

Mass movements have a big influence on water conditions of mountainous areas. Landslides which usually cut water horizon are abundant in springs, swamps and streams which control local river system. Landslide peat bog accumulates water and regulates its distribution being a very important water reservoir.

Some biotopes developed within slid masses are significant for land use in mountain and foothill regions. Flat colluvial areas of old steady landslides, abounding with springs and water bodies were attractive as settlements since prehistorical times. Now, they are also cultivated or used as pastures or orchards. However they have recently become depopulated owing to hard conditions of living and in consequence they are being transformed by plant succession and geomorphic processes.

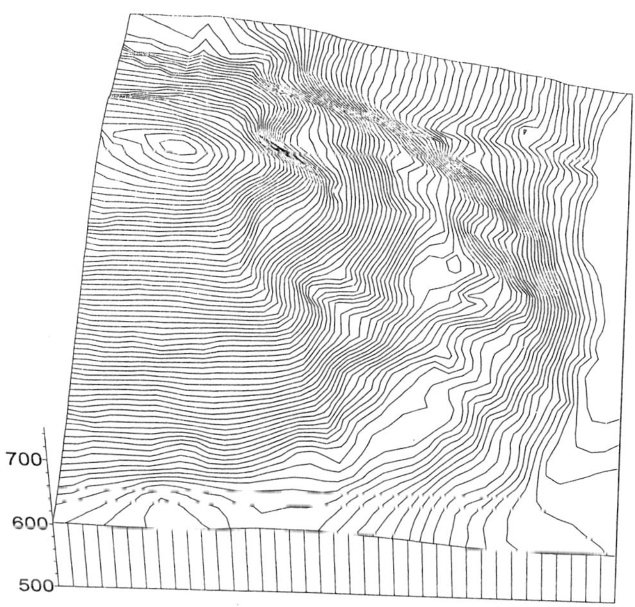

Figure 2. Stereogram of landslide relief within park of Krynica health resort (after Alexandrowicz & Alexandrowicz 1998). view from north (azimuth 175°)

CONCLUSIONS

Differentiated relief of the Carpathian landslides and connected with them biotopes diversity reflect transformation of mountain environment controlled by both geological structure and climate. The activity of mass movements depends not only on natural factors but also on the human impact. Sliding processes are less intensive within forest mountains mainly due to annual precipitation and heavy rainfalls. They are particularly hazardous in deforested and overbuilt areas as well as on unstable slopes cut by stream/rivers or routes, quarries.

REFERENCES

Alexandrowicz S.W. 1997. Holocene dated landslides in the Polish Carpathians. Paläoklimaforschung Bd 19. Special Issue: ESF Project. European Palaeoclimate and Man 12: 75-83.

Alexandrowicz S.W. & Alexandrowicz Z. 1998. Landslide at the Parkowa Góra Mt in Krynica (Polish Carpathians) - sediments, malacofauna and phases of the development. (In Polish with English summary). Kwartalnik AGH, Geologia 24,4: 271-331.

Alexandrowicz S.W. & Alexandrowicz Z. 1999. Recurrent Holocene landslides: a case study of the Krynica landslide in the Polish Carpathians. The Holocene 9,1: 91-99.

Alexandrowicz Z. 1978. Sandstone tors of the western flysch Carpathians. (In Polish with English summary). Prace Geologiczne PAN 113.

Alexandrowicz Z. & Alexandrowicz S.W. 1988. Ridge top trenches and rifts in the Polish Outer Carpathians. Annales Societas Geologorum Poloniae 58: 207-228.

Alexandrowicz Z. ed., Margielewski W., Urban J., Gonera M. 1996. Geoconservation of the Beskid Sądecki and the Sacz Basin, Polish Carpathians. (In Polish with English summary). Studia Naturae 42.

Alexandrowicz Z. & Poprawa D. eds., Margielewski W., Urban J., Gonera M., Otęska-Budzyn J., Rączkowski W. (in print). Geodiversity conservation of the Polish Carpathians. (In Polish with English summary). Polish Geological Institute. Warszawa.

Baumgart-Kotarba M. 1974. Development of mountain ridges in the flysch Carpathians. (In Polish with English summary). Prace Geograficzne IG PAN 106.

Margielewski W. 1997a. Landslide forms of the Jaworzyna Krynicka and their connection with the geological structure of the region. (In Polish with English summary). Kwartalnik AGH, Geologia 23: 45-102.

Margielewski W. 1997b. Dated landslides of the Jaworzyna Krynicka Range (Polish Outer Carpathians) and their relation to climatic phases of the Holocene. Annales Societatis Geologorum Poloniae 67: 83-92.

Margielewski W. 1998. Landslide phases in the Polish Outer Carpathians and their relation to climatic changes in the Late Glacial and the Holocene. Quat. Studies in Poland 15: 37-53.

Starkel L. 1997. Mass movements during the Holocene: the Carpathian example and the European perspective. Paläoklimaforschung Bd 19. Special Issue: ESF Project. European Palaeoclimate and Man 12: 385-400.

Ziętara T. 1988. Landslides area in the Polish Flysch Carpathians. Folia Geographica, ser. Geogr.-Phys. 20: 21-31.

Risk Assessment of Landslide Using Fuzzy Theory

S. M. ALI JAWAID, Civil Eng. Dept., M. M. M. Eng. College, Gorakhpur, India

INTRODUCTION

Natural events such as landslides can not be prevented from occurring, but their impact can be reduced if effective measures are taken to reduce their severity, frequency and possible size. The hazard coping measures include planning for disaster preparedness based on the estimation of the magnitude of the risk in the past and those anticipated in the future. This study is intended to prepare a low cost quantitative evaluation scheme using "Fuzzy Set Theory" analysis for evaluating such a potential in terms of vulnerability and risk.

METHOD PROPOSED

The main thrust of the proposed methodology is decision tree analysis. On each level or branch of the decision tree, the knowledge and information expressed in terms of the selected evaluation criteria, the weight among them and the rating of the risk potential according to the adopted criteria may be processed with the following equation:

$$R = [(r_1)^{w1} (r_2)^{w2} \ldots\ldots\ldots (r_j)^{wj} \ldots\ldots (r_n)^{wn}]^{\beta} \quad \ldots\ldots\ldots\ldots (1)$$

Where

R = Combined rating of the risk potential based on all criteria on a given branch of the decision tree.

r_j = Risk potential rating of area according to criteria j.

w_j = Weight(or relative importance) of the criteria j as compared with other criteria on the same branch of the decision tree.

n = Criteria adopted in each branch of the decision tree.

$$\beta = [1.1\Sigma_{j=1}^{n} w_j]^{-1}$$

Equ. 1 is a deterministic model, often used for combining evidence obtained from various knowledge sources. In this study, it is used to determine the overall rating of the risk potential based on all the criteria. For evaluation and mapping of risk at a regional or an area scale, quantitative evaluation approach is involved, as it does not need large amount of precise data regarding parameters, but a rating in linguistic terms such as low, moderate, high, etc which can be obtained from a single study of the area under consideration, supplemented by the information obtained from interviews conducted among local inhabitants. The weight among parameters can be determined by the frequency of occurrence of certain hazard based on that particular factor and its effects. By repeating equ. 1 step by step through various branches of decision tree, a final rating that represents the overall assessment of the risk associated with an element. The final rating, thus, obtained are again represented in linguistic terms using Table 1.

Landslides in research, theory and practice, Thomas Telford, London, 2000.

Table 1. Rating in Linguistic Terms

Letter assigned	Scale of weight	Description of rating	Range of values	Average value
A	Highly important	Very High	0.85 – 1.00	0.93
B	Very important	High	0.70 – 0.85	0.78
C	Important	Medium	0.45 – 0.70	0.58
D	Moderately important	Low	0.25 – 0.45	0.35
E	Little important	Very Low	0.00 – 0.25	0.13

RISK ASSESSMENT OF LANDSLIDE HAZARDS

The factors affecting the stability (or failure potential) of a slope were identified and examined, primarily through a comprehensive review of the literature on the subject. Four categories of factors namely topography, geology, environment and metrology were considered. Each category has two to five sub-factors. Thus, a total of thirteen factors were adopted for slope failure potential. These factors and weightings among them are shown in Table 2.

A pairwise comparison (Saaty, 1977) is done for every pairing of two classes, and score is assigned each time to the two entries. At the end of this process, the scores for each class are added up. Each class is then assigned a linguistic rating grade such as very high, high and so on, based on final total scores.

PRACTICAL APPLICATION

Case Study of Kaliyasur Landslide located at 147 km on Rishikesh-Joshimath Road:
(Data extracted from Proc of National Seminar on Construction of Roads in Hill Areas, Nainital, 1985)

Description of landslide:

Geology:

1. Open folded white and purple red gritty quartzite with minor bands of purple and leaf green shale's varying in thickness from a few cm to a maximum of 1.5 m.
2. Two prominent bedding shear zones (or fault zones). In shear zones, the rock have practically lost all their hardness and cohesion.

Soils

Usually thick unconsolidated overburden, prominently clay, along the slope. The clay content is evidently derived from the underlying argillaceous bed rock / shale and its predominance in the overburden material impedes latter's free drainage resulting in build-up of excessive pore pressure. Such material lying on inclined base is most susceptible to sliding under saturated conditions.

Highly fractured and disintegrated bedrock exposed in a few sections; also constitute potentially unstable slopes. Open joints and fractures facilitate deep percolation of surface runoff. The hydrostatic pressure developed along such fractures initiates slides in the vicinity of slopes.

TABLE 2. Factors and Weight for Evaluation of Risk of Landslide Hazards

Major category of factors	Weight among major categories	Secondary level factors	Criteria for rating of (3)	Weight among all factors within each major category
1	2	3	4	5
Topography	A	Gradient		D
			A[1](deg)> 40	A
			A (deg)= 30 – 40	B
			A (deg)= 25 - 30	C
			A (deg)= 20 – 25	D
			A (deg)< 20	E
		Horizontal Profile		B
			Straight	A
			Slightly convex	B
			Slightly concave	C
			Salient concave and convex	D
		Vertical Profile		A
			Straight	A
			Up slope	B
			Complex	C
			Down slope	D
		Location		B
			Shoulder	A
			Central or Upper	B
			Toe	C
			Others	D
		Height		A
			>20 m	A
			10 - 20 m	B
			05 - 10 m	C
			02 - 05 m	D
			00 - 02 m	E
Geology	B	Geological Origin		D
			Plutonic Soil	A
			Pyroclastic rock or Sedimentary rock	B
			Entrusive rock or metamorphic rock	C
			Others	D
		Soil Texture		A
			Top Soil	A

			Weathered rock	B
			Alluvium	C
			Terrace deposit	D
		Depth of weathering		B
			> 03 m	A
			02 – 03 m	B
			01 – 02 m	C
			0.5 – 01 m	D
			< 0.5 m	E
		Direction of Slope		C
			Dip Slope	A
			Cross Slope	B
			Escarped Slope	C
			Non-bending	D
Environment	C	Vegetation		B
			Broad leaved tree	A
			Grass covered needle leaved tree	B
			No cover	C
			Bamboo	D
			Others	E
		Land use		C
			Recreation zone/swimming zone	A
			Forest/roadway grass land	B
			Residue	C
			Shrub	D
			Other	E
Metrology	B	Maximum daily precipitation		A
			> 150 mm	A
			100 - 150 mm	B
			50 - 100 mm	C
			20 - 50 mm	D
			00 - 20 mm	E
		Maximum hourly Precipitation		A
			> 30 mm	A
			20 – 30 mm	B
			10 – 20 mm	C
			05 – 10 mm	D
			00 – 05 mm	E

[1] A refers to slope angle.

Physiography
Tension cracks varying in length from about a few meters to a few hundred meters were observed.

Climate and Vegetation
Annual rainfall of about 1500 mm spread over three months of monsoon and nearly one and half month of winter. A heavy rainfall of 100mm to 150mm continuously in a day has been recorded. Area around the slide has thick forest. Part of the slide area covered by bush/grass.

Hill slope

Slope	Before slide	After Slide
Uphill side	$60^0 - 70^0$	45^0
Downhill side	$60^0 - 70^0$	45^0

Description of slide
Length of slide = 300 m; Height of crown of slide above road level = 100m; Vertically above and 160 m along the slope.

Based on the above information, the assigned linguistic and numerical gradings are given in Table 3, for the landslide hazard assessment. Processing the numerical data by decision tree analysis, using Equ. 1.

Topographical rating = $[(0.93)^{0.35} (0.58)^{0.78} (0.78)^{0.93} (0.78)^{0.78} (0.93)^{0.93}]^{(1/4.14)} = 0.79$
Similarly, Geological rating = 0.50; Environmental rating = 0.87;
Meteorological rating = 0.79
The combined rating of cell under consideration
$= [(0.80)^{0.93} (0.50)^{0.78} (0.87)^{0.35} (0.79)^{0.78}]^{(1/3.12)} = 0.72$

Now, using the Table 1, the linguistic rating for the cell under consideration may termed as **high**, which conforms the actual field condition of recurring landslide in that zone.

CONCLUDING REMARKS

Present study has an edge over many methods proposed earlier. Claims of its being perfect, has not been made, because prominence of any factor varies from site to site. Thus, before application of this method, weights of various factors may be added in check list prepared and according to site conditions, few factors can be deleted as per its prominence. Such flexibility in application makes it more practical and handy, which is not available in many methods.

REFERENCES

Juang, C. H, Lee, D.H and Sheu, C. (1992), "Mapping slope failure potential using fuzzy sets", Journal of Geotechnical Engineering, ASCE, Vol. 118, No. 03, pp. 475 – 493.

Kandel, A. (1986), "Fuzzy mathematical techniques with application", Wesley Publishing Company, New York.

Pant, M., Pachouri, A. K. (1989), "Land hazard mapping – a step towards the preparedness and mitigation of land hazards", Proc. symposium of preparedness and mitigation of natural disasters, New Delhi, pp. 230 – 245.

Saaty, T.L. (1977), "A scaling method for priorities in hierarchical structures", Journal of mathematical psychology, Vol. 15, No. 03, pp. 234 – 281.

Sato, T. and Terada, M. (1985), "Data processing and prediction of slope failure based on fuzzy set theory", Annual report of natural hazard mitigation disaster prevention research institute, Kyoto university, No. 28(B-2), pp. 63 – 86.

Seshasri, D. N., Lakshmikantam, C. B., Upendran, R. and Subramanium, K. (1982), "Landslide zonation in Nilgiri plateau, Tamilnadu, India", Proc. IV congress of intl. association of engineering geologist, New Delhi, pp. 379 – 389.

Table 3. Linguistic and numerical grading assigned

Major category factor	Weight among major category	Corresponding numerical values	Secondary level factors	Weight among each category	Corresponding numerical value	Rating based on field condition	Corresponding numerical values
Topography	A	0.93	Gradient	D	0.35	A	0.93
			Horizontal Profile	B	0.78	C	0.58
			Vertical Profile	A	0.93	B	0.78
			Location	B	0.78	B	0.78
			Height	A	0.93	A	0.93
Geology	B	0.78	Geological Origin	D	0.35	B	0.78
			Soil texture	A	0.93	A	0.93
			Depth of weathering	B	0.78	C	0.58
			Direction of slope	C	0.58	C	0.58
Environment	D	0.35	Vegetation use	B	0.78	A	0.93
			Land use	C	0.58	B	0.78
Metrology	B	0.78	Maximum daily precipitation	A	0.93	B	0.78
			Maximum hourly precipitation	A	0.93	B	0.78

Measuring of Slope Surface Displacement by Using Digital Image Processing

H. G. B. ALLERSMA, Delft University of Technology, Netherlands

ABSTRACT
A measurement based on digital image processing has been developed in order to determine the displacement field of a slope during the failure of a dyke. The dyke was brought to failure by water infiltration, which was supposed to be a simulation of wave overtopping. It was found that the measuring technique could successfully be applied to digitize the deformation of the slope during the field test. Thanks to the measurement a good comparison could be made between small-scale tests in a geotechnical centrifuge and field tests.

INTRODUCTION
A real scale tests has been performed under the supervision of the Dutch Ministry of Public Work in order to examine the mechanism of failure of dykes during water infiltration as a result of wave overtopping. The interest in the failure mechanism dates from 1953. In that year a large area in the Southwest part of the Netherlands was flooded by the sea during a heavy storm in combination with spring tide. It appeared that several dykes were not damaged at the first instance by mechanical forces at the seaside, but on the landside. The water table of the sea had reached such a high level that it flowed over the crest of the dyke thanks to wave action. The amount of water itself was not very critical because a polder has sufficient capacity to store this water. The point was, however, that the water infiltrates into the dyke. This was possible because the top of the dyke and the slope at the landside were not covered with a water impermeable layer. The water infiltration in the soil dramatically reduces the stability of the slope at the landside, which finally causes failure of the dyke.

In order to estimate the potential danger of the actual dykes, several theoretical analyses have been made for determining the stability of dykes when wave overtopping occurs. It appeared that the steepness of the slope and the appearance of different soil layers are important parameters. The theoretical approaches, however, did not explain the behaviour satisfactory, so that tests are required to get more insight in the contribution of the different parameters. Since a real scale tests is expensive and time consuming centrifuge tests are performed in the firsts instance to gain more insight in the problem. Several different profiles of dykes were tested in order to examine the effect of water flow through the dyke. It appeared that the centrifuge tests show in some cases a different failure mechanism than assumed in the theoretical analyses and in previous field observations. For that reason special attention was paid to examine the failure mechanism in more detail in the field test, which was performed after the centrifuge test. The results of the centrifuge tests appeared to be helpful in the design of the setup and measuring plan of the real scale test. An old dyke located in the province of Zeeland was chosen as a location for the field tests. The water infiltration by wave

Landslides in research, theory and practice, Thomas Telford, London, 2000.

overtopping was simulated by running water over the slope of the dyke, which was supplied by a tube with holes located at the crest of the dyke. The configuration of the dyke agrees with the dyke simulated in the centrifuge, namely: a sand core based on a impermeable layer and covered with a clay layer. Several sensors are installed in the soil in order to monitor the behaviour of the dyke. However, a problem arose to find a method to monitor the surface deformation of the slope of the dyke. This information was important in order to compare the field tests with centrifuge tests. It appeared that no standard equipment was available to measure the surface deformation. Therefore the Geotechnical Laboratory of the University of Delft was invited to develop a measuring method to detect the surface displacement of the slope of the dyke. It appeared soon that the measurement could be performed by image processing techniques. A measuring system based on image processing was developed by Allersma (1990) in order to measure surface deformations during centrifuge tests. It was already found that this measuring technique could be applied for several other purposes i.e. field measurements (Allersma et al., 1994a)

A B

Figure1. a) Photograph of the dyke section, prepared for performing a water infiltration test. b) Photograph of the housing of the video camera.

DESCRIPTION OF THE FIELD TEST

A dyke with a height of 8.7 m and a slope angle of 60 degrees (length slope ca. 10m) was flooded by water. Globally the dyke profile consists of a sand body, which was covered by a clay layer with a thickness of 0.3m. The water supply system consists of a tube with a diameter of 200mm, which was located at the crest of the dyke (Figure 1a). To be sure that the settlement of the dyke does not influence the distribution of the water, holes are made in the tube at regular distance. The water flow through the holes was made completely dependent of the pressure in the tube. To prevent that erosion disturbs the course of the test in the first stage folio was used to protect the slope close at the outlet of the water supply tube. The water supply was kept constant during the tests. Three tests were carried out. In the first tests no measuring equipment was installed. These tests were used to adjust the water supply system and an indication was obtained how the tests proceeds and how much time it takes before failure occurs. The second tests were for performing a water infiltration test. The second tests were equipped fully. The third tests were performed additionally. Since the measurement by image processing requires little preparation only this measuring technique was used to monitor the third test.

DEFORMATION MEASUREMENT BY IMAGE PROCESSING

Equipment

Two image-processing stations were used to monitor the behaviour of the slope. One system was programmed to store an image at the hard disk at every three seconds. The second system was used to perform real time operations in order to monitor the tests in more detail. A simple but effective operation was for example subtraction of two stages. This operation shows the changes in two images taken at different time intervals.

Each image processing station comprises a PC with a frame grabber and a CCD video camera. The frame grabber is the interface between the PC and the video camera. Several frame grabbers are available on the market. In this test a rather simple device (PFGPLUS-512-3-Hz-A from Imaging Technology, USA) was used. An industrial black-and-white video camera (Sony, model AVC-D7CE) with a lens with automatic diaphragm (25mm, F1.3 from Ernitec, Denmark) was connected to the frame grabber. A video mixer was used to make time continuously visible in the image. To enable manipulation of the images a software toolkit (TIM from Difa Measuring System, the Netherlands) was loaded.

Additionally a video recorder was connected to one of the stations, in order to have an extra guarantee that the test was monitored. In principle it is possible to elaborate the images from video tape.

The camera was located at the slope of the dyke at a distance of approximately 30m of the center of the monitored area (Figure1b). Special attention was paid to fix the position of the camera. In the first instance a metal shaft was blown into the soil. The metal shaft was supported by three beams, which are also fixed to the soil by shafts. The whole camera support was confined in a small wooden house, which was fixed by ropes and sticks. This house protects the camera from rain and reduces the influence of wind. Afterwards this protection was very useful, because there was a strong wind at the day of the tests.

Labels

In order to digitize the deformation of the slope markers are required, which can be separated from the background by processing the image. Since water is running over the slope simple labels are preferred. Therefore it was chosen not to use an active label e.g. a light emitting source. By trial and error it was found that a black spot on a white plate was an object which could rather easily be recognized by digital image processing. Plastic sheets with a size of 250x250 mm with a circular black spot with a diameter of 100 mm were fixed to aluminum tubes (dia. 25mm) with a length of 300mm. The tubes were driven into the soil over a length of 100mm. Three arrays of six labels were installed: one array in the centre line of the test section and two arrays 2.5m out of the center (Figure 1a).

Practical conditions

Several points have to be taken into account in order to obtain a good result. The location of the dyke was East- West, where the camera was viewing in the East direction. This means that the visible side of the labels were in the shadow in the morning, where the background was in the sun shine. This would not led to sufficient contrast. Therefore the tests had to be carried out in the afternoon, with the sun in the back.

A large part of the sky was visible at the image. Since the camera had an automatic black and white balans control the dyke at the foreground become very dark. This problem was solved by covering a part of the lens, so that the sky was not longer visible. The camera was equipped with an automatic diaphragm. This was a very useful option, because due to partly clouded weather in combination with a strong wind there was a large variation in light intensity.

Trial and error exercises had shown that he contrast between the labels and the grass was low when the grass was just mowed (the grass has a brown color). For that reason it was arranged that the grass was mowed a week before the tests was carried out.

ELABORATION IMAGES

Digital image processing can be considered as a universal tool to visualize, digitize and analyze parameters during geotechnical tests. By means of image processing objects in video images can be isolated, digitized and analyzed. Firstly relevant objects have to be extracted from the images and next the coordinates of the object have to be digitized. The principle of the measurement in the field tests is simple. In order to follow the behaviour of the slope a series of labels were placed at the slope surface. The labels are isolated from the background and the coordinated are determined. The coordinates for each label can be digitized in time, which finally gives the displacement field of the slope of the dyke.

The image frame consist of a matrix of 512 x 512 pixels, where each pixel represents 255 different gray levels. By calibration, the distance between the pixel can be transformed to field conditions. In the x-direction one pixel represents a distance of 15.6mm, where the distance is 11.4mm in the y-direction. The coordinates of a label can now be calculated from the row and column number of the pixel which is located at the center of the label. If there is a significant difference in gray value between the labels and the background the isolation of the labels can be performed simply by selecting objects with has a specific range of gray values. This selection can be carried out with a threshold command which is supplied by the toolkit. The command is like VAL a b, which means that only pixels are left in the image which have gray values in the range of a to b. If the gray value of the background is larger than 100, and the gray level of the labels are for example lower than 20, than a selection is obtained with e.g. VAL 0 30. Objects which are not connected to each other are treated as separate objects. The objects can be counted and linked to a serial number by a single command (LABEL). The serial number can be used to recall a specific object (MARK a), where the pixel number in x- and y-direction of the center are reflected. The procedure can be defined in a script file, so that the elaboration of the image proceeds automatically, where the coordinated are stored in a file for later use.

In practice, however, the elaboration of the image is much more complicated. Several disturbing effects have to be eliminated by digital manipulation of the image. Disturbing effects are generated by e.g. other objects, differences in light intensity, people who were walking around and wind. Several tricks can be applied to eliminate unwanted objects. Information which can help in the separation procedure is e.g. the amount of labels, the size of the labels and the approximate position of the labels. In spite of the disturbing effects it appeared to be possible to digitize the labels completely automatically in more than 3000 images.

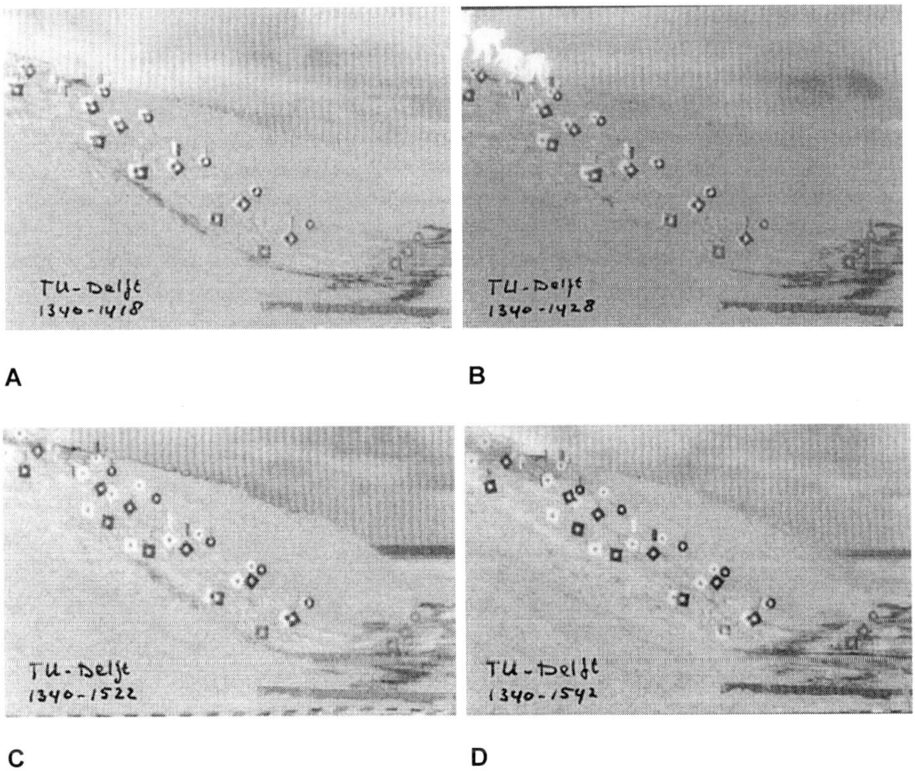

Figure 2. Visualization of the deformation of the slope by subtracting two images after different time intervals.

RESULTS

During the tests 3180 images were stored on the hard disk. The time interval between the images was 2.9 seconds. In each image the time was written. Since the course of the test was not known as many frames were captured as possible. During the test the course of the test could be followed in more detail by subtracting images, taken at different time intervals. An example of different time intervals are shown in Figure 2. Displacements of 5mm could be made visible in this way. Subtraction is a nice tool to visualize the deformation in the case that the displacement proceeds with a slow speed. It is clear visible that the displacements are largest in the area close to the crest. Further it can be observed that the direction of displacement is parallel to the slope close to the crest, where the direction is almost horizontal in the region close to the toe. This is caused by elevation of the covering clay layer thanks to overburden pore water pressure. It is visible that the labels beyond the toe did not show any movement. It appeared that the motion of comparable labels in the different arrays are very similar. This means that the test conditions in the measuring area can be considered to be two dimensional.

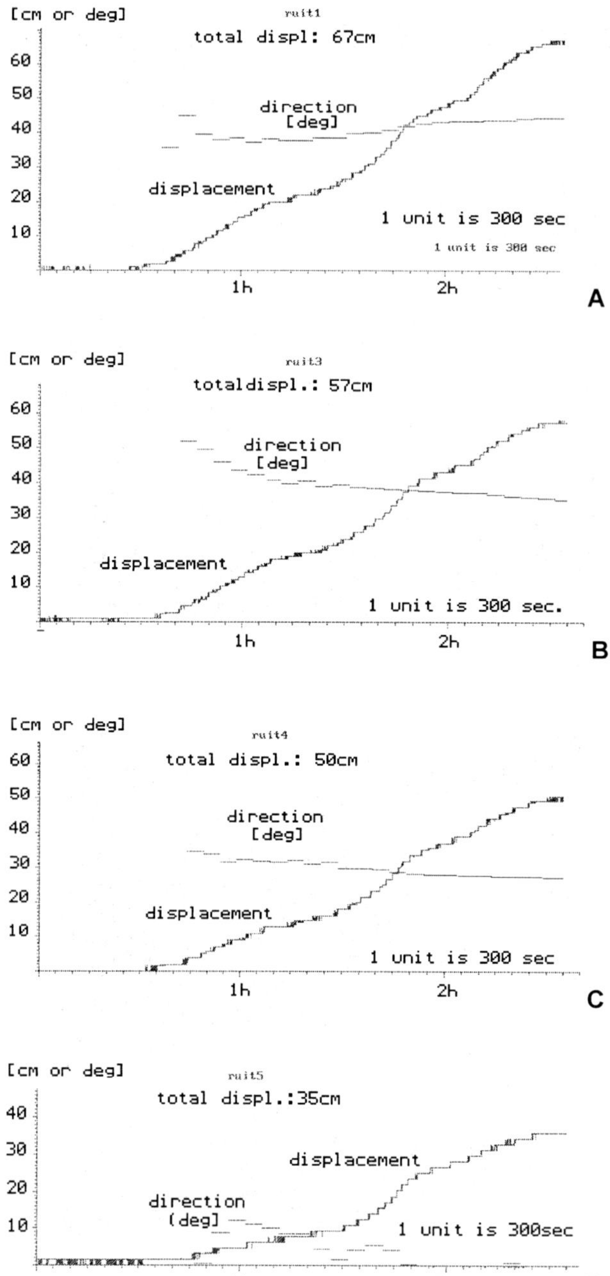

Figure 3. Graphical presentation of the displacement of some diamond shaped labels during the course of the test

Figure 4. Addition of the measured locations of the labels to a photograph of the initial stage of the dyke

Figure 5. Photograph of dyke failure in a small scale test, where water infiltration was simulated in a centrifuge

In Figure 3 the displacement of some diamond shaped labels (number 1, 3, 4 and 5, starting from the crest in Figure 2a) is shown in time. The digitization is performed each 2.9 seconds, beginning from the start of the test. Also the direction of the displacement with respect to the horizontal axis is presented. It can be seen here also, that closer to the toe the direction becomes more horizontal. The displacement is largest close to the crest, probably because the clay is softer because infiltration starts at that location. It seems that the displacement is not linear in time. During two time intervals of approximately 15 minutes the displacements of all the labels are smaller. Probably more pore water pressure has to be build up in order to increase the speed. The well-synchronized behaviour indicates that the whole clay layer moves uniform during the test.

In Figure 4 the position of the labels during some stages of the tests are added to the photograph of the initial stage of the dyke. This addition can be performed easily using commands of the image processing toolkit. The difference in direction of the displacement at the crest and at the toe is clear visible in this presentation.

The behaviour of the field test was close the same as a similar test in a geotechnical centrifuge (Figure 5) (Allersma et al., 1994b). In this test a stiff clay dyke was simulated, which was made higher by a sand cover. To prevent erosion by wind the sand surface is usually covered by a clay layer. Also in this test the clay layer moves parallel to the slope, where there was horizontal displacement at the toe causing a bubble. An important observation was that there was not a shear band failure mechanism as is supposed in many calculations model. Instead of that initially the clay layer moves down thank to reduction of friction caused by overburden pore water pressure. This finally causes cracks in the clay layer, so that further failure is caused by gradual erosion of the sand body.

CONCLUSIONS

Digital image processing was found to be a valuable method to observe the behaviour of a dyke in field tests, simulating water infiltration by wave overtopping. In relation to the large scale of the experiment the measurement was sufficiently accurate. The accuracy can be improved by using more cameras or by optimizing the digital processing of the image. The method is simple, because only labels have to be installed at the slope of the dyke, and only a video camera is needed as sensor. Locations with poor assess can be monitored in this way.

Thanks to the measurement a better insight was obtained in the failure mechanism of the dyke, and a better comparison was possible with small scale tests preformed in a geotechnical centrifuge. It was found in the field test that the dyke did not fail along shear bands. Instead of that a mechanism of gradual erosion could be observed. A very similar mechanism could be observed in small-scale centrifuge tests.

REFERENCES

Allersma, H.G, B., 1990. On line measurement of soil deformation in centrifuge tests by image processing. 9th Int. Conf. on Experimental Mechanics, Copenhagen, 1739-1748.

Allersma H.G.B. H.G. Stuit P. Holscher, 1994a. Using image processing in soil mechanics. Thirteenth Int. Conference on Soil Mechanics and Foundation Engineering, New Delhi, p.1341-1344.

Allersma H.G.B. I.A.G. Ligtenberg & B.A.N. Koehorst, 1994b. Simulation of failure of dykes by water infiltration by waves. Int. Conf. Centrifuge94, Singapore, 289-294.

Neural Networks: a Solution for the Factor of Safety Problems in Slopes

A. A. AL-TUHAMI, Department of Structural Engineering, Faculty of Engineering, Zagazig University, Zagazig, Egypt

ABSTRACT

A back-propagation neural network model for calculating the values of the factor of safety in slopes is proposed. A large number of factors of safety calculated using nonlinear finite element method is identified as input attributes. These input data include combination of different slope angles, heights and different soil properties. Outputs of this neural network model are compared with the calculated results of other models. Results indicate that neural network models can successfully and satisfactorily be used in obtaining the safety factors.

1. INTRODUCTION

Nowadays many methods for the calculation of the values of the safety factor in slopes are available. The degree of accuracy in many of these methods is close. Unfortunately, the determination of the safety factor in the traditional Limit Equilibrium Methods requires complicated procedures. Moreover, the factor of safety obtained by these methods is not uniquely defined.

The finite element method proved effective in finding the factor of safety in slope analysis (Salem & Al-Tuhami, 1998; Smith, 1994). But in this method, the large number of degrees of freedom and non-linearity present in the problem make it computationally expansive. In addition, the presence of a many parameters adds more complexity to the already complex problem in all of the above mentioned techniques.

Any new technique significantly reduce the computer time required to solve this problem would be beneficial. One promising technique is the back-propagation neural network model. This model is used in the current study to predict and obtain the values of the safety factor.

The neural network refers to a method of information processing using a network of processing elements modeled after the structure of human brain (Rosenblatt 1958; Rumelhart, 1986).

Neural networks are capable of self-organization and learning. They can be introduced with a series of examples, and from these examples the networks learn the relationships between the inputs and the outputs.

Landslides in research, theory and practice, Thomas Telford, London, 2000.

Neural networks have recently found many applications in civil engineering. For example, in geotechnical analysis, neural networks were used to predict pile capacity (Teh et al., 1997), and diagnose pre-stressed concrete pile problems during driving (Yeh et al. 1993).

2. DEVELOPMENT OF THE NEURAL NETWORK MODEL

Many types of neural networks have been proposed by changing the network topology, the node characteristics, and the learning procedures. The input layer is the means by which data are presented to the network. In this study, the input layer data include slope angle; slope height; soil angle of internal friction; and the soil cohesion. It may also include other factors such as, the effect of seepage force, external applied load and the effect of dynamic loads such as earthquakes. The output layer holds the response of the network to the input signal. It is the hidden layer(s) and the connecting weights that enable the network to represent and process the complicated association between the input and the output. In the present study the output layer has the safety factor only. A typical back-propagation neural networks model for the problem presented in this study is shown in Figure 1.

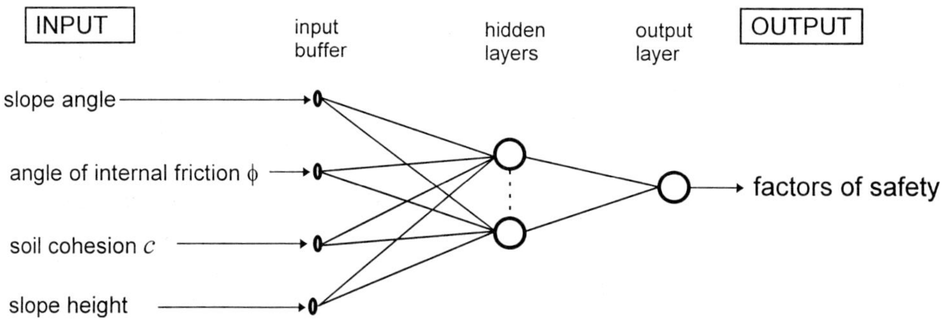

Figure 1. Structure of neural network for obtaining the values of the factor of safety in slope stability problems.

The neural network paradigm used in this work may be summarized as follows:
(1) Present an input; (2) Calculate the actual output; (3) Alter the weights to reinforce correct decisions i.e. reduce the error; (4) Present the next input; and (5) Repeat the whole process until the overall error is less than a prescribed value.

2.1 Implementation

As summarized in Elazouni et al., 1997, the implementation phase involves three main aspects: (1) acquiring the knowledge; (2) selecting the network configuration; and (3) training and testing the network.

2.1.1 Knowledge acquiring

In this paper, a nonlinear finite element model is used to obtain the required training set of the minimum safety factor in slopes. A visco-plastic Mohr-Coulomb soil model with a non-associated flow rule is adopted. The factor of safety obtained using the finite element method is achieved when the solution of the problem diverged. This method of determining the factor of safety is closer to the real meaning of the failure in which a permanent deformations (unrecoverable) took place, (Zienkiewicz, 1971; Smith & Griffiths, 1988). The finite element model is verified against a full-scale trial embankment and other traditional solution techniques as presented in Salem & Al-Tuhami, 1998.

2.1.2 Normalization of input and output data

Neural networks usually provide improved performance when the data lie within the range (0,1). To avoid the slow rate of learning near the end points of the output range, the soil cohesion, the slope height, the slope angle, the soil angle of internal friction, and the associated outputs are transformed to values ranging from 0.1 to 0.9 according to the following equation:

$$XT = 0.1 + 0.8\left(\frac{X - X_{\min}}{X_{\max} - X_{\min}}\right) \qquad (1)$$

where XT = normalized value; X = original value; $X_{\max}$ = maximum value of the attribute or output; and $X_{\min}$ = minimum value of the attribute or output.

2.1.3 Selecting network configuration

The neural network algorithm for obtaining the factor of safety in slopes is coded and used in the present study. The neural network model is then configured, trained, and tested. Setting the configuration of a neural network that gives the best performance is a highly problem dependent task. How well the network can learn the training examples and produce satisfactory response when presented with patterns outside the training examples measures the performance. Many variables need to be specified in the back propagation algorithm to define the network configuration. The present network model is configured as having a generalized delta rule learning algorithm to reduce the error as rapidly as possible. Also the model contain, Sigmoid transfer function to account for the continuous inputs.

The training is conducted using over 98 examples in different network tasks with different degrees of refinement. The height of the embankment was varied from 8 m to 30 m for the following slope angles and soil strength parameters; slope angle (β) ranging from 25° - 60°; soil cohesion (c) ranging from 10 to 50 KPa, and soil angle of internal friction (ϕ) ranging from 5° to 25°.

Many tasks are conducted with different neural network configurations, to choose the best performance of this method. The configurations of each task include number of hidden layers, number of processing elements in each hidden layer, maximum training and individual errors, the momentum η rate coefficient and α learning rate coefficient. These taskes are implemented for η value of o.8 and α of 0.6 which showed the best results during testing the networks model. Some of these tasks configuration are shown in the following table:

Table 1. Configuration of different tasks.

Task number / Model Configuration	Task1	task2	task3	task4	task5
Maximum total normalized error	8.0×10^{-7}	1.0×10^{-5}	1.0×10^{-5}	1.0×10^{-6}	1.0×10^{-5}
Maximum individual error	4.0×10^{-7}	5.0×10^{-6}	5.0×10^{-7}	5.0×10^{-7}	5.0×10^{-6}
Number of hidden layers	1	1	1	1	2
Number of processing elements in each hidden layer	5	15	25	15	15
Input iterations	30000	30000	30000	30000	30000

2.1.4 Network training and testing

In order to check the validity of the used technique, the author compared the obtained results with traditional methods (Huang & Yamasaki, 1993; Liang et al. 1997; and finite element method, Salem & Al-Tuhami, 1998). The two slope configurations are 2 : 1 (horizontal : vertical) slope and 1 : 1 slope and the slope height is 8 m. The soil is assumed to be homogeneous and the depth to the underlying rigid layer boundary is 6 m from the toe of the slope. The ground water table is assumed to be below the rigid layer.

As presented by Huang (1993), different pairs of soil strength parameters were selected to evaluate the sensitivity of strength parameters on the calculated slope stability. Cohesion values ranging from 2.5 KPa to 30 KPa and friction angles ranging from 10 to 30 degrees were used. The soil is assumed to be homogeneous and the depth to the underlying rigid layer boundary is 6 m from the toe of the slope. The ground water table is assumed to be below the rigid layer. It should be noted that Liang et al., 1997; in Huang and Yamasaki's method considering the inter-slice forces. Table 2 shows some of the results obtained by these tasks. By comparing the tasks result with that obtained with the finite element method, it can be noticed that task 3 gave the best results of all the tasks.

The results also showed that increasing the number of processing elements in hidden layers from 15 to 25 sharply increased the time consumed for the run and giving overestimated values for the safety factors. Increasing the number of hidden layers from 1 to 2 with 15 processing elements in each hidden layer gave inaccurate results, task 4 of Table 2.

Table 2. Comparison of safety factors between FEM and neural networks, for (2: 1 slope)

Technique		**Neural Network Results**				**FEM**
Task Number		Task1	task2	task3	task4	
Reached error Iterations reached Hidden layers PEs		0.000001 30000 1 5	0.000001 30000 1 15	0.00001 22669 1 25	0.00001 1848 2 15	
c (Kpa)	(ϕ^o)	**Factor of Safety**				
20	20	2.1274	2.071	2.128	1.6958	2.069
15	20	1.8852	1.8135	1.8835	1.4425	1.801
10	20	1.6244	1.5334	1.6121	1.1826	1.517
5	20	1.3621	1.2452	1.3251	0.9394	1.198
2.5	20	1.2371	1.106	1.1831	0.8332	1.01
25	15	2.0395	2.0363	2.0842	2.0346	2.028
20	15	1.829	1.8078	1.8683	1.7953	1.787
15	15	1.5993	1.5602	1.63	1.54	1.535
10	15	1.3624	1.3007	1.3742	1.2740	1.257
5	15	1.1341	1.0458	1.1157	1.0176	0.967
15	10	1.3111	1.3016	1.3728	1.6405	1.264
10	10	1.1109	1.0726	1.1448	1.3701	1.012

In which: c is the soil cohesion; ϕ is the soil angle of internal friction; PEs: is the number of processing elements in each hidden layer; and FEM is the solution using the finite element.

Tables (3) show the comparison of safety factors between the traditional and finite element methods and the results obtained using neural networks technique, task3. These traditional techniques include: (1) Huang and Yamasaki, 1993; (2) Huang and Yamasaki, 1993, considering the inter-slice forces [Liang et al., 1997]; (3) Bishop's method, considering the inter-slice forces [Liang et al., 1997]; and (4) Bishop's method (1955).

Results show a very good agreement between the neural network results and those of the finite element method. Although no training is applied for the neural networks in the case of c = 5 and 2.5 KPa, results are still in good agreement with other solution techniques, especially Huang and Yamasaki 1993.

Table 3. Comparison of safety factors between traditional methods and neural network, task 3, for (2: 1 slope)

c (KPa)	(ϕ^o)	(1)	(2)	(3)	(4)	(5)	(6)
20	20	2.05	2.07	2.08	2.09	2.069	2.071
15	20	1.85	1.81	1.83	1.82	1.801	1.8135
10	20	1.60	1.55	1.59	1.54	1.517	1.5334
5	20	1.23	1.30	1.25	1.21	1.198	1.2452
2.5	20	1.00	1.19	1.20	1.06	1.01	1.106
25	15	1.87	2.06	2.05	2.05	2.028	2.0363
20	15	1.72	1.79	1.79	1.78	1.787	1.8078
15	15	1.54	1.53	1.54	1.53	1.535	1.5602
10	15	1.29	1.28	1.29	1.29	1.257	1.3007
5	15	1.00	1.03	1.05	0.99	0.967	1.0458
15	10	1.19	1.28	1.27	1.27	1.264	1.3016
10	10	1.00	1.02	1.02	1.03	1.012	1.0726

Table 4. Comparison of safety factors between traditional methods and neural networks, for (1: 1 slope)

c (KPa)	(ϕ^o)	(1)	(2)	(3)	(4)	(5)	(6)
25	20	1.87	1.72	1.73	1.71	1.708	1.8568
20	20	1.68	1.51	1.51	1.50	1.493	1.6198
15	20	1.46	1.30	1.30	1.29	1.266	1.3701
10	20	1.00	1.10	1.09	1.05	1.029	1.1185
30	15	1.85	1.76	1.78	1.75	1.735	1.7711
25	15	1.65	1.54	1.56	1.53	1.532	1.5615
20	15	1.45	1.33	1.34	1.32	1.323	1.3445
15	15	1.24	1.12	1.12	1.11	1.109	1.1255
10	15	1.00	0.92	0.91	0.89	0.880	.91531
25	10	1.42	1.38	1.40	1.35	1.351	1.3038
20	10	1.23	1.17	1.18	1.15	1.150	1.1134
15	10	1.00	0.96	0.97	0.97	0.956	.92866

(1) Huang and Yamasaki, 1993; (2) Huang and Yamasaki, 1993, considering the inter-slice forces [Liang et al., 1997]; (3) Bishop's method, considering the inter-slice forces [Liang et al., 1997]; (4) Bishop's method (1955); (5) Solution uses the finite element (Salem & Al-Tuhami 1998); and (6) Third task of neural networks technique.

In table 4, the comparison of safety factors between traditional and finite element methods and the results obtained using neural networks technique for slope 1:1 is shown. In the present comparison, only one task is performed with the following configuration: the setting normalized error is 0.000001, individual error is 0.000005. One hidden layer and 15 processing elements in this hidden layer are chosen. The reached normalized error is 0.000004 in 30000 iterations. Based on the result of this task, it can be seen that, the first two values of the safety factors (column 6 in Table 4) with corresponding c = 25 and 20 KPa, ϕ=20°, the results close to Huang and Yamasaki, 1993. All the other results lie between the values of other traditional methods (1-5). In general, good agreement between the neural networks technique and the other traditional methods. The numbers from (1)-(6) in table 4 indicate the same definitions as that in table 3.

4. CONCLUSIONS

The safety factor in slope stability problems usually takes exhaustive effort and not uniquely defined using the finite element and other traditional methods. Back-propagation neural networks showed high suitability for finding the safety factors to slope stability problems. This method has good capability to learn complex associations between input and output patterns. Best performance of the neural network is highly dependent on the setting configuration and the learning examples. The neural networks technique is a direct way for finding the safety factor and does not depend on the shape of the slip surface. Based on the results of this research, the neural networks can provide a very good estimate of the safety factor.

REFERENCES

Arai, K. & K. Tagyo 1985. Determination of noncircular slip surface giving the minimum factor of safety in slope stability analysis. *Soils and Foundations, Vol.25, No.1, pp.43-51.*

Elazouni, A.M., I.A. Nosair, Y.A. Mohieldin & A.G. Mohamed 1997. Estimating resource requirements at conceptual design stage using neural networks. *ASCE J. comp. civil engg. 11(4): 217-223.*

Huang, S.L. K. Yamasaki 1993. Slope failure analysis using local minimum factor of safety approach. *ASCEJ. Geotech. Engg. Vol.119, No.12, pp.1974-1987.*

Indraratna, B., A. Balasubramaniam & S. Balachandran 1992. Performance of test embankment constructed to failure on soft marine clay. *ASCEJ. Geotech. Engg. Vol.118, No.1, pp.12-33.*

Kim, J.Y. & S.R. Lee 1998. An improved search strategy for the critical slip surface using finite element stress fields. *Computers and Geotechnics, Vol.21, No.4, pp.295-313.*

Liang, R.Y., J. Zhao & S. Vitton 1997. Determination of interslice force in slope stability analysis. *Soils and Foundations, Vol.37, No.1, pp.65-72.*

Rosenblatt, F., 1958. The perception: A probabilistic model for information storage and organization in the brain. *Psychology Rev., 65(6), 386-408.*

Rumelhart, D.E., & J.L. McClelland 1986. Parallel distributed processing, *Vols 1 and 2, MIT Press, Cambridge, Mass.*

Salem, T. & A.A. Al-Tuhami. 1998. Finite element analysis of slopes. *J. Egyptian Soc. Geotech. Engg.* 13: 40-49.

Smith, I.M. 1994. Validation of finite element solutions. *XIII ICSMFE, New Delhi, India, 229-231.*

Smith, I.M. & D.V. Griffiths 1988. Programming the finite element method. *John Wiley & Sons, London.*

Teh, C.I., K.S. Wong, A.T.C. Goh & S. Jaritngam, 1997. Prediction of pile capacity using neural networks. *ASCE J. comp. in civ. engg. 11(2): 129-138.*

Yeh, Y.c., Y.J. Kuo & D.S. Hsu 1993. Building KBES for diagnosing PC pile with artificial neural network. *ASCE J. comp. in civ. engg. 7(1): 71-86.*

Zienkiewicz, O.C. 1971. The finite element method in engineering science. *McGraw Hill, New York.*

Slope Stability Studies of Vyasi Dam Reservoir Area, Garhwal Himalaya, U. P. India

R. ANBALAGAN, N. C. N. SRIVASTAVA, Department of Earth Science, University of Roorkee, Roorkee, India and VIKRANT JAIN, Engineering Geology Group, Department of Civil Engineering, Indian Institute of Technology, Kanpur, India

ABSTRACT

The Vyasi dam in Garhwal Himalaya, India envisages the construction of a 61m high concrete dam. The stability studies of slopes around the dam reservoir have been taken up to identify hazard prone slopes. For that purpose geological mapping and Landslide Hazard Zonation mapping of the reservoir area were carried out on 1:50, 000 scale. Percentage distribution of various sub-categories of individual causative factors in the study area have also been studied. Landslide Hazard Zonation map shows seven high hazard facets.

Key words : dam, landslide, reservoir, stability

INTRODUCTION

The stability of hill slopes forming the reservoir rim is an important consideration in evaluating the life and performance of dam reservoirs. Since landslide in the reservoir rim area would reduce the effective storage capacity, stability analysis of slopes in the vicinity of reservoir rim has been carried out. Initial investigations have been done on regional scale to cover large areas in a relatively short time. For that purpose, Landslide Hazard Zonation (LHZ) mapping has been carried out using an empirical approach.

GELOGICAL SETTING OF THE AREA

The reservoir and its vicinity lies in Garhwal-Kumaun Lesser Himalaya. The reservoir area has been mapped on 1:50000 scale to establish the lithological and structural setting. The rocks of Chandpur Formation and Mandali Formation of Jaunsar Group (Precambrian) are exposed in the study area, which are intruded by basic intrusives. Mandali Formation, the lower most unit consists of the massive quartzites and phyllites with thin band of quartizite. Basic intrusives show sharp contact with Chandpur Formation which consists of phyllites, quartzitic phyllites and massive quartzite intercalated with calcareous slates.

General strike of beds is in N40^0 to 75^0W-S40^0 to 75^0E direction with a dip of 70^0-80^0 in NE direction. This trend has been changed locally at some places due to presence of small scale folds in the area.

LHZ MAPPING OF THE RESERVOIR AREA BASED ON LANDSLIDE HAZARD EVALUATION FACTOR (LHEF) RATING SCHEME

Landslide Hazard Zonation mapping is a macro-zonation numerical approach done on 1:50,000 or 1:25,000 scale and numerical values depend on factors of slope instability. A Landslide Hazard Zonation (LHZ) map depicts division of land surface into zones of varying degree of stability based on the estimated significance of the causative factors in inducing instability (Anbalagan, 1992). It is done using a numerical rating scheme called LHEF rating scheme .

LHEF Rating Scheme

The Landslide Hazard Evaluation Factor (LHEF) rating scheme (Anbalagan, 1992) is based on major inherent causative factors of slope instability namely lithology, structure, slope morphometry, relative relife, land use and land cove and hydrogeological conditions. The maximum LHEF rating for different categories are determined on the basis of their estimated significance in causing instability. The number of 10 indicates the maximum value of the Total Estimated Hazard (TEHD). The detailed rating scheme includes various sub-categories of major causative factors (Anbalagan, 1992). These rating values have been tested in a number of localities in parts of Kumaun and Garhwal Himalaya of India (Anbalagan, 1992; Anbalagan et al., 1993). The LHZ map is prepared on the basis of distribution of TEHD.

Landslide Hazard Zonation (LHZ) Mapping of Vyasi Dam Reservoir Area

The Landslide Hazard Zonation (LHZ) map of the area has been prepared based on Landslide Hazard Evaluation Factor (LHEF) rating scheme (Anbalagan, 1992). For the preparation of LHZ maps, the whole area is divided into a number of segments called facets. Initially, a facet map (Fig. 1) has been prepared, which has been used as the base map for thematic maps. In order to evaluate the status of causative factors, other thematic maps such as lithological map, structural map, slope morphometry map, relative relief map, landuse and landcover map, hydrogeological condition map have been prepared using the detailed LHEF rating scheme.

Lithology

Slate-phyllite interbedded with quartzite covers most of the area. These rocks are moderate to highly weathered. Calcareous slate, which occurs in small portion of reservoir area are slightly to moderately weathered. Their resistance is comparatively higher. Pinkish-brown and white coloured massive quartzite occurs in a narrow band. It is compacted and hard in nature. Dolerite intrusives occur in eastern part of area(Fig.2). The soil type of the area is debris in nature, containing finer materials mixed with rock blocks of varying size. The soil is about 5 to 10m thick. The percent distribution of different rock types is given in Table 1 and facet wise distribution of LHEF ratings for lithology is given in Table 2.

Structure

The structure map has been prepared after plotting the attitude of observed discontinuities

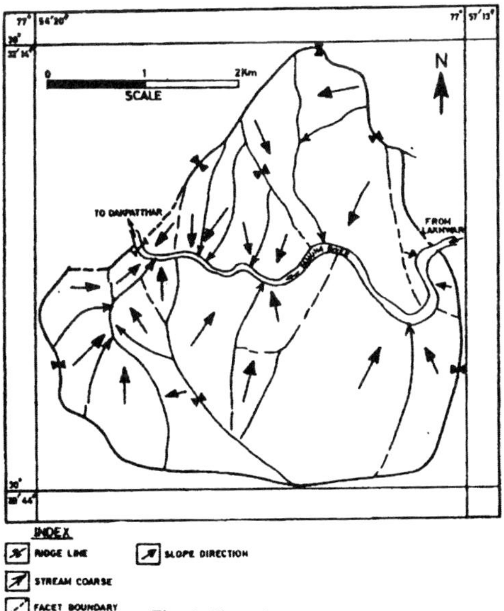

Fig. 1 Slope facet map

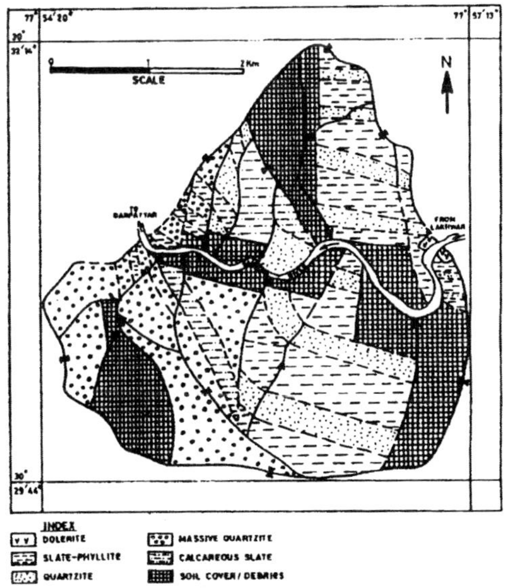

Fig. 2 Lithological map

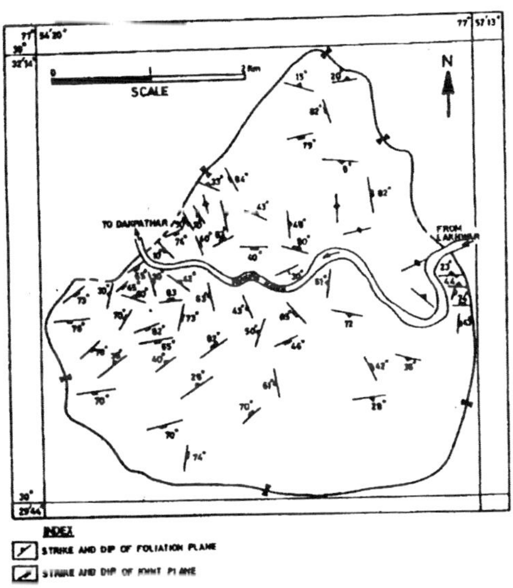

Fig. 3 Structural map

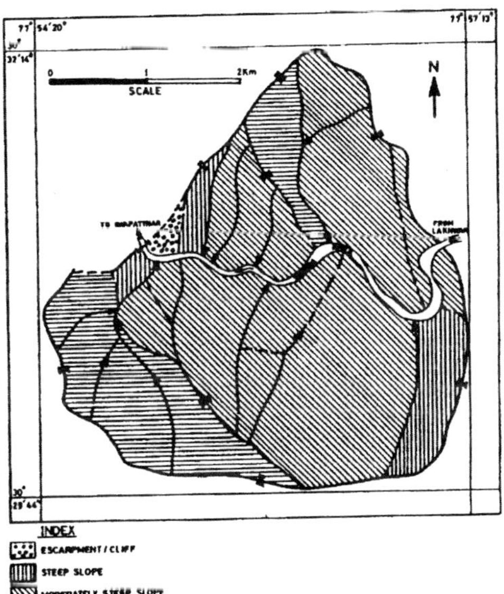

Fig. 4 Slope morphometry map

Table 1 : The percent distribution of the various sub-categories of individual causative factors in the study area

Causative Factors	Categories	Percentage of area in (%age)
Lithology	Soil Cover/Debris	30.36
	Dolerite	0.52
	Slate-phyllite	25.72
	Quartzite	14.82
	Massive quartzite	22.23
	Calcareous slate	6.69
Slope	Very gentle slope	-
	Gentle slope	15.60
	Moderately steep slope	73.94
	Steep slope	9.34
	Escarpment/Cliff	1.02
Relative Relief	Low	-
	Medium	13.17
	High	86.83
Landuse & Landcover	Cultivated/Populated land	5.63
	Dense forest	40.15
	Moderately vegetated area	30.09
	Sparsely vegetated area	12.15
	Barren land	0.38
Hydrogeology	Dry	17.14
	Damp	75.36
	Wet	6.39
	Dripping	1.02
	Flowing	0.06

Table 2 : Facetwise distribution of LHEF rating for various causative factors and total estimated hazard (TEHD) values

Facet No.	Slope Morpho-metry	Litho-logy	Relative Relief	Land-use & land-cover	SR*	Hydro-geology	THED value	Hazard zone
1.	1.2	0.91	0.6	0.8	1.2	0.6	5.71	MH
2.	1.7	1.0	1.0	0.8	0.85	0.2	5.55	MH
3.	1.2	0.96	1.0	0.76	1.3	0.2	5.42	MH
4.	1.2	1.24	1.0	1.04	1.3	0.32	6.10	HH
5.	1.2	1.2	1.0	0.8	1.0	0.0	5.20	MH
6.	1.2	0.96	1.0	0.8	0.65	0.16	4.77	LH
7.	1.2	0.96	1.0	1.2	1.65	0.2	6.21	HH
8.	1.2	0.5	1.0	1.2	1.0	0.18	5.08	MH
9.	0.8	0.2	0.6	0.8	1.0	0.05	3.55	LH
10.	0.8	1.0	1.0	0.8	0.85	0.1	4.55	LH
11.	0.8	0.44	1.0	1.0	1.45	0.2	4.89	LH
12.	0.8	0.2	0.6	1.2	1.0	0.2	4.00	LH
13.	1.7	1.0	0.6	1.2	1.9	0.4	6.80	HH
14.	2.0	1.0	1.0	1.7	1.0	0.2	6.90	HH
15.	1.7	1.0	1.0	1.5	1.6	0.2	7.00	HH
16.	1.2	1.0	1.0	1.2	1.0	0.2	5.60	MH
17.	1.2	1.0	1.0	1.2	1.4	0.2	6.00	HH
18.	1.2	1.0	1.0	1.5	1.6	0.2	6.50	HH
19.	0.8	1.0	1.0	1.2	0.85	0.2	5.05	MH
20.	1.2	1.02	1.0	1.2	0.8	0.3	4.72	LH
21.	1.2	0.91	1.0	1.4	1.0	0.1	5.61	MH
22.	1.2	0.5	0.6	1.2	1.0	0.1	4.60	LH

SR* = Structural Relation with Slope

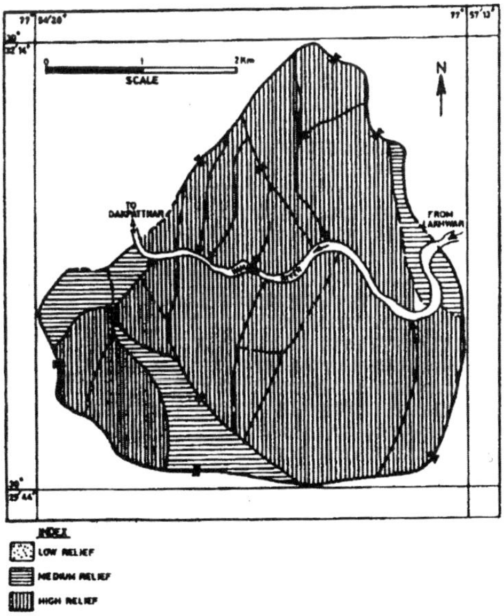

Fig. 5 Relative relief map

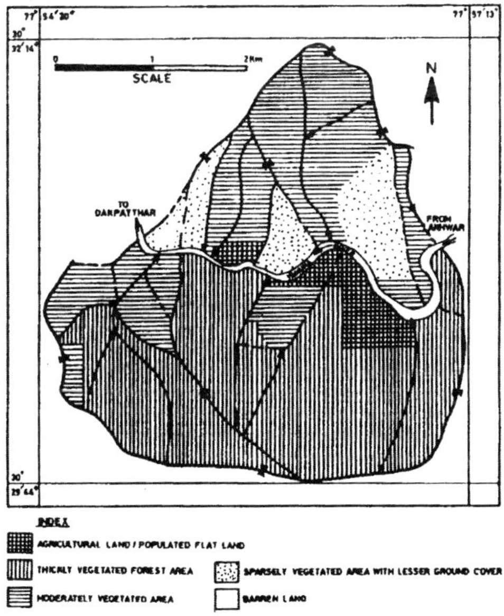

Fig. 6 Landuse and landcover map

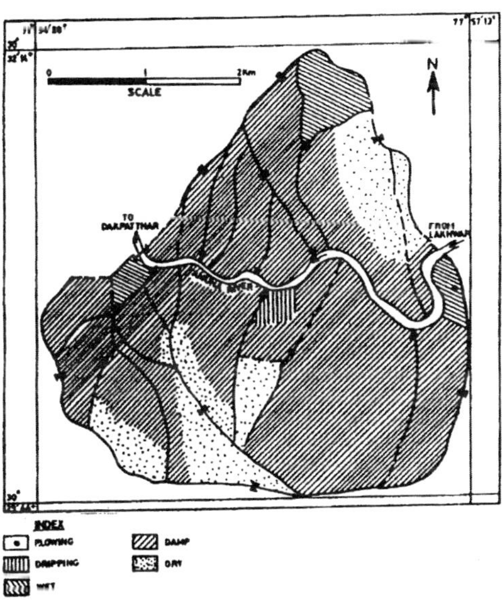

Fig. 7 Surface moisture map

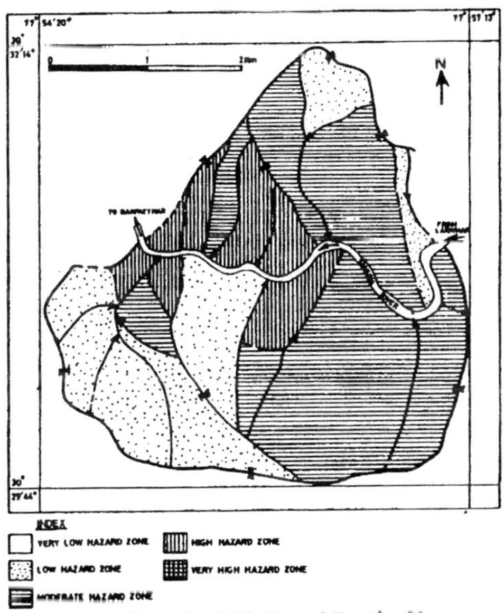

Fig. 8 Landslide Hazard Zonation Map

in the reservoir area (Fig. 3). The LHEF ratings for structure are awarded based on the relationship of slopes with attitudes of discontinuities (Anbalagan, 1992). In case of soil, the rating has been provided based on the depth inferred.

The rocks are highly jointed. The joint data is variable from place to place due to folding. The bedding and joint data has been collected facet wise and then plotted on the equal angle stereonet for obtaining preferred orientation of joint sets. The most unfavrouable joint(s) for individual facets has been determined and analysed with a steronet to find out the LHEF rating for structure.

Other Contributory Factors

Various maps pertaining to different contributory factors namely slope morphometry, relative relief, landuse and landcover and hydrogeological condition were prepared and LHEF ratings were awarded as per site conditions.

Landslide Hazard Zonation

The sum of all causative factors within an individual facet gives the Total Estimated Hazard (TEHD) for a facet. The TEHD indicates the net probability of instability within an individual facet . Based on the TEHD value,'facets are divided into different categories of hazard.

CONCLUSION

The three categories of hazards mamely, low hazard (LH), moderate hazard (MH), high hazard (HH) are found to be present in the area of study (Fig. 8; Table 2). Low hazard faces constitute the 31.46% of the study area. These facets are mainly concentrated towards the S-W part of the reservoir area. Moderate hazard facets which cover the 52.56% of the study area and well distributed over the region. Remaining facets (15.98%) are high hazard facets which are concentrated towards the downstream side of reservoir area. High hazard facets are potential sites for landslides, which should be further studied in detail for evolving proper remedial measures. .

REFERENCE

Anbalagan, R. : 1992, Landslide Hazard evaluation and zonation mapping in mountainous terrain, Engineering Geology, 32, 269-277.

Anbalagan,R., Singh, B. and Sharma, S. : 1993, Some aspects of Environmental impacts of dam reservoir in Himalaya region, Proc. Int. conference on Environmental Management Geo-Water and Engineering Aspects, Wollongong, Australia, pp. 735-739.

Sentinel Landslide, Dam Breach, and Road Reconstruction Zion National Park, Utah

S. A. ANDERSON, J. L. WILLIAMS, URS Greiner Woodward Clyde, Landslide Technology Group, Denver and S. GREENWELL, Federal Highway Administration – CFLHD, Denver, USA

INTRODUCTION

Zion National Park is located in southwestern Utah. The park contains deep canyons with steep walls and cliffs of sandstone, the largest of which is Zion Canyon. In April of 1995, the Sentinel landslide caused damming of the river and flooding along the Valley Floor Highway. The flooding caused the complete erosion of 180 m (590 ft) of the highway and damage to underground utilities buried beneath the road. The damage left the central part of the Park and Zion Lodge inaccessible, with 450 people isolated. Road restoration has involved several stages and has proved to be very difficult because of the physical constraints of the site, the need to restore public access quickly, and the interest to maintain a "natural" appearance in the National Park.

GEOLOGIC CONDITIONS

Zion Canyon is a canyon about 900 m (3000 ft) deep, cut through the relatively flat lying Navajo Sandstone, Kayenta Formation and Moenave Formation by the North Fork of the Virgin River. Landsliding has been an important part of the erosional process of the canyon and landslide deposits cover many parts of the lower valley slopes. A large prehistoric landslide, identified as a Toreva block by Wieczorek and Schuster (1995), originated from the Sentinel, on the west side of the valley about 4000 years ago (Hamilton, 1992). The prehistoric landslide deposit has been mapped for more than 2000 m (6000 ft) along the valley, and for 200 m (600 ft) or more above the current valley bottom on both sides of the valley (Grater, 1945; Hamilton, 1987, 1992). The prehistoric landslide deposits include a wide variety of materials, from clayey silt to large blocks of Navajo Sandstone, shown at the top of the lower slope in Figure 1.

The landslide dammed the river, forming a large lake, before the dam was breached by the river about 3600 ± 400 years ago (Hamilton 1992). Lake sediments of more than 25 m (75 ft) thickness have been located upstream of the dam (Grater, 1945). The eventual breach of the landslide dam and the draining of the lake eroded the landslide into two parts that are now separated by the river (Grater, 1945; Hamilton, 1987).

The Sentinel landslide is located in the middle of the prehistoric Toreva block and is shown on Figure 1. An outcrop of the Moenave Formation immediately downstream of the Sentinel

Landslides in research, theory and practice, Thomas Telford, London, 2000.

Figure 1. Southerly view of 1995 Sentinel landslide (Photograph taken by D. Grimes)

landslide is structurally in place, and appears to have been overridden by this prehistoric landslide. The Moenave Formation contains mudstone and thin gray lake clays, which have lower strength than the sandstone and siltstone, tend to perch water, and could contribute to instability of the valley slopes. Beds of this type outcrop on the east side of the road, upstream of the site, and they may have influenced the prehistoric and recent landslides. Portions of the Moenave Formation may have been part of the prehistoric landslide block and may have been transported to an elevation beneath the current valley floor during sliding.

The Moenave Formation outcrop separates two areas where the prehistoric landslide debris has been remobilized and it has created a seep on the north (down dip) side of the outcrop. The seep apparently flows year round, as indicated by wet ground throughout the year. The seep likely indicates the approximate location of groundwater within the portion of the slope that failed in 1995.

SENTINEL LANDSLIDE HISTORY

Landslides have occurred from within the prehistoric landslide debris in 1923, 1941 and most recently in 1995. Each of these landslides moved a large volume of material rapidly enough to have dammed the river and caused overtopping and damage to the park access road. The 1941 landslide blocked the Virgin River with a dam approximately 6 m high for 17 minutes before the dam was breached (Wieczorek and Schuster, 1995). The subsequent floodwater washed away a 150 m section of highway, eroded the highway subgrade to a depth of 9 m, tore out a water line, and damaged an irrigation dam downstream.

On April 12, 1995, during spring flooding of the North Fork of the Virgin River, another landslide dammed the river and created a temporary lake upstream of the landslide. Photos from Grater (1945) show that the 1995 landslide occurred in the same location as the 1941 landslide. Cumulative precipitation measured at Zion Park Headquarters for the months of January, February and March of 1995 totalled 386 mm, which is well above average (Hamilton, 1992). The 1995 landslide is believed to be a remobilization of most of the 1941 landslide mass resulting from erosion of the toe by the river and elevated water pressure. In general, the debris moved across the valley and impacted a mortared rock retaining wall that was across the river, between the river and road. A wedge of material moved only partially and is now isolated from the main landslide mass by an intermediate scarp. The toe of the landslide abutted the wall where the river used to be and formed a dam.

The lake that was created by the landslide backed up about 365 m upstream and was not more than about 25 m (80 ft) wide. Based on hydrograph readings, Wieczorek and Schuster (1995) estimated partial failure of the landslide dam occurred in about 1.5 hours, when 3 m (10 ft) of the approximately 6 m (20 ft) high dam was eroded. The landslide debris adjacent to the wall was higher than the road surface, so the river was directed over the road and the resultant flooding of the area caused about 180 m (590 ft) of the Valley Floor Highway to be washed out along with buried power, telephone, sewer and water lines.

RESTORATION OF VALLEY FLOOR HIGHWAY

Three phases of road reconstruction have been undertaken since the 1995 landslide. The work has been directed by the Federal Highway Administration and its consultants, in coordination with the National Park and the state of Utah. Due to the economic importance of the park to this region of Utah and the significance of this road in the park, a temporary road was constructed within about one month. This allowed the park to open the route for tourist traffic through the summer. After the peak tourist season the road was reduced to one lane for public access, and construction of the permanent road commenced. The reconstructed road was completed in 1996 but did not last long. Flooding in 1998 caused no damage to the landslide or its toe buttress, but caused failure of the reconstructed road and indicated the need for further measures of protection. These measures were implemented in 1999.

Temporary Road Construction

The temporary road was opened by cutting the toe of the east valley slope to provide access for two narrow lanes of traffic. The cut width had to be limited because the east valley slope is comprised of prehistoric landslide debris, of generally low strength. A concrete barrier and narrow ditch were used to catch rockfall ravelling from the slope. The slope was adequately stable for the short term but not for the long term without providing increased width for the ditch and shoulder.

1996 Road Reconstruction

In the fall of 1995 work commenced on a permanent repair. The design of the permanent repair was based on many criteria, some of them conflicting. For example, the desire to preserve the landslide mass as much as possible in its current form and let the river erode its toe naturally and the desire to protect and control the landslide so that catastrophic events, such as the 1995 landslide, do not occur here in the future. Access to the landslide for subsurface investigation would be difficult and would cause lasting impact to the landslide mass, so no subsurface investigation was conducted on the landslide (west) side of the river at this time.

Test holes were drilled from the temporary road surface on the east side of the river. The test holes showed that the subsurface conditions of the east side of the river (part of the prehistoric Toreva block) are highly variable and include silt, sand, cobbles and boulders. It appeared that boulders might be several meters across, possibly consisting of ledges of bedrock that remained intact during the prehistoric Toreva block landslide.

Several design alternatives were considered, including a no action plan, excavating a portion of the landslide toe with a near vertical cut and restraining the landslide material with a large retaining wall, constructing a viaduct to span the site, realigning the road farther to the east, and draining the landslide mass. Although each of these alternatives satisfied some of the design criteria, they were dismissed due to the importance of the criteria they did not address. Without the benefit of subsurface investigation during the design phase, it was decided the repair would impact the landslide only enough to armour the toe with a rock buttress, and reconstruction would focus on measures on the east side of the valley.

The selected repair consisted of a rock buttress at the landslide toe and about 145 m (475 ft) retaining wall on the east side of the valley to support the roadway about 6 to 8 m (20 to 25 ft) above the river. With special attention to construction scheduling, this option satisfies the majority of the design criteria, with the exception of significantly improving the hydraulics of the river. The rock buttress provides a resisting mass against landslide movement and protects the slope from river scour. The selected retaining wall consists of pre-cast concrete T shaped units. It resembles a crib type wall, but its design and function are based on mechanically stabilized earth (MSE) principles.

The pre-cast concrete panels could be installed quickly and could withstand impact of high flow velocities and bed load. To protect against scour at the toe, the wall was embedded approximately 1.5 m (5 ft) below the channel bottom and the toe lined with riprap boulders ranging 1 to 1.2 m (3.3 to 4 ft) in diameter. To ensure free draining conditions following flash floods, clean granular backfill was used. The lowest course of pre-cast units was founded on a non-structural concrete levelling pad at the wall face and on native materials at excavation grade (generally, sand with cobbles and boulders), elsewhere.

Figure 2. Soil nail wall construction

Site conditions and the need for maintaining public access to Zion Canyon and Lodge called for an engineered temporary shoring system during construction. A soil nail wall consisting of approximately 345 soil nails of 5.8m (19ft) length, installed at 1.5m (5ft) spacings, was used to maintain one lane of safe public access during excavation and construction of the permanent wall. The soil nail wall facing consisted of wire mesh and shotcrete, as shown on Figure 2.

All major construction activities were completed in time to open the two lane roadway to public traffic by the Memorial Day holiday weekend, 1996, with paving and aesthetic restorations completed immediately thereafter. The project was generally considered a success. The design and construction of the repair was completed on schedule through coordination by both the private and public sectors including Zion National Park, National Park Service, United States Geological Survey, Utah State Historic Preservation Office and Utah State Engineer's Office (Mentz et.al., 1997).

ROAD COLLAPSE

On September 11, 1998, flooding of the river caused damage to the two year old retaining wall. The flooding, which involved a peak discharge of about 120 to 130 cubic meters per second (4,200 to 4,500 cubic feet per second), scoured riprap from the toe of the wall and, apparently, scoured the wall foundation. At certain locations, the wall backfill collapsed into the scoured foundation and voids progressed upwards to the road surface, which then collapsed, as shown on Figure 3. The process of scour and backfill collapse may have happened entirely during the September flood or it may have started during the smaller floods that preceded the September event, only to become evident during this flood.

Figure 3. Road damage from flood on September 11, 1998

The extent of scour and voids in the wall backfill was investigated by drilling from the road surface and probing along the toe of the wall, where accessible. A large portion of the wall was not reachable because of deep, swift current, even during low flow. The investigation showed additional voids in the wall backfill that had not yet propagated to the ground surface and caused failure of the pavement, and the investigation also showed the presence of sediment material with little resistance to scour for at least 3 to 5 m (10 to 15 ft) below the base of the wall. Based on the level of damage to the wall and uncertainty regarding the ability to stabilize the foundation and backfill in-situ, it was decided to deconstruct the majority of the wall as part of the repair.

Alternatives for providing further protection to the wall were considered. These alternatives included providing a scour protection wall at the base of the retaining wall, constructing grade control structures within the channel, and widening the channel to avoid directing flow directly at the wall and decrease flood depth and velocity. Once again, the selected design had to meet the criteria of the National Park, the Federal Highway Administration, the State of Utah, and the environmental compliance process. These criteria included completing construction by the Memorial Day holiday (within 8 months) and not crossing or working substantially within the river prior to completing the environmental permitting process.

To satisfy the design criteria, a secant pile wall was constructed at the toe of the retaining wall. The construction sequence was as follows: (1) deconstruct about 90% of the retaining wall, stockpiling elements for later use, (2) construct a steel rein-forced secant pile wall at approximately the river level, along the toe of the retaining wall (Figure 4), (3) construct foundation support piles in the foundation of the retaining wall, (4) construct an apron of riprap (0.5 to 1.25m (2 to 4ft) size diameter) in front of and covering the top of the secant pile wall, and (5) reconstruct the retaining wall. The reconstructed wall was de-signed to withstand 4m (12ft) of scour and was completed on schedule, before the Memorial Day holiday, 1999.

Figure 4 Soil nail wall construction

In early September, 1999 the river flooded again, reaching a peak discharge of about 65 cubic meters per second (2,300 cubic feet per second) (Sharrow, 1999). This flood washed out some newly placed riprap along the toe of the T wall and exposed a maximum of 1.5 to 2 m (4 to 6 ft) of the secant pile wall. The wall performed as designed, but the apparent frequency of flood events causing significant scour of riprap along the wall and scour of channel in front of the wall suggests that without further measures, or considerable periodic maintenance, the life of the wall could be limited. In anticipation of this, the environmental permitting process was initiated during the design and construction of the secant pile wall to permit additional measures such as widening the channel by excavation of a portion of the landslide toe and/or installation of natural (riprap) grade control structures. It is anticipated that some combination of these measures will be constructed in the Spring of 2000.

CONCLUSIONS

Scour at the toe of the retaining wall was a design issue recognized as requiring attention during the landslide repair design in 1995 (Mentz et.al., 1997). However, floods with return intervals of less than 10 years have caused significant damage and it is now apparent sufficient measures were not taken in the original design to protect the retaining wall foundation from scour and undercutting. The active scour and downcutting of the channel through this reach, as compared to adjacent reaches of the river is believed to be not only due to the narrow, steep channel, but also due to the process of landslide dam breach. When the 1995 landslide dam was breached, the channel grade was left far from equilibrium. Mitigation measures such as the wall designed here, rock buttresses, or other measures impacted by the channel grade need to consider the potential for active scour during relatively frequent events, when employed in the breach of a landslide dam.

REFERENCES

Grater, R.K. 1945. Landslide in Zion Canyon, Zion National Park, Utah: Journal of Geology, V. 53, pp. 116-124

Hamilton, W.L. 1987. Geological Map of Zion National Park, Utah: Zion Natural History Association, Zion National Park, Springdale, Utah, Scale 1:31,680

Hamilton, W.L. 1992. The Sculpturing of Zion – Guide to the Geology of Zion National Park: Zion Natural History Association, Zion National Park, Springdale, Utah, p. 132

Mentz, K.M., Worrell, E., Zanetell, F.D. 1997. Park Project is a Paragon of Partnership: Public Roads, U.S. Department of Transportation, Washington D.C., Vol. 60, No. 4, pp. 13-18

Sharrow, D. 1999. Zion National Park Hydrologist, personal communication

Wieczorek, G.F., Schuster, R.L. 1995. The 1995 Landslide in Zion Canyon, Zion National Park, Utah, U.S.A.: Landslide News, Japan Landslide Society, Tokyo, Japan, No. 9, pp. 28-31

The Landslide at Vagnhärad in Sweden

H. ANDERSSON, P-E. BENGTSSON, C. BERGLUND, R. LARSSON, Swedish Geotechnical Institute, Linköping, Sweden, G. SÄLLFORS and A-L. ÖBERG-HÖGSTA, Chalmers University of Technology, Göteborg, Sweden

INTRODUCTION

In the middle of the night on May 23, 1997, a large landslide occurred in a residential area in Vagnhärad about 70 km south-west of Stockholm. The slide took place in a clay slope and finally encompassed a 200 m long stretch along a small river, Trosaån, reaching about 60 m up the slope. During the slide seven one family houses were destroyed or undermined. The large slide had been preceded by observed slide movements within a smaller area at the toe of the slope and a smaller initial slide covering an area of meadow land of 50*30 m^2 close to Trosaån, see Fig. 1.

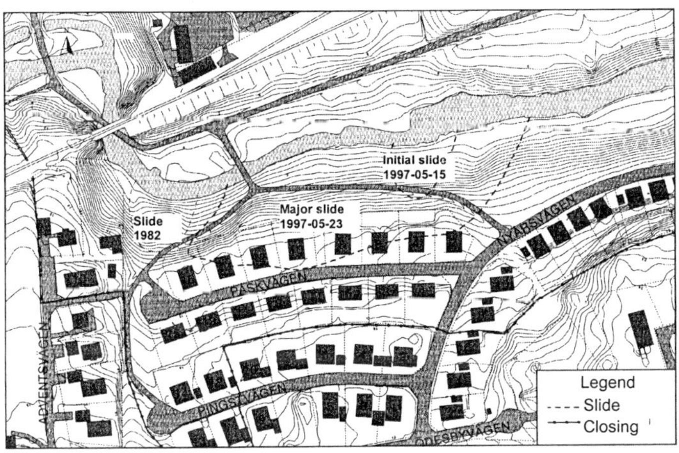

Figure 1. Plan over the residential area in Vagnhärad with the major landslide and the initial slide (Andersson et al. 1997).

After the initial slide the area was kept under observation and a few houses in the immediate surroundings had been evacuated. However, as an ultimate consequence of the main slide a total of 29 houses had to be abandoned and finally demolished.

Slope stability is a great problem in Sweden and each year numerous slides occur both in clayey and silty soils. The result of a stability analysis is normally a factor of safety, which in general is the base for decisions for further action. It is therefore of utmost importance that the geotechnical investigations give the right information for making the analysis as relevant as possible and that the uncertainty in the calculated factor of safety is minimised. In 1995 the Swedish Commission on Slope Stability within the Royal Academy of Engineering Sciences put forward recommendations for how to investigate the stability of a slope (IVA Skredkommission, 1995). These national recommendations are the result of ongoing research in Sweden during the last two decades. The guidelines given are structured in such a way that the investigations should be carried out step by step with increasing extent.

GEOTECHNICAL INVESTIGATIONS AND PROPERTIES

Topography and Stratigraphy

The slide area consisted of a long and fairly steep slope leading towards a watercourse. It had an inclination of about 1:15 in the upper part and about 1:5 closer to Trosaån. The total difference in elevation from the crest to the ground level at the toe of the slope was about 15 m and the erosion of Trosaån increased this difference by another 2 m.

The soil profile consisted of a dry crust of about 1-2 m followed by post glacial clay and glacial clay with varves and layers of silt to a depth of about 10 m. The silt layers increased with depth and gradually the soil transformed into silt with clay layers. The fine-grained soil layers were underlain by moraine and coarse silt and sand on bedrock, see Fig. 3. The clay was high plastic and medium sensitive with values of 15-25. The liquid limit varied between 50-80%, where the lower value represents the varved clay with silt. The bulk density was about 1.6 t/m^3 in the upper layers and increased with depth to 1.8 t/m^3.

Ground Water Pressures

The ground water situation in the area was rather complex. Hydrogeological investigations showed that the moraine at the bottom constituted a confined aquifer. Behind the clay covered slope, there were areas with outcropping rock or a covering of moraine at higher elevations, which constituted areas for infiltration by precipitation. In the slide area and in the adjacent areas, the clay acted as an almost impervious lid on top of the coarse soil, which created conditions for high water pressures in the coarse bottom layers. Consequently artesian pore pressures were measured in these layers close to Trosaån with an excess pore pressure head of 2-3 m above the ground surface. Unfortunately, no measurements were made in the upper part of the slope in the slide area before the slide took place. However, at the back of the slope the groundwater table probably fluctuated and varied between 1-3 m below the ground surface. The pore pressure situation was not obviously extreme at the time of the slide. However, during the week prior to the first observations of ground movements there was an unusually high rainfall for the season. Normally, the ground water levels are high at this time of the year in this region, and start to drop off in May. This year, the ground water remained high and even increased somewhat. This indicated that the aquifer was saturated and probably had a ground water level just below the ground surface in the upper part of the slope.

Stress History and Over Consolidation Ratio

The geological development of the area with a gradual erosion of the river through the soft sediments indicates that the clay layers at the bottom of the slope had earlier been exposed to a higher stress level than the present. The clay close to Trosaån had also been changed owing to time effects and creep. Results from constant rate of strain oedometer tests show over consolidation ratios of 4 -5 at depth of 2 and 4 m respectively. Further up in the slope, the clay was also overconsolidated, with OCR equal to 3.5-4. This was probably mainly caused by time effects and variations of the ground water table. In Fig. 3 an outline of the principal stress history is given.

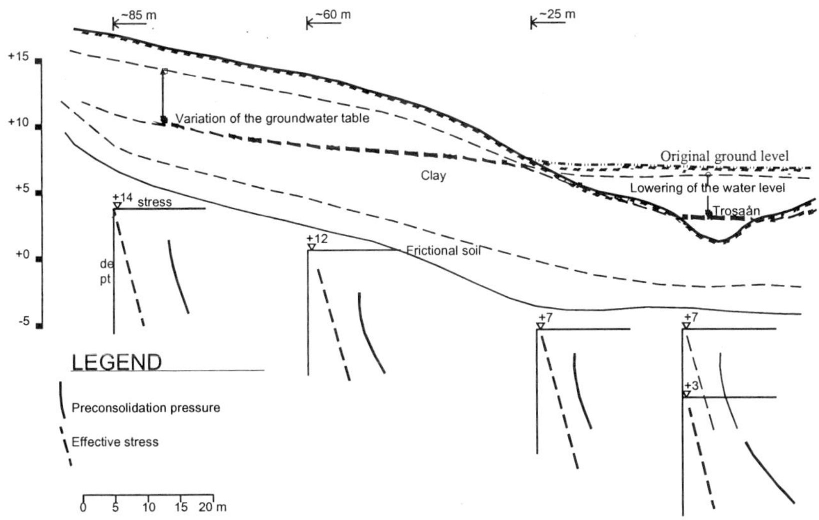

Figure 3. Typical outline of stratigraphy and probable stress history for the slope at Vagnhärad.

SHEAR STRENGTH

Undrained Shear Strength

The undrained shear strength has been determined with field vane tests and cone penetration tests *in situ* and in the laboratory with both fall cone tests, direct shear tests and triaxial tests. The shear strengths obtained from the fall cone and field vane tests have been corrected with respect to liquid limit (SGF, 1993). The undrained shear strength has also been estimated using empirical relations based on liquid limit and over consolidation ratio (Larsson et al., 1984). The stratigraphy in the area with varved clay with silt layers has greatly affected the evaluation of the shear strength for the different methods. Therefore, great care was taken when evaluating the strength parameters to account for this fact.

The undrained shear strength varies rather logically according to the stress history, which can be seen in Fig. 4. The results from CPT-, field vane- and direct shear tests showed a relatively good agreement, whereas the results from the fall cone tests generally showed lower values. The investigations with various methods and the comparison with empirical experience, illustrate the importance of an evaluation of the applicability of the various methods to the particular type of soil as well as the importance of the stress history when evaluating the shear strength of a slope. The shear strength, below the dry crust can approximately be expressed as 11 kPa at a depth of 2 m increasing with 2.1 kPa/m, but the variation in the stress history alters this general picture to some extent

Drained Shear Strength

The drained shear strength parameters can for Swedish soft clays, empirically be estimated as $\phi'=30^o$ and $c'=0.1\tau_{fu}$ or $0.03\ \sigma'_c$. Using the empirical relation $0.1\ \tau_{fu}$, the value of c' would be 1.5 to 2 kPa. The drained shear strength of the clay was determined with drained triaxial tests. The results varied somewhat, which was probably due to the varying silt content. In Fig. 5 the evaluated results from the triaxial test are shown together with the empirical relation. The test results indicate an average friction angle of ϕ′=31° and c′=3 kPa which is in fairly good agreement with the empirically estimated parameters.

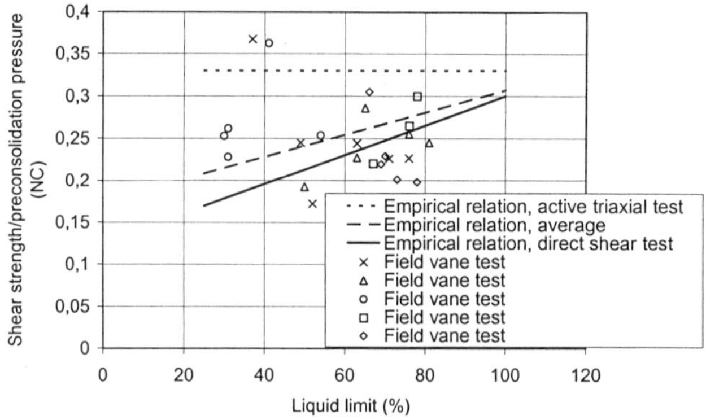

Figure 4. Comparison between shear strength determined with field vane tests and empirical relations.

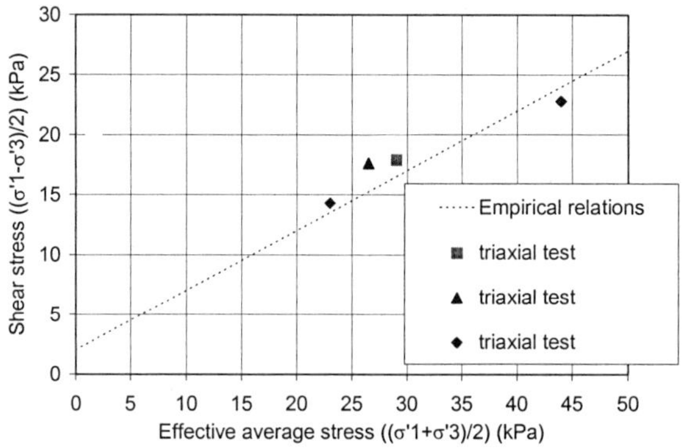

Figure 5. Results from drained triaxial tests compared to the estimated strength parameters according to the empirical relations.

SLOPE STABILITY ANALYSES

The calculations of the stability of the slope have been performed according to the recommendations given by the Swedish Commission on Slope Stability.

The shear strength used in the calculations was chosen according to the discussion in the previous section, that is in general about 11 kPa at a depth of 2 m and thereafter increasing with 2.1 kPa/m. However, consideration was also taken to the overconsolidation and this is reflected in the chosen variation in shear across the slope. Consequently, the shear strength versus depth is slightly higher than the general values close to Trosaån and at the top of the slope where the OCR is highest and slightly lower in the mid part of the slope where the OCR is lower. The drained shear strength of the clay has been chosen according to the empirical relations to ϕ′=30° and c′=1.5-2 kPa. The friction angle of the silt has been assumed to ϕ′=34° based on triaxial tests.

Since the pore pressure situation was rather complex, simulations were made with the computer program SEEP/W. The simulations were done in order to find a probable pore pressure distribution in the slope. They were made with different assumptions accounting for high artesian pressures close to Trosaån and in the bottom layers and varying pressure profile in the clay and at the top of the slope. The estimated most probable pressure contours are shown in Fig. 6. Most of the slope stability analyses have been performed using the computer program Slope/W. The undrained analysis indicated a factor of safety of 1.3 and the combined analyses in which the pore pressures are considered, gave factors of safety between 0.9-1.1, see Fig 6. The calculated probable slip surfaces coincide well with the observed initial slide.

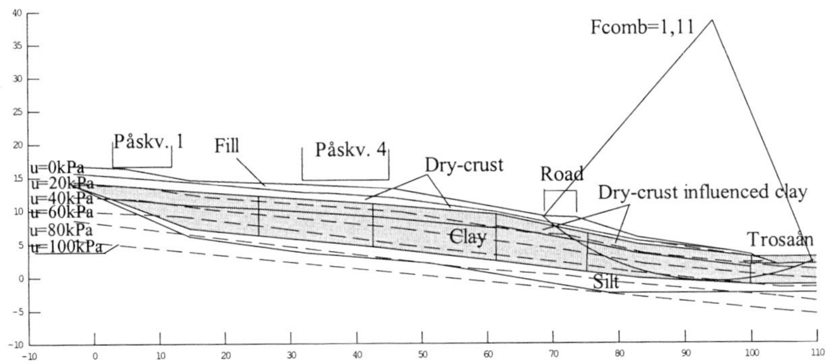

Figure 6 Calculated critical slip surface for the initial slide with combined analysis.

The calculations showed that the pore pressure situation at the river was decisive for the stability of the slope. Extensive analyses were therefore made with varying pore pressure profiles both in the upper and lower part of the slope. The ranges of chosen pore pressures were within the limits of what could be expected to have existed. The results given in Fig. 7 indicate that only minor changes in the conditions were sufficient to cause failure of the slope.

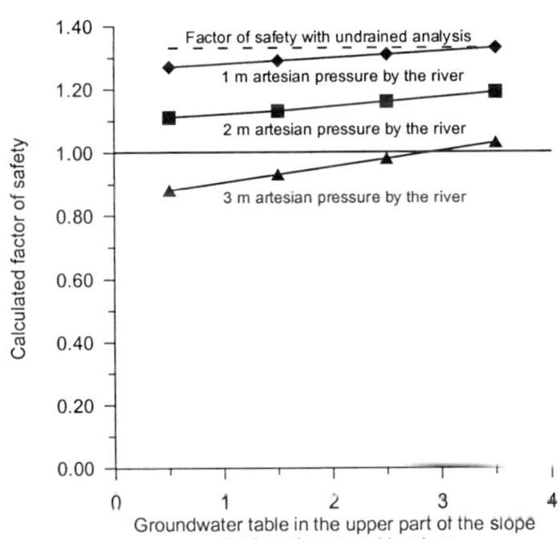

Figure 7. Variation of factor of safety with assumed ground water pressures both in the upper and lower part of the slope.

The stability of the soil masses in the upper part of a slope is dependent on the support given by the soil masses in the lower parts. Schematic calculations show that only a small reduction in the ability of the masses involved in the initial slide, to support the forces from the soil above, is required for the slide to be able to retrogress and reach its final extent.

The guidelines given by the Swedish Commission on Slope Stability require that certain criteria must be fulfilled in order to classify a slope as stable. Also after an extensive investigation, the factor of safety in an undrained analysis, F_c must be ≥1.3 á 1.4 and at the same time the factor of safety in a combined analysis must be ≥1.2 á 1.3. If these minimum values are to be accepted, restrictions for the use of the area have to be applied. These requirements were not fulfilled, but the safety factors calculated after the careful and extensive modelling of the geotechnical conditions and pore pressure situation were close to unity for the slide area. It can thus be concluded that the stability of the slope was very poor in its normal conditions and that only marginal changes, particularly in the pore pressures were sufficient to cause its failure.

CONCLUSIONS AND RECOMMENDATIONS

The careful investigation and modelling of the geotechnical conditions in the area, such as soil stratigraphy, shear strength, ground water situation and pore pressures, have resulted in calculated factors of safety close to unity for the slide area. It can be established that the main cause for the slide was that the slope had a poor stability and that only minor changes in the conditions were sufficient to cause its failure. Possible factors, which may have caused the failure on this particular occasion, are increased pore water pressures, due to the unusually heavy rain for the season. Other possible factors are that erosion and minor local slides had occurred in the river banks, that recurrent movements in the soil masses in the slope had brought a reduction in the shear strength and that the water level in the river was low for the period of time. The most probable cause for the initiation of the slide is a combination of two or more of these factors.

If the recommendations from the Commission on Slope Stability had been used, the slope would never have been built upon. However, the recommendations did not appear until 1995 and therefore several of the developments on clay slopes in Sweden may have been based on geotechnical investigations which do not correspond to the demand on quality and extent presented by the Commission. There is thus a considerable risk that there are other built-up slopes which, based on previous practice have been considered as safe but which in reality may have an unsatisfactory stability. Consequently, these investigations must be re-examined and supplemented. This mainly involves comparison of evaluated shear strength parameters with existing empirical knowledge, control of possible high pore pressures and artesian ground water pressures and calculation of stability with combined analysis. Also more attention should be paid to observations in connection with slopes, which may indicate instability or beginning ground movements.

REFERENCES

Andersson H., Bengtsson P-E., Berglund C., Larsson R., Sällfors G. & Öberg-Högsta A-L., 1998. The landslide at Vagnhärad. *Report No 56*, The Swedish Geotechnical Institute.

IVA Skredkommission, 1995. Recommendations for Slope Stability Analysis. *Report No 3:95*.

Larsson R., Bergdahl U & Eriksson L., 1984. Evaluation of Undrained Shear Strength in Cohesive Soils. *Information No 56*, The Swedish Geotechnical Institute.

SGF, 1993. Recommended Standard for Field vane Shear tests. SGF report 2:93.

An Analytical Approach to Obtain the Actual Groundwater Response Using Open Standpipe Piezometers

M-G. ANGELI, National Research Council, IRPI Institute, Perugia and P. GASPARETTO
IQT Consulting S.r.l., Rovigo, Italy

ABSTRACT

The paper deals with the interpretation of piezometric data recorded in open standpipes. Any piezometric change in slopes induces an instantaneous measurement error, the intensity of which depends on the rate of change and is inversely proportional to the soil permeability. This error shows itself in two different ways in any piezometric apparatus working with large transfer of water volumes. The first way is displayed with a time lag response of groundwater fluctuations; the second is showed by a flattening of the piezometric peaks. Starting from the static analysis on time lag responses in piezometers installed in different soil conditions, done by Hvorslev (1951), the Authors of this paper developed a numerical model based on the solution of the differential equation governing the groundwater inflow from the soil toward the open standpipes piezometers. The final result is the prediction of the actual transient groundwater conditions working in landslide bodies of homogeneous soils. The paper presents also a successfully application of this methodology in a real landslide case of sudden collapse, through the interpretation of data recorded in an open standpipe piezometer.

INTRODUCTION

The study of the behaviour of large landslides makes the extensive use of electrical pore pressure transducers, directly installed into the ground and non-removable, difficult and sometimes impossible. In fact, non-removable electrical piezometers guarantee a high degree of precision and punctual pore pressure readings, on the other hand, they present problems from a practical point of view. They are relatively expensive and not recoverable; it is practically impossible to check and correct possible instrumental drift; any instrument failure involves a definitive loss of the data recording at the selected location.

Hence, the advisability of using open standpipe piezometers equipped with electrical pore pressure transducers arises. This makes it possible to avoid the above disadvantages. But, in this last case, the measurements are affected by a significant time lag, that presents itself as a delayed equalisation of the piezometric level inside the pipes with respect to the actual water table change in the soil surrounding the open standpipe piezometers. In other words, the lag is in the time required for the equalisation of the pressure difference existing between the hole and the surrounding soil when the surrounding pore pressure increases or decreases. It is inversely proportional to the hydraulic conductivity of the soil and varies with the size and type of pressure measuring device. In practice, if the volume of water that is required to register a head fluctuation in a piezometer standpipe is large to the rate of entry at the intake, there will be a time lag introduced into piezometer readings (Figure 1 and 2).

Landslides in research, theory and practice, Thomas Telford, London, 2000.

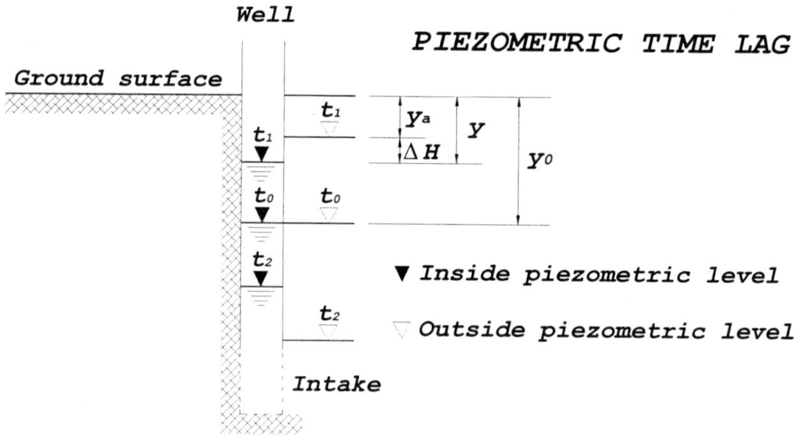

Figure 1. Scheme of an open standpipe piezometer

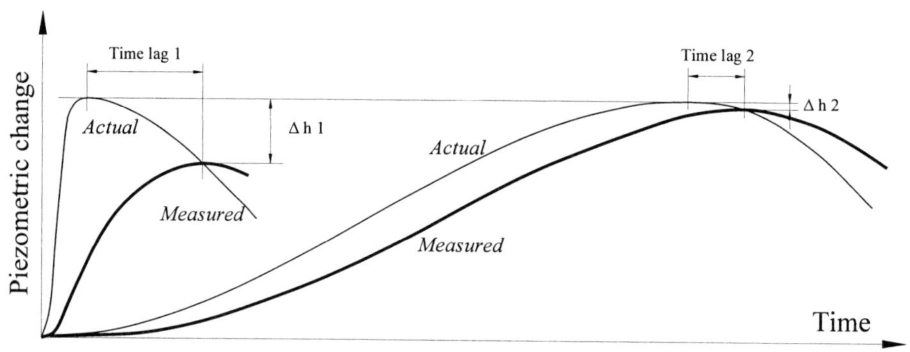

Figure 2. Time lag errors: time lag response and flattening of piezometric peaks

The larger the diameter of the open standpipe is, the larger is the volume of water required to reach the equalisation: in other words, the time lag is higher. This phenomenon is well known since many years (Hvorslev, 1951) and, thus any measurement of groundwater flow must take this into account, especially when very rapid transient flow phenomena occur (Figure 2). With the aim of interpreting these transient flow phenomena in slopes, which often cause sudden collapses, an analytical approach to the problem was used. Starting from the recorded values of piezometric levels in the open standpipes, a computer routine, which takes into account mainly the permeability of the soil and the geometry of the open standpipe, provided the necessary corrections of the data. The final result gave us the actual level and the trend of the piezometric fluctuations in slopes (Figure 2). The paper, besides dealing with the numerical model of data correction, presents a case history in which the Authors were able to fully understand the actual landslide mechanism at work during a collapse event.

THE NUMERICAL MODEL

Any device for the measuring of the pore pressure distribution based on the principle of transfer of water from the surrounding soil needs a finite time to operate. This finite time depends on the geometry of the measuring apparatus, the active head and the soil permeability. The only value of soil permeability (*k*) that must be taken into account is the one coming from in situ permeability tests. This choice guarantees that the utilised value will be representative of the permeability of the finest soil matrix and of the possible presence of cracks crossing through the soil mass.

The volume of water transferred in a infinitesimal time interval *dt*, from and toward the soil surrounding the piezometer (cross section area equal to *A*), can be expressed by the following equation:

$$q * dt = A * dy$$

where *q* represents the flow rate and *dy* is the water level change inside the open standpipe piezometer (Figure 1). According to Hvorslev (1951), the flow rate, for any shape of piezometer in an incompressible soil, may be expressed as follows:

$$q = - F * k * \Delta H$$

where *F* is the piezometer shape factor, *k* is the soil permeability and *ΔH* is the active head (Figure 1). The *F* factor depends on the kind and the geometry of the piezometric intake. Combining the two above equations, we get to the following differential equation:

$$\frac{dy}{y - y_a} = \frac{- F * k * dt}{A}$$

where y_a represents the actual piezometric level and *y* the water level inside the standpipe. The integration of the above differential equation, under the boundary conditions ($y = y_0$ when $t = t_0$), gives the actual groundwater table y_a:

$$y_a = \frac{y - y_o * e^{\frac{- F * k * (t - t_o)}{A}}}{1 - e^{\frac{- F * k * (t - t_o)}{A}}}$$

This equation leads to the actual value of the piezometric level, by inputting the values of the water level measured inside the open standpipe piezometer in a finite time interval. The accuracy of the results is strictly dependent on the time frequency of the sampling. The higher is the frequency of the sampling, the higher is the accuracy of the results.

THE CASE HISTORY

Hydrogeological and geotechnical investigations (Angeli, 1987; Angeli, 1992; Angeli et al., 1984; Angeli & Silvano, 1987; Angeli et al., 1987; Angeli et al., 1988; Angeli et al., 1989; Angeli et al., 1990) have been carried out since 1982 on a landslide located in Northern Italy, in the Dolomite Mountains, near Giau Pass, about 30 kilometres from Cortina d'Ampezzo. The landslide affected a steep slope formed of morainic deposits overlying a marl bedrock and limited by two streams.

The climatic conditions are the ones typical of high mountain climate, characterised by two distinct seasons with the temperature respectively keeping under 0°C or over 0°C. Two maximum values of rainfall depth can be observed in late spring or in early autumn. The maximum yearly value of rainfall depth can reach 1,200 mm, whereas the maximum yearly value of total precipitation depth can overcome 1,600 mm. The maximum snow precipitation mainly occurs in January and February, and the daily mean value of temperature stays under 0°C from December to March.

The evolution scheme of the slope is represented in Figure 3. The initial triggering condition was evidenced by the erosion occurring at the base of the slope by the deepening of the Codalonga stream, after the major 1966 flood which affected all of North and Central Italy. The final triggering condition that caused the definitive collapse was represented by the extremely rapid snow melting and the consequent rapid water filling of the natural tank located at the rear of the landslide body (the graben area).

The automatic instrumentation installed in the landslide body allowed the detailed recording of two critical events against the stability of the slope (Figure 4). In the first case (April 1987), displacements of a few centimetres were recorded, whereas in the second case (April 1988), the complete collapse of the landslide mass occurred with displacements up to 20 m. The recording of the only hydraulic and kinematic features of this last event (over a period of about 100 hours) is shown in Figures 4 and 5. The piezometric curve refers to an electric pressure transducer installed in open standpipe, whereas the displacement curves refers to a steel wire extensometric device installed in another equipped borehole.

The comparison between the two above critical events, especially as regards the recorded piezometric graphs (Figure 4), is not sufficient to justify the collapse event of April 1988. But after the application of the numerical method proposed in this paper (inputting a k = 10-7 m/s as the value obtained by in situ permeability tests in the morainic deposits, a shape factor F = 0.022 according to the Hvorslev theory and the measured piezometric data) it was possible to obtain the actual piezometric data that looked very different from the observed ones. As shown in figure 4, although the measured values of 1988 event appeared much lower than the measured values of 1987 event, the calculated piezometric levels followed a complete opposite trend. In fact, the calculated graph of 1988 rose far above the graph calculated for the event 1987. The reason of this different behaviour can be attributed to the faster rising of the observed piezometric graph of 1988 event. In other words, as mentioned before, the higher the observed velocity of water rising in open standpipes, the higher are the time lag and the flattening effects on the predicted (or actual) groundwater levels (Figure 2).

CONCLUSIONS

The usage of open standpipe piezometers is widespread all through the world, due to their simple way of working and their lower cost of management. In spite of that, although they often give acceptable results, sometimes they lead to a complete misleading of the transient groundwater phenomena occurring in natural slopes. However, it is always possible to obtain data leading to erroneous conclusions as regards the understanding of the mechanism at work in the landslide, even utilising sophisticated instrumentation.

With the method proposed, the Authors want to underline the importance of a thorough examination of the data obtained even by means of automatic data collectors. In fact, in the case of a trend of recorded piezometric data showing an extremely rapid change, special attention must be paid and the corrections proposed in this paper must be taken into account. Therefore, in special landslide cases (i.e., affecting inhabited areas), a dedicated numerical routine (resident inside the data collector system) could be implemented, in order to obtain automatic recordings of the actual hydraulic condition in slopes and to prevent sudden soil mass collapses.

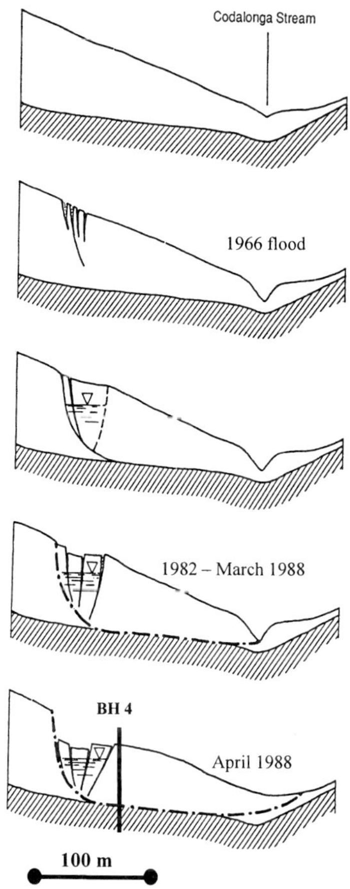

Figure 3. Evolution scheme of the landslide from the initial conditions to the definitive collapse (April 1988)

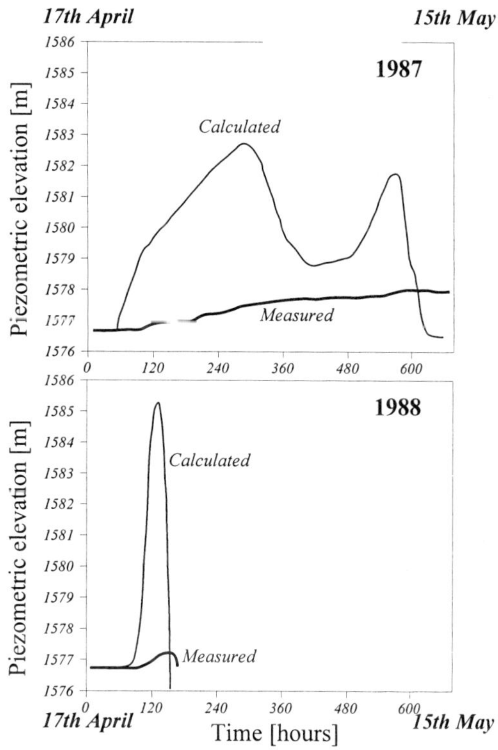

Figure 4. The critical events of 1987 and 1988: measured and calculated piezometric response

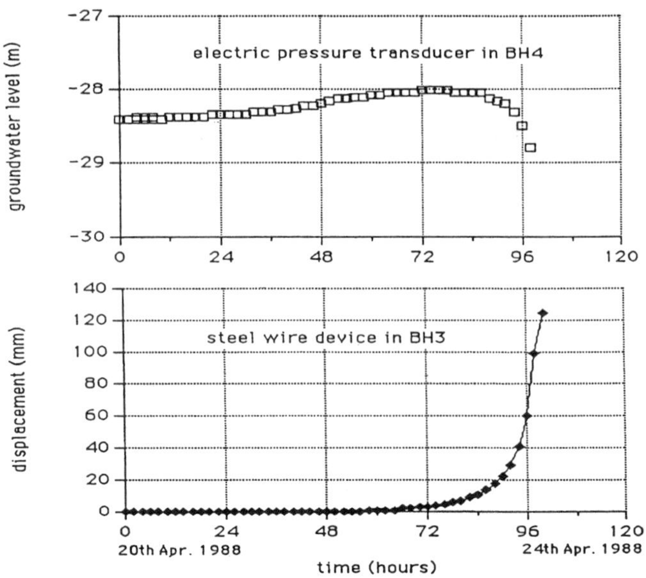

Figure 5. Piezometric and kinematic data automatically recorded during the collapse event (20th-24th April 1988)

REFERENCES

Hvorslev, M. J. (1951). Time lag and soil permeability in groundwater observations. *Bulletin No. 36, Waterways Experiment Station, Corps of Engineers, Vicksburg, Miss., USA.*

Angeli M.-G., Carampin R., Mearini G., Silvano S. (1984). An example of periodic reactivation of a landslide in morainic materials. *Proc. of CNR-PAN Meeting on Progress in Mass Movement and Sediment Transport Studies, Torino, pp. 137-150.*

Angeli M.-G. (1987). Strumentazione automatica nel controllo dei pendii instabili. *Workshop su: 'Applicazioni di tecnologie nella conoscenza, nella previsione e nel pronto intervento in caso di movimenti franosi', Atti della Fondazione "Giorgio Ronchi", Firenze, vol. 62, pp. 99-113, (in Italian).*

Angeli, M.-G., Silvano, S. (1987). Groundwater flow effects on a landslide body in morainic deposits. *Proc. IX Europ. Conf. on Soil Mechanics and Foundation Engineering, Dublin, vol.1, pp.363-365.*

Angeli M.-G., Gasparetto P., Silvano S. (1987). Automatic recording systems in landslide surveying: recent experiences. *Oral contribution in: Proc. IX ECSMFE, Dublin, vol. 3, pp. 1075.*

Angeli M.-G., Gasparetto P., Silvano S., Tonnetti G. (1988). An automatic recording system to detect the critical stability conditions in slopes. *Proc. V Int. Symposium on landslides, Lausanne, vol.1, pp. 375-378.*

Angeli M.-G., Gasparetto P., Pasuto A., Silvano S. (1989). Examples of landslide instrumentation (ITALY). *Proc. XII ICSMFE, Rio de Janeiro, vol. 3, pp. 1531-1534.*

Angeli M.-G., Gasparetto P., Pasuto A., Silvano S. (1990). Analisi di un caso di frana in materiali morenici (Val Fiorentina, Belluno). *Memorie di Scienze Geologiche di Padova, vol. XLII, pp. 129-154, Padova (in Italian).*

Angeli M.-G. (1992). Observed and Predicted Critical Hydraulic Conditions in Natural Inhomogeneous Slopes. *In "Geomechanics and Water Engineering in Enviromental Management", Ed. by R.N. Chowdhury, A.A. Balkema, Rotterdam, pp. 103-169.*

Slopes affected by soil slips: validation of an evolutive model

L. ANTRONICO, G. GULLÀ, CNR-IRPI, Rende (CS), Italy

INTRODUCTION

In a study area in Calabria (Southern Italy) a multidisciplinary study outlined several characteristic evolutive schemes on the behaviour of slopes affected by shallow instability (Antronico et al., 1998; in press). This type of instability, usually indicated in the literature as soil slip, involves small amounts of soil and can endanger people and damage property. The hazard which soil slips represent is a consequence of their kinematic mechanisms and their spreading over wide areas during heavy rainfall (Brabb, 1989; Campbell, 1975; Cancelli & Nova, 1985).

With the aim of acquiring a detailed knowledge of these phenomena five zones, representative of larger parts of the study area where soil slips occur, were chosen. For three of these five zones the evolutive models have been sufficiently well defined to proceed with their validation and then to construct their geotechnical models. Studies are being conducted to achieve this end both on site and in the laboratory and the results so far attained are illustrated in this paper.

SETTING

The study area, in the middle of Calabria (Southern Italy), lies inside the Plio-Pleistocene Catanzaro Graben, bounded by fault sets trending from WNW to ESE (Fig. 1). This graben, that represents the transition zone between the Sila–Catena Costiera range and the Aspromonte-Serre range, comes from the Upper Pliocene. The north and the south rims of the study area, are mainly composed of allochtonous paleozoic crystalline units emplaced during Middle-Upper Miocene (Alpine Unit). They include gneiss, slates of a high metamorphic grade and phyllites (Amodio-Morelli et al., 1976). Lying uncomfortably on the crystalline units are neogene deposits which fill the tectonic depression. These, grouped in lithological units are represented, from the bottom to top, by: conglomerates, evaporitic levels, clay and sandstone. Quaternary is represented by post-orogenic units, consisting of marine and continental sands and conglomerates (Pleistocene) and by alluvial deposits of the Amato and Corace rivers and their tributaries (Olocene) (Gullà et al., 1998).

As a result of its geological history (all the metamorphic and igneous rocks experienced intense folding, faulting, jointing), quaternary regional uplift and climatic conditions this sector of Calabria, as is the case for the region as a whole, is prone to slope instability. Detailed studies of the area have brought out the fact that different types of slope instability are present; there are landslides, some of which are large-scale, and widespread examples of moderate to intense processes of erosion (Sorriso-Valvo & Tansi, 1996; Gullà et al., 1998). This can be seen from Figure 1 which shows the phenomena, mapped at the 1:50,000 scale, with at least one side equal to or exceeding 250m. Soil slips most commonly occur on sedimentary terrain and, to a lesser degree, on weathering profiles of crystalline rocks.

Calabria has a mediterranean climate with precipitation and temperature strongly dependent on altitude. Along the coastal zones and the slopes which face the sea the climate is characterized by mild winters and hot and dry summers; with increasing altitude and in the more inland areas the winters are cold and the summers relatively wet. In the study area a first indicator as regards climate

Landslides in research, theory and practice, Thomas Telford, London, 2000.

are the temperatures recorded at the Catanzaro rain gauge for the period 1951-1987, showing daily values ranging between 41° C and –6° C.

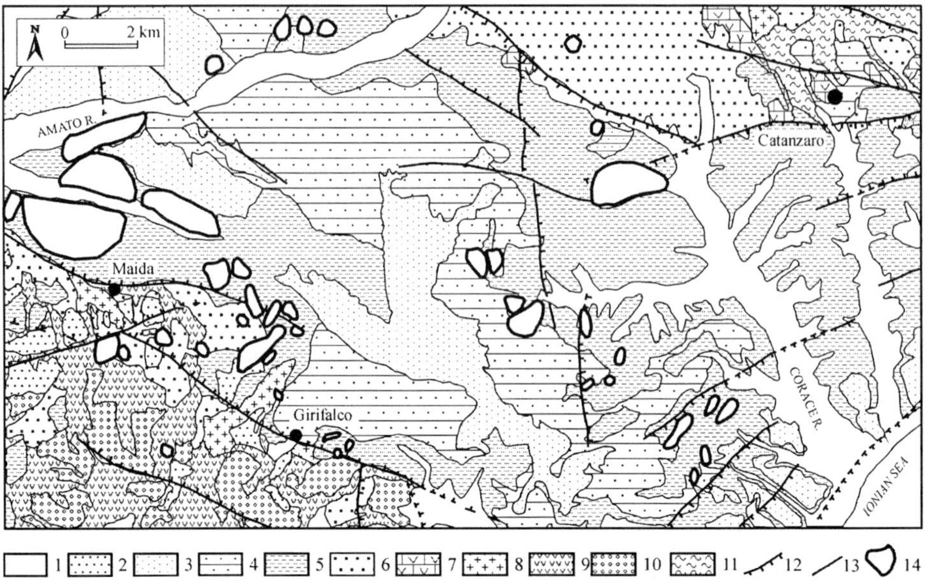

Figure 1. Geological sketch of the study area. 1) Alluvial deposits (Olocene); Post-orogenic unit (Pleistocene): 2) continental conglomerates and sand; 3) marine conglomerates and sands; Trasgressive litological unit (Neogene): 4) mostly sandstone deposits; 5) mostly clay deposits; 6) mostly conglomerate deposits; 7) evaporitic level; Alpine Unit (Paleozoico): 8) Castagna Unit (two mica gneiss); 9) Polia-Copanello Unit (garnet and sillimanite-bearing gneiss); 10) sedimentary deposits which overlay Polia-Copanello Unit; 11) Stilo Unit (phyllites); 12) nornal fault; 13) transcurrent fault; 14) body of landslide. Modified from Gulla et al., 1998.

From a pluviometric point of view, Versace et al. (1989) subdivided Calabria into three areas, with the same types of rain regimes, according to statistical elaboration of maximum depths of daily rainfall over a number of years; the first zone (T in Fig. 2) characterized by very frequent rainfall, the second (I in Fig. 2) by extremely intense, clustered rainfall and, finally, the third zone (C in Fig. 2) characterized by frequent and intense rains. The study zones examined in this paper fall inside the third zone.

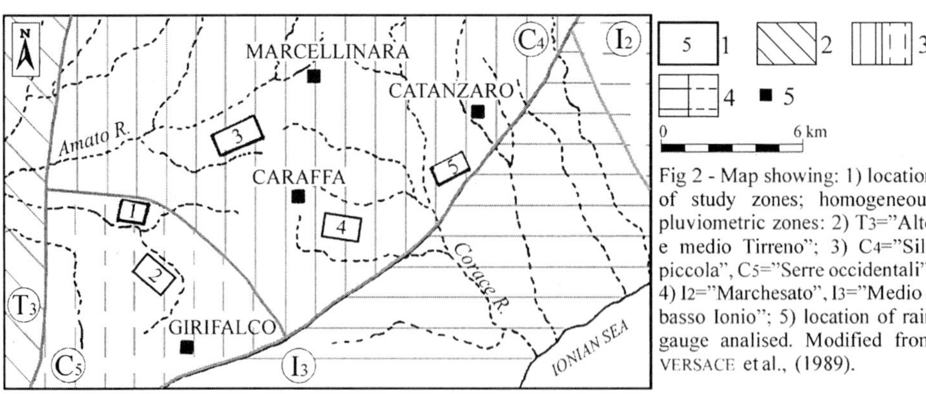

Fig 2 - Map showing: 1) location of study zones; homogeneous pluviometric zones: 2) T3="Alto e medio Tirreno"; 3) C4="Sila piccola", C5="Serre occidentali"; 4) I2="Marchesato", I3="Medio e basso Ionio"; 5) location of rain gauge analised. Modified from VERSACE et al., (1989).

SOIL SLIPS AND THE EVOLUTIVE MODEL

The angle of the slopes affected by soil slips estimated on the basis of available different scale maps and topographical surveys are shown in Table 1.

Zone	Slope (°)		
	1:10,000 Scale	1:5,000 Scale	Topographical Surveys
Zone 2	29°	25°	37°
Zone 3	21°	25°	39°
Zone 5	34°	40°	-

Table. 1 – Angle of the slopes affected by soil slips deduced from available maps and topographical surveys.

From this table it can be seen that there is a substantial difference between the gradients according to the maps and those found by the topographical surveys. Further surveys will lead to a better understanding of these disparities and this will enable us to specify the gradient intervals to be utilized in order to identify, at a preliminary level, areas potentially susceptible to soil slips from the available maps.

As regards the other morphometric characteristics direct evidence, gained from a study of most recent soil slips, show a similar pattern for the three zones studied. In particular, in Zone 5, where miocenic conglomerates crop out, the width of the crown is about 10m and its length is about 20m; in Zone 3, where clay crop out, the width is about 8m and the length is about 10m; finally, in Zone 2, where clay crops out, the width is between 3-10m and the length is between 10-20m.

The thickness of the superficial cover involved, at least in the most recent episodes, in the zones considered are all of an order of 0.5-1m. Evidence, albeit of a general nature, allows us, however, to hypothesize in particular environmental conditions the occurrence of higher values but not above 2m. The area of each phenomenon is limited to not more than 200m^2; however, it has been observed that critical pluviometric conditions can produce contiguous single phenomena involving much larger portions of slope (10,000m^2).

Photographic documentation, over ten years of observation, analysed together with detailed geo-morphological relieves, allowed us to make a general evolutive scheme of the slopes affected by soil slips. Zones 2 and 3 are the areas studied most thoroughly up to now. Soil slips in these zones mainly involve the superficial cover which is composed substantially of degraded clays affected by pedogenosis to a depth usually no greater than one metre, above the intact clay. Subsequently, where

DATE			March 13-22 1985	March 6 1988	December 17 1990	March 15 1996
IC	$P^c_{apr-jun}$ [mm]	CZ	122.8	134.6	-----	79.4
		MA	108.0	168.0	128.4	-----
		CA	-----	141.3	-----	-----
		GI	210.5	-----	-----	-----
	$P^c_{jul-sept}$ [mm]	CZ	96.8	47.6	-----	191.2
		MA	135.0	68.0	49.8	-----
		CA	-----	106.6	-----	-----
		GI	105.4	-----	-----	-----
PP	$P^c_{oct-trig}$ [mm]	CZ	808.8	789.6	-----	964.2
		MA	1477.0	668.5	411.7	-----
		CA	-----	725.6	-----	------
		GI	1275.8	-----	-----	-----
TP	P^g_{trig} [mm/g]	CZ	45.6	143.4	-----	50.8
		MA	64.0	156.3	44.0	-----
		CA	-----	127.8	-----	-----
		GI	76.0	-----	-----	-----
			Zone 2 & 4	Zone 1 & 2	Zone 5	Zone 1 & 2

Tab. 2 – Values of the pluviometric parameters that have given rise to the triggering of soil slips.

IC = Initial condition
PP = Preparation phase
TI = Triggering phase

CZ→Catanzaro Rain Gauge
MA→Marcellinara Rain Gauge
CA→Caraffa Rain Gauge
GI→Girifalco Rain Gauge

the slope is without degraded cover as a result of soil slips, the intact clay itself begins to degrade and crack and vegetation starts to take root. On the same slope, the occurrence of heavy rainfall can trigger successive soil slips involving both the material accumulated on less steep parts from previous slips, and the newly degraded surface layer.

RAINFALL AND SOIL SLIPS

The study zones lie in an area which, on a regional scale, has widely differing rainfall patterns (Versace et al., 1989), as can be seen from Figure 2.

As regards the relationship between rainfall and shallow instability in the study area, certain preliminary indications are given by Antronico et al. (1998) based solely on rainfall data recorded by the Catanzaro rain gauge. Further data from other rain gauges lieing inside the study area was acquired in order to gain a better understanding of this relationship (Fig. 2).

Referring to the available photographic documentation, which shows that soil slips have always occurred in the period from December to March, we selected specific date on the triggering of soil slips, focusing on the periods which were covered by rainfall data (Table 2). Using these dates and referring to the general guidelines outlined by Clarizia et al. (1996) we defined values which are representative of rainfall patterns triggering soil slips.

Taking the days when triggering of soil slips was a probability as our temporal point of reference, we calculated the following parameters:

- $P^c_{apr\text{-}jun}$ rain cumulated in the months of April, May and June prior to the soil slips;
- $P^c_{jul\text{-}sep}$; rain cumulated in the months of July, August and September prior to the soil slips;
- $P^c_{oct\text{-}trig}$ rain cumulated from the month of October to the day prior to the soil slips;
- P^g_{trig} rain on the day in which the soil slips occurred.
-

The values assumed by $P^c_{apr\text{-}jun}$ and $P^c_{jul\text{-}sep}$ are connected to the necessary initial condition for triggering the processes which give rise to soil slips. These values are indicative of the lowering of the water table compared to its maximum winter level and the effect of the dry season on the distribution of pore pressures respectively.

$P^c_{oct\text{-}trig}$ is correlated to the preparation phase which can predispose the superficial layers to soil slips and it is indicative of the reconfiguration of the distribution of pore pressures after the dry season. Obviously, the ways in which rainfall cumulates play an important role on this aspect. Finally, P^g_{trig} represents the necessary amount of rainfall, given the right conditions, to trigger soil slips (triggering phase).

From Table 2 the following ranges of variation can be seen: for $P^c_{apr\text{-}jun}$ from 79.4 to 210.5mm (average 136.6mm); for $P^c_{jul\text{-}sep}$ from 47.6 to 191.2mm (average 100.1mm); for $P^c_{oct\text{-}trig}$ from 411.7 to 1477.0mm (average 890.2mm); finally for P^g_{trig} from 44.0 to 156.3mm/day (average 88.5mm/day). There are quite wide ranges of variation for all the defined parameters. Nevertheless, the high value for $P^c_{jul\text{-}sep}$ shows that the initial condition, determined by a dry season with bouts of heavy rainfall, is an important element in the preparatory phases which lead to triggering. This element, if confirmed by subsequent research, could well be of crucial importance as regards prediction.

These results supply an initial element in validating the evolutive models of the slopes in the study zones. Moreover, they allow us to analyse the time series of available rainfall data, with respect to the values in Table 2, with the aim to choose the periods to concentrate historical and documentary research for this type of instability in the study area.

GEOTECHNICAL CHARACTERIZATION

At the present time, the geotechnical investigations are being concentrated in Zones 2, 3 and 5. Tests are being conducted on samples taken up to a depth of one metre from ground level, as these soils are the most frequently involved in soil slips.

In the zones where research is being undertaken, the grain size distribution curves are available and the relative envelopes in Figure 3a show that, moving from Zone 2 to Zone 5, the grain size distribution passes from clay with silt to slightly silty gravel with sand; the grain size envelopes for Zones 2 and 3 is partly overlayed. In some of samples taken from Zones 2 and 3 a significantly percentage of sand is present; instead, in some samples of Zone 5 levels with high percentage of fine fraction was found (Antronico et al., 1998; in press). These levels mightwell play a crucial role in the instability mechanisms connected with soil slips and for this reason further research is being undertaken to learn more.

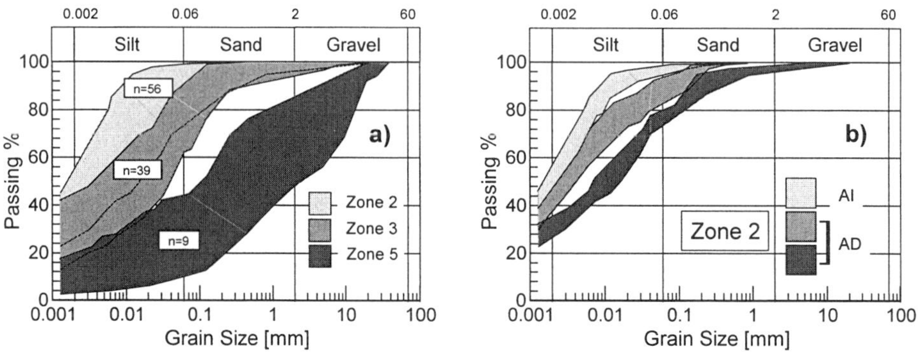

Figure 3. Grain size envelopes. a) Soils tested for Zones 2, 3 and 5; b) AI and AD soil types tested for Zone 2.

If attention is focused on the soils in Zone 2, one can see from Figure 3b that the overall grain size envelope is broken down into three envelopes which, as the lower bounds show, move from slightly sandy clay with silt to sandy silt with clay and, finally, to silt with clay and sand. Inspection of the samples underline the fact that the soils denominated AD in Figure 3b show an initial condition of degradation (Antronico et al., 1998; in press), as a result of daily and seasonal climatic variations. As is brought out in the definition of the evolutive model, the effects of small-scale instabilities- small sods of earth being displaced breaking up the structure of the soil- join with the processes of degradation.

The plasticity and activity of the soils in Zones 2 and 3 is shown in Figure 4a. The soils of both zones are inactive and, in particular, the points relating to Zone 3 are placed predominantly on the line separating inactive soils from normally active ones. As regards plasticity, the soils in Zone 3 all have medium plasticity, while for those in Zone 2 the plasticity ranges from medium to high. This differentiation for the soils in Zone 2, which to a lesser extent is also shown in the activity chart (Fig. 4b), can be explained by their granulometric differentiation (Fig. 3b).

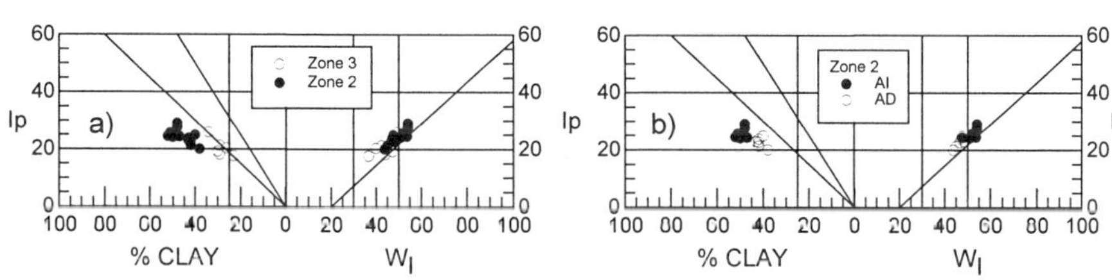

Figure 4. Plasticity and activity chart. a) Soils tested for Zones 2 and 3; b) AI and AD soil types tested for Zone 2.

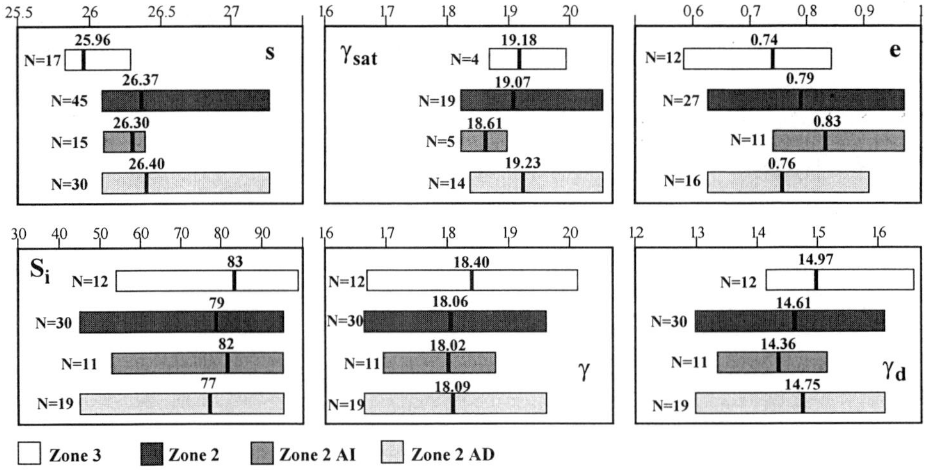

Figure 5. Minimum, maximum and medium values of some index properties relative to soils tested.

In Figure 5 the minimum, maximum and medium values are compared for various index properties. The available data, does not reveal any substantial differences between the soils in Zone 2 and those in Zone 3. This is also the case for AI and AD soils in Zone 2.

To sum up we can assume that the average values indicated in Figure 5 are representative of the index properties of the soils under study.

Turning to saturated permeability, the information available comes from Antronico et al. (1998; in press): for Zone 2 from 1.5×10^{-5} cm/s to 7.4×10^{-5} cm/s; for Zone 3 from 6.0×10^{-5} cm/s to 5.2×10^{-4} cm/s. As regards this aspect of the study, further investigation is planned to gain indications on the permeability characteristics in conditions of partial saturation.

With the same objective in mind in situ suction measures were carried out in Zone 2 at the end of November 1998 (Ferraina, 1998), and the average value for all the tests undertaken, to a depth varying from 5 to 20 cm, is about 9kPa.

Next, in Figure 6 are displayed the preliminary results of the direct shear tests carried out in order to define failure envelope for the soils in Zone 2. All the samples taken up to a depth of one metre, and Figure 6 shows that the completely saturated AI and AD soils have a failure envelope whit a pronounced curvature to lower stress levels: $\varphi=41°$ and c=0 up till $\sigma_v \cong 80$kPa, $\varphi=26°$ and c=41kPa for $\sigma_v > 80$kPa. Several specimens, taken once again from AI and AD soils, tested with an average saturation degree of 50%, with values of $\sigma_v < 50$kPa, give rise to a failure envelope characterized by $\varphi=45°$ and c=100kPa.

Finally, two reconstituted and artificially degraded groups of specimens were tested (Gullà et al., in press.); the relative failure envelope are very close and characterized by c=0, $\varphi=18°$ and $\varphi=20°$ respectively (Fig. 6).

To sum up, the findings reveal that the stress level, the degree of saturation and the state of degradation have a strong influence on shear strength.

These indications supply further elements in the validation of the evolutive model. In particular, they show that on parts of the slopes with lower gradients soil slips only occur when the soils are highly degraded.

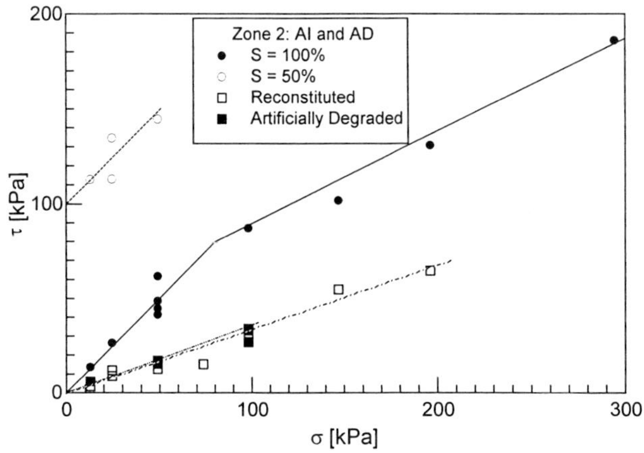

Figure 6. Shear strength envelopes for soils in Zone 2.

CONCLUSIONS

The serious economic and social impact of the effects of soil slips has lead to the setting up of a comprehensive study in a part of Calabria where such phenomena are widespread. Their characteristics and the way in which these are manifest in association with heavy rainfall directly and indirectly produce dangerous situations. An object of the research is to draw up guidelines which will enable us to optimize stabilization and remedial works.

To this end, after having selected several zones representative of larger areas where soil slips occur in Calabria, surveys and tests were carried out in order to define and to validate the evolutive model of the slopes and then to construct a geotechnical model.

Investigations have reached different stages in the various zones. The evolutive model proposed for Zones 2 and 3 can be substantially extended to other zones where clay crop out. The present paper has illustrated some elements which validate this model.

In particular, as regards the rainfall, this paper has also identified combinations of values that have given rise to the triggering of soil slips in the study zones. These values form an essential element in the calibration of parameters to utilize in the geotechnical model.

Turning to shear strength parameters, the on-going study, which still needs further work in a number of zones, has built up a comprehensive framework, which brings into focus the role played by processes of degradation and instability processes themselves in the evolutive model.

In fact, soil slips which occur on the portion of the slopes with steeper gradients are generally of modest proportions, and involve soil degraded by climatic variations. This kind of instability mobilizes sods of earth, which then come to rest on slopes where the gradients are less steep, forming over time a new layer superimposed on already degraded soil. Along these tracts, probably in tandem with periods of critical intense rainfall, soil slips of much more significant dimensions involving wider areas are triggered.

ACKNOWLEDGEMENTS
The authors wish to express their appreciation to L. Aceto for help in supplying geotechnical tests, D. Niceforo for data analysis, and A. Guido for graphic elaboration. This study has been carried out using the financial budget provided by Calabrian Regional Authority (European Community - P.O.P. 1994-99), and CNR-GNDCI.

REFERENCES

AMODIO-MORELLI L., BONARDI G., COLONNA V., DIETRICH D., GIUNTA G., IPPOLITO F., LIGUORI V., LORENZONI S., PAGLIONICO A., PERRONE V., PICCARRETA G., RUSSO M., SCANDONE P., ZANETTIN-LORENZONI E. & ZUPPETTA A. (1976) - *L'arco Calabro-Peloritano nell'orogene appenninico-maghrebide*. Mem. Soc. Geol. It., 17, 1-60.

ANTRONICO L., GULLA G. & ODDO B. (1998) – *A methodological approach to characterized soil slips: preliminary results*. Proc. of the 8th Congress of Int. Assoc. of Eng. Geology and Environment, Vancouver, Canada 21-25 September, 1577-1584.

ANTRONICO L., GULLA G. & ODDO B. (in press) – *Considerazioni preliminari sui risultati relativi alla caratterizzazione geotecnica di soil slips in alcune zone di studio della stretta di Catanzaro (Calabria)*. Boll. Soc. Geol. It.

BRABB E.E. (1989) - *Landslides: Extent and economic significance in the United States*. In E. E. Brabb & B. L. Harrod (eds), Landslides: Extent and Economic Significance, Proc. of the 28th Int. Geol. Congr.: Symposium on Landslides, Washington D.C., 17 July 1989, 25-50.

CAMPBELL R.H. (1975) - *Soil slips, debris flows, and rainstorms in the Santa Monica Mountains and vicinity, Southern California*. U.S. Geological Survey Professional Paper 851, 1-51.

CANCELLI A. & NOVA R. (1985) - *Landslides in soil debris cover triggered by rainstorms in Valtellina (Central Alps - Italy)*. Proc. 4th International Conference and Field Workshop on Landslides, Tokyo, 1985, 267-272.

CLARIZIA M., GULLÀ G. & SORBINO G. (1998) - *Sui meccanismi di innesco dei 'soil slips'*. Proc. Int. Conf. 'Prevention of hydrological hazards: the role of scientific research'. CNR-GNDCI, CNR-IRPI (TO), Alba (CN), 5-7 November 1996. Vol. II, 585-597.

FERRAINA G. (1998) – *Caratterizzazione geotecnica dei soil slips nella stretta di Catanzaro*. Unpublished Graduation Dissertation, Università della Calabria, pp. 160.

GULLÀ G., ACETO L., ANTRONICO L., FERRARI E., SORRISO-VALVO M., TANSI C. & TERRANOVA O. (1998) – *Primo stato di avanzamento del Progetto "Linee guida per interventi di stabilizzazione di pendii in aree urbane da riqualificare"*. P.O.P. 1994/99 Regione Calabria- Fondi UE, Misura 4.4 "Ricerca Scientifica e Tecnologica.

GULLÀ G., ACETO L. & NICEFORO D. (in press) – *Caratterizzazione geotecnica delle argille della Stretta di Catanzaro coinvolte in soil slip. Risultati Preliminari*. Rapporto Interno, CNR-IRPI, Cosenza.

SORRISO-VALVO M. & TANSI C. (1996) - *Grandi frane e deformazioni gravitative profonde di versante della Calabria. Note illustrative della carta al 250.000*. Geog. Fisica e Din. Quater., 19, 395-408.

VERSACE L., FERRARI E., FIORENTINO M., GABRIELE S. & ROSSI F. (1989) - *Valutazione delle piene in Calabria*. Geodata, 30, CNR-IRPI, Cosenza.

Coastal Protection in a Proposed World Heritage Site

D. ATHERTON and K. BURBIDGE, Peter Brett Associates, Reading, UK

INTRODUCTION

Continued erosion of the cliffs along the Dorset coastline has led to the development of large scale landslides. As a result of this process, several properties in the small village of Seatown (Figure 1) were at risk of becoming damaged or lost to the sea. In 1994 West Dorset District Council (WDDC) commissioned Peter Brett Associates to carry out an investigation to identify the failure modes and, subject to MAFF grant aid approval, to progress to detailed design and construction of a coastal protection scheme.

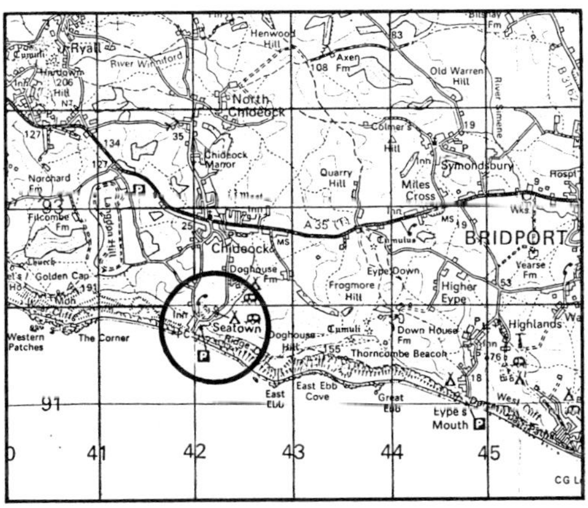

Figure 1 Location Plan

The design of the investigation phase and subsequent remediation scheme had to take into account the fact that the study area is within a proposed World Heritage Site, an Area of Outstanding Natural Beauty and a Site of Special Scientific Interest. Careful planning was required to take into account the often conflicting requirements of, for example, the planners, English Nature, local residents and local geological societies together with statutory bodies at District and County Council level. The final scheme was ultimately designed to be a 'soft engineering' solution and for the remediation to be as unobtrusive as possible on completion of the works.

Landslides in research, theory and practice, Thomas Telford, London, 2000.

GEOLOGY

The study area is underlain by Recent Head Deposits overlying in turn the Green Ammonite and the Belemnite Beds which are of Lower Lias Age.

The composition of the Head Deposits varies as they are derived from local solid and drift formations. In the study area, they tend to mainly comprise orange brown silty clays with varying proportions of gravel. These could be seen exposed in the backscar of the landslip.

The Green Ammonite Beds are described as a series of some 15m to 30m of clays with irregular limestone bands developed at three principal horizons, known as the Lower Limestone, the Red Band and the Upper Limestone. The Belemnite Beds are described as some 22m of mainly pale grey marls, with darker marls at the top and base of the sequence. In contrast with other divisions of the Lower Lias, there are no definite bands of limestone except at the extreme top and base of the sequence; nevertheless the pale marls, being more calcareous and harder than the dark marls, do, to a marked degree, behave like limestones as regards resistance to weathering.

METHOD OF INVESTIGATION

When carrying out analyses of slope failures it is important to identify the geometry of the failed mass (shape, depth, etc). Identifying the location of failure planes along which movement has taken place is of critical importance. In order to maximise the possibility of identifying failure planes, a number of complimentary techniques were adopted as described below.

As part of an initial feasibility study, a number of sources were contacted regarding any relevant information that they may have concerning the area around Seatown. The sources contacted provided aerial photographs, historical Ordnance Survey maps and photographs held by local residents.

The field work was carried out in a number of phases. Initially, a geomorphological survey was carried out to accurately determine the visible surface features related to past failures at the site and to allow failure models to be developed. A number of surface survey stations were established to allow vector movements to be determined and to provide further information on failure modes.

The results of the geomorphological survey and vector monitoring allowed the scope of the intrusive investigation to be planned and agreed with the relevant parties. Trial pits were dug at optimum locations to allow detailed information to be collected and to provide a minimum of disruption. Standpipes and slip indicators were installed within boreholes. These were installed using hand operated drilling techniques, using window sampling and dynamic probing techniques, to minimise any disruption at optimal locations indicated by the geomorphological survey.

FAILURE PROFILE

The results of the investigation indicated that failure was occurring as translational movement along a deep seated (up to 7m depth) failure plane. Several smaller scale multiple retrogressive failure planes are present within the main failed mass, with a postulated common basal failure surface near the toe (see Figure 2).

The toe of the main failure plane appeared to be coincident with the junction between the Green Ammonite Beds and the underlying Belemnite Beds which marks a distinct lithological boundary. The results of the laboratory testing indicate the Belemnite Beds to be a calcareous silty clay of high plasticity. Where exposed in exploratory holes, the Belemnite Beds were stiff to very stiff in consistency. The Belemnite Beds are exposed in the cliff section to the west of the study area. There they can be seen to form near vertical seacliffs with the sheared, slumped and disturbed Ammonite Beds resting upon them. It was therefore considered that the mineralogy of the Belemnite Beds produces more competent strata that are more resistant to erosion and shearing than the Ammonite Beds and this is controlling the position of the base of the main translational failure surface.

The results of the vector monitoring tended to confirm the failure was being structurally controlled to a certain degree. Vector movements of 2° to 3° east-south-east were recorded which is consistent with the regional dip direction from the geological memoir.

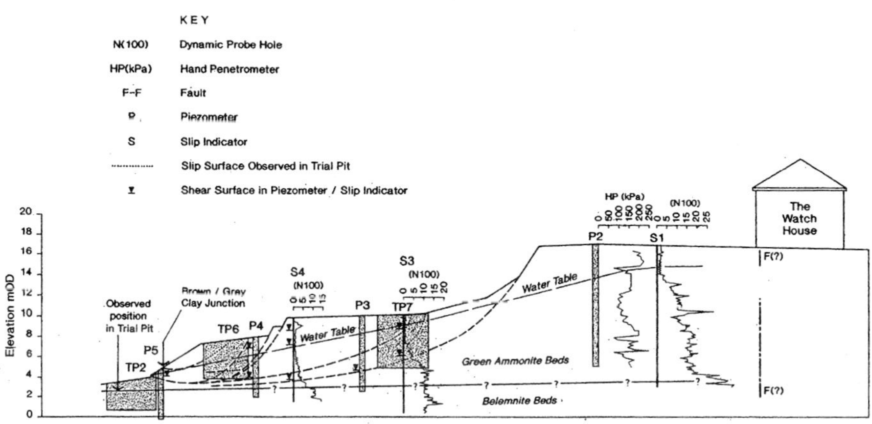

Figure 2. Typical section through failure

The results of the geomorphological investigation indicate that relatively flat benches are present within the main failed mass, ranging from around 1m wide to up to 15m wide. These reflect historical or retrogressive failure surfaces that gradually decrease in size as they are transported towards the seacliffs where continued erosion results in their breakdown into smaller scale failure blocks. It was therefore postulated that when a large-scale failure takes place it occurs as cyclic movement with failure blocks of the order of 10m to 20m wide. Historical evidence suggests that the large-scale failures occur every one hundred years or so.

SLOPE STABILITY ANALYSES

The results of the modelling indicated that, for the main failed mass, the overall factor of safety could be significantly improved by the installation of drainage measures. Toe protection measures would also be required to ensure the long-term stability of the slope and effectiveness of the drainage scheme. As well as the main failed mass, remedial measures were also required for the main backscar. Continued movement and erosion had resulted in a near vertical face up to 4m high comprising mainly Head Deposits. Failure of the backscar could therefore occur independent of the main failed mass, resulting in a loss of residential gardens.

REMEDIAL MEASURES

Counterfort drains were considered to be the most appropriate drainage scheme for the site. They comprise deep trenches backfilled with geotextile wrapped granular fill and a collector pipe at the base. Counterfort drains have the advantage of intercepting a relatively large surface area and will penetrate secondary slip surfaces together with the main basal failure thus increasing the shear strength parameters throughout the whole of the failed mass.

The results of the computer modelling indicated that, with continued toe erosion, the effectiveness of a drainage only scheme would reduce with time. The adoption of rock armouring was environmentally the most contentious because of its visual impact. Low maintenance and acceptable risk were prerequisites to satisfy MAFF and WDDC. A compromise was reached by minimising the amount of armour to reduce the visual impact but accepting this would not entirely prevent wave overtopping in extreme events. The independent Environmental Impact Assessment confirmed the acceptability of the approach. A general layout of the scheme is given in Figure 3.

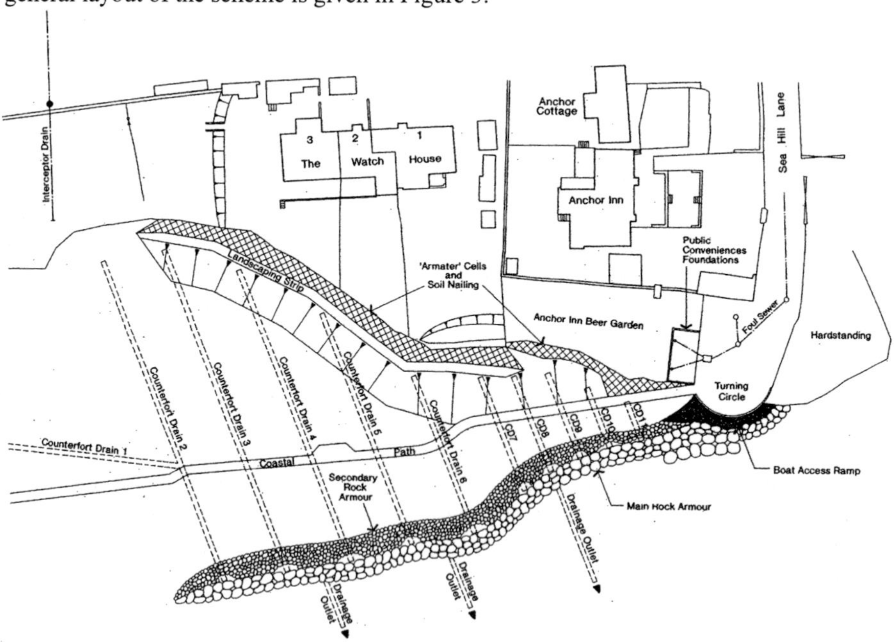

Figure 3. General layout of the remediation scheme

A soil nailing scheme was preferred to stabilise the main backscar as this would best meet the requirements of environmentalists and local residents. The main disadvantage of a soil nailing scheme is that the nail heads and face plates are visible. This was not only undesirable from an aesthetic point of view but were also a health and safety issue as the site is popular with tourists. A facing scheme for the main backscar was therefore devised which comprised the fixing of a PVC coated polyester geogrid to the backscar using the soil nails and attaching 'Armater' (honeycomb cells), supplied by MMG Civil Engineering Systems to the geogrid. Within the 'Armater', indigenous topsoil could be placed in the cell to allow local vegetation to establish and completely hide the nail heads (Figure 4).

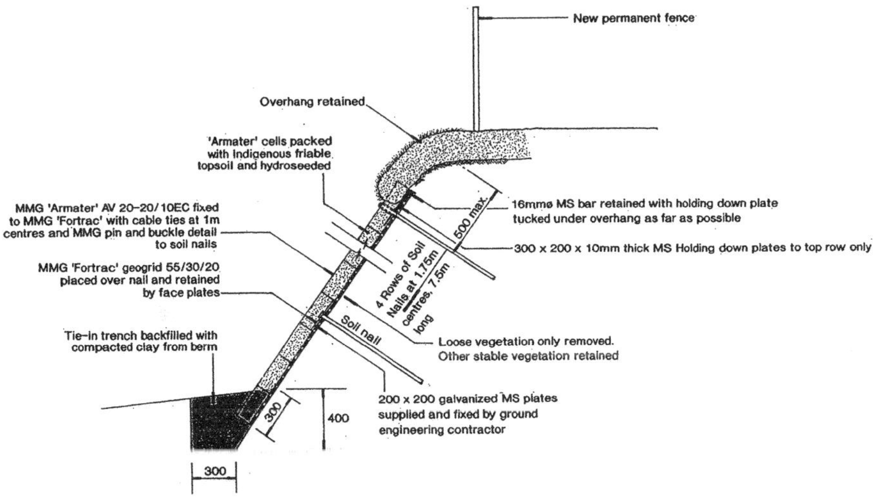

Figure 4. Typical section through remediated backscar

CONSTRUCTION

Construction of the remediation scheme commenced in September 1997, following the award of a MAFF grant. The main Contractor was P Trant Limited, with AMEC Piling carrying out the soil nailing. During construction and after a period of prolonged heavy rainfall, continued movement of the slip occurred along the eastern end of the scheme. This resulted in the generation of a new backscar some 10m in land within the Anchor Inn beer garden and gardens of No1 The Watch House. Severe damage also occurred to public conveniences, requiring the building to be demolished. At the time of the failure, work was ongoing to install the drains and rock armouring to this area and hence the remedial measures were not effective.

As a result of the failure in November 1997, the profile of the failed mass had altered from the original scheme design, with an additional mass now driving the main slide at the crest, as indicated in Figure 5.

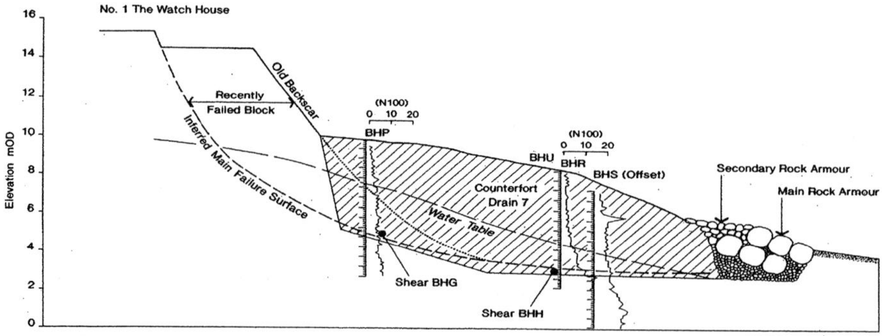

Figure 5. Typical section through recent failure

Consideration was given to extending the drainage scheme further inland to encompass the recently failed material. This was however discounted as it would have required drains to be installed to depths of the order of 8 to 9m in residential gardens and the beer garden. Instead, the soil nailing scheme was extended to include the recently failed area. A temporary remedial measure, following initial failure, was the removal of 1m of soil from the crest of the beer garden. To reinstate the beer garden to its original condition prior to failure, soil nails were also installed in this temporary bench and then buried with fill material (Figure 6). For the failed section in the garden of No1 The Watch House, additional soil nails were also installed but hidden by regrading of the slope.

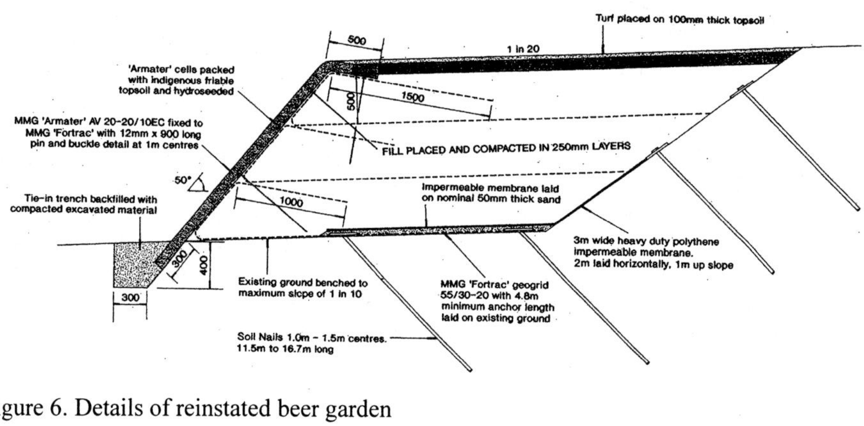

Figure 6. Details of reinstated beer garden

CONCLUSION

Following completion of all of the remedial measures, a monitoring regime was established. This included the monitoring of groundwater levels and vector movements at specific locations within the scheme. The results of the monitoring have not revealed any significant movements and general steady reductions in groundwater levels have been evidenced as the drains became effective. The use of a soft engineering remediation scheme has therefore been effective in stabilising the coastal section whilst remaining unobtrusive on a highly designated and environmentally sensitive coastline.

Space and Time Distribution of Landslides in a Mediterranean River Basin: Trigno River Valley (S. Italy)

P. P. C. AUCELLI, Università degli Studi del Molise, Facoltà di Scienze MM. FF. NN., Isernia, Italy, A. CINQUE, G. ROBUSTELLI, Università Federico II, Dipartimento di Scienze della Terra, Napoli and C. ROSSKOPF, Università degli Studi del Molise, Italy

INTRODUCTION

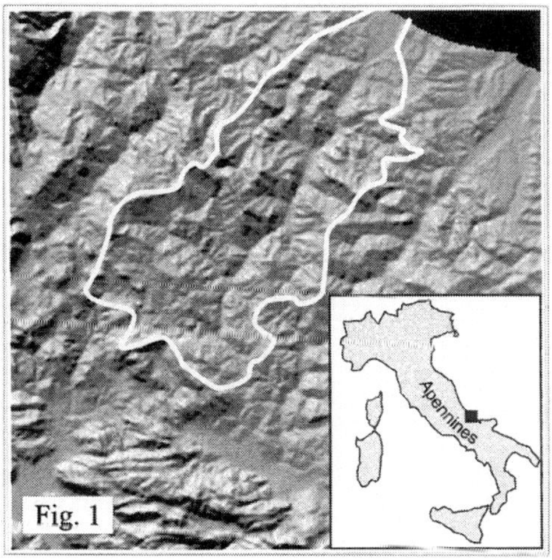

Fig. 1

The Trigno river basin is a major catchment of the Molise segment of the Southern Apennines (Figure 1), a NE verging orogenic belt interested by compression and orogenic transport at least until the Late Pliocene, then uplifted and fragmented by high angle faults (both normal and oblique slip) during Quaternary times (Di Bucci et Al., 1999). The most advanced and recent thrust fronts are found offshore the Adriatic coast, toward which all the major drainage lines of the region converge following the general slope of the outer flank of the accretionary wedge. The drainage system main trunks cut thus the thrust and fold structures of the chain perpendicularly, while many tributary valleys (of subsequent type) follow the geological structures. Almost the whole drainage system shows a marked tendency to downcutting and accelerated slope dynamics due both to the effect of the recent uplift and the weak nature of most the relief forming rock units. Aim of this study is to examine the role of landslide phenomena in the evolution of small catchments and, conversely, the influence of fluvial processes on landslide distribution and characteristics.

GEOMORPHOLOGY OF THE TRIGNO VALLEY

The Trigno river is a 120 km long, 7th order stream, which drains an area of about 1200 km^2, with an average elevation of about 600 m and peaks up to 1400 m high. The valley, elongated from SW to NE, is cut into thrusted and folded terrains of Tertiary age (Molise sequences), mainly represented by transitional to basinal sequences (alternations of marls and marly limestones) and flysch deposits (pelitic and arenaceous siliciclastic formations). The lower Trigno also cuts some less severely deformed terrains of the last Apennines Foredeep (clayey to sandy/arenaceous formations of Plio-Pleistocene age.

Our investigation focused mostly on the medium reach of the Trigno valley because it discloses a higher density and a greater diversity of landslide features, often in clear genetic

Landslides in research, theory and practice, Thomas Telford, London, 2000.

relation with landforms in different stages of evolution. This reach, comprised between 710 and 60m a.s.l., is characterised by a fluvial pattern ranging from a single confined channel to a braided (single to multi-threated) channel system. The valley side slopes of both the main trunk and the major tributaries are characterised by composed profiles.

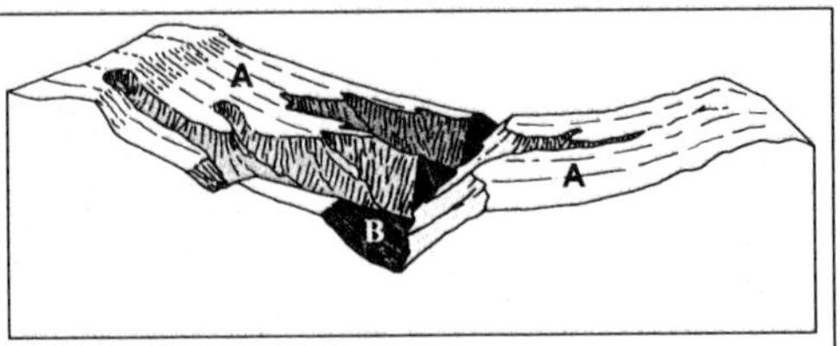

Fig. 2

In fact, the upper part of the slopes (**A** in Figure 2) is characterised by more mature landforms than the lower one. In particular, it is dominated by gentle convex-concave hillslopes grading downward to relics of almost planar surfaces dipping toward the valley axes of 3 to 8 degrees. Said gentle planar elements are interpreted as erosion glacis (hereinafter *hanging valley-side glacis*, or HVSG) shaped during a period of decreased downcutting rates (decreased water discharge and/or increased detrital supply) and prevailing slope decline. The deposits covering the basal portions of some HVSG remnants witness with their facies the importance of diffuse erosion and landslide processes (mainly slides and flows) in the genesis of those ancient surfaces. Some portions of the Trigno valley show more generations of glacis hanging at different elevations above the thalwegs. The lowest one (which is also the most frequent and best preserved) is tentatively ascribed to the dry and cold climatic conditions of the Last Glacial.

The lower portion of valley side slopes is typically made of a steep fluvial scarp (herein after FS; **B** in Figure 2) representing the effect of a phase of deepening which affected the whole study area after the shaping of the last generation of HVSG. These scarps are a few to several tens of meters high, but their low degree of evolution (i.e. degree of decline, retreat and/or dissection) coupled with considerations about the high erodibility and instability of the local rocks types, makes reliable the hypothesis that they appeared after the end of the Last Glacial, when linear dissection was favoured by a return to warmer and moister climatic conditions.

LANDSLIDE TYPES AND DISTRIBUTION

Presently the FS of the Trigno middle reach are affected by deep to medium seated landslides which are the main factors of decline and retreat of those scarps. The former mode of evolution is active in places where the valley floor is broad enough to permit a lateral shift of the riverbed when a landslide accumulation comes to obstruct its original course. The mode of scarp retreat, on the contrary, is common in places where the Trigno actively undermines the footslope, either because confined in a V-shaped valley or because running at the margin of a wider valley floor.

The FS are also deeply affected by accelerated linear erosion. The resulting gullies are, on one hand, able to determine new landslides on their flanks and heads; on the other hand the evolution of gully systems appear clearly influenced by landslide phenomena. The interference between fluvial and slope processes is also evident at a greater scale. In fact, the shape of the valley side slopes and their present dynamics are strongly linked with the morphology of the valley bottom. In this respect, the medium reach of the Trigno valley may be subdivided in three sectors.

The Upper Sector is characterised by a confined channel system and a longitudinal gradient of about 2,5 percent. It shows a general tendency to valley deepening, locally interrupted by phases of aggradation related to channel obstruction by landslides and fast growing alluvial fans. An important detrital supply comes from the FS located inside 4th and 5th order tributary basins where the post-HVSG dissection wave is still very active.

The Middle Sector discloses a stepped longitudinal profile in which steep and short elements (corresponding to gorge-like reaches cut in hard rocks) alternate with gentler and longer elements where the valley is broader and flat floored. The reaches of the second type show evidence that the current phase of dissection was preceded by a recent (historical) phase of valley floor aggradation and widening. This phase (probably triggered by a climatic fluctuation) was partially controlled by the arrival of very large earth and debris flow generated within 3rd and 4th order tributary catchments as a consequence of fast dissection of previous landslide deposits.

The Lower Sector has a longitudinal gradient of about 0.6% and shows a broad valley floor (0.6 to 1.0 km wide) occupied by recent alluvial deposits. The Trigno river channel has entrenched 3 to 4 m in this *alluvium* during the last 50 years. The existence of a broad plain implied that, at least all along the Holocene, the river undermined the FS only rarely and diachronically. Consequently, the FS bordering the main valley have rather low relief energy and mature landslide landforms. On the contrary, high slope instability is observed in the medium and upper portions of the main (4th and 5th order) tributary valleys which are in course of deepening and headward extension.

Within the medium reach of the Trigno valley, slope instability is represented mainly by slides and flows (sensu Cruden & Varnes, 1996) in the areas exposing chaotic clayey units (Argille Varicolori *Auct.*) and/or the most thinly stratified portions of formations composed of alternating calcareous and marly beds (e.g. Tufillo and Daunia Units), or arenaceous and clayey ones (e.g. Frosolone and Agnone Flysch). Falls and topples are restricted to places where thick calcareous or arenaceous beds of the above listed formations form free-faces along the slopes. Similar phenomena also affect the portions of the fluvial scarps which are currently undermined by a stream. With few exceptions of landslides of small dimensions, the observed phenomena evidence a complex origin and activity. Particularly frequent are complex slide-flows involving rock or soil, and multiple landslides (slides, above all) which generally are retrogressive or enlarging (*sensu* Cruden & Varnes, 1996). By analysing the spatial distribution of landslides and their characters it was noted that they correlate with the stage of evolution of the fluvial landscape in which they are inscribed, more than to the local lithological and structural conditions. In particular, four fundamental types of landslide controlled catchments were recognised in association with specific geomorphological contexts.

Along the steep FS which characterises the Upper Sector the most typical features of slope instability are landslides in a relatively young stage of evolution (i. e with steep profiles and incipient signs of scar branching) whose style of activity ranges from single slides and flows to complex slide-flow phenomena. The dissection and removal actions performed by the river on the displaced materials, trigger new failures in the toe part of the landslide, thus favouring the tendency to deepening and retrogradation of the landslide catchment (Type 1 in Figure 3).

Also common along the FS of in that Sector are some wider landslide catchments shaped by multiple to complex slide-flow phenomena in co-action with the linear erosion performed by consequent streams of the 1st and 2nd order. In cases of this type, only the toe portion of the landslide accumulation is eventually removed by the main river, while large parts of the landslide catchment remain generally occupied by a debris/earth tongue (Type 2 in Figure 3 and example a in Figure 4).

In the reaches of the valley where downcutting was followed by a period of valley floor broadening (parts of both the Middle and the Lower Sectors), the FS are transversally cut by landslide catchments which are similar to the previous ones, but larger (between 1 and 2 km^2), less elongated and with a better developed drainage network reaching generally the 3rd order. The materials displaced by the most recent landslides rest in the upper portion of the catchments, but more ancient landslide deposits can be found also in their middle and lower portions. The older deposits are sometimes in course of dissection (Type 3 in Figure 3 and example b in Figure 4) and the resulting detrital material accumulates as an alluvial cone at the mouth of the catchment. In other cases, the mouth zone shows a prominent lobe of mixed origin (alluvial plus flow deposits) which is the terminal part of a tongue extending up to the upper catchment.

Landslide catchments even larger and gentler that the previous ones are found along the flanks of valley reaches which had a longer phase of riverbed stability after that of downcutting. Their longitudinal profiles evidence a concave basal element followed upstream by a convex element and finally by a more gentle concave head portion. The latter one corresponds to the sector of catchment which is intensely enlarged by shallow landslides (slides and flows) operating with low threshold angles. The convex element is a zone of substantial balance between accumulation and removal of landslide deposits. The lower concavity represents the portion of the catchment where the action of running waters prevails, in the long term, on landslides. The prominent bodies forming at the catchment mouths are, in this case, fans with a strong alluvial component. (Type 4 in Figure 3 and example c of Figure 4).

THE LONG TERM EVOLUTION OF THE LANDSLIDE CATCHMENTS

The described four types of landslide related catchments are quite widespread along the middle Trigno valley and its tributaries. It was observed that the higher the hierarchical order of a valley, the greater the maturity of its flanks and also the diffusion of catchments of Types 3 and 4. On the contrary, cases of the first and second types are particularly common in low order valleys, because they are still in a stage of deepening. This distribution leads us to retain that the observed four types of landslide controlled landforms form an evolutionary sequence that started with the dissection of the HVSG and reached different evolution stages in different parts of the drainage system. In Figure 5 we schematise a conceptual model of evolution for the landslide catchments at issue. The model starts from the moment when the river bed stabilises its elevation (local end of the post-HVSG dissection phase) and starts forming a flat valley floor (VF). The very initial phases of FS degradation by both linear erosion and type 1 landslides are not shown.

The longitudinal profile 2 and the cross-section II refer to a Type 2 catchment. Its evolution is almost entirely due to landslides occurring along its flanks and head. It gradually evolves to a Type 3 catchment (profile 3 and cross-sections IIIa and IIIb) of gentler mean gradient. While both the slope failures and the landslide deposition migrate upslope, the gradual increase of the catchment area leads to a progressive growth of stream power in the lower reach of the catchment itself. This permits the dissection of the older landslide deposits and, occasionally, new failures from the incision flanks. As the longitudinal profile has still high gradients, earth/debris flows generated in the middle-upper reach may eventually reach up to the VF contributing to the construction of a terminal cone.

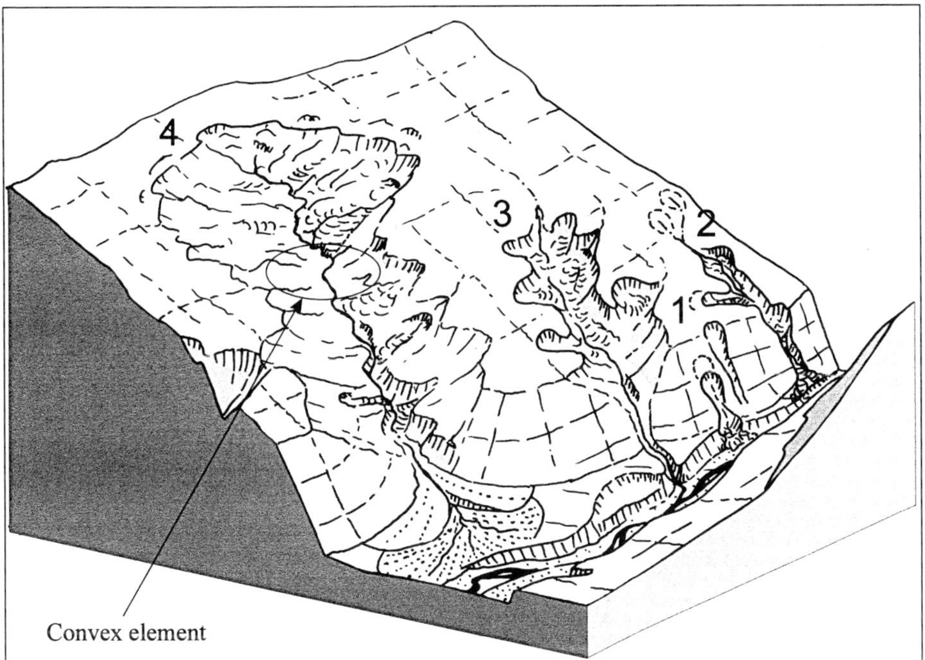
4
3
2
1
Convex element

Fig. 3

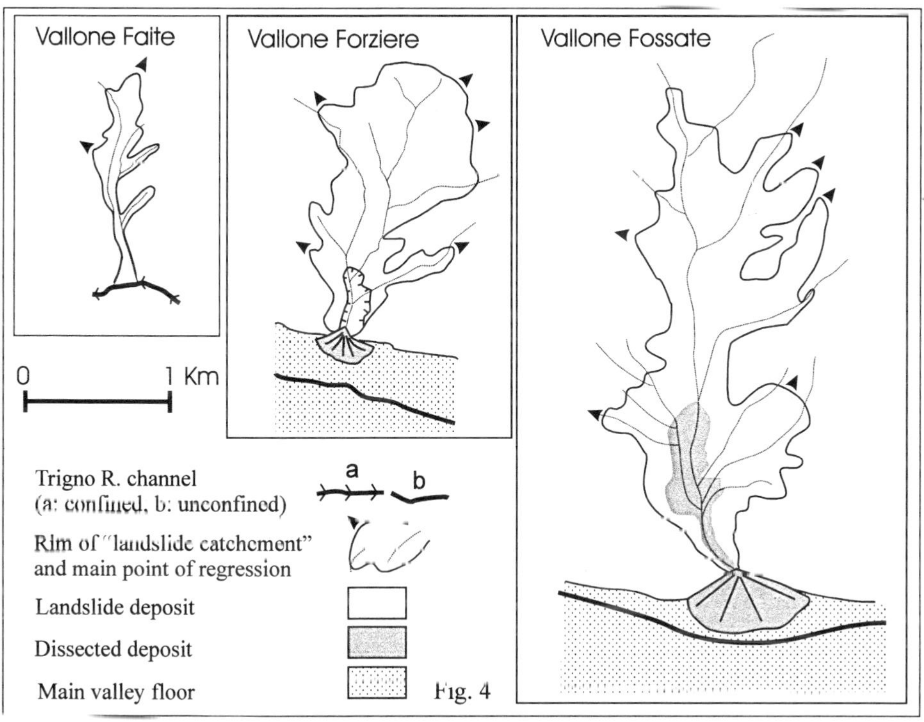
Vallone Faite
Vallone Forziere
Vallone Fossate
0
1 Km
Trigno R. channel
(a: confined, b: unconfined)
a
b
Rim of "landslide catchment"
and main point of regression
Landslide deposit
Dissected deposit
Main valley floor

Fig. 4

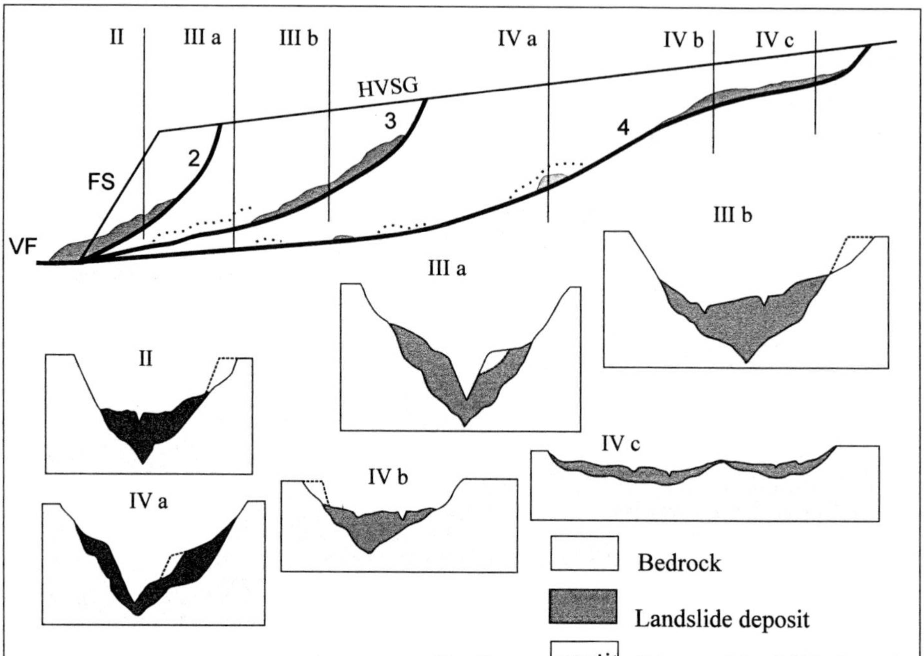

Fig. 5

The most advanced stage of catchment evolution (profile 4 and cross-sections IVa, b, c) is characterised by the appearance of a zone of convexity in the upper part of the longitudinal profile. While landslides of noticeable depth continue to occur in the middle reach of the catchment (both from retreating heads of the tributary branches and from the shoulders of incised deposits), the zone resting above the convex element is widened only by shallow landslides and slow mass wasting processes (creep and solifluction). Two hypotheses can be made as regards the origin of the convexity. The first one assumes that, at a certain moment of the catchment evolution, a point of its longitudinal profile starts receiving displaced materials at a rate that almost equals the capacity of removal by stream erosion and mass wasting, thus becoming an almost stable base level for the further evolution of the upper catchment. Most probably this happens in a zone where numerous unstable branches converge and/or a narrowing of the valley is present. The other hypothesis assumes that the longitudinal profile breaks at the point where it intercepts the passage between the fresh bedrock and the weakened mantle developed on the glacis, which has lower shear-strength and higher permeability than the underlying rocks.

REFERENCES

Cruden D. M. & Varnes D. J. (1996) *Landslide types and processes*. In: A. K. Turner and R. L. Schuster Eds. "Landslides, investigation and mitigation". Special report **247**.

Di Bucci D., Corrado S., Naso G., Parotto M. & Praturlon A. (1999) – *Evoluzione tettonica neogenico-quaternaria dell'area molisana*. Boll. Soc. Geol. It. **118**, 13-30, 12 ff.

Sorriso Valvo M. (1985) Mass movement and slope evolution in Calabria. Proceed. IV I.C.F.L., 23-30.

Sorriso Valvo M. (1988) Landslide related fans in Calabria. Catena Suppl. **13**, 109-121.

Varnes D.J. (1978) *Slope movement types and processes*. In: Schuster R. L. & Krizek R. T. (Eds.), Landslides Analysis and Control. National Acad. of Sciences, Washington, D.C.

Some Problems Linked With Landsliding in Albania

S. AXHEMI, University "Luigj Gurakuqi", Department of Geography, Shkoder, Albania

INTRODUCTION

Some of the most characteristic relief forms of Albania result from landslides. Their widespread extent is closely linked with the natural conditions of our country. To study them is of great importance, because in many territories they play a major role in creating the relief. Studies of the landslides have great practical importance. Many Albanian territories touched by landslides are important habitation centres for the local population, and in special sectors are built some important economic works. It is therefore important that landslides are studied with the intention of taking some technical and practical engineering precautions to prevent them or to eliminate their wider negative consequences. It is important to closely observe and continuously study particular landslides to discover their peculiarities with which they are accompanied during their development, not forgetting here that the phase of reactivation, may have considerable consequences. In reference to the damage they bring, different scholars are working (geologists, geomorphologists and hydrographers) in order to determine the most active zones for landslide activity, and constructing geomorphological maps (hazard maps) of Albanian territory where landslides will be evidenced. We must stress that landslides are evidenced and put in map for particular zones and soon the geomorphological map will be ready, including here the other forms of relief in Albania.

THE FACTORS THAT INFLUENCE THE OCCURRENCE OF LANDSLIDING

Landslides are widespread in Albania and this is closely linked with many factors where the most important ones are the relief, climatic factors, vegetation cover and the artificial elements (as a result of human intervention in the landscape).

A primary and important role is played by the relief of the surface of earth: where the considerable extent of hilly and mountainous territories, the high values of horizontal and vertical parcelling of these territories, especially the stressed inclination of relief and the orientation, and the development of valley slopes, have influenced the presence of landslides in many surfaces of Albanian territory. Numerous scholars evidence the fact that about 76.6% of Albania territory lies above the height of 200m, where about 28.5% of Albanian territory is above the height of about 700m. Such indications show clearly the predominance of hilly and mountainous relief, creating the potential for the occurrence of landslides in these regions. To this considerable height is conjoined another important geomorphological element, that has to do with the considerable inclination of Albanian relief, which is one of the most important factors that influence in the occurrence of landslides.

It must be stressed that a considerable surface of arable agricultural land is found on these slopes. The most frequent landslides are seen in the height of 1000-1500m and this is linked not only with the influence of human activity. In these heights a bigger extent of the terrain has soft formations such which are susceptible to the process of landsliding. In the same time the human activity at these heights is very expressed, for the fact of the concentration of the

Landslides in research, theory and practice, Thomas Telford, London, 2000.

living territories of the population. An important role in the occurrence of landslides in Albanian territory is played by the soft composition. The considerable extent of the weak formations (which in the reported to take up about 58% of territory) has clearly influenced the occurrence and extent of landslides. Such weak formations, not being resistant to the operations of external factors, are immediately given to creating interesting forms and aspects in the relief.

Climatic factors have considerable importance in the occurrence and extent of landslides. A very important role is also played by the quantity of rain, its intensity, and the irregular pattern of rainfall during the seasons of the year. From the climatic point of view, Albania is included in the belt of Mediterranean subtropical climate, with a mild winter and humidity, while the summer is long, dry and a very hot one. The large quantity of rain (from 1500-2000mm in a year) has greatly influenced the occurrence of landslides. Even a lower intensity (average of 100mm till 200mm) makes the material saturated temporarily, giving the opportunity to shear away from the mother formations and to begin to move (creating the probability of occurrence of the landslides) so that with the continuation of the rain for a long time, formations are saturated with water, creating landslides.

An important place in the occurrence of landslides is played by even the vegetation cover. In those territories where exist a pure vegetation, then the formations are mostly given to the landslides. Even the artificial activity and the human activity plays a great role in the occurrence of landslides in the Albanian territory. Construction, development and the opening and tillage of agricultural land is very often accompanied with negative effects in the environment that have to do with the perversion of natural balances and parallel in the occurrence of landslides. As a conclusion we can say that the factors that influence in the occurrence of landslides are numerous, and from these factors results the scale of the influence and of types and sorts of landslides.

THE EXTENT OF LANDSLIDING IN ALBANIA.

All territory in Albania is under the influence of erosional factors, so consequently the widespread occurrence of landslides is considerable. We can find them everywhere in the hilly and mountainous territories of our country, including here all the physical and geographical regions. Among the bigger landslides seen in the Albanian territory are those of:

- Moglica,
- Zoll Bastari,
- Kaptina e Martaneshit,
- Fierza.

One of the most tangible zones of landslides is the valley of Osumi. The rocks contain soft formations, creating the opportunity for the occurrence of landslides. Also the quantity of rain has played a role, as well as the inclination of territory. These landslides are seen especially in the sector of Mican. They are present along both sides of valleys, although their dimensions are not large. But they are noted for their frequency. Numerous landslides are seen in the south of Albania. The wide occurrence of soft formations, especially of clayey ones has influenced the development of landslides. They are present especially in the zone of Lunxheri, where we can find the landslides:

- Lekli,
- Hormova.

Also landslides are noted in the zone of Pogoni as are those of Polica, Sopiku, etc. A considerable territory includes the big sliding of Gramsh of the year 1988. That included a surface of 100 hectares causing difficult economic problems. Also a lot of erosional processes are present (where are included in landsliding) in the zones of valleys of:

- Tomorrica,
- Voskopoja,
- Molishti.

Though with a fragmental character, landslides are present even in the central part of the Luras mountains. Among the most important landslides are those of:

- Krej Lura,
- Grykegjuraj,
- Majes se Madhe.

We must stress that these landslides have a peculiarity that differentiates them from each other. The landslide of Krej Lura is presented a stabilised territory and where vegetation appears everywhere the integrity of cone of the depositing in both other landslides though it may be created these are in the phase of stability their territory is presented relatively calm. Especially this relativity is presented flexible in the season of autumn and winter with the beginning of rain. They are made in activity in a periodical manner with annual cycle, but moving scale of the material is small.

Landslides occur in the highland of Puke – Mirdite, especially in the sectors as in those of

- Mirdite,
- Bugjon,
- Gomina.

Even though these are not noted for big measures, they are clearly shown in the relief with their expressed morphological elements An important role in the occurrence of landslides in this territory has been played especially fact that the highland Puke - Mirdite is one of the most ancient zone of Albania where the average rainfall is about 2000mm. One of the most interesting zones with the presence of landslides is the south east of Albania. Here is seen the biggest slide of all Albanian territory, which is the landslide of Moglica.

CONCLUSIONS

Beginning from the considerable widespread extent of landslides and their problems, authors and geomorphological scholars have introduced a synthesis study upon the negative effects of these landslides in relief. Here is seen their important role in the creation of degraded territories. Also is realised over a particular classification of degraded territories in Albania where are evidenced some levels:

1-completely degraded
2-medium degradation
3-limited degradation
4-possibility of future degradation.

In these levels are integrated different territories of Albania depending on their report with the degradation phase in which they are. It must be stressed that typical zones and sectors are in alarming degraded situations, which require urgent studies to be initiated and to eliminate the negative consequence that may happen.

REFERENCES

Axhemi Sokol, 1995 Some fisico-geographic characteristics of Luras mountains in Albania. Bulletine Geographic Studies, Nr.6.

Biancotti, A, 1982 L' erosione e i problemi della conservazione del suolo. Aspetti e problemi della geographia, Marzarati Editore.

Qiriazi, Penkli, Sala , Skender, 1992. The map of Erosion in Albania. Tirana, Albania.

Sala, Skender, 1993. The development of actual processes in Mati territories. Bulletine Geographic Studies, Nr.4.

Factors Affecting Slope Stability in the Blue Nile Basin

L. AYALEW, Technical University of Clausthal, Department of Engineering Geology, Clausthal-Zellerfeld, Germany

INTRODUCTION

Landslides and rock falls are two of the most damaging natural hazards in Ethiopia. Many active and inactive units of mass movements are recently observed in various gorges, mountain chains, towns and road sections of the country damaging dwelling houses, infrastructures and farmlands. Indeed, as the population increases in an alarming rate and construction and agricultural activities are extending towards valleys and unstable grounds, the chance for humans to be part of a tragedy linked with slope instabilities is becoming high. This issue gains high curiosity at present and a circumstance is created in which responsible professionals and government officials are starting to give attention to the geological aspects of construction sites.

In this paper, an attempt was made to present the nature and distribution of landslides in the Blue Nile Basin of northern part of Ethiopia. The study specially treats the problem of slope instability along the main asphalt road, covering some distance further to the east and the west. It is observed that many factors are involved in generating landslides and rock falls throughout the basin. The contribution of these factors to the problem of instability is discussed at first before relevant variables are selected, combined and integrated in a GIS platform to divide the area into different zones of slope instability.

GEOLOGY AND GEOMORPHOLOGY

The Blue Nile River crossed a variety of geological formations and units along its entire course from the starting point around Lake Tana, in Ethiopia, to the Mediterranean Sea. The first 150km of its course lies on lava flows of Tertiary age, after which the river flows through Mesozoic and Palaeozoic sediments until the Precambrian is reached approximately 330km from Lake Tana (Jespen and Athearn, 1961). The study area, which is part of the Blue Nile Basin, is covered by Mesozoic sedimentary rocks and a Trap Series basalt (Figure 1).

The oldest recognisable rock unit in the area is a cliff forming, thickly bedded and cross-laminated sandstone exposed along the main river channel and around its junctions with several tributaries. It grades upward to a unit which is composed of fine sandstone, siltstone, mudstone and shale on both sides of the valley. Members of the Gohatsion Formation, which include gypsum, mudstone, shale and claystone (Getaneh, 1981), are exposed along the road in the southern part of the gorge. In the northern part, however, only a thick accumulation of colluvium and residual soils with rough, entombed boulders of basalt are present at the same elevation, with some isolated hills of the Gohatsion Formation exposed a few hundred meters away from the road towards the east and the west.

A lower cliff-forming limestone intercalated with thin layers of marl overlies the Gohatsion Formation. The marl content increases upward due to the increase in percentage of clay

Landslides in research, theory and practice, Thomas Telford, London, 2000.

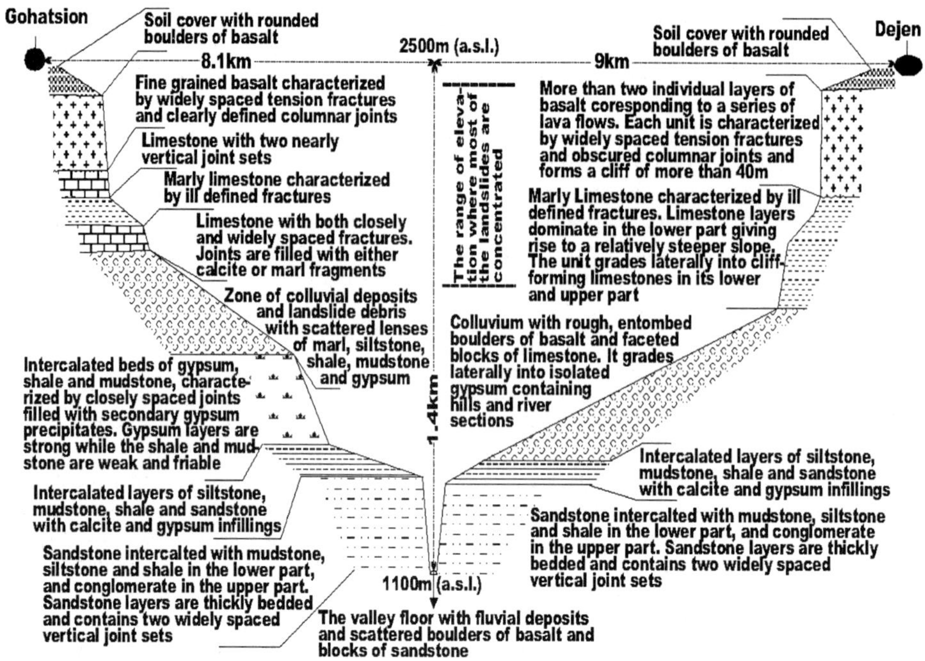

Figure 1. A geological cross-section along the main road and the physiography of the gorge.

compare to calcite. At an approximate elevation of 2000m, individual layers of marl, although not distinct, dominates limestone beds and forms the marly limestone unit. The top part of the sedimentary sequence in the area is marked by an upper cliff forming, thickly bedded sandy limestone. Both the upper and lower cliff-forming limestones are missing along the road in the northern part of the gorge either due to erosion alone or combined with minor scale faulting. Horizontal flows of basalt cover the sedimentary series in an extensive area on both sides of the gorge. The general characteristics of these flows indicate that they are products of fissural eruption poured out sequentially from long fractures and accumulated in the form of thick volcanic deposits (Jespen and Athearn, 1961).

The degree of fracturing of rock masses varies throughout the area in terms of joint frequency, orientation, and persistence. The cliff forming sandstone and limestone units and some parts of gypsum are characterised by two nearly vertical joints parallel and perpendicular to the cliff face. While the line of fracturing becomes less apparent in soft intercalation units, individual basaltic rock màsses are highly dissected by tension fractures and obscured columnar joints.

The topography of the Blue Nile Basin in general is a result of the nature and structure of bedrocks and the denudational processes operating in the geologic past. From the cross section shown in Figure 1, it is apparently clear that the gorge is characterised by a very rough morphology. Generally, steep cliffs and low mesas are results of strongly bedded sandstone, limestone and voluminous out-pouring of basalt, whereas the soft intercalated units of shale, gypsum, mudstone and siltstone yield a relatively gentler slope. The slight asymmetrical appearance of the entire gorge, as it is shown in the figure, is a result of the variation in the response of the slope to river incision owing to the cyclic nature of basal erosion on one side of the valley and accompanying deposition on the other side throughout the history of basin development. The presence of frequent rapids, water falls and channels cutting into the rock units are all indicative of the modification of the old hydrographic pattern.

RAINFALL AND GROUNDWATER

During field operations, a marked climatic variation both in terms of the amount of temperature and the degree and extent of rainfall has been experienced in the entire basin. The least elevated portions of the area around the main river channel are very hot especially in dry seasons whereas the surrounding highlands are relatively cool. Recording stations both in Gohatsion and Dejen indicated that throughout the gorge, rainfall is high from June to September. While the yearly average precipitation in both stations is nearly 1250mm, the mean annual rainfall in a small town called Filikilik located in the middle section of the southern part of the gorge is roughly 1100mm. An amount less than this is normally expected around the lowest portion of the valley, although hard data are currently not available.

The determination of the condition of groundwater flow and distribution is difficult in the area. The existing boreholes are few in number to justify whether there is regional hydraulic continuity between rock masses. In general, rocks and soils exhibit high spatial variability in their hydraulic characteristics, and the occurrence of alternate layers of permeable and impermeable materials might yield perched water tables at various depths. A series of springs and seepage lines observed along much of the gorge would be the result of secondary permeability owing to the presence of master joints and minor karstfication in limestones.

THE ROLE OF DIFFERENT FACTORS ON INSTABILITY

Slope instability in the form of landslides and rock falls was a frequent problem in the entire basin even in early periods, although its effect was much limited to scattered, small villages. The magnitude of events has increased in the last few decades. At present large-scale landslides in highly populated areas and rock falls along lateral cliffs and road sections are common. This is especially true in localities where adverse geological and topographical conditions are coupled with unusual climatic influences and unplanned artificial activities, which make demands on the ground to adjust itself into a new equilibrium. To give a detail description on specific roles of all factors contributing to the slope failure in this paper is difficult. Accordingly, discussions are more integral, and they are based on regional analyses.

The effect of geology and geomorphic configuration on the stability of materials of the gorge was highly varied. The lithological succession, the geological structures, weathering and river erosion had an influence on initiating landslides and rock falls in the area. The occurrence of bands and lenses of highly fissile, slightly dipping, carbonaceous mudstones, claystones and siltstones within the thickly bedded, relatively competent main sandstone unit, and the presence of marl intercalated with layers of limestone make high material strength of hard rocks little relevant to cliff stability especially along the road. In localities where the upper cliff forming limestone is missing, the thick accumulation of lava flows directly overly the marl sub-unit creating a much favourable condition to the initiation of landslides particularly in the rainy season. In these areas, water from gullies and direct precipitation enters tension fractures and columnar joints of basalt from the top of the hill and moves downward into the marly limestone unit. Since marl is relatively impermeable, it absorbs much of the water and gets wet easily. With an increase in moisture, the marl looses strength, and hence, its bearing capacity. The tall, thick columns of basalt, which are at places only due to slight shear strength at their touch with the relatively stable rock, dislodge and fall in the form of large slabs owing to the decrease in the bearing capacity of the marl and the high water pressure built around the contact (Figure 2a). These slabs of rocks accumulate at the base of the cliff until they finally transported down the slope with the help of runoff.

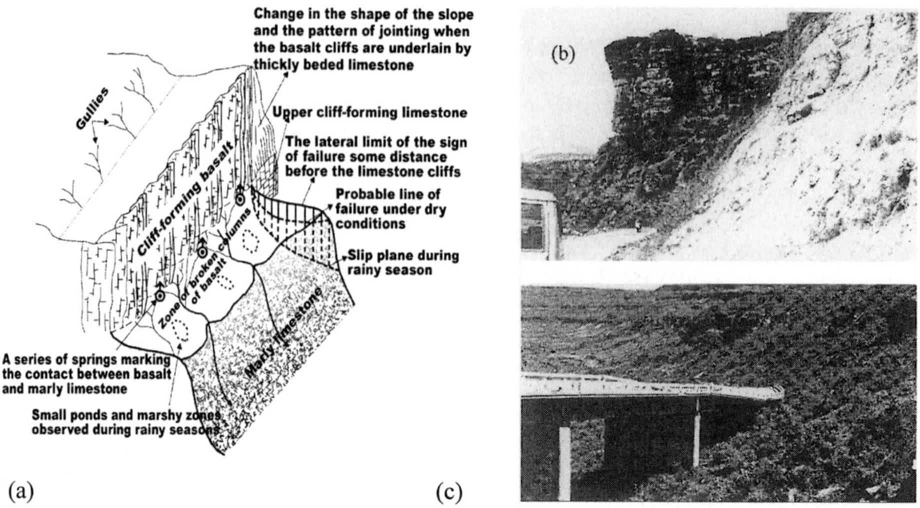

Figure 2. Graphical illustration and some examples on how landslides and rock falls are initiated in the gorge; (a) the mechanism of slope instability at the marly limestone-basalt contact; (b) long columns of sandstone treating the normal service of the road; (c) a bridge damaged because of old and newly rejuvenated planar landslides provoked by river erosion.

The intersection of various joints was also a major geologic constraint mainly for the presence of rock falls along the entire road and in the lateral cliffs of basalt, sandstone and limestone. Of greater importance were specially those master joints oriented parallel to the cliff face. These joints coincident with the slightly inclined rock beds and basal erosion yield long columns of sandstone that are top-heavy and likely to fall over to the main road (Figure 2b), alarming that the factor of safety against rock fall and rock toppling is only marginal.

In addition to the lithological make up and the structure of rocks, weathering and erosion played a significant role in initiating landslides and rock falls. The action of weathering was differential in most places owing to the presence of both soft and hard rocks. The effect of erosion was not also regular and depends on the slope angle and shape, and the nature of rock masses. Deep in river channels where large isolated columns of sandstone are present and along limestone and basalt cliffs, erosion was active around the base and in the mid height benches. At times, failure of a few columns of rock had a large impact on the slope profile and led to extensive cliff recession. The problem of cliff retreat places the service of the road and some villages on the cliff top at great risk. In the middle section of the gorge where the gradient is low, the main effect of river erosion was merely to reduce the slope angle further and to destroy bridges with the help of running water and mudflow (Figure 2c).

Mass movements in the form of landslides are frequently mobilised during periods of prolonged rainfall in summer and accompanying variation in groundwater storage. The unusual high amount of precipitation in 1993 and 1994, were specially responsible for rapid transformation of small scale non destructive landslides into those which were characterised by a mesa and butte rim collapse and large destruction (Lulseged and Berhanu, 1995). The effect of groundwater was specially high in concave topographic segments in the upper and middle section of the gorge where large amount of moisture from springs reduces the shear strength of soils and initiates semi-circular slope failures.

The concentration of landslides and rock falls along road sections and settlement zones is a prove to blame human beings for their role to facilitate the problem of slope instabilities through the action of extensive land clearing for agricultural purposes. In addition, old road sections which had been trimmed back to natural joints largely by hand in early periods, and which maintained almost vertical faces for more than half a century were disturbed in the last few years because of unplanned quarrying activities and poor system of road utilisation.

SLOPE INSTABILITY CLASSIFICATION

In this study, a GIS technique is adopted to classify the area into different zones on the basis of the degree of instability. The database used for mapping was a constitute of a comprehensive inventory of those landslides recorded from 1992 to 1996, the geology of the area, rainfall and groundwater condition, data on the geotechnical properties of rocks and soils, the topographic map and the hydrographic network. A GIS platform was used to integrate the information coming from different sources and was helpful in making rational decisions to divide the area into different zones of instability. The slope stability map shown in Figure 3, which later was modified with some graphic tools, was then the end result of the whole combination and integration process.

At first, those areas which were assumed to be zones of rock falls were demarcated based on objective and subjective information. Geological boundaries were found to be useful in this case as they usually mark sharp changes in engineering properties and are, therefore, significant for slope stability analyses and classification. Then, the rest part of the valley where mass movements in the form of landslides are expected, was divided into three zones: stable, semi-stable and unstable. In the stable zone, except for an evidence of creep in areas where the slope angle is relatively high, both active and inactive landslides are unlikely. The semi-stable zone is characterised by inactive landslides, and could be distinguished with the help of "lobes" and "arcuate" scarps, hummocky surfaces, and signs of old mass movements. Recent active slope failures, relatively fresh erosion channels, and unconsolidated superficial deposits are main features of the unstable zone. A few number of slope analyses at selected localities in the unstable zone showed that the factor of safety in these regions goes down to 0.73 specially under wet conditions.

CONCLUSIONS

The problem of slope instability in the Blue Nile Basin is accredited to various factors. The geological properties of rock masses, the rugged topography and the high amount of rainfall had a greater contribution. Human interference was also crucial specially concerning the recent landslides in the highlands and along road sections. All the information collected were combined using a GIS technique to divide the area into different zones of slope instability.

ACKNOWLEDGEMENT

The preliminary phase of this study was financed by the Ethiopian Science and Technology Commission. The author, therefore, owe a lot to the geologists of the commission, Mulugeta Amha, Abebe Mekuriaw and Fanta Demissie.

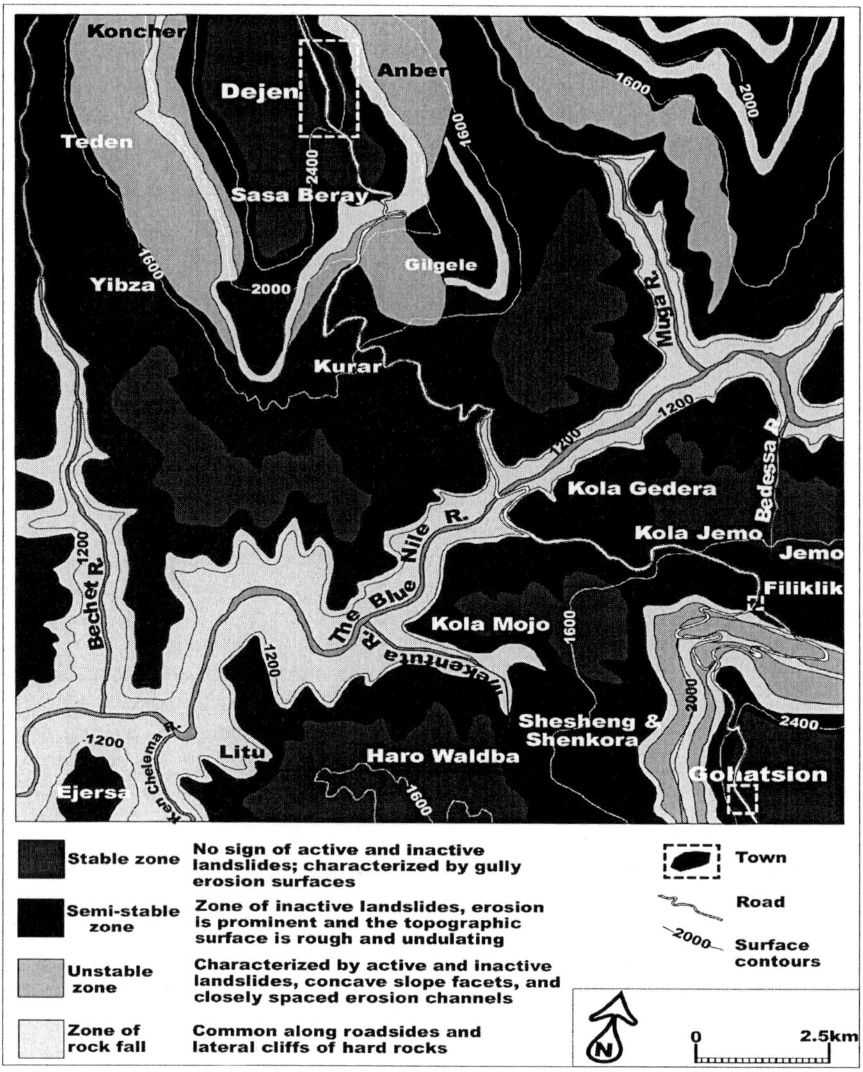

Figure 3. Slope stability map of the Gohatsion-Dejen stretch in the Blue Nile Basin.

REFERENCES

Getaneh Assefa (1981). Gohatsion Formation, a new Lias-Malm lithostratigraphic units from the Abay river basin, Ethiopia. Geoscience J., 2:63-88.

Jespen D. and Athearn M. (1961). General geological map of the Blue Nile river basin, Ethiopia. Water resources department, Ethiopia-United States co-operative program, Addis Ababa.

Lulseged Ayalew and Berhanu Temesgen (1995). Assessment of slope movements from Gohatsion to Dejen, Abay (Blue Nile) gorge, Ethiopia. A report submitted to the Ethiopian Science and Technology Commission, Addis Ababa.

New Typology of Low Environmental Impact Consolidation for Rockfall Prone Cliffs Through Interventions from the Underground

S. BANDIS, University of Thessaloniki, Greece, V. COLOMBINI, University of Basilicata, Potenza, Italy, G. DELMONACO and C. MARGOTTINI, ENEA CR Casaccia, Ministry of Environment, Rome, Italy

INRODUCTION

The problem of unstable towns in Italy is well represented by Civita di Bagnoregio, which appears as a typical medieval town, built on the top of a volcanic cliff and joint with the municipality of Bagnoregio through a modern bridge, founded on a narrow morphological clayey crest. Landslides involving the cliff-slope system have caused the disruption or endangering of many buildings founded close to the border, so that Civita is commonly known as "the dying town". This situation is largely common to other many historical towns of the surrounding area (i.e. Orvieto), sited on soft rock plateaux of variable thickness, surmounting sloping deformable clay formations and in a wider context to most of municipalities, along the Alpine and Apennine chains, in territories characterised by lithological complexity, high seismic and geomorphological activities. In this paper a new typology of low environmental impact reinforcement works for consolidating cliffs prone to rock fall, is discussed.

GEOLOGICAL AND GEOMORPHOLOGICAL SETTING

The village of Civita di Bagnoregio lies in a hilly area in the northern part of Latium (central Italy), surrounded by steep valleys intensively eroded by two torrents presenting an E-W direction. The cliff is composed of a 20-25 m top layer of jointed ignimbrite, underlain by intensively stratified pyroclastic formation; the Quaternary tuffs rest on a bedrock of a Plio-Pleistocene clayey succession which form the valley in the major area. The clays depth can be estimated in some hundreds meters, while the outcropping portion is about 150 and 200 m in the northern and southern valleys respectively.

Due to the particular geological and geomorphological features, the town of Civita di Bagnoregio has suffered from large and frequent slope instability phenomena: in the clayey formation landslides are represented by mudflows, debris-flows and rotational slides, while in the upper portion of the volcanic cliff, due to a retrogressive mechanism of erosion, rock-falls, toppling and block-slides are the common landslide typologies.

Landslide mechanism can be explicated by complex and integrated causes that, starting from the valleys erosion water-courses, is transferred to the slopes and, then, to the cliff. Such causes can be summarised as follows:

1. deepening of bottom valleys due to runoff;
2. decaying of mechanical properties of the clays due to weathering (swelling and softening) up to an average depth of 5-10 m; in the first 0.5-1.0 m this phenomenon is more evident;

Landslides in research, theory and practice, Thomas Telford, London, 2000.

3. occurrence of mudflows with a typical 0.5-1.0 depth, after heavy rainfall, with a continuous erosion of materials and exposition of underlain layers to weathering;
4. intense superficial erosion due to weathering (some centimetres/year);
5. deformation of the clayey bedrock due to the decaying of geotechnical parameters and induction of deformation processes to the overlying tuff rocks;
6. opening of pre-existing joints in the basal stratified tuffs, due to the significant increasing of deviatory stress associated with lateral unloading, and deformation of clays;
7. opening of pre-existing failures in the upper part of the ignimbrite due to thermoclastic and cryoclastic phenomena as well as to pore pressure increase along the joints;
8. increasing of deformations in the upper part of the cliff with a progressive failure mechanism from the pre-existing joints of upper tuffs to the stratified pyroclastic materials; in the latter the straining process produce rotational failures in correspondence of plastic layers;
9. rock-falls in the upper portion of the cliff, in correspondence of massive tuffs.

HISTORICAL AND PRESENT-DAY MORPHOLOGICAL EVOLUTION

A great deal of historical records, starting since 1373, on landslides occurred and stabilisation works, has been recovered and analysed in order to reconstruct the evolution of the urban setting. The town, of Etruscan or Villanovian origins (VII century BC), had a great expansion from the Roman Age to the late Middle Age, when the quarters Ponte and Carcere, actually disappeared, where added to the original urban nucleus. Topographical and cadastral maps, dating back to from the beginning of the 18th century as well as other historical maps and documents, have verified the progressive shortening of the cliff due to landslides that have caused, in different times, the disappearance of portions of the town sited on the border. The most important phenomena have affected: the Ponte quarter (in the western side) where many houses, surveyed in the map of 1705, are gradually disappeared in the cadastral maps of 1869 and 1937; the southern side, where the Church-House of S. Bonaventura, still existing in 1849, is definitively disappeared in 1937; the Carcere quarter, in the eastern side, which was almost completely ruined by the strong 1695 earthquake and by a large landslide (few buildings still remain in the 1869 map); the northern side, where some rock-falls destroy a portion of the cliff.

In the last decade some landslides have affected the northern border of the cliff of Civita. The rock-fall of February 1992, caused a retrogression of the scarp of 2 m, along a crown surface of about 40 m; the displaced blocks covered, in the cliff foot, a pre-existing rock-fall accumulation. Moreover, along the cliff border of new formation, in correspondence of *Piazza del Vescovado*, an opened joint, sub-parallel to the scarp direction, was observed. The joint, in W direction, intersected a building, heavily damaged, where this discontinuity appeared as widely opened and lowered. The rock-fall of August 1993 interested an area located 20 m W of the previous landslide with a scarp length of ca. 60 m. The fall involved both massive tuffs and part of the stratified pyroclastic materials, producing an average cliff retrogression of 6 m. During the previous months, in correspondence of the western side of the recently formed cliff border, an opened and lowered failure as well as local sinking were observed. Such failure represents the prosecution of the existing joint in *Piazza del Vescovado*, prolonged in W direction, under the damaged building. The displaced materials accumulated in the pre-existing valley below, reaching, with some isolated blocks, the bottom valley of the torrent Cireneo, at a distance of about 300 m from the cliff. The debris-flow of December 1996 re-mobilised the materials of the 1993 landslide, a small rock-fall occurred on August 1998 and a 1000 m^3 ca of a rock-fall on April 1999, that involved the stratified pyroclastic complex.

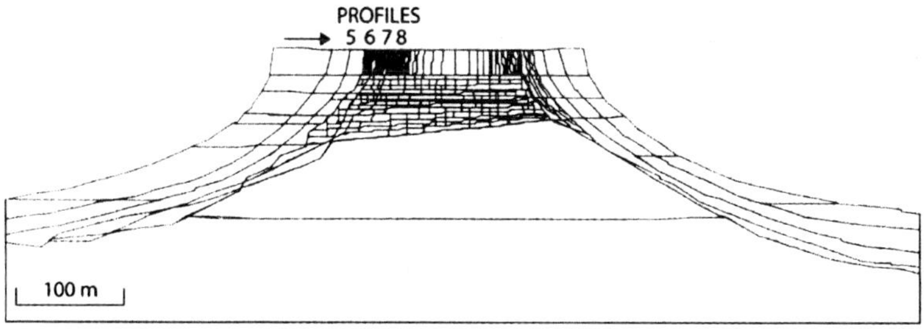

Fig. 1. Numerical representation (UDEC model) of 1600 ground profile with representation of geological setting and predominant rock mass structure (Bandis, 1999).

NUMERICAL MODELLING

A numerica modelling study has been undertaken to attempt an investigation of the progressive complex stability phenomena through the reconstruction, based on long term historical data, of the valley erosion and slope forming processes and monitoring of the induced changes in the stress regimes close to the critical zones. The practical implications of the back-analyses concerned the selection of type and the dimensioning of the reinforcement measures. A 2-D model of a representative section was developed using the program UDEC (Universal Distinct Element Code). A stepwise process of valley erosional unloading was implemented according to records and the induced principal stresses in the tuff formations were monitored at all stages of the simulated landscape development. The traces of known failure surfaces were introduced as fictitious cracks with properties equivalent to those of the tuff formations (Fig. 1).

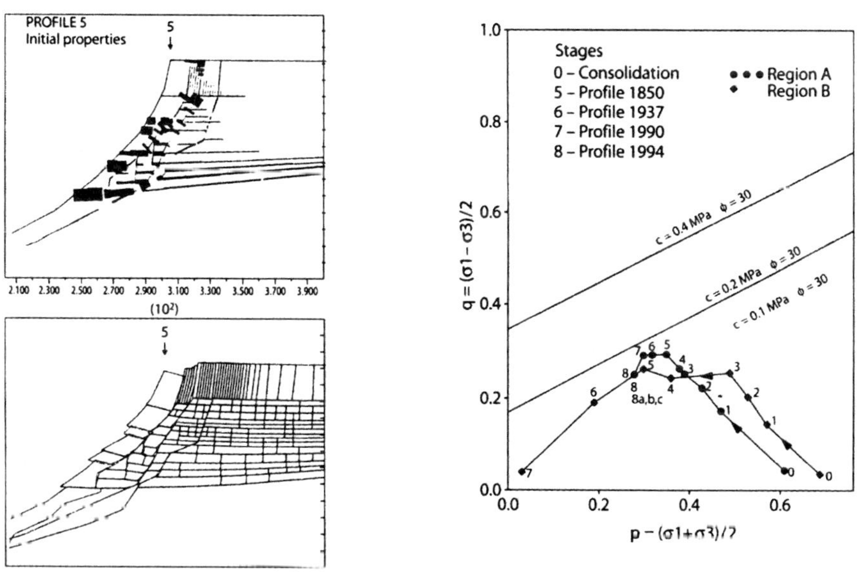

Fig. 2. UDEC modelling of shear stress development and progressive failure (left diagrams) and predicted stress path with respect to valley profiles reconstructed from historical records (right diagram).

Numerical analyses indicate that rock slope instability is related with significant build-up of differential shear stresses in the stratified tuff substrate. The increase in the shear stresses is the result of lateral unloading (stress relief) imposed by the erosion of the valley materials. Long-term overstressing of the materials in the outer slope zones and weathering effects both causing material weakening also appear as an additional contributing factor. Shear overstress development along potential failure planes and excessive shear strains (upper series of Fig. 2) may lead to progressive failure mechanisms (lower series of Fig. 2). The process has been applied repeatedly to several of the slope profiles 5, 6, 7, 8. Stress path within the stratified tuff associated with the simulated erosional processes (lateral unloading) as predicted from the UDEC model. A small decrease in cohesion apparently suffices to induce failure in the strata underlying the jointed massive tuff. Rock cliff failures may then develop due to loss of foundation reactions or excessive straining, promoting rotational failures of the rock columns rather then of a toppling mode. Confirming evidence on the latter was provided from field observations and monitoring data.

RESTORING WORKS OF THE CLIFF

As mentioned previously, in the northern part of the town recently some rock-falls occurred causing a decreasing of the cliff perimeter, in terms of several meters. After these landslide events the Civil Protection Department funded the restoration of a limited portion of the cliff, mainly affected by the 1992 rock fall. A pilot project was then developed according to the following purposes:

1. congruence with the limited available financial resources (ca. 500,000 Ecu);
2. to solve definitively the problem of slope stability, although for a limited area of the cliff;
3. to safeguard the medium-term stability of the surrounding portions of the cliff, not directly interested by restoration works;
4. to favourite the execution of future restoration in adjacent areas;
5. to preserve the external aspect of the cliff, in respect of the landscape and historical and cultural heritage of the town;
6. to facilitate a possible future restoration in the hazardous area of *Cavon grande*, involved by the rock-fall event of 1993, which is actually not restorable with the available funds.

The selected area is actually interested by a strong residual risk, as testified by an opened fracture isolating for a ca. 1 m thickness, the external portion of the cliff. The restoration of this area is indispensable for a successive intervention in the landslide area of *Cavon grande*, that can be realised under security conditions. This intervention is limited to the upper massive ignimbrite since the underlain stratified tuffs are completely covered and protected by a stable ancient landslide and recent debris materials; moreover the clayey bedrock does not evidence any deep deformation processes which may progressively involve the cliff.

Restoration works (Colombini et al., 1994) have been projected to stabilise the upper volcanic formation of massive tuffs through the realisation of wells located on the back with respect to the potential landslide scarp; the wells, hand-excavated in order to minimise vibrations, represent the link between a system of active anchorages and passive nailing of the external cliff border and a system of active anchorages oriented toward the internal side of the tuff masses (Fig. 3). The anchorages provide a lateral reinforcement to the cliff walls, preventing, as well, toppling, detachment or fall of rock masses; moreover their function is to reduce the vertical load component in the contact between massive and stratified tuffs, characterised by poor mechanical properties. The injections provide a better stability of medium-small rock masses, prone to sliding, toppling or fall as well as an increase of mechanical properties of rock masses exposed to weathering. A system of micropiles have been realised in the bottom of wells to link the structure with the bedrock, increasing the shear strength of masses potentially involved in landsliding.

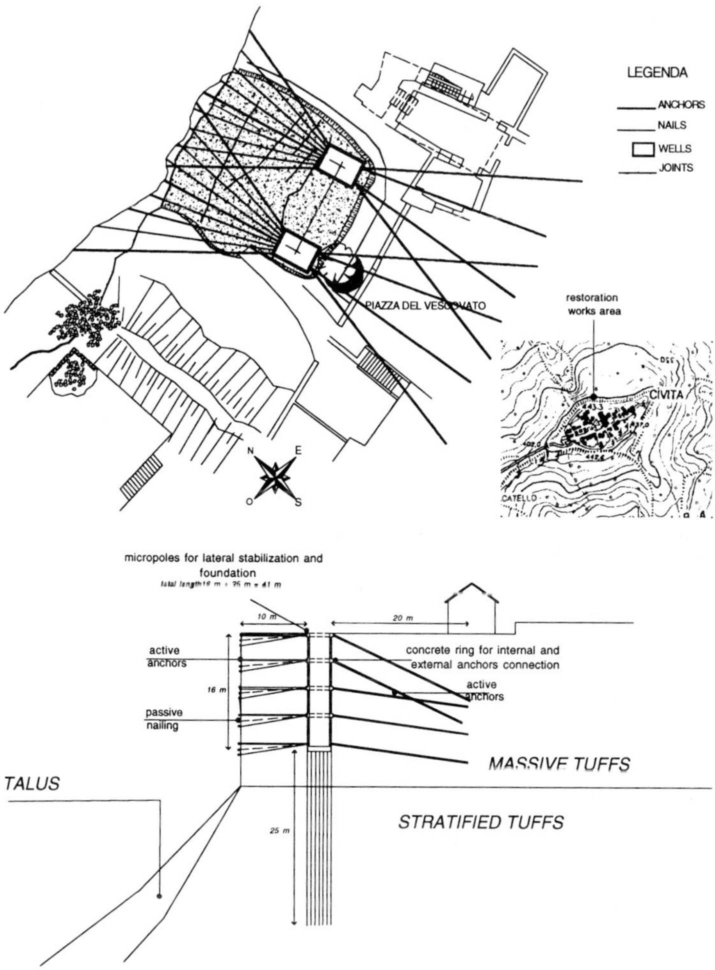

Fig. 3. Planimetry (above) and section (below) of the remedial works carried out in the northern border of the cliff in Civita di Bagnoregio.

This pilot reinforcement project has permitted to:

1. realise the works in the town in safety conditions for operators, not from the external side of the cliff, where a danger of landslides or rock-falls exist;
2. annul the environmental impact due to scaffolds or service roads;
3. unburden the landslide scarp, decreasing the load of rock masses in the basal lithotypes;
4. realise a structure permitting a direct control of the structural ground conditions during the excavation and working also as a draining system;
5. allow a monitoring of deformations and an eventual stressing of anchorages;

6. create a semi-rigid structure inside the tuff masses;
7. undertake a network on different horizontal planes both in perpendicular to the landslide scarp (nails and tension bars) and in parallel to the cliff (anchorage of wells) in order to diffuse and absorb possible deformations and overloads;
8. maximise the yard operators security realising the wells anchorage toward the internal side first, and then toward the landslide scarp. Execution time was reduced to 50%.

Fig. 4. Civita di Bagnoregio with, superimposed, the experimental restoration works.

CONCLUSIONS

The pilot project in Civita di Bagnoregio, although still far from exhaustive for the resolution of the landslide problems which affect the cliff/slope system of the town, demonstrates that an effective and correct restoration may be realised in the local ecosystem, contributing to the preservation of the landscape without disfiguring the natural environment. At the same time, the conceptual solution adopted for the stabilisation of rock-falls, represents an innovative tool for the consolidation of unstable towns, mainly where geomorphological conditions do not permit any intervention from the external portion of cliffs. Finally, this experience has proved the importance to transform the comprehension of geomorphological processes into engineering projects at low environmental impact which take into account the effectiveness of works and the respect of places.

REFERENCES

Bandis S.C., 1999. Numerical modelling techniques as predictive tools of ground instability. In Casale and Margottini (eds), Floods and Landslides, Springer, 1999, pp. 101-118.

Colombini S.I., Colombini V., Margottini C., Soccodato F.M., 1994. Progetto di risanamento di un settore settentrionale della rupe di Civita di Bagnoregio. Executive Project.

Margottini C. & Serafini S. (eds.), 1990. Civita di Bagnoregio: osservazioni geologiche e monitoraggio storico dell'ambiente. Una ricerca ENEA. 174 p.

Slope Movement Patterns in Young Valley Slopes in Northern Alberta, Canada

J. P. BARLOW, AGRA Earth & Environmental Ltd., Edmonton, Canada.

ABSTRACT
The river valleys in the plains of Western Canada are relatively young in a geological time scale and often have extensive landslide activity on the valley slopes. Many of these landslides have very low movement rates and show little or no visible signs of slope movement. However, pipelines, roads and other facilities installed in gradually moving landslide terrain can be subject to some risk of damage, depending on the details of the installation and the level of activity of the slide. Subsoil data collected at 17 investigated slopes has provided an opportunity to gain insight into the slide mechanisms in this terrain and into long term movement patterns. The effectiveness of drainage improvements to reduce slope movement rates is also assessed.

INTRODUCTION

Long term movement data have been synthesised from case histories on 17 valley slopes in Northeastern Alberta, Canada. The location of this region is indicated on Figure 1. This is a non-agricultural area where the two main industries are logging and oil and gas production. In addition to conventional oil and gas fields, the massive Athabasca oil sands deposit is located in this area, which is currently being exploited by two of the largest open pit mining operations in the world. The rivers and creeks in this area have extensive landslide terrain along the valley slopes which presents a challenge to resource development. This paper examines data that has been compiled from 17 slopes where a drilling and monitoring program was implemented because ground movements were affecting road or pipeline installations.

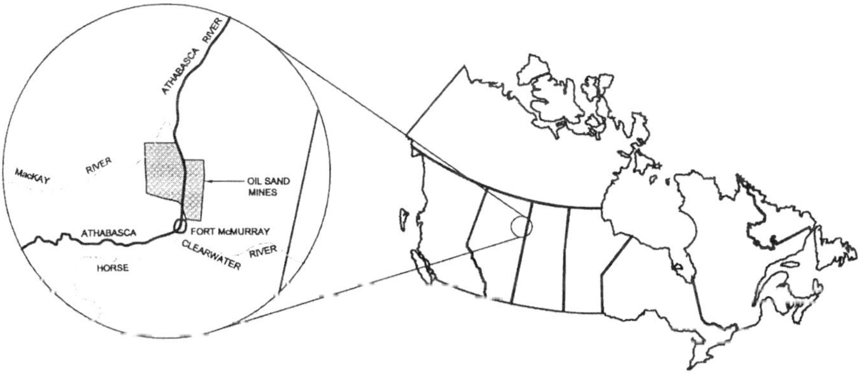

Figure 1: General Location of Case Histories in Northeastern Alberta, Canada

Landslides in research, theory and practice, Thomas Telford, London, 2000.

OVERVIEW OF GEOLOGY

The geology of the area consists of glacial deposits overlying sedimentary bedrock of Cretaceous age. The ice retreat from the last glaciation was complete approximately 12,000 years before present (Christiansen, 1979). Since that time the rivers re-established themselves either by incising new channels through the surficial deposits and into bedrock or by incising through glacial drift and re-entering previously excavated bedrock channels. Most of the case histories included in this paper relate to new drainage courses that do not follow previously established pre-glacial valleys.

The landslides in the area of interest in Northeastern Alberta are typically associated with the marine shales in the Cretaceous bedrock of the area. The bedrock geology generally consists in descending order of Clearwater Formation shale, McMurray Formation oil sand and limestone and shale of Devonian age (Carrigy, 1959).

The Clearwater Formation comprises uncemented clay shales, clay-silts and fine grained sands with thin indurated siltstone layers. The clay shales dominate the slope stability characteristics of the Clearwater Formation. The more critical subunits have very high plasticity, with liquid limits in excess of 150 percent, and residual angles of friction in the range of 6 to 9 degrees (Isaac, et al., 1982). The term “clay shale” is used to describe this material because it is classified geologically as a marine shale, but is more soil-like in behaviour relative to its potential to soften and break down on exposure to water in an unconfined state (Morgenstern and Eigenbrod, 1974).

The McMurray Formation oil sand tends to form steep slopes when undercut by stream erosion and is not typically associated with instability. The oil sand consists of bituminous sand with an interlocking sand structure (Barnes and Dusseault, 1982), with interbeds of clay and indurated siltstone. The interlocking sand structure and the interbedded nature of the deposit makes it a relatively strong, stable component of the bedrock sequence relative to valley slope stability. In some cases, clays at the base of the McMurray deposit can exhibit low shear strength and be involved in slope failures.

The Devonian interbedded limestone and shale that are encountered below the oil sand, form a competent base in the valleys relative to both slope stability and stream erosion.

VALLEY GEOMORPHOLOGY

The 17 case histories presented in this paper are from slopes along 8 separate rivers or creeks in the Athabasca drainage basin. Two of the case histories are from slopes of the Athabasca River, and the remaining cases are from tributary rivers and creeks that range in size from small creeks with valleys less than 5 km long to rivers that have overall valley lengths exceeding 80 km. The 7 tributary rivers and creeks are in active phases of down-cutting or lateral migration and have virtually continuous landslide terrain along the valley slopes, except in the upper reaches of the valleys that are not incised into the Clearwater Formation bedrock. The roads and pipelines traversing this terrain were generally located along what were perceived at the time to be the best slopes, yet every one of the 17 slopes discussed in this paper was moving. It could therefore be inferred that virtually all of the slopes along these valleys are moving at some rate.

The case histories can be divided into two reasonably consistent characteristic landslide profiles, depending on how deeply incised the stream is into the bedrock. If the stream is incised only into the Clearwater Formation (valleys less than about 60 m deep), the valley will be V-notched in shape with gently inclined slopes ranging from 8 to 11 degrees, and have a general stratigraphic profile as shown on Figure 2. The slopes in this category have a layer of colluvium of reasonably consistent thickness gradually translating down-slope.

Figure 2: Characteristic Slope Movement Mechanism for Streams in Clearwater Shale

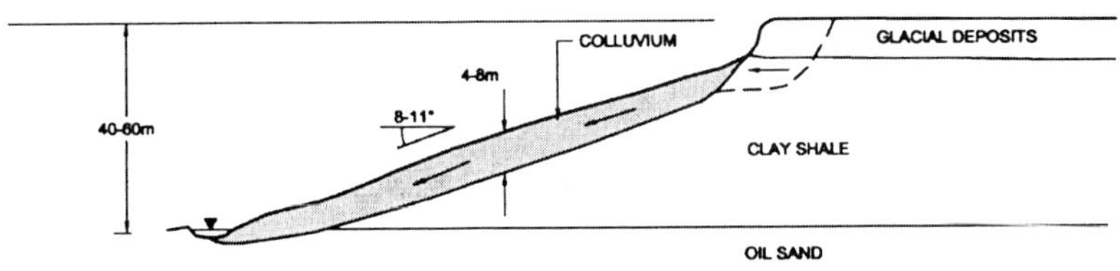

If the stream is incised into the McMurray oil sand, there is a distinct change in the subsurface profile that can be seen on Figure 3. While the river is under-cutting the base of slope (Figure 3a), the McMurray oil sand forms a steep erosional face in the order of 50 degrees. The failure process in the overlying Clearwater clay shale continues with the slide debris largely being transported downstream by the river. Once the river shifts, or if the volume of slide debris exceeds the transport capacity of the stream, an accumulation of deeper colluvium develops in the lower slope as illustrated on Figure 3b.

Figure 3: Characteristic Slope Movement Mechanism for Deeper Valleys

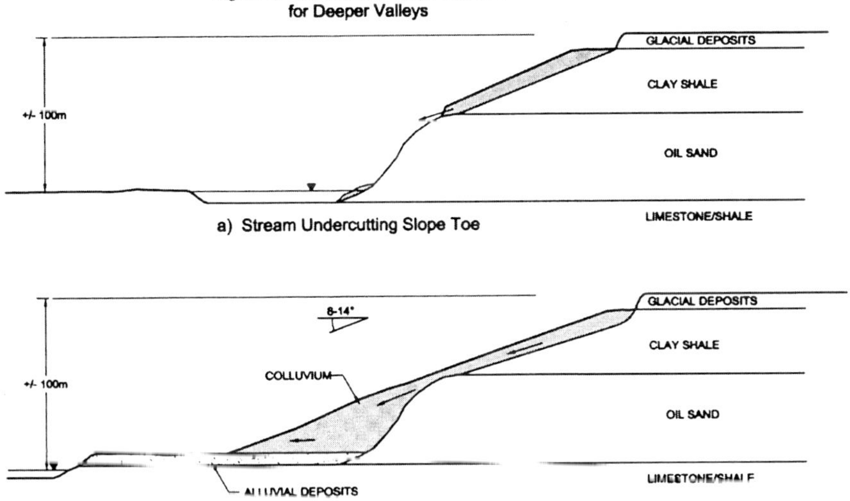

a) Stream Undercutting Slope Toe

b) Once Stream has Shifted Away from Slope Toe

Once the streams reach the base of the McMurray oil sand, the rate of down-cutting is much lower due to the erosion resistant harder limestone beds in the Devonian strata. The streams exhibit more active lateral migration and terrace and flood plain development begins to occur as the valleys widen.

Table 1. Slope movement data from 17 case histories

SLOPE	INCLIN-ATION	VALLEY DEPTH	MOVEMENT		INDICATIONS OF ACTIVE MOVEMENT AT SLOPE SURFACE
			DEPTH (m)	RATE (mm/yr)	
1. Athabasca Tributary Creek No. 1, North Slope	8	38	5	73	None
2. Athabasca Tributary Creek No. 1, South Slope	9	38	4	33	None
3. Athabasca Tributary Creek No. 2, North Slope	11	47	3	8	None
4. Athabasca Tributary Creek No. 2, South Slope	11	50	8	1	None
5. Poplar Creek South Slope	11	56	6	40	None
6. Athabasca Tributary Creek No. 3, South Slope	9	62	5	131	Minor cracks, small scarps
7. Goat Head Creek Slope A	11.5	66	6	75	Small tension cracks
8. Goat Head Creek Slope B	12	66	3.5	50	Small tension cracks
9. Goat Head Creek Slope C	8	66	8	44	None
10. House River Slope Section A	9.5	80	16	31	None
11. House River Slope Section A in 1975	9.5	80	16	188	Tension cracks, small scarps
12. House River Slope Section B	10	80	11	2	None
13. Horse River North Slope Section A	13	102	6	15	None
14. Horse River North Slope Section B	8.5	102	13	23	None
15. Horse River South Slope	12	102	4.5	180	Minor cracking
16. Athabasca River North Slope – Upper Slope	11.8	170	5	5	None
17. Athabasca River North Slope – Lower Slope	14	170	3	25	None

MOVEMENT DATA

A summary of the movement data for the 17 slopes is provided in Table 1. The table includes the overall valley depth, average slope inclination, long term movement rate, and the depth to the slide plane. Comments on the level of slope movement activity that is visible at the slope surface are also included.

The first 6 case histories shown on Table 1 are from shallower valley sections, where the streams are incised partway into the Cretaceous bedrock, as deep as the base of the Clearwater Formation shale. These valley slopes range from 38 to 62 m in height and have a relatively consistent inclination, varying from 8 to 11 degrees. The layer of colluvium overlying the in situ strata on these slopes varies between 3 and 8 m in thickness. There is a wide range in long term movement rates on these slopes from 1 to 131 mm per year, with an average of about 50 mm per year. The estimates of long term movement rates are either based on slope inclinometer data, or on physical observations of movement where pipeline installations were buried in the layer of colluvium. In several cases, the slowly moving landslides adhere to the pipeline coating and gradually pull the coating down-slope along the pipeline, providing a lower-bound measure of accumulated displacement of the slide.

Once the streams are incised into the McMurray Formation oil sand, a different characteristic subsurface profile develops as discussed earlier and is illustrated on Figure 3. Case histories 7 through 17 on Table 1 are from the deeper valleys that are incised part-way, or to the bottom of the oil sand. The only exception is the House River, which is excavating the in-filled glacial deposits in a pre-glacial valley. The slopes in these deeper valleys exceeded 60 m in height and ranged in inclination between 8 and 14 degrees. Most of the movement data shown in Table 1 is taken from the upper sections of these slopes where the layer of colluvium is sliding over the Clearwater Formation shale. The long term movement rates for these slopes ranged between 2 and 188 mm per year, with an average of 58 mm per year. A much lower rate of movement was generally observed in the lower slope section where the colluvium is considerably deeper, which would be expected.

SLIDE MECHANISMS

Based on the observations at these slopes and other slopes in the area, the following process of instability has been identified:

- The colluvium blanketing the slope is derived from the failure of the clay shale and overlying glacial clay till in the upper section of the stratigraphy.
- Due to the high plasticity of the clay shale, and how readily its structure breaks down on exposure, the colluvium becomes a low permeability remoulded clay that tends to block the seepage from the more pervious beds in the underlying in situ strata.
- This creates a build-up of pore pressures at the base of the colluvium, which promotes further movement of the colluvial layer.
- With continued movement, the slip plane at the contact between the colluvium and the underlying shale trends toward the residual angle of friction, which can be as low as 6 degrees for the shale as stated earlier.
- Further destabilisation is promoted as on-going movement then disrupts surface drainage on the slope, producing ponding and high levels of infiltration of surface water into cracks and into the more pervious portions of the colluvium
- As the colluvium continues to translate down the slope, an over-steepened scarp at the slope crest develops that will eventually lead to a translational failure based in the upper section of the clay shale that will add material and further loading to the colluvial layer at the top of the slope, allowing the process to continue.

A plot showing movement rate versus slope inclination is shown on Figure 4 for the case histories discussed above. As can be seen on the figure, there is no correlation between movement rate and slope inclination. In fact, the slopes with the higher movement rates are more gently inclined. There is also no appreciable difference in the rate of erosion at the toe of any of these slopes. The key difference between the cases is the surface and groundwater conditions, with the more mobile slopes all having higher groundwater levels. While slope instability is initially triggered by the under-cutting of the stream, the process of instability discussed above can lead to such extensive slide activity that the stream's ability to tranport the debris away can be overwhelmed. In some of the cases, the colluvium extended well below the base of the stream.

The data have also been divided into two categories based on whether or not there were any surficial indications of active slope movement. In 11 of the cases, where the movement was generally less than 50 mm per year, there were no visible indications of any active slope movements. The 6 cases with movement rates between 50 and 188 mm per year showed some minor indications of active movement. This was manifest as small unvegetated soil scarps, and tension cracks that were generally less than 5 cm wide. However, even for the highest movement rates of over 180 mm per year, the tree cover was generally not affected, and cracking or fresh scarp development could easily be missed under the forest litter. In some of these cases, the total accumulated slope movement as evidenced by pipeline coating losses exceeded 2.5 m over a 32 year period, with no obvious distress to the tree cover, which included 80 year conifers that were standing vertically.

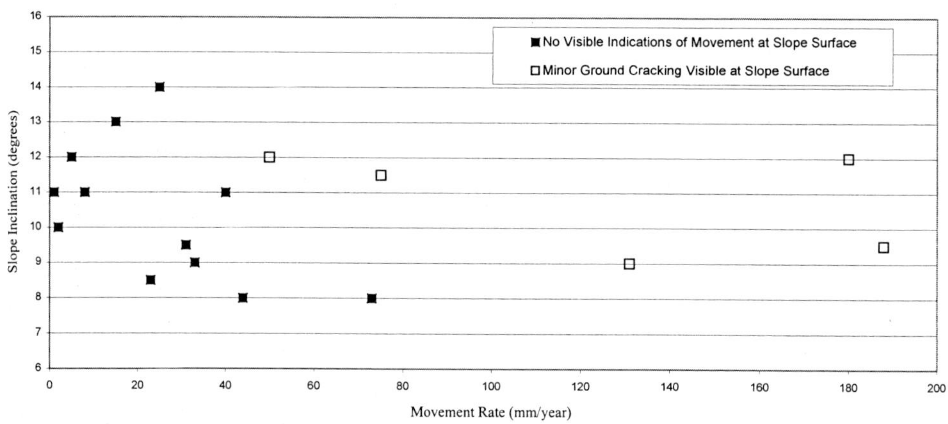

Figure 4. Movement rate data for valley slopes

Slope inclinometer data consistently show that virtually all of the movement occurs at the slip plane at the base of the colluvium, with little straining within the translating layer of colluvium. This is consistent with the lack of surface disruption for these slopes as discussed above.

INFLUENCE OF SLOPE DRAINAGE IMPROVEMENTS

Given that the groundwater levels are the key driving factor for these instabilities, drainage improvements are typically the most efficient way of abating slope movements. Various forms of drainage improvement were implemented at 10 of the 17 slopes in an effort to reduce slope movement rates to mitigate the impact on pipelines, roads or power lines. In each of the cases, an investigative program was done to establish the groundwater flow patterns and identify target zones for de-watering. At each of the 10 slopes, one or more of the following drainage measures were adopted:

- Construction of ditches and re-contouring of the slope surface to more effectively drain surface water from the irregular, broken topography of the landslide terrain;
- French drains tied into pervious zones in the strata underlying the colluvium where the colluvium was thin enough to excavate a trench through; and
- Horizontal wells to intercept deeper zones of seepage below the sliding layer of colluvium, where colluvium depths were larger.

Table 2. Influence of drainage measures on slope movements

SLOPE	MOVEMENT RATE (mm/yr)		REDUCTION IN MOVEMENT RATE	TYPE OF DRAINAGE IMPROVEMENT
	BEFORE DRAINAGE	AFTER DRAINAGE		
1. Athabasca Tributary Creek No. 1, North Slope	73	4	95%	French Drains
2. Athabasca Tributary Creek No. 1, South Slope	33	3	91%	French Drains
6. Athabasca Tributary Creek No. 3, South Slope	131	3	98%	Surface and French Drains
7. Goat Head Creek Slope A	75	2	97%	Horizontal Wells
9. Goat Head Creek Slope C	44	4	91%	Horizontal Wells
13. Horse River North Slope Section A	15	0	100%	Surface and French Drains
14. Horse River North Slope Section B	23	6	74%	Surface and French Drains
15. Horse River South Slope	180	5	97%	Surface and French Drains
16. Athabasca River North Slope – Upper Slope	5	0	100%	Horizontal Wells
17. Athabasca River North Slope – Lower Slope	25	0	100%	Horizontal Wells, Surface and French drains

A summary of the slope movement data before and after implementation of drainage measures is presented on Table 2. As can be seen on the Table 2, slope drainage had a significant positive effect on reducing movement rates. The average reduction in movement rate was 94%. In many cases, the actual quantity of water extracted from the subsurface was relatively small. However, when seepage discharge is blocked by a low permeability layer of clay-rich colluvium, significant pressures can develop, even with small flow quantities.

In all of the above projects where drainage improvements were implemented, the cost of drainage was a small fraction of the cost of other solutions being considered, such as toe berms, retaining walls, or methods to isolate the facilities from the effects of the moving ground. However, the site investigation required to adequately identify the pervious strata and groundwater flow patterns to design the de-watering system was often more extensive than would have been required to design other stabilisation techniques.

CONCLUSIONS

1. Streams incised into the Clearwater Formation clay shales produce extensive slide activity on the valley slopes. Most of the valleys examined in this paper had landslides with some level of active movement covering nearly 100 per cent of the valley slopes.
2. The gradually moving slides from 17 valley slopes showed little to no indication of active slope movement visible at the slope surface. Up to 50 mm/year of movement, no indications were found. For the slopes with 50 to 188 mm/year of movement, minor scarps and tension cracks were found, but the tree cover was generally undisturbed.
3. Implementation of surface and subsurface drainage measures was a highly effective means of reducing or eliminating movements, with an average reduction in movement rates of 94 percent measured at 10 remediated slopes.
4. The overall cost of the drainage improvements was typically a fraction of the cost of other stabilisation methods, but a detailed investigation was crucial to successful implementation.

REFERENCES

Barnes, D.J., and Dusseault, M.B., 1982. The influence of diagenetic microfabric on oil sands behaviour. Canadian Journal of Earth Sciences, 19, pp. 804-818.

Carrigy, M.A. 1959. Geology of the McMurray Formation. Part III General Geology of the McMurray Area. Research Council of Alberta. Geological Division. Memoir 1.

Christiansen, E.A. 1979. The Wisconsinan deglaciation of southern Saskatchewan and adjacent areas. Canadian Journal of Earth Science, Volume 16, pp. 913-938.

Isaac, B.A., Dusseault, M.B., Lobb, G.D., and Root, J.D. 1982. Characterization of the Lower Cretaceous overburden for oil sands surface mining within Syncrude Canada Ltd. Leases 17 and 22. Proceedings, 4th International Congress of the International Association of Engineering Geology, New Delhi, India.

Morgenstern, Norbert R., M. ASCE and Eigenbrod, Kurt D. 1974. Classification of Argillaceous Soils and Rocks. ASCE Geotechnical Engineering. Vol. 100. Nos. 7-12, pp. 1137-1156.

Stability of a Slope in Perm with a Building Complex Located on it, Analysis and Slope Stabilization

A. A. BARTOLOMEY, L. A. BARTOLOMEY, Perm State Technical University, Perm, Russia, and S.V. IRUNDIN, FBB PSTU Perm, Russia

INTRODUCTION

The city Perm is located in the piedmonts of the western Ural mountains. The central part of city is separated into parts by two old broad gullies. The intensity of building in a city with a population of one million has encroached on the banks of these broad gullies. On the slope of the broad gully containing the river Egoshiha, the construction of buildings and structures associated with road traffic management in the Perm area was begun. In geomorphological terms, the attitude the construction platform is located on terrace IV above the flood plain of the river Egoshiha. The height of the platform above the thalweg of the broad gully was 11-12m. The slope angles of the banks of the broad gully are around 18-25°.

Before the start of construction, the platform was excavated to in-situ soil by benches cut into the slope. The in-situ soil is represented by stiff loams with inclusion of shingle, gravel, crushed rock, building debris and brick rubble. The natural slope is combined alluvial deposits and quarternary systems by the way of stiff loams with silty partings and sand with gravel up to 10 %, further clays from half-firm up to soft-plastic consistency, inhomogeneous with frequent silts, sand with shingle and gravel up to 20 %. Beneath this lies a lower bed of coarse gravel, which is saturated, varies in thickness from 2.8-15.8m, and is formed from weathered argillites and sandstones of the Perm system below.

Ground water is met in the natural soils at a depth 2.2-4.7m from the surface and in quarternary deposits at a depth of 10.2-16.6m. Water-saturated soils are the loams, sandy loams and gravel-pebble soils. Ground water was encountered at a depth of 14.4-15.8m within the limits of the platform, and at a depth of 2.8-3.7m in the broad gully. The ground water has a hydraulic connection with water in the alluvial deposits.

During a period of 10 years on this slope there has been constructed a whole complex of buildings with a 125m transmitting antenna. Then at the bottom of a slope a two storied building was constructed. As a result of such building, the slope has appeared to be overloaded by the weight of constructed buildings, the drainage lines of underground waters were impeded, some parts of the buildings have deformed, and there was a vital need for the prognosis of stability of this slope and, if necessary, its strengthening.

CHOICE OF THE DESIGN SCHEME, ANALYSIS OF LOADING HISTORY

For the prognosis of stability of a slope by us, were used the program and method, designed by Bartolomey L.A. (1999). Before drawing up of the design scheme, the numerous reports on geologic surveys were investigated and the summary section below was composed. In the most dangerous zone of possible movement, the double-storied building at the bottom of the slope and 125m transmitting antenna were located.

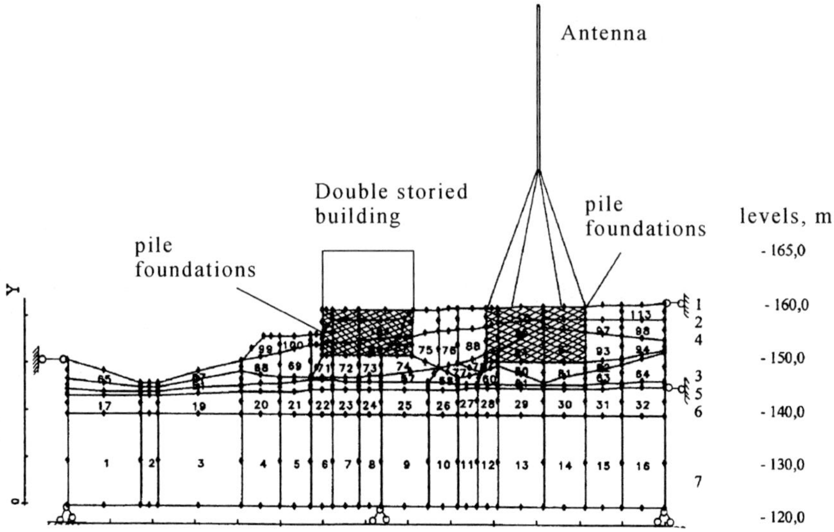

Figure 1. Design scheme for the prognosis of stability of a slope

In Figure 1 the design scheme for the prognosis of stability of a slope is represented. All elements in the considered area are eight-noded parabolic Serendipity-family finite elements. The condition of a plane deformation was considered. The mass of soil was represented with a viscoelastic-plastic soil model. The function of fluidity was set in an exponential kind, with the flow rule associated with the Mohr-Coulomb yield function. The properties of soils that are taken into account in the report, were obtained from geological investigations and laboratory researches. In the Table 1 the properties of each stratum of a soil are represented.

Table 1. Properties of soils that are taken into account in design

N	Name of the Soil	Modulus of deformation E, MPa	Poisson's ratio ν	Specific soil weight, kN/m^3	Cohesion c, MPa	Angle of friction ϕ'^{o}	Parameter of viscosity
1	Bulk soil	1.25	0.4	18.3	0.0212	9.0	0.002
2	Loam soft-plastic	2.5	0.38	19.1	0.013	17.0	0.0035
3	Clay hard-plastic	5.0	0.3	19.0	0.0415	20.0	0.0015
4	Loam hard-plastic	6.0	0.3	20.3	0.039	24.0	0.0016
5	Gravel-pebble soil	20.0	0.27	21.5	0.31	33.0	-
6	Argillite	40.0	0.23	22.0	0.6	35.0	0.001
7	Sandstones	80.0	0.23	23.0	1.0	37.0	-

For ferro-concretes pile foundations of property were nominated on normative data.

The analysis of stability of this slope was conducted stage by stage. An initial body-weight of soil and pile foundations of the transmitting antenna was applied. The extraction of soil from foundation pit and system pile foundations of the two-storied building was then modelled. At the last stage vertical and horizontal loads acting on the foundations of the antenna and the double storied building were incorporated in the model.

OUTCOMES OF ANALYSES OF STABILITY OF THE SLOPE

The outcomes of the finite element analyses are represented on Figures 2 & 3 by the way of isolines of horizontal displacements, isolines of horizontal stresses and area of viscoplastic deformations.

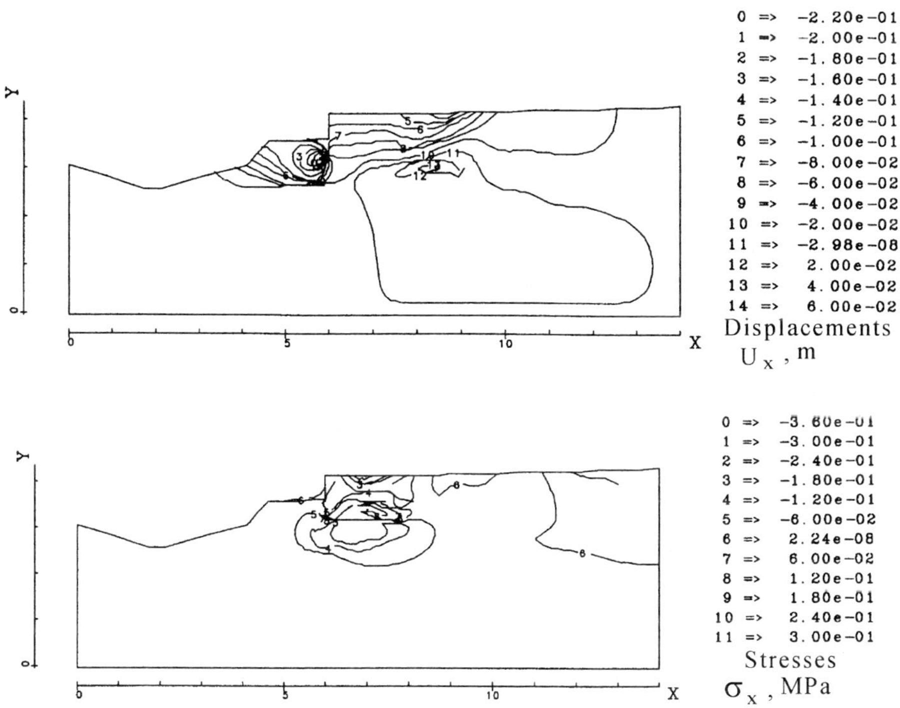

b

Figure 2. Isolines of horizontal displacements (a) and isolines of horizontal stresses (b)

The analysis of horizontal displacements shows, that at a design load on the foundations of a double storied building there are local displacements of in-situ soil on the slope behind the building. Building receives horizontal displacements from 4.0 up to 8.0cm. Viscoplastic deformation of the soil happens in a zone immediately adjoining to edge of the foundations of a building on the bottom part of the slope (Figure 4).

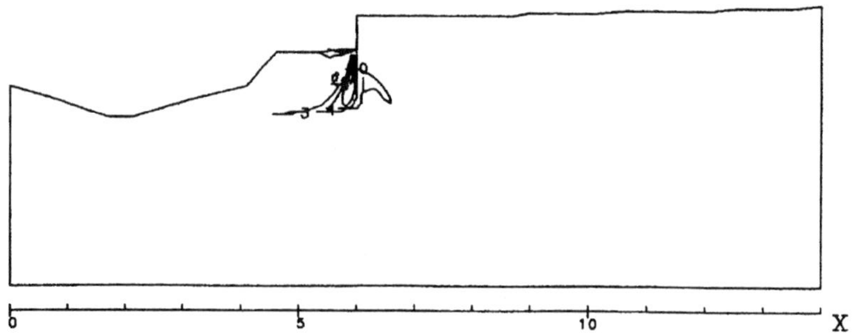

Figure 3. Area of viscoplastic of deformations

As the foundations under a building, the six-meter prismatic piles were arranged which fully penetrated the strata of in-situ soils and penetrated a further distance of 1.5-2. into sloping rockhead beneath. The geodetic surveying measured vertical non-uniform settlings of a building of about 2.0-3.0 cm and horizontal displacements on the slope of from 3.0-5.0cm. The survey was executed after the erection of the building but prior to the start of fitting-out.

The account of factor of a stability margin of a slope has shown, that as a whole slope is in a stable state: F=1.13-1.8. Further overload of a slope could result in the development of a more extensive zone of deformation and decrease of stability margin, that would cause sustained deformations both in situ soils and bedrock.

PROGNOSIS OF STABILITY OF A SLOPE IN TIME

On our long-term observations, the building programme in an urban area results in raising the level of ground water on the average by 0.5m annually. The considered slope at the given state will be saturated intensively by atmospheric and floodwater at the expense of filtering last through the bulk soil. Besides this, a considerable volume of underground waters will arrive from overlying territory. Thus strength and the deformation properties of water-saturated soils will decrease with time.

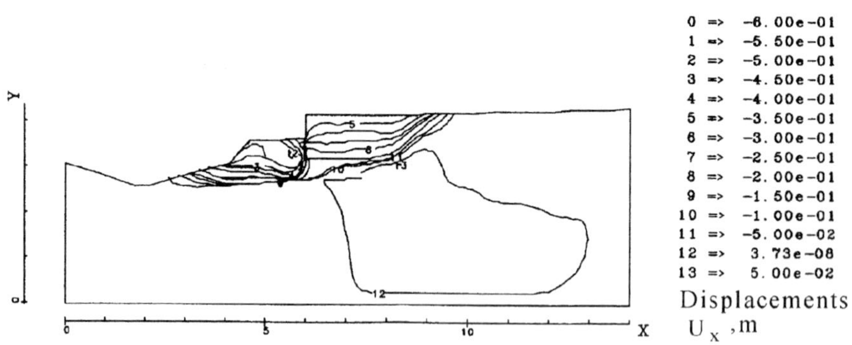

Figure 4. Isolines of horizontal displacements

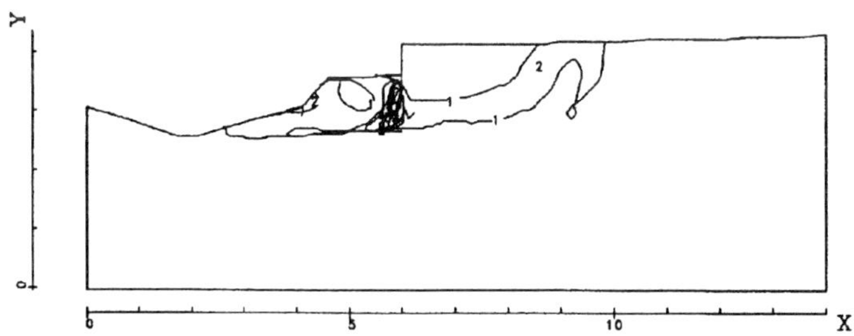

Figure 5. Area of development viscoplastic of deformations

Calculations were made of the stability of a slope in a condition of saturation by water in 6-10 years timescale of an in-situ soil (1) and soft-plastic loam (2). Strength and the deformation properties of 1-st and 2-nd stratums were reduced progresssively with time. The outcomes of calculations are represented by the way of isolines of horizontal displacements (Figure 4) and area of development of viscoplastic deformations (Figure 5). As it is visible from these figures, there is an extensive development of horizontal deformations and areas of viscoplastic deformation of a soil. The slope loses stability. Factor of safety is computed to be F=0.8.

FIXING OF THE BOTTOM OF THE SLOPE BY CHISEL PILES

It was accepted a solution to anchor a dangerous site of a slope with the help of a construction from chisel piles with a diameter 50cm and length 12m, arranged in casing pipes. In this connection the following account was executed. Immediately behind the double storied building, the construction from chisel piles incorporating a monolithic foundation mat was provided. The properties of soils and loads acting on the foundations of structures were taken as the same as in the first analysis. The construction from chisel piles is not loaded.

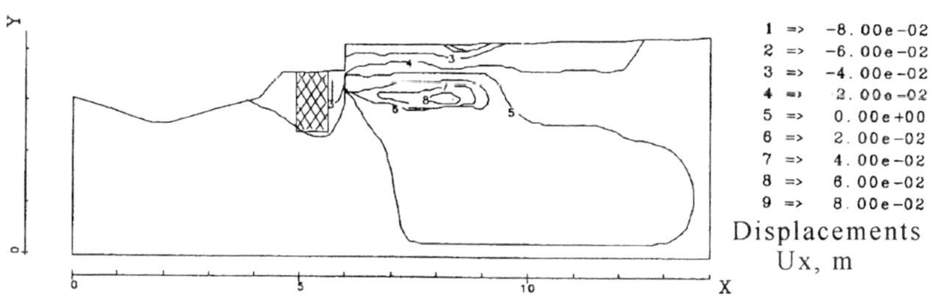

Figure 6. Isolines of horizontal displacements after fixing

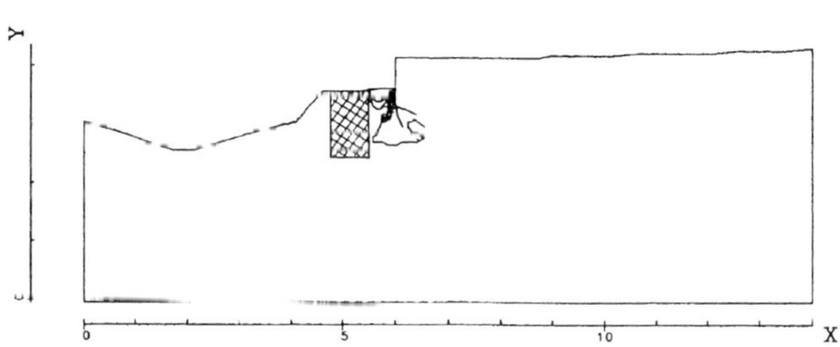

Figure 7. Area viscoplastic of deformations after fixing

As a result of the analysis, the displacements, stresses and area of development of viscoplastic deformations were determined. On Figure 6 the isolines of horizontal displacements, on Figure 7 the areas of viscoplastic deformations, are represented. As it is visible from Figure 7, viscoplastic deformation arises only in a zone between the building and chisel pile wall, occurring in the bulk soil and gripping 2-nd stratum - soft-plastic loam. All the remaining mass of the soil is only deformed elastically.

The horizontal displacements of a double storied building have decreased and make from 0.0 up to 4.0cm in comparison with 4.0 and 8.0cm in the analysis without fixing the slope by piles. Factor of a stability margin of a slope F is in between the limits F=1.32-1.8. Thus, the wall from chisel piles essentially augments stability of the slope at its bottom.

CONCLUSIONS

Hereinafter we executed geotechnical studies and drainage on the building platform. On the basis of numerical analyses and geophysical researches, the scheme to protection this sloping site from moistening, and stabilizing works consisting of chisel piles in casing were executed to treat a dangerous site on a slope.

During 1998 the system of 0.5m diameter vertical drains for interception of underground waters to transfer them to the underlying underlying gravel-pebble deposit was executed. At the bottom of the slope, chisel piles (diameter 0.5m and length 12m) in metal casings incorporating a monolithic foundation mat were constructed. On this foundation, the one-storey building is constructed. For July 1999 of adjustments of a slope and deformations of buildings is not fixed.

REFERENCE

Bartolomey L.A. 1999. The subsidence forecast of buildings taking into account the joint work of bedding, foundation and overground constructions/ Perm State Technical University. 147p.

Inclinometer and Tiltmeter Monitoring of a High Chalk Cliff

M. E. BARTON and A. M. McCOSKER, Department of Civil & Environmental Engineering, University of Southampton, UK

INTRODUCTION

The southern coastal road (A 3055) at Afton Down, Isle of Wight (UK) approaches to within 15 to 30m of the edge of a steep chalk cliff of 70m height. The threat posed to the security of the traffic by cliff falls and landward recession of the cliff edge was reinforced by the discovery of open joints beneath the road surfacing.

To minimise the hazard to traffic, a tiltmeter-based early warning system was installed in 1981. Tilt angles going beyond a pre-set upper limit actuate traffic signals located at each end of the threatened section of road. A total of 5 tiltmeters were installed as shown on the site plan (Fig. 1). Readings of tilt angles have been recorded manually from the date of installation and provide a record of mainly steady creep but with some accelerated movements (Barton and McInnes, 1988).

Further investigation of the displacements taking place within the cliff were made in 1987 by installing two inclinometer casings. One of these (Borehole No. 2) was located within the area covered by the tiltmeters (as shown on Fig. 1). After an initially fairly frequent set of readings, the inclinometers have been monitored annually. The site thus provides what is believed to be a useful (at least in terms of what has been published in the literature) case record of where both inclinometer and tiltmeter measurements have been carried out and hence affords an opportunity to compare their respective records of ground movements.

GEOLOGY

Afton Down (and the area immediately east, known as Compton Down) is on the southern side of the Isle of Wight monocline . At this location, the beds are dipping at about 55° to the north (inland direction). The cliff with its highest elevation of 73m above O.D. consists wholly of the Chalk, with the Lower Chalk just present near the toe, the Middle chalk occupying the lower half of the cliff and the upper half being in the more massively bedded Upper Chalk. The cliff face shows a marked break of slope with a slope angle of 53° in the Middle Chalk increasing to 71° in the Upper Chalk. Between the road and the cliff edge, there are a series of large, open vents with widths up to several metres and depths of 1½ m. The vents are the surface expression of tension cracks developed along joint planes by the toppling mode of failure of the Upper Chalk (Barton 1990). A variable thickness, up to several metres at maximum, of soliflucted chalk overlies the in-situ Chalk.

Landslides in research, theory and practice, Thomas Telford, London, 2000.

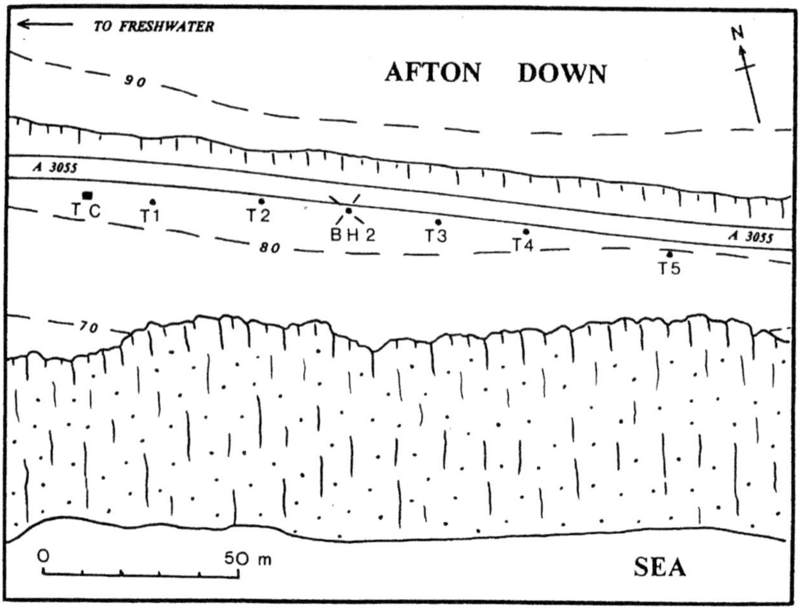

FIGURE 1. Site Plan. T (1 - 5) = tiltmeters. TC = tiltmeter control box. BH 2 = Inclinometer

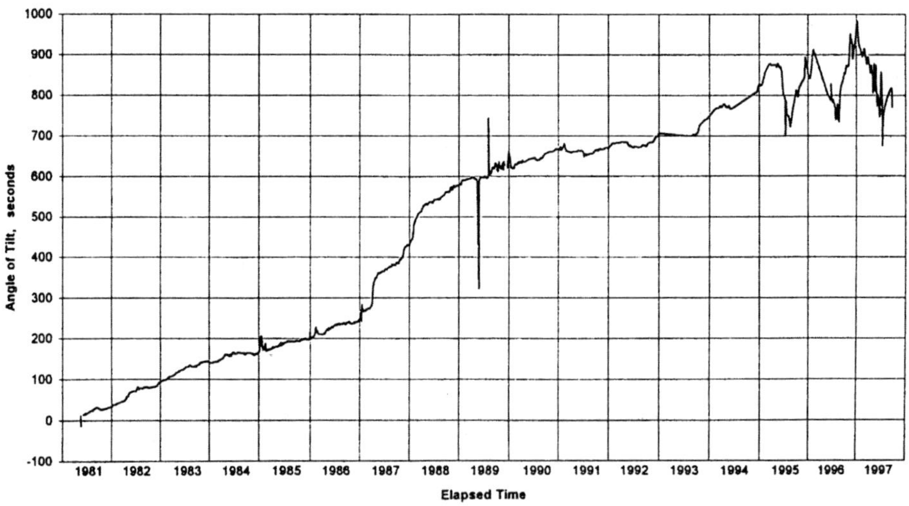

FIGURE 2. Tiltmeter No. 3: true tilt angle versus time.

TILTMETERS

Each tiltmeter incorporates a pair of orthogonally orientated electrolevels, one vial being in the seawards/landwards direction. These are fixed to the top of a tube, of 1.24m length, concreted into a borehole with a base at 2m below ground level. Each electrolevel has a range of 1,999 arc seconds with a resolution of ± 1 arc second.

The general trend for all tiltmeters is an increasing tilt but at a varying rate. The tilt is mainly directly seawards but in some cases has a time dependant easterly or westerly component. The total range of tilting varies between 240 seconds (in Tiltmeter No. 2 between 1981 and 1997) and 1,090 seconds (in Tiltmeter No. 4 which ceased to function during 1991). Characteristic features of all the tiltmeter records are fluctuations, which in the case of Tiltmeter No. 2 are clearly seasonal in origin, and 'spikes' which are very short term fluctuations (probably due to problems with the power supply, frost action or possibly, in some cases, due to sudden ground movements).

The record from Tiltmeter No. 3 over the period from 1981 to 1997 is shown in Figure 2 The tilt record from this instrument is virtually confined to the seawards direction only. This plot shows the marked increase in the rate of tilting during years 1987 - 1988 which is a feature characteristic of all the tiltmeter records.

INCLINOMETER

The inclinometer tubing in Borehole 2 extends to a depth of 30.5 m below ground level. The tubing was firmly grouted into the Chalk except from a depth of about 3.5m to 7.5m below ground level where the borehole passed through an open joint. Although there is a lack of stiffness in this section, as shown by the readings taking a long time to settle down, nevertheless the actual readings are very consistent and conform to the trends shown by the rest of the borehole. Records over a two-yearly interval since 1987 are given in Figure 3

The inclinometer displacements are of particular interest in providing a record of movements within a well jointed rock (the Chalk) which is without any well defined shear surface. Inclinometer installations are more usually located in locations where a slip surface acts as the boundary between two rigid, or fairly rigid, sliding masses.

Figure 3 shows a pattern of movement where the displacements increase upwards (in the seawards direction) but without the presence of a single identifiable slip surface. This is in keeping with the expected pattern of increasing displacements between individual jointed bounded blocks. Figure 3 also shows an inland movement at depth, particularly from 21 to 26m, which may represent either the natural pattern of joint displacements or the thrust developed by the tubing in response to the bending strains imposed higher up.

The displacements taking place at a depth of 1m are plotted versus time in Figure 4 The general trend of increasing displacement in the seawards direction is in keeping with the trends shown by the tiltmeters. Also in keeping with the tiltmeter records is the marked increase in the rate of displacement which took place during the years 1987 - 1988. This is believed to be related to the notably stormy winters of these two years (R. G. McInnes, personal communication).

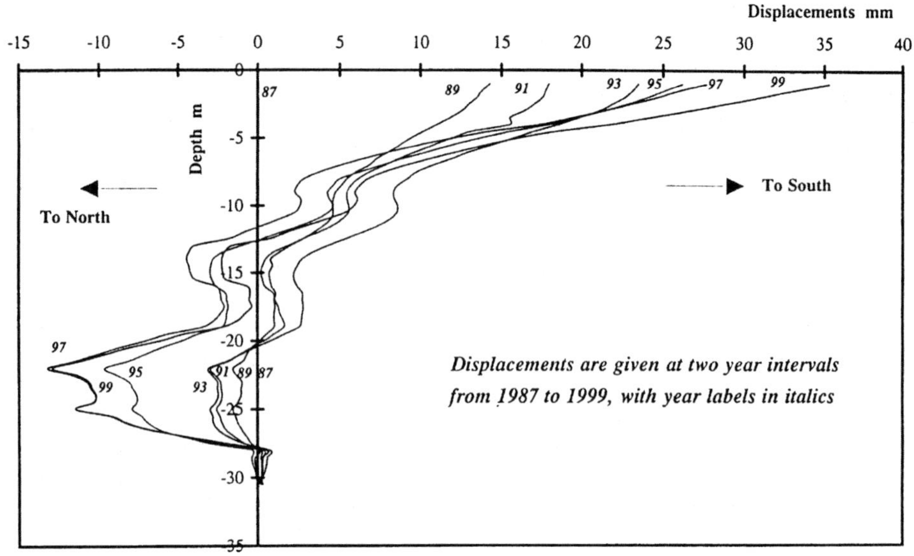

FIGURE 3. Inclinometer, Borehole 2, N - S Keyways: Displacement versus Time.

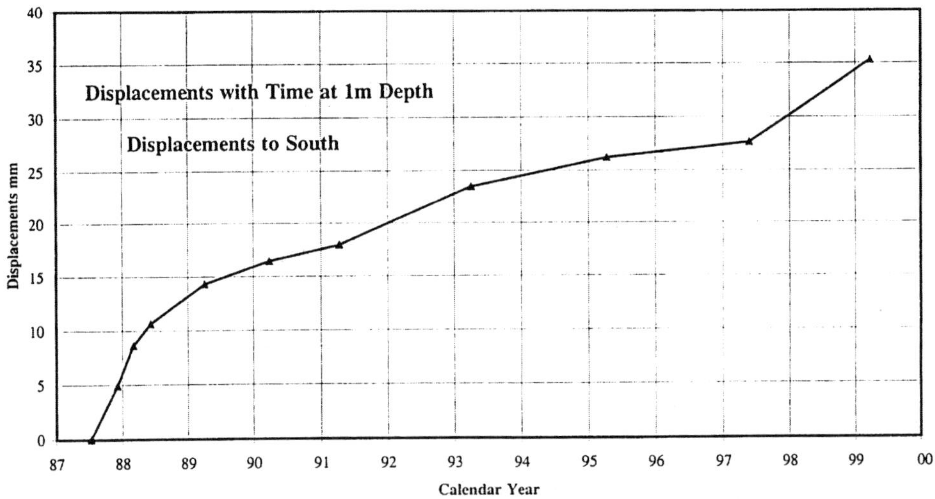

FIGURE 4. Displacements at 1 m depth for Inclinometer, Borehole 2, N - S Keyways.

TILT ANGLES FROM INCLINOMETERS

Inclinometers are used to measure relative displacements at depth (being "*transverse deformation meters*", Dunnicliff and Green 1988). However, they do this by measuring the inclination of the torpedo and hence the readout provides an angular measurement. The readout meter used has a resolution of ± 0.1 mm which, with a torpedo length of ½ m, gives an angular precision of about ± 41 seconds. Although this is satisfactory for the principal function of the instrument and far less precise than the tiltmeter, nevertheless, it affords an opportunity to compare their respective outputs.

Table 1 gives a comparison of tilt angles for seaward tilting over the period from July 1987 to May 1997. The reading given for the inclinometer is based on that taken at 2 m depth on the assumption that this is similar to that obtained from the tiltmeters which are rigidly fixed to a base at the same depth. It is seen that although there is considerable scatter, the output from the inclinometer is not unreasonable in comparison to that from the tiltmeters. The wide variation shown by the tiltmeters accords with the independent behaviour expected from the joint bounded blocks on which they are supported. The inclinometer is located close to the site where open joints were discovered beneath the road surface and hence the somewhat larger degree of tilt at this location is in conformity with the geological observations.

INSTRUMENT	TILT (arc seconds)
Tiltmeter No. 1	492
Tiltmeter No. 2	124
Inclinometer (Borehole 2)	660
Tiltmeter No. 3	503
Tiltmeter No. 4	not functioning
Tiltmeter No. 5	409

TABLE 1. Comparison of change in seawards tilting over the period from July 1987 to May 1997 and assumed to apply at a depth of 2 m.

DISPLACEMENTS VERSUS TIME

The initially frequent readings of the inclinometer gave some scatter (perhaps only just above the limit of instrumental accuracy) but the later readings, as plotted on a two year basis in Fig. 3, show definite trends. The creep rates recorded by the inclinometer at 1 m depth (shown in Fig. 4) are listed in Table 2

PERIOD	CREEP RATE mm/year
July 1987 to May 1988	11.7
May 1988 to May 1997	1.9
May 1997 to March 1999	4.2

TABLE 2. Creep rates as shown by the inclinometer displacements at 1 m depth as plotted in Fig. 4

In geomorphological terms, these measurements belong to the category of *depth creep* (Selby 1993) but most of the published values for depth creep relate to landsliding with one or more shear surfaces. In comparison to the few published cases where toppling is a significant mechanism, the above values are quite small. Thus for a sandstone canyon wall, Zvelebil (1991) records short term rates up to 365 mm / year while for a limestone pinnacle, Glawe and Lotter (1996) record an overall rate of 78 mm / year over an 8 year period. For both these examples, the authors record short term fluctuations as experienced at Afton Down. The major use of these rate determinations must be in terms of investigating the history and geotechnical character of the slope; use of the values for predictive purposes is not possible at present, for as Zvelebil has shown, the rates can decrease towards the time of a sudden failure.

The values obtained at Afton Down relate to a point some 29 m from the cliff edge. Some variation in rate with distance from the edge is to be expected, not only with the dissipation of strain energy but also with the influence of the vents acting as open tension cracks. Such an investigation of displacements with proximity to a cliff edge would provide a very useful understanding of the mechanisms at work within the rock body. In the case of Afton Down, it has been proposed (Anon, 1998) to undertake a stabilisation scheme with a piled wall support for the highway and incorporating an extended system of geotechnical instrumentation and monitoring.

CONCLUSIONS

1. The inclinometer and tiltmeter angular rotations are in broad agreement, allowing for the natural variation to be expected in well jointed rocks.
2. The pattern of rotation versus time for the tiltmeters and the displacements versus time for the inclinometer show similar trends with accelerated movements in wet winters.
3. The much shorter reading interval for the tiltmeters has picked up short term fluctuations and surges. This phenomenon is common to other slopes subject to toppling and there is evidence that this is also true for the inclinometer as shown by the early series of frequent readings. A longer time interval between readings shows more definite trends.

ACKNOWLEDGEMENTS

Grateful thanks are recorded to Mr. Robin McInnes, Coastal Manager, and other the staff of the Isle of Wight County Council for provision of the tiltmeter records. An especial thank you is also due to Mr. Steve Wake for his untiring assistance with the field measurements and for maintaining the inclinometer in a serviceable state.

REFERENCES

Anon 1998. Piled wall slows cliff road erosion. Ground Engineering. 21, 6.

Barton, M. E. 1990 The stability and recession of the chalk cliffs at Compton Down, Isle of Wight. In Chalk (Burland, J. B. et al, eds). Thomas Telford, 541 -544.

Barton, M. E. and McInnes, R. G. 1988. Experience with a tiltmeter-based early warning system on the Isle of Wight. Proceedings of the 5th International Symposium on Landslides, Lausanne. A. A. Balkema, Rotterdam. 1, 379 -382.

Dunnicliff, J. and Green, G. E. 1988. Geotechnical Instrumentation for Monitoring Field Performance. John Wiley & Sons, New York.

Glawe, U. and Lotter, M. 1996. Time prediction of rock slope failures based on monitoring results. Proceedings 7th International Symposium on Landslides, Trondheim. A. A. Balkema, Rotterdam. 3, 1551 - 1555.

Selby, M. J. 1993. Hillslope Materials and Processes. 2nd. Edition. Oxford University Press.

Zvelebil, J 1991. Some problems in the interpretation of monitoring records of rock slope movements. In Field Measurements in Geomechanics (G. Sørum, ed.), Oslo. A. A. Balkema, Rotterdam. 1, 387 - 393.

Causes of Landslides in the Darjiling Town of the Eastern Himalaya

S. R. BASU, Department of Geography, Calcutta University, 35, Ballygunge Circular Road, Calcutta 700019, India

INTRODUCTION

Landslides are perhaps the most rampant environmental hazard threatening the Darjiling town itself. During or after every monsoon, landslides create havoc in and around the Darjiling township area. Numerous slips have occurred in the past; however, the intensity, cause and severity of the slides have only been recorded since 1899. From the records of the recent landslide events, it is clearly revealed that the frequency of landslips is increasing year after year. In order to have an insight into the probable causes of such increased vulnerability of the Darjiling town it is necessary to trace the course of events from the very inception of the town.

GEOMORPHOLOGICAL BACKGROUND OF THE STUDY AREA

Darjiling is situated at the nothern extremity of a ridge, which extends southwards through Ghum and is bifurcated by numerous spurs extending at all angles. The main ridge and the spurs are narrow and steep and are separated by water channels of steep gradient locally known as *Kholas* and *Jhoras*. Within the confines of Darjiling town the foliation dips of the gneissic rocks are generally towards the east, ranging from 20^0 to 40^0. There are two prominent sets of joints, one running roughly NW to SW and the other NNW to SSE. Both the joints have steep westerly dips varying from 40^0 to 70^0 (Figure 1). The direction of the hill spurs agreeing with the joint directions indicates that blocks of rocks loosen easily along these joints.

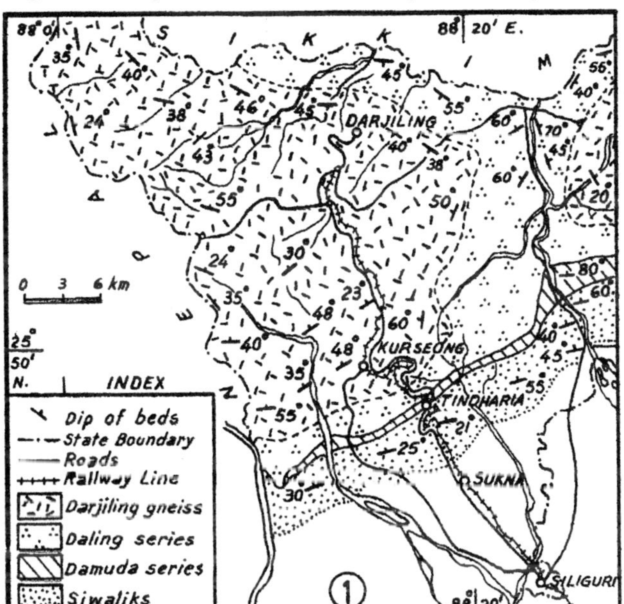

Figure 1. Geological Formations

In Darjiling town, there is a topsoil of about 3 to 5m near the crest of the ridges, but the thickness decreases downwards and lower slopes are generally covered with boulders rolled down from the top. The soil everywhere is residual as it is derived from the weathering of gnessic rocks. Weathering is selective in these rocks and proceeds along some suceptible bands – mica rich bands in preference to quartzose bands and also along joints and shear planes. Consequently, blocks of fresh rocks are generally found encircled on all sides by highly weathered rocks of the nature of clay. The clay is found mixed up with grains of quartz, feldspar and flakes of mica. This has an important bearing on the slips, as the air, filling up the voids in clayey soil, gives it an apparent cohesion, which is lost as soon as the pores get filled with water during rains.

The reason for formation of the soil-covered (1.5 to 3m) western slopes contrasting with the boulder-covered eastern slope is probably to be attributed to the characteristic structural features of the underlying rocks (joints, foliation, dip etc.) in the respective area. On the eastern slope, foliation plane and joints are favourable for loosening of blocks to form talus while on the western slope, the foliation dip being inward, the looseness of blocks is retarded and the rocks get time to weather into soil.

DEVELOPMENT OF DARJILING TOWN

On the 1st of February 1835 by a deed of grant by the King of Sikkim, Darjiling was ceded to the East India Company. At the onset, the population consisted only of a few hundred of people. With the eastablishment of a sanatorium for ailing British soldiers, purchase of land by Europeans for residential purposes, opening up of barracks for soldiers and a *bazaar* (Market), population grew up to 10,000 in the following 50 years. When the tea gardens were being opened out (1873) and the communication with the plains became easier with the construction of the Hill Cart Road and Darjiling Himalayan Railway, labourers poured in and a phenomenal growth of population (16.924 persons) resulted in 1901. Success of tea plantation, partition of Bengal, introduction of tourism and the resultant massive construction works (e.g. expansion of roads and construction of new urban structures) ushered in more and more immigrants from the neighbouring countries and thereby almost doubling the population of Darjiling (33,605 people) in 1951 and by 1991 the town recorded a population of about 71,471 persons. The density figure changed from 1603 persons/km^2 in 1901 to 6,769 persons/km^2 in 1991 with a consequent changes in percentage of increase by 322 (Table 1).

Table 1. Decrease in Forest and Increase in the Built-up Area and Population in Darjiling Town (1800 – 2001)

Year	Forest (%)	Built–up Area (%)	Population
1800	95	Nil	100
1850	87	3.0	10,000
1901	65	7.5	16,924
1951	40	39.0	33,605
1991	15	52.0	71,479
*2001	10	65.0	85,000

* Projected. (Source: Census reports, Forest census and other documents)

Needless and reckless obliteration of forests, with unscientific use of slopes (especially in construction works), coupled with geological, rainfall and slope characteristics have changed the local scenario completely. As a result Darjiling, one of the most densely populated tourist-centres in a comparable environment, exists on the verge of an environmemtal catastrophe as with just one concentrated shower of 50mm.h^{-1} would initiate numerous landslides endangering the lives and properties of the local inhabitants.

LANDSLIDES EVENTS

From the available records, it is found that the disastrous landslip occurred on the 24th September 1899, in Darjiling town. Unprecedented rainfall of 1065.5mm during 23rd to 25th September (4 days) triggered about fifteen landslides in different parts of Darjiling town killing 72 persons and causing enormous loss to land and property (Griesbach, 1899 – 1900). Most of these slips were confined to the soil-cap overlying the gneissic rocks. The immediate cause of the slips was attributed to excessive rainfall. The instability of the hillside gradually increased due to increased saturation caused by absorption of moisture and the cutting of hillslopes both for natural and artificial needs increased this instability further. Defective drainage was also considered an important factor in the absorption of moisture. The second major event of landslips in the town took place on the 15th January 1934 due to Bihar-Nepal earthquake, which was responsible for widespread destruction though not of equal magnitude as was experienced in 1899. On the 11th and 12th June, 1950, the hill slopes in and around Darjiling town was affected by another disastrous event of landslips causing several deaths and heavy damage to roads, houses and public works, due to heavy shower (834.1mm) from 10th to 13th June, 1950. Darjiling town and its environs were again eclipsed with large scale landslips owing mainly to very heavy and concentrated rainfall that continued from 2nd to 5th October, 1968 with total of 1121.40mm. Such landslides caused widespread damage to human lives, properties, roads, railways and cut off the town from the rest of the world for about a week. The period between 1969 and 1979 was relatively undisturbed. But heavy and continuous rain on the 27th August and again during 3rd to 4th September 1980 triggered off widespread landslips in and around Darjiling. Since 1980, it has been observed that almost in every monsoon some parts of the town have been suffering from major or minor landslips causing loss of life and properties. The years 1991, 1993, 1995 and 1998 are the latest cases of landslips in Darjiling town (Figure 2).

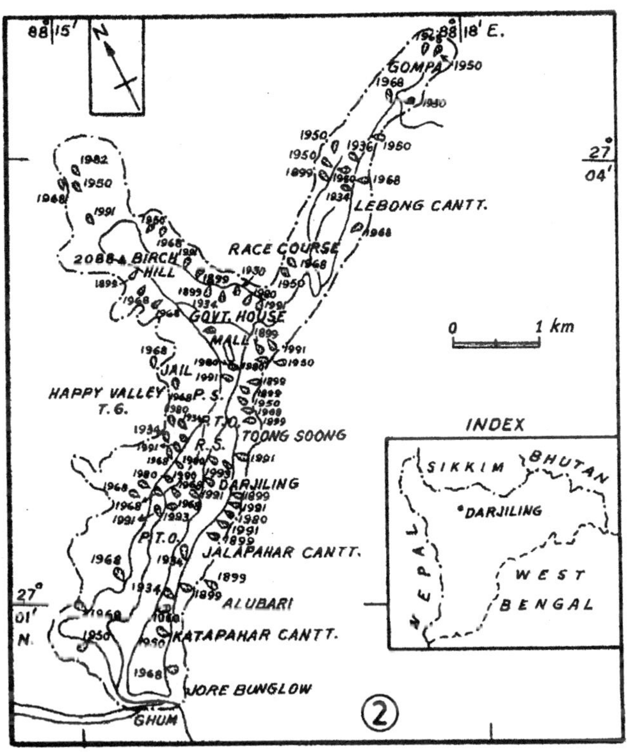

Figure 2. Chronology of landslides (1899-1993)

Plate 1. A landslide on the Toy-Train of Darjiling

LANDSLIDE-PRONE AREAS

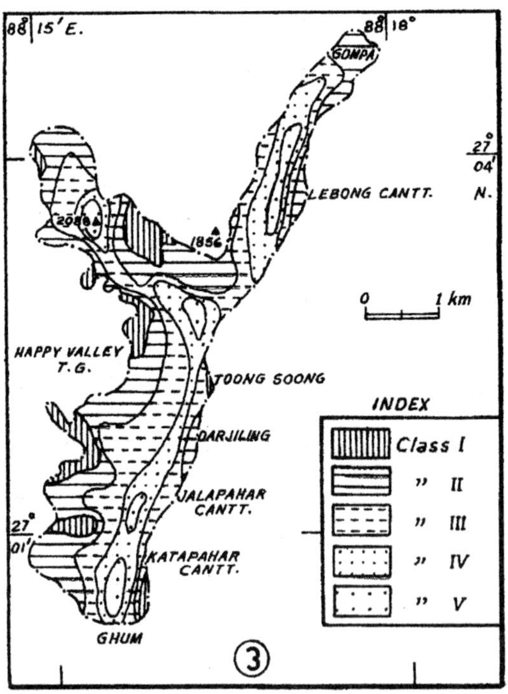

For better understanding of the geographical distribution of landslip-prone areas in Darjiling town, a map (Figure 3) has been produced with the help of a 'check list', topographical sheets (78 A/4 & 78 A/8) and direct field observations at 35 sample sites. The following 5 categories of the susceptibility zone have been put forward:

Figure 3. Landslide – prone areas

Class-I: Extremely high slip-prone zone: Almost after every torrential rain, these tracts experience slips. They are mostly found on the eastern slopes of Jalapahar-Katapahar ridge mainly covering the areas like Alubari, Manpari *busti*, Toongsoong, Pandam tea garden, Bhutia *busti* and Hermitage, eastern slope of Lebong spur around Ging and Bannockburn tea gardens and in small pockets on the western slope of the Lebong spur i.e. Pattabomg and Rangit tea gardens. It can also be observed along the western part of the town below Batasia.

Class – II: Very high slip-prone zone: Areas where slips have occurred more than 5 times in 10 years are found along both the eastern and western slopes of the ridge i.e. upper Alubari, upper Toongsoong, along the Tenzing Norge Road, C. R. Das Road, eastern slopes of the Mall, below Raj Bhavan; also on both sides of the Lebong spur mainly in the tea gardens of Bannockburn, Rangit and Pattabong. On the western slopes of the ridge it covers Rajbari *busti*, Kagjhora, Victoria Falls, Dr. Zakir Hussain *busti,* Dhobitala, around the jail, below the railway station, Lochanger, Haridas *hatta* and Singamari.

Class – III: High slip-prone zone: It covers the western spur of the town along the Hill-Cart Road, Gandhi Road, Nimkidara, Police line, Mary Villa, Mayapuri, upper Kagjhora, below the convent cemetery, Dr. Zakir Hussain Road, along the Birch Hill spur and the Lebong spur. Here landslips occur 2-5 times in ten years.

Class – IV: Moderate to low slip-prone zone: Landslips occured once or twice in the last decade mostly along the ridges of Jalapahar-Katapahar up to the Mall, including the *bazaar* area and also along the Lebong spur including the Lebong Cart Road.

Class – V: None to negligible: It is found in pockets on the ridge tops of the Jalapahar-Katapahar ridge, the Lebong ridge (Military Cantonments) and the Observatory hill and on the top of Birch Hill ridge where slips occur rarely.

RECOMMENDATIONS

Identification of landslide prone areas and the study of individual slips within the town have helped in producing a map on soil conservation (Figure 4) and this indicates that there is no portion, excepting only the very top of the ridges which is stable against slips. However, the western slopes of the ridges and spurs are comparatively more stable than the eastern slopes. In the western slopes, the debris-cover is fairly thick below the present town limit, i.e., below the Cart Road and Victoria Road where the gradient is also steeper. So, construction of buildings on the debris-covered slopes should be restricted. In soil slopes, building may be constructed but they should be provided with proper revetment walls with adequate weep-holes.

Narrow drains extending only a few metres down the hillside are found to have helped in the seepage into the hill-slope below causing slips. *Pucca* (cemented or plastered) drains wide enough to drain out the storm discharges should be constructed from the top to below, the lower level of the town.

In the *busti* areas (slums), especially Bhutia, Toongsoong, Alubari, Manpari, Butcher, Kagjhora etc., the greater incidence of slips is mainly due to bad drainage, from the densely packed houses. Here, in the absence of proper outlets, water gets absorbed in the soil. Unscientific and badly managed terrace cultivation adjoining the houses aggravates soil erosion. Low, badly designed and poorly constructed revetment walls also assist the slip instead of providing protection against them. Hence, measures should be taken to ensure wider spacing of the houses with proper drainage and suitable revetment walls. Here in the *busti* areas the bearing capacity of the soils and the debris rich in sands

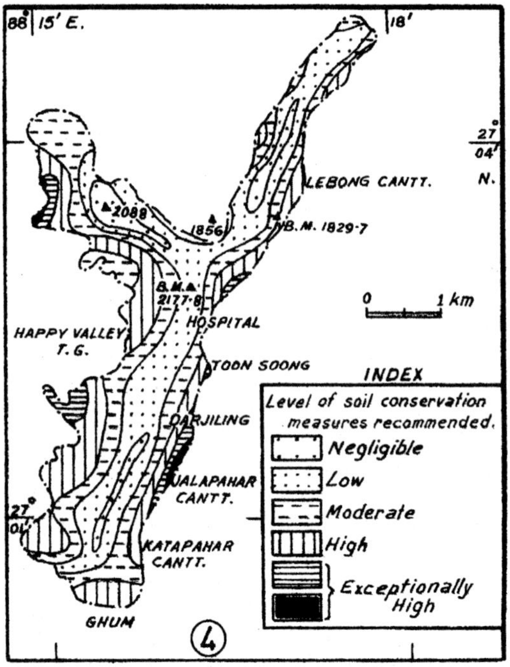

Figure 4. Proposed soil conservation scheme for Darjiling town

particles ranges between 4.5 to 6.6kg.cm^{-2}, corresponding to the strength suitable for the construction of three-storeyed buildings only (Bandyopadhyay, 1980). So any construction beyond this limit must be prohibited. Moreover, along the *nullah* (unlined drain) courses in the *busti*, spaces should be left on either side at an angle of 45^0 with the *nullah* bank, which are unstable zones for construction as there are chances of slope failure within that zone.

CONCLUSIONS

In view of the ever-increasing problems of landslides in Darjiling town, man must be made aware of the possible dangers that he is inviting due to his careless dealing with nature. It is true that one has to make room for the growing population and in this pursuit he has to utilise every piece of land available. But the precautions that have to be adopted should not be neglected. In the town, the revetments are not maintained properly, the weep-holes are choked and the drains are dumped with garbage restricting free drainage of water. Moreover, the present land-use system should be properly evaluated. The construction of high rise buildings should be stopped immediately. The people should be provided with some alternate sources of energy through construction of mini-hydel power projects utilising the springs, which can be an option to prevent them from cutting down more trees. But above all, it should be of utmost priority to develop mass-awareness among both local people and the tourists, so that they become aware of the possible dangers they are inviting by interfering with natural laws.

REFERENCES

Bandyopadhyay, M (1980). *Progress on slope stability study of Toongsoong Area, Darjiling Town.* Unpublished Report, Geological Survey of India.

Griesbach. G. L. (1899 – 1900): *General Report*, Geological Survey of India.

Ground Water Isolation by Low-Permeability Clays in Landslide Shear Zones

R. L. BAUM, U.S. Geological Survey, Denver, Colorado, USA, and M. E. REID, U.S. Geological Survey, Menlo Park, California, USA

INTRODUCTION

Observations at several sites in the western United States indicate that the persistent movement of large translational slides is related to entrapment of water in the landslide masses. The structure and stratigraphy of many large slow landslides create the conditions that allow them to retain water perennially. Weak, low-permeability clays are widespread in basal and lateral shear zones of such slides in various climates (Figure 1; Baum and Reid, 1995; Baum and Fleming, 1996). The strength and hydraulic conductivity of the clays are markedly lower than the bulk of the materials in the landslide masses. Clay layers tend to isolate a landslide both mechanically and hydrologically from adjacent, unfailed materials, effectively destabilizing the slide by both weakening its shear surfaces and allowing the slide to fill with water like a bathtub. These effects tend to perpetuate movement in the slide areas rather than in adjacent otherwise similar ground; even when boundaries shift, they tend to remain inside previously failed materials. In many cases the clay layers probably formed by shear at the base of the slide; however, in at least one case, existing clay layers provided a preferred surface for sliding.

Our field observations and measurements from central Utah, southwestern Colorado, and the island of Oahu, Hawaii, U.S.A. document the causes of persistent movement in translational landslides in three different climatic and geologic settings. Simple hydrologic calculations illustrate how water collects in slides under present-day climatic conditions and how the accumulation of water and formation of weak clay layers affects the stability of the slides.

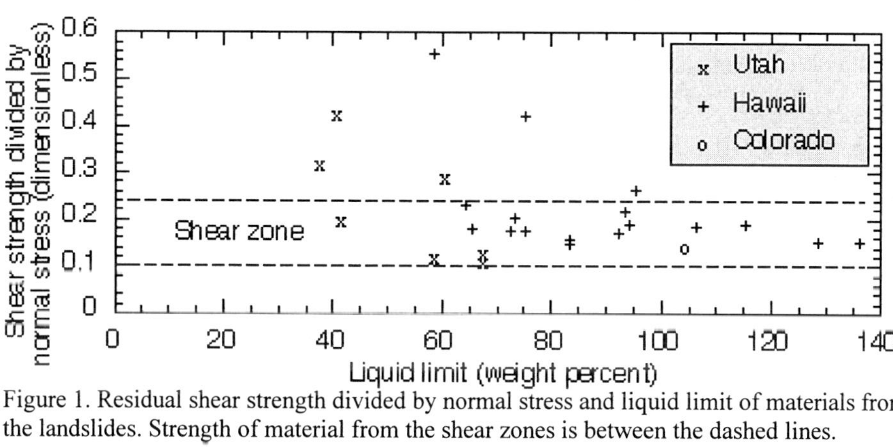

Figure 1. Residual shear strength divided by normal stress and liquid limit of materials from the landslides. Strength of material from the shear zones is between the dashed lines.

Landslides in research, theory and practice, Thomas Telford, London, 2000.

LANDSLIDES OF CENTRAL UTAH

On the Wasatch plateau of central Utah, borings and cross-sectional exposures have revealed highly plastic clay in the basal and lateral shear zones of at least four slides in colluvium derived from sedimentary rocks. The landslide material typically consists of rocky debris in a matrix of sandy clay. Plastic clay appears to have been injected during sliding into the rocky landslide debris adjacent to shear zones of the Twin Lake landslide near Mayfield, Utah (Fleming and Johnson, 1989) and at Cottonwood Creek near Ephraim, Utah (Baum and Johnson, 1993, Figure D3). Residual shear strength determined by ring shear testing of the low plasticity clay (LL=42, PI=25) from Cottonwood Creek was relatively low with an angle of residual friction, ϕ'_r=9° and residual cohesion, c'_r=12.2 kPa, both for effective stress.

Piezometers in the Aspen Grove landslide about 7km east-southeast of Ephraim, Utah, documented accumulation of water in one of these slides. The landslide was 30-100 m wide, 6-8m thick and 1000m long (Baum and Fleming, 1989). The average slope of the landslide surface was 11°. Infiltration of snowmelt in May 1983 triggered the slide, reactivating 3.6 x 10^5 m^3 of previously failed landslide debris. The slide moved for several weeks each spring from 1983 to 1986. Water levels in the slide peaked near the ground surface during April in response to snowmelt and gradually fell through the summer (Figure 2). Movement slowed gradually in response to the falling water levels. In contrast, maximum water levels outside the slide peaked 3 m below the ground surface in April and dropped 2m by the time movement stopped in August.

The Aspen Grove landslide consists of sandy clay that contains abundant boulders and blocks of sandstone. High-plasticity clay (59≤LL≤68, 43≤PI≤49) in the basal shear zone of the landslide is extremely weak (ϕ'_r =5°, 5.2 kPa≤c'r≤10.2 kPa, ring shear) and has low saturated hydraulic conductivity, K_s=9.8 x 10^{-13} m/s. Low-plasticity sandy clay representing the bulk of the landslide debris has residual strength parameters two-three times greater than material from the shear zone (ϕ'_r =15°, c'r=17kPa, ring shear).

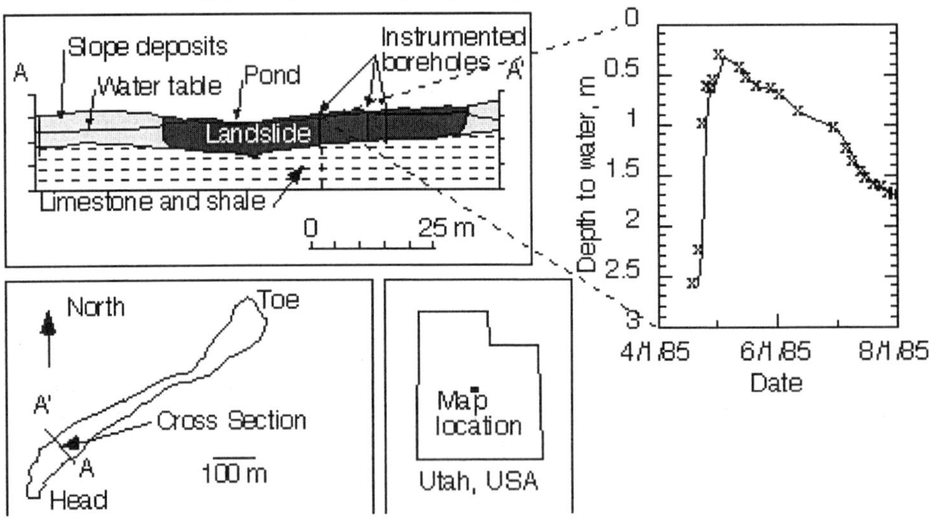

Figure 2. Map and cross section of the Aspen Grove landslide, Sanpete County, Central Utah, showing water table inferred from piezometric measurements (elevated within slide). Piezometer hydrograph shows water level changes related to spring snow melt.

A simple 3-dimensional stability analysis of the upper part of the slide indicated failure was consistent with the measured strength parameters, landslide dimensions, and pore pressures ($FS = 1$ assuming $\phi'_r=5°$ and $c'_r=8.5$ kPa, or $\phi'_r=12.5°$ with no cohesion, Baum, 1988).

THE SLUMGULLION LANDSLIDE, HINSDALE COUNTY, COLORADO

The Slumgullion landslide in Colorado consists of rocky debris in a fine matrix derived from hydrothermally altered volcanic rocks. The active part of the slide is 3.7km long, 120-420m wide, and probably 12-30m thick, based on the observed height of the toe, 12m. The average slope of the active part is about 9°. The slide has moved steadily at rates that have varied over the surface of the landslide from 0.5 to 6m annually during a 30-yr period of observation (Baum and Fleming, 1996). Thin, dike-like masses of weak ($\phi'_r=9.7°$, $c'_r=0$, ring shear) highly plastic clay ($105 \leq LL \leq 116$, $74 \leq PI \leq 77$) coat the outboard faces of flank ridges. The clay has been extruded to the surface along lateral shear zones (Baum and Fleming, 1996). Iron oxides in adjacent debris within the slide indicate past presence of ground water above the clay. The low hydraulic conductivity of the clay ($K_s=1.4 \times 10^{-11}$ m/s, consolidometer) indicates that the clay layer forms an effective barrier to water percolating downward through the slide. The slide evidently is a conduit for ground water as indicated by springs and perennial ponds that are abundant on the slide and rare outside of it, but few observations of ground water in the slide have been attempted because of its size and remoteness. Seismic surveys indicated that the inactive part of the slide directly downslope from the active toe is water saturated a few meters below the ground surface (Williams and Pratt, 1996).

THE ALANI-PATY LANDSLIDE, OAHU, HAWAII

In subtropical Hawaii, the very slow-moving Alani-Paty landslide has damaged houses, streets, and utilities on 60 lots of a residential neighborhood built on a debris apron near Honolulu (Figure 3). The slide is similar to several others near Honolulu that have caused millions of dollars in property damage. The slide showed signs of movement in the mid or late 1970s and residents reported damage to houses and streets during and immediately following rainy periods, especially during the winters of 1987 and 1988. Maximum net displacement by June 1989 was 4.3 m. The 250,000-m^3 slide formed in weathered, bouldery, smectite-rich deposits of a debris apron and consists of two displaced masses that move somewhat independently: the larger mass is 280 m long, 110m wide, and slopes 9°; the smaller mass is 150 m long, 60 m wide, and slopes 12° (Figure 3). The bouldery deposits have a crudely stratified matrix of silty clay and clayey silt. A perched water table fluctuates seasonally in the upslope part of the slide; water-table fluctuation elsewhere is insignificant.

Subsurface exploration of the landslide revealed a clay-rich layer ($74 \leq LL \leq 129$, $40 < PI < 55$) of low hydraulic conductivity near the basal slip surface. Mean hydraulic conductivity near the slip surface was two orders of magnitude lower than conductivity of the overlying landslide material (Figure 4). Such conductivity contrasts can lead to locally elevated pore pressures (Reid, 1997). The clay-rich layer appeared to be a buried paleo-vertisol formed in debris derived from weathered basalt. Shearing may have enhanced clay layer continuity and low permeability. Measurements at 68 piezometers over a period of 2 yr revealed that water was perching on the clay. Water levels outside the slide were several meters lower than water levels in the slide.

Monitoring from October 1989 to April 1991 indicated that infiltration of surface water during extended rainy periods caused the pressure head at the shear surface to increase 1-3m. Periods of intense rainfall further the increased pressure head 0.2-1.1m, enough to trigger

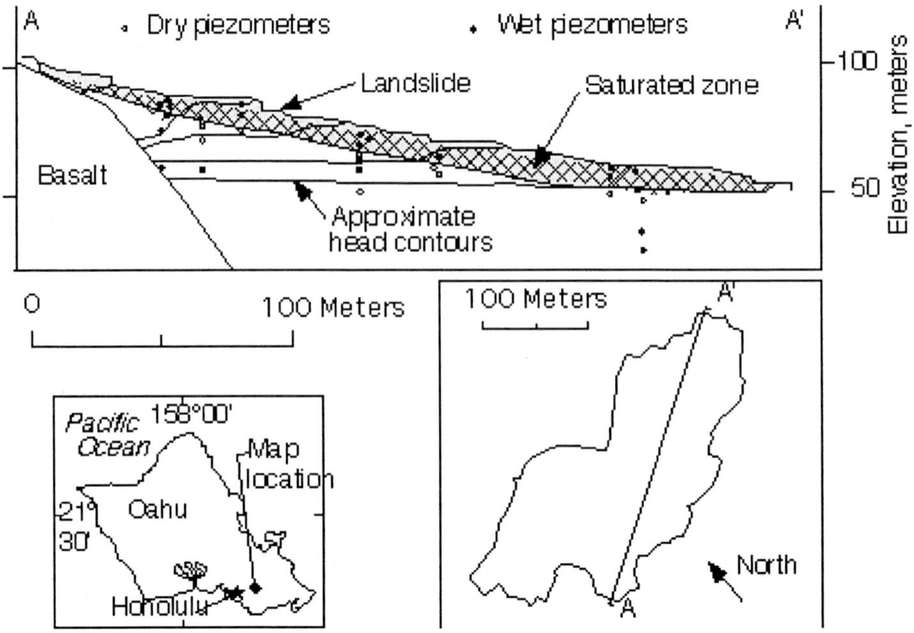

Figure 3. Map and cross section of the Alani-Paty landslide, Honolulu, Hawaii showing lines of equal ground-water head. Average thickness of the saturated zone (hatched) is 7-8m

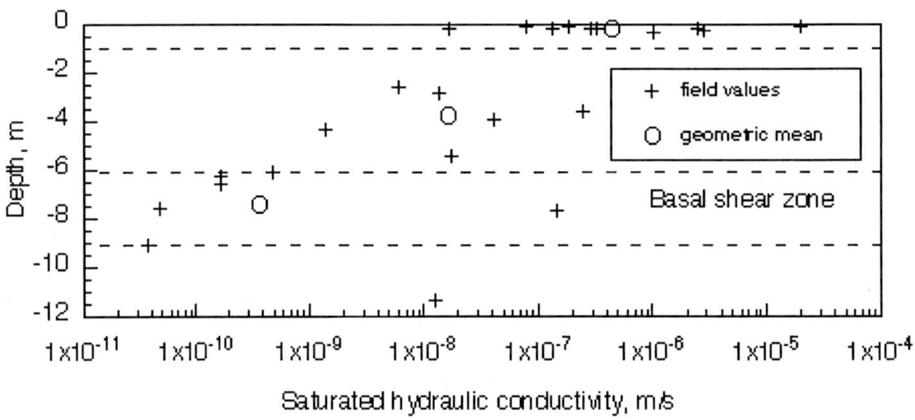

Figure 4. Variation of saturated hydraulic conductivity (determined by field tests) with depth in the Alani Paty landslide, Honolulu, Hawaii (after Baum and Reid, 1995)

small movements during January 1990 and March 1991 (Baum and Reid, 1995). A limit-equilibrium stability analysis using wet-season water pressures with ϕ'_r=8° and c'_r=6.1 kPa, or ϕ'_r=11° if c'_r=0, gave FS=1, consistent with strength parameters measured in direct shear tests of materials from the basal shear zone (6°≤ϕ'_r≤11° and 1.9 kPa ≤c'_r≤6.7 kPa, Figure 1). Substituting the shear strength of materials away from the slip surface into the analysis increased the *FS* to 1.5 or greater. Reducing the pore pressure to zero and using strength of the basal shear zone increased the *FS* to 1.3 (Baum and Reid, 1995).

CONDITIONS REQUIRED TO MAINTAIN PERENNIAL SATURATION

Maintaining a perennial saturated zone requires that inflow from precipitation (minus runoff, evapotranspiration, and other losses), *r*, exceeds outflow from drainage. Steady, saturated drainage through a landslide can be estimated by separating vertical discharge through the landslide base, q_v, from horizontal discharge, q_h (Figure 5). To estimate the range of drainage rates, we computed saturated hydraulic conductivity normal and parallel to the layers according to standard formulas for isotopic layers (Freeze and Cherry, 1979, p.34) and our measurements. We transformed these values into the vertical and horizontal components using the formula for the hydraulic conductivity ellipse (Freeze and Cherry, 1979, p.34). Assuming conditions shown in Figure 5, the vertical hydraulic gradient is –1 and the horizontal hydraulic gradient is $\tan\theta$, where θ is the slope angle. We computed drainage rates per unit area using Darcy's law and reduced the downslope drainage rate by the factor, $(d_1 + d_2)/L$, to account for the smaller drainage area through the toe compared with drainage through the base of the slide.

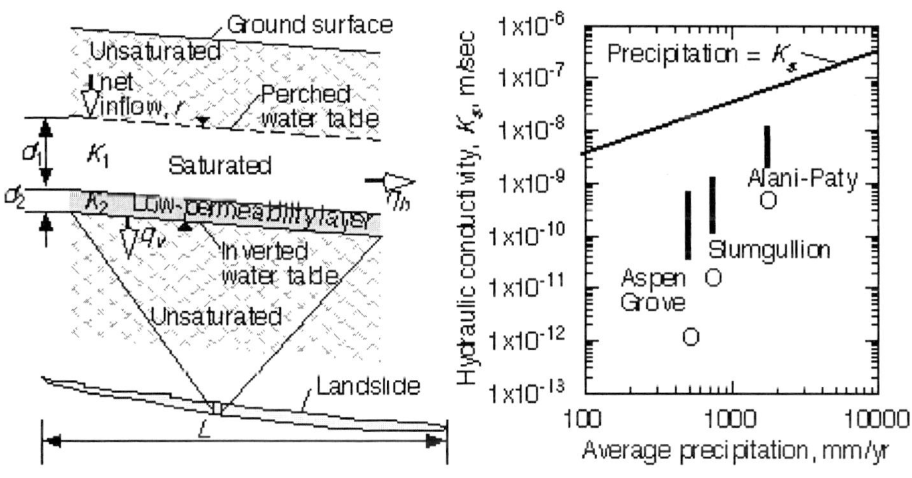

Figure 5. Model for drainage of perched water for a gently sloping landslide, assuming steady flow in isotropic, homogeneous layers ($K_1 \geq 10K_2$). Graph shows measured K_s of basal shear zone materials (squares) and range of effective drainage rates for the three landslides (vertical bars) estimated from the model assuming measured ranges of dimensions, pore pressures, and hydraulic conductivity.

In the slides we have studied, the estimated total drainage rate is roughly half to a tenth or less of the annual precipitation rate, consistent with probable rates of flux into shallow saturated zones in the climates represented by these slides (Figure 5). The hydraulic conductivity of the basal shear zone is two-four orders of magnitude less than the annual precipitation (Figure 5). For example, at the Alani-Paty landslide, the mean K_s of the slip surface, 12 mm/yr (3.9 x 10^{-10} m/sec), was significantly less than K_s of the overlying materials, 540 mm/yr (1.7 x 10^{-8} m/sec), and the average annual precipitation rate, 1800 mm/yr. Inflow amounting to one fifth (350mm) of the annual precipitation is sufficient to maintain the observed 8-m-thick saturated layer in the slide even if the if the basal shear zone is only 0.1 m thick. If the basal shear zone is 1 m thick, then only 90 mm of inflow is needed to maintain the saturated layer. Even small variations in K_s are sufficient to impede ground-water flow (Reid, 1997).

Transient effects related to alternating wet and dry periods complicate the process of perennial perching of ground water. The perched water table typically rises following periods of rain or snowmelt and then falls to pre-storm levels over a period of days or weeks following the rise. These transient rises initiate or accelerate movement. Increased head following storms drives transient increases in drainage through the base of the landslide and increases in seepage at the ground surface directly above the toe of the slide.

DISCUSSION AND CONCLUSIONS

Weak, low-permeability clay layers occur in the basal and lateral shear zones of large translational landslides in diverse climates and geologic settings. Whether the layers were deposited during formation of the hillside or formed later during sliding, the layers have significant influence on slope stability. The layers impede drainage of water from the slides, keeping their shear zones perennially saturated and intermittently allowing pore pressures to increase sufficiently to destabilize the slides. Decimeters-thick clay layers with $K_s \leq 10^{-9}$ m/sec in the basal and lateral shear zones of landslides are capable of trapping water and keeping the slides partially saturated throughout the year in many climates. K_s of the basal clay layer/shear zone is 2-4 orders of magnitude less than the annual precipitation rate. The clay layers also have a lower shear strength than surrounding materials, making it possible for slides to move on slopes flatter than 12°. Both the hydraulic and mechanical effects of clay seem to focus and perpetuate movement in previously failed areas.

REFERENCES

Baum, R.L., 1988, The Aspen Grove landslide, Ephraim Canyon, central Utah: Cincinnati, Ohio University of Cincinnati, PhD dissertation, p. 363.

Baum, R.L., and Fleming, R.W., 1989, Landslides and debris flows in Ephraim Canyon, central Utah: U.S. Geological Survey Bulletin 1842, p. C1-C12.

Baum, R.L., and Fleming, R.W., 1996, Kinematic studies of the Slumgullion landslide in Varnes, D.J., and Savage, W.Z., editors, The Slumgullion earth flow--A large-scale natural laboratory: U.S. Geological Survey Bulletin 2130, ch. 2, p. 9-12.

Baum, R.L., and Johnson, A.M., 1993, Steady movement of landslides in fine-grained soils--A model for sliding over an irregular slip surface: U.S. Geological Survey Bulletin 1842, p. D1-D28.

Baum, R.L., and Reid, M.E., 1995, Geology, hydrology, and mechanics of a slow-moving, clay-rich landslide, Honolulu, Hawaii: Geological Society of America Reviews in Engineering Geology 10, p. 79-105.

Fleming, R.W. and Johnson, A.M., 1989, Structures associated with strike-slip faults that bound landslide elements, Engineering Geology, v. 27, no. 1-4, p. 39-114.

Freeze, R.A., and Cherry, J.A., 1979, Groundwater, Prentice-Hall, 604 p.

Reid, M.E., 1997, Slope instability caused by small variations in hydraulic conductivity: American Society of Civil Engineers Journal of Geotechnical and Geoenvironmental Engineering, v 123, p. 717-725.

Williams, R.A., and Pratt, T.L., 1996, Detection of the base of Slumgullion landslide, Colorado, by seismic reflection and refraction methods, in Varnes, D. J. and Savage, W. Z., eds. The Slumgullion Earth Flow--A Large Scale Natural Laboratory: U.S. Geological Survey Bulletin 2130, p. 77-83.

Monitoring Principles of Civil Works located in Landslide-prone Areas

Ch. BONNARD, F. MAYORAZ and F. NOVERRAZ, Swiss Federal Institute of Technology, Lausanne, Switzerland

ABSTRACT

Numerous road and railway works are founded on slightly active unstable slopes, inducing long-term maintenance and safety problems. Therefore, research was undertaken for the Swiss Union of Road Professionals (VSS), aiming at the definition of technical control and management processes, which has lead to the formulation of general principles to be followed in monitoring campaigns.

On the basis of the behaviour of 17 typical road works (bridges, viaducts, anchored walls), general monitoring strategies and data interpretation procedures have been formulated. The 8 expressed monitoring principles will be included in a new code of the Federal Highway Office in order to ease the preparation of monitoring campaigns and gather reliable data which may be used in predictive models based for instance on neural network techniques.

INTRODUCTION

Due to the constant development of roads and railway lines which have to cross unfavourable land like hilly areas or forest slopes (in order to preserve flat agricultural land and also to avoid bends around topographical obstacles), civil works like bridges, viaducts and anchored walls are often founded on landslide-prone areas which cannot be totally stabilised. On the other hand the existence of substabilised slopes liable to experience acceleration phases following unfavourable climatic conditions is more and more recognised; indeed it has been established then more than 8% of the Swiss territory has been affected by large-scale slope stability phenomena (Noverraz and Bonnard, 1990).

Furthermore, the stability requirements for the transportation infrastructures are constantly increasing in order to ensure safe and high speed tracks on carriage ways and to limit long-term maintenance costs. But in various regions, significant change in climatic conditions (rainfall intensity or annual precipitation) induce a sudden or progressive increase of landslide movements, which may prove to be detrimental to the works if it is not properly investigated and monitored.

Therefore, it is absolutely necessary in such situations to undertake and maintain an adequate monitoring program. Such an action can be divided into two components, the objectives of which are clearly different:

- The **monitoring campaign** strictly speaking consists in analyzing the behaviour of the foundation ground and of the civil work along with time, so as to understand the deformation mechanism of the soil-structure system and to be able to assess its evolution as well as its impact on the serviceability of the transportation way.

Landslides in research, theory and practice, Thomas Telford, London, 2000.

- The **surveillance** campaign consists of observing systematically and maybe automatically some characteristic parameters on the basis of which urgent decisions can be taken in relation with the functionality of the transportation works.

In this study the features of the monitoring campaign are mainly analysed as it is the basis for the assessment of the long-term safety of the works.

OBJECTIVES

The main objectives of the research can be summarised as follows (Bonnard, 1996):

- To examine the technical problems related to the monitoring of several civil works founded on landslide-prone areas.
- To define the general principles which are applicable to all monitoring systems so as to adapt the design of the civil works which may be affected by landslides.
- To determine relevant organisation and management technical procedures in order to ensure a continuous control of the studied civil work from its design phase to its operation, so as to be able to assess its safety and reliability.
- To establish recommendations allowing the relevant administrative services to co-ordinate the control of their various civil works and to precise the tasks of the different partners involved.

The final aim is to organise rationally a monitoring campaign so as to follow the evolution of the civil structures at any time and to avoid any critical situation occurring due to a lack of pertinent data concerning the unstable foundation ground and the work itself.

APPLIED METHODOLOGY

In order to meet the objectives, the research has included three stages:

- a selection of 9 typical works, namely motorway bridges, viaducts, anchored walls and cut slopes in the Canton of Vaud that had been previously monitored for 7 years in a preliminary phase (ISRF & al., 1992), plus 8 bridges and anchored walls concerning motorways and cantonal roads in 6 different Swiss cantons and in France, for some of which monitoring data had been available for 10 to 20 years. All these data have allowed the formulation of a monitoring strategy (see case study below).
- a detailed analysis of the monitoring procedures and of the most appropriate interpretation systems, so as to formulate recommendations on the use of such data and on the corresponding assessment methods (Bonnard & Despond, 1988).
- a global assessment of the behaviour of the studied works and of the required maintenance works induced by the slope instability problems.

The main conclusions gathered have allowed the preparation of a code on monitoring for works laid out in landslide-prone areas, by a VSS expert commission.

PRELIMINARY QUESTIONS TO BE ADDRESSED

In order to select properly the relevant monitoring methods, 9 preliminary questions have to be solved:

a) Is the considered site stable, unstable or potentially unstable?
b) On which bases (documents, maps, data) is it possible to determine the past and present movements and their spatial distribution?
c) Which is the long-term average displacement velocity of the slide?
d) To which extent the determined velocities may vary with time?
e) Which is the depth of the moving mass?

f) By which appropriate means can we observe the slope movements, at the surface or in depth, in order to understand the landslide mechanism?
g) How is it possible to determine adequately and then to record the hydrogeological conditions?
h) What is the significance and importance of the planned works?
i) Which can be the impact of the slope movements on these works?

The detailed and justified answers to these questions will significantly help to plan adequately the monitoring devices, which can be classified into three main categories: displacement / piezometry / forces and stresses. The most common systems with their performance and accuracy are summed up in an annex in the new Federal Highway Office (code SN 570.200) and can be found in the literature (Dunnicliff, 1988 or LCPC, 1994, for instance).

GENERAL MONITORING PRINCIPLES

In order to duly implement a monitoring system, some general principles listed in the table 1 have to be respected in all circumstances. Although some of them may look basic, it has been proved that by ignoring any of them difficulties had occurred in the long-term acquisition and interpretation of data.

Table 1. General monitoring principles

I	To plan the monitoring system already at the preliminary design stage with lasting devices in order to allow a long-term control.
II	To take the probable behaviour of the works and of the site into account when designing the monitoring system.
III	To provide for discrete and/or continuous measurements with time, according to the magnitude of the movements and of their velocity, as well as of the piezometric variations.
IV	To use redundant measurement means, implying several techniques or different specialists, in order to confirm the obtained information and to avoid a diagnosis based on a single type of monitoring data.
V	To take up solid, long-lasting, reliable and easily accessible measurement systems which have to be rather numerous and simple than isolated and sophisticated.
VI	To analyse the monitoring system regularly and critically within the data interpretation process, according to the behaviour of the site and/or the works and considering the quality of the measurements.
VII	To carry on a synthesis of data in time and space within the work management processes so as to identify the global evolution mechanism of the movements, taking the reliability and the accuracy of the measurements into account in such an analysis.
VIII	To determine evolution models of the slope and works on the basis of measured data in order to assess their possible behaviour in the future and to implement maintenance programs.

MONITORING STRATEGIES

Paradoxically, the planning of the monitoring system of a work founded on a landslide-prone area begins even before its preliminary design. Indeed, during the layout selection phase, a first general investigation determines the limits of main and secondary unstable zones and the probable evolution of their movements, taking the existing data into account, like previous survey works.

A second investigation campaign indicated precisely the characteristics of the unstable zones at the sites where parts of the works will be founded.

The monitoring program includes closely repeated measurements for some specific devices and more occasional measurements concerning the entire site area. During construction, the program takes the building stages into account as well as the occurrence of events that might affect the quality of the measurements, in order to be able to adapt the monitoring system at any time.

Near completion of the works, a complete measurement campaign allows the establishment of a reference state for the structure and its foundations at the beginning of commissioning. It is followed by a surveillance program of limited duration (5 years for instance) with regular measurements, which will describe in detail the behaviour of the works and of the slope.

At the end of this first phase a multidisciplinary committee will assess the behaviour of the works and establish a long-term monitoring program. This program must enclose the maintenance of the measurement systems so as to ensure their perennity and allow thus a constant control of the safety of the works.

A CASE STUDY: MONITORING OF POLMENGO BRIDGE

Some of the previous statements will be illustrated by the monitoring data at Polmengo bridge, crossing the Ticino River near Faido (TI) (Figure 1). This cantilever arch bridge made of stone masonry, built in 1941-43, has its east abutment founded at the toe of a very large slide (area of 11 km2), which had not been identified at that time. Topographic measurements on the whole slope initiated in 1929 and even 1895 for one point and repeated 5 times between 1958 and 1990 for most of the points allowed however the determination of general movements varying between 1 and 3 m/year, and even more in the upper part of the slide (6cm/year between 1919 and 1958) (Noverraz et al., 1998).

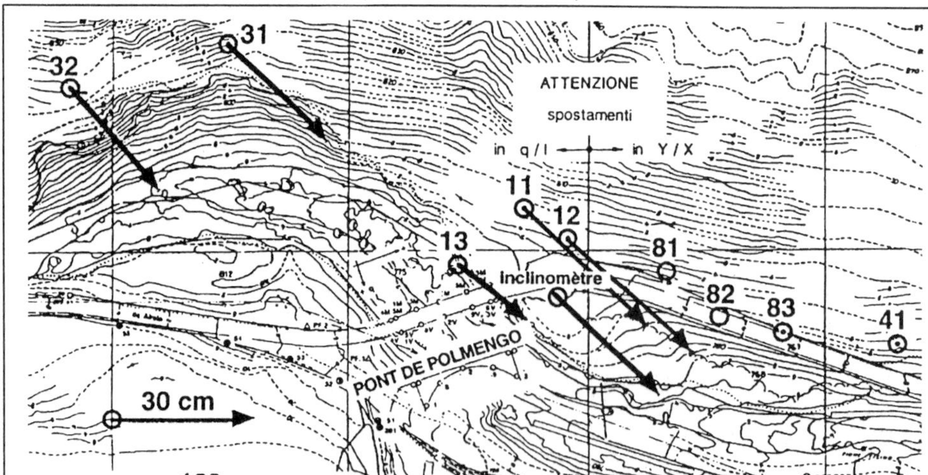

Figure 1. Location of Polmengo bridge on Ticino River at the toe of Faido landslide, with layout of topographic points and horizontal displacements for the period 1989-1996

The bridge affected by an oblique thrust on its east abutment seems to have behaved fairly well until 1987, but following a very high flood which eroded the riverbed over a depth of 6m, the movements of the structure increased causing numerous cracks. Thus, in 1988, the initial span of the arch (41m) had been reduced by some 25cm and the elevation of the crown had raised of 30cm.

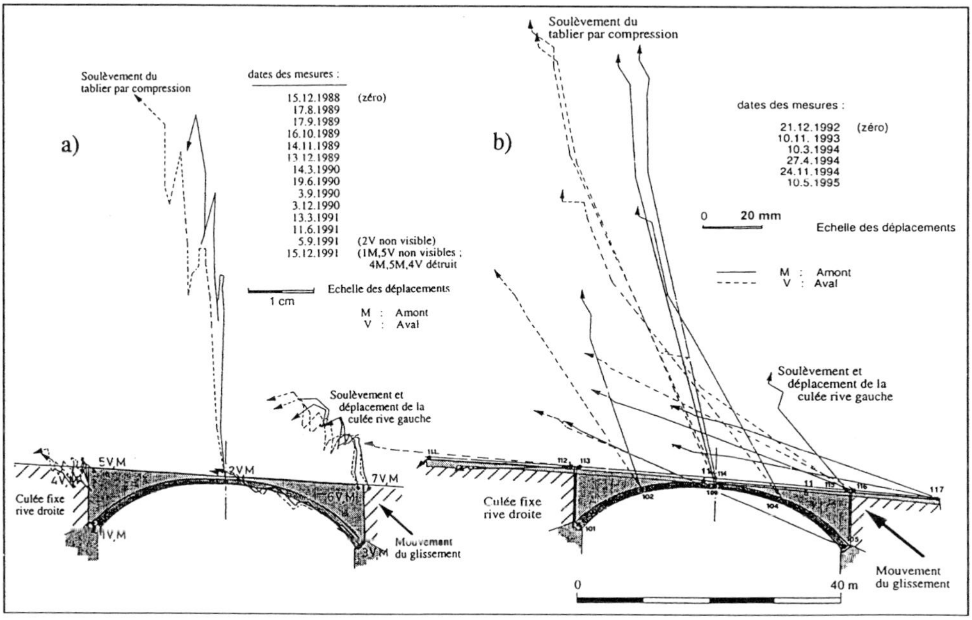

Figure 2. Displacements of the various parts of the Polmengo bridge (a) from Dec. 15, 1988 to Dec. 15, 1991; (b) from Dec. 21, 1992 to May 5, 1995 (Note the different scales for the displacements) (M are measurement points on the upstream face of the bridge, V are measurement points on its downstream face)

The movements of the structure and of the nearby slope (20 points) were monitored in detail by triangulation between December 1988 and December 1991 (Figure 2a), confirming this tendency, so that it was decided to strengthen the bridge by a new concrete slab. An inclinometer was also installed near the left abutment in August 1990, but showed small but unclear movements until March 1992. Unfortunately no topographic observations were done in 1992 due to the execution of the repair works.

During this period two new inclinometer readings put forward very important movements between 35 and 45m below the ground level (i.e. 25m below the river bed): the total displacement between June and September 1992 seems to have reached 34cm, so that no further readings could be done later, but this value seems to be doubtful if compared with the surface readings which were started again in December 1992 (Figure 2b).

During one and a half years (1993 to mid-1994) the displacements were impressive, as the crown was raised by 200 mm and as the left abutment, at the level of the road, moved horizontally by the same amount. But progressively, during the next 12 months, the movements were greatly reduced, so that, after May 1995 no continuous monitoring of the bridge was carried out.

The consideration of the mechanism of the whole slide and of its sensitivity to high rainfall (2178mm in 1993 which is the highest annual value ever recorded), as well as its tendency to raise the left abutment of the bridge (namely a consequence of the magnitude of the versant slide), show that the repair works do not guarantee a long-term safety of the bridge. It is thus necessary to increase the monitoring activities in order to understand more precisely the movements at the toe of the slope and their causes, for instance by a neural network analysis (Mayoraz et al., 1996) in order to determine if a new bridge may be planned at the same site in the future or if a new layout of the road has to be studied (Bonnard, 1996).

CONCLUSIONS

Detailed and long-term monitoring of landslide prone areas in which civil works are located is essential to ensure their safety and reliability and also to reduce the maintenance or repair costs. The measurements have to include discrete and/or continuous observations of movements, at the surface and in depth, and of piezometric data. The lack of redundant information or an interruption of the monitoring program may have severe consequences by inducing a false feeling of safety or by deciding unappropriate repair works. For all major civil works, it can be stated that the monitoring costs are always economically justified in the long-term.

ACKNOWLEDGEMENT

The work of the Expert Commission 5.07 of the Swiss Union of Road Professionals (VSS) to prepare a code based on the presented research was highly appreciated.

REFERENCES

Bonnard Ch. (1996). Principes d'auscultation des ouvrages routiers soumis à des instabilités de terrain. Rapport de recherche No 384 de l'Office Fédéral des Routes, Berne (125 p.)

Bonnard Ch., Despond J.-M. (1988). Principes d'auscultation des ouvrages soumis à des instabilité de pente. Pub. Soc. Suisse de Méc. des Sols et des Roches No 118, pp. 23-32

Dunnicliff J. (1993). Geotechnical instrumentation for monitoring field performance. John Wiley & Sons Inc., New York

Institut des sols, roches et fondations de l'EPFL (ISRF), Bureau de construction des autoroutes de Lausanne, Bureau J.-C. Ott à Thônex (1992). Comportement des fondations d'ouvrages d'art, rapport final. Rapport de recherche de l'Office Fédéral des Routes, Berne

Laboratoire Central des Ponts et Chaussées (LCPC) (1994). Surveillance des pentes instables. Guide technique LCPC, ISSN 1151-1516

Mayoraz F., Cornu T., Vulliet L. (1996). Using neural network, to predict slope movements. Proc. 7th Int. Symp. on Landslides, Trondheim, Vol. I, pp. 295-300

Noverraz F., Bonnard Ch. (1990). Mapping Methodology of Landslides and Rockfalls in Switzerland. Proc. 6th Int. Conf. and Field Workshop on Landslides, Milano, pp. 43-53

Noverraz F., Bonnard Ch., Dupraz H., Huguenin L. (1998). Grands glissements de versant et climat. Rapport final d'un projet de recherche PNR 31, FNRS. v/d/f, Zurich (314 p.)

Measured Response to a Drainage Adit in a Deep Creeping Slide Mass

L. BONZANIGO, Bonzanigo Geologo Consulente, Bellinzona, Switzerland,
E. EBERHARDT and S. LOEW, Engineering Geology, Swiss Federal Institute of Technology, ETH Zürich, Switzerland

ABSTRACT

Down slope deformations of the Campo Vallemaggia landslide in the southern Swiss Alps have been reported for well over 200 years. The average rate of deformation has been measured as being approximately 30 cm/year, with a minimum of 1 cm/year and a peak of 5 cm/day (reported in 1940). Although these rates could be described as creep deformations, other mechanisms relating to multiple slip surfaces are viewed as being equally critical. Following an in-depth study between 1984 and 1991, a decision to stabilize the slope through the construction of a drainage adit was undertaken. This paper reports the findings and the interpretation of an extensive *in situ* monitoring program with respect to the measured response of the slope before and after implementation of the drainage adit system.

INTRODUCTION

The Campo Vallemaggia landslide is located in the crystalline penninic nappes of Ticino, in southern Switzerland, 50 km NW of Lugano. Two small villages, Campo Vallemaggia and Cimalmotto, are located on the toe of the slide mass from which surface displacements have been geodetically measured for over 100 years. Surface and borehole investigations of the Campo Vallemaggia landslide have shown that the unstable mass incorporates approximately 800,000,000 m^3 of weathered and intact rock.

The geology of the slide mass can be described as a metamorphic series consisting of amphibolite and micaceous schists, several types of gneiss, ultramafic metaperidotites and metacarbonates. The geological structure is characterized by isoclinal folding, dipping 30° SSE, with amphibolitic formations positioned at the core of the fold. A brittle fault system crosses the region in a NNW-SSE direction and divides the landslide body into distinct blocks (Figure 1). These longitudinal blocks can be described as displaced clusters of rock which are primarily intact higher up near the head of the slope and pass gradually to a highly weathered material towards the toe.

The unstable mass reaches depths of approximately 300 m, as delineated through seismic surveys and inclinometer measurements of slip and creep. Slide masses of this scale are considered rare (Petley, 1996) and, as in the case of Campo Vallemaggia, are almost always restricted to crystalline rock masses. Examples of deep rock slides analogous to the Campo Vallemaggia slide include 'La Clapière' in southeast France (Julian and Anthony, 1996), the Rosone slide in northern Italy (Forlati *et al.*, 1991), and multiple slides at the Clyde Dam in New Zeland (e.g. Beetham *et al.*, 1992).

Landslides in research, theory and practice, Thomas Telford, London, 2000.

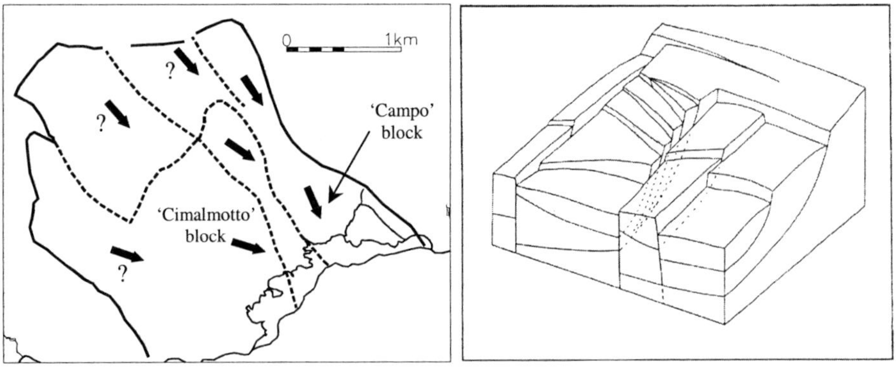

Figure 1. Plan view (left) and three-dimensional block model (right) showing the moving elements and complex subdivision of the Campo Vallemaggia creeping slide mass.

In deep sliding masses, high pore water pressures are a significant contributor to instability. One of the more notable characteristics of the Campo Vallemaggia slide is the presence of deep artesian water pressures. Hydraulic conductivity within the slide mass is highly anisotropic owing to the fractured and weathered state of the rock mass. The distribution of potential heads, as measured in several boreholes, indicate a high vertical component likely relating to the presence of a subvertical fracture system below the sliding mass. These upward moving seepage forces, together with the strong reduction in effective stress, were viewed as the instigating factors for the deep mass movements and instability. Given the potential threat to local communities, mitigative measures were subsequently taken. In the following sections, this paper will describe the acting slide mechanisms based on surface and borehole observations, the actions taken to decrease slope movements (i.e. construction of a drainage adit), and the monitored effect of these mitigative measures.

SLIDE MASS MECHANISMS

Several key factors have been examined in an attempt to better understand the acting mechanisms responsible for the slope movements/instability observed. These include:

- the presence of deep artesian overpressures;
- slope movements which do not directly correspond to precipitation;
- the pulsing behavior of the slide mass;
- the massive but unbalanced erosion at the slope's toe; and
- several hydrogeological anomalies which are still not completely understood (e.g. eruptive source behaviour and the presence of hydrogen).

The slow but consistent displacement rate of 30 cm/year, the depth of disturbed material and the low slope dip angle (approximately 30°), suggest that the Campo Vallemaggia slide can be classified as a 'deep-seated creeping landslide' as defined by Hutchinson (1988).

Inclinometer measurements reveal that displacements are distributed unevenly along the slide profile and concentrate along zones of several meters in thickness. Displacements in these zones indicate visco-plastic type deformations as described by Desai *et al.* (1995). Further analysis shows that this behaviour is highly sensitive to precipitation and accumulated pore pressures. Figure 2 demonstrates that pore pressure values exceeding an apparent threshold value coincide with sudden accelerations of the slide mass. Velocities then return to background levels as pore pressures dissipate and drop below the threshold.

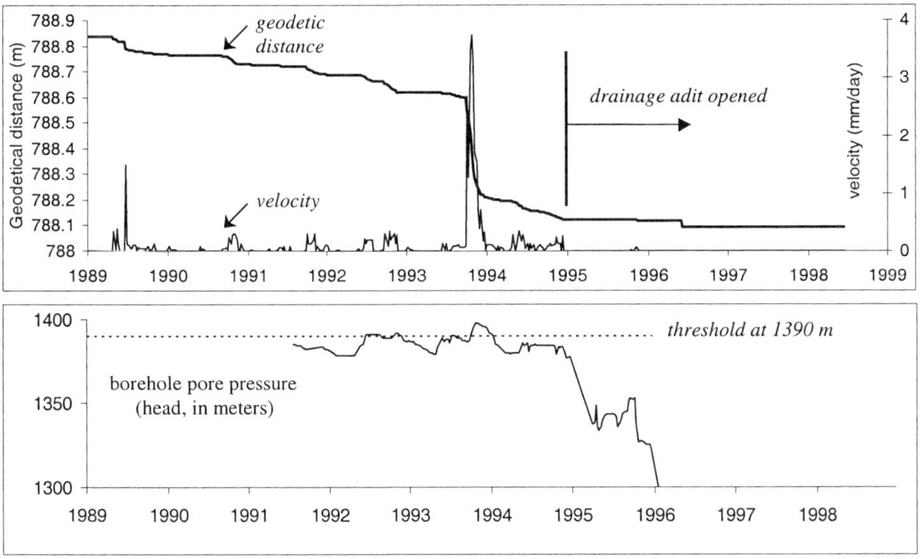

Figure 2. Plots of slide block movement and velocity versus time (top), as measured using an automated geodetic station, and borehole pore pressure versus time (bottom).

Hydrogeology

The hydrogeology of the region is characteristic of crystalline environments, with strongly anisotropic flow occurring within discontinuous and fractured aquifer systems. Flow paths are largely controlled by jointing in the undisturbed rock mass with higher hydraulic conductivities being found along fault zones. Discontinuities associated with these fault zones produce a strong subvertical flow component (i.e. hydraulic conductivity much greater vertically). These conditions are similar to those found below the slide mass and combine with regional topographical effects to produce a reversed equipotential gradient and upward acting seepage forces, which in turn contribute to the overall destabilization of the slide mass.

Within the slide body, multiple zones of large deformation and deep weathering result in a predominantly horizontal anisotropy. Hydrogeological data collected at the site (i.e. piezometers, borehole pressure transducers, etc.) was subsequently used to develop models of the pore pressure distributions and groundwater flow patterns within the unstable slope mass (Figure 3).

IMPLEMENTATION OF DRAINAGE ADIT SYSTEM

The successful use of drainage tunnels/adits to lower pore waters pressures and stabilize hazardous slopes has been widely documented in a number of case studies. For example, Moore and Imrie (1995) report mitigative measures at Dutchman' s Ridge where the piezometric head was lowered more than 150 m through one drainage gallery. Other examples include the Taren slide, UK (Martin and Warren, 1992) and the Pacific Palisades area, USA (Krohn, 1992).

Similarly, the decision was made to construct an 1800 m long drainage adit in the undisturbed rock below the creeping Campo Vallemaggia slide mass. The design, shown in Figure 4, included 30 drainage boreholes drilled from the adit into the transition zone forming the slides base. The total discharge upon completion of the drainage adit system (at the end of 1995) was approximately 50 l/s, decreasing to 30 l/s by 1998. It should be noted that drainage into the adit alone (i.e. before the auxillary boreholes were drilled) was 9 l/s.

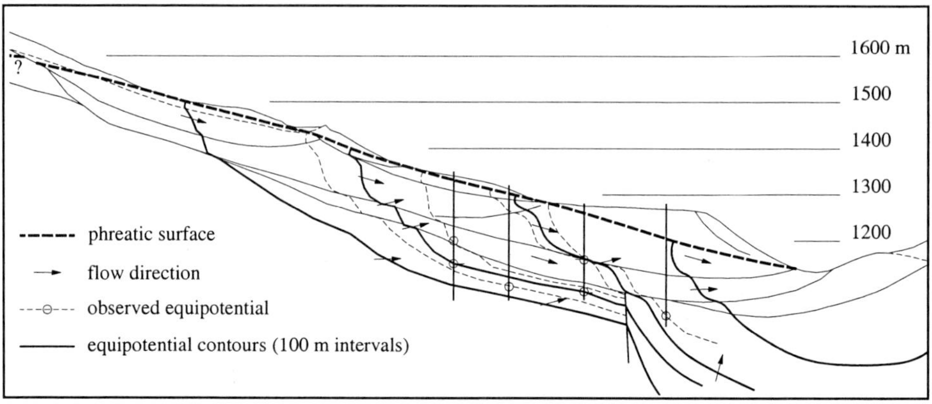

Figure 3. Cross-section of the slide mass showing groundwater flow vectors and equipotential contours *before* drainage adit construction.

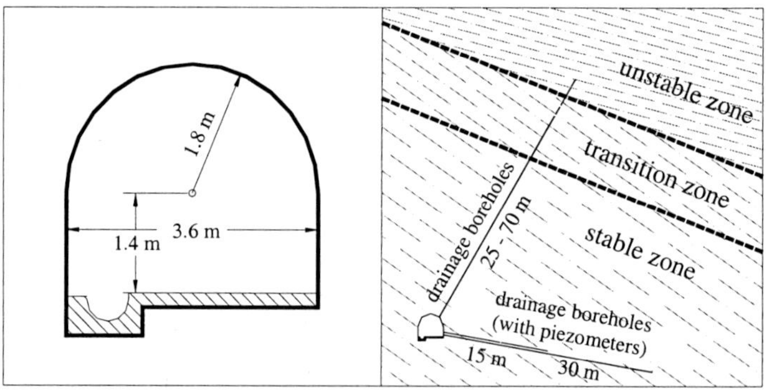

Figure 4. Drainage adit profile (left) and adit with perforated drainage boreholes (right).

Drainage, Ground Settlements and Post-Adit Behaviour of the Slide Mass

Following the completion of the drainage adit system, an immediate drop in water head was detected and surface settlements of up to 40 cm were measured (Figure 5). Slope movements decreased significantly and in some cases, upslope displacements were recorded (relating to the development of a subsidence cone).

The effect of drainage on pore pressure distributions, as verified through borehole pressure transducer measurements, is shown in Figure 6. From this figure, it can be seen that the upward moving seepage forces within the unstable transition zone (as shown in Figure 3) are inverted, thus suggesting slope stabilization through a subsequent increase in effective stress and resisting forces. Steady state conditions have yet to be reached but are expected in the next few years. Meanwhile, monitoring of the drainage adit performance continues.

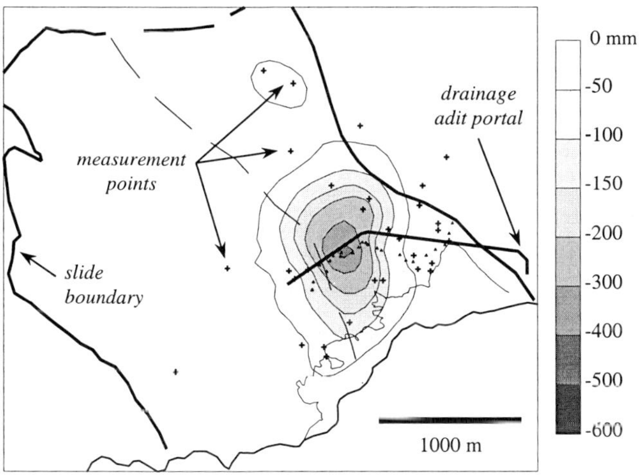

Figure 5. Total surface settlements above drainage adit between 1995 and 1998 (settlement contours in mm).

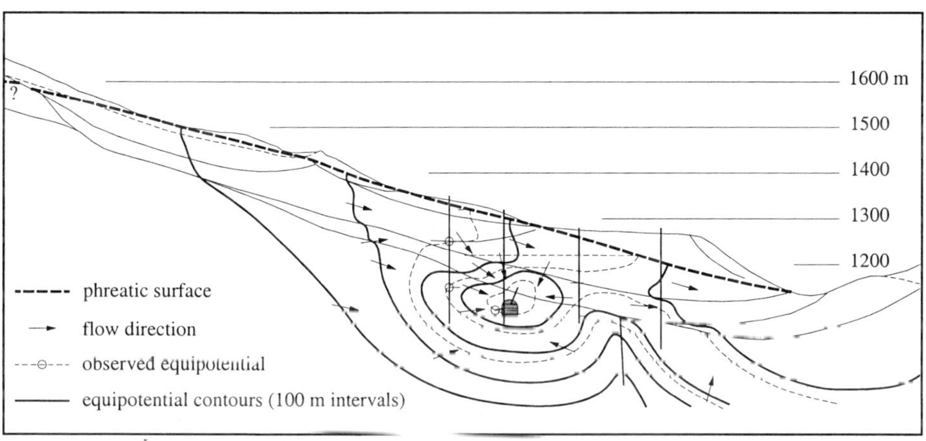

Figure 6. Cross-section of the slide mass showing groundwater flow vectors and equipotential contours *after* drainage adit construction.

CONCLUSIONS

The Campo Vallemaggia landslide represents a creeping mass movement of 800,000,000 m^3 of metamorphic rocks broken into a complex assemblage of blocks divided by tectonic elements. The unstable mass reaches depths of approximately 300 m, within which viscous rock deformations and artesian water pressures have been measured.

Slope movements were attributed to these high pore water pressures, for which an 1800 m long drainage adit was constructed to reduce pore pressures and to stop or decelerate slope movements. In addition, a series of boreholes were driven from the adit through the artesian aquifers to increase drainage and to speed up the mitigative effort.

The measured response of the slope to the drainage adit system shows that pore water pressures have been greatly reduced. These results are reflected in geodesy measurements which indicate that downslope movements have nearly ceased and in some cases have begun to move back upslope due to settlement and consolidation effects.

REFERENCES

Beetham, R.D., Smith, G., Jennings, D.N. and Newton, C.J. (1992). The geology of Nine Mile Creek schist landslide complex. *In* Proc. 6th Int. Symp. on Landslides, Christchurch. *Ed.*. D.H. Bell, A.A. Balkema, 25-31.

Desai C.S., Samtani N.C. and Vuillet L. (1995). Constitutive modeling and analysis of creeping slopes. Journal of Geotechnical Engineering, **121 (1)**: 43-56.

Forlati, F., Ramasco, M., Susella, G., Barla, G., Marino, P. and Mortara, G. (1991). La deformazione gravitativa profonda di Rosone; Un approccio conoscitivo per la definizione di una metodologia di studio. Studi Trentini di Scienze Naturali: Acta Geologica, **68**: 71-108.

Hutchinson J.N. (1988). General report: Morphological and geotechnical parameters of landslides in relation to geology and hydrogeology. *In* Proc. 5th Int. Symp. on Landslides, Lausanne. *Ed.* C. Bonnard, A.A. Balkema, 3-35.

Julian, M. and Anthony, E. (1996). Aspects of landslide activity in the Mercantour Massif and the French Riviera, southeastern France. Geomorphology, **15**: 275-289.

Krohn, J.P. (1992). Landslide mitigation using horizontal drains, Pacific Palisades area, Los Angeles, California. *In* Landslides/Landslide Mitigation. *Eds.* J.E. Slosson, A.G. Keene and J.A. Johnson, Geological Society of America, 63-68.

Martin, P.L. and Warren, C.D. (1992). The design and performance of drainage measures installed for the stabilization of Taren Landslide, South Wales, UK. *In* Proc. 6th Int. Symp. on Landslides, Christchurch. *Ed.*. D.H. Bell, A.A. Balkema, 777-784.

Moore D.P. and Imrie A.S. (1995). Stabilization of Dutchman's Ridge. *In* Proc. 6th Int. Symp. on Landslides, Christchurch. *Ed.*. D.H. Bell, A.A. Balkema, 1783-1788.

Petley, D. (1996). The mechanics and landforms of deep-seated landslides. *In* Advances in Hillslope Processes. *Eds.* M.G. Anderson and S.M. Brooks, John Wiley & Sons, 823-835.

Mass Movements on Granitic Slopes and the Question of Human Interference

J. A. BOURNE and C. R. TWIDALE, Department of Geology and Geophysics, University of Adelaide, Adelaide, South Australia

INTRODUCTION

Granite outcrops occupy about 15% of the continents. Granitic rocks in contact with water are readily altered, and as shallow groundwaters are ubiquitous, most exposures have been weathered and eroded. Nevertheless granite underlies massifs and isolated hills or inselbergs. On the slopes of such uplands mass movements develop, triggered by earthquakes, by heavy rains and by the steepening of basal slopes, either naturally or artificially.

EARTH MOVEMENTS AND INSTABILITY OF SLOPES

Major earthquake impacts cause granite terrains to be drastically changed, with widespread mass movements of rock and soil, as exemplified by the Mt Rokko area after the Hanshin Earthquake of 1995. Small shakes cause minor dislocations, of which the earth tremor which affected Minnipa Hill on northwestern Eyre Peninsula, South Australia, in January, 1999, is a recent example, though similar disruptions manifested as fault scarps, A-tents or pop-ups, and displaced slabs, could be cited from many other granite hills in southern Australia (Figure 1).

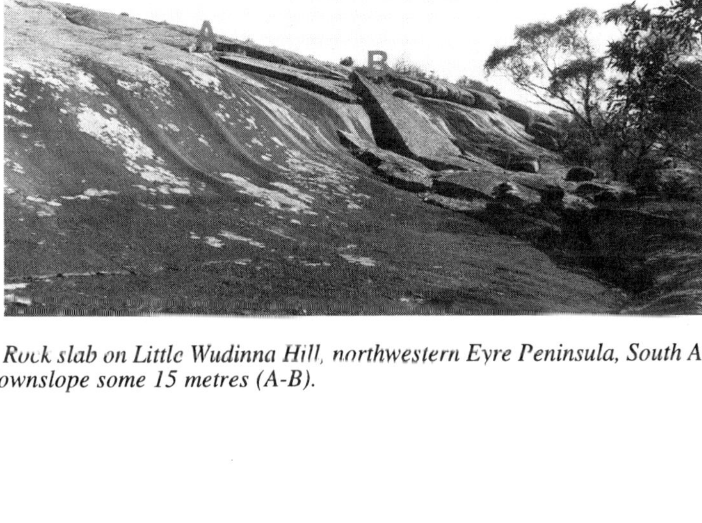

Figure 1. Rock slab on Little Wudinna Hill, northwestern Eyre Peninsula, South Australia has slipped downslope some 15 metres (A-B).

Mt Rokko and the Great Hanshin Earthquake of 1995 (Ikeda 1996)

On 17 January 1995 an earthquake struck the Kobe area of central Honshu, Japan causing great loss of life and property damage. It measured 7.2 on the Richter Scale. It was manifested in the slippage of a fault on the south side of granitic Mt Rokko. Slopes are steep. The regolith failed at many sites causing slides of weathered granite, blocks and boulders. Ikeda reports rock bursts with blocks jumping, probably because they were squeezed between adjacent blocks and boulders (Vidal Romani and Twidale 1998, pp. 347 and 369), as well as the splitting of blocks and slides of mixed debris. Many rock surfaces suffered flaking or spalling.

The pattern of destruction in an earthquake-initiated dislocation differs from that associated with heavy rains, for instead of water accumulating at midslope and saturation and instability extending under gravity toward the valley axis, the impacts of large energy release on midslope extend upwards (Ikeda 1996). This is because the crests of hills and ranges are less confined than are lower slopes, so that the effects of energy dissipation are more evident there.

Minnipa Hill and other minor seismic events (Twidale and Bourne 2000)

At 1137 CDST on 19 January 1999, a small earth tremor, estimated at about 2.3 on the Richter Scale, struck part of Minnipa Hill, a low granitic bornhardt located near Minnipa on northwestern Eyre Peninsula, South Australia. It generated fault scarps, a low amplitude horst, overthrust blocks, A-tents, and also rock bursts involving the ejection of blocks weighing almost 90 kg and the dislocation of a group of slabs together weighing over 3000 kg. Surficial flaking was widely developed. These mass movements are small-scale, but they illustrate the effect a minor seismic event can have on a rock mass already in substantial compression.

Most bornhardts are upstanding not because they are of a rock type more resistant than that in which the surrounding plains have been eroded, but because they are in compression and massive (e.g. Twidale 1982). The event which affected Minnipa Hill on 19 January 1999 was but the latest in a series of low intensity disturbances, for there is evidence of similar rock bursts and flaking in the recent past. Prior to the January tremor the quarrymen working on the eastern side of the hill reported flaking adjacent to cut slots, random fracture propagation from and distortion of drill holes. The quarry was abandoned following the January event.

GRANITE AND WATER

Characteristics of granite weathering

Like all rocks, granite is stable when dry but water causes rapid alteration by solution, hydration and hydrolysis. Fresh granite is of low porosity, but once weathered, both porosity and permeability increase dramatically. Granite is also pervious, and in particular is subdivided by orthogonal fracture systems into cubic or quadrangular blocks, and by arcuate, commonly (though not everywhere) convex-upward sheet fractures into massive slabs or sheets.

Granite consists of quartz, feldspar (typically orthoclase or microcline) and mica. Feldspar and mica are altered to clay, the character of which depends on the nature of the original minerals, drainage and time. Quartz is less reactive and many regoliths resulting from the weathering of granite consist of a stiff clay with fragments of quartz and possibly also large feldspars. When of the consistency of a fine gravel or coarse sand, the material is known as grus.

Thus, granite tends to be either fresh and cohesive or well weathered and friable according to whether it is dry or in contact with water. The junction between the weathered mantle or regolith, and the intrinsically fresh rock is most commonly abrupt (Figure 2), and is known as

Figure 2. Weathering front seen in Tapah Quarry, Perak, West Malaysia.

the weathering front (Mabbutt 1961). Water infiltrates cohesive crystalline granite only slowly but provided the partings are open enough it readily infiltrates along joints. Because the corners and edges of joint blocks are attacked more rapidly than plane faces, angular blocks of fresh rock are converted into more-or-less spherical masses or core-boulders (Scrivenor 1931) set in a matrix of grus (Figure 3).

These fracture and weathering patterns are critical to an understanding of the behaviour of the regolith. Grus is permeable and shallow groundwaters flow through it and along various connected fracture systems to emerge at the weathering front. Springs and seepages are common at the base of slopes. If augmented by heavy rains or artificial diversion and concentration of meteoric waters, subterranean drainage within, and especially at the base of the regolith can flush out sediment, and create voids or low density zones. Not only does the surface subside but the grus becomes saturated, liquifies and loses strength. Such slope sectors may flow downhill.

Figure 3. Core-boulders near Pine Creek, Northern Territory, Australia. Note intrusive vein.

Basal waterlogging can cause slope sectors to become effectively unbuttressed and move downslope en masse. The clayey products of weathering lubricate downslope movement of detritus. Corestones, some of considerable volume and mass, are carried downslope with the finer constituents of the regolith.

Pulau Penang, West Malaysia, 28 November 1998

On Pulau Penang, West Malaysia, in a suburb known as Paya Terubong, tall apartment blocks (Sun Moon City) were constructed on a granite platform cut into the base of a high, steep-sided and wooded granite ridge. On 28 November 1998 a major landslide occurred on the slope above one of the apartment towers (Figure 4). Sand, clay and huge core-boulders (one some 10 m

Figure 4. Landslide related to construction, Pulau Penang, West Malaysia.

diameter) were carried downslope, burying and crushing a number of parked cars (all fortunately unoccupied) and reaching within about 10 metres of one of the tower blocks.

Why the landslide developed at the time it did is not known, but the basic cause of the mass movement was the unbuttressing of the base of the slope combined with the weathering characteristics of the local bedrock and the consistent heavy rainfall of the area. Over the previous two years there were warning signs in the shape of small blocks falling down the slope and the development of small springs. Water may have run along the weathering front and caused the regolith to become unstable over a considerable sector of the slope. With the base cut away the mass was unsupported and slipped downslope.

Thredbo, New South Wales, 30 July 1997

Thredbo is a popular ski resort in the Kosciusko region of southeastern New South Wales. It is a region of deep, steep-sided valleys eroded into a massif in which Palaeozoic granite batholiths are prominent. At Thredbo Village the slopes to the north rise steeply some 1100 m from the floor of the valley to The Ramshead (2194 m) and to the south, the hillslope rises almost 800 m. Many of the lodges are perched on the valley sides. Just before midnight (2340 EST) on 30 July 1997 a mass of slope material collapsed, taking trees, two ski lodges and several cars with it to form a chaotic pile of debris. Eighteen people lost their lives.

The landslide was initiated just above a major road, the Alpine Way, located upslope from the lodges. No heavy rains or snow melt had occurred prior to the landslide. The basic cause of the mass movement of debris was the undermining of the granitic slope by a subterranean stream. Runoff, possibly concentrated by the upslope roadworks, flowed both at the base of the regolith and through it. Fines were evacuated causing volume loss and eventual collapse of the slope. Huge boulders were carried downslope with the finer debris.

Regoliths consisting of core-boulders set in a matrix of grus are exposed on steep hillslopes throughout the region. Landslides had been reported in the area for several decades previous to the 1997 disaster and it was noted that the Alpine Way had been sited on some of the steepest slopes in the vicinity.

HUMAN ACTIVITIES AND MASS MOVEMENT

The conventional wisdom is that vegetation stabilises the soil surface. Clearance of vegetation renders the soil susceptible to erosion by water and wind. There are many well documented cases of vegetation clearance leading to accelerated soil erosion, mainly in the form of gullying and sheetwash but also by way of mass movements (e.g. Van Deur 1978). But on the basis of the distribution of landslides following the rainstorms of early June 1966 in Hong Kong, So (1971) disagrees. Is there evidence of mass movements prior to the advent of humans in any given area, and is there any general reason to suggest that human activities and mass movements of debris are inextricably and genetically linked? It is difficult to find sites which have not in greater or lesser degree been affected by human activities *sensu lato*. Certainly many mass movements develop on engineered slopes or on slopes which have been changed by human activities; and notably vegetation clearance. But earthquakes and volcanic eruptions also trigger mass movements and have occurred throughout geological history. The Blackhawk Landslide of southern California, USA (Shreve 1968) is considered to be prehistoric. It is an air fluidisation flow probably triggered by an earthquake. Such tectonically induced mass movements are essentially independent of human activity. nevertheless, changes wrought by humans have greatly increased the instability of slopes and regoliths.

IMPLICATIONS

Granite hills commonly carry a scatter of slabs, blocks and boulders resulting from the disintegration of the thick shells or sheet structures that are developed within the granite mass. Many of the granite hills are in compression and earthquakes and tremors readily give rise to slippage of the disconnected blocks. This susceptibility of mass movements ought to be taken into account, not only during construction works but also when directing tourist traffic, for example, on granite rocks. Granite in contact with water is weathered readily and in particular ways, but the primary product of granite alteration, grus, consisting of quartz and clay, is not only permeable but the clays also provide lubrication. The common production of a bimodal mass, namely core boulders set in a matrix of grus, is not only potentially unstable but the large contained blocks and boulders render it especially hazardous.

REFERENCES

Ikeda H. 1996. A strong earthquake and the changes it caused in granitic landscape: the case of Mt Rokko due to the Great Hanshin Earthquake of 1995. *Mem. Nara Uni.* 24, 105-116.

Mabbutt J.A. 1961. 'Basal surface' or 'weathering front'. *Proc. Geol. Assoc. (London)* 72, 357-358.

Scrivenor J.B. 1931. *The Geology of Malaya.* Macmillan, London, 217 p.

Shreve R.L. 1968. The Blackhawk Landslide. *Geol. Soc. Amer. Spec. Paper* 108.

So C.L. 1971. Mass movements associated with the rainstorm of June 1966 in Hong Kong. *Trans. Inst. Brit. Geog.* 53, 55-65.

Twidale C.R. 1982. The evolution of bornhardts. *Amer. Sci.* 70, 268-276.

Twidale C.R. and Bourne J.A. 2000. Rock bursts and associated neotectonic forms at Minnipa Hill, northwestern Eyre Peninsula, South Australia. *Env. Eng. Geosci.* In press.

Van Deur W.J. 1978. Earthflows in the Yankalilla area of South Australia: significance of rainfall, soil properties and Man's activities. *Trans. Roy. Soc. S. Aust.* 102, 159-167.

Vidal Romani J.R. and Twidale C.R. 1998. *Formas y Paisajes Graníticos.* Servicio de Publicacións da Universidade da Coruña, Serie Monografías 55, A Coruña, 411 p.

Regularities of Spreading and Forming of Landslides in Armenia

V. R. BOYNAGRYAN, B. V. BOYNAGRYAN, A. V. BOYNAGRYAN, Yerevan State University, Yerevan, Armenia

INTRODUCTION

Dangerous geological processes in Armenia were active in last 10-15 years in connection with total activation in the region of endogenous geodynamic situation and resulted in catastrophic earthquakes in Armenia, Georgia, Iran and Turkey, as well as in connection with a sharp increase of anthropogenic activity. Landslides cause damage to the Armenian economics in some tens of millions of US$ yearly. Therefore, one of the actual problems is detailed studying of regularities, spreading and forming of dangerous geological processes (including landslides) in Armenia and working out protection measures.

DISTRIBUTION OF LANDSLIDES

There are more than 3500 landslides of various sizes in Armenia – from small surface dislocations and small landslides along river beds in erosion districts or on steep and comparatively sloping moistened slopes to large landslide-rockfall, landslide-blocks and flowslides of huge size (complex tectono-seismogravitational displaced bodies). The territory of Armenia, situated in the Alpine-Himalayas belt of high tectonic activity, is characterized by complicated geological-tectonic and geomorphologic structures, differential vertical displacements of blocks of earth crust and high seismicity (strong earthquakes of magnitude 7-8 and more happen here quite often: they induce to dip slope failures, break massive rocks, redistribute natural stresses, and generate forces of inertia). Therefore, dangerous geological processes are widely spread here. Landslides are one of these varieties.

More than 90% of all landslides in Armenia are situated at high-altitude interval of 2000-1500m and below. Only separate landslides are met on altitudes of 2200-2400m and else rarely – 2500-2600m and more. Landslides are widely spread in the basins of the rivers Aghstev, Getic, Arpa, Pambak, Vedi (Figure 1). Armenian landslides are subdivided according to their formation in time into: 1) ancient Pliocene-Upper Quaternary; 2) young Holocene and recent ones.

Ancient landslides are large block slides and flowslides with length of 5-8km, width of 1-2km, and thickness of 100-170m. These huge landslides are named by us "complex tectono-seismogravitational displaced bodies". They are met in the basins of the rivers Aghstev, Getic, Vedi, Arpa (Figure 1). The amplitude of horizontal displacement of these landslides is equal 0,5-1 km; riverbeds in their tongue part are heavily deviated to the opposite side (Figure 2). Vertical displacements are relatively small. Most of large tectono-seismogravitational displaced bodies coincide with fault zones (tectonic active districts with high seismicity). Rocks are often altered to clay conditions and moistened in these zones. Young Holocene and recent landslides quite often complicate concluding parts of majority

Landslides in research, theory and practice, Thomas Telford, London, 2000.

ancient landslides, as well as develop independently in the districts that are favourable for their forming. In slope relief these landslides are identified clearly. Scarps, tension cracks, the landslide body and landslide steps can be seen very well.

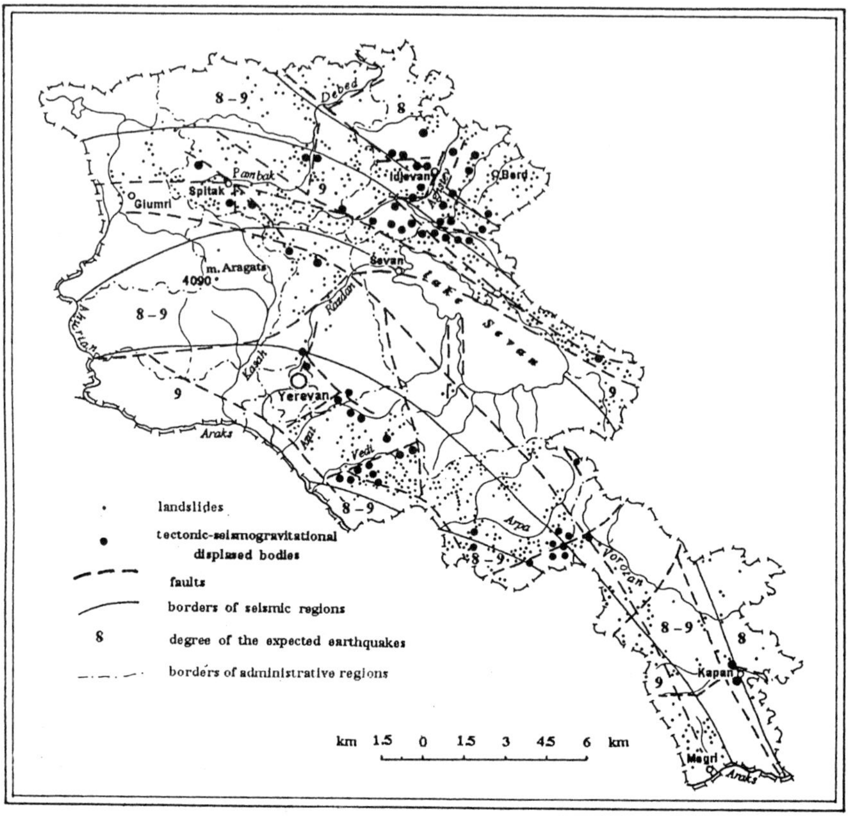

Figure 1. Map of the spread of landslides and tectonic-seismo-gravitational displaced bodies in Armenia

FORMING OF LANDSLIDES

Landslides in Armenia are formed by the geological-geographical conditions of the region, as well as by development of this territory in Pliocene-Quaternary time. Hydrothermal altered (frequently to clay condition), very fissured and broken rocks, gypsipherous clays, clays, various tuff-aleurolites, tuff-breccias, tuff-sandstones, aleurolites, porphyrites, are met frequently here on the slopes. These rocks are reduced to clay by weathering and slip themselves or are good smearing for slipping of rocks, situated above them. Wide prevalence of landslides in Armenia will be clear if we also add a great steepness of slopes and quite often their convex profile; wide development of weakened zones; differential vertical displacements of earth crust separate blocks, which violate condition of slope stability by changes of base level of denudation, their steepness and height alterations. Man's economic activities exercises great influence on forming of new and activating existing landslides. Cutting and overloading of slopes, their remoistening in the time of water leakages from irrigation canals, water-supplies and in the time of excessive watering of cultivating districts, slope vibrations from motor transport, etc. are incomplete list of anthropogenic factors, which destabilise slopes and cause deformations.

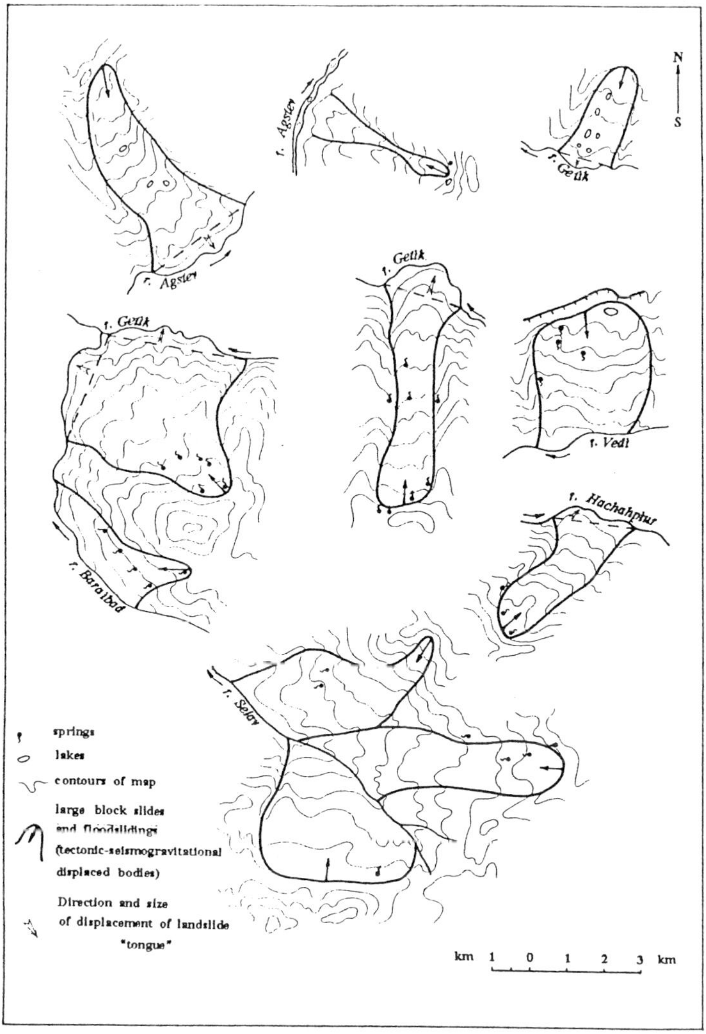

Figure 2. Examples of large block slides and flowslide forms in plan

The large ancient landslides are formed in the time of differential vertical dislocations of tectonic blocks along faults and strong earthquakes, accompanying these dislocations. It is underlined by large size of horizontal and small size of vertical displacements. That kind of displacement character is inherent to landslides, in forming which earthquakes have a great significance. The large amplitude of horizontal displacement is connected with surplus horizontal acceleration during earthquakes. Forming of young and recent landslides, as a rule, is connected with exogenic and anthropogenic factors. Only part of them is formed in the time of earthquakes (for example, 19 new landslides appeared just in North Armenia and many old landslides were activated in the time of the catastrophic Spitak earthquake).

EXAMPLES OF LANDSLIDES

The Dilijan landslide block is the largest and most studied one. Differential displacements of neotectonic blocks along faults, complex geomorphological and geological structure of district; variegated lithological composition of rocks; their intensive weathering, bentonitization, strong fracturing and disunity; wide spreading of underground waters; increase of weight of rocks because of their moistening by atmospheric precipitations and surface waters assisted in forming of this block. Surface parts of this landslide block are active. It is stipulated by cutting and overloading of slopes because of construction of roads, heavy buildings, leakages of water from various water supplies and excessive watering of cultivating districts. The Jermanis (Kanligel) landslide is one of the large landslide bodies in the basin of river Vedi. Here, near village Jermanis, in fault zone, effusive series (andesites and spreading tuffs and tuff-breccias) with covering slope deposits, creeping over surface of ancient clay weathering crust. Wall of scarp of landslide is well traced; suffosion funnels and pits, filled in water are spread at foot of scarp. Some of them reach of large size, as, for example, lake Kanligel (length of lake is 350m, width – 150m). Length of landslide is 3-3,5km, width – 1,5-2km.

Near the village of Vohchaberd a landslide block (size of 2,5x1,3km^2) is displaced along of reversed fault line. Amplitude of vertical displacement of this block is of the order of 200-250m. At the present time the Vohchaberd block shows signs of activity. Filtration of surface and underground waters in the landslide body, unstable clay soils, repeated earthquakes, and economical activity of man, are causes of it. Here, rate of landslide displacements are, now and then, some millimeters per day. Separate districts of Large Yerevan, where Sarmatian stage clays (canyon of river Razdan) and gypsiferous clay (Nubarashen, Shorbulah) are spread, are strongly suffered from landslides too. In canyon of river Razdan lava creep over spreading Sarmatian stage clays and these clays. Here some large landslide blocks of 0,8x0,7km^2 sizes are distinguished in fault zone. These landslide blocks drove back bed of river Razdan to opposite board of canyon till 150-200m. Vertical displacement is 25-50m, in some places – till 75m.

In Nubarashen gravels of the same name plateau overlie gypsiferous clays and clays themselves, including large territories affected by landslide movements. Individual, more active, landslides are distinguished here on general background of landslide slope with evidently expressed signs of landslide movements. One of landslides reached Nubarashen cemetery and destroyed many graves. This landslide formed in 1980 because of moistening of clay by underground and filtration waters, as well as in consequence of slope cutting by construction of high road. Thickness of landslides is 16-18m in central part. Surface of landslide is covered by numerous cracks. Displacement of landslides in Armenia happens with various intensity. If some landslides are stabilized at present, others show weak activity (for example, separate landslides in Dilijan are displaced with average rate of 5-6 cm/year. Displacement is in addition interrupted in time). In other districts rate of displacement of landslides can reach 20-36 cm/year or have a catastrophic nature.

CONCLUSIONS

Landslide processes are widely spread in Armenia and introduce essential discomfort into natural environment, determine instability of relief, increase degree of danger and risk by engineering and economic development of separate districts. At the same time regularities of spreading of large landslides can help in determination and verification of degree of seismic danger, and in making of corresponding forecast maps.

Landslide-Rockfalls in Valley of River Baidak

V. R. BOYNAGRYAN, S. P. BALYAN, R. H. GAGINYAN, A. A. AIRIYANTS, A. V. BOYNAGRYAN, A. S. MKRTCHYAN, K. R. CHOBANYAN, Yerevan State University, Yerevan, Armenia

INTRODUCTION

All types of landslides occur in Armenia, however landslide-rockfalls are rare. In the valley of the river Baidak, there are two large landslide-rockfalls which formed in 1946 and 1968, is exceptional.

COMMON NATURE OF VALLEY OF RIVER BAIDAK

The valley of the river Baidak occupies the northern slope of the Pambak range, where the amount of precipitation is 600-700 mm/year. Here distribution of precipitation by season is irregular, reaching a maximum in July.

Valley of river Baidak is produced in Middle Eocene porphyrites, tuffites, tuff-sandstones, tuff-breccias, which lie transgressively over eroded surface of Lower Eocene and Upper Cretaceous. Very dislocated sedimentary and volcanic rocks, broken granodiorites and quarzdiorites, as well as hydrothermally-altered rocks are exposed on slopes of valley. The transverse profile of valley is V-shaped, its bottom is deeply cut into, and large declivities are marked. Block deposits, stone taluses, and sandy-clay deposits are widely spread on the side-slopes of the valley. Slopes are not covered. The valley of river Baidak is situated a little diagonally respect to the strike of tectonic structures.

Left slope of valley is steep (40-50°), it is complicated by faults, along of which granitoid intrusions and hydrothermally-altered porphyrites are situated. Springs with constant discharge occur at a height of 50-60m above bed of river. Landslides on this slope coincide with altered rocks and springs (figure 1). The right slope is steady, as here hydrothermal altered rocks are absent, and dip of layers is opposite in respect to the slope.

LANDSLIDE-ROCKFALL OF 1946 YEAR

This landslide-rockfall is situated in 1500m farther south of village Kishlag (figure 1) and occupies area of 0,36km^2. Height of wall of scarp is 30-35m, steepness is 80-85°. Landslide body occupies slope with steepness of 45° and height of 400m above bed of river. The thickness of landslide in head part is 10-25m, in the middle part it is 30-35m, and near the foot it is more than 60m. Total volume of landslide-rockfall is 600 000m^3. A large fault passes along the right failure plane of this landslide

Our researches showed that there is landslide-rockfall of packets and blocks of hydrothermally-altered rocks near the village of Kishlag. Repeated displacements have caused a stepped structure in this landslide. At present surface of landslide-rockfall is covered by grass, new displacements no longer occur.

Landslides in research, theory and practice, Thomas Telford, London, 2000.

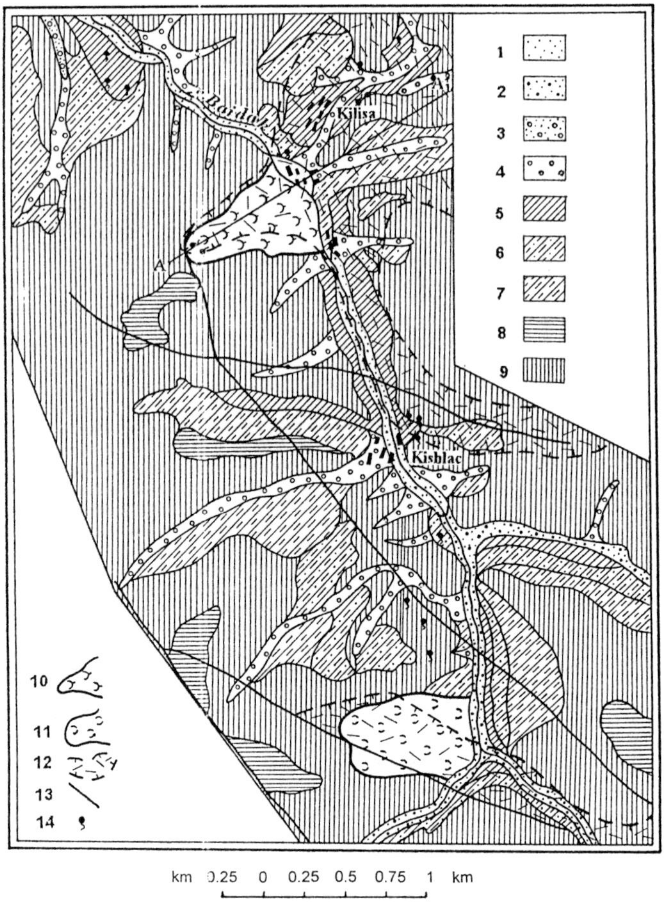

Figure 1. Engineering-geological map of part of basin of river Baidak:
1 – alluvium of flood-plain and terraces; 2 – alluvium-alluvial fan deposits; 3 – alluvial fan deposits; 4 – deluvium-alluvial fan deposits; 5 – deluvium of foot of slope (thickness more than 5 m); 6 – deluvium of middle part of slope (thickness of 3-5 m); 7 – deluvium of high slope (thickness of 1-2 m); 8 – eluvium-deluvium; 9 – rock on steep slopes; 10 – landslide of 1968; 11 – landslide of 1946; 12 – hydrothermally altered rocks; 13 – faults; 14 – springs

LANDSLIDE-ROCKFALL OF 1968 YEAR

This landslide is formed on the steep left slope of valley of river Baidak near village Kilisa (figure 1) and in process of its displacement, it went out to the right slope, re-covered the fluvial terrace above flood-plain with village Kilisa. The landslide buried 80 buildings and 34 persons. A lake with depth of 10m, length of 600-660m and width by the landslide dam of 150-200m was formed behind of landslide in a very short time.

The landslide-rockfall is situated on a steep slope and has in plan the form of an isosceles triangle with a base of 400m and a height of 500m (it is the same for the length of rockfall). The first signs of landslide forming (of fracture type) appeared here else in 1967 year, but signs of the start of movement of the weak rock massif down-dip were marked some days before disaster. The landslide-rockfall happened very quickly, in some minutes, but with short interval, during which the children were taken out of the school building. But 34 persons had no time to escape.

The landslide-rockfall consists of four large levels (figure 2). The lowest level has 75-80m height above bed of river and it is wide ground with smooth surface. Upper three levels have 40-50m height above the first and one above the other. Their surfaces (except the fourth level) are slightly inclined to the bed of river. Thickness of slipped masses in the bed of river is more than 35m, but in the middle part it is 20-25m.

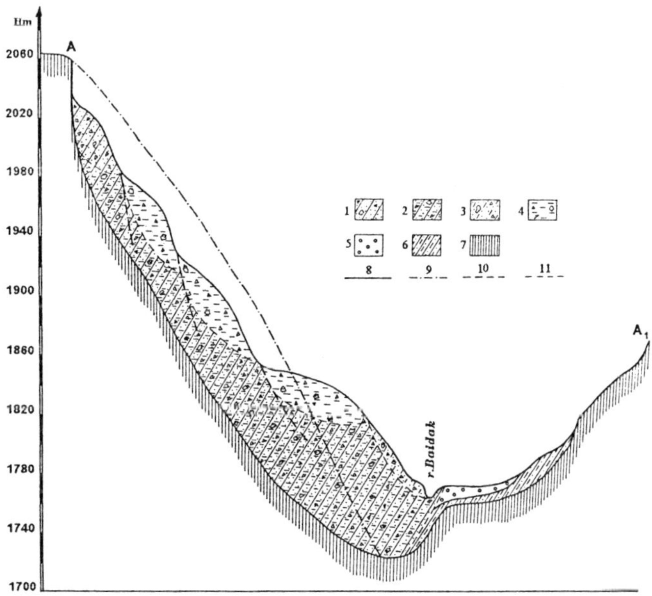

Figure 2. Geological profile of valley of river Baidak by village Kilisa:
1 – large detritus with loamy sand and loam; 2 – loams and clays with detritus and block; 3 – large detritus with blocks and loamy sand; 4 – clay with detritus; 5 – alluvium of terrace; 6 – deluvium; 7 – rocks; 8 – recent surface of landslide; 9 – slope before landslide displacement; 10 – boundaries of landslide steps; 11 – lithological boundaries

The slipped masses of landslide-rockfall consist of two groups: 1) yellow loamy – loamy sandy deposits with rock debris skeleton (they are 2/3 of landslide), 2) gray rock debris – block deposits with loamy sand and loam (1/3 of landslide).

The first group is formed as result of displacement and following destruction of hydrothermal altered very pyritinized and kaolinized effusive rocks.

The second group is formed as result of interception by landslide of exposing on steep slope part of steady unweathered Eocene porphyrites.

Forming of this landslide-rockfall is result of constant penetration of water from two springs and melted waters into fractures of rocks. It weakened rock massif.

Rapid melting of snow in warm April of 1968 on slope and re-moistening of soil are immediate causes of landslide-rockfall, as well as slope erosion by Baidak River in the time of spring rise of water level.

CONCLUSION

Landslide-rockfalls in valley of river Baidak are connected with fault zones, where rocks are hydrothermally altered, broken by numerous fractures on separate blocks and massives, and much weakened. Moistening of that kind rocks by melted, pluvial and spring waters, as well as concussion of steep slopes during strong earthquakes can assist in appearance of new landslides and landslide-rockfalls with dangerous consequences for populated locations in this valley. Therefore, the valley of the river Baidak coninues to be a potentially dangerous district for the occurrence of catastrophic landslides.

An Interpretation of Slope Dynamics in Pisticci (Southern Italy)

F. BOZZANO, M. FLORIS, Dipartimento di Scienze della Terra, Università degli Studi di Roma "La Sapienza", Rome, Italy, and M. POLEMIO, CNR CERIST, Bari, Italy

INTRODUCTION

The flat top of the Pisticci hill (392m a.s.l.) is totally occupied by the town (12,000 inhabitants), and steep slopes deeply eroded and involved in slope movements surround it. Access roads and external part of the built-up area have been frequently damaged, sometimes very seriously, by landslides. For that reason the hill has been largely investigated by some authors (to be mentioned Guerricchio & Melidoro, 1979): some remedial works had carried out and other remedial measures had been adopted. In this paper slope movements occurred in Pisticci hill have been re-analysed in order to focus into rainfall as last trigger factor. Moreover the combination of historical landslides series and statistical hydrological models as a tool to assess landslide hazard has been dealt.

GELOGICAL AND HYDROGEOLOGICAL OUTLINE

Pisticci hill is located in the SE section of the Apennine Foredeep, a structural depression that had been filled up by Pliocene and Pleistocene deposits. In Figure 1 geological and hydrogeological setting has been synthesised. Groundwater paths have been inferred by the definition of the main hydrogeological structures, each one characterised by recharge areas, no flow limits and springs. From the bottom to the top the succession is (Figure 1): 1) Grey-blue clay (80 - 350m a.s.l.): stiff and jointed marine marl silty clay (lower Pleistocene) with silt or sand thin horizons and rare volcanic fall levels. Spectacular badlands morphology shows that intense erosive processes involve this formation. Respect to the remaining lythologies, the blue clay acts as an impervious boundary, with a continuous permeability limit, and, wherever it does not outcrop, it underlies those with which it is in contact. 2) Marine terraced deposits (350 - 392m a.s.l.): silty sand with thin gravel levels, yellow clayey silt, uniform fine sand, yellow sand with sandstone levels. These succession (middle Pleistocene) is transgressive on the grey-blue clay and shapes a steep scarp (Figs. 2, 3). It is noteworthy that at present the Pisticci plateau (dated 650,000yrs BP) is an isolated relict of an originally more extended marine terrace progressively destroyed by erosive processes and landslides. The marine terraced deposits exhibit the highest relative hydraulic conductivity, owing to original porosity and to jointing. Due to strata with different hydraulic conductivity, they constitute a layered aquifer; it is the main aquifer in this area. The groundwater is sustained by the impervious blue clay and it partially drains through some springs (n. 3, 5, 7, 10, 11, 12, 13, 15, 16, in Figure 1) that are located along the impervious boundary with the blue clay. The spring discharge Q of the n. 3 spring is very low ($Q \cong 0.01$ l/s); this spring is periodic as the discharge is null every summer; as it regards the others no discharge data are available as tapped and no accessible. Some collected spot data on water head referred to the base of the aquifer suggest values ranging from 8 to 20m and exceptionally to 30m. Where perched water exists it is narrow and drains directly hillside or it recharge the groundwater through some joints. At present, losses of sewers and aqueduct have been thought to be the main recharge of this layered aquifer since rain water infiltration is cancelled by urbanisation and paving of the hill top. 3) Continental deposits: irregularly shaped bodies of reddish sand or conglomerate covering the marine terraced deposits.

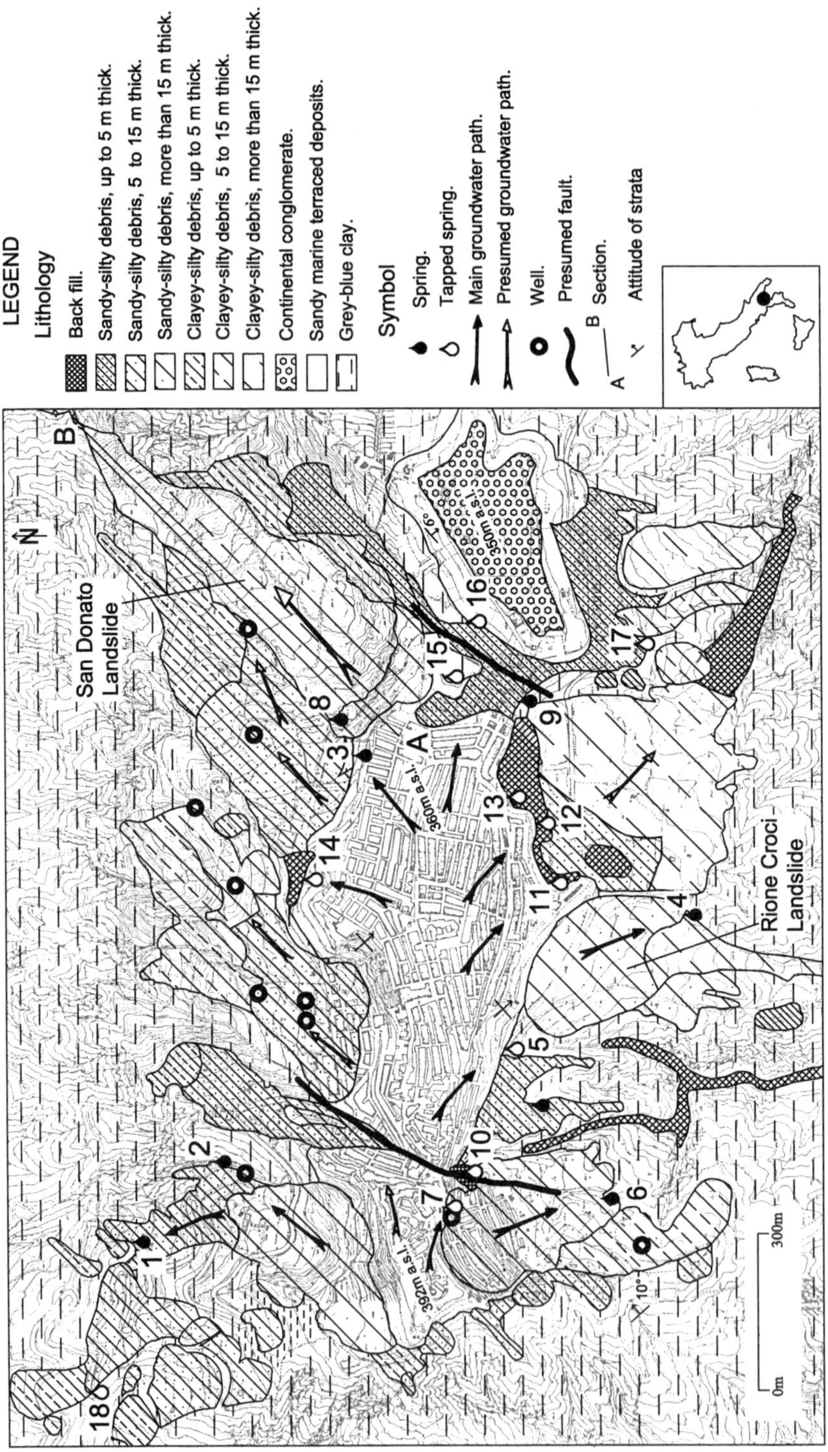

Figure 1. Pisticci hill: geological map with hydrologic elements.

Debris deposits mantle the slopes surrounding the top surface of Pisticci hill (Figs. 1, 2, 3). They can be regarded as colluvium mainly originated by landslides involving either marine terraced deposits or grey-blue clay. On the upslope side, they are hydrogeologically in contact with the main aquifer and they make up a secondary aquifer. 4) Sandy and silty debris: from reddish to yellow in colour mainly outcrop in the upper part of the slopes and in some areas in the medium part too; large volumes blocks constituted by well-stratified sand dipping into the slope are embedded in this type of debris. In sandy silty debris bodies direct rainwater infiltration and groundwater flow are possible. Moreover, in the cases of lateral hydraulic continuity with the marine terraced deposits the above mentioned layered aquifer is drained by them. In Figure 2 it is shown how slope movements could cause either this hydraulic continuity condition or the displacement of a spring along the slope as it is for springs 1, 2, 4, 6, 8, 9, 17 and 18 (Figure 1). These springs are perennial (Q in the range 0.01 - 0.33 l/s) and are located where sandy silty debris is in contact with grey-blue clay formation or clayey - silty debris. 5) Clayey and silty debris from grey to yellowish in colour outcrop in the medium part of the slopes and in topographic depression of the upper part. The presence of some wells and other information coming from S. Donato area (Figs. 1, 3) show that the role of this type of debris could be that one of a secondary aquifer. If groundwater paths, springs location and hill morphology are considered, they suggest that the NNW-SSE trending tectonic discontinuities (Figure 1) and the dipping of marine terraced deposits contact with grey-blue clay control groundwater flow (Figure 1) and that debris bodies, mainly originated by landslides, constitute significant hydrogeological complexes. These complexes and the role of impervious limit played by blue clay are very relevant to slope stability conditions as it is discussed in the next paragraph.

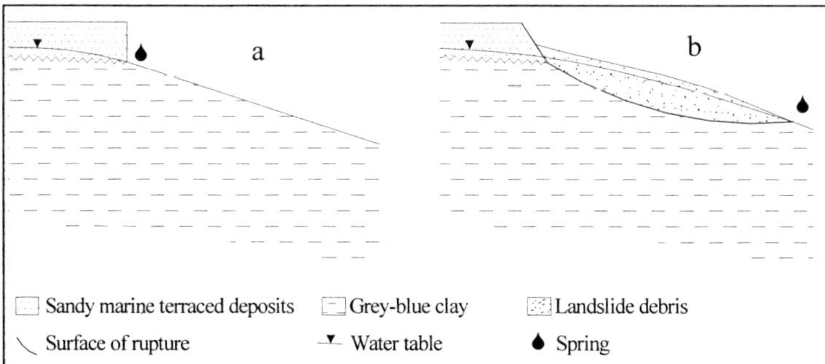

Figure 2. Sketch of spring displacement induced by slope movements.

ANALYSIS OF RAINFALL AS TRIGGER FACTOR OF LANDSLIDE

Based on field survey, interpretation of aerial photographs, consulting of historical and technical documentation, two main landslide types have been identified. First type is retrogressive rotational complex rapid earth slide-earth flow (Cruden & Varnes, 1996) involving the marine terraced deposits and the grey-blue clays. A good fitting example is represented by the Rione Croci landslide (Figure 1) dated 11/21/1976: since 1952 numerous previous instability marks in

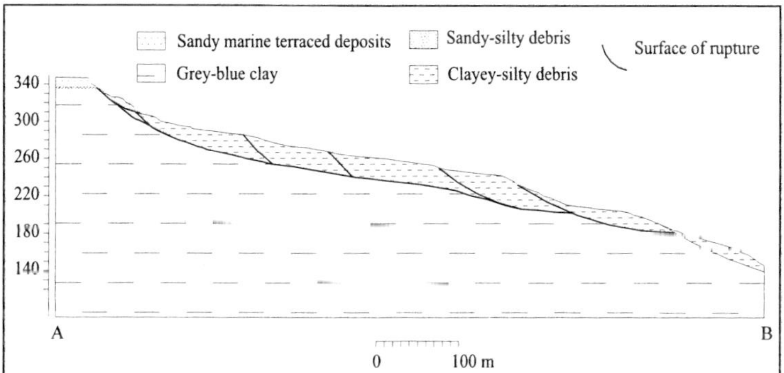

Figure 3. Cross section (trace A B in figure 1) along San Donato landslide.

the crown area predicted the 1976 collapse (Guerricchio & Melidoro, 1979). This event produced a main scarp 25m height in the marine terraced deposits and maximum retrogression of the top surface up to 30-40 m; the slip surface is from 30 to 40m deep. The width of surface of rupture is 300m, the length of displaced mass is 200m and the total volume is over 3 millions m^3; an earth flow is located in the zone of accumulation. As it regards slope failure mechanism we made reference to Bozzano et al. (1996). Supposed geological and hydrogeological setting before the 1976 collapse is shown in sketch of Figure 2a. Spring n. 4 (Figure 1) originated after the landslide as hypothesised in Figure 2b. Second type is classified as translational slides involving debris deposits and retrogressively and successively inducing rotational slides in marine terraced deposits. It has been well recognised in San Donato area (Figs. 1, 3) where in the time interval 1972-'73 four landslides have been recorded. Aerial photograph dated July '72 shows the first event: some debris masses have been displaced down-slope along slip surfaces following the contact debris deposits - grey-blue clay. In the period August '72 - January '73 landslide area shifted up-slope: 2 rotational slides occurred in the built up area involving marine terraced deposits and producing maximum retrogression of the top surface up to 8 m. In the sketch of Figure 2b the geological and hydrogeological setting before the '72 landslide sequence has been hypothesised as it was interpreted by means of air photos taken in 1946, '54, '67. In these photos debris deposits show evident marks of slope movements before '72 demonstrating their high susceptibility to landslide. Numerical modelling performed by FLAC (Fast Lagrangian Analysis of Continua; ITASCA 1996) on the section of Figure 4A reinforces our reconstruction of landslide sequence. In fact if water table level 8 m b.g.l. is supposed in clayey-silty debris deposits ("ordinary seepage conditions") the distribution of the yield points shows an instability area in the lower part of the slope (Figure 4B). Moreover, if water table level 5 m deep b.g.l. in marine terraced deposits and in clayey-silty debris a water table corresponding to ground level ("extraordinary seepage conditions") are supposed, the instability area shifts up-slope and involves the terraced deposits (Figure 4C). The transition between ordinary and extraordinary seepage conditions could be related to the recharge of aquifers by rainfall as discussed in the next lines. A hydrologic-statistical method has been applied as a means of investigating, through the study of the selected hydrological variable, to what extent the rainfall event believed to be associated with a given landslide had been exceptional. Cascini and Versace (1986) indicate that for deep-seated landslides, the only significant variables are the ones, which take into account rainfalls lasting more than one day. Some hydrologic variables have been introduced, such as the so-called cumulated or cumulative rainfall, that is the summation of daily rainfall occurred over a given number of consecutive days, prior to the considered day. The identification of the most suitable variable for the correlation between rainfalls and landslides results from essays run on more than one cumulative rainfall. By means of these models, the exceptional character of landslide-related rainfall events is expressed in terms of return period (T) (methodology is reported in Polemio & Sdao, 1999). The Figure 5 and Tab. 1 synthesise the hydrologic-statistical results. The Rione Croci landslide was activated on November 1959 by a very exceptional storm that hit seriously the region: the return period (T) of cumulative rainfall was more than 50 years for duration (D) less than 55 days. The slope stability was temporary; two reactivations followed after subsequent four years (during negligible rainfall: T< 4 years).

Piezometric water pressure was monitored from May 1993 to February 1995, using two couples of electric piezometers which were installed upslope the scarp in two boreholes from 15 to 45 m b.g.l.. The shallow piezometric devices were in the main aquifer; the deep ones were in a transition level from the aquifer to the impervious blue clay. The monitoring data show that the lag between daily rainfall events and piezometric variations was 40 to 60 days for all piezometers. So this physical approach is coherent with the more significant duration, as a result of the empirical hydrologic-statistical analysis. The first San Donato landslide occurred on March 1972, due to significant rainfall. The critical duration is 80 days and T is about 18 years. If we consider also the Rione Croci landslide, the higher duration for the San Donato landslide is due to lower permeability of the landslide body but also and principally to the groundwater inflow very probably coming from the main aquifer. In fact, there is a lateral hydrogeological continuity between main aquifer and the San Donato landslide as a

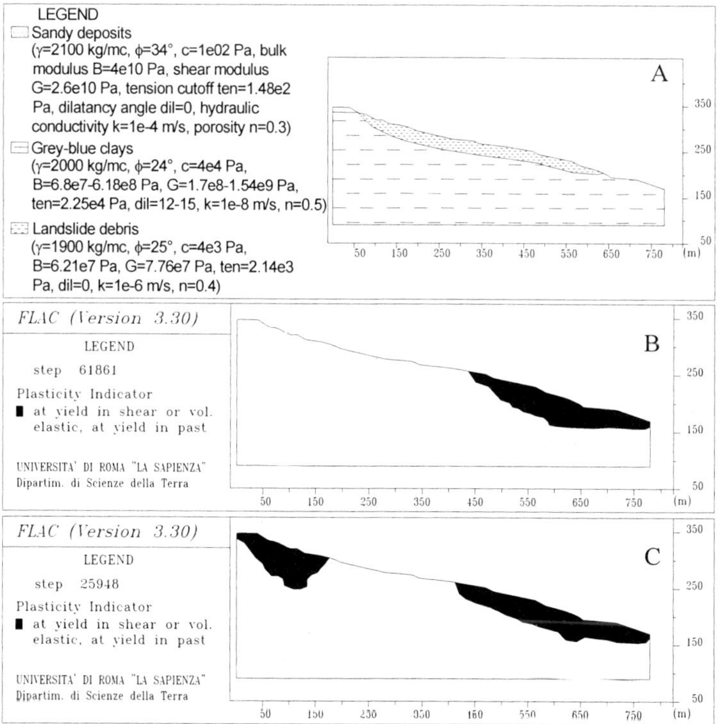

Figure 4. Results of numerical modelling of the San Donato slope. For A, B, C see text.

perennial n. 8 spring, confirms. The overlapping effect of direct rainfall infiltration and groundwater inflow can be also observed from a hydrologic-statistical point of view. The San Donato landslide was followed by 3 reactivations during 10 months for a complex of 4 known movements in the last forty years. In any case, significant 15 days were observed up to 48 days before each landslide movement (Figure 5); this lag agrees quite well with the lag of main aquifer piezometric response.

CONCLUDING REMARKS ON LANDSLIDE HAZARD EVALUATION

An important output of the hydrologic-statistical study of the S. Donato landslide sequence is that in this case the classical "threshold model" applied to the relationships rainfall- landslidereactivations doesn't work and that a modified concept of rainfall threshold has to be introduced. When a threshold [A] has been fixed from the historical data on reactivations and rainfall, it needs to associate to it the probability to trigger a landslide reactivation [B] named conditional probability (P [B/A]). If the period 1950-'87 is chosen, 4 reactivations have been recorded in S. Donato landslide complex. Their trigger have been associate to the overcoming of a rainfall threshold of 150 mm in 15 days occurred up to 48 days before each landslide date (Figure 5). Return period of this rainfall is 2.6 yrs (probability of occurrence 0.38 named P[A]); when it occurs there is a probability of 0.29 to initiate a landslide reactivation inside this area (P[B/A]) as it is computed from the observation that in the chosen period 14 times the threshold had been overcome and 4 times it has been associate to a landslide. Finally, the landslide hazard evaluation (H) derives from the computation of the intersection probability:

$$H = P[A \cap B] \equiv P[A]P[B|A] = 0.11$$

That means a return period between 9-10 years, a very high hazard according to Fell (1994).

It has to be noted that in this exemplifying numerical evaluation other observations, such as that one regarding the stage and dynamics of the slope, aren't considered.

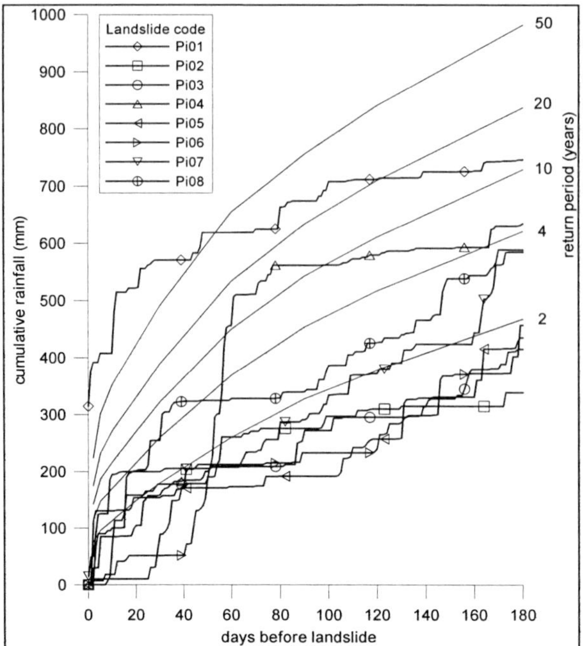

Figure 5. Cumulative rainfall curves of assigned return period (2, 4, 10, 20 and 50 years) and cumulative rainfall curves of the analysed landslide events.

Table 1. Code, name and date of the selected landslides and hydrologic-statistical characteristics of cumulative rainfall. (D) Critical duration of cumulative rainfall (day) and (T) return period (year).

Code - Name	Date	D	T
Pi01-Croci	11/24/59	29	>50
Pi02-Croci	01/17/61	4	>4
Pi03-Croci	10/19/63	30	2-4
Pi04-S. Donato	03/15/72	78	10-20
Pi05-S. Donato	08/21/72	180	<2
Pi06-S. Donato	09/05/72	180	<2
Pi07-S. Donato	01/02/73	14	4-10
Pi08-Croci	11/21/76	37	4-10

REFERENCES

Bozzano F., De Pari P. & G. Scarascia Mugnozza 1996. Historical data in evaluating landslide hazard in some village in southern Italy. Proc. of the 7th Int. Symp. on Landslides (ISL '96), 1, 159-164.

Cascini, L. & Versace, P. 1986. Eventi pluviometrici e movimenti franosi. Atti XVI Convegno Nazionale di Geotecnica, Bologna, Italy, 3, 171-190, AGI, CLEUP.

Cruden, D.M. & D.J. Varnes 1996. Landslide type and processes. In: A. Keith Turner & R.L. Schuster (eds.), *Landslide investigation and mitigation,* Trasportation Research Board, Special Report 247, 36-75.

Fell R. (1994): Landslide risk assessment and acceptable risk. Can. Geotech. J., 31(2), 261-272.

Guerricchio A. & Melidoro G. 1979. Fenomeni franosi e neotettonici nelle argille grigio-azzurre calabriane di Pisticci (Lucania) con saggio di cartografia. Geol. App. e Idrogeol., XIV, 105-135.

ITASCA 1996. FLAC version 3.3, User's Manual. Itasca Consulting Group.

Polemio M. & Sdao F. 1999. The role of rainfall in the landslide hazard: the case of the Avigliano urban area (Southern Apennines, Italy), Eng. Geol. (53), 3-4, 297-309.

Numerical Modelling of Slopes Surrounding Ignimbrite Plateaux

F. BOZZANO, M. GAETA, G. SCARASCIA MUGNOZZA and G. VALENTINI,
Dipartimento di Scienze della Terra, Università "La Sapienza", Roma, Italy

INTRODUCTION

The adoption of a representative geological model is the key-step in slope stability analysis aimed at designing effective remedial works. Examples come from computer modelling of state of stress and failure mechanisms involving tuffaceous cliffs, tens of meters high, overlying a more deformable susbtratum (Lembo-Fazio et al., 1984). Such a geological and geomorphological setting is typical of a wide area north of Rome (Italy), where since Etruscan times many small towns (e.g. Orvieto, Civita di Bagnoregio) were founded on the top of hills and spurs which represent the relics of large plateaux of extremely heterogeneous quaternary volcanic products (Martinetti, 1980; Campobasso et al., 1994). In this paper criteria are reviewed about effective transferring from geological formations and their relative volcanic eruptive and depositional units into engineering geological units in order to achieve simplified though representative geological model. This operation is mainly based on the assessment of lithological and geomechanical features of the investigated rock mass in view of Distinct Element analysis to highlight the principal factors bearing on the failure mechanisms observed along the cliffs surrounding Calcata and Faleria, two small towns located 50 km north of Rome. Plateaux of quaternary volcanic rocks of Northern Latium and Southern Umbria and Tuscany overlay Pliocene and Pleistocene clastic successions composed of silty-clay, sand and conglomerate that were deposited in marine and continental basins whose origin is related to the opening of the Tyrrhenian sea in a tectonic extensional regime. Volcanic rocks belong to three volcanic districts, namely Vulsino, Cimino-Vicano and Sabatino (Figure 1), whose

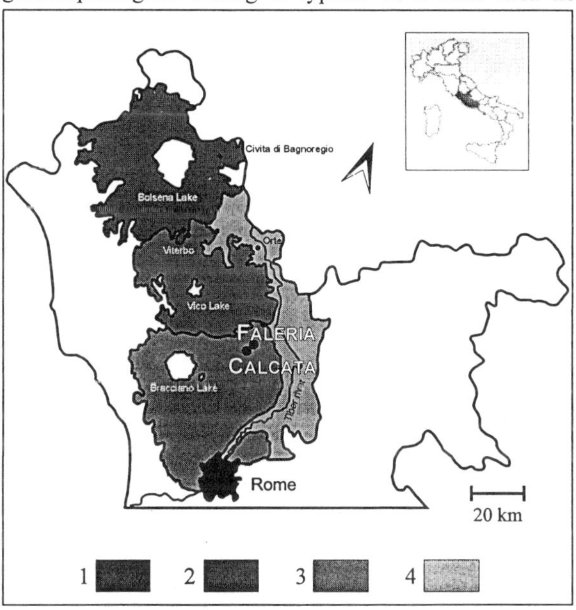

Figure 1. Volcanic districts in northern Latium and location of Calcata and Faleria. 1 - Vulsino district ; 2 – Cimino-Vicano district; 3 - Sabatino district; 3 - alluvial deposits of the Tiber river

Landslides in research, theory and practice, Thomas Telford, London, 2000.

activity spans between 1.31 and 0.09 My. Vulsino and Sabatino districts are characterised only by alkali-K products, while the Cimino-Vicano district by quartzlatitic products also. The outcropping rocks are mainly lavas, pyroclastic flows, fall and surge deposits whose thickness is widely variable from tens up to hundreds of meters (Campobasso et al., 1994). The geometrical relationships between the products of the volcanic districts are very complex due to their depositional mechanisms, previous morphology of the area and erosion that they underwent in time intervals between successive eruptive phases.

GEOLOGICAL MODEL

In addition to the outcropping sedimentary substratum, the geological survey allowed to distinguish some eruptive units within the volcanic rocks, each one characterised by different depositional units. Among the depositional units, the dominant one and consequently its lithotype and geomechanical behaviour has been considered as representative of the engineering geological unit labeled with the name of the geological formation including it. From the bottom to the top they are (Figure 2):

- Conglomerates, constituted by pebbles in not much matrix. They have been interpreted as deposited by the PaleoTiber course (Pliocene and Pleistocene) – (labeled as Cgl).
- "Tufo Giallo della Via Tiberina" (0.52 - 0.60 My). Pyroclastic flows deposits composed of the superposition of a thick sequence of depositional units. It shows a lithoid behaviour but some more erodible accretionary lapilli rich levels are present (TGVT). Heteropically they pass to "Epiclastic deposits" mainly represented by thin lacustrine clay levels (Epic.).
- "Tufi stratificati varicolori di Sacrofano". This formation is formed by loose fallout levels of tephra, ashes and scoria widely argillified (TSVS).
- "Tufo rosso a scorie nere sabatino" (about 0.54 My). Massive pyroclastic flow deposits with a typical reddish and lithoid facies made up of fine ash and pumiceous lapilli matrix. Very porphyritic black scoriae and abundant lithics constitute the framework of this facies. At the bottom a loose pozzolanic facies with black fiamme is locally present (TRS).
- "Tufi stratificati varicolori di La Storta". Very similar to the above described "Tufi stratificati varicolori di Sacrofano" (TSVL).
- "Tufo rosso a scorie nere vicano" (0.15 My). Similar to the above mentioned "Tufo rosso a scorie nere sabatino" except for the lack of the bottom pozzolanic facies. Its geometry and the unconformity on the previous listed unitss show that it had been deposited along the existing valleys (TRV).

Very large volcanic plateaux gently sloping and deeply eroded by gullies are the main morphological feature of the study area; volcanic and tectono-volcanic depressions, in some cases occupied by lakes, and volcanic relief are present. The valley slopes are sub-vertical or steps carved while the valley floors are flat. Fluvial erosion of the volcanic plateaux caused the outcropping of the underlaying sedimentary substratum along the valley flanks. On the western side of the geological map (Figure 2) the Treja creek represents the paleo-course of the Tiber river before its damming and deviation caused by a pyroclastic flow ascribed to the TGVT unit (Alvarez, 1973). The ancient small town of Calcata (area in box A of Figure 2) is built on the top of a hill (172 m a.s.l.) surrounded by very steep slopes. Respect to the complete lithostratigraphic succession shown in Figure 2, Epiclastic deposits, TRS and TSVL units are absent (Figure 3).

Figure 2. Geological and geomorphological map of the area between Calcata and Faleria and stratigraphical relationships scheme (front page). 1 - Alluvial deposits; 2 - Debris deposits; 3 - TRV; 4 - TSVL; 5 - TRS; 6 - TSVS; 7 - TGVT; 8 - Epiclastic deposits; 9 - Conglomerates; 10 - Edge of quarry scarp; 11 - Edge of degradation and/or landslide scarp; 12 - Alluvial fan; 13 – Flat valley floor; 14 - Scree slope; 15 - Landslide body due to fall or toppling; 16 - Landslide body due to complex slope movement. Box A and box B surround Calcata and Faleria cliffs respectively

Faleria
Calcata
Treja River
F.so della Mola
A
B
N
0
0.5 km
1
2
3
4
5
6
7
8
9
10
11
12
13
14
15
16
0
20 m
3
4
5
6
7
8
9

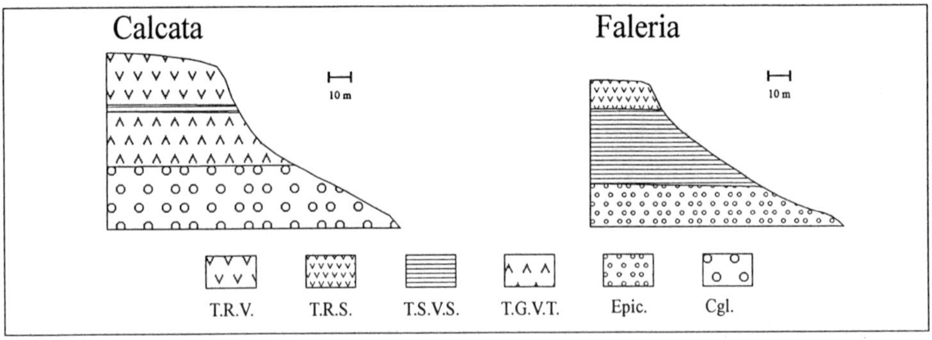

Figure 3. Geological sections used for the numerical models. Location of the corresponding slopes is shown by arrows in box A and B of Figure 2

Debris fan characterises the NW slope. Houses are founded on the TRV and face the about 40 m high scarp bordering the cliff. This formation is frequently involved in slope movements classified as falls and topples that induce a progressive retreat of the edge of the scarp. In the period 1990 –1992 the SE part of the Calcata hill was damaged by fall and topples but the small dimensions of the blocks don't allow their mapping in Figure 2. Faleria is built (210 m a.s.l.) on a 13 m thick tuffaceous slab of the TRS overlaying TSVS and Epiclastic deposits 50 m thick (Figure 3); a 20 m high scarp surrounds the perimeter of the hill. The ancient part of the town (area in box B of Figure 2) located on the northern part of the hill was abandoned 40 years ago as strongly damaged by falls, topples and complex slides involving the TRS and TSVS units and Epiclastic deposits (Figure 2). At present, the tuff cliffs bordering the two villages are considered susceptible to slope movements, as shown by the events recorded in the last 50 years and the presence of landslides masses and fresh scars on the slopes. Jointing of the tuff rocks isolates blocks which are prone to fall or topple; moreover tree and shrub roots along joints, man made caves and unpiped rain and domestic grey waters draining onto the scarps below the town worsen the local slope stability conditions.

Discontinuity pattern mapping of the tuffaceous cliffs has been performed systematically at various locations, both in horizontal and vertical direction, with particular reference to the sections that have been successively analysed through numerical modelling. The statistical analysis of orientation data points out the presence of two discontinuity sets (K1: 83°/85°; K2: 132°/80°) along the southern face of Calcata cliff; along the NW slope of Faleria two sets have been detected as well (K3, K4), whose orientation is 66°/87° and 315°/74° respectively. Deformability and strength parameters for the rock mass and discontinuities have been determined from rock characterisation studies. For the rock mass, reference was mainly made to previous laboratory testing (e.g. Nappi et al., 1979) and empirical relationships. For the discontinuities, the geometrical and mechanical parameters were defined on the basis of shear tests and geomechanical survey. The obtained results are summarised in Table1.

NUMERICAL MODEL

Computer modelling was based on the distinct element method, with completely deformable blocks, using the UDEC code (ITASCA, 1992) and combining both data on intact rock properties and rock mass discontinuities. Such method of analysis can be regarded as particularly effective when dealing with jointed rock mass mainly involved in fall and topple phenomena (e.g. Barla et al., 1995; Tommasi, 1996). In the studied sections of the cliffs surrounding Calcata and Faleria (Figure 4) are included traces of the discontinuity sets, respectively K1-K2 and K3-K4, that affect the rock mass behaviour.

	D	E	ν	K	G	Coh	ϕ	σ_τ	JKn	JKs	Jco	Jfric	JRC	JCS	Jv	Spac
	kg/m^3	Pa		Pa	Pa	Pa	°	Pa	Pa/m	Pa/m	Pa	°		MPa	1/m^3	m
FALERIA																
TRS	1450	2e^8	0.2	1.3e^9	8.1e^8	x	x	6e^5	7.1e^8	4.9e^8	0	30	17	23	0.2	5.1
TSVS	1080	3e^7	0.17	1e^7	0.9e^7	3.4e^4	30	5e^4	7.1e^5	5.7e^5	3.4e^4	27	x	X	x	x
EPIC	1080	3e^7	0.13	1e^7	0.9e^7	3.4e^4	30	5e^4	1.4e^6	1.1e^6	3.4e^4	27	x	X	x	x
CALCATA																
TRV	1600	2e^8	0.24	1.3e^9	8.1e^8	X	x	6e^5	6.9e^8	4.8e^8	0	30	17	33	0.33	2.5;3.5
TSVS	1080	3e^7	0.17	1e^7	0.9e^7	3.4e^4	30	5e^4	7.6e^8	3.8e^8	3.4e^4	27	x	X	x	x
TGVT	1250	1.1e^8	0.22	6.7e^8	4.6e^8	X	x	6e^5	6.4e^8	4.6e^8	0	30	17	21	0.66	1.5
CGL	2100	2.6e^8	0.26	1.8e^9	1e^9	X	40	X	1.6e^8	1e^8	x	40	x	X	x	x

Table 1. Geomechanical parameters introduced in the model for each unit

Figure 4. View of the Calcata southern slope (left) and of the Faleria north-western slope (right) analysed by numerical modelling (for location see arrows in Figure 2)

The vertical sides of the models have been assumed to move vertically only and the horizontal ones only horizontally. The rock material and the discontinuities are assumed to be elastic-perfectly plastic when the Mohr-Coulomb failure envelope or tensile failure are reached. The models run over a number of iterations for each site after the initial equilibrium conditions were attained and the initial in situ state of stress was determined (Figure 5).

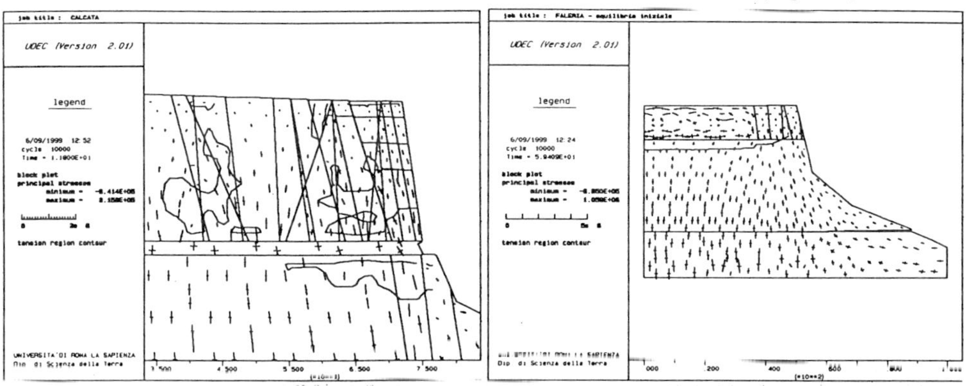

Figure 5. In situ state of stress in initial equilibrium conditions: (left) Calcata numerical model of the upper section of profile A in Figure 3; (right) Faleria numerical model of profile B in Figure 3

Next step was a parametric analysis to highlight the most important factor bearing on the stability conditions, e.g. water table within joints, softening processes of the TSVS units, dip angle of joint sets, toe erosion. In the case of Calcata cliff, introducing dip angle values consistent with the determined standard deviation range of K2 set, the model reached unbalanced conditions and topples were generated (Figure 6). As regards Faleria cliff, the results point out the significant role played by the softening processes which can involve the TSVS unit. In this case displacement vectors show a rotational slide as the most representative failure mechanisms within the Epiclastic and TSVS units, causing high tensile stress and fracturing in the overlaying TRS unit (Figure 6).

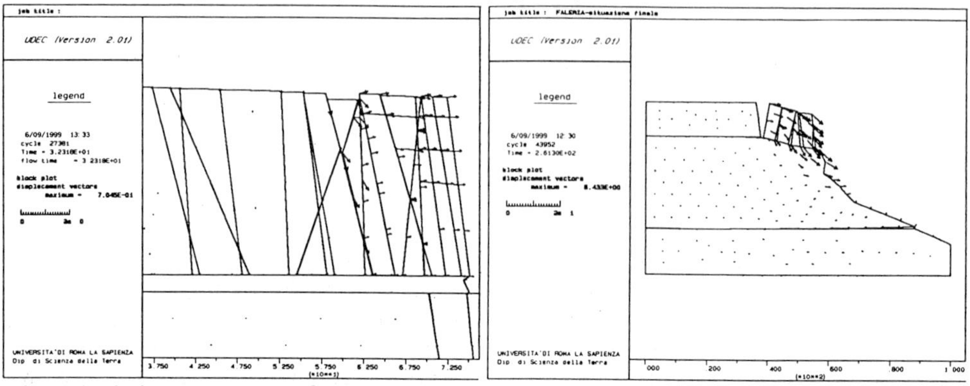

Figure 6. Displacement vectors after a large number of computation cycles: Calcata and Faleria

Calcata model shows the importance of discontinuity pattern data when treating soft rocks like pyroclastic flow deposits tens of meters thick (TRV and TGVT); for Faleria model, the significant thickness of deformable tuffs (TSVS) below the stiff cap rock (TRS) is determinant in inducing complex slide. Further refinement of both geological models, based on the coincidence between each depositional unit and the engineering geological one, could lead to significant improvement in the numerical model outputs? Within a rough cost-benefit assessment, mainly including geomechanical characterisation and computing-time, the two case studies indicate that the simplification made in the adopted approach yields failure mechanisms and involved volumes consistent with the observed ones, thus showing that the assumed models can be regarded as representative.

ACKNOWLEDGEMENTS

This research has been financially supported by CNR-GNDCI grants (CT97 00065.PF42).

REFERENCES

Alvarez W. (1973) – Ancient course of the Tiber River near Rome: an introduction to the Middle Pleistocene volcanic stratigraphy of Central Italy. Geol. Soc. Am. Bull., 84, 749-758.

Barla G., Borri-Brunetto M., Devin P. & Zaninetti A. (1995) – Validation of a distinct element model for toppling rock slopes. Proc. 8th ICRM, 1, 417-423.

Campobasso C., Cioni R., Salvati L. & Sbrana A. (1994) - Geology and paleogeographic evolution of a peripheral sector of the Vico and Sabatini volcanic complex, between Civita Castellana and Mazzano Romano (Latium, Italy), Mem. Descr. Carta Geol. d'It., 49, 277-290.

Lembo-Fazio A., Manfredini G., Ribacchi R. & Sciotti M. (1984) – Slope failure and cliff instability in the Orvieto Hill. Proc. 4th Int. Symp. on Landslides, 2, 115-120.

ITASCA Consulting Group Inc. (1992) – UDEC, Universal Distinct Element Code, ver 2.01.

Martinetti S. (1981) – Saving old towns raising on hill tops. Proc. 10th ICSMFE, 4, 841-846.

Nappi G., De Casa G. & Volponi E. (1979) – Geologia e caratteristiche tecniche del "tufo giallo della Via Tiberina". Boll. Soc. Geol. It., 98, 431-445.

Tommasi P. (1996) – Stabilità di versanti naturali ed artificiali soggetti a fenomeni di ribaltamento. Riv. It. Geotecnica, 30, n. 4, 5-34.

Shallow Landslides in Wadhurst Clay at Robertsbridge, Sussex, UK

E. N. BROMHEAD, Kingston University, M. HUGGINS, Fugro UK Ltd, M-L. IBSEN, Kingston University, UK

INTRODUCTION

The outcrop of the Wadhurst Clay, an overconsolidated clay shale of Lower Cretaceous age, in Kent and Sussex, although quite extensive, is very irregular. Its outcrop is often fault-bounded into small areas and has no coastal slopes. Due to this limit of extent it has, therefore, not received the intensity of geotechnical attention that has been devoted to other strata in this area. The lack of published case records on slides involving the Wadhurst clays, is particularly surprising when one reads, for instance "These soils have achieved a notoriety in a district which abounds in landslips" (Cracknell, 1968). More data has since become available (Pugh, Weeks and Hutchinson 1991), but in principle the sentiment is still correct.

The outcrops of the Wadhurst, as with many of the other clayey Wealden strata, are often the site of shallow earth movements, frequently the reactivation of earlier slips or solifluxion sheets. Movements of this kind are usually only of significant engineering interest when they also involve highways or housing, but their influence on agriculture can be quite severely damaging. This paper describes one particular instance of such a movement at a site to the East of the A21 Hastings to London road just north of Robertsbridge (Figure 1).

The site contains three slides that have developed in sloping agricultural land, which have been under observation since 1975. None of the slides have shown any tendency to enlarge significantly or move rapidly in that time, although they have clearly move from periodically. They appear to be local reactivations of an extensive sheet coverage of ancient landslides in the area. This paper describes some of subsurface investigations that were carried out on the site, sufficient to show the general form of the landslides. In the discussion it is then demonstrated that the morphology of the slides could have resulted from the breakdown of field drainage.

OUTLINE GEOLOGY

The Wadhurst Clay is a deposit within the Hastings Beds and lies between the Tunbridge Wells Sand, which is above the Weald Clay, and the Ashdown Sand. It is comprised of dark grey clay shales, mudstones with subordinate beds of silt, shelly limestone and ironstone totalling between 60 and 70m in thickness. The shales are often banded due to rhythmic variations in the coarseness of sediment. The formation outcrops in a South-Easterly direction from East Grinstead to Hastings, but the continuity is largely broken by faults so that the clay rarely covers a wide and continuous expanse. Outliers of Tunbridge Wells Sand together with valleys eroded through to the underlying Ashdown Sand, contribute to the irregularity of the outcrop.

Landslides in research, theory and practice, Thomas Telford, London, 2000.

Much of the iron ore which formed the basis of the Wealden iron industry in East Sussex and Kent was obtained from the Wadhurst Clay, in which it occurs both as nodules and as tabular masses, particularly near the base of the formation. Hence, outcrops are characteristically studded with small ponds, the degraded remains of pits that were dug for ironstone or sometimes as brick pits. Overspill from these "hammer ponds" may be responsible for local mudslide activity.

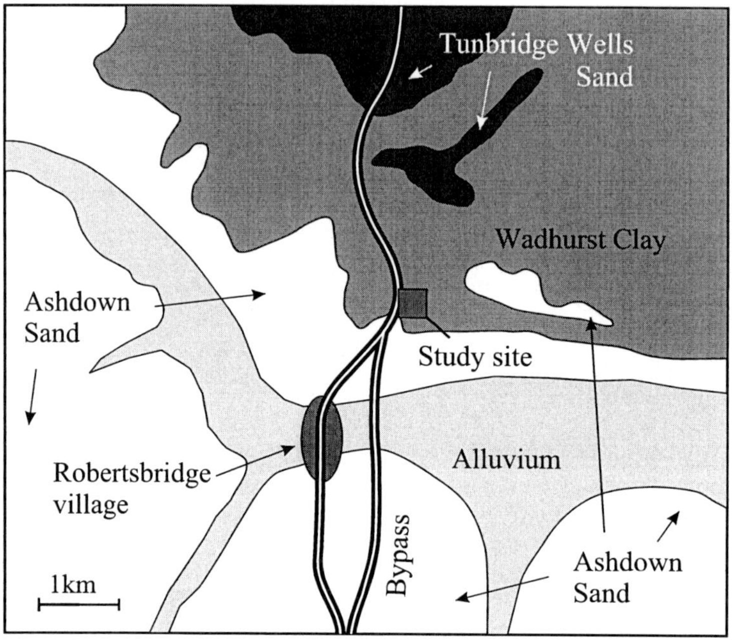

Figure 1. Location of the study area and the surrounding geology

GEOMORPHOLOGY OF THE SITE

The landslides that are the subject of this investigation occur in two adjacent fields that extend over an area of approximately four hectares. In the larger field, there are two slips (slips 1 and 2) which are lobate in plan but coalesce at their lower part to form a single sheet type of slide, and in the adjacent, more northerly and smaller, field there is another sheet type slip - slip 3. From the evidence of the ground undulations and a 5m displacement of a hedge, this latter slip appears to have been subject to at least two separate components of movement. These directions of movement are shown on a plan of the slides (Figure 2) which have been drawn from a survey made in the winter of 1975/6, and updated periodically since. The present day morphology is almost unchanged. Generally, however, the surface features indicate that the movements are of a very shallow nature. Sheet piling has been installed at the lower extremity of slip 2, possibly positioned to prevent closure of the ditch at the toe of the slope in which a stream runs.

The present slips 1 and 2 can only be dated roughly in the decade 1950-60 since they have occurred subsequent to the return of the land to pasture after the war during which they were ploughed and used for arable crops. Slip 3, on the other hand, may be more accurately dated since the landowner recollects that it first moved in the winter of 1963, probably as a result of the high rainfall that year.

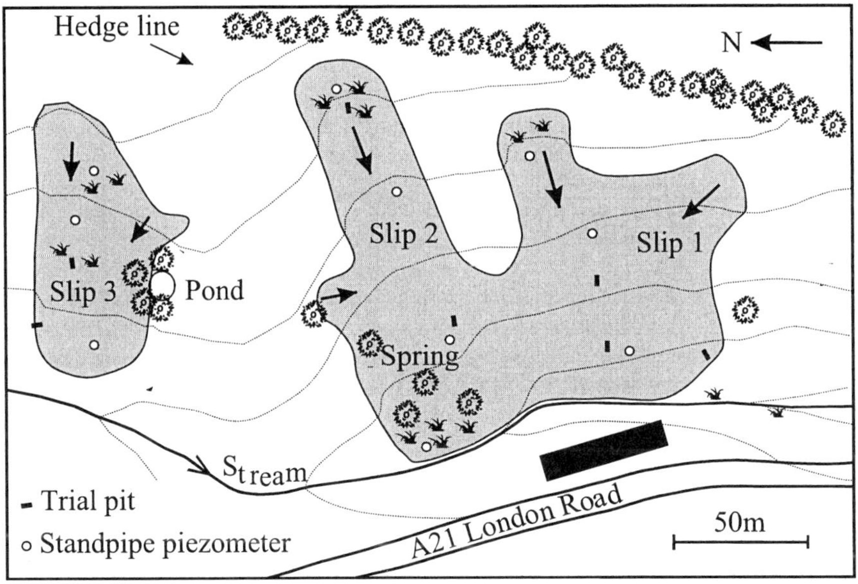

Figure 2. Survey map of the three slides which was drawn in the winter 1975/6

Evidence of past movements on the site is indicated from the form of a distorted fence and hedge lines shown on large scale (25") Ordnance Survey maps. Where such a fence or hedge line deflects between consecutive series of maps, or where a line was initially straight, movements can safely be inferred. For example, a bulge in a fence line is shown on the 1st Edition of the County Series (1873) at the position of slip 1, but unfortunately this fence or hedge was removed before the second edition. The bulge in the hedge involved in slip 3 is shown only on the latest map. It is believed that traces of former slips may have been ploughed out, and that the present slide morphology has resulted from fresh subsequent activity.

In the immediate vicinity of the slips there were several indications of the ground water hydrology, notably a series of large ponds above the fields, and a spring near the bottom of slip 2 which feeds the stream running along the bottom of the slips. A small pond is also found alongside slip 3. This has a water level which agrees closely with the piezometric observations (0.5 to 0.7m) below ground level during the winter 1975-76, but closer to ground level during the winter 1976-77. It is believed that the ponds are in fact degraded clay-pits which were dug between 1841 (the approximate date of the Parish Tithe Map on which they do not appear) and 1870. The upper ponds probably feed the spring, and indicate the presence of a drainage path under the slip.

FIELDWORK

Seven machine-dug trial pits were excavated to a maximum depth of 4m. These generally showed a succession of 300mm of topsoil overlying 1 to 1.2m of firm light brown clay which with depth became increasingly mottled with blue grey clay. Beneath this was a stiff blue grey clay, the transition from the overlying mottled clay being quite distinct. Within the blue grey clay, zones of up to a metre thick were found to contain several shear surfaces. It was originally expected that a single well-marked shear surface or at least a thin zone of shear surfaces would be found. The presence of such wide zones was, in fact, the subject of some concern. Most of the slip surfaces were slickensided and striated, but otherwise not easily distinguishable from the material in which they were found. Except, in Pit 5 where there was an extensive 1mm thick zone of a light blue material within the mottled brown clay at a depth of 1.5m, which was also of a much higher moisture content (36% as against 22%) than the surrounding material. This was thought to be a "fossil" shear zone.

Two of the trial pits showed a slightly different succession. In pit 6 (which was near the middle of slip 3) the shear zone was much shallower than elsewhere (between 0.5 and 1m deep), and in pit 5 (where slip 3 overrides the unslipped ground) there were three repetitions of the basic succession dipping sharply into the slip. This is believed to clearly indicate that successive phases of overriding have taken place although the appearance of the surface feature, a 1.5m high toe heave, was more consistent with only one such phase. The basic succession was also observed in the part of pit 5 which penetrated the unslipped ground.

In all of the pits a number of narrow bands of grey shale fragments and occasionally pockets of dark brown ironstone were found. These did not appear to be related to the shear features. It was, however, quite noticeable that the fragments in pit 2 were orientated with their major axis in a vertical position, a feature which suggests that the soils have been soliflucted, since fragments of shale and ironstone will tend to become oriented as they settle through the active layers.

Along with the trial pits, ten shallow 150mm boreholes were hand augered in early 1975, three in slip 1, four in slip 2 and a further three in slip 3. Their depths varied, but were up to a maximum of 4.5m depending on their position within the landslides. In most cases boring was curtailed when layers of hard material were encountered. These later proved to be bands of shale and ironstone about 25mm or more in thickness which were impenetrable with the hand auger.

A Casagrande type piezometer was installed in each bore hole and left for about three weeks to settle. Readings of the piezometers have been made sporadically and although all instruments have now been lost, some survived for over 2 decades. In general, however, the piezometric readings were mirrored by water levels in the ponds. These overflow when the water level is 300-400mm below ground level. The piezometric conditions at the site almost certainly rise to ground level, as is evidenced by standing water in hollows on the slides, and the presence of reeds over much of the area.

LABORATORY RESULTS

The mean index properties of the soils on site are given in Table 1. A noteworthy feature is the reversal of the usual distribution of plasticity between more and less weathered clays. Ordinarily, weathering might be expected to increase the liquid limit of a soil, but here in the Wadhurst the reverse appears to be true. The blue-grey soil has a markedly higher liquid limit than the more extensively weathered brown soil.

Direct shear tests were carried out on samples trimmed from block samples containing shear surfaces taken from the trial pits. Further residual strength testing was undertaken using the Bromhead ring shear machine and the range of results was confirmed by conducting a back analysis over several sections of the site. The laboratory tests show a lot of scatter, with the shear box tests generally being higher than the ring shear test results. Between the three methods, there was some agreement at about 13 - 14° for the residual angle of shearing resistance. There is little published data for comparison, the single value of 12° which is quoted by Weeks (1969), as the result of direct shear tests on slip surfaces taken from the base of solifluxion sheets at Vauxhall Lane and Quarry Lane, Tonbridge.

Table 1. Index properties of Wadhurst Clay from Robertsbridge

Soil description	**Liquid limit**	**Plastic limit**	**Natural water content %**
Firm, reddish brown mottled with blue grey, clays	43	19	21-25
Firm to stiff blue grey clays	56	19	21-25

DISCUSSION

The morphology of the slips is intriguing, with several channels retrogressing upslope, especially slip 2. This sort of failure is not amenable to a simple analysis on a plane section. Use of the infinite slope method (Skempton & Delory, 1957) underestimates the overall stability of the slope, reporting failure at water levels often exceeded on this site. It is found that within the possible (and indeed actual) slide lengths possible on the site, the toe restraint adds appreciably to stability.

An approximate 3-D analysis of this type of slide shows that major contributions to overall shear resistance come from side shear and the passive resistance in the toe zone. The relative proportions of each in contrast to the shearing resistance of the basal shear zone are functions of width, length and depth of the slide. The side shear contribution is lessened in a wide slide, and the passive resistance at the toe is lessened in a long one. In this case, stream erosion at the toe of the slope may be responsible for movements of the lower section of slips 1 and 2, as there may have been some removal of passive restraint, but slip 3 breaks out above the stream.

The side shear is overcome in the analyses by locally high pore water pressures. This can account for the localisation of the shear. Often, mudslides have this localisation where a large catchment feeds water and debris through a narrow neck into a channel through which the slide progresses. In such cases (but not here), there are often additional pre-existing morphological factors which control channelization. Frequent small failures not only enlarge the catchment bowl, but add to load and pore pressure through the process of *undrained loading* (Hutchinson & Bhandari, 1971).

A local source of underground water may cause the retrogressive behaviour of the slip. Watson and Bromhead (2000, this volume) cite the role of soakaway (foul) drainage. In the present case, the land is not used for this purpose.

Land drainage has been extensively used in Britain to improve agricultural land for well over 150 years. Shallow open-jointed pipes or shallow tile drains were originally used and now modern systems rely on geofabric protected perforated plastic piping. There are often no records of the work having been carried out, although government grant aid was provided for many years in the nineteenth century, and large areas of sloping agricultural land are known to have been treated. Where the flow in such drain systems is impeded, for example by crushing the pipe under agricultural machinery, removal of a section of pipe in a pit or trench and failing to reinstate the pipe run, or other means, the collected water cannot run away. However, it is still collected by the section of undamaged drainage upslope, and when surcharged in this way, provides a source of water at a greater-than-hydrostatic pressure. This source is precisely the right shape in plan to provide the resulting slide shape, because the pore water pressures fall off rapidly laterally, and diminish upslope. Severance of the drainage system at the head scarp of the slip moves the problem uphill, and the process may recommence. It will only stop when the remaining section of drain has too small a catchment to permit further slippage.

ACKNOWLEDGEMENTS

Permission to carry out the fieldwork on his land was kindly given by Mr Powell. The early work was done by Mr S. D. Ralph in 1975/6 and numerous students have contributed to the ongoing observation of the site over 25 years.

REFERENCES

Cracknell, D. W. (1968). Building on a Hillside. *The Structural Engineer, **46,** No. 3., 73-82.*

Hutchinson, J. N. & Bhandari, R. K.(1971). Undrained loading, a fundamental mechanism of mudflows and other mass movements. *Geotechnique, **21**, 353-358.*

Hutchinson, J. N. & Gostelow, T. P. (1976). The development of an abandoned cliff in London Clay at Hadleigh, Essex. *Phil. Trans. Roy. Soc. **Series A, 283,** No. 1315., 557-604.*

Pugh, R. S., Weeks, A. G. & Hutchinson, D. E. (1991) Landslide and remedial works in Wadhurst Clay. *Slope Stability Engineering: developments & applications, ed. R. J. Chandler, Thomas Telford, 377-382.*

Skempton, A. W. & Delory, F. A. (1957). Stability of natural slopes in London Clay. *Proc. 4th International Conf. Soil Mech., London, **2,** 378-381.*

Weeks, A. G. (1969). The stability of slopes in South-East England as affected by periglacial activity. *Q. Jnl. Eng. Geol., **2,** No. 1., 49-61.*

Analysis of Engineering Geological Efficiency of Anti-Landslide Measures in Odessa, Ukraine

B. V. BUDKIN, Centre of Engineering Research, Odessa, Ukraine and E. A. CHERKEZ, Odessa State University, Ukraine

ABSTRACT
The experience of long-term supervision of slope stability after realization of antilandslide measures permits to evaluate their engineering-geological efficiency and to check up objectively the conformity to actual results of theoretical assumptions serving as the basis for design decisions.

1. INTRODUCTION.

The geological section of Odessa coast to the depth of possible influence of large landslides (60-70m from the surface plateau, absolute marks of which make 40-50m) is submitted by stratigraphic-genetic complexes of the Meotian, Pontian, Middle and Upper Pliocene and Pleistocene Ages. General fall of formation is monoclinal, weakly inclined with azimuths to the south. In Odessa bay region, the Meotian clay deposits, the Pontian limestones, the Upper Pliocene sand-clay deposits, and Pleistocene loess-like rocks are outcropped above the sea level.

According to geological and historical data, approximately 250 various types of landslides were fixed on Odessa coast from 1797 till 1964 (by the beginning of construction of the antilandslide complex). The diversity of landslide types is caused by geological conditions of slope structure, rock parameters of strength and intensity of abrasion, and other geological processes (Zelinsky et al. 1993). Four principal types of landslides were formed:

I - landslides-flows in the Quaternary loess-like rock, caused by overhumidity; II - block landslide displacements of the Quaternary loess-like rocks along the Lower Quaternary and Pliocene clay, in transition to landslides-flows, in their further development; III - block landslides which are concerned with the Neogene-Quaternary formation occurring above the sea level; IV - block landslide displacements deforming the Meotian rock deeply, with the surface of displacement located considerably below the modern sea level.

As a result of data analysis on availability of factors and their influence on landslide development, it is possible to assign as main factors the following ones.

1. Structural-geological features of rock massifs (availability of a strong limestone layer, lithogenetic weakness zones in the basic deformable horizon (the Meotian clay), tectonic blocking of rock massifs).
2. Abrasion processes resulting in constant increase of slope steepness and, connected with it, redistribution of pressure in rock massifs.
3. Underground waters (availability of three aquifers: in the loess-like rock formation, in the Pontian limestone, in sand lenses locked in the Meotian clay).
4. Construction and economic activity.

2. ANTILANDSLIDE CONSTRUCTION EXPERIENCE, AND TYPES OF MEASURES

First approaches to realization of antilandslide measures in Odessa relate to 1883, when a gallery for interception of the Pontian aquifer was constructed in the area of Cape Large Fountain. However, construction of this gallery and all subsequent ones on other sites of the coast have not given desired result. These constructions were destroyed by landslides. On-going landslide activity had shown that the interception of a part of the Pontian aquifer is an unsufficient measure for termination of landslide displacements. A similar situation took place with shore protective constructions (boons) constructed on a few sites of the coast, which were deformed immediately after construction.

The downtown area of Odessa bay can be a good example of reliable strengthening of the coast. There, in early 19th century were erected sea-port constructions loading the slope in its basis (an embankment 4-6m high was lead in the sea for over 100m), as well as outward wave-suppressing constructions - piers. That resulted in loading the slope and liquidation of abrasion, after which deep block landslides were stopped.

The cited examples show that the experience of research and prevention of landslides on separate sites is not always acceptable on the whole length of coast, as separate sites differ in landslide specific features. This completely excludes the standard approach to evaluation of conditions and factors of landslide occurrence, and requires development of projects of antilandslide measures on the basis of comprehensive account of natural factors role, and assumes the complex character of antilandslide measures.

In 1968, antilandslide constructions of the 1st section were completed on the coast area 6.2 km long. At present, construction of the 2nd section and design of the 3rd section have been completed. Taking into consideration the natural conditions of Odessa coast, the antilandslide measures were directed on (1) reduction or liquidation of harmful influence of all main slide forming factors, and (2) maintenance of necessary slope stability. These measures are following:

- liquidation of wash-out of coastal cliff by making artificial beaches and beach prevention constructions;
- reduction of shearing stress values in the landslide slope rock by cutting the vertical part of the coastal precipice, laying out the slope, and construction of quays;
- interception and organized removal of underground waters of the Quaternary (using drainage wells) and the Pontian (construction of a complete drainage gallery and water-removing adits) aquifers into the sea;
- organization of surface runoff, fastening of the slope by turf covering.

3. GEODETIC SUPERVISION IN UNDERGROUND DRAINAGE CONSTRCTIONS

Large amount of data has been collected on natural supervision (hydrological, hydro-geological, geophysical, geodetic, etc.) of the efficiency of the 1st section of antilandslide constructions within Odessa coast. These constructions have been used and observed since 1964. Geodetic observations carried out by Odessa antilandslide management include:

1) geodetic supervision of surface bench-marks network on plateaus and slopes;

2) geodetic supervision in underground drainage constructions: levelling of bench-marks, measurement of longitudinal deformations of adits on sites between bench-marks, measurement of radial deformations of workings;

3) geodetic supervision for hydraulic-engineering structure (wavebreakers, traverses).

It is important to emphasize that geodetic supervision in underground constructions, in comparison with surface ones, is characterized by greater accuracy, since fluctuations of air temperatures in galleries and adits are insignificant; workings are straight; there is no mechanical damage of bench-marks caused by external effects; surface transportation and communications have no influence. The underground bench-mark network was laid within fastening elements of derivation adits driven in the formation of landslide accumulation and in the bed-rock at a distance of under 150-200 meters from the slope edge (fig.1).Adits are subhorizontal workings 350-400m long. They are located on the coast approximately 1 km apart from one another. Deformations of these adits characterize the condition of stability of the corresponding site. In the bed-rock, adits are connected by the gallery intended for draining of the Pontian aquifer. Adit deformations were determined through measuring bench-marks' vertical and horizontal displacement. There was completed a total 18-20 cycles of measurements in each adit for the whole period of supervision.

The experience of geodetic supervision shows that of all bench-marks, even the most remote from the landslide slope, are moving in vertical and horizontal directions. In this connection, while processing data of each levelling cycle we determined relative changes of marks in reference to the mark which is nearest to the bed-rock. It has allowed to reveal the deformations of adit "body" in the vertical plane, which were accumulated for whole period of operation of the constructions.

The accumulated deformations represent wavy bends with waves varying in different adits from 60 to 120 meters long (fig.1). The vertical deformations amplitude of adits within the bed-rock usually does not exceed 40-50mm, with the tendencies of increase to 100-200mm in landslide accumulations.

The analysis of distribution of horizontal deformations along adits shows that slope undergo some expansion. The most significant increase of distances between bench-marks occurs on the site of contiguity of landslide body to the bed-rock. These time-stable zones of stretch are revealed on sites of adits where the bench-marks vertical displacements are negative. Within the bed-rock, the characteristic distance between the stretch sites of adit makes 60-120m.The longitudinal deformations within the landslide slope limits are displayed synchronously with the bed-rock deformations, exceeding them usually several times.

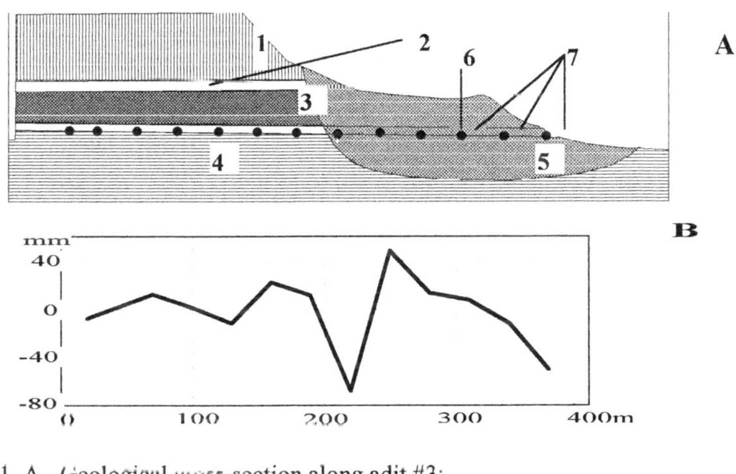

Figure 1. A - Geological cross-section along adit #3:
1 - Pleistocene loess; 2 - Upper Pliocene red clay; 3 - Pontian limestone, 4 - Meotian clay, 5- landslide deposits, 6 - adit #3, 7 - bench-marks.
B - Graphs of accumulated deformation (mm) of the 'body' of adit #3 in the vertical plane.

As a result of data processing, it has been found out that values of the horizontal displacement of bench-marks within the limits of the bed-rock vary slightly (Cherkez et al. 1991). In 1992 these displacements varied from 200 to 300mm (for example, adit 2 and 3, fig.2). In recent decades, the speed of bench-marks displacement within the bed-rock does not exceed 3-5mm/yr, with a general tendency of decrease. Distinctions in horizontal displacement of bench-marks are more significant on sites where adits are driven within landslide accumulations.

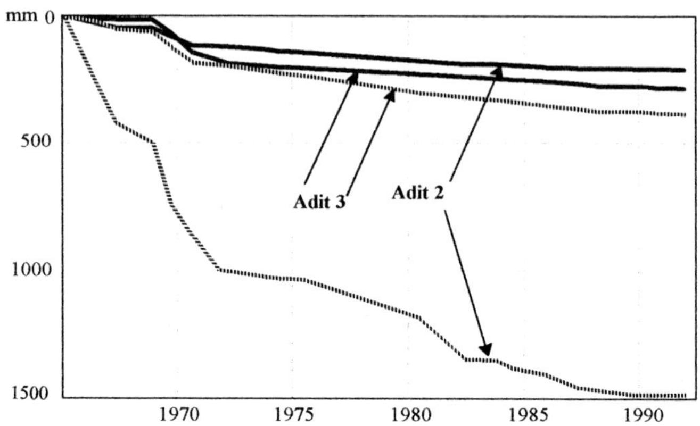

Figure 2. Horizontal displacements of bench-marks in adits #2 and #3 (mm). Solid lines — in the bed-rock, dotted lines — in landslide accumulations.

For example, in adit #2 values of displacement in 1992 made up 1,480mm, and in adit #3 - 370mm (fig.2). The speed of rock displacement of landslide slope on sites of coast nearest to adit #2 till 1971 made an average of 200-250 mm/yr, and they exceeded the speed of displacement established in other adits by 5-10 times.

4. ENGINEERING-GEOLOGICAL ANALYSIS OF THE ANTILANDSLIDE MEASURES COMPLEX EFFICIENCY

The standard quantitative criterion of general slope stability is the factor of slope safety. By the value of this factor, the probability of loss of slope stability is determined; the relative role of factors, the influence of change of engineering-geological conditions, and efficiency of antilandslide measures are evaluated (Zelinsky et al. 1983).

At the same time, analysis of modern condition of theory and practice of calculation of rock stability and simulation of rock intense condition of the landslide slope supports the idea that the use of these methods for eestimation of efficiency of types of antilandslide measures gives correct results but does not allow to take into account the summary antilandslide effect of the complex.

The reason is that in a separately calculated or experimental model it is not practical to reproduce all types of measures. Therefore, estimation of efficiency received through different models (for example, simulation of drainage constructions function, of intense condition of slope rocks, of the function of shore-protective constructions) can not be totalled. Besides, the absence of supervision of changes in physical-mechanical soil properties after termination of construction and during the operation of antilandslide constructions excludes the opportunity of estimation of slope stability at the required time.

The most objective criterion takes integrally into account both the summary efficiency of the whole antilandslide measures complex functioning and the change of natural conditions. It is presented by the values of slope rocks displacement and deformations, established by data of geodetic supervision. The bed-rock untouched by landslide processes is in steady condition and has rather small deformation speed, while the deformation speed of landslide accumulation within the slope limits is considerably larger.

The agreement of deformation speeds of landslide accumulations and the bed-rocks, with their general tendency to decrease in time, is the criterion of slope stability and maximum efficiency of carried-out antilandslide measures; this corresponds to the behaviour of the slope as a uniform kinematic system.

While conducting the antilandslide measures complex of the 1st section, construction was carried out on slopes at various stages of landslide cycles and having different degrees of stability: 1) unstable slopes, where the landslide deformations just before beginning of construction actively developed (for example, the site around adit #3); 2) temporarily steady slopes, where the landslide displacement has taken place before construction (for example, the site around adit #2).

In the area of adit #2, a deep squeeze-out landslide occurred in 1953, stretching for 2 km along the sea-coast. Geodetic supervision for deformations of the landslide slope started after the landslide; then, the speed of the slope vertical deformations made on the average 400-500 mm/yr. At beginning of making the underground drainage constructions (1964) deformations decreased to 50-100 mm/yr. On this site of coast complex, the antilandslide measures were not completed at beginning 1969.

There were no transverse buttress and artificial beach; in connection with high construction density on the plateau, cutting and levelling were not carried in total volume (average slope steepness decreased by just 1 degree). Therefore, landslide displacement activity by this period had not decreased (fig.2) and resulted in local destructions of the adit on the site of contiguity of landslide accumulation to the bed-rock. The decrease of speed of horizontal landslide accumulation displacement to 25-30 mm/yr was achieved as a result of placing an additional load in the slope basis of 100 m^3 in volume on 1 m of the coast.

Geodetic supervision over the slope area in the region of construction of adit #3 from 1956 till 1965 revealed large landslide motion. For this time a broken-away block 350 m along the coast and 40 m wide sank 2-3 m down. On the 40 m distance from the coast, a squeeze-out bank was generated. The speed of vertical deformations changed from 2500 mm/yr at the time of activity to 320 mm/yr in 1963.

Only by 1966 the speed had decreased to 30 mm/yr and was close to this value in 1969, when the antilandslide measures were carried out on this site. As a result of the slope cutting and levelling, its steepness decreased by 4°30′; the load volume in the slope basis made 120 m^3 on 1 m of the coast. This allowed to stop one of the largest landslides developed on the coast in the decade.Geodetic supervision over horizontal displacement of bench-marks in adit #3 showed that the bed-rock and the landslide slope behave as a uniform kinematic system (fig.2).

Deformations of the landslide slope rocks all along the coast attenuate gradually after completion of antilandslide measures. In all cases, except for site of coast in the area of adit #2 where the necessary measures carrying-out was delayed, values and speeds of bench-marks' horizontal displacement in adits differ slightly, both within the limits of the bed-rock and within landslide accumulations. This testifies about high efficiency of the antilandslide measures complex.

The analysis of geodetic supervision results in adits shows that in case of insufficient volume of slope cutting, when the rock displacement reaches its utmost values, for increase of slope stability and prevention of local destructions it is necessary to increase load volume to 200-230 m^3 on 1 m of coast in the slope bottom through creation of beaches and transverse buttresses.

This conclusion is confirmed by the experience of port construction in the early 19th century. On the city territory contiguous to the sea- port, the historical centre was formed; it is uniqueness for the concentration of numerous monuments and architectural ensembles within a small area. It is important to emphasize that architectural projects were of a complex character and provided both prevention of landslides and construction development of the area.

For example, consruction and design of Primorsky boulevard included the Potyomkinskaya stairs construction (1841); the stairs connected the boulevard with the quay and were an antilandslide construction. 400 piles driven under the stairs basement fixed the friable accumulation of the landslide terrace and the top leaky layers of the Meotian clay.

While building the sea-port constructions for communications roads, there was made a soil embankment up to the absolute levels of +6 m for the distance of approximately 100 m into the sea. This corresponded to the load volume of the slope basis about 300-500 m^3 on 1 m of coast. The antiwaves walls, retaining walls and linear drainages were also constructed, the levelling of slope was made. For over 100 years this complex has provided protection of the port constructions and adjoining territory of city from deep landslides.

At the same time, the geodetic supervision data show that the slope deformations occur within this part of the coast. They are caused by strengthening of engineering overhumidity of rocks within edge parts of the plateau and landslide accumulation.

5. CONCLUSION

1. Efficiency of the Odessa coast antilandslide measures complex was evaluated in comparison to the bed-rock and landslide accumulations deformation speed. Coincidence of these speeds — at the general tendency of its decrease in time — corresponds to the behaviour of the slope as a uniform kinematic system and testifies about high efficiency of carried out measures.

2. After completion of antilandslide measures — as a result of elimination or considerable decrease of the negative influence of slide factors - the character of landslide processes, its dynamics and quantitative parameters changed essentially. The antilandslide constructions of Odessa coast provide long-term slope stability. The slope deformations are of attenuated character on a long-term scale.

3.Local dislocations within the slope after the antilandslide complex construction are caused by an insufficient load volume in the slope basis.

REFERENCES

Cherkez, E.A., Pilipenko, M.N., Gutkovsky, V.N. 1991. Geodynamic phenomena study on Odessa territories with the help of complex geophysical and engineering-geological regime supervision. *In Work of Conf. on geophysical researches in hydrogeology and engineering geology. Part 1:* 137-142. Tashkent: SAIGIMS (in Russian).

Zelinsky, I.P., Cherkez, E.A. Guzenko, A.V. 1983. *Engineering-geological prognosis and simulation. Educational allowance.* Odessa: Odessa State University (in Russian).

Zelinsky, I.P., Korzhenevsky, B.A., Cherkez E.A. et al. 1993. *Landslides of Black sea north-western coast, their study and prognosis.* Kiev: Naukova dumka (in Russian).

Investigation and Monitoring of Landslides with Torrential Significance in the Bavarian Alps

G. BUNZA, Bavarian Water Management Agency, Munich, Germany and University of Innsbruck, Tyrol, Austria

PREFACE

Landslides and their displaced material represent huge potential debris sources for torrents. During flood events debris flows can develop and endanger the settlement areas on the alluvial fan. These processes were made particularly clear on the occasion of the intensive rainfalls on 21st and 22nd of May 1999. Therefore the investigation and observation of landslides situated in the areas of torrents became an important objective of the geological-torrential service department at the Bavarian Water Management Agency, not only for the understanding of these processes but also for the planning of control measures. This will be demonstrated by two examples in this paper.

THE LANDSLIDE FALKENBACH

For several years the village of Tiefenbach near Oberstdorf has been threatened by the debris flows of the torrent Falkenbach, due to a landslide in the upper reach of this torrent, which developed a huge debris source (Fig. 1, Fig. 2).

Geology

The catchment basin of the Falkenbach covers an area of 1.2km² and is completely composed of cretaceous sediments (thrust sheet of „Säntis“). They form a north-verging fold which is spreading E-W. Karstic limestones and siliceous sandstones build up the anticlines in the North and the South. Waterproof marls and marl schists are building the syncline. On the left valley side these sediments are overlaine by moraines, in which the landslide is situated (Fig. 1).

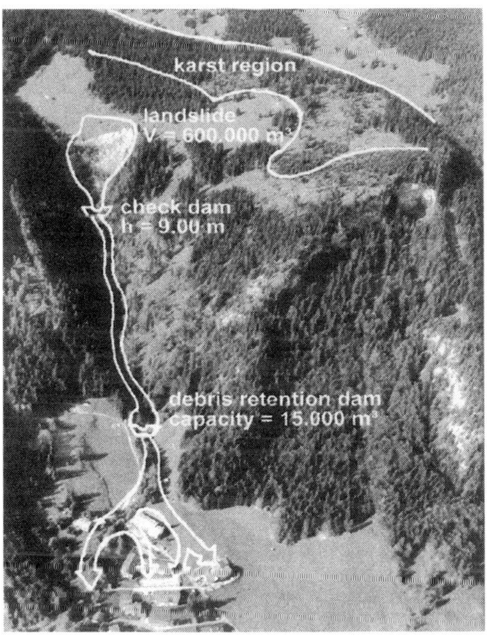

Figure 1. The landslide Falkenbach with the debris flow endangered houses of Tiefenbach

Landslides in research, theory and practice, Thomas Telford, London, 2000.

The landslide

The landslide developed in the sixties and increased between 1973 and 1988. In 1984 therefore a check dam was built at the foot of the landslide (Fig. 1). This measure became inefficient very soon because of the continuous activities. During the following year's extensive geological and geotechnical investigations (seismic, sounding, etc.) were carried out.

The landslide, an enlarging rotational slide with a dimension of 30.000m², extends from an altitude of 1210m about two hundred metres downwards to the valley bottom. The main scarp is about ten metres high. The volume of the slide amounts to 600.000m³, of which the displaced masses are some 300.000m³. About 95.000m³ became loose and are now ready to slide. Within the immediate surroundings 200.000m³ are unstable.

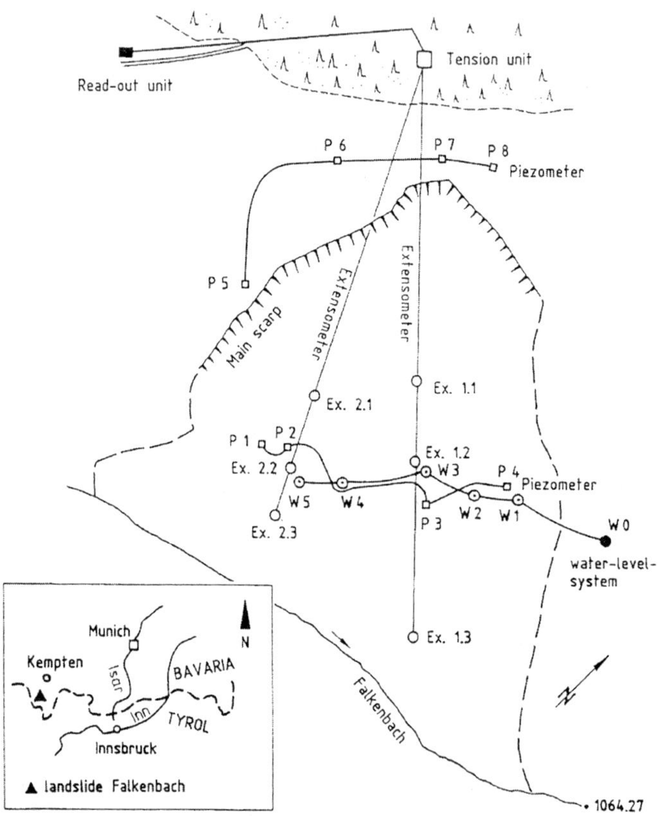

Figure 2. Automatically monitoring system at the landslide and site of the catchment area.

The material could be described as a widespread not cohesive gravel (50%), sand (25-30%), silt (5-30%) mixture with a water content between 4 and 16 weight percent. The shear strength, the angle of inner friction [φ] was determined between 27° - 35°, those with a silty matrix between 22,5° - 27,5° (cohesion c' = 0,01 - 0,02 MN/m²; Ulrich, 1991).

On the one hand the most likely cause of the slide was the erosive forces of the torrent affecting the bottom of the slope, especially during floods. This is a permanent process. But, on the other hand the water content plays an important part. In addition to the seepage of rainfall and snow, water from the karst area above (Fig.1) can also stream in using subterranean channels which are marked by lineaments. Due to the saturation, the angle of friction can decrease to less than 20° and the silty matrix becomes viscous. Subsequently the bracing by the gravel detritus gets lost and slides can occur.

Protective measures

In the year 1990 a debris retention dam with a height of 15m, and, at the top a width of 54m was built at the valley mouth. For the reinforcement 180 tons of steel were used. In spite of the best exploitation of the valley section only a removable retention-volume of approx. 15.000m³ could be realized (Fig.1). Considering the remaining risk an automatic geotechnical measurement system was installed within the landslide area in 1994 (Fig. 2). It is the aim of this device to identify increasing landslide activities at a very early stage, in order to be prepared for disasters and to provide for appropriate measures. The movements of the displaced material are measured and recorded once a day by two different systems, working irrespectively of each other: cable extensometers measure horizontal movements (in the inclination of the slope), and a water-level records the vertical movements.

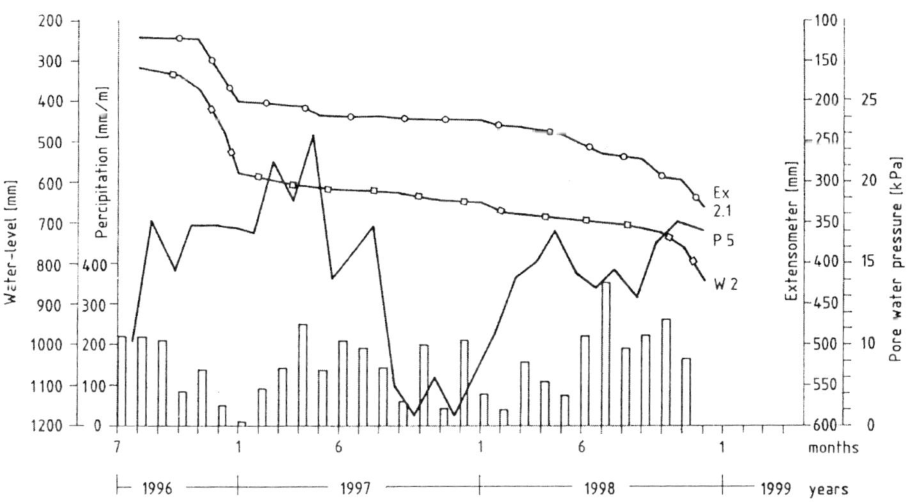

Figure 3. A comparison between the monthly precipitation amounts from July 1996 until Nov. 1998 and the recorded data by piezometer P5, extensometer 2.1 and water level W2.

In 1995 eight piezometers were installed at a depth of about 6m in the active and the potential slide areas to measure the pore water pressures (Fig. 2). Additionally a rain gauge was installed. All the results are recorded by a read-out unit which is provided by a laptop. The energy demand of this equipment is met by solar-energy.

Accordingly the results of the measurement period between August 1996 until November 1998, which are represented in Figure 3 help to understand the landslide kinematics. From this it appears that the pore water pressures react very fast to a water input by continuous intensive rainfalls. This was the case in the summers of 1996 and 1998. The triggering of movements, however, could only be registered in the following winters (Nov.-Feb.). This means that the movements will increase but with delay and be dependence on the seepage capacity of the material. With this connection it seems, that the subterranean tributaries are very important.

Conclusions

It is necessary to pay attention to the times of movements during and after continuous and intensive rainfall periods to be able to recognize dangerous temporal situations, especially when there were wet periods before, for example, snowmelt. So far 1.5 Million DM were spent on protective measures. Beyond that for a successful stabilisation of the slope by French drains 500.000DM can be made available.

THE LANDSLIDE IN THE TORRENT OF JENBACH

The torrent of Jenbach near the village Bad Feilnbach (Fig.4) has a catchment area of 13 km^2. Already in 1897 thirty check dams were built for torrent control as protection against debris flows. In this century these check dams were replaced by concrete dams with a height of 7-8m. In the year 1981 a check dam was destroyed in the area of „Brechries“ in the valley of Jenbach. Afterwards the geological investigations brought to light, that the damage was caused by deep-seated movements (saggings) in the bedrock of the mountain front at the right hand side of the valley.

Geology

The slope is built by marls, lime marls and limestones of the flysch-facies. The sediments which dip against the slope (s = 230/35 SE) have a high degree of jointing and a disintegrated rock bonding caused by the thrust fault Alpine facies / Flysch facies in the neighbourhood. Their weathered materials behaviour is distinctly plastic due to the great fractions of silt and clay. The recent failures and movements are controlled by two joint sets (principal dip surfaces) which dip outward from the slope (j1 = 210/80; j2 = 280/55). The movements cover an area of 16.5ha within a huge relict slip area of 26ha.

The rock flow

The slope deformation occurs as a sagging according to a non-shearing model with a depth of 70-100m. Shearing in the shallow zone causes secondary slides and falls. The total volume of the sagging masses could be estimated at approximately 7 Million m^3. The rock flow was and is produced by slope undercutting, by residual stresses due to older failures and by gravitational stresses due to the young valley formation.

Within the sagging area four zones with different directions and intensities of movements could be identified (Fig.6). The monitoring of the slope since 1982 by a cable extensometer, by terrestrial geodetic surveying and by GPS recorded movements in the order of magnitude of an average 10-20mm/month in the lower parts, and of an average 2-3mm/month in the upper parts of the slope.

In the years 1995/96 increasing movements in zone c were measured and observed. They lead to huge rock slides and rock falls which continued until spring 1999. Behind the main scarp the bedrock moved 2-4m downwards. All together masses with a volume of about 100.000m³ were deposited in the valley bottom at a length of 80m and a height of about 30m (Fig.5). Due to the blocky displaced material the torrent generally could run off beneath the dam, causing backwater only temporarily.

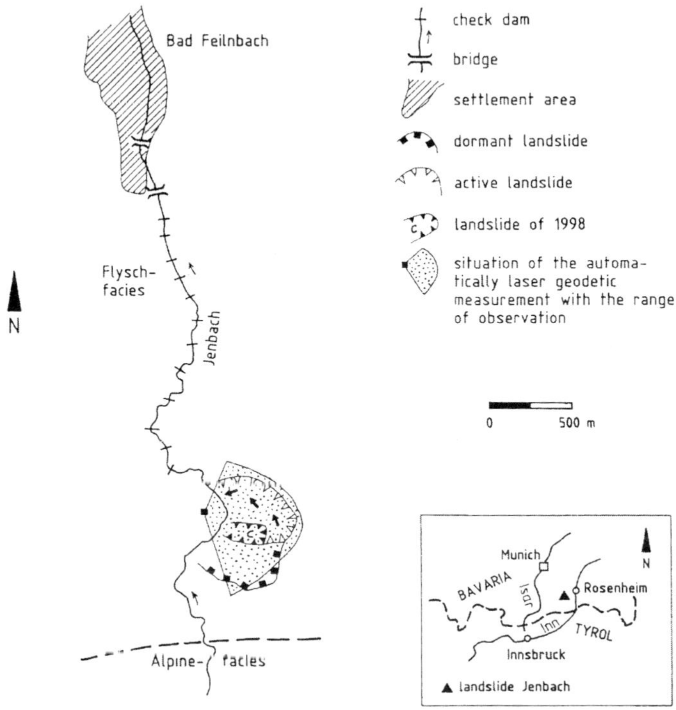

Figure 4. Site of the torrent Jenbach and the landslide (rock flow) in the flysch-fazies.

The movements can be interpreted as secondary failures in the brittle zone of the sagging area. The movements were triggered partly by joint and pore water pressures due to rainfalls and snowmelt in the autumn of 1997. Springs, mud flows and debris flows in a lateral zone of the failure zone during the first months of the movements demonstrated the intensive participation of water.

Protective measures

Triggered by the current movements and the massive material input into the valley bottom of the Jenbach several protective measures were initiated by the Bavarian Water Management Agency:

- With regard to enlarging movements scenarios were produced to estimate the potential height of a dam and the volume of the dammed lake. A complete failure of zone c would have built a dam in the valley bottom with a height of about 57m and with a possible lake with a volume of about 600.000m³ (Fig.6). Debris flows triggered by overtopping or piping could endanger parts of the village of Bad Feilnbach.
- The measures by torrent control were intensified. Several check dams were built in addition to the existing ones, sedimentation basins were cleaned up and a disaster plan was made up by the local authorities.
- Beyond that a comprehensive investigation programme was started in co-operation with the Bavarian Geological Survey. It includes the determination of the soil mechanical indices, shearing tests, geoseismic investigations, two drills to a depth of 80 - 100m, inclinometer measurements, dilatometer tests, mineralogical investigations and the preparation of a slope model by the finite-difference method. The results of this investigation programme will be presented in autumn 1999.
- At present an automatically laser-geodetic measurement is installed at the opposite slope (Fig.4). The measurements will start in September 1999.

Figure 5. In 1998 the collapse in the brittle zone (c) of the sagging “Brechries” lead to rockmasses of about 100.000m³ in the valley bottom.

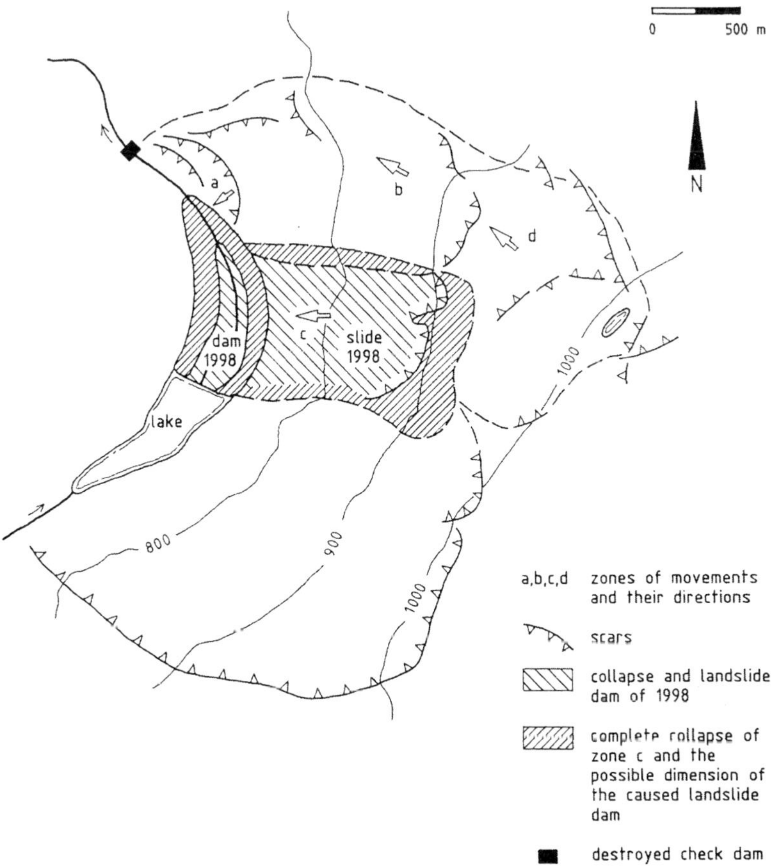

Figure 6. The possible dimension of a landslide dam and ist lake (V ≅ 600.000 m³) due to a complete collapse of zone c.

Results from the first drill to a depth of 80m verified the supposition of a deep seated movement (rock flow) without distinct shear planes. The drill core and the dilatometer tests showed over the whole thickness jointed rocks and no water slips or water bearing strata. The movements caused by saturation and joint water pressures after rainfalls and snow melting in front of the rock flow are secondary and shallow. At present the zone c seems cleaned out and in spite of the intensive rainfalls on 21st and 22nd of May 1999 (180mm/24h) movements were rare. On the other hand those rainfalls triggered a flood and the overtopping of the dam. 30 000m³ of the displaced material were eroded and small debris flows were triggered. The processes formed a canyon in the displaced material with a width of 20 - 40m and a height of 25 - 30m. The eroded material was deposited after several hundred metres and destroyed check dams, the forest road and a dredger.

CONCLUSION

This paper was designed to explain the work of the Bavarian Torrent Control Agency in handling such non-common problems, especially the difficulties of risk assessments and their defence by applying justifiable methods.

Investigations in the past few years have shown, that debris flows are influenced definitively by different types of landslides in many catchment areas of the Bavarian Alps. The increasing building activities of settlements and roads augment the hazards caused by these phenomena. Therefore monitoring and investigations are planned for further landslide areas with torrential significance. Last but not least it is also necessary to investigate other hazard zones of mass movement in order to develop protection measures in time. With this in mind a method is being developed at present within a joint EU project (Joint Pilot Action Programme under Art. 10 EFRE - Project EGAR) with the Austrian and Italian Torrent Control Agencies which started in autumn 1998.

REFERENCES

Bunza, G. und Glogau, A., 1992: Rutschung und Murgangsperre am Falkentobel. - Int. Symp. Interpraevent 1992 - Bern, Tagungspubl. Bd. 2, S. 113-130.

Ulrich, G., 1991: Standsicherheitsuntersuchung Rutschung Falkentobel. - Ingenieurgeol. Gutachten, unveröff. Kempten.

A Dynamic Model of Shallow Slope Motion Driven by Fluctuating Ground Water Levels

R. BUTTERFIELD, Department of Civil and Environmental Engineering, University of Southampton, UK

INTRODUCTION

In many areas of the world, shallow slopes in clays undergo progressive downhill motion which accelerates and decelerates in response to fluctuations in pore-water pressure brought about by seasonal changes in ground water levels. Whereas typical movements are a few mm/year, relative to a plane of sliding at a specific depth in the hillside, they can, and frequently do, speed-up by an order of magnitude sometimes accelerating to failure. The paper presents a new, simple, lumped- mass, dynamic model of such landslide processes.

The parameters used in the model are empirical, back-calculated from in-situ data comprising time-series records of piezometric levels and downhill displacements of the sliding mass. The key parameters can be determined directly from a small number of such measurements - possibly from as few as five or six records but preferably from a more extensive data-set. The model, so calibrated, can then be used to predict the future response of the hillside to any predicted pattern of ground-water level variation, without recourse to any laboratory determined soil parameters. There is, as far as I am aware, no reliable, generally accepted method for making such predictions at present.

SUMMARY OF THE MODEL

The non-linear differential equation of motion underlying the model is of the form

$$\ddot{X} + G(\dot{X})^n = H(t) \qquad (1)$$

i.e. a single-degree-of-freedom system driven by a known, time-dependent force H(t), the source of H being gravity and the fluctuating pore-water pressures. The resisting forces are inertia and soil strength, augmented by a non-linear, velocity-dependent damping term $G(\dot{X})^n$. The parameters G and n are to be determined by fitting the time-dependent velocity solution $\dot{X}(t)$ of this equation to field-data. All the parameters and variables used are dimensionless.

Field records of the downhill motion of such slopes show (Figure 2) static phases followed by a step change of velocity. It is usually possible to identify quite clearly a frequently occurring minimum velocity ($\dot{x}_0 = v_0$) which accompanies a specific measurement of a characterising pore-water pressure distribution in the slope (see below). This value of velocity is used to create the following dimensionless variables.

$$X = xg/(v_o)^2 : \dot{X} = (dX/dT) = \dot{x}/v_o : \ddot{X} = (d\dot{X}/dT) = \ddot{x}/g : T = gt/v_o \qquad (2)$$

Landslides in research, theory and practice, Thomas Telford, London, 2000.

Analytical solutions of equation (1) are well known for n =1 and n =2 . It is easily shown that, as expected, the complementary functions of these solutions go to zero at the very high values of T which are usually of interest in slope dynamics. Therefore the particular integral, which remains, provides the long-term, slope response. We assume that this conclusion holds for all values of n (for which analytical solutions are not available). The particular integral of equation (1) is simply $X = T(H/G)^{1/n}$ and the solution, which we will use throughout the analysis, becomes,

$$X = T\,(H/G\,)^{1/n} : \dot{X} = (H/G\,)^{1/n} : \ddot{X} = 0 : \text{with } X(0) = 0,\ \dot{X}(0) = v_o \qquad (3)$$

G and n are to be found by fitting this equation to the field data using, in particular, the downhill velocity of the sliding mass $\dot{X}$ and the associated value of H, representing the piezometric level distribution in the hillside.

Once G and n have been determined, equation (3) can be used to predict X versus T (or x versus t) for any assumed, or known, time series values of H. Such a diagram will clearly comprise a series of straight-line, constant-velocity, segments (since $\ddot{X} = 0$ and $\dot{X}$ = constant in each of them). Whilst this is an approximation it is also the form in which in-situ data will usually be available and therefore a more elaborate interpretation of it is not warranted.

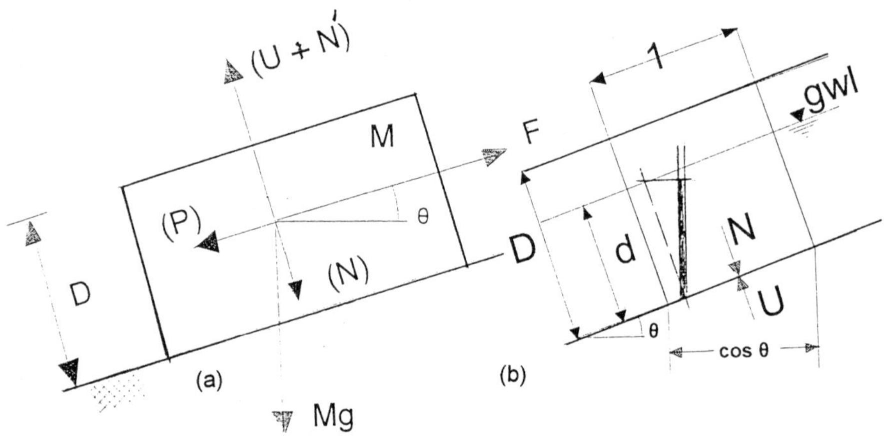

Figure 1. (a) Free-body diagram: (b) Uplift diagram

DEVELOPMENT OF THE MODEL

Figure1a shows a free-body diagram representing unit length of a slab-sliding slope, mass M, inclined at angle θ which is less than the residual friction angle of the slope φ'_r. If a mean effective friction angle φ' and a non-linear damping force Cx^n are mobilised then the equation of motion for the slope is,

$$Mx + Cx^n = (P\text{-}F) = N(\tan\theta - \beta\tan\varphi') \qquad (4)$$

Where $P = Mg\sin\theta$: $F = N'\tan\varphi'$: $N = Mg\cos\theta$: $N' = N - U$ and $\beta = N'/N$ with U the uplift force arising from the piezometric water level in the hillside.

Figure1b shows the phreatic surface in a slab-slide parallel to (and at depth d below) the surface of the sliding mass, thickness D, whence,

$$N' = N\text{-}U = \gamma_s D \cos\theta - \gamma_w d/\cos\theta$$
$$\beta = N'/N = (N\text{-}U)/N = 1 - (\gamma_w/\gamma_s)(d/D)(1/\cos\theta)^2 \quad (5)$$

in which γ_w and γ_s are the unit weights of water and the soil respectively. A model rather similar in principle to the above, but incorporating arbitrary upper and lower pore-water pressure thresholds and restricted to n= 1, was published by Angeli *et al* (1996,1998). Their fitted displacement-time curve for the Alverà data is shown in Figure 2.

Adequate field records will comprise time-series measurements of characteristic piezometric levels d and corresponding displacements x (yielding $\dot{x}$). Displacements are usually obtained by integrating inclinometer measurements, which also identify D. The corresponding series of β values can then be calculated from equation (5). From such a set of (x,β) values the unique, and usually recurring, pair, (v_o, β_o) say, which identify the step change from a static slope to a sliding one, can be selected. Substituting these values into equation (4), together with $\ddot{x} = 0$, leads to the value of $\varphi' = \varphi'_o$ which is mobilised at such instances. i.e.

$$\beta_o \tan\varphi'_o = \tan\theta - C(v_o)^n / N \quad (6)$$

The simplest version of the model follows if we assume that φ' = constant = φ'_o throughout the dynamic history of the hillside. Substituting $\varphi' = \varphi'_o$ from equation (6) in to equation (2) yields,

$$(\ddot{x}/g) + C(\dot{x})^n/Mg = \sin\theta\,(1-\beta/\beta_o) + (\beta/\beta_o)(v_o)^n/Mg$$

or, when dimensionless variables are introduced from equation (3)

$$\ddot{X} + G(\dot{X})^n = (B + (\beta/\beta_o)G) = H \quad (7)$$

wherein, $B = \sin\theta(1-\beta/\beta_o)$ and $G = C(v_o)^n/Mg$. G therefore becomes an empirical parameter, to be determined, which subsumes the two unknown quantities (C/M). Both G and n are assumed constant for a specific slope regime whereby, the model and its solution, equation (3), are fully defined. Substituting G into equation (6) generates the value of φ'_o via

$$\tan(\varphi'_o) = (\sin\theta - G)/(\beta_o\cos\theta) \quad (8)$$

Equation (1) has no meaning when $\beta > \beta_o$, since $v = v_o$ when $\beta = \beta o$ and sliding stops when $\beta > \beta_o$. The velocity only increases when $\beta < \beta_o$, driven by increasing pore-water pressures in the slope. Thereafter, the velocity response to changes in β (increasing or decreasing) is an immediate step change; the velocity cannot, of courrse, become negative.

The particular integral of equation (7), which, together with the field data, is to be used to determine the unknown parameters (n,G), is that already given in equation (3). In particular that,

$$\dot{X}^n = H/G : \quad \text{i.e. } n\log(\dot{X}) + \log(G) = \log(H) \quad (9)$$

For lists of ($\dot{X}$,H) values it is a standard linear-regression operation to find 'best-fit' values of (n,G) to the data. However, since $(\beta/\beta_o) < 1$ and G turns out to be small, $B > (\beta/\beta_o)G$ and we can either: (i) simply approximate H=B in equation (7) and determine (n,G) directly from equation (9), or, (ii) retain $H = (B + (\beta/\beta_o)G)$ in which case equation (9) has to be used iteratively to arrive at the best-fit (n,G) values.

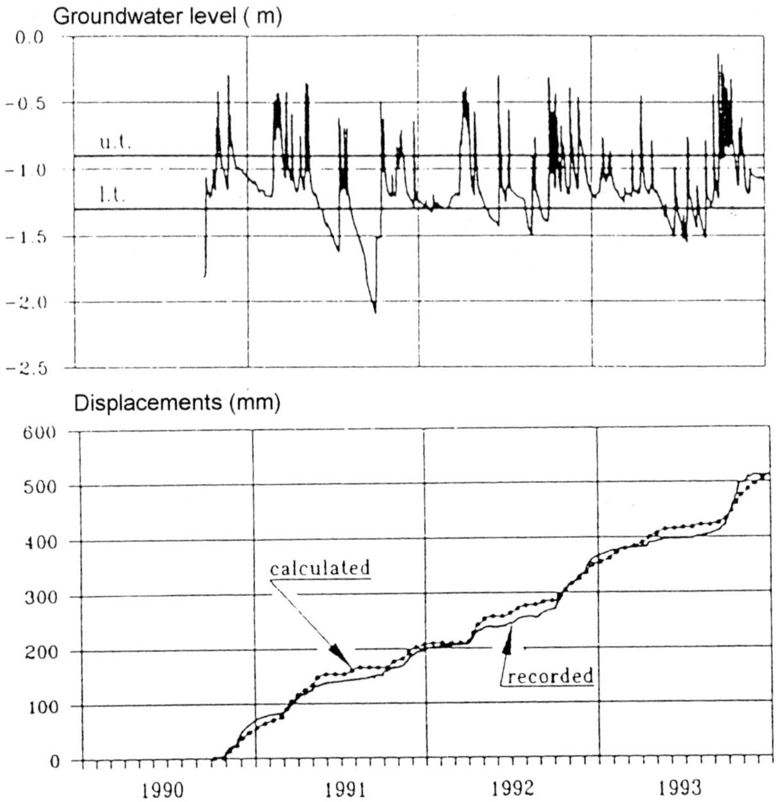

Figure 2. Recorded groundwater levels and downhill displacements at Alverà

CASE HISTORY:

The 'mudslide' at Cortina D'Ampezzo (Alverà) in the eastern Dolomites was described by Deganutti & Gasparetto (1992). Extensive data on its downhill movement and the ambient piezometric levels has been recorded (Angeli *et al*, 1989), presented and analysed (Angeli *et al*, 1996,1998). The total slide is about 1700m long with the central, more extensively studied section, some 800m long by 80m wide, sloping at about 8° on average, with a well-defined slip surface at approximately 5m depth. The material, within which the sliding surface developed, "consists of irregular, poorly sorted blocks of the original rock dispersed in an argillaceous matrix". Although meaningful soil properties are therefore not easily established Angeli *et al* (1996) reported tests on two samples recovered at the slip surface. These provided average values of: clay fraction 70%, $w_L = 0.95$, $I_p = 0.48$, $\gamma_{sat} = 18.73$ kN/m^3 and $\varphi'_r = 15.9°$ (ring shear). From Lupini *et al* (1981) and the above classification data the quoted value of φ'_r is unexpectedly high. Since 1989, Angeli *et al* have recorded piezometric levels in this slope (from 11 transducers) and adjacent soil mass displacements (from 11 steel-wire extensometers and 4 inclinometers) together with detailed rainfall data. They have studied the period 1990-1993 in particular and published the measurements shown in Figure 2 (Angeli *et al,* 1996). In order to use the graphical data the diagrams were enlarged (to A3 size) and discretised at 30-day intervals (38 data points) and 15-day intervals (78 data points). The latter data set (Pignagnoli, 1998) is shown in Figure 3. Whereas discretising the displacement data (and thereby the velocity) is straightforward, care is needed to ensure that representative discretisation of the piezometric level data is achieved

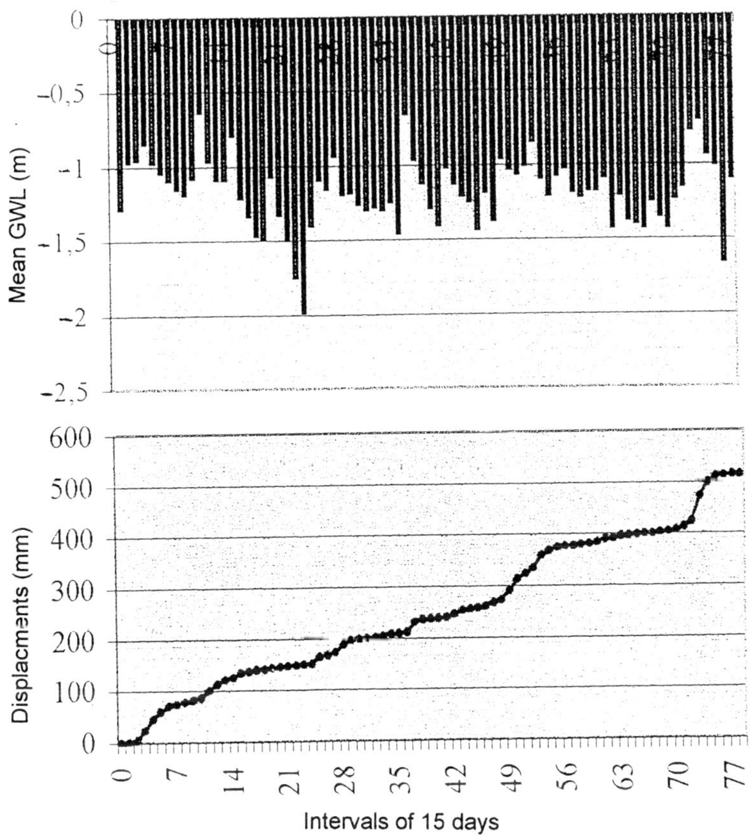

Figure 3. Discretised groundwater and displacement data for Alverà

RESULTS OF ANALYSIS

The data so obtained can be transferred to a spread-sheet and best-fit values of (n,G) evaluated, either directly (by assuming H=B) or iteratively, as explained previously The fitting procedure turns out to be sensitive to the value selected for β_o. The precision can be improved substantially by introducing a second level of iteration in order to allow the initial, graphically assessed, value of β_o to be adjusted to optimise the overall fit to the data. Figure 4 shows an example of the quality of fit that can be obtained. The values of (v_0, β_0) used were

(40mm/yr, 0.605) which led, using the fully iterated model, to (n,G) = (0.2474, 0.0123) when β_o was modified to 0.609. From equation (8) the corresponding value of $\varphi'_o = 11.9°$.

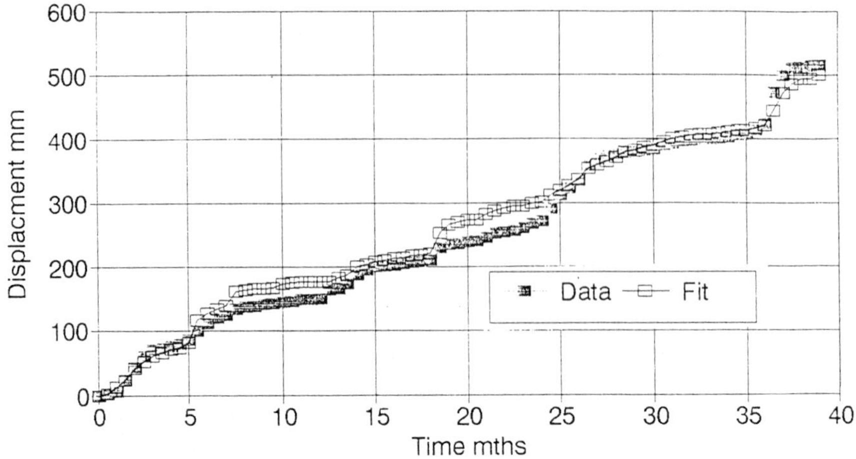

Figure 4. Measured and predicted downhill displacements at Alverà

CONCLUSIONS

A simple, well defined, one-degree-of-freedom, non-linear damped dynamic model has been presented. The two, initially unknown, dimensionless parameters which, together with the basic geometry of the slope and the ground water regime, define the model can be determined from a time sequence of slope movements and associated piezometric-level measurements. The model has been correlated using an extensive data set from a mudslide at Cortina d' Ampezzo. Excellent overall agreement between prediction and reality was obtained for data spanning more than 3 years. This implies that downhill displacements and velocities of the sliding mass, due to future variations in typical, mean piezometric levels, can now be predicted with confidence in this specific hillside

REFERENCES

Angeli, M.G., Gasparetto, P., Pasuto, A.& Silvano, S. (1989). Examples of landslide instrumentation (Italy). *Proc. 12th ISSMFE*, **3**, 1531-1534. Rio de Janeiro, Brazil.

Angeli, M.G., Menotti, R. M., Pasuto, A. & Silvano, S. (1991). Landslide studies in the eastern Dolomite Mountains, Italy. *Landslides*, Ed. Bell, Balkema Rotterdam.

Angeli, M.G., Gasparetto, P., Menotti, R. M., Pasuto, A.& Silvano, S. (1996). A visco-plastic model for slope analysis applied to a mudslide in Cortina d'Ampezzo, Italy. *Quart. Jour. Eng. Geology*, **29**, 233-240.

Angeli, M.G., Gasparetto, P., Pasuto, A. (1998). A combined hillslope hydrology/stability model for low-gradient clay slopes in the Italian Dolomites. *Eng. Geology*, **49**, 1-13.

Deganuti, A. & Gasparetto, P. (1992). Some aspects of mudslide in Cortina, Italy. *Proc. 6th. Int Symp. on Landslides*, **1**, 373-378. Christchurch , NZ.

Lupini J.F., Skinner, A.E. & Vaughan, P.R. (1981). The drained residual strength of cohesive soils. Géotechnique **31**, No.2, 181-213.

Pignagnoli, G., (1998). Modellazioni di movimenti franosi soggetti a spostamenti lenti, Facolta di Ingegneria, Universita' degli Studi di Bologna (unpublished UG project).

The Influence of Meteoric Events in Triggering Shallow Landslides in Pyroclastic Deposits of Campania, Italy

D. CALCATERRA, Department of Geotechnical Engineering, Section of Applied Geology, Federico II University of Napoli, M. PARISE, CERIST, National Research Council, Bari, B. PALMA, Vico Equense (Napoli) and L. PELELLA, Somma Vesuviana (Napoli), Italy

INTRODUCTION

The Campania region of southern Italy is widely characterised by outcropping of pyroclastic materials deriving from historic volcanic activity of the Phlegraean Fields and Mount Vesuvius. These deposits mantle with thickness usually ranging from a few centimetres to some metres most of the carbonate mountains in the region, covering their original morphology, which had been mostly shaped by tectonics and karst processes during the Quaternary. Similar deposits crop out at Napoli and in the Phlegraean Fields, where, on the other hand, the loose pyroclastics overlie lithified volcanic tuffs. Due to poor consolidation of the pyroclastites, and high slope steepness, the risk posed by detachment of these surficial deposits, and the following mobilisation of the material as rapid to very rapid debris flows is very high. The strong permeability contrast between the surficial pyroclastic cover and the underlying carbonate bedrock, and the vertical and lateral differences in permeability within the pyroclastic cover itself, play a fundamental role in the trigger of mass movements.

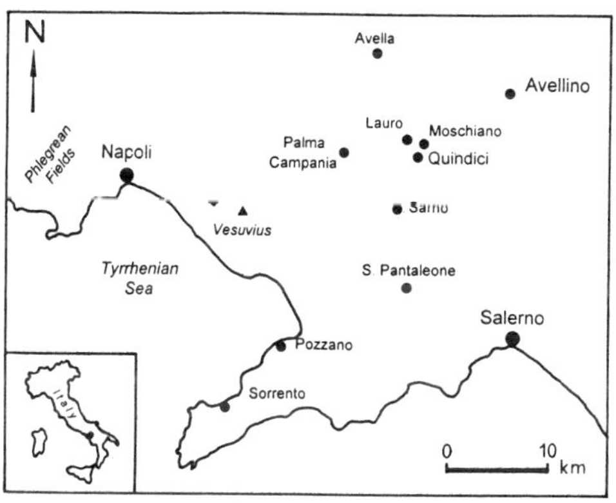

Figure 1. Location of the study areas and of the main localities cited in the text

Shallow landsliding in pyroclastic deposits mantling carbonate bedrock is probably the most frequent mass movement typology in Campania. Two recent events tragically confirmed this statement: on January 9-11, 1997, a severe storm in the Sorrento Peninsula and the Napoli area (Figure 1) triggered several hundreds of mass movements (mostly soil slide-debris flows), causing 5 victims (Calcaterra et al., 1999a; Calcaterra & Guarino, 1999). On May 4-5, 1998, another prolonged rainfall event struck a wide area between the provinces of Salerno and Avellino: some hundreds of shallow soil slides evolving into channelised debris- mud-flows, which travelled up to 4 km, reached several inhabited areas, killing 160 victims. Eleven of them lived at Quindici, a small town in the Avellino province (Figure 1). In this

Landslides in research, theory and practice, Thomas Telford, London, 2000.

paper, the main characters of the 1998 rainfall event at Quindici are analysed; Quindici is a case of great interest, since it was involved, even if marginally, by the January 1997 event, too. In addition, a research carried out in local archives and other sources provided very interesting information on the historical recurrence of landslides and floods in the area.

THE MAY 4-5, 1998 LANDSLIDING EVENT AT QUINDICI

The May 1998 pluviometric event originated from a barometric depression located over northern Africa on May 2, which reached the Campania region two days later. In the Quindici area the rainfall started between 0:00 and 1:00 of May 4 (Fig. 2) and, after a 11 hour break, resumed in the afternoon, and was continuous until the late evening of May 5. At the rain gauge nearest to Quindici (Lauro) a total rainfall of 173 mm was measured, while the highest 2-day value (275mm) was recorded some 40 km east.

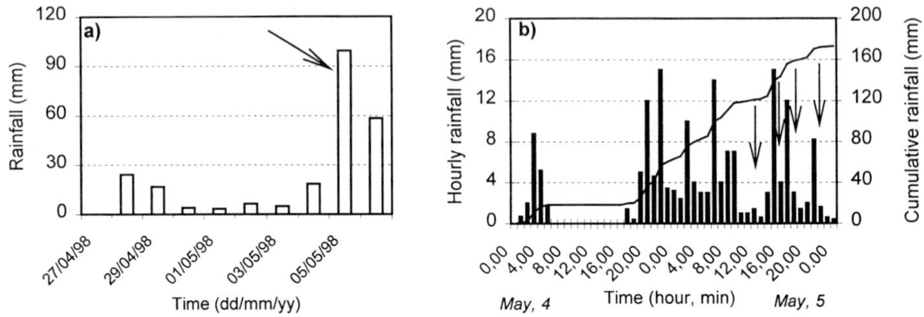

Figure 2. a) Daily rainfall in late April - early May 1998 at Lauro rain gauge (192m a.s.l. – mean annual precipitation 1,122mm). b) May 4-5, 1998 hourly and cumulative rainfall (arrows indicate the main landslides at Quindici)

Eyewitness accounts allowed the identification of the temporal distribution of the landsliding events at Quindici. The first slope movements occurred between 12:00 and 12:30 on May 5, when about 122 mm of the total rainfall had already accumulated. After a second cluster of failures (from 14:00 to 15:30), between 18:00 and 18:30 the main landslide occurred, which was responsible for invasion of the town and the consequent casualties. The last documented events were observed at 20:30-21:00. Figure 2b shows that timing of the main landsliding event well relates to the period of peak rainfall intensity, which reached the value of 15 mm/hr at 15:00. The average hourly rainfall, over a period of 32 hours (from 17:00 of May 4 to 0:00 of May 6), was 4.8 mm/hr, while, prior to the occurrence of the first landslides, it was slightly higher (4.92 mm/hr).

The heavy rainfall of May 4-5 triggered more than 300 landslides (Calcaterra et al., 1999b), over an areal extent of about 10km^2. The great majority were shallow (from 0.5 to 3m), planar soil slides, which started from steep (highest concentration: 31°-43°; mean: 36°), vegetated slopes, and involved fine- to coarse-grained loose pyroclastics. Most of the detached material reached the steep channels and transformed into channelised debris flows, which rapidly travelled downslope up to about 3.5 km. Volumes involved in the slides ranged from a few to about 16,000m^3; the overall volume of landslide material was estimated in about 1,500,000m^3.

Given the distribution of mass movements over a time span of 8-9 hours, the antecedent rainfall here considered refers to the first documented events, that is those observed at 12:00-

12:30. All the rainfall data come from the Italian Hydrologic and Oceanographic Service (SIMN), which publish the hydrologic yearbooks and, for the May '98 event, diffused the main data through the Internet (dstn.pcm.it/simn/FranaSarno).

The long-term antecedent rainfall at Lauro is unfortunately not available; however, taking into account the data recorded at Avella (198 m a.s.l., 1,119mm Mean Annual Precipitation - MAP) and Sarno (36 m a.s.l., 987 mm MAP), which are the other closest rain gauges, the 10-30-90 days antecedent rainfall showed values constantly lower than both the equivalent seasonal data and the yearly maximum.

a)

Duration (hrs)	Mean rainfall (mm)	Max rainfall (mm)	May '98 rainfall (mm)	Return period (yrs)
1	28.6	50.0	15	1
3	39.3	94.2	31.6	1.6
6	50.2	131.4	43.2	1.8
12	64.7	156.0	79.6	4.2
24	84.2	172.6	103.4	4.9
48	100.6	203.2	121.8	4.9

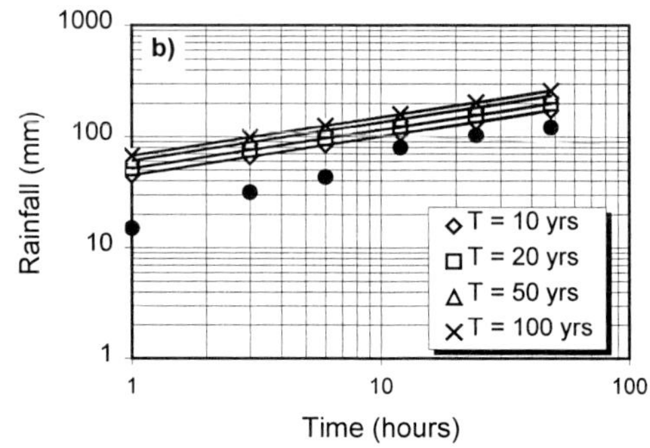

Figure 3. a) Return periods of the May 4-5, 1998 rainfall; b) Climatic curves (Gumbel distribution) for rainfall at Lauro rain gauge. Black dots indicate the May 4-5, 1998 rainfall values

In Figure 3a, the 1- to 48-hours rainfall for the May 4-5, 1998 event is reported: the comparison with the mean and maximum values recorded at the Lauro rain gauge station shows that the 1998 data are lower than historical maximum and, in some case, lower than mean data, too. Aiming at a preliminary definition of the recurrence interval of the May 4-5 antecedent rainfall, the historical set of data recorded at Lauro has been statistically analysed, as regards the maximum rainfall for duration of 1, 3, 6, 12, 24, and 48-hours, by means of climatic curves computed for different return periods. Data sets have been elaborated using the Gumbel probability function. The May 1998 short-term precipitation (Fig. 3b) falls well below the considered recurrence interval, with probability of occurrence between 1 and 5 years (Fig. 3a); this result implies that the rainfall delivered on May 1998 at Lauro can not be regarded as an "extreme" pluviometric event.

HISTORICAL HYDROGEOLOGICAL DISASTERS IN THE QUINDICI AREA

The May 1998 event is to be considered as unprecedented for magnitude, areal extent and overall damage caused. Nevertheless, an accurate historical research, carried out mostly through local sources, led to identify 24 documented past instabilities, which affected Quindici, Lauro and Moschiano (Fig. 1). The oldest of these phenomena dates back to 1632, when Quindici was struck by a flood, some months later a huge explosive Vesuvian eruption, which occurred on December 16-18, 1631. On October 10, 1640, 40 lives were lost due to another flood which devastated Quindici and Lauro. Again in the 17th century (1660), at Lauro some monumental buildings and churches were destroyed or abandoned due to a flood. In more recent times, 3 events at least deserve to be mentioned. The first of them is the flood which invaded Moschiano on August 27, 1990 (Fig. 4a), thoroughly documented by citizens and local Authorities. The peculiar feature of this event was the very high intensity and concentration of the rainfall: 30.8 mm were discharged in about 1 hour. On 1997 two other events were recorded: on January 10-11, when a vast portion of Campania was involved in a rainstorm, significant damages were recorded in the study area: at Quindici, a spring tap was completely destroyed and several minor roads were damaged by landslides. On the night between November 12 and 13, a woman at Lauro was killed by a flood, which followed a long rainfall period (Fig. 4b); on the same occasion, several landslides occurred on the slopes surrounding Moschiano.

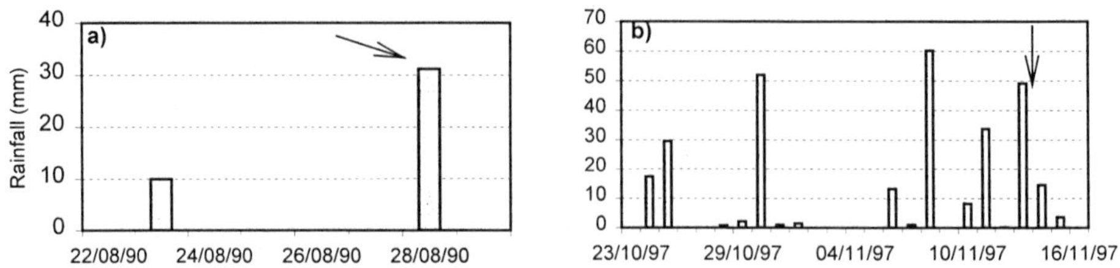

Figure 4. Daily rainfall related to some of the instability events in the Quindici area (Lauro rain gauge)

Six of the studied past events have been selected, based on availability of the rainfall data, in order to be compared with the May 1998 phenomena (Table 1). The absolute amount of rainfall varies over a very wide range, for each event and for the different durations as well. The same values, when "normalized" with respect to MAP (Cannon & Ellen, 1985; Govi et al., 1985; Wieczorek, 1987), allow some interesting considerations: in the short-time durations (24- and 48-hours), antecedent rainfall is generally below 10%. However, the majority of the events show values above a common threshold, if the long-term antecedent rainfall is considered. In fact, in four cases the 30-days precipitation is equal or higher than 20% of MAP, which turns to more than 35% when considering the 120-days duration. The May, 1998, event falls within this group, although the long-term data at Lauro rain gauge were not available. Nevertheless, taking into account the Avella station, normalized values of 21.6% and 36.9% are obtained, respectively, for 30- and 120-days.

DISCUSSION AND CONCLUSIONS

Recognition of rainfall thresholds is a fundamental aim to be achieved in the mitigation of the risk posed by rainfall-triggered landslides. Antecedent seasonal rainfall is usually thought to be important over a period as long as 7 to 30-60 days, as regards fine-grained, low permeability materials such as the loose pyroclastics from Campania. In recent papers, after the last major geomorphic crises in the region, the working hypothesis of considering a longer antecedent rainfall (120-240 days) has been proposed (De Falco et al., 1999).

An alternative approach to rainfall thresholds emphasises the relationships between two main parameters of the storm rainfall, namely the intensity (I) and the duration (D): the May 1998 rainstorm in the Quindici area was characterised (prior to the first landslides) by a 21-hours continuous rainfall, whose average intensity was 4.9 mm/hr. This event falls in the heavy rainfall field of Moser & Hohensinn (1983) of relatively long duration (1-2 days or more) and low average intensities (0.1÷10 mm/hr). In the broader regional context, many recent rainfall-induced landslides in Campania fall within this category. On the other hand, the major events on January 1997 (de Riso et al., 1999; Calcaterra & Guarino, 1999) and May 1998 can be defined as triggered by a prolonged but low-medium intensity rainfall (Fig. 5). Moreover, two other events significantly differ from the two categories above: they are the well known

Salerno flood of October 1954 (Lazzari, 1954), when 504 mm of rain accumulated in about 16 hours, and the August 1990, Moschiano event, which was characterised by more than 30 mm of rain in about 1 hour.

Event	24 hrs	48 hrs	10 days	30 days	60 days	120 days
May 4-5, 1998	103.4	121.8	200	n.m.	n.m.	n.m.
	9.2%	*10.9%*	*17.8%*	-	-	-
Nov. 12-13, 1997	49.2	49.4	165.8	277.4	302.9	441.9
	4.4%	*4.4%*	*14.8%*	*24.7%*	*27.0%*	*39.4%*
Jan. 10-11, 1997	31	75.6	147	230.8	452.4	517.2
	2.8%	*6.7%*	*13.1%*	*20.6%*	*40.3%*	*46.1%*
May 4, 1993	16.8	30.4	38.8	50.8	127.6	230.6
	1.5%	*2.7%*	*3.5%*	*4.5%*	*11.4%*	*20.6%*
Dec. 26, 1993	38	67.2	98.6	213.6	371	645.4
	3.4%	*6.0%*	*8.8%*	*19.0%*	*33.1%*	*57.5%*
Aug. 27, 1990	31.2	31.2	41.2	83.6	92.2	135.6
	2.8%	*2.8%*	*3.7%*	*7.5%*	*8.2%*	*12.1%*
April 7-10, 1978	46	92.2	148	215.4	302.2	456
	4.1%	*8.2%*	*13.2%*	*19.2%*	*26.9%*	*40.6%*

n.m. = not measured.

Table 1. Antecedent rainfall of selected instability events at Quindici and surrounding areas, also expressed as percentage of mean annual precipitation. January 1997 data were recorded at the Avella station

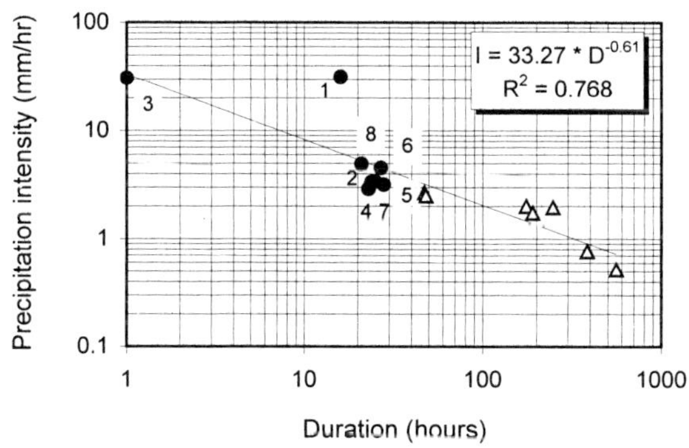

Figure 5. Antecedent rainfall duration and intensity for landslides in the Campanian pyroclastics. 1) Salerno (26.10.54); 2) Quindici (7.04.78); 3) Moschiano (27.08.90); 4) Napoli (10.01.97); 5) S. Pantaleone (10.01.97); 6) Pozzano (10.01.97); 7) Sarno (5.05.98); 8) Quindici (5.05.98). Triangles indicate data from Guadagno (1991)

Even with these somehow marked differences, and considering also other historical data collected for the same areas (Guadagno, 1991), a tentative relationship between duration and intensity of known case studies has been drawn (Fig. 5):

$$I = 33.27 * D^{-0.61} \quad (1)$$

This equation is quite similar to the lower bound associated with several landsliding events in the Alps (Moser & Hohensinn, 1983, Cancelli & Nova, 1985). In addition to the regression line proposed, a lower bound for the case histories has also been computed, which resulted as follows:

$$I = 28.1 * D^{-0.74} \quad (2)$$

In conclusion, some key points of the research so far carried out can be highlighted:

- The May 5, 1998, landsliding event at Quindici was triggered by a long duration, low intensity rainstorm. The 1- to 48-hours antecedent rainfall has return periods, referred to the first landslides, lower than 5 years.
- More than 20 past flood and landsliding events were documented in the Quindici area through historical and archive research.
- Antecedent rainfall for the May 1998 and the other previous events tend to cluster around values higher than 20% and 35% of the MAP for, respectively, 30 and 120 days.
- Soil slide-debris flows in pyroclastic deposits of Campania appear to belong to the same group in terms of intensity and duration of preceding rainstorm. A linear relationships between I and D was found, together with a lower bound equation, consistent with some of the well known equations reported in literature.

ACKNOWLEDGMENTS
Research partly financed by MURST-PRIN (National Interest Research Projects) funds, assigned to prof. R. de Riso (Univ. of Napoli). The Authors gratefully acknowledge prof. G. de Marinis (Univ. of Cassino) and dr. eng. C. Farina, for helpful comments and suggestions.

REFERENCES

Calcaterra D., Guarino P.M. (1999) - Morphodynamics and recent landslides in western hillslopes of Napoli. Geologia Tecnica ed Ambientale, 2/99: 11-17. (In Italian)

Calcaterra D., Santo A., de Riso R., Budetta P. & co-workers (1999a) - January, 1997 intense rainfall related landslides in Sorrentine Peninsula - Lattari Mts.: first contribution. Proc. 9th Nat. Congr. of Geologists, Rome, 17-20 April 1997: 223-231. (In Italian).

Calcaterra D., Parise M., Palma B., Pelella L. (1999b) - The May 5th 1998, landsliding event in Campania (southern Italy): inventory of slope movements in the Quindici area. Proc. Int. Symp. on Slope Stability Engineering, IS - Shikoku '99, Matsuyama, Japan, Nov. 8-11. In press

Cancelli A., Nova R. (1985) - Landslides in soil debris cover triggered by rainstorms in Valtellina (Central Alps–Italy). Proc. 4th Int. Conf. Field Work. on Landsl., Tokio: 267-272.

Cannon S.H., Ellen S.E. (1985) - Rainfall conditions for abundant debris avalanches, San Francisco Bay region, California. California Geology, dec. 1985: 267-272.

De Falco M., de Riso R., Ducci D. (1999) - The pluviometric regime of the Sorrentine Peninsula-Lattari Mts. area related to the January 1997 heavy rainfall. Proc. 9th Nat. Congr. of Geologists, Rome, 17-20 April 1997: 321-329. (In Italian).

de Riso R., Budetta P., Calcaterra D., Santo A. (1999) - Debris flows in the pyroclastic deposits of Campania. Proc. Nat. Symp. "Prediction and prevention of rapid mass movements", Trento, 17-19 june 1999: 133-150, GEAM. (In Italian)

Govi M., Mortara G., Sorzana P. F. (1985) - Hydrological events and landslides. Geol. Appl. e Idrog., 20(2): 359-375. (In Italian)

Guadagno F.M. (1991) - Debris flows in the Campanian volcaniclastic soils (Southern Italy). Proc. Int. Conf. on Slope stability eng. develop. and applications, Isle of Wight: 109-114.

Lazzari A. (1954) - Geological aspects of events occurred in the Salerno area as a consequence of the October 25-26, 1954 storm. Boll. Soc. Natur. in Napoli 63: 131-142. (In Italian)

Moser M., Hohensinn F. (1983) - Geotechnical aspects of soil slips in Alpine regions. Eng. Geol., 19: 185-211.

Wieczorek G.F. (1987) - Effect of rainfall intensity and duration on debris flows in central Santa Cruz Mountains, California. Rev. in Eng. Geol., Geol. Soc. of America, 7: 93-104.

Preliminary Study for Landslide Hazard Assessment: GIS Techniques and a Multivariate Statistical Approach

S. CAMPUS, F. FORLATI, Regione Piemonte, Direzione Servizi Tecnici di Prevenzione, Settore Studi e Ricerche Geologiche, Torino, Italy and C. SCAVIA, Politecnico di Torino, Structural Engineering and Geotechnique, Torino, Italy

FOREWORD

In connection with the program *Geological Cartography of Italy* (*CARG Project*), the Piedmont Region has stipulated a specific program agreement with the Department for National Technical Services-Geological Service of the Chair of the Council of Ministers. The purpose of this agreement is the computer-based production of some Geological Maps on a 1:50,000 scale. A special project has also been started in parallel, called *"Alluvial Events in Piedmont"*, which includes, among other things, the production of a 1:50,000-scale map concerning landslides hazard.

A salient feature of the project is the interdisciplinary character of the working group set up to deal with the stages of definition of the methodology and data control, from all points of view. The group is staffed by representatives from all the disciplines involved, i.e. professionals who, in most cases, unfortunately, do their job separately and in a non co-ordinated fashion: the geologists, first of all, who contribute to the knowledge of the geological processes and the way such processes affect the territory; the engineers, who study the mechanical implications; the mathematicians/statisticians, who analyse the data from the mathematical viewpoint; the computer analyst formalises everything in conceptual models for the design and development of an integrated Geographical Information System (GIS).

INTRODUCTION

The assessment of geological hazard is strictly connected with the characterisation of the non-deterministic predictability associated with the occurrence of landslides. This is a complex process requiring an estimate of the probability of occurrence of such events in both *spatial* and *temporal* terms, and it also requires an evaluation of the probability of the *intensity* and the *evolution* of the phenomena.

The definition of landslide hazard in its current acceptation uses the concept of probability of occurrence of a potentially damaging landslide within a specified time period and in a specific area. Hence, it follows that the term *hazard* cannot be applied to landslides *tout court,* as each specific, recognised type of landslide has its own hazard attributes, which cannot be extended to other phenomena having distinctive characteristics in terms of kinematics and evolution.

Landslides in research, theory and practice, Thomas Telford, London, 2000.

We begin focusing on a typical landslide: the planar landslides that occur in the Langhe area of Piedmont, as illustrated in Figure 1. This type of phenomenon has been chosen on account of its frequent occurrence over a vast area, in typical forms, which make it suitable to an analysis, also in quantitative terms. The territory of the Langhe represents a vast hilly region of southern Piedmont, in north-western Italy. The general morphology of the area is clearly affected by the structural situation, giving rise to long slopes dipping in the NW direction with down slope stratification, and to short, steep slopes dipping in the SE direction with counter slope stratification (figure 2). The landslides that took place in the Langhe region concomitantly with extreme rainfall events are mainly of soil slip type along the SE facing slopes and rock-block slide type along NW facing slopes. During the violent rainstorm of 4-6/11/1994, about 800 cases of rock-block slide were recorded.

Figure 1. Location of the Langhe region

In rock-block slides, the unstable mass slides along a surface coinciding with a bedding plane, with dip direction sub-parallel to the slope one and dipping from 8° to 14°.

Figure 2. Morphology of the area

The tools employed to process the data are mostly the application tools associated with GIS technology, as well as software programs developed both for the management of databases and for the execution of statistical analyses.

GENERAL CRITERIA

For the evaluation of slope instability, recent studies (Carrara *et al*, 1995), (Baeza *et al.*, 1996), (Leroi, 1996), (Amanti *et al.*, 1997), (Pegoraro, 1998) have stressed the potential of a statistical methodology relying on application tools related to the GIS. A quantitative estimate of hazard implies the knowledge of a series of parameters, concerning both the landslide phenomenon in itself and the boundary conditions, in order to create a model for the prediction of such phenomena in space and time. The approaches most commonly adopted are of a heuristic, deterministic or probabilistic nature. The methodology adopted in this study is based on the integration of several interdisciplinary aspects which make it possible to minimise the errors. On the basis of these considerations, it was decided to take advantage of some of the typical features of a heuristic model within a statistical framework.

Thus, on the one hand, each *"domain expert"* guided the choice of the parameters to be used in the statistical analyses, based on his knowledge and experience of the territory and of the phenomena. On the other hand, the statistical techniques used (univariate and multivariate) made it possible to capture in a more objective way the diversified characteristics of landslides. The employed and outlined model is based on the well-known and widely applied principle *"the past and present are keys to the future"* implies that slope failure in the future will be more likely to occur under those conditions which led to past and present instability.

WORKING STAGES

Briefly, the steps that have led to a first attempt assessment have been: i) choice of the area in which to carry out the study; ii) identification and mapping of a set of parameters, i.e., the instability factors deemed to be directly or indirectly correlated with the landslides; iii) organisation of the existing parameters and analysis of newly developed parameters, so as to be able to define the statistical variables; iv) discretisation of the chosen area into elementary terrain units; v) allocation of the chosen parameters to the elementary terrain units and vi) statistical analysis.

Area of study and available data

Consequently to the devastating meteorological event of November 1994, an accurate and systematic landslide investigation was carried out. This investigation lead to realisation of a detailed data-base related to over 300 rock-block slides and landslide inventory maps (1:10,000 scale) of the whole territory involved. Further information on slope instability was derived from Geological Information System, implemented by the Piedmont Region-Geological Studies and Research Sector, since 1990. It contains a lot of records (more than 2000) concerning landslides affecting this area occurred before 1994. The area of study (over 60 km^2) was chosen on the basis of the availability of a considerable quantity of significant data. In particular, the acquisition of data was conducted in form of vector data and data-bases. The vector data encompassed information of a general nature and concerning specific aspects, such as for instance special landslide themes: i) main and secondary hydrography of the Piedmont Region; ii) land cover according to the *Land Cover Corine* project; iii) map of geological formation; iv) landslides occurred during the extreme rainfall of 4-6 November 1994; v) slope sectors involved in the landslides during 1972, 1951–1972 and 1972–1974, period and vi) measured points (25-50 m sided square mesh) for the creation of the DTM.

Parameter identification

The role of the expert at this testing stage was decisive. Thanks to the knowledge of the landslide phenomena and of the boundary conditions, in fact, it proved possible to select the parameters to be subjected to the subsequent statistical analysis. This was not done solely through the expert judgement; in the selection process, in fact, the parameters which, based on an initial analysis, had been rated as fundamental for the interpretation of the landslide phenomena, were also subjected to a thorough univariate statistical analysis, to pin-point their specific characteristics. In view of the fact that the studies undertaken in connection with the Geological Cartography project (understood as a *summa* of geological, geo-mechanical, hydro-geological analyses, etc.) have not yet been concluded, to enable this study to attain a certain degree of completeness (and an albeit partial physical significance), the parameters already available were taken into account

Variable structuring and domain discretisation

The organisation of the existing parameters and the acquisition of new parameters were carried out in the GIS environment through the ArcView® module, with its Spatial Analyst extension, in order to:

- create, in automatic way, the *digital terrain model* (DTM) starting from the measured points, and therefore be able to process the functions relating to *level curves* and slope *dip* and *dip direction*;

- use the hydrological functions, and obtain in automatic mode new themes concerning the *accumulation* and the *preferential direction* of the flow, starting from the respective functions, and finally, the *hydrographic basins*.

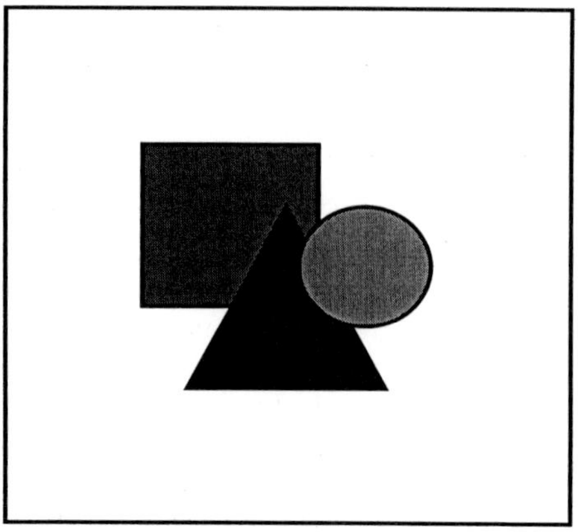

Figure 3. Slope units (left) and landslide mapping (right)

At this point, it proved necessary to proceed with the identification of the terrain unit, or *slope unit*, understood as an elementary cell of the area of interest, i.e., the entity with which all the data processed previously must be associated and entered into the statistical processing module. This *slope unit* is obtained from the automatic definition of the hydrographic basins. In order to provide reliable results, a subsequent discretisation has made by the expert. After this step, thanks to suitable processing operations, each *slope-unit* was assigned all calculated or measured parameters obtained from the GIS application, to be employed in the statistical analysis (figure 3).

MULTIVARIATE STATISTICAL ANALYSIS TECHNIQUES

At this point, in order to define a value of significance to distinguish between slope instability and stability, univariate and multivariate type statistical tests were performed. The methodology mainly used tests aimed at the determination, in automatic mode, of a discriminant function, i.e., a function able to produce a first attempt classification of the different terrain units. Multivariate statistical analysis is a technique for the processing of statistical cases through the simultaneous use of several variables (as opposed to univariate analysis) whose aim is to identify multiple correlations between variables and to establish the statistical weight of the latter. The outcome is a mathematical function (in particular, a linear function) which makes it possible to evaluate the ability of a certain number of variables to sort out the cases into pre-defined groups. The value of this function can be determined for each *slope unit*, that is to say, for each statistical case.

In parallel with this, another linear function, called *classification function*, can also be determined; its purpose is to re-classify a statistical sample into two or more distinct groups, having assigned *a priori* each individual case to a specific group. Each classification group is associated with a different classification function having a twofold task: a) to generate the classification matrix, containing the success percentage of the method, i.e., the number of cases correctly re-classified and b) to work in a predictive manner, that is to say, to classify new cases (which have not been employed in the construction of the function itself) by assigning them to the different groups.

The pre-defined classes, or *grouping variables*, to be used for an *a priori* classification of *slope units* as *stable* or *unstable*, were identified by computing the ratio between the landslide area and the total area included in each individual *slope unit*. If this ratio is greater than a threshold value (according with mapping errors) of 2 %, then the slope unit is defined as **unstable**, whilst, for a value lower than or equal to 2 %, the *slope unit* is defined as **stable**.

This resulted in the formation of two samples, one made up of stable slope units and the other of unstable slope units, each characterised by a distinct value of the set of variables defined. At this point, the classification attempts through the discriminant analysis method were finally performed. The first attempt took into consideration 57 variables concerning: i) *slope-unit* morphology; ii) overall percentage of landslide area contained in each terrain unit; iii) geological formations; iv) land cover; v) slope dip and dip direction and vi) number of rainfall events triggering landslide in each terrain unit.

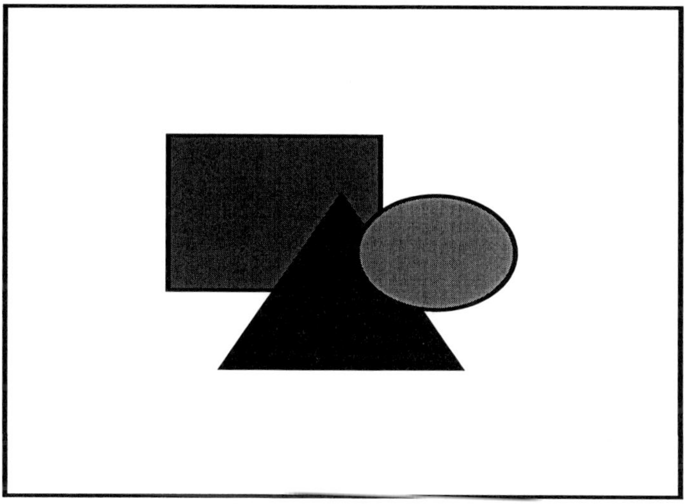

Figure 4. Synthetic report of the first attempt results.

The success percentage, in this case, was 97 %, and the number of variables rated as significant in making the classification was 22. In other words, the statistical model took into consideration 22 variables (out of an initial choice of 57) to classify the slope units as either stable or unstable, and was able to re-formulate the classification defined *a priori* in 97% of the cases. Figure 4 shows both the statistical distribution and the success percentages obtained on this analysis. The discriminant function showed that the most influencing parameters governing slope stability are: slope dip direction, geological formations, land use and number of rainfall events triggering landslide in the last 40 years, watershed area.

The second attempt took into account only the 53 variables of a, so to speak, *environmental* nature, i.e., those concerning the geological formations, land use, the 36 dip classes identified for the terrain units, and neglected the variables concerning the landslide areas. The *a priori* classification was still maintained as a function of the percentage of *slope unit* involved in the landslides, as discussed above. In this case, the success percentage decreased to 93 % and the variables rated as significant in making the classification reduced to 17. The initial results seem promising, especially if we consider the success percentages of the multivariate analysis and the number of statistical cases incorrectly classified. This might be partly due to the fact that the choice of the slope units had been very accurate and specific in relation to the pre-established objective.

In order to verify the model prediction capability, the whole slope units (73) of the study area was randomly split into two parts: one part, a selection of 56 slope units, was used to estimate the model while the remaining case (17) were reserved for testing the goodness of fit. This operation yielded to these results: a) a success percentage on the reclassification of 56 cases equal to 96 % and b) a success percentage on the reclassification of 17 cases equal to 90 % (obtained using the classification function determined on the 56 cases).

CONCLUSIONS AND FUTURE DEVELOPMENTS

The operational advantages made available by multivariate statistical analysis to the entire operating procedure are emphasised: a) it proves possible to quantify, in a clear and concise manner, the data obtained from a simple observations analysis and b) after converting the available parameters into statistical variables, it proves possible to evaluate objectively the weight of each individual variable in the re-classification of the statistical cases, so providing an useful tool in the choice of influencing parameters. The results obtained display a clear-cut discrimination between the stable group and the unstable group; in terms of superposition of distributions, we find that there is an interference between the tails, and it is precisely on these cases falling astride the two groups that we should concentrate our attention, in terms of both the statistical calculation and the phenomenon in itself. It should be noted, however, that the methodology and the specific parameters to be employed should be re-defined for each specific type of landslide. Another interesting observation is the fact that, in actual practice, the efforts made in coping with the problems associated with data processing by means of GIS techniques have been decisive, despite the fact that this stage consisted solely of preparatory work for an accurate statistical analysis. It is felt, in fact, that it is indispensable that the data to serve as input for the statistical analysis be as accurate as possible. Finally, it should be pointed out that there is much to be done before this research project can come to its conclusion. Further and more complex processing techniques (main components analysis, logistic regression, time series) are still being studied in order to assess their degree of effectiveness in terms of their applicability to the problem of the overall assessment of geological hazard, that is to say, their ability to take into account all the aspects to do with the timing, intensity and evolution of the phenomena.

REFERENCES

Baeza C. & Corominas J., (1996), Assessment of shallow landslide susceptibility by means of statistical techniques, Proc. 7th Int. Symp. on Landslides, Trondheim, vol. 1, 147-152;

Berardi R., Passalacqua R., Scavia C. & F. Forlati, (1994), Translational slope instabilities in weak rock, Proc. 1st North American Rock Mech. Symp, Austin, 859-865;

Carrara A., Cardinali M., Detti R., Guzzetti F., Pasqui V. & Reichenbach P., (1991), GIS techniques and statistical models in evaluating landslide hazard, Earth Surface processes and landforms, vol.5, 427-445;

Carrara A., Cardinali M., Guzzetti F. & Reichenbach P., (1995), GIS-based techniques for mapping landslide hazard, G.I.S. in assessing natural hazard, Academic Pub., Dordrecht;

Forlati F., Lancellotta R., Scavia C., & Simeoni L., (1998), Swelling processes in sliding marly layers in the Langhe region, Italy, Proc. 2nd Int. Symp. on Hard Soil-Soft Rocks, Napoli, 1089-1099;

Leroi E., (1996), Landslide hazard–Risk maps at different scales: objectives, tools and developments, Proc. 7th Int. Symp. on Landslides, Trondheim, vol. 1, 35-51;

Pegoraro C., (1998) Studio propedeutico alla valutazione della pericolosità geologica inerente all'instabilità dei versanti mediante tecniche GIS ed approccio statistico multivariato, Corso di Specializzazione Post Lauream in Sistemi Informativi Territoriali, COREP-GISFORM, Torino.

Long-term Stability of Overburden Dump Slopes in Recultivated Lignite Mines of Central Germany - Prediction of Maximum Pore Water Pressures

A. CARSTENSEN and W. POHL, Institute of Geosciences, Technical University, D-38023 Braunschweig, Germany

INTRODUCTION

Large-scale lignite mining in eastern and central Germany left hundreds of opencut pits, which are to be restored to public use. For this, they will be filled with water to about the pre-mining water table. The slopes of the future lakes will have to be stabilised to a high degree of safety. More than 50% of the length of these slopes, totalling several hundred km, will border spoil dumps. One of the most important factors controlling slope stability are pore pressure maxima caused by extreme precipitation events. The slopes have to be designed to withstand pressures at least due to rains with a recurrence period of one hundred years. The hydraulic characteristics of the large spoil dump bodies were little understood at the outset of this project (1995). Therefore, we undertook an investigation comprising *in situ* observations over two years (1996-98) as a base for predicting possible pore pressure maxima.

The observation field of approximately 0.5 km² was situated in the lignite opencut mine at Zwenkau, about 10 km south of Leipzig (Fig. 1a). Here, the interior spoil dump is 20-30 years old. It has an average thickness of 50m and comprises three different bodies (Fig. 1c). The lowermost (AFB-dump) consists of rather well homogenized clayey and sandy silts with minor coarse material (Fig. 2), which have been dumped on the former pit floor. Compaction is highly variable, mainly due to changing dumping depth and to varying water contents during dumping. The (buried) surface of the AFB-dump forms ridges and depressions with slope angles of 38 ± 2° (Fig. 1b). The upper dump (AK I) was placed to create a plane surface for recultivation. It consists of silt and fine sand with layers of sand and gravel, and with fragments of kaolinitic clay and lignite. There was much less homogenisation during dumping, resulting in a pronounced heterogeneous and anisotropic internal structure. The scale of individual soil type bodies is so small and variable, however, that it is not possible to determine economically the 3D internal structure of the AK I-dump by drilling or geophysics. At the crown of the slope to the pit bottom, a third dump (PUK) was built up from waste produced by cleaning railway wagons used for spoil transport. The waste is dumped over the edge, resulting in a slope angle of about 38 ± 2°.

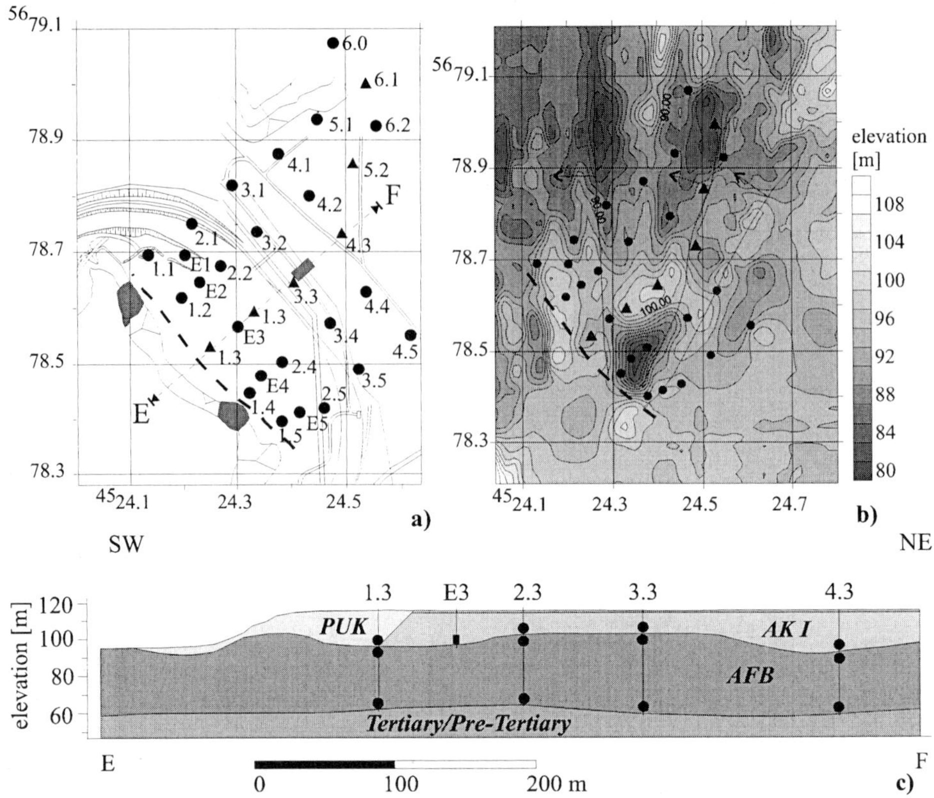

Figure 1. a) Field observation area with measuring sites (triangles: 6 deep drillholes with pore pressure gauges; black dots: groundwater observation wells in AK I). b) Buried relief of the AFB-dump. c) Section E-F through centre of field (black dots are pore pressure gauges).

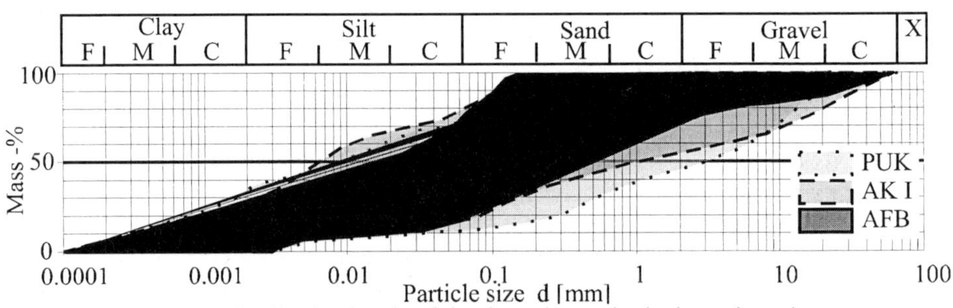

Figure 2. Particle size distribution bands of the three dumps in the investigated areas.

The mechanical parameters of the soils forming the dumps vary considerably due to dumping technology, to nearly dry versus waterlogged deposition, and to the conditions of consequent consolidation. Shear parameters have been determined by laboratory tests of undisturbed samples (direct shear, triaxial tests), and by charts correlating index test results with the angle of shearing resistance (Fig. 3, Table 1).

Table 1. Main mechanical parameters of spoil dumps at project site.

Dump	γ_r [kN/m³]	ϕ' [°]	c' [kN/m²]
PUK	19	34 [29-41]	0.0
AK I	18	30 [25-39]	0.0
AFB	24	31 [25-31]	14.6

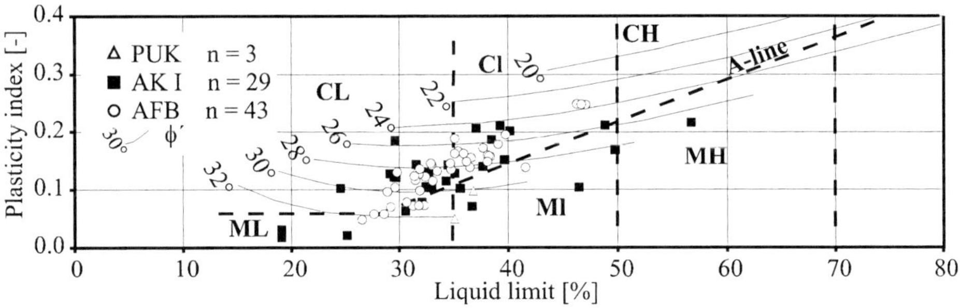

Figure 3. Plasticity chart with data from overburden dump soils as a base for determination of the effective angle of shearing resistance (Reuter et al. 1992).

HYDRAULIC REGIME IN THE OVERBURDEN DUMPS

The infiltration of precipitation into the dumps during 20-30 years caused nearly total saturation, and the formation of a ground water body with a free surface situated a few meters (1.5-3.0 m) below the ground surface. Towards the slope to the pit bottom, the distance between groundwater surface and ground level increases. Springs at the base of PUK form the only visible drainage. Generally, ground water flow is towards the pit in the southwest, with an increasing hydraulic gradient in the same direction (Fig. 1b). At least for the higher and more mobile part of the groundwater (in AK I) this is favoured by the SW-orientation of the AFB-ridges in this part of the area, while a more complicated pattern occurs in the northern part. Back calculation of pore water pressure into water column heights and observations during drilling both confirm that pressures in the AFB-dump may deviate from purely hydrostatic to higher as well as lower values. This is due to its low permeability, and to variable compaction and water content during dumping operations. The consistency of the soils varies between very soft and stiff.

Field and laboratory tests to determine the hydraulic permeability confirm the heterogeneous and anisotropic character of both upper dump bodies. The "best" estimates for overall hydraulic characterisation were k_f = 5E-7 to 5E-8 m/s for PUK, and k_f = 1E-7 - 1E-9 m/s for AK I. Hydraulic modelling as well as field observations indicate, however, that the hydraulic behaviour of these dumps may locally be dominated by more permeable soil lamellae (up to 1E 5). Only laboratory-tests have been carried out for determination of the permeability of the AFB-dump, resulting in a mean value of k_f = 2E-8 m/s with small variance.

PORE WATER PRESSURES

Predicting pore pressures for future high rainfall scenarios is not possible without an experimental base. Therefore, a systematic data collection effort was undertaken over a period of two years. A measuring field was established adjoining the slope to the Zwenkau open pit, comprising a weather station (precipitation, air temperature and pressure, humidity), 5 tensiometers in the unsaturated zone and 18 pneumatic pore water pressure gauges in 6 deep drillholes, surrounded by 24 water level observation wells sited in AK I (Fig. 1a). Three pressure gauges were installed in each drillhole, at the base of AK I, in the top of AFB, and near the base of AFB at about 45m below ground (Fig. 1c). Pore pressures were measured and recorded every two hours, the other data with even higher frequency. The influence of the air pressure on the pore water pressure measurements was eliminated by frequency-filtering (Kuehl, 1998). Various methods were used to estimate actual infiltration as opposed to total precipitation.

There are various possibilities to analyse the covariance of precipitation/infiltration and pore water response. One approach has been developed by Tilch (1999), exploring the linear correlation between infiltration rate, the duration of infiltration/rainfall, and the resulting pore pressure rise. In this project, we chose to analyse the data in the time and frequency domains. Long-term trends were quantified (ranging from -54.2 and +66.9 hPa for different gauges) and then subtracted from the data series. These results were related to specific rainfall events (Fig. 4), as expected showing that pore pressure rises after rainfall occur with a time lag, and that this is larger with increasing depth. The actual duration of the time lag and the intensity of the pore pressure response vary widely at similar depth, however, obviously caused by the hydraulic heterogeneity of the dumps. Cross correlation of precipitation with pore water pressure indicate a temporal shift in the upper dump of 10 to 20 days; at the top of the AFB-dump about 30 days, and approximately 50 days at its base.

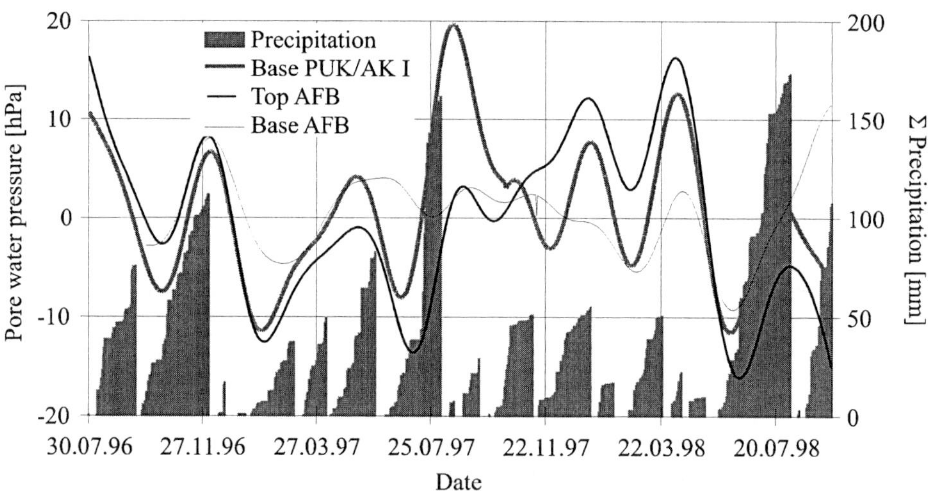

Figure 4. Precipitation phases and pore water pressures at different depth levels.

Fluctuating pore water pressures can be modelled as sine waves due to flow in a porous medium (Iverson & Major 1987). They are initiated by recharge of groundwater. Their propagation velocity depends mainly on the flow resistance of the medium. In the simplest case, the flow is vertically downward. This can be described by the diffusion equation

(Gambolati 1973), which allows calculating the transient flow of an incompressible homogeneous liquid, moving in a linearly elastic porous medium:

$$\frac{\delta p}{\delta t} = C \cdot \left(\frac{\delta^2 p}{\delta y^2} \right) \tag{1}$$

Expressing this relation as a sine wave, attenuation and phase shift towards depth can be deduced from the following equation (Iverson & Major 1987):

$$\frac{p}{p_0} = \exp\left[-y \cdot \sqrt{\frac{n \cdot \pi}{C}} \right] \cdot \cos\left[2\pi \cdot nt - y \sqrt{\frac{n \cdot \pi}{C}} \right] \tag{2}$$

p_0: amplitude [m]; n: frequency [1/s]; p: pore water pressure [Pa]; t: time [s]; y: vertical depth under ground surface [m]; C: hydraulic diffusivity [$m^2 s^{-1}$].

While phase lag and attenuation of the pore pressures towards depth are clearly exhibited by the data, calculations based on average hydraulic conductivity of the dumps according to (2) give results that deviate by 30-200% from the measured values (Fig. 5). For most of the gauges, substitution of higher conductivity leads to calculated wave amplitudes quite near measured ones. The resulting phase lag, however, is much smaller than measured. Of course, this is due to the heterogeneous and anisotropic nature of the dumps.

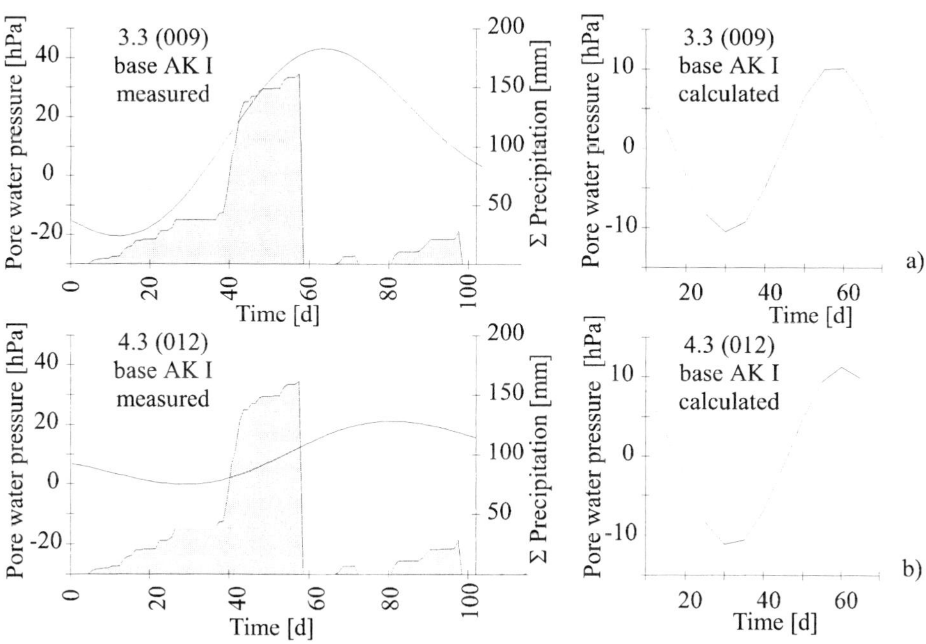

Figure 5. Measured (left) and calculated (right) pore water pressure wave in AK I due to one precipitation event. a) gauge 009 b) gauge 012.

Analysis of the pore pressure data in the frequency domain produces two significant periods [81 and 121 d] indicating that cyclical climatic influences are predominant. In the FOURIER-spectra, shorter precipitation cycles are not expressed (Fig. 6). Pore pressure waves with lower frequency penetrate further and attenuate less than shorter waves. The phase difference between rainfall and pore pressure waves is between 23 and 43 days for the 120 d period, and 8-21 days for the 80 d period. These results help to predict normal pore pressure cycles as a background for possible catastrophic events.

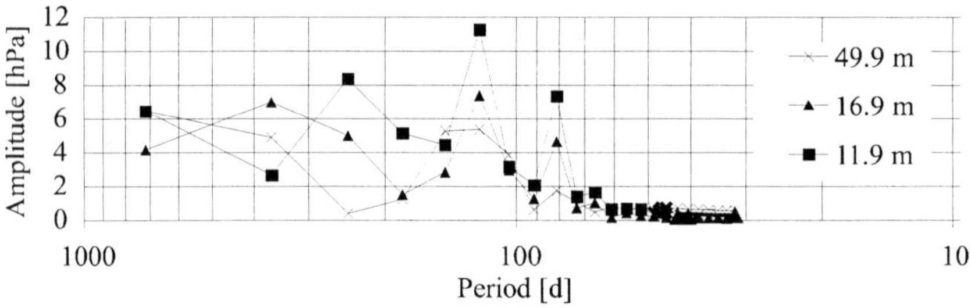

Figure 6. FOURIER-spectra of pore water pressure in deep drillhole 3.3 at different depth.

CONCLUSIONS

Analysis and modelling of measured pore water pressure response to precipitation/infiltration confirm generally accepted physical concepts. A satisfying reproduction of the observed relations by back-calculation, especially of the pore pressure amplitudes at a mechanically significant depth of 10-30 m can be obtained only partially. This is due to the highly heterogeneous and anisotropic nature of the spoil dumps, which cannot be economically investigated. For an „average“ situation representing about 80% of the slope length, maximum pore pressures produced by extreme rainfall can be calculated using the sine wave model. The remaining length is characterised by a much stronger pore pressure response, undoubtedly due to special hydraulic conditions. For these critical sections, only „best estimates“ are possible. Nevertheless, the resulting projections for extreme precipitation with a recurrence period of 100 years must be used for the design of the spoil dump slopes.

Acknowledgements: This work was supported by the German Ministry of Education and Research (BMBF). We thank MIBRAG mbH for considerable technical support. The German Meteorological Service (DWD) provided extensive weather data.

REFERENCES

Gambolati, G., 1973, Equation for one-dimensional vertical flow of groundwater: I. The rigorous theory. Water Resources Res. 9 (4), 1022-1028.

Iverson, R.M. & Major, J.J., 1987, Rainfall, groundwater flow, and seasonal movement at Minor Creek landslide, northwestern California: Physical interpretation of empirical relations. Geol. Soc. America Bull. 99, 579-594.

Kuehl, H., 1999, Bereitstellung und Test ausgewählter digitaler Filter unter Einbeziehung der „Schnellen Hartley-Transformation“. (diploma thesis), 128 pp, 56 Abb., 5 Tab.

Reuter, F., Klengel, K.J. & Pasek, J.P., 1992, Ingenieurgeologie. 3. Ed., 604 pp, Dt. Verlag für Grundstoffindustrie.

Tilch, N., 1999, Rutschungssuszeptibilität im südlichen Niedersachsen - von der Anatomie der Rutschung bei Brunkensen/Alfeld zur Prognose instabiler Hanglagen. Braunschweiger Geowiss. Arbeiten 22, 184 pp, 97 Abb., 22 Tab., 8 Tafeln.

Interaction Between Dams and Landslides: Three Case Histories

A. CATALANO, V. CHIEPPA and C. RUSSO, Geotechnical Bureau of the Italian National Dam Service, Italy

INTRODUCTION

Italian dam history is marked by the huge catastrophic event of the Vajont slide, whose reactivation in 1963 produced a 100m high wave that overtopped the dam and killed thousands of people in the valley below. Since then, some other landslides have been detected on reservoir slopes: some of them are not a threat to public safety and in many cases just required some monitoring equipment to evaluate their evolution, others require large remedial works or severe limitations to the reservoir to be securely operated.

With the aim of avoiding the problems connected with safe operation of the basin in similar cases, in 1982, a very strict rule was introduced in the Italian technical regulations concerning new reservoirs: "*dams can not be built if on the dam abutments, even above the maximum water level, a predictable risk of landslides exists in normal conditions or as an effect of an earthquake*" (D.M., 1992). No other advice is provided for existing dams. The Italian National Dam Service, whose main task is to watch over the correct design, construction, operation and maintenance of large dams, is facing the problem of setting up operative procedures to manage the risk connected with existing landslides on reservoir slopes.

Among the detected landslides affecting reservoir slopes in Italy, three case histories have been selected whose common features are the location in the vicinity of the dam abutments and the quiescent state of activity modified by huge rainfalls or water impounding in the reservoir. In two cases the reactivation is non-catastrophic, also thanks to the restraint exerted by the embankment at the landslide toe, in one case severe restraints have been imposed on reservoir operations. The mechanical aspects of the landslide response to boundary condition modifications will be discussed through the description of results from monitoring; some information about set up procedures for the operations are also provided.

DESCRIPTION OF SELECTED CASE-HISTORIES

Trinità Dam

The reservoir delimited by Trinità dam is located in Southern Italy, in the centre of Sicily. The involved formation is a clayey formation dated Upper Miocene, dislocated and covered by a transgressive calcarenite formation, probably dated Pleistocene, with several block falls and toppling failures (Figures 1 and 2). The contact between the two formations is irregular. The superficial layer of the argillaceous formation, 7-8m thick, is weathered and altered, while the underneath formation is intact; above the clayey soil, an almost uniform layer of debris covers the slopes. The subsoil in the abutment area is composed of a relevant clay percentage, with silt percentage varying between 20 and 40%; several sandy layers are present, sometimes of relevant thickness, with sand percentage up to 70%. Natural water content measured in the laboratory is close to the plastic limit or slightly in excess of it. Friction angle is on average of about 15°, with values up to 32° due to the presence of sand. Cohesion is very low.

From the hydrogeological point of view, the water stored in the highly permeable calcarenites emerges at the contact with the clayey formation and partly flows towards the lake along the slope and partly infiltrates in the superficial layers of the clayey soils.

The whole area is affected by earthflows, many of them in a quiescent state of activity. In the following, the attention will be focused on the left abutment of the dam, where an earthflow is present, about 8m thick, whose cyclic reactivations created several problems in the past.

The zoned earth dam was completed in 1959; it is 28.5m high and 322m long, the hydrographic basin is 200km^2 and the water volume stored is 18million cubic metres. The reservoir is used to provide water for civil use and for irrigation.

The existence of the landslide was known since the dam construction and was roughly described in the geological study dated 1953: as a consequence, the spillway was designed to be founded on piles. The landslide is first reported in official documents in 1963: the first reactivated portion was located about 150m upstream from the dam and was about 150m long and 50m wide. It progressively enlarged in the following year, reaching the base of the calcareous formation. Measured maximum displacement was 1.4m. Several trees were then planted in the area but, before they could really become effective, a new reactivation took place in 1965 that severely damaged the electric cabin (later demolished) and the guardians' house. The landslide enlarged downhill towards the dam abutment and the spillway. Small depth trenches were dug along the water paths and filled with calcareous blocks in order to reduce the progressive excavation due to overland water flow. The landslide displacement rate reduced in the following years.

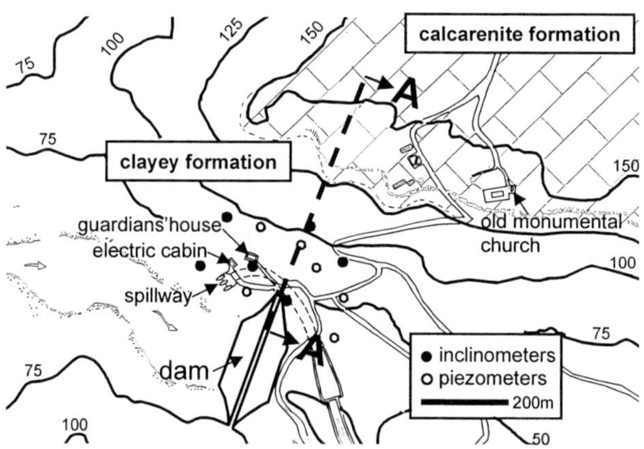

Figure 1. Plan of the dam area

In 1981 further damages were detected at the guardians' house which required its demolition and new reconstruction a few years later using a piling foundation. Other piles have been built in the same period to protect the deposit located uphill from the spillway.

Detailed geotechnical studies have been performed in 1990 that allowed the description of the landslide. Inclinometers and piezometers were installed in the following year. Measured displacements are depicted in Figure 2: the average displacement rate was of 1cm/yr (Ente Di Sviluppo Agricolo - Palermo, 1997).

Neither visible deformations of the dam due to the landslide load have been reported nor serious damage to appurtenant structures relevant for public safety (spillway and bottom outlets). Nevertheless signals of progressive movements are clearly visible uphill from the guardians' house, where the wall of the backyard has been displaced: landslide pressure on the concrete pavement determined its lifting and damage to the house bricks.

Future monitoring of the landslide will include installation of new instruments and continuous data acquisition aimed, on the one side, at evaluating the load increment on the spillway structure, on the other side, at defining the possibility of introducing remedial measures to reduce the displacement rate.

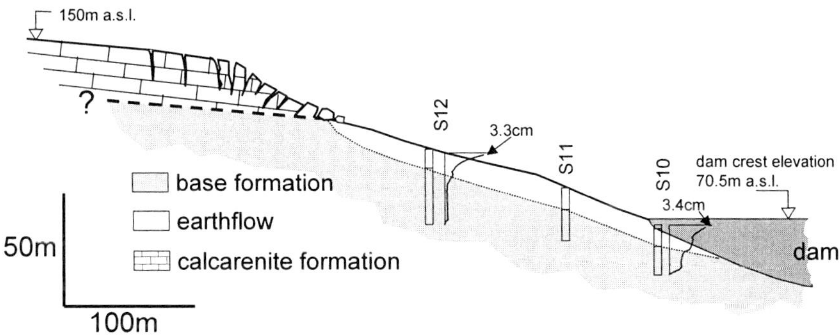

Figure 2. Section A-A corresponding to the dam abutment; measured displacements between March 1991 and January 1996.

Rossella Dam

Rossella dam is a zoned dam with inclined core, located in Sicily and founded on a flysch formation dated Miocene, locally covered by argillaceous and sandy deposits of the Tortonian period. In the basin area Mesozoic calcareous rocks outcrop.

The dam is about 30m high and 350m long and was completed in 1965. The hydrographic basin is 20km^2 and the storable water volume is of about 20millions of cubic metres. The reservoir is used to provide water for civil use and for irrigation. At present the maximum allowed water level has not been attained. The landslide on the left abutment of the dam was reactivated during the excavations performed for the core in 1961. Remedial measures for the slope consisted in plantation of trees. Very slow movements probably continued in the following years as witnessed by the actual evident deformation of the pipeline located uphill from the dam.

During winter 1997-1998 a sudden reactivation occurred with a larger displacement rate, witnessed by the formation of cracks on the cut-off wall at the dam toe and displacements measured on the dam crest. A new monitoring equipment was installed, consisting in inclinometers, piezometers and topographic monuments. Measured displacements are depicted in Figure 4 from December 1997 to September 1998 (Ente Acquedotti Siciliani, 1999): it is important to notice that displacement entity is also influenced by the dam settlements still in progress.

The reactivation is related to the attainment and to the maintenance for two months of a critical water level (524m a.s.l.). Remedial measures consisted in water level lowering down to a safe position and set up of a program for a new controlled water rising in the reservoir.

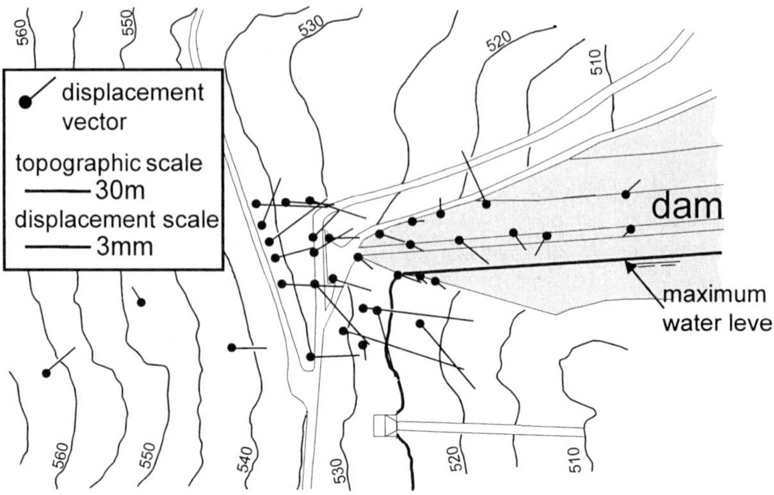

Investigations on soils and monitoring program are supervised by the University of Palermo.

Figure 4. Displacement vectors in the abutment area obtained through topographic measurements between December 1997 and September 1998

Casanuova Dam

Casanuova dam is a earth zoned dam located in Central Italy. The base formation is constituted by marls and clayey marls, with interposed strata of limestones, sandstones and calcarenites. Clay shales locally outcrop on the reservoir slopes.

The dam is about 90m high, 450m long (Figure 5) and was built between 1981 and 1993. The original designed volume of water to be impounded was more than 200millions of cubic metres. First monitored filling of the reservoir occurred in 1991, during assembly of the bottom outlet's gates, nevertheless the presence of the foredam determined some antecedent fillings as a consequence of heavy rainfalls. In 1984, fissures and cracks had appeared on the withdrawal tunnel: these were the first visible effects of the slow movements of the landslide affecting the mountainous relief near the right abutment of the dam. Local interventions did not stop the evolution of cracks. Since 1990 several investigations have been performed aimed at defining the main features of the landslide affecting the reservoir.

The movement is of translational style, involving the upper destructured portion of the arenaceous and marly formation, for an average thickness of 70m (Figure 6). The sliding surface was directly observed from the withdrawal tunnel and consists in a slickensided softened clayey layer, between 10 and 50cm thick, with $c'=0$, $\varphi'=16°$ and $\varphi'_{res}=8°$. The displacing soil mass is probably an old quiescent landslide reactivated as an effect of water level increase in the reservoir and pore pressures rising along the shear surface (Ente Irrigazione Umbro Toscano – Arezzo, 1993).

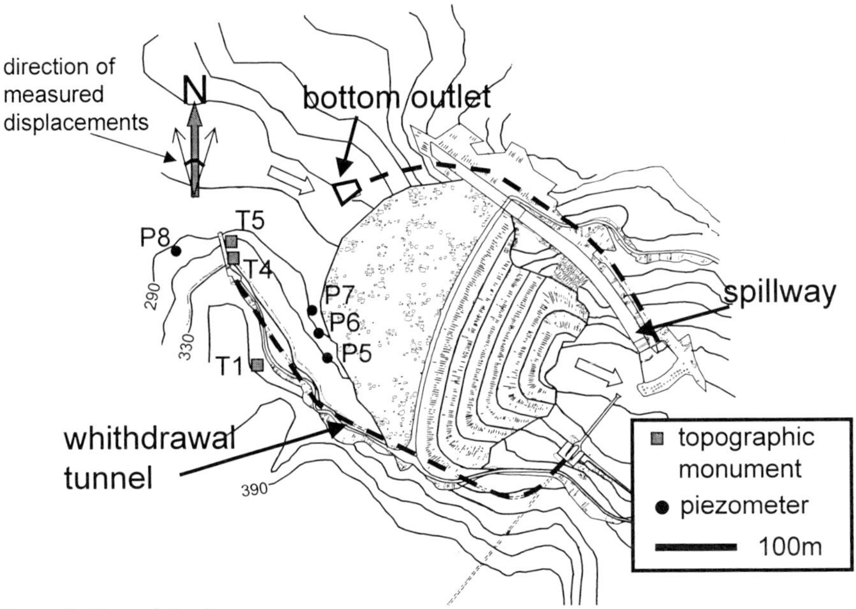

Figure 5. Plan of the dam area

Monitoring of the landslide showed a sudden response of piezometers to water level fluctuations in the reservoir (the upper level attained was the top of the foredam and corresponding impounded volume was 15millions of cubic metres) and concurrent increase of the displacement rate. Data reported in Figure 7 have been collected during the reservoir filling in 1991. Maximum measured displacement rate on that occasion was of 1-2mm/day, five times larger than displacement rates measured before. Moreover, displacement rates measured during successive attainment of identified critical water levels, were far smaller than those exhibited during first water level rising, in agreement with what observed in similar cases (Vajont slide, Ragoleto reservoir). Until now data are not sufficient to clearly distinguish effects related to rainfalls from those connected to the reservoir.

The landslide thickness is quite relevant in comparison to the dam size and the movements are affected by water level fluctuations in the reservoir. Therefore, safety procedures were required to stop the filling programme in order to study the feasibility of landslide stabilisation and set up remedial works to guarantee the possibility of emptying the reservoir in case of an unexpected evolution of the landslide. In the meantime, existing monitoring instruments are going to be integrated.

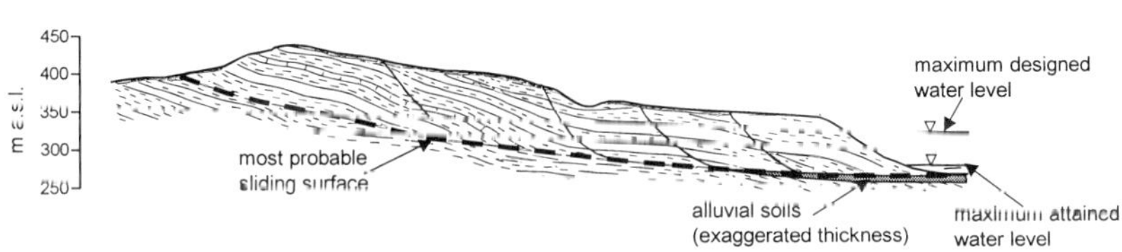

Figure 6. Section of the landslide

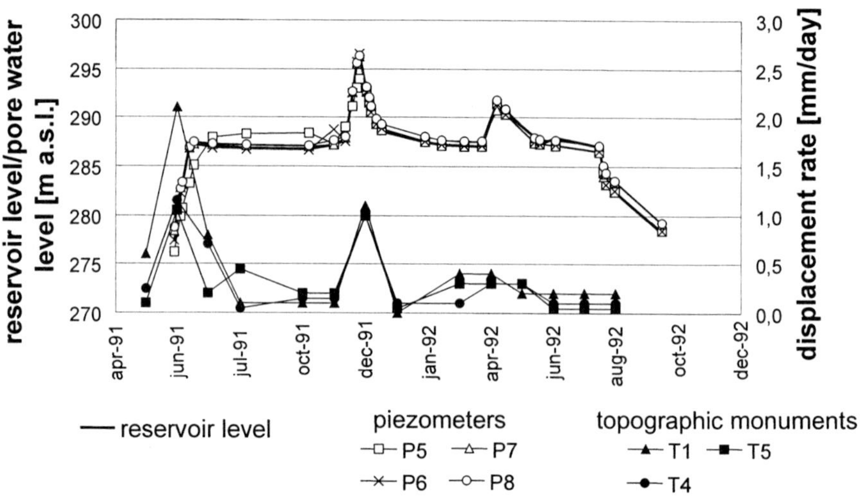

Figure 7. Evolution of water levels, pore pressures and displacement rates

FINAL COMMENTS

The management of the risk connected to the presence of landslides on the reservoir slopes must be seriously considered. Together with potentially catastrophic events which would require massive stabilisation works or the drainage of the reservoir, there are several other cases that, although less relevant as far as public safety is concerned, require consideration in view of a correct management of the reservoir. No clear indication and experience is provided on how to deal with such cases when, although the potential for a catastrophic failure is low, particular care must be put in to controlling dam leakage and operation of the overflow spillway, the bottom outlet and other appurtenant structures.

In the Authors opinion, there is a strong need for simple models capable of drawing the risk scenarios, following boundary conditions modifications, and helpful in establishing safety procedures and evaluating the efficacy of remedial works.

In the end, it is important not to focus only on the dam abutments, since potentially moving landslides can be present even on other reservoir slopes. Analyses of similar cases showed that this circumstance frequently takes place and particular care should then be placed when dealing with these phenomena in controlling the stability of the whole area of the basin.

REFERENCES

D.M. 24 marzo 1982: “Norme tecniche per la progettazione e la costruzione delle dighe di sbarramento”.

Ente Acquedotti Siciliani (1999). Diga Rossella: rilievi topografici per il controllo dell’opera di sbarramento. Archive of the Italian National Dam Service.

Ente Di Sviluppo Agricolo - Palermo (1997). Diga Trinità: controlli strumentazione in opera in sponda sinistra. Archive of the Italian National Dam Service.

Ente Irrigazione Umbro Toscano – Arezzo (1993). Diga di Casanuova sul fiume Chiascio. Geological and geotechnical studies by prof. S. Olivero. Archive of the Italian National Dam Service.

A Landslide along Gently Dipping Bed Surfaces

BYUNG-GON CHAE, YONG-CHAN CHO, WON-YOUNG KIM, KYEONG-SU KIM
Korea Institute of Geology, Mining and Materials (KIGAM), 30, Kajung-dong, Yusung-gu, Taejon, 305-350, South Korea

ABSTRACT
A landslide which occurred within gently dipping sedimentary rocks in southeastern Korea was investigated to find the factors that caused the failure. The landslide is probably one of the biggest rock slides ever reported in Korea with dimensions up to 100 m long, 40 m wide and 10 m thick. The landslide occurred on a hill slope with an inclination of only about 25 degrees. Narrowly spaced tension cracks and vertical scars developed perpendicular to the slope indicating that the landslide developed slowly as rock creep associated with block tilting.

One of the contributing factors for the landslide may have been a small roadcut passing through the middle part of the hill. The roadcut was constructed with a slope angle of 60 degrees. The landslide terminated at the road cut and did not extend downslope of the road. Slope stability analysis shows that the roadcut would allow the hill slope to be triggered. These facts imply that the road cut would cause tension cracks on the hill slope.

The area consists primarily of sandstones, but a few centimeter thick clay layers are interbedded within the sandstones. Moreover, several sheets of shalely mudstones and coaly shales are also developed within the landslide area. Heavy rainfall during the summer of 1998 may have infiltrated along tension cracks developed at an early stage due to stress relief on formation of the roadcut. The thinly interbedded mudstones and coaly shales probably acted as sliding surfaces. Since the mudstones and coaly shales have lower permeability than sandstones, perched water probably developed above the mudstone beds. The water pressure would have decreased the effective stress, thereby, reducing the shear resistance. This mechanism would contribute to the rock sliding along the mudstone beds with very slow velocity for a long time like a rock creep type.

INTRODUCTION
Most of the landslides which occur along mountain slopes in Korea are typically small in scale and debris flow in type. These kinds of debris flows generally occur during periods of intense rainfall from July to August when more than half of annual rainfall occurs (Kim and Chae, 1998). But, the studied landslide is one of the largest reported rock slides which have ever occurred in gently dipping bed surfaces in Korea. The failure was also unusual in that, according to the inhabitants near the study area, the landslide seems to have begun in late September 1998. The landslide continued to develop over three months until the gentle mountain slope collapsed totally. For the study, a field survey was carried out together with physical property tests for rocks, XRD analysis and direct shear tests of samples to help identifying the triggering factors of the landslide.

Landslides in research, theory and practice, Thomas Telford, London, 2000.

SITE DESCRIPTION

The study area is located on a gentle slope of a hill with a height of about 140 m. The width of slope containing the landslide was about 55m and the overall slope angle about 25 degrees. A small road passes through middle part of the mountain. The size of landslide reaches 100 m long and about 40 m wide (Figure 1). The thickness of the collapsed block is about 10m. Head of the landslide is located close to the top of the hill. The general sliding direction is about N110 degrees.

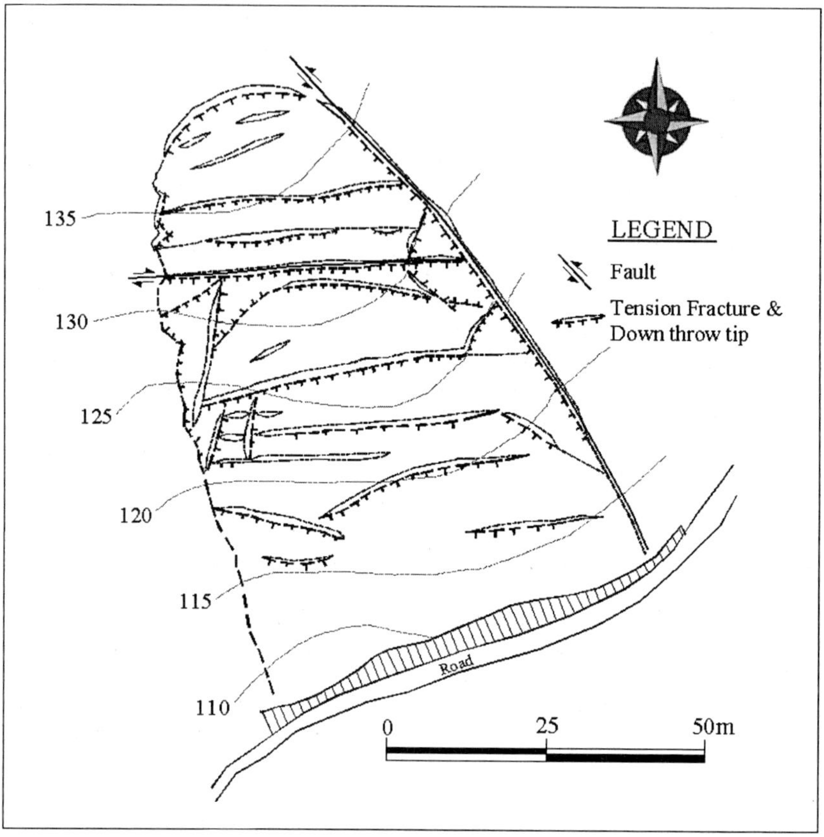

Figure 1. Plan view of the landslide. The magnitude of the slide reaches about 100m long, 40m wide and 10m thick. A small road passes through the middle part of the hill slope.

The study area is composed of medium grained sandstones, siltstones and mudstones of the lower Cretaceous. The sandstones are the most dominant rocks in the area whose thickness reaches between 30 and 40cm. The sandstones of the landslide area are moderately weathered and relatively weak compared with other areas of the same stratigraphy. The siltstones and mudstones are thinly interbedded in the sandstone beds. The sedimentary rocks have gentle dipping angles of about 15 ~ 25 degrees. It implies that the sedimentary rocks have not experienced severe tectonic deformation since deposition.

Geologic structures in the study area are mainly composed of faults and joints. Faults are developed in two sets oriented N38~54W, 85~90SW and N70~88E, 80~90SE. The eastern boundary of the landslide coincides with the N38~54W fault that has a sinistral sense of strike slip movement. The N88E and the N54W faults intersect as conjugate faults at the eastern part of the landslide. At the intersection, the eastern part has large-scale collapse of rock blocks. Joints within the area have developed parallel to the two sets of faults. The joints oriented N20~40W and N70~80E intersect each other and formed saw tooth features at the western boundary of the landslide.

Different parts of the slide have displaced by different amounts. Divided as a quadrant, the upper right part, the first quadrant, has the largest horizontal and vertical displacement. Although rock blocks contact each other near the ground surface, there are wedge-shaped spaces below the surface. This feature seems to be due to block tilting. In other words, sliding of lower part of the rock blocks is thought to have opened up the wedge-shaped space. At the upper left part of the landslide, the second quadrant, large horizontal displacements and small vertical displacements are observed. The joints make large displacement of horizontal component with small vertical one in this quadrant. The third, lower left quadrant, shows small displacement of horizontal and vertical components. The fourth quadrant is covered with big rock blocks with relatively larger vertical displacement than horizontal component.

ANALYSIS OF TRIGGERING FACTORS

Several triggering factors may have contributed to the landslide. One of the triggering factors was probably the construction of a small road passing through the middle part of the hill. The road cut was constructed with a slope angle of 60 degrees in 1994. The slope was cut again in early 1998 with widening of the mountain road. On detailed observation of the landslide, it was revealed that the landslide terminated at the crest of the road cut and did not extend down slope of the road. Due to the recovery works at the site, it is impossible to find the exact toe of the landslide. But, a lot of rock blocks that are piled up above and near the road cut indicate that there must have been some displacement of rock masses.

Graphical slope stability analysis was performed to verify the relationship between road cut and slope failure. The slope stability analysis demonstrates that the construction of road cut would allow the hill slope to be triggered (Figure 2), because of the gently dipping bed surfaces plotted in the unstable daylighting area for the road cut. In forming the road cut, the equilibrium of stress condition was probably lost, which led to the development of small tension cracks on the hill slope.

By two boring surveys, however, it revealed that several thin sheets of coaly shales intercalated in the sedimentary layers (Kim et al, 1999). One shale bed with thickness of 40~60 cm is also observed at the head of the slope. This shale bed appears to be continuous through the landslide area at a depth of about 9 ~ 10 m below the surface. The sequence also has intercalations of clays as filling material between bedding surfaces. XRD analysis of the mudstones shows that they contain more than 20~30 percents of clay minerals including muscovite and montmorillonite.

Laboratory tests were performed to identify rock properties. Four samples of completely weathered material between sandstone beds were collected. According to results of the tests, all of the materials are silty sands. The range of permeability is between 1.38×10^{-5} m/sec and 6.28×10^{-6} m/sec. Although they are sandy materials, the permeability has a relatively low value. It may be due to that some clay minerals are contained in the sands.

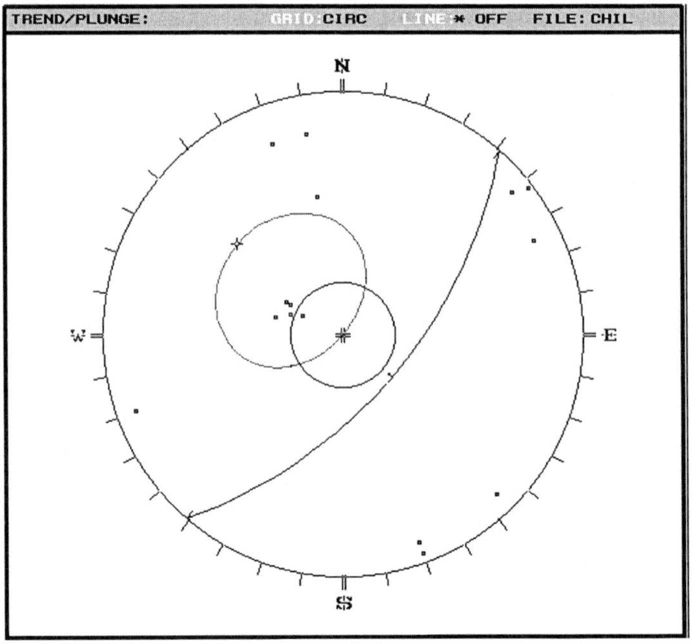

Figure 2. Graphical analysis of slope stability. Gently dipping bed surfaces are plotted in an unstable area of roadcut whose dip angle is 60 degrees. Pole points of high angles are faults and joints in the study area. The internal friction angle is 20 degrees.

Precipitation records from Korea Meteorological Bureau show that there was intense rainfall of more than 400 mm during August and 300 mm during September. Several debris flows were reported in the area at that time. The large amount of precipitation would infiltrate to the underground along tension cracks, which were formed by roadcut. The mudstones, coaly shales and clay material between bed surfaces might be expected to have acted as aquicludes or aquitards to infiltrating water during the heavy rainfall in August and September 1999. The water pressure in the perched table would decrease the effective stress, leading to lower shear resistance. Water may also have reduced shear strength fundamentally (Peres and Sousa, 1981). As a result, it is envisaged that the rock masses slide along mudstone beds with very slow velocity. As the rock masses slide, tension cracks would have propagated and the apertures opened considerably. This would have accelerated larger displacement of the rock slope during subsequent rainfall event. Sliding surfaces were actually observed on shale beds by core samples.

DISCUSSION

In order to know the shear strength of the rock masses, direct shear tests for bedding planes were performed. All of the three samples are moderately to highly weathered sandstones with completely weathered muds or clays intercalated between the sandstone beds. As a result, all of the samples showed relatively low shear strengths and internal friction angles (Table 1). The result is thought to be due to weathering of the rocks. If the rocks were fully saturated, the shear strength would have decreased much more. This softening may act as one of triggering factors of landsliding.

Table 1. Results of direct shear tests for bedding planes

Sample No.	Area (cm^2)	Cohesion (MPa)	Internal Friction Angle (degree)	Kn (GPa/m)	Ks (GPa/m)
CH-1	53.82	0.07	35.6	2.30	4.63
CH-2	48.92	0.04	28.1	1.80	2.32
CH-3	64.13	0.05	21.7	0.78	0.50

It is thought that the landslide is a kind of multiple rotational slide of each block (Figure 3). It is impossible to observe the major sliding plane. But, the rock blocks were disturbed and rotated with different amount of displacement. During the sliding, the pre-existing tectonic faults and joints led the rock masses to be collapsed with large displacement. This big collapse at the middle part induced relatively large displacement and backward tilting of rock masses near top of the hill. But, the rock masses near the road cut are not disturbed severely. The small displacement near the road cut indicates that the slide would initiate from the road cut and proceed towards upslope of the hill. If the slope failure had started from top of the hill, toe of the landslide must have large displacement.

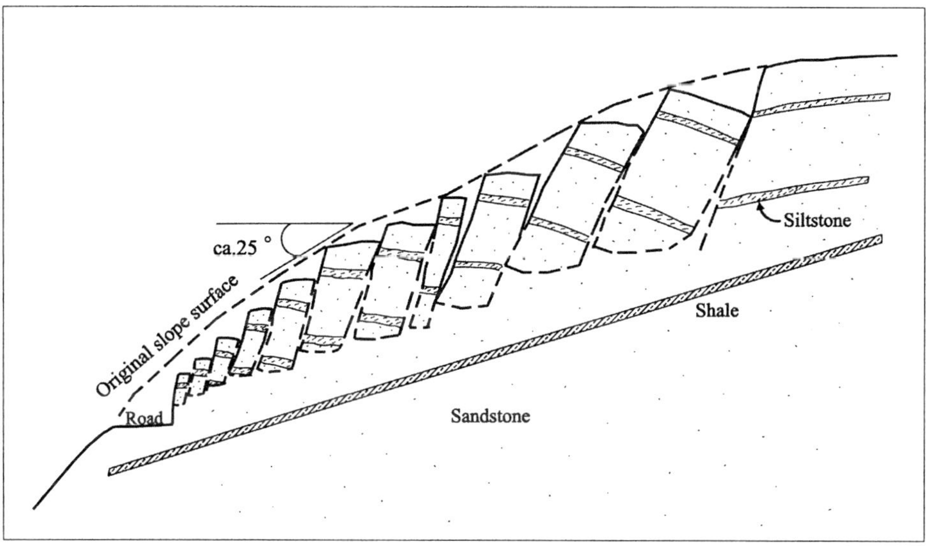

Figure 3. Conceptual schematic cross section of the landslide. Top and the middle part of the hill show large displacement and block tilting, whilst bottom of the slide has small disturbance.

CONCLUSION

The landslide occurred on gently dipping bedding surfaces and it extended with dimensions up to 100 m long and 40 m wide. The failed material was composed of moderately to highly weathered sandstones, mudstones and coaly shales. Triggering factors of the landslide were roadcut, heavy rainfall and low shear strength of mudstone beds. The construction of the roadcut led to the formation of some tension cracks. The heavy rainfall infiltrated to underground along the tension cracks and perched above of low permeable rocks such as mudstones and coaly shales leading to decrease in shear strength. Initiated rock sliding along the gently dipping mudstones and shales would have been at a very low velocity. The sliding initiated above the road cut, and then, propagated towards the top of mountain. This is supported by the small displacement near the road cut, no extension of landslide to the downslope of the road cut.

ACKNOWLEDGEMENT

This research was performed for the Natural Hazards Prevention Research Project, one of the Critical Technology-21 Programs, funded by the Ministry of Science and Technology of Korea.

REFERENCE

Kim, W.Y., and Chae, B.G., 1998, Landslide Characteristics in Korea, Proceedings of Korea-France Joint Seminar, Seoul, Korea.

Kim, G.Y., Roh, B.D., and Doe, K.Y., 1999, Report on Research Method of Slope Stability Assessment, Kyungpook National Univ. p.123.

Peres-Rodrigues, F and Reis E Sousa, M., 1981, A Contribution to the Shear Test Techniques, Proceedings of the International Symposium on Weak Rock, Tokyo, A. A. Balkema, pp.27-32.

The Stabilisation of the A4061 at Foel Goch Landslide

S. CHAMBERS, Halcrow Group Ltd, Cardiff, UK

ABSTRACT

As a result of carriageway disruption to the A4061 at Foel Goch landslide, a series of studies and investigations was carried out to determine the cause of the movements and to enable the design of remedial works. Landslipping within an ancient complex rock failure and mudslide was confirmed. Stabilisation measures were installed comprising ground anchorages and soil nails, and monitoring has confirmed their effectiveness in arresting movements of the highway. The ground and groundwater conditions, the design principles and the construction of the scheme are described.

INTRODUCTION

Foel Goch landslide is situated some 1.6 km north west of the village of Treherbert near the head of the Rhondda Fawr Valley (Figure 1). Since opening of the A4061 in 1929, ground movements had caused damage to the carriageway where it crosses the head of the landslide. The highway is of strategic importance to the area, providing the only link from the Rhondda to the A465 trunk road, a key route from South Wales to the Midlands.

The location and extent of the landslide at Foel Goch is shown on Figure 2. The section of the A4061 affected by landslipping was primarily on embankment up to 3m high constructed on sidelong ground and some 170m in length. Uphill of the road, the ground slopes up at about 1 in 2 or steeper. Downhill, the valley side slopes down at an overall gradient of about 1 in 4. The hillside comprises rough grazing land, the nearest inhabited properties being some 400m to the west in the valley floor.

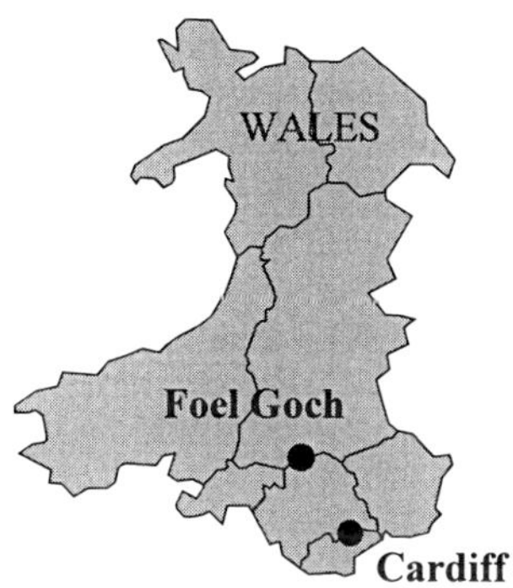

Figure 1. Location Plan

GEOLOGY AND GEOMORPHOLOGY

The solid strata at the site consist of the Llynfi Beds and the overlying Rhondda Beds, both sequences belonging to the Upper Coal Measures. A generalised geological section is shown as Figure 3. The boundary between the generally arenaceous Rhondda Beds and the essentially argillaceous Llynfi Beds is the No. 2 Rhondda Seam.

The natural superficial deposits at the site comprise scree materials overlying glacial till. Post-glacial, and recent landsliding at the site has displaced some of this material with varying degrees of disturbance. The scree deposits exposed immediately adjacent to the

Landslides in research, theory and practice, Thomas Telford, London, 2000.

A4061 comprise slightly silty sand with angular gravel and cobbles of sandstone, consistent with the parent Rhondda Beds. Made ground is represented by localised deposits of colliery spoil and quarry waste uphill of the road, and by locally derived embankment fill beneath and directly downslope of the road.

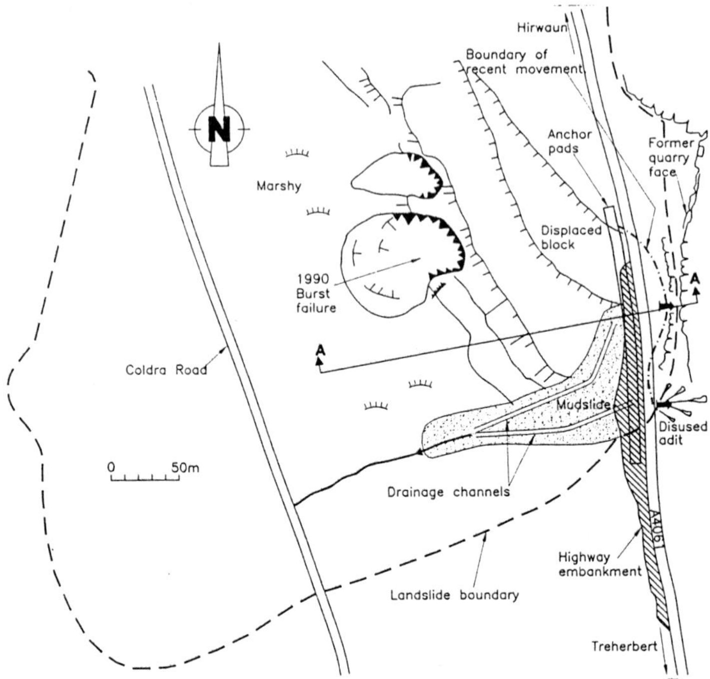

Figure 2. Site Plan

The landslide at Foel Goch extends from a backscar just above the A4061 down to the valley floor below Coldra Road (Figure 2), and was referenced by the Institute of Geological Sciences Landslip Survey (IGS, 1980) as Landslip RR1. It comprises two distinct components, namely an ancient rotational failure in glacial till downslope of the A4061, over-ridden by a later deep-seated rock failure, over which the road passes. The rock failure is also rotational, but involved the detachment of blocks of sandstone from a rear scarp upslope of the present A4061. The scarp has subsequently been exploited and enlarged by quarrying, Figure 2. A shallow mudslide is present at the southern margin of the displaced block, post-dating the original failure, resulting from large quantities of groundwater which would have been diverted to that area by the rock movement.

RECENT INSTABILITY

The earliest reports of ground movements affecting the carriageway at Foel Goch were in 1954, when cracks up to two inches wide and one hundred yards long were reported. The south-eastern cracking corresponded to the rear of the mudslide and the north-western to the inferred upslope edge of the displaced block. Repairs to the road took the form of limited patching, which was repeated approximately twice a year thereafter.

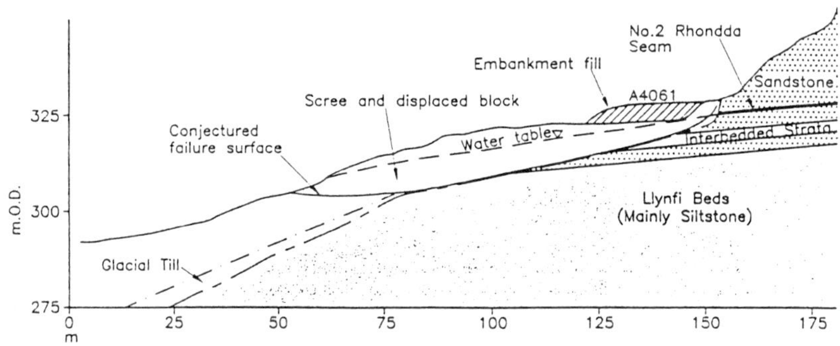

Figure 3. Geological Section

This repair strategy was continued until 1982, when larger movements necessitated further action. Small-scale drainage improvements adjacent to the carriageway were carried out by Mid Glamorgan County Council (MGCC) in 1983. However, inclinometer and surface movement monitoring showed that significant landslide movements were continuing to occur throughout the winter of 1983/4.

Halcrow was appointed by MGCC in 1984 and remedial works were commenced in 1985 with improvements to the drainage from disused trial levels in the No.2 Rhondda Seam adjacent to the A4061 (Figure 2). Survey, monitoring revealed a reduction in rates of movement, however cracking and deformation of the carriageway continued.

HAZARD STUDY AND OUTLINE DESIGN

In 1987, Halcrow carried out a geomorphological hazard study of some 9.5km of the A4061 between Treherbert and Tower Colliery. Investigations took the form of aerial photograph interpretation, assessment of archival records of past landslide events and geomorphological mapping in a corridor enclosing the route. The study confirmed Foel Goch as the most significant active feature on that section.

Following the adit drainage works, further reductions in groundwater levels were impracticable. Therefore, earthworks and structural solutions were considered to stabilise the road including excavation and replacement of landslipped material, stabilisation by ground anchorages and soil nailing, or a structure spanning the landslide. Key factors involved in the final choice of option included cost, period of road closure required for construction, ease of construction and future maintenance requirements. The preferred option was stabilisation by ground anchorages and soil nailing. In February 1990, the A4061 was temporarily closed due to severe cracking, and a large burst failure took place immediately downslope. The westbound lane remained closed and traffic light controls were put in place.

STABILISATION DESIGN

A detailed ground investigation was carried out in summer 1990 comprising 12 boreholes sunk by cable percussion and rotary techniques with associated sampling and in situ and laboratory testing. Casagrande type piezometers were installed in the boreholes, and the instruments were monitored during and following the site operations. Slope stability analyses were undertaken using the laboratory test results from the ground investigations. Incorporating back-analyses and sensitivity studies, these confirmed that the ground movements which were affecting the A4061 resulted from the ancient mudslide, and were taking place along a pre-existing non-circular slip surface in highly weathered argillaceous strata at rockhead.

From the stability analyses, anchorage designs were carried out in accordance with BS8081 (BSI, 1989). For overall stability, a design minimum factor of safety of 1.2 was to be achieved using anchorages fixed into the interbedded Llynfi sandstones and siltstones. The theoretical anchorage requirements were developed into a design solution incorporating 149 anchorages each of 1100kN working load. These transferred the load into anchor pads generally measuring 3.6m wide by 4.6m high and inclined at 60 degrees to the horizontal (Figure 4). Anchorage spacings were therefore 3.7m horizontally and 2.7m vertically. In order to optimise the anchorage lengths and to minimise any problems from drillhole convergence, anchorage inclination was varied, the upper rows being at 25 degrees below horizontal, and the lower at 35 degrees.

In order to prevent undermining of the anchor pads, a reinforced block of soil was designed using soil nails beneath the lowest anchorages. A grid of 390 soil nails at approximately 1.5m spacings on a triangular grid was the final design solution.

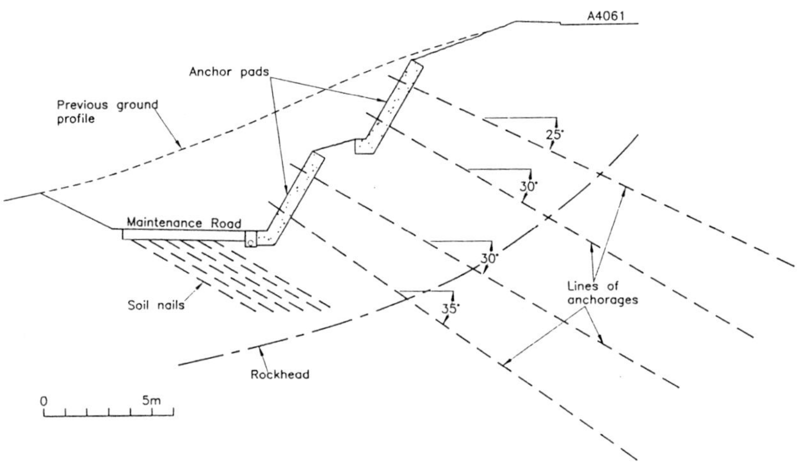

Figure 4. Typical Section through Stabilisation Measures

CONSTRUCTION

The works commenced on site in July 1991, and the initial phase of the geotechnical works comprised the installation, testing and monitoring of three test anchorages. To obtain a permanent record of the strata into which these were installed, continuous coring was employed over the intended free and fixed anchorage lengths.

Two of the test anchorages withstood the full testing cycles up to the maximum load of 3300kN with minimal non-recoverable deformation. The remaining test anchorage failed at a load of 2100kN. Back analysis of the test results indicated ultimate bond stresses in excess of 868kN/m^2 in two anchorages, and 552kN/m^2 in the remaining one. The difference in developed stress was interpreted as being due to poorer quality siltstones beneath part of the site than had previously been indicated by the investigations. On this basis, a conservative ultimate bond stress of 550kN/m^2 was adopted, using holes of 220mm diameter. The required fixed anchor length using this combination was some 8.7m.

The 149 'production' anchorages comprised 8 tendons of 15.2mm drawn strand to BS5896, with specified characteristic strength of 300kN per strand, and ranged in length from 20m to 44m. Double corrosion protection was provided by two layers of HDPE encapsulation and grout, the details being a proprietary system. A removable, galvanised steel cap, filled with grease covered the anchorage heads.

Following grouting and curing, each anchorage was subjected to an acceptance test. Load was applied in cycles in accordance with BS8081 (BSI, 1989) up to a maximum of 1.5 times the working load. Deflections of the anchorage head and anchor pad were measured and the elastic responses of each anchorage calculated. A typical test plot is included as Figure 5. All the anchorages passed the acceptance tests with no significant permanent deformations. Following the acceptance test, each anchorage was locked-off at 110% of working load to allow for some loss of prestress and creep. Further lift-off tests were carried out at 7 and 28 days and on substantial completion of the contract. Losses of prestress were generally well within the specified limits, only a small number requiring re-stressing

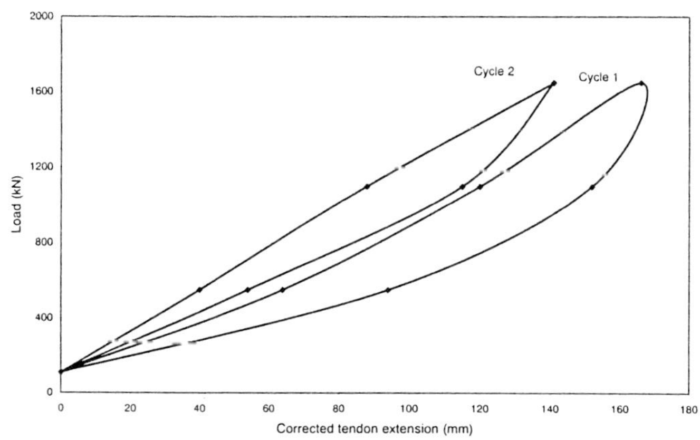

Figure 5. Typical anchorage acceptance test.

Similar testing and installation of the soil nails followed the anchorage works, and on completion of these and landscaping works, the A4061 was fully re-opened in spring 1992 (Plate 1)

CONSTRUCTION AND LONG TERM MONITORING

As the works proceeded, drilling rates and grout-take were recorded, to enable any changes in ground conditions to be detected to allow assessment of the design against the actual conditions encountered. No changes were actually made, the design being robust enough to cope with the ground conditions across the whole site.

Following completion of the stabilisation works, a series of movement markers was installed in the anchor pads and in the kerbs of the A4061, and in the landslide downslope. These were monitored up until August 1993, and showed no significant movements affecting the road or the embankment. Further downslope, landslide movements continued at some 6-12mm/year.

Plate 1. View of completed Works

CONCLUSIONS

The installation of a system of ground anchorages and soil nails has provided an efficient means of stabilising some 170m length of the A4061 where it crosses Foel Goch landslide. Detailed mapping and sub-surface investigation of the landslide confirmed the failure mechanism and provided the geotechnical design parameters. Construction was carried out without closing the road, and landscaping has enhanced the final appearance of the site.

Monitoring of anchorage loads and movement pins has confirmed the effectiveness of the stabilisation measures, and to the present day the road remains open and free of deformation at the site.

ACKNOWLEDGEMENTS

The Client for this project was Mid Glamorgan County Council, and the successor Authority is Rhondda Cynon Taff County Borough Council. The stabilisation measures were designed by Halcrow Group Ltd, together with PSC Freyssinet who carried out detailed design and installation of the anchorages and soil nails. The main contractor was Taylor Woodrow Civil Engineering Ltd. The author expresses his thanks to colleagues in those organisations.

REFERENCES

Institute of Geological Sciences, 1980, Report EG80/4 South Wales Coalfield Landslip Survey

British Standards Institution, 1980, BS5896 Specification for high tensile steel wire and strand for the pre-stressing of concrete

British Standards Institution, 1989, BS8081 Code of Practice for Ground Anchorages

Remediation of the Watawala Landslide, Sri Lanka

K. R. CHANDLER and M. BROISE, Douglas Partners, Sydney & Melbourne, Australia

INTRODUCTION

This paper is a case history describing the design and construction carried out to stabilise a landslide at Watawala in the southern highlands of Sri Lanka. From 1980 the slide disrupted an important rail link from Colombo to Bodalla situated in the south east of the country. Stabilisation was achieved using directional drilling techniques to install a drainage system beneath the slide to permanently lower the water table. This was supplemented by the use of deep wells and an eductor system for rapid draw down and dewatering. The paper includes results of monitoring of the slide following construction. As far as is known this is the first application of directional drilling techniques to landslide stabilisation on a relatively large scale.

THE WATAWALA LANDSLIDE SITE

Site Location and Description

Watawala is located approximately 160 km by rail and 126 km by road from Colombo in the central southern highlands of Sri Lanka. Average annual rainfall is about 5500 mm most of which falls during the south west monsoon from may to september. The slide was situated in a colluvium filled valley and extended over a length of some 500 m with a width up to 60 m The Colombo - Bodalla rail line crossed the slide about halfway up.

The slide area was formerly a tea plantation which was abandoned. The first investigations of slide movements took place in 1957. Movement of some 89 m occurred between 1976 and 1992 when an event took place that precipitated a decision by Sri Lanka Rail to to stabilise the slide. In the early hours of 2 June, after 200 mm of rain and submergence of the track, a 97 tonne locomotive sank into the softened embankment. Further movements of about 8 m occurred in the same month.

When visited by the first author in November 1993, the slope was in a highly degraded state with pools of surface water and mud, extensive areas of bare ground, a hummocky surface and transverse cracks up to 100 mm wide. Sliding of saturated earth slopes along the flanks of the main slide was common. The rail line was supported on an embankment continuously extended by maintenance crews to keep pace with loss of elevation and displacement resulting from slide movements.

Geology and Hydrogeology

The slide is located in a prominent 'V' shaped valley in an area of high grade metamorphic faulting and fracturing. The metamorphic rocks include quartzites, marbles and Charnockitic (a coarse grained rock, mainly quartz, feldspar and hypersthene, named after Job Charnock, founder of Calcutta) gneisses. The site overlies a fault zone which is deeply weathered, eroded and filled with colluvium comprising mostly sandy silts but also zones of clayey silt up to 6 m thick. Depths of colluvium increased gradually from the top end, to a maximum of

26 m a short distance above the toe. Cavernous limestone was reported in two test bores at depths of about 66 m.

Water levels recorded (ref 1) over many months indicated artesian conditions at the top of the slide and groundwater levels varying up to 14 m from wet to "dry" season. Groundwater level was therefore very sensitive to rainfall.

Pump tests in two bores, 60 m apart and drilled through the colluvium and into the gneiss, showed good hydraulic connectivity between the two. Falling head permeability tests in bores indicated hydraulic conductivities averaging 8x10-6 cm/sec for the colluvium but 50 times higher for the fractured gneiss.

TENDERS FOR STABILISATION

Tenders were called on the basis of a design concept included in reference 1. This concept involved the sinking of two 5 m diameter caissons to rock at depths up to 22 m, from which inclined drainage holes were to be drilled into the colluvium.

At the request of one of the tenderers, John Holland International (JHI), Douglas Partners (DP) evolved an alternative solution based on placing drains beneath the colluvium to lower the water table. The concept was based on:

- a correlation (ref 1), between prolonged rainfall (and therefore groundwater level) and slope movement.
- the observed natural lowering of groundwater tables during the dryer months of the year,
- the rapid response demonstrated in the pump tests of water table drawdown of some metres at a point 60 m from the pumped well,
- the significantly higher permeabilty of the gneiss compared with the colluvium.
- use of directional drilling to install drains up to 480 m long in the stable, upper zones of the gneiss underlying the colluvium

Simplified, three dimensional analysis of the slope by Hutchinson (quoted ref 1) had indicated an angle of residual shearing resistance (ϕ'_R) of 14°. This value was used in a comparative study of the effects of water table lowering on stability, using conventional two dimensional analysis. That study clearly demonstrated that for an idealised slope, and assuming the water table was at the surface during failure, a factor of safety of 1.3 could be achieved with a drawdown of 5 m. Since the dry season water levels were generally lower than this, the drainage concept was aimed at minimising the rise in water table during the wet season, to give an acceptable factor of safety.

The JHI alternative tender was accepted and construction began in November 1994, to be completed the following April.

ELEMENTS OF THE DRAINAGE SOLUTION

Deep Wells

The first operation undertaken on site was the installation of six wells, 200 mm diameter, drilled to about 60 m depth into stable ground along the eastern side of the colluvium. 150 mm diameter slotted steel casing was installed to support potentially unstable ground. An electrical submersible pump was permanently installed in each well and pumping began immediately the well was completed. As the slide surface was too soft to install appropriate piezometers at the time pumping began, it was not possible to measure the drawdown within the colluvium. Observations between the wells themselves clearly indicated a significant response from one to the other. Although the deep wells were designed primarily to lower the water table and to depressurise the artesian aquifers, the pumps have been left in the wells in case of temporary problems with the underdrainage system.

Eductor System

An eductor system was installed consisting of four independent groups of eductors, each group comprising about 20 individual eductor wells connected by a single delivery and return line. Two pumps were used, each pump serving two eductor groups. The eductors were installed near the bottom of wells which were drilled to the level of the underdrains. The purpose of the eductor wells was to rapidly dewater the colluvium, and also to hydraulically connect the colluvium above and below the clay layers, to the gneiss. Discharges from the wells rapidly declined after pumping began, and intermittent pumping was then adopted until after about six weeks, pumping from three of the four groups was terminated.

Underdrainage

Directional drilling in civil engineering is an adaptation of oil well technology. It has been used routinely for installation of infrastructure such as pipelines, cables etc, beneath rivers, or to void geological hazards such as escarpments. More recently, it has been applied to major geotechnical site investigation for long tunnels, particularly in mountainous terrain.

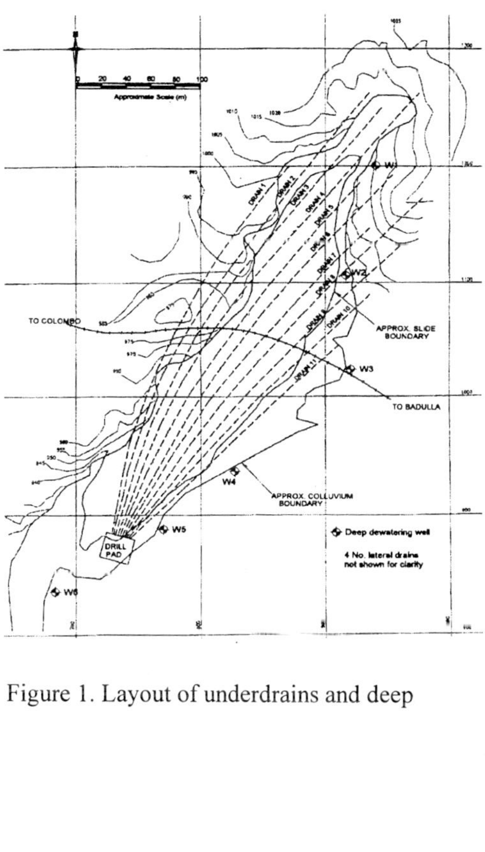

Figure 1. Layout of underdrains and deep

Eleven holes were drilled from a drilling pad located close to the toe of the slide as indicated in Figure 1. The holes were profiled vertically to lie within the upper weathered zone of the gneiss, a few metres beneath the base of the colluvium. The levels of the interface surface between the gneiss and the colluvium was estimated by computer interpolation from bore data. Each hole, therefore, was a unique, complex, three dimensional shape, (see Figure 2). Limited options for siting the drilling pad meant that the first 60-80 m of each drain had to pass through the colluvium. A high density polyethylene (HDPE) pipe, was chosen for its ability to stretch (800% at failure load), and therefore withstand any movements which might occur during and after construction. A further four short, straight drains were installed from the drill pad, to drain the area just above the toe of the slide The method of installation was to drill an open 215 mm diameter hole using 130 mm diameter drill pipe and to daylight the hole at the upper end.

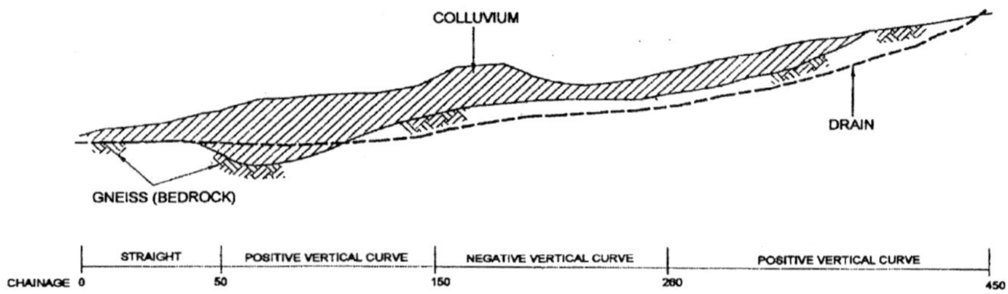

Figure 2. Elevation showing typical underdrain profile

The drain pipe was pulled through the drill pipe which was then withdrawn over the top of the drain. This restricted the maximum size of the drain pipe to 63 mm OD. The drain pipe was supplied in 250 m rolls and the joints butt welded. Water entry points into the drain pipe were provided through filtered openings spaced 1 m apart. The drains were completed in February 1995.

Surface Drainage

Improved surface drainage was vital to control surface water flow velocities and to ensure that water was removed from the slide as quickly as possible. Components of the surface drainage system included: installation of stone pitched drains up slope of the colluvium to intercept water in the upper catchment and divert it into the adjoining valleys; construction of 'contour' drains or ditches across the slide at a grade of 2% to discharge into a new collector drain on the eastern side of the slide (above the rail track) and to a natural water course (below the rail track); use of trench drains up to 3 m deep along the western side of the slide to dewater unstable slopes, and also at the toe to lower the water table.

Surface Regrading

An early priority was to regrade the slide surface as part of run off control, but also to create berms along the lines of the eductors for access of the eductor drilling and installation equipment. At the same time as this regrading was carried out, the contour drains were cut and erosion gullies were filled under controlled compaction. Ridges impeding lateral surface flows were removed.

Revegetation

Revegetation of the regraded surface was essential to prevent erosion. Advice was obtained from the University of Kandy and a selection of indigenous grasses, shrubs and trees were planted. Growth was rapid.

Scarp Stabilisation

An old near vertical slip scarp which extended to within one metre of the rail track was stabilised by 'soil' nailing, and by shotcreting the face. This was a precautionary measure as the scarp face showed little sign of instability or deterioration since the scarp was formed in1986.

Realignment of the Rail Track

It was considered that the existing embankment supporting the rail track required reconstruction but to do this whilst maintaining rail traffic was impracticable. A new alignment was, therefore, chosen just up slope from the existing track and a new embankment constructed. Sri Lanka Rail chose to defer the track laying for some 12 months after stabilisation was completed in spite of the somewhat precarious state of the existing embankment. They reported, however, that no movements in the embankment were detected in this period.

MONITORING

Monitoring Programme

A monitoring programme was implemented to measure post construction performance of the stabilisation works. The main elements comprised:

- Groundwater level monitoring at three different levels in the colluvium and in the underlying undisturbed gneiss, at 16 stations on the slide.
- Measurement of discharges from the 15 inclined underdrains, flows in the eastern surface collector and northern diversion catchment drains using automatic solar powered gauging equipment, which also recorded site rainfall.
- Installation of approximately 60 surface movement monitoring survey points to measure vertical and horizontal displacements along the length and breadth of the slide, including monitoring of the rail track.
- Monitoring of discharges from the eductors and deep wells.
- A physical condition and surveillance survey of the stabilisation works.

Unfortunately, persistent problems were encountered with the lower gauging station, hence data pertaining to the surface water discharge were incomplete. The monitoring data were initially collected by the contractor, then the responsibility was assumed by the principal. A complete monitoring dataset is available for the first nine months of 1995. It is regrettable that no longer term measurements are available.

Underdrain Discharges

The relationship between site rainfall and the total discharges from the 15 underdrains is shown in Figure 3. The annual rainfall for 1995, which includes the post construction period indicated on Figure 3, was equal to the previous maximum encountered during 1960 - 1992. Key aspects to note from this Figure are:

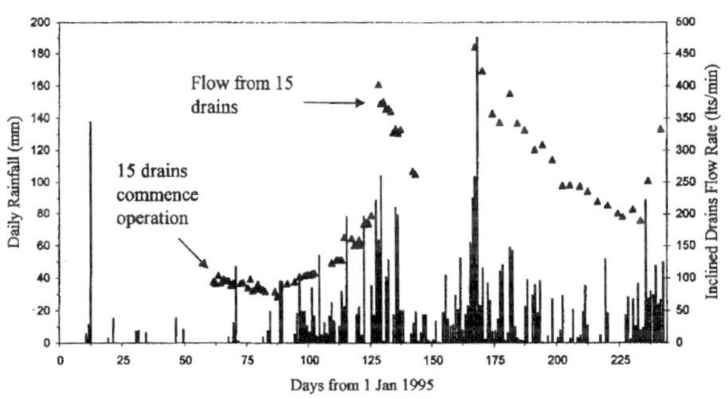

Figure 3. Watawala landslide rainfall & underdrains flow rate

- The base flow rate from the underdrains during relatively low rainfall in March 1995 is slightly less than 100 litres/min.
- The peak flow rates coincide with periods of heavy rain over several days. Significantly, isolated storms have a noticeable but small effect on the overall flow rates, unless followed by successive days of heavy rainfall.
- response of the underdrains to heavy rainfall is rapid, within about one day. This suggests that the colluvium is more permeable than indicated from permeability tests. This observation is encouraging in that there appears to be good hydraulic connection between the colluvium and the underdrains.

By adopting a coarse approximation of the landslide hydrology, an analysis of the water balance indicates a surface infiltration of 63%. This is considerably greater than would normally be assumed, implying recharge from the gneiss, flow into the slide from surrounding strata and from artesian groundwater above track level.

Surface Movements

With regard to the movement monument points on the slide, the cumulative displacements since commencement of readings indicate horizontal downslope movements typically in the range 10-50 mm, and vertical movements typically in the range 15-55 mm. The vertical settlement experienced is directly related to the stabilisation works in that consolidation will naturally occur as a consequence of lowering the groundwater table.

Piezometric Levels and Slope Stability

The piezometric levels measured in the colluvium and the underlying gneiss pre and post installation of the underdrains are shown in Figure 4. At around the peak rainfall events (see Figure 3), the groundwater levels in the colluvium in the middle and toe regions of the slide were maintained at 15 m or greater, and 3 m to 5 m below ground level, respectively. At the head of the slide, in the artesian area, drawdowns of one metre or so were achieved.

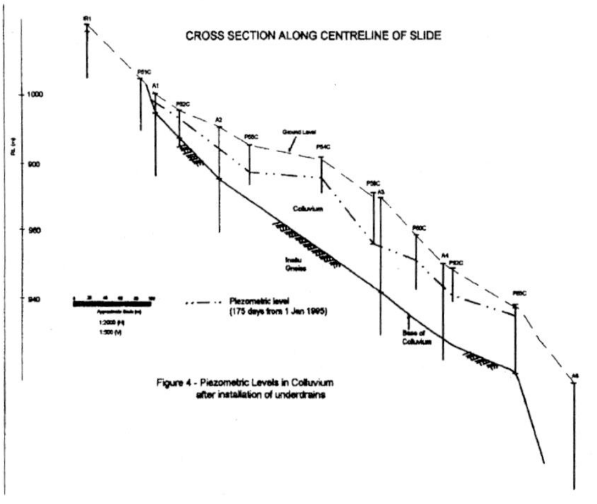

Slope analyses were performed using a 2D model with the original soil parameters, and the 3D CLARA program with colluvium properties of $\phi' = 18^{\circ}$ and $c' = 5$kPa and post stabilisation groundwater levels. The reasons for using both the latter values are not examined here, but were derived from further studies. Both approaches gave factors of safety greater than 1.5. The maximum theoretical factor of safety assuming fully efficient underdrains (i.e., drawdown to underdrain level) is 1.8.

Figure 4. Piezometer levels in colluvium after installation of underdrains.

CONCLUSIONS

As with the great proportion of landslides, water played a dominant role in destabilising the Watawala slide. The work has succeeded in stabilising the slide, and has demonstrated the feasibility of installing long underdrains using directional drilling technology to provide a "drainage blanket" beneath the slide. At Watawala, the high permeability of the upper zones of jointed gneiss enabled the drains to be installed, for the most part, in stable rock. This may not be the case at other sites. Nevertheless where hydrological conditions are suitable, directional drilling offers a powerful technique to install drainage, with few restrictions only, on length, diameter, and drain profile.

REFERENCE

1. "History of Earthslides and Investigations" Report issued by Sri Lanka Rail as part of tender documents.

Landslides in the North-Western Black Sea Region

Y. CHERKEZ, T. KOZLOVA, V. SHMOURATKO, I. I. Mechnikov Odessa State University, Odessa, Ukraine, V. KHARITONOV and A. KARAVAN, Black Sea State Regional Geological Enterprise, Odessa, Ukraine

LANDSLIDES IN THE NORTH-WESTERN BLACK SEA REGION

Research of regional landslide processes at the northern Black Sea coast (territory between the Dniester and the Yuzhny Boug) has been considerably long. Landslides at the Odessa coastline have been researched for over 150 years; there are a number of publication on this topic. A landslide research station was established in Odessa in 1932. Development of a 12-km long anti-landslide construction complex was finished at the Odessa coastline in 1968. The complex provides reliable protection of the city coastline territory against deep squeeze-out landslides.

Landslide processes developed beyond the Black Sea coastline are researched insufficiently. As a rule, there are no key stationary observation points, thus the characteristics of landslide mechanism and dynamics are not clear. To analyse conditions and factors of landslide forming beyond the sea coastline, the Black Sea State Regional Geological Enterprise has started a programme of landslide research. The purpose of the programme has been to form a database for the development of regional models of anti-landslide protection of Odessa (33.3 thousand km^2) and Nikolaev (24.6 thousand km^2) regions (45 administrative districts). There have been used two groups of methods: annual aero-visual observations to reveal newly formed landslides, and annual land field observations to create a landslide catalogue.

The main objective of aero-visual observations was to reveal the number of newly formed and activated landslides. The work was mapped on topographic map of the 1:50000 scale. Single landslides were indicated according to the map scale or with as non-scale signs. Where possible, the interrelationships of landslides and other exogenic geological processes, and the elements of relief, outputs of underground water; draft drawings were made. At this stage of research, parameters on more than 6,000 landslides in the database were generalised according to administrative districts (Fig. 1).

Based on the results of land field observations, a landslide catalogue was designed comprising such parameters and characteristics of a landslide slope as lithological-generic characteristics of rocks, gypsometric elevation of a landslide on a slope, generic landslide type, landslide form in a plan, landslide measurements (length, width), size, height of displacement surface, and geological age of the main deformed horizon (MDH).

Landslide slopes in the region are formed by Sarmatian, Meotian, Pontian, Middle and Upper Pliocene, and Pleistocene deposits. Clay deposits prevail in the geological section, while sand and cemented carbonaceous deposits are less developed. General layer bedding takes the monoclinal weakly inclined fall with azimuths to the south and south-east. The erosion-hydrographic network creates horizontal and vertical relief dissection; it decreases monoclinally from the north-west to the south-ast, according to the general inclination of surface. The annual atmospheric precipitation (520-440 mm) decreases in the same direction. Clay deposits of the Neogene and Quaternary ages act as the MDH. Generically, these are continental, marine, alluvial, delta, eluvial, and loess deposits.

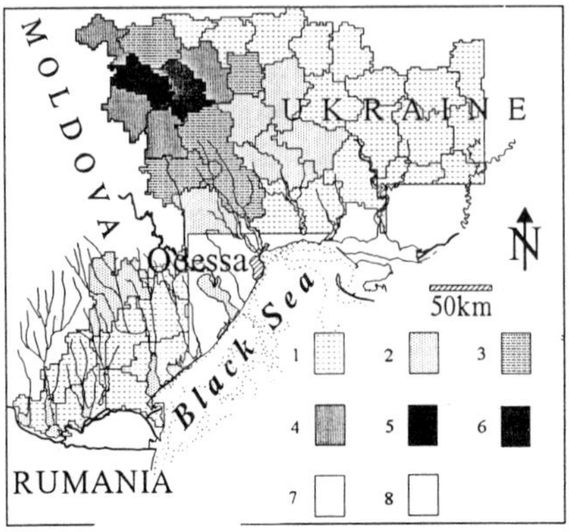

Figure 1. Number of landslides per 1 km^2 (%): 1 – 0-0.1; 2 – 0.1-0.2; 3 – 0.2-0.3; 4 – 0.3-0.4; 5 – 0.4-0.5; 6 – 0.5-0.6.

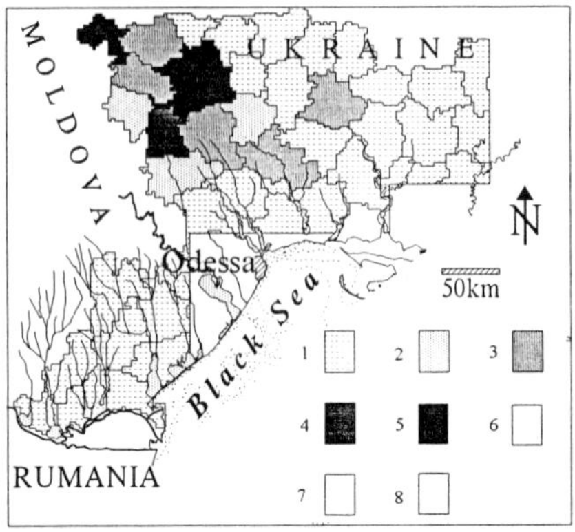

Figure 2. Number of landslides in the Pleistocene loesses: 1 – 1-10; 2 – 11-20; 3 – 21-30; 4 – 31-40; 5 – 40-60.

The number of landslides, depending on the MDH age, changes on a large scale (Table 1). Most landslides is related to main deformable horizons developed in clay deposits of the Neogene. Considerably fewer landslides are related to the loess rocks of the Quaternary age.

Table 1. Quantitative distribution of landslides depending on age of the main deformed horizon.

MDH	Percentage
DpQ_{IV}	0.19
th-Q_{IV}	0.11
Vd-Q_{III-IV}	0.74
d Q_{III-IV}	3.03
Vd-Q_{I-III}	7.16
N_2^3	1.07
N_2^{2-3}	0.08
N_2kj	0.50
N_2nv	10.42
$N_{1-2}bl$	32.72
N_1m	32.11
N_1S_3	11.24
N_1S_2	0.61

Depending on the MDH age, landslides are classified into two main types: 1) landslides in the loess rocks (A type), 2) landslides in clay deposits of the Neogene (B type).

General statistical characteristics of parameters and distribution of landslides of A and B types are somewhat different.

Most landslides of type A are related to the north-western part of the territory, to the districts with high gypsometric relief marks (180-270 m) and high factor of horizontal dissection (0.5-0.9 km/km^2).

The length of landslides (L) varies from 5 to 250 m (average 45 m); width — 3 – 285 m (average 48 m); steepness — 5 – 45 m (average 14.6 m); height — 3 – 100 m (average 41.8 m); size — 18 – 46,000 m^2 (average 2,100 m^2). The factor I_l equal to the ratio of maximum length (L_{max}) of a landslide (measured in the displacement direction) to its maximum width (W_{max}) varies from 0.1 to 7 (average 1.21). The value of this factor characterises the form of landslides on the map, this form reflecting indirectly the level of the technogenic charge, besides the combination of various natural influences. According to the values of factor I_l, landslides were subdivided into two types. Type one (P1) comprised landslides with $I_l > 1$, type two (P2) comprised landslides with $I_l < 1$. Fig. 3 shows the map of distribution of parameter $P = N_{P1} / N_{P2}$, where N is the number of landslides of the mentioned types. It is evident that the region of distribution of landslides with P>1 forms a narrow band from the north-west to south-east.

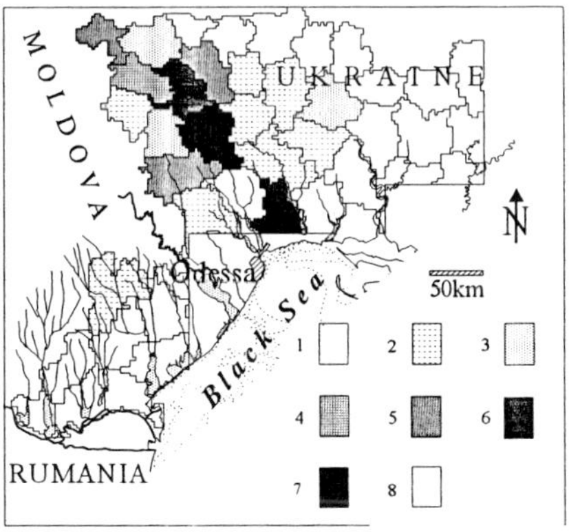

Figure 3. Map of parameter P for landslides in the Pleistocene loesses: 1 – no landslides, 2 – P=0...0.5; 3 – P=0.6...1.0; 4 – P=1.1...1.5; 5 – P=1.6...2.0; 6 – P=2.1...2.5; 7 – P>2.5.

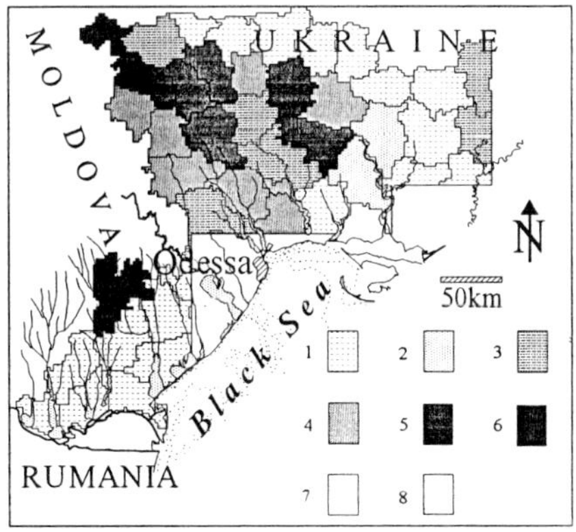

Figure 4. Number of landslides in the Neogene deposits: 1 – 1-25; 2 – 26-50; 3 – 51-100; 4 – 101-150; 5 – 151 200; 6 – >200.

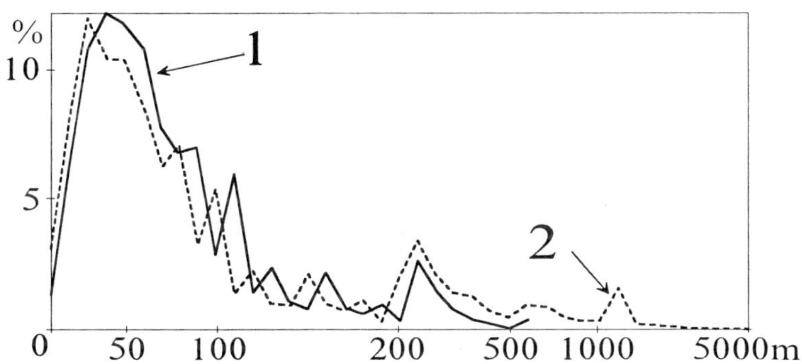

Figure 5. Variation curves of distribution of length (1) and width (2) of active landslides in the northern Black Sea coast.

The maximum quantities of landslides of B type are related to districts of the deepest erosion dissection (main valleys of the region) (Fig. 4). Landslides of B type are developed on slopes 3 to 120 m high (average 39.4 m). The steepness of landslide slopes changes from 4 to 60 degrees (average 16.8); length (L) — 3 – 550 m (average 74.3 m); width (W) — 1 – 4,600 m (average 141.26 m); size — 15 – 1,600,000 m^2 (average 14,433 m^2); factor $I_{л}$ — 0.01 – 18 (average 1.06). The map of the spatial distribution of parameter P for the landslides of B type is shown in Fig. 6. The comparative analysis of Figs. 3 and 6 shows that landslides with the MDH of different ages but the same form coincide spatially.

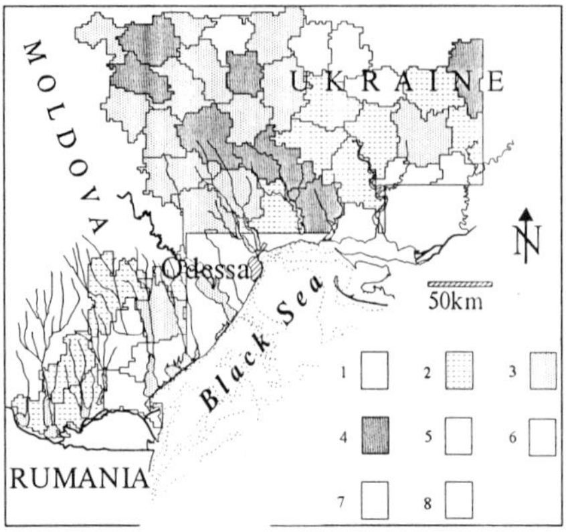

Figure 6. Map of parameter P for landslides in the Neogene deposits: 1 – no landslides; 2 – P=0...0.5; 3 – P=0.6...1.0; 4 – P=1.1...1.5

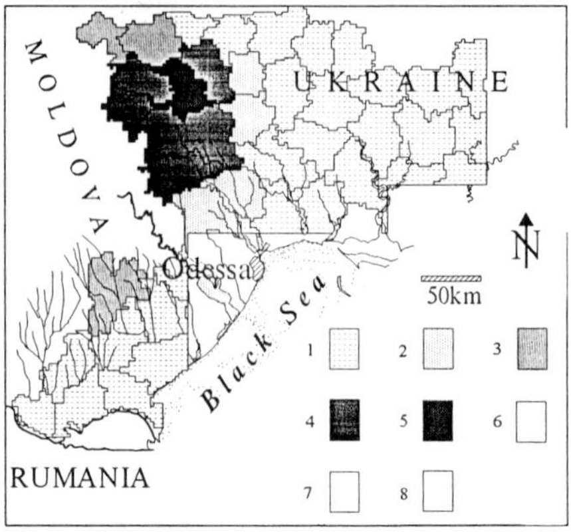

Figure 7. Average long-term velocity of increase of number of landslides (%): 1 – 0-5; 2 – 6-10; 3 – 11-15; 4 – 16-20; 5 – 21-25.

Concerning landslide sizes, the statistical analysis shows that geometrical parameters (length and width) of most landslides developed at the northern Black Sea coast do not exceed 100 m. Fig. 5 shows distribution curves of length (L) and width (W) (the length being measured in the displacement direction, and width — parallel to the slope foot) of active landslides of the northern Black Sea coast. Two modes can be reliably revealed — 20...30 m and 200...250 м. the above-mentioned geological parameters of landslides nearly coincide with characteristic measurements of tectonic micro-blocks on the territory of the northern Black Sea coast.

The dynamics of increase of the number of landslides is shown in Fig. 7. The most landslide activation in the latest decades is characteristic of administrative districts situated in the north-western part of the region, i. e., districts with largest vertical and horizontal dissection of the relief.

Presently, research is performed to reveal regularities of formation and development of landslides. A network of stationary regime observations in the most landslide-damaged districts is being planned and equipped. Projects of the territory anti-landslide protection are being developed.

Study of Planar Sliding Instability Phenomena in the Langhe area (Piedmont Region, Italy)

A. CHIAPPONE and C. SCAVIA, Department of Structural and Geotechnical Engineering, Politecnico di Torino, Torino, Italy

ABSTRACT

This paper illustrates the results of a research carried out on the planar sliding instability phenomena that occur periodically, following abundant rainfall, in the marly-arenaceous formations of the Langhe area (Piedmont Region, North-Western Italy). First of all, the importance of meteorological events in triggering instability phenomena is borne out through a specific analysis of rainfall data with regard to the event of 2-6 November 1994. Then, the results of back-analyses conducted on well-documented landslides occurred in november 1994 are given. Finally the results of the tests performed on the marly materials involved in the landslides are discussed, with special regard to the application of the methylene blue technique, which was used to identify the marly levels containing the greatest concentrations of active clay minerals, i.e., those exhibiting the most significant swelling behaviour in the presence of water.

INTRODUCTION

The hilly territory of the Langhe area (Piedmont Region, North-Western Italy, Figure 1) is characterised by the periodical occurrence of landslides. Planar sliding instability phenomena (Figure 2) represent the prevailing type of landslide in terms of frequency and spatial distribution.

Figure 1. Location of the Langhe area (dotted area) (Simeoni, 1998)

Figure 2. Typical planar sliding instability phenomenon, Murazzano (CN), November 1994 (photograph: courtesy of the Piedmont Region)

These phenomena take place periodically, following abundant rainfall (Govi et al., 1985), in geological formations of the Oligo-Miocene consisting of marly, silty-marly, and arenaceous-sandy lithotypes. Landslides are triggered in planes with a dip of between 8 and 15 degrees, virtually coinciding with the inclination of the slopes. The sliding planes need not always coincide with a layer joint; they are often situated inside a marly level (Forlati et al., 1995) and correspond to a sedimentation plane. Studies conducted on the marly materials involved in the planar sliding phenomena have revealed (Simeoni, 1998) the highly swelling behaviour of these materials. The investigation described below, which is still underway at the Department of Structural and Geotechnical Engineering of the Politecnico of Turin, focuses in particular on the landslides that took place in connection with the floods of November 1994.

THE EVENT OF 2-6 NOVEMBER 1994

The flood that hit the basin of the Tanaro River, and hence the Langhe area, on 2÷6 November 1994 was caused by abundant rainfall and was followed by numerous landslides. In particular, the Management of the Technical Prevention Services of the Piedmont Region identified about 800 instability phenomena due to planar sliding, most of them concentrated in an area of about 400 km^2. Different types of planar sliding instability phenomena have been identified (AA.VV., 1998), as schematised in Figure 3.

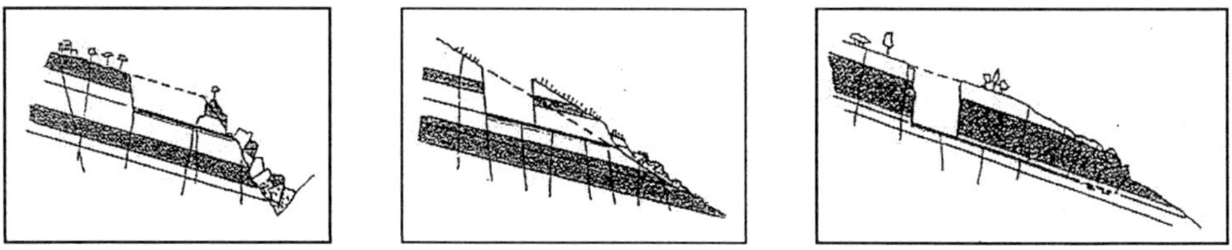

a), b): Sliding free at the toe c): Sliding confined at the toe

Figure 3. Types of planar sliding instability phenomena (AA.VV., 1998)

In order to confirm the "rainfall-landslide" correlation established by other authors (Govi et al., 1985) to account for other events recorded in the same area before 1994, the data given in the Meteorological Annals of the Piedmont Region for 1994 and the hourly rainfall data (relating to the event of 2-6 November) supplied by the Meteohydrographic Sector and Monitoring Networks of the Management of the Technical Prevention Services of the Piedmont Region were analysed.

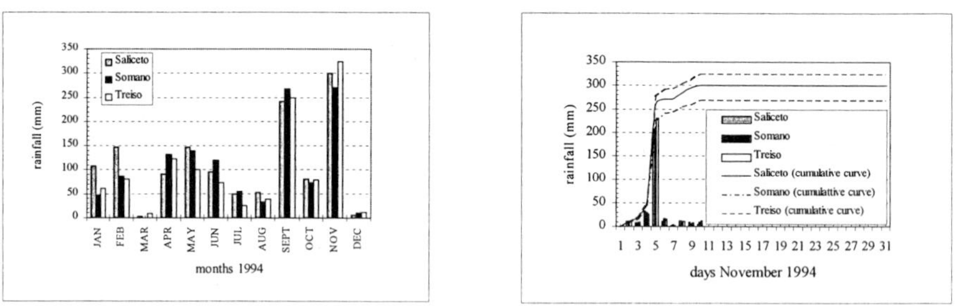

Figure 4. Analysis of rainfall data

The analysis was conducted on the data (Figure 4) collected at three pluviometric stations (Saliceto, Somano and Treiso), deemed significant on account on the exhaustiveness of the data and the location of such stations. All of them are included in an area where several planar sliding phenomena have occurred (56 slides recorded), for which detailed information was available concerning the time of triggering. The results of the analyses performed have been plotted in a diagram given by Govi et al. (1985) (Figure 5) in terms of cumulative rain level for the 60 days preceding the event, as well as cumulative rain level for the 60 days preceding the event plus the cumulative quantity of rain that fell in the course of the event (in particular, the minimum and maximum values worked out from the analysis are shown). The analysis shows that, during the event of 2÷6 November 1994, the threshold values indicated by Govi et al. (1985) as the minimum values beyond which instability phenomena are triggered were amply exceeded, thereby confirming the "rainfall-landslide" correlation already pointed out in investigations concerning earlier events (Figure 5).

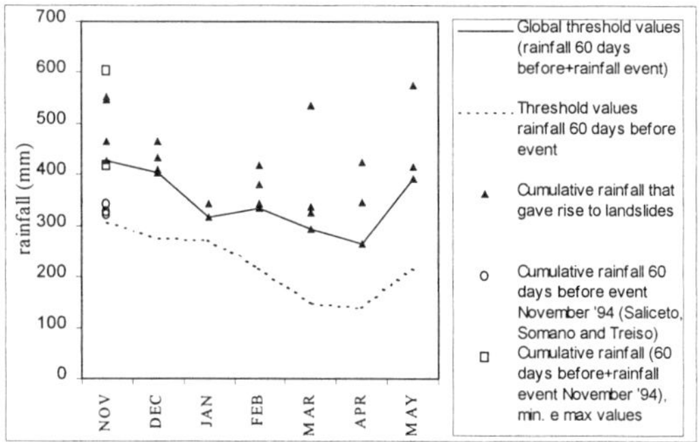

Figure 5. Results of the analysis performed on the rainfall data relating to the event of 2÷6 November 1994 plotted in the "rainfall-landslide" correlation diagram given by Govi et al. (1985, modified)

BACK ANALYSES

In order to quantify the instability phenomena being considered in terms of the shear strength parameters mobilised on the sliding surface at the time of collapse, back analyses were conducted on real, well-documented cases recorded in November 1994 (the data processed were supplied by the Management of the Technical Prevention Services of the Piedmont Region)

The analyses were carried out according to the Limit Equilibrium method on 26 cases of slides freely evolving at the toe, by taking into account the hydraulic pressure distribution scheme illustrated in Figure 6. Assuming cohesion along the sliding surfaces to be nil, the friction angle mobilised at the time of collapse on the same surfaces was determined. The results are given in Table 1. Account taken of the shear strength data concerning the marly materials of the Langhe area contained in the literature (Table 2), the results of the analyses show that only by considering the minimum values of the residual strength parameters and extremely severe hydraulic conditions (degree of saturation: S=75%-100%) it proves possible to account for nearly all the landslides recorded. Cases of confined landslides are currently being analysed through the stress methods, and, needless to say, the explanation of these phenomena on the basis of literature shear strength data is even more difficult. Thus it may be concluded that, in order to work out a satisfactory explanation of the phenomena being considered, it is necessary to investigate the following aspects: geological-structural characterization of the area to determine whether any post-sedimentary tectonic phenomena are involved, which might account for the possible mobilisation of residual values of the strength parameters; hydrogeological characterisation of the slopes to define the groundwater flow conditions and the evolution in time of hydraulic pressures; laboratory tests for the characterisation of the swelling marly levels already examined in earlier studies (Simeoni,

Table 1. Back analysis. Friction angle (φ) mobilised on the sliding surface (cohesion on the sliding surface=0; S=degree of saturation)

SLIDES CODE	φ(°) S=0	φ(°) S=25%	φ(°) S=50%	φ(°) S=75%	φ(°) S=100%
37	10	12	15	21	30
64	19	22	27	34	44
72	9	11	13	17	24
101	11	13	16	20	27
120	8	10	12	17	26
193B	7	9	14	23	37
194	11	13	15	19	24
235	11	13	16	21	30
237	10	12	14	18	25
245	5	6	8	10	15
246	11	13	16	21	29
249	9	11	14	19	27
273	11	13	16	21	29
274	10	11	13	16	22
277	9	11	13	16	21
282	10	12	15	20	28
300	5	6	8	13	20
305	11	13	17	23	34
329	8	9	11	15	20
339A	11	13	16	22	31
339B	10	12	15	20	29
341	11	12	15	18	25
92	10	12	15	21	30
219	8	10	16	27	43
244	11	13	16	20	27
343	10	12	15	20	28

Table 2. Strength parameters of the marly materials of the Langhe area (data from the literature: AA.VV., 1998)

Cohesion (c)		Friction angle (φ)	
peak: c_p (kPa)	residual: c_r (kPa)	peak: φ_p (°)	residual: φ_r (°)
18÷30	0÷15	29÷30	19÷24

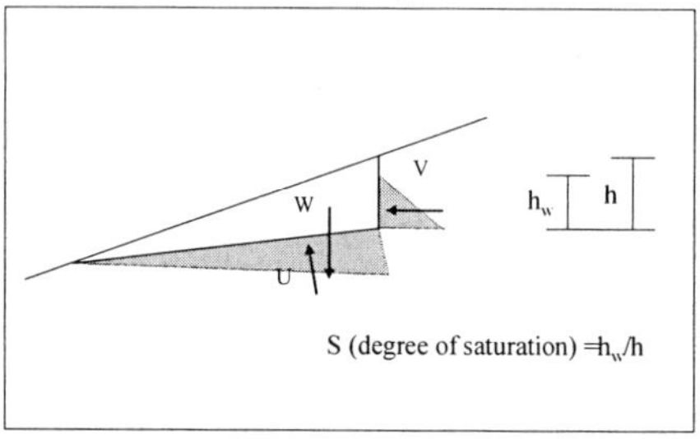

Figure 6. Hydraulic pressure distribution scheme

1998). Let us now briefly examine some of the most important results of the testing program carried out so far to assess the behaviour of the material.

EXPERIMENTAL INVESTIGATIONS

The behaviour of the marly materials involved in the planar slides of the Langhe area has been studied through tests for several years (Forlati et al., 1995; AA.VV., 1998; Simeoni, 1998). These investigations have revealed the highly swelling behaviour of these materials (average values of swelling pressure of between 1 and 3 MPa) (Simeoni, 1998). This behaviour is motivated by the presence of active clay minerals, and, in particular, of smectite, with percentages of up to 50%. It has also been observed that the highest concentrations of smectite are located within millimetric levels. In order to be able to identify such "critical levels" quite readily, the methylene blue technique was employed. This test (Lautrin, 1987; Tran, 1980; Tourenq and Tran, 1989) consists of quantifying the ionic absorption capacity of soil by measuring the quantity of methylene blue (an organic dye which is a cationic electrolyte) necessary to cover the total surface (external + internal) of the particles making up the soils. This is based on a chemical phenomenon of absorption caused by the negative charges of the particles which tend to balance out through the attraction of methylene blue cations following the separation of the latter from the anions (Cl^-) in water. Based on the total quantity of methylene blue used up in the test to balance the negative charges of a known quantity (Ps) of dry material, it proves possible to calculate the quantity corresponding to a 100 g sample, referred to a "blue index of soil" (V_B), which is expressed in grams of methylene blue per 100 g of dry soil:

$$V_B = V \bullet 0.01 \bullet 100/Ps$$

where: V (in cm^3) is the volume of the methylene blue solution employed
0.01 (in g/cm^3) is the concentration of the solution (i.e. 10 g/l)
Ps (in g) is the weight of the dry sample
with reference to 100 gr of material.

The blue index of the fraction of the elements with grain size smaller than 2 μm is defined as the "activity index of clay fraction of the soil" (A_{CB}) (Lautrin, 1987) and is expressed in grams of methylene blue per 100 gr of the fraction smaller than 2 μm:

$$A_{CB} = 100 \bullet V_B/C_2$$

where: V_B is the blue index of the sample (g/100 g)
C_2 is the clay percentage (particles smaller than 2 μm).

Table 3 lists the results obtained in terms of methylene blue index on samples taken from the same marly layer at the sliding surface (sample SS), and at the top (sample T) and bottom of it (sample L), from two sites involved in the planar slides observed in November 1994.

Table 3. Methylene blue index (V_B), in g/100g, determined on marly samples taken from two landslide sites

Sample	S. Benedetto Belbo slide	Somano slide
T	1.6	2.0
SS	2.7	2.9
L	2.2	2.6

Table 4. Results of the tests performed on marly samples (San Benedetto Belbo landslide)

Sample	C_2 (%) (grain size <2 μm (¹)) (³)	C_5 (%) (grain size <5 μm (²) (³)	Plasticity index: IP (%) (³)	Skempton index: A_C (-) (³)	Smectite (%) (³)	Blue index: V_B (g/100g)	Activity blue index: A_{CB} (g/100g)
T	20	37	20	1.00	20	1.6	8.0
SS	24	46	25	1.04	32	2.7	11.3
L	19	38	18	0.95	31	2.2	11.6

Table 4 compares the results of the methylene blue tests listed in Table 3 with the results obtained from grain size analyses, determination of the Atterberg limits and X-ray diffractometry (percentage of smectite). It can be seen (Tables 3 and 4) that the blue index reflects the presence of more active levels at the sliding surfaces and is in good agreement with the results of X-ray difractometry analyses. In Figure 7 the blue index (V_B) is plotted as a function of the swelling pressures measured on the same materials (Simeoni, 1998); the correlation obtained is satisfactory.

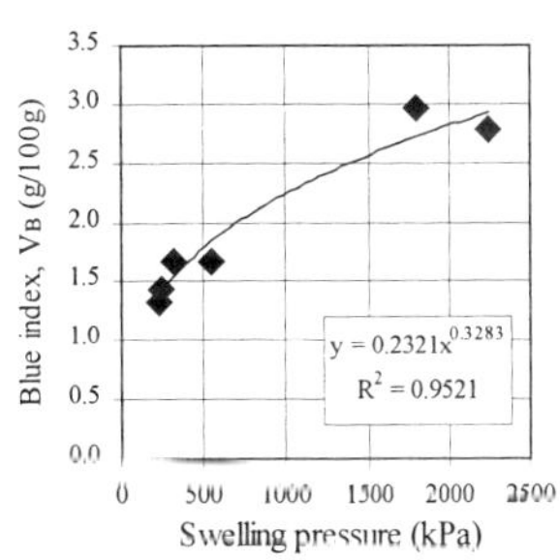

Figure 7. Blue index as a function of swelling pressure

(¹) British Std. & M.I.T.
(²) ASTM.
(³) From Simeoni, (1998).

CONCLUSIONS AND FUTURE DEVELOPMENT OF THE RESEARCH

The studies conducted so far ber out the complexity of the phenomena in question; in particular, the following aspects should be pointed out: i) rainfall plays a decisive role in triggering the landslides; ii) the strength parameters mobilised on the sliding surfaces at the time of collapse are very low (lower than or equal to the residual parameters); iii) "critical" levels exist within the marly layers being studied, which are characterised by the presence of active clay minerals; the latter can be identified quite readily through the methylene blue testing technique. In order to acquire more information and to improve the understanding of the phenomena, the research is being developed through the following stages: i) new stability analyses on landslides with confined evolution at the toe; ii) direct shear tests (in drained and undrained conditions) on the marly materials (with shear applied in a direction parallel to the sedimentation planes) in order to assess the decreasing of the strength parameters in the presence of swelling; iii) further study of the geological-structural features of the area in order to identify possible signs of post-sedimentary tectonic phenomena which might explain the mobilisation of the residual values of the strength parameters and iv) hydrogeological characterisation of the slopes being considered by setting up suitable in situ measurements through the automatic acquisition of piezometric data in order to define the groundwater flow conditions and the evolution in time of hydraulic pressures.

ACKNOWLEDGEMENTS

The authors wish to thank: the Management of the Technical Prevention Services of the Piedmont Region that supplied the data necessary for the analyses, Ing. Monica Ferraro who helped develop the computation models, Prof. Pierre Antoine (LIRIGM - Laboratoire Interdisciplinaire de Recherche Implicant la Geologie et la Mecanique - Université Joseph Fourier - Grenoble) who provided the contacts with the French Institutions of Grenoble, Ing. Paul Camporota and Mr. Laurent Sorita (Direction Départementale de l'Equipement - Isère - Service des Routes" – Grenoble) who made available the laboratory for the initial stage of the testing program with methylene blue.

REFERENCES

AA.VV. (1998) – *Eventi alluvionali in Piemonte: 2-6 novembre 1994, 8 luglio 1996, 7-10 ottobre 1996.* Direzione Sevizi Tecnici di Prevenzione – Regione Piemonte. Torino, 1998.

Forlati F., Lancellotta R., Osella A., Scavia C., Veniale F. (1995) – *Analisi dei fenomeni di scivolamento planare nelle Langhe.* Difesa del suolo, GEAM, dicembre 1995.

Govi M., Mortara G., Sorzana P. F. (1985) – *Eventi idrologici e frane.* Geol. Applicata e Idrogeol., 20, 2, pp. 359-375.

Lautrin D. (1987) - *Une procédure rapide d'identification des argiles.* Bulletin de Liaison des Laboratoire des Ponts et Chaussées, No. 152, nevembre-décembre, 1987, pp. 75-84.

Simeoni L. (1998) - *Fenomeni di scivolamento planare nelle Langhe.* Politecnico di Torino, Dipartimento di Ingegneria Strutturale e Geotecnica. Dottorato di Ricerca in Ingegneria Geotecnica. X Ciclo, febbraio 1998, 159 pp..

Tourenq C. and Tran N. L. (1989) - *Mise en évidence des argiles par l'essai au bleu de méthylène. Application aux sols, roches et granulat.* Bulletin de Liaison des Laboratoire des Ponts et Chaussées, No.159, janv.-févr. 1989, pp. 3-16.

Tran N. L. (1980) - *L'essai au bleu méthylène, un progrès dans la mesure et le controle de la propreté des granulats.* Bulletin de Liaison des Laboratoire des Ponts et Chaussées, No. 107, mai-juin, 1980, pp. 130-135.

Improving Foundation Conditions on an Unstable Slope

A. CHIRICA, A. OLTEANU and M. S. SERBULEA, Technical University of Civil Engineering Bucharest, Romania

INTRODUCTION

The city of Nehoiu is located in the eastern part of Romania at an altitude of 350-500m above Black Sea level. A slope on which a future building is to be erected has a slope of 1:3 on the E-W direction and a 1:12 slope on the S-N one. The building in question is a bank and, on its above mentioned location there used to be the ex City Council, heavily damaged by earthquakes and heavy traffic on the national road in the area (Figure 1).

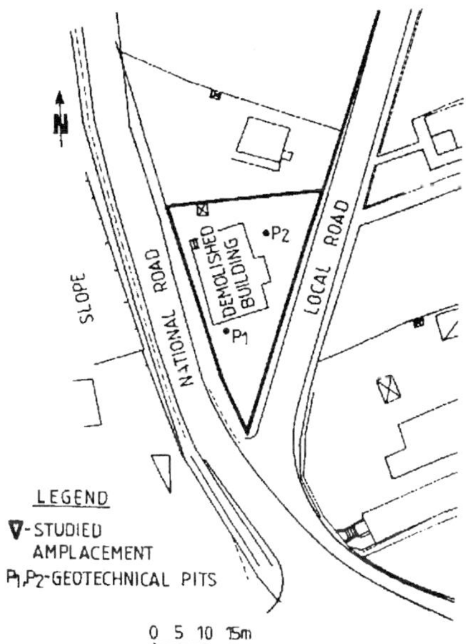

Figure 1. Location of the studied area

It is bounded, both upstream and downstream, by retaining walls damaged due to improper execution and bad backside drainage. The trees on the slope have the "drunken forest" appearance, being inclined towards the west (Fig. 2), pointing out the existence of some general instability phenomena. Besides, it should be mentioned that the location in question belongs to the hill area of the Carpathian Mountains, with frequent slow flow or slide phenomena. The difference between the two (Terzaghi, 1950) can be made considering the displacement velocity, but first of all the character of the displacement should be taken into

Landslides in research, theory and practice, Thomas Telford, London, 2000.

account – slow flow is a mass displacement, while slide is a displacement located on already existing discontinuous surfaces. The two can transform into one another.

Figure 2. "Drunken forest" on the landslide area

GEOLOGICAL AND GEOTECHNICAL ASSESMENT

Between July 1995 and February 1996 two pits (P_1 and P_2 with a diameter of 1m; Fig. 1), respectively 4 drillings (F_0 to F_3) located as in the layout plan which can be seen in Fig. 3 were performed. Borehole F_1 was provided with the necessary tubes in order to carry out inclinometric measurements before, during and after the building construction. Three of the four drillings (F_0, F_1 and F_2) were placed on a line in order to establish calculation profile I-I, on the most likely direction of the instability phenomena (Fig. 4). In all cases, silty clay deposits placed over the bedrock – poorly cemented sandstone – were found below the top layer.

In the studied area the sandstone was found at 6,5m depth. When drilling F_2, ground water under pressure (25kPa) was found. In order to determine the lateral extension of this confined aquifer, drilling F_3 was performed, but no ground water was found in it. Thus may be concluded that either a lens shaped aquifer exists in the F_2 drilling zone, or there is a ground water circulation on the preferred direction $F_1 - F_2$, below the road. Whatever the case, it is necessary to have the bottom of the foundation raft of the new building in the F_2 area at least 1m above level 387,5m where ground water was found.

Laboratory tests performed on samples taken during the drillings – consisting of silty clay through which failure can occur – gave the following results:

- natural water content w = 15,6-19,8%,
- plasticity index PI = 35,5-40,3%,
- liquid limit w_L = 53,1-58,6%,
- consistency index CI = 0,91÷1,00,
- unit weight γ = 18,7-20,2kN/m^3,
- porosity n = 37-39%,

- degree of saturation S_r = 0,62÷0,86%,
- shear strength parameters in terms of effective stresses ϕ' = 14°÷17°, c' 5÷30kPa.

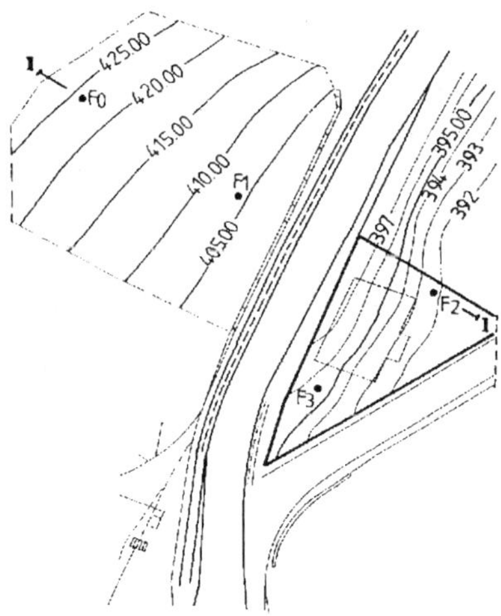

Figure 3. Landslide zone morphology

Based on these results we can assume that a high activity index, and consequently, a plastic behaviour in a wide range of water content characterise studied soil. Studying the variation of saturation degree with depth in F_1 and F_2 drillings we found the position of a possible failure surface, where the water content pointed out the maximum values (Fig. 4).

The parameters selected for the stability analysis were ϕ' = 14° and c' = 5kPa. As a matter of fact from the relationship between plasticity index and residual shear strength (Voight, 1973) for PI = 40% results ϕ_r = 14° (Fig. 5).

STABILITY ANALYSIS

Figure 4 shows the calculation profile on which two possible sliding surfaces were considered. First one, (S_1), corresponds to the maximum values of water content, found in the studied boreholes and the second one, (S_2), was considered the limit between the shallow deposits and the bedrock itself.

The mass of unstable ground was divided into slices, following the ground surface morphology and the failure surfaces geometry. A computer program based on the Fellenius method was used (Chirică, 1982). According to this method, for a predetermined failure surface, the earth thrust E_i at the downstream slice limit was computed using the relation:

$$E_i = \sum_{j=1}^{n} G_j \cdot \sin\alpha_j \cdot \cos\alpha_j - tg\phi' \cdot \sum_{j=1}^{n} G_j \cdot \cos^2\alpha_j - c' \cdot \sum_{j=1}^{n} b_j$$

where G_j is the weight of the considered slice, b_j is the horizontal width of slice j (between the upstream and the downstream limits), ϕ' and c' are the shear strength parameters in terms of effective stresses.

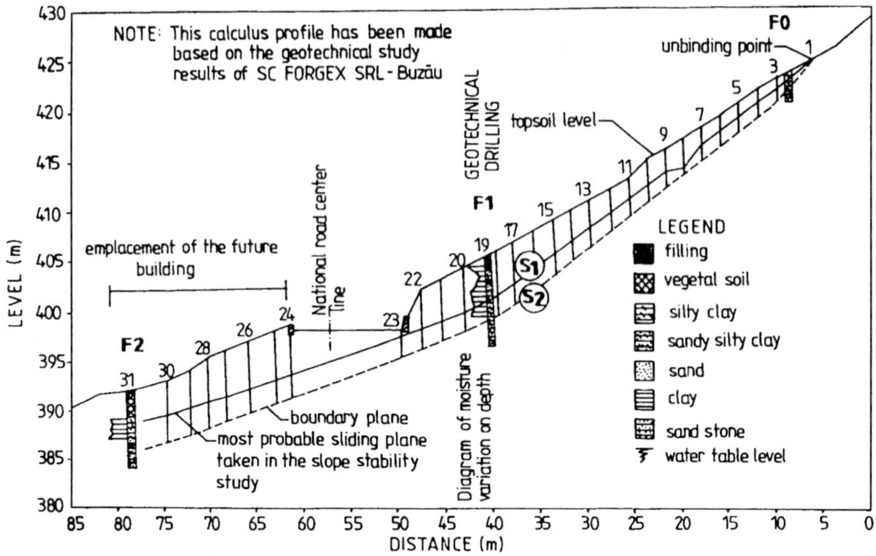

Figure 4. Cross-section of the landslide area

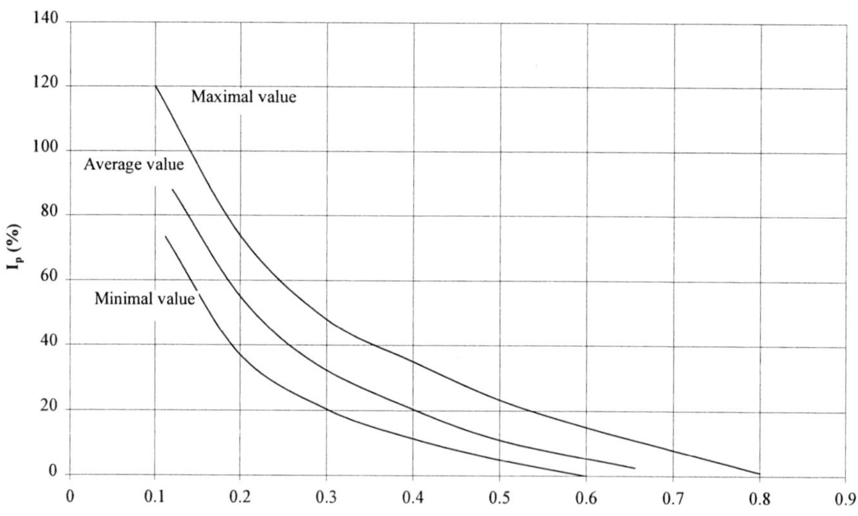

Figure 5. Relationship between the plasticity index and the residual internal friction angle (Voight, 1973)

Stability analysis pointed out that for the S_1 failure surface minimum value for safety factor corresponded. Considering the actual ground morphology, for the S_1 failure surface the calculated earth thrust value was E = 311,6kN/m, in the vertical section passing through point 24 (Fig. 4). Taking into account the seismic action this value increases 2-2,5 times in the same vertical section. Consequently it is necessary that retaining works or/and the foundation of the future bank building should be made out of bored piles with large diameter (> 0,6m) rigidly fixed on the bedrock.

STUDY OF THE FOUNDATION CONDITIONS

As a result of the previous considerations, we can assume that the investigated location is placed in an area that is susceptible to landslides. Therefore, the following strategies may be adopted:

- to separate the slope stabilisation works from the foundation system of the building, or
- the foundation of the designed building to be constructed in such way to support the actions induced by thc slopc instability.

As it can be noticed, for both strategies there is the issue of a discontinuous support on piles made on site. To calculate the reactive horizontal load, R, for that the piles will be designed, the Ito – Matsui method (Ito & Matsui, 1975) was used. This is based on the plastic equilibrium of the volume of earth that has a tendency to flow between two adjacent piles (Fig. 6).

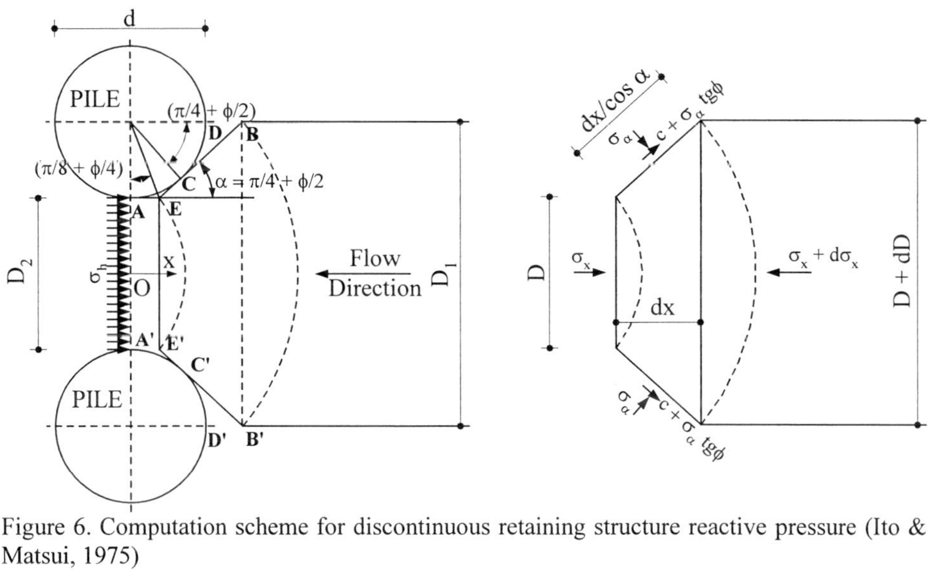

Figure 6. Computation scheme for discontinuous retaining structure reactive pressure (Ito & Matsui, 1975)

The calculation of the effective distance between piles resulted from the condition that the soil thrust E_i is equal the reactive load R determined with the above-mentioned method.

Two possible thicknesses of the sliding mass H were considered (Fig. 7).

- for supporting work madc by piles situated closely down the existing national road (Fig. 2), H = 4,5m,
- for piles as components of the foundation of the designed building, H = 2,5m.

According to the graph in figure 7, for earth thrust value E = 312kN/m, optimal distances D = 2m for the first case and D = 1,5m for the second case were obtained.

It is the opinion of the authors of this paper that in the given conditions the supporting works should be made closely down to the national road. This would ensure – on one hand – the stability of the national road and on the other – a higher security during the new building foundation works.

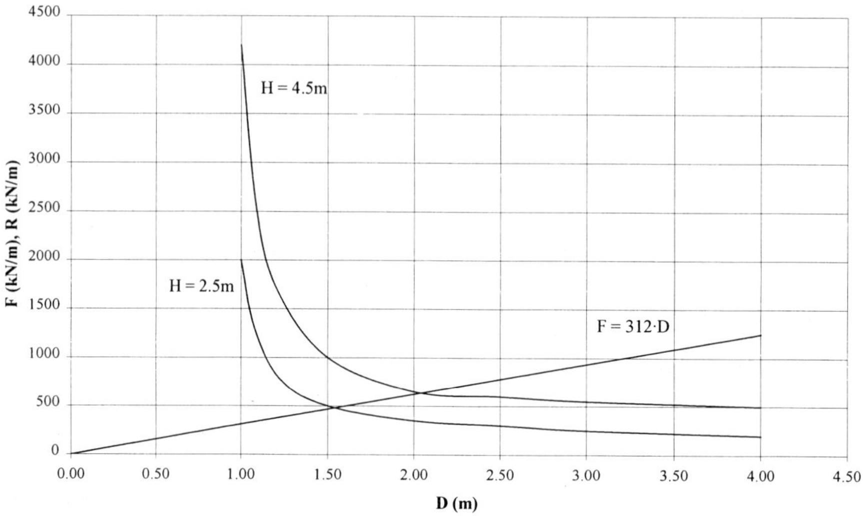

Figure 7. The reactive load values as function of slide surface depth and the distance between the piles

REFERENCES

Chirica, A. 1982 – Computer programs for slope stability analysis. Proc. of the 3rd Symposium of Computer Sciences in Civil Engineering, Sibiu, Romania.

Ito, T. and T. Matsui 1975 – A method to estimate lateral force acting on stabilising piles. Soils and Foundations, vol. 15, No. 4.

Terzaghi, K. 1950 – Mechanics of Landslides. Application of Geology to Soil Mechanics, G.A.S., Harvard S.M. series.

Voight, B. 1973 – Correlations between Atterberg Plasticity Limits and Residual Shear Strength of Natural Soils. Geotechnique XXIII.

Nonlinear Features and Prognosis of a Landslide

C. Y. CHOU, Department of Geosciences, Zhongshan University, Guangzhou, China

INTRODUCTION

Dynamic prognosis of landslide is an essential problem in the fields of landslide research. For about several decades, the development of the dynamic prediction of landslides have developed slowly because of the limits of theoretical methods and the testing methods. In the early 1950s', the dynamic prediction of landslide, concerning about the time when the slope developed to a critical sliding stage based on the monitoring of displacement ,was carried out in Japan. Later, studies, concentrated on the dynamic prognoses of landslides by astrogeological dynamics or by dynamic geology, presented that the happened time of landslides coincided well with the active periods of some planets, e.g.80-90years or 11 years of sunspot and 30years of Saturn(A.I. Sheko ,1976; E.Fussganger,1976).Besides, many researchers sensitively realized the influence of rainfall on landslides, and variety kinds of relationships between rainfall and landslides were carried out to express the dynamic regularities of landslides(S.Wang,1987;G.Guidicini,1976; T.Endo,1970;T.Z.Yan, 1988). More recently, the research has focused on the sliding velocity, distance (A.E.Schiedegger, 1974, J.Z.Pan,1988; T.Z.Yan,1985)and any other dynamic features of a single landslide, of which the theoretical basis was the exchange of potential energy and kinetic energy. At the same time, the growth model of living things and GM(1,1) model of grey system has introduced to build the evaluating equation of a landslide so as to found the *maturity point* which corresponded to the occurrence of landslide(T.Z.Yan,1988,M.D.Chen,L.S.Wang,1985). The grey prediction method, based on the monitored data series, aiming at giving the test prognoses of landslides at the critical sliding time, included two certainty solution; which one is reliable was very difficult to fix on. A simulation experiment of a sandy slope had carried out to give the form of deformation equation and the dynamic prognosis by Fukuzono(1989): $d^2x/dt^2 = a(dx/dt)^m$. Where x was the surface displacement, a, m were the integral constants. t represented time. Another dynamic equation had also built for describing the deformation process of landslide by distortion deformation theory of a cantilever beam (Z.Q.Cui1990) . However, the occurrence of landslide is the re-sliding process on about the same place periodically or aperiodically(Ф•Л•Щавалинский, И•В•Лполов, А•Д•Нисантов,1957). So, We thought that landslide, caused by the interaction of geological environment and the extra factors such as rainfall, must be presented as a connection of certainty and uncertainty chaotic dynamic process; The occurrence of landslide is a self-organization process comprising spatial, temporal, functional and structural organization of every factors within a landslide system by a series nonequilibrium and unstable action. In fact, the rock mechanical test, comprising deformation measure, wave velocity measure and micro-sampling-analysis etc. were all shown the nonlinear features of the deformation process of rocks(V.I.Keilis-borok,1987; Y.Chen,1987; X.W.Yang,1987). In the study described in this paper, a new approach was developed to set the nonlinear dynamic method of analysis into a dynamic research framework and provides a reasonable and a new way to landslide prediction. The

Landslides in research, theory and practice, Thomas Telford, London, 2000.

author here was also making great efforts to give the idea that the landslide system was an open and complexity one, it should be studied by way of the nonlinear dynamics.

THEORITICAL BACKGROUD OF DATA PROCESS—THE BASIS FOR BUILDING THE NONLINEAR DYNAMIC EQUATION

Corralation Fractal Dimension

It is supposed that we have some small balls as the amount of $N(\varepsilon)$, and the diameter of each ball is denoted as r. Screening them with a sieve, denote the number of small balls which pass through the holes of the sieve as $N_d(\varepsilon)$. Thus, we obtain $C(\varepsilon)$ (Li and Cheng,1990): $C(\varepsilon) = N_d(\varepsilon)/N(\varepsilon)$. Evaluate the area of $C(\varepsilon)$ by changing the length of ε properly, then we have: $C(\varepsilon) \propto \varepsilon^r$, where r is a kind of approach to correlation dimension. Thus, the correlation dimension can be defined as D_c: $D_c = \lim_{\varepsilon \to 0} \lg C(\varepsilon)/\lg(\varepsilon)$. Cmmonly, with the aims of calculating the value D_c conveniently, we may give the curve of $\lg N(\varepsilon) - \lg(1/\varepsilon)$ or $\lg C(\varepsilon) - \lg \varepsilon$ first , and then, find out the slope from the non-scale area in the curve. That slope is called correlation dimension D_c.

Saturated Correlation Dimension -Theoretical Background for Data Processing

Give a time-dependent sequence as $x_0(t)$, from which we may build the n dimensional phase space. Suppose a certain time interval $\tau(\tau = m\Delta t$, where m is an integer ; Δt is the sampling time interval), develop the original time-dependent sequence gradually according to the integral times of τ .Besides, choose n interval points from the data series, then, obtain the dispersed value as follows:

$$x_0 : x_0(t_1), x_0(t_2), \cdots, x_0(t_n)$$
$$x_1 : x_0(t_1+\tau), x_0(t_2+\tau), \cdots, x_0(t_n+\tau)$$
$$\vdots \qquad \vdots$$
$$x_{n-1} : x_0(t_1+(n-1)\tau), x_0(t_2+(n-1)\tau), \cdots, x_0(t_n+(n-1)\tau)$$

Introduce a vector symbol x_i , take an arbitrary point from the phase space :

$$\bar{x}_i\{(x_0(t_i), \cdots, x_0(t_i+(n-1)\tau)\}$$

chose an arbitrary data point from the data sequence $\bar{x}_i$, calculate the distance from the chosen point to other (n-1) points ($\bar{x}_j$) : $\bar{x}_i - \bar{x}_j$, thus, obtain the data points in phase space with a center of $\bar{x}_i$, and the diameter of r. Repeat this process to all the numbers of i, we have:

$$C(r) = \frac{1}{n^2} \sum_{\substack{i,j=1 \\ j \neq i}}^{n} \theta(r - |\bar{x}_i - \bar{x}_j|)$$

where, θ is called Heaviside function, $\theta(x) = \begin{cases} 1 & x > 0 \\ 0 & x \le 0 \end{cases}$, $|x_i - x_j|$ shows the distance of the two arbitrary points from the chosen time-dependent sequence; C (r) is the integral correlation function of the attractor.

Give a small value ε as a scale to measure the structure of the attractor. If the structure is a line,
then the number of data points within a given certain distance should be in proportion to r/ ε ; If the structure is a surface, then the number of data points should be in proportion to $(r/\varepsilon)^2$, and generally, if the attractor is with a dimension of d, then the number of data points should be in proportion to $(r/\varepsilon)^d$. So, the small C (r) which coincided with r may be changed according to the estimation: $C(r) = r^d$, where, d is the dimension of the evaluating shape in a subsidiary phase space. Consider the relationship between d and n, if value d is saturated when

value n is over a certain value, then, the saturated value d, shows the trends of the time-dependent sequence, is the dimension of the attractor. The critical value, just when the saturated value appeared, is provided the least needed values for the modeling of a system with the properties which the attractor described.

DYNAMIC FEATURES OF LANDSLIDE SYSTEM

Xintan landslide, took placed in 1986, in Xilingxia Gorge of the three gorges, Yangtze River, China, has important significant to the Three Gorges Dam. Here, taking Xintan landslide as an example to give the analysis of dynamic features of landslides. Although, several groups of variables which related to landslide, should be needed for well describing the dynamic features of a landslide system, we should give the analysis to landslide system by a variable which contains all the features of other variables and can reflect the features of the whole system from the view point of nonlinear dynamics(G.Nicolis, I.Prigogine, 1976, H. Harken, 1977). Commonly, displacement was considered as the represented value. Figure 1 have shown the monitored time-dependent displacement curve at point A′ of Xintan landslide, from which we could get the original data and developed them by two dimensional process. According to the method we introduced above, take τ =9 days , obtain the data series $x_i(t) - x_i(t+\tau)$; calculate the curve showing the relationship between $C(r)$ and r, and moreover the statistics between the dimension D and the phase space n(see fig. 2) was turned out. From figure 2 we knew that the saturated correlation dimension is 2.2; and the corresponded phase space number is 3. It means that at least 3 variables needed for describing the dynamic process of Xintan landslide.

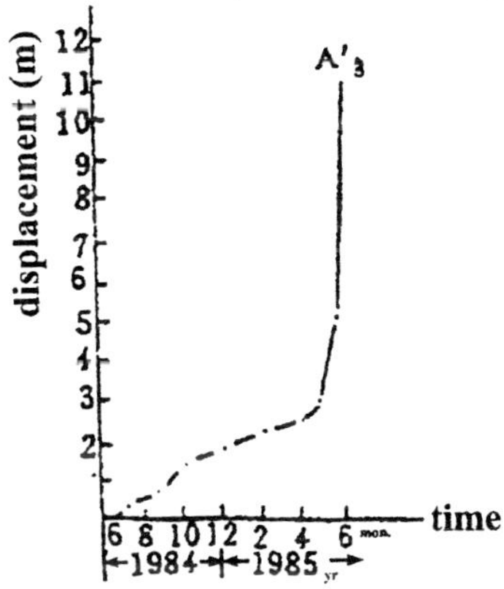

Figure 1 Dynamic curve of displacement at point A_3' of Xintan landslide from May,1984 to June,1985

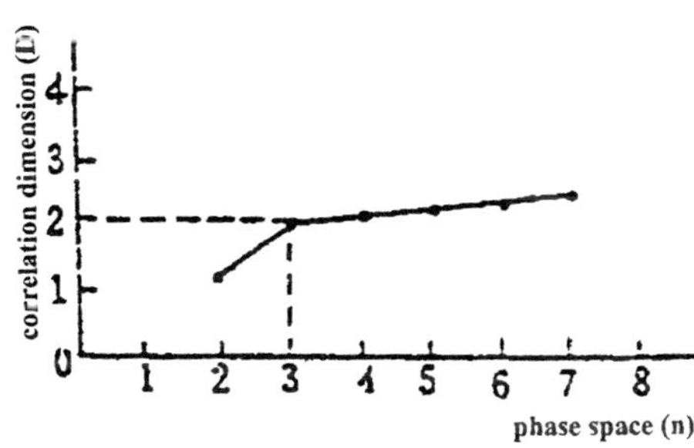

Figure 2 Correlation curve of $\lg C(r) \sim \lg r$ of Xintan landslide (n- phase space number)

THE BUILDING OF NONLINEAR DYNAMIC MODEL AND PROGNOSIS OF LANDSLIDE SYSTEM

The Basic Inverse Theoretical Method(Y.H.Pei,1991)

Suppose the physical regularity of the state of a system varying with time is:

$$\frac{dq_i}{dt} = f_i(q_1, q_2, \cdots, q_n), i = 1,2,\cdots,n \qquad (1)$$

where, f_i is a nonlinear function. Change equation(1) into a deviation form as :

$$\frac{q_i(z+1)\Delta t - q_i^{(z-1)\Delta t}}{2\Delta t} = f_i(q_1^{j\Delta t}, q_2^{j\Delta t}, \cdots, q_n^{j\Delta t}) \qquad (2)$$

representative, means it can reflect the kinetic features directly or be the main factors to influence the stability of the slope; (2) the sum of all the variables should be reflect the dynamic features of the whole system; (3) consider the feasibility and reliability of the monitored data. Chose the time-dependent sequence of the three variables from the second half of 1987 to the first half of 1989 of Xintan landslide as the special solution of the theoretical equation through the uniformity and analysis of the original data of the three variables.

The Building of Nonlinear Dynamic Model of Landslide System

By coupling the previous studies and the property analysis, the dynamic equation of a landslide system should be given as:

$$\left.\begin{aligned} dx/dt &= a_1x + a_2y + a_3z + a_4x^2 + a_5y^2 + a_6z^2 + a_7xy + a_8yz + a_9zx \\ dy/dt &= b_1x + b_2y + b_3z + b_4x^2 + b_5y^2 + b_6z^2 + b_7xy + b_8yz + b_9zx \\ dz/dt &= c_1x + c_2y + c_3z + c_4x^2 + c_5y^2 + c_6z^2 + c_7xy + c_8yz + c_9zx \end{aligned}\right\} \quad (6)$$

According to the inverse theory introduced above, taken the time sequence of displacement, discharge and stress of Xintan landslide into the equation(6), we can obtain matrix D and G as well as each coefficient of equation (6). Then, we have:

$$\left.\begin{aligned} dx/dt = {} & -1.125x - 1.7099\times10^{-3}y + 11.2020z - 1.8059\times10^{-2}x^2 - 5.0835\times10^{-6}y^2 \\ & - 24.9744z^2 + 1.1349\times10^{-6}xy + 1.305\times10^{-2}yz + 2.5898zx \\ dy/dt = {} & -0.2517x - 2.7583\times10^{-2}y - 6.9807z + 1.8296\times10^{-2}x^2 + 1.8043\times10^{-5}y^2 \\ & - 9.5541z^2 - 3.1987\times10^{-4}xy + 4.6088\times10^{-2}yz + 2.5647\times10^{-2}zx \\ dz/dt = {} & -0.2957x - 6.7141\times10^{-3}y + 5.7673z + 1.3374\times10^{-2}x^2 - 6.7756\times10^{-6}y^2 \\ & - 8.9592z^2 - 5.8942\times10^{-4}xy + 1.7245\times10^{-3}yz + 0.6494zx \end{aligned}\right\} \quad (7)$$

where x denotes displacement, y denotes discharge, z denotes stress. This is the nonlinear dynamic equation of Xintan landslide.

Dynamic Prognosis of Landslide System

After numerical integration for equation (7)and the returning to the original dimension of each variable, we obtained the time sequence of x_i, y_i, z_i at the time of t_i ($i=1,2,3,\cdots$). Fig.3 shown the comparison of the predicted curve and the monitored one, from which we know the trends of the two curves were coincided well within a certain period of time. The mean absolute dispersion or error rate (ER)of displacement, discharge and stress of Xintan landslide in this period are 0.078, 0.045 and 0.020 separately.

TIME SCALE OF CERTAINTY PREDICTION AND ERROR ANALYSIS OF XINTAN LANDSLIDE

Calculate the Jacobian matrix of the right term by equation (7), we have:

$$A_{ij} = \frac{\partial f_i(x_1, x_2, \cdots, x_{(n)})}{\partial x_j}\Big|_{(x_1, x_2, \cdots, x_n)(t)} \quad (8)$$

Calculate the specific values in a certain period of time and give the average of these values in this period, then we obtain lyapunov exponent (LE):

$$LE_1 = 0.1379 > LE_2 = -0.0096 > LE_3 = -0.6529 \quad (9)$$

Lyapunov exponent gives the exponential increasing ratio of the systematic error in the phase space. If the error of a system is continuously increased with the time developed, obviously, the long term behavior of the system shouldn't be predicted. The sum of all the positive LE is defined as Kolmogorov (entropy K)entropy in chaotic dynamics:

$$K = \sum_{LE_i > 0} LE_i \quad (10)$$

Entropy K gives the way to measure the average productive ratio of the information of a system, representative, means it can reflect the kinetic features directly or be the main factors to influence the stability of the slope; (2) the sum of all the variables should be reflect the dynamic features of the whole system; (3) consider the feasibility and reliability of the monitored data. Chose the time-dependent sequence of the three variables from the second half of 1987 to the first half of 1989 of Xintan landslide as the special solution of the theoretical equation through the uniformity and analysis of the original data of the three variables.

The Building of Nonlinear Dynamic Model of Landslide System

By coupling the previous studies and the property analysis, the dynamic equation of a landslide system should be given as:

$$\left.\begin{aligned} dx/dt &= a_1x + a_2y + a_3z + a_4x^2 + a_5y^2 + a_6z^2 + a_7xy + a_8yz + a_9zx \\ dy/dt &= b_1x + b_2y + b_3z + b_4x^2 + b_5y^2 + b_6z^2 + b_7xy + b_8yz + b_9zx \\ dz/dt &= c_1x + c_2y + c_3z + c_4x^2 + c_5y^2 + c_6z^2 + c_7xy + c_8yz + c_9zx \end{aligned}\right\} \quad (6)$$

According to the inverse theory introduced above, taken the time sequence of displacement, discharge and stress of Xintan landslide into the equation(6), we can obtain matrix D and G as well as each coefficient of equation (6). Then, we have:

$$\left.\begin{aligned} dx/dt &= -1.125x - 1.7099\times10^{-3}y + 11.2020z - 1.8059\times10^{-2}x^2 - 5.0835\times10^{-6}y^2 \\ &\quad - 24.9744z^2 + 1.1349\times10^{-6}xy + 1.305\times10^{-2}yz + 2.5898zx \\ dy/dt &= -0.2517x - 2.7583\times10^{-2}y - 6.9807z + 1.8296\times10^{-2}x^2 + 1.8043\times10^{-5}y^2 \\ &\quad - 9.5541z^2 - 3.1987\times10^{-4}xy + 4.6088\times10^{-2}yz + 2.5647\times10^{-2}zx \\ dz/dt &= -0.2957x - 6.7141\times10^{-3}y + 5.7673z + 1.3374\times10^{-2}x^2 - 6.7756\times10^{-6}y^2 \\ &\quad - 8.9592z^2 - 5.8942\times10^{-4}xy + 1.7245\times10^{-3}yz + 0.6494zx \end{aligned}\right\} \quad (7)$$

where x denotes displacement, y denotes discharge, z denotes stress. This is the nonlinear dynamic equation of Xintan landslide.

Dynamic Prognosis of Landslide System

After numerical integration for equation (7)and the returning to the original dimension of each variable, we obtained the time sequence of x_i, y_i, z_i at the time of t_i $(i=1,2,3,\cdots)$. Fig.3 shown the comparison of the predicted curve and the monitored one, from which we know the trends of the two curves were coincided well within a certain period of time. The mean absolute dispersion or error rate (ER)of displacement, discharge and stress of Xintan landslide in this period are 0.078, 0.045 and 0.020 separately.

TIME SCALE OF CERTAINTY PREDICTION AND ERROR ANALYSIS OF XINTAN LANDSLIDE

Calculate the Jacobian matrix of the right term by equation (7), we have:

$$A_{ij} = \frac{\partial f_1(x_1, x_2, \cdots, x_{(n)})}{\partial x_j}\Big|_{(x_1, x_2, \cdots, x_n)_{(t)}} \quad (8)$$

Calculate the specific values in a certain period of time and give the average of these values in this period, then we obtain lyapunov exponent (LE):

$$LE_1 = 0.1379 > LE_2 = -0.0096 > LE_3 = -0.6529 \quad (9)$$

Lyapunov exponent gives the exponential increasing ratio of the systematic error in the phase space. If the error of a system is continuously increased with the time developed, obviously, the long term behavior of the system shouldn't be predicted. The sum of all the positive LE is defined as Kolmogorov (entropy K)entropy in chaotic dynamics:

$$K = \sum_{LE>0} LE_I \quad (10)$$

Entropy K gives the way to measure the average productive ratio of the information of a system, shows the average exponential increasing ratio of a micro-volume in a certain direction in phase space. $1/K$ shows the needed time for increasing a certain times of error of a system, i.e. the time scale to give the certainty prediction of a dynamic system in the attractor distributed area. Here, we can obtain the certainty predicted time scale of Xintan landslide by calculating the Kolmogorov entropy by equations (8), (9)and (10):

$$K_{熵}=0.1379 \qquad (11)$$

thus,

$$\frac{1}{K_{熵}}=\frac{1}{0.1379}=7.2516 \qquad (12)$$

In the procedure of monitoring, one time-dependent point (x, y, z) was given out every one month. Let the error is limited within 0.3, means, when the time increased from 0.1month to 0.3month, the error increased 2 times, then, the needed time when the error increased two times was:

$$\frac{1}{K}\times\frac{0.3}{0.1}=7.2516\times 3=21.7557m=1.8\text{years} \qquad (13)$$

Hence, under this error condition, the certainty predicted time scale is 1.8 years.

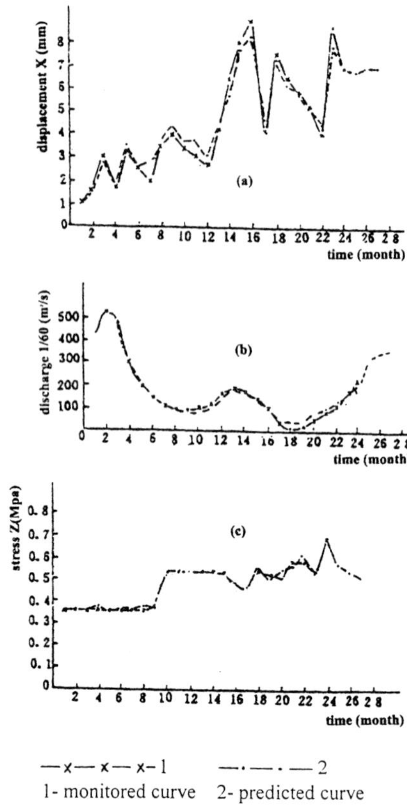

Figure 3 Time-dependent comparison curve of the prognostic results and the monitored ones of displacement (x), discharge (y) and stress (z) of Xintan landslide

CONCLUSIONS

Landslide system is an open and complexity one. As illustrated in this study, The nonlinear dynamic method is a coupling of certainty and uncertainty method, provides a new way to landslide research. The numbers of variables to be used to build the dynamic equation can be determined by the analysis of correlation dimension and then, the nonlinear dynamic model was build in this paper. The developed solution shows how the nonlinear dynamic method can be utilized to describe the whole dynamic process of landslide system and to determine the time scale of the certainty prediction of the system.

REFERENCES

T.Fukuzono, Rrecent studies on time prediction of slope failure, Landslide News, No.6,1990

V.I.Keilis-Borok,chaos and self-organization in the lithosphere and earthquake prediction, a report in the 19th International Conference of IUGG,1987

Y.Chen, Applications of fractal and chaos to Geosciences, Academic Journal Press, Beijing,1989

G.Nicolis, I.Prigogine, Exploration of complexity (in Chinese , translated by J.L.Luo et. al.), Sichuan Educational Press, 1986

S.D.Liu, Simulating of earth system and the chaotic time sequence, Journal of Geophysics, Vol.33, No.2,1990

J.P.Huang, Y.H.Yi, Inverse of nonlinear dynamic model by monitored data, Chinese Science(B), No.3, 1991

The Dramatic Cluster of Pyroclastic Debris Flows which Ocurred on 5th and 6th May 1998 in the Sarno Mountains (Vesuvius Region, Southern Italy): a Geomorphological Perspective

A. CINQUE, G. ROBUSTELLI, F. SCARCIGLIA, Dip.to di Scienze della Terra – Università "Federico II", Napoli and F. TERRIBILE, DISCA – Università "Federico II", Napoli, Italy

DEBRIS FLOWS IN THE SARNO MOUNTAINS

In the evening of May 5th, 1998 and the following morning, some tens of debris flows mobilised more than one million m^3 of slope materials from the flanks of the Sarno Mountains (Guadagno et Al., 1999) and caused 161 fatalities and severe damages to the towns of Sarno, Quindici, Siano and Bracigliano, them all settled at the footslope of the massif. The Sarno Mts. cover an area of 70 km^2 and rise from the Campana coastal plain up to 1133 m a.s.l. (Fig.1). The massif top is characterised by a Late Tertiary mature landscape (hereinafter Summit Palaeosurface, or SP) with gentle residual relieves and flat karstic plains. It is bounded by steep and high scarps (Marginal Scarps, or MS) which reflect Quaternary phases block faulting and appear dissected by consequent incisions having steep lateral and longitudinal gradients. A gentle accumulation glacis (Piedmont Glacis, or PG), made of coalescing alluvial fans, connects the massif to the plain (Fig.2).

The bulk of the massif is made of Mesozoic limestones on which a prone to slide, loose pyroclastic cover occurs. The thickness of the latter is up to 2-3 m on the MS and up to many metres on both the SP and the PG. It is made of layers of ashes and/or pumice fragments deriving from different Late-Quaternary explosive eruptions of the nearby Somma-Vesuvius volcano and the farther Phlegrean Fields (Fig 1). The alternation of fall-out events (one every $n \cdot 10^3$ yrs) and pedogenetic phases is recorded on the mountain slopes by stratigraphic sequences made of fresh pyroclastics units and typical andosols (Terribile et Al., 1999) (Fig. 3) both of them often truncated by erosion surfaces. The depositional record of these erosional events is found in the sequences of the PG's fans as water laid and debris flow deposits that intercalate with gravel beds and *in situ* pyroclastic layers (Cinque et Al., 1999).

Fig. 1 - Location of the Sarno Mts. massif (3) with respect to the Somma-Vesuvius (2) and the Phlegrean Fields (1) volcanoes.

The study area has a typical upland Mediterranean climate, with mild wet winters and warm dry summers. Annual rainfall varies with elevation between 1000 and 1500 mm and shows two peaks: one in the autumn and one in the early spring. The natural vegetation cover (mixed oakland) is preserved only in small, scattered patches, because it was gradually replaced with plantations of wild and edible chestnut-trees and, more recently, of hazel-trees. This transformation was accompanied by the cut of many new trackways and by the diffusion

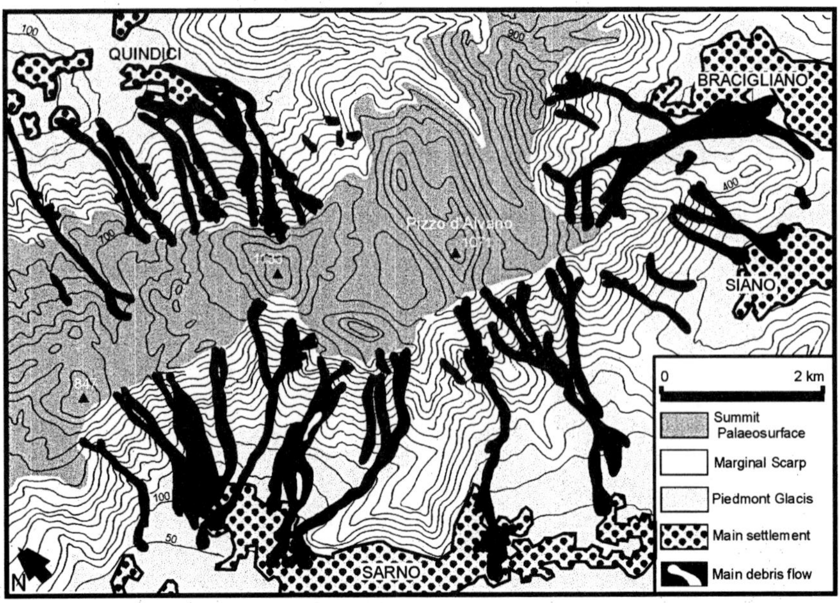

Fig. 2 - A threefold geomorphological classification of the Sarno Mts. area and the main landslides of 5-6 May 1998 (slope failures that did not evolve to long flows are not mapped).

of the custom to clear the undergrowth by fire. Moreover, in the last decades, forest fires and overgrazing produced many almost treeless spots along the slopes, especially along the south-western side of the massif.

Landslides like those occurred in the Sarno Mts. on May 1998 are quite common all over the belt of calcareous mountains that surrounds the Campana Plain (Fig. 1). They are debris flows (*sensu* Varnes, 1978) of exceptionally high fluidity because derived from water saturated pyroclastic materials having an abundant fine-textured matrix and light coarse elements (pumice fragments with Ø ≤ 3cm). Some of them, after an accelerating run along steep slopes and/or valleys, reach the densely populated footslope areas with velocities that may exceed 100 km/h. Isolated slides of that type occur almost every year, while dense clusters of slides (tens of events within spots of $n \cdot 10^2$ km^2) show recurrence times which are of decennial order if referred to the whole peri-volcanic mountain belt and probably pluri-centennial for restricted parts of it. Recurrence times are scale dependent because the slides at issue are strictly correlated with rain events, whose distribution in the region is controlled by small cells and by orography. Debris flows are generally triggered by one day or two of concentrated rainfall occurring after a week or more of frequent rains (not necessarily intense) able to saturate the deepest part of the pyroclastic cover. In the case of May 1998 there were 100 to 150 mm of rain on May the 4th and 5th and 150 to 300 mm in the preceding 30 days (50 to 120 of which in the ten days before the slides).

The role of rainwater is related to the great water retention capacity at saturation (over 100% of their weight) of the pyroclastic-andosolic sequences, which are rich in allophane and proto-halloysite (Fig. 3). Moreover, the high thixotropy of these materials expose them to liquefaction when remoulded upon falling or sliding, when impacted by materials coming from upslope and/or when shocked by vibrations.

It is evident in the field that the soil failures phenomena initiating the debris flows were sometimes facilitated by the pedo and tefro-stratigraphical discontinuities that characterise the cover sequences.

Classification Soil Taxonomy (1998):	Vitric Hapludands, ashy over pumiceous mesic	**Aspect:**	225° N
Date:	13-06-1998	**Elevation:**	800 m
Location:	Quindici	**Stoniness:**	Absent
Position Site:	Alvano	**Parent Material:**	Volcanic ashes and pumice.
Physiography:	Slope	**Natural Drainage:**	Well Drained
Slope:	140%	**Land use:**	Copse

Horizon	Depth (cm)	Description
A	0-25	moist; very dark graysh brown (10YR3/2); common (5-15 %) not weathered irregular medium gravel (5-20 mm) pumice and few (<5%) not weathered angular fine gravel (2-5 mm) pumice; many (25-200) very fine (<1 mm) roots and few (1-10) fine (1-2 mm) roots; carbonate reaction null; Al(ox) + 0.5Fe (ox): 4.12; clear and wavy boundary to
Bw1	25-47	moist; olive brown (2.5Y4/4); few (<5%) not weathered angular fine gravel (2-5 mm) pumice; few (1-10) very fine (<1 mm) roots and few (1-10) fine (1-2 mm) roots; carbonate reaction null; Al(ox) + 0.5Fe (ox): 5; clear and wavy boundary to
Bw2	47-70	moist; brown (10YR4/3); common (5-15 %) not weathered angular medium gravel (5-20 mm) pumice and few (<5%) not weathered angular coarse gravel (20-75 mm) pumice; few (1-10) very fine (<1 mm) roots; carbonate reaction null; Al(ox) + 0.5Fe (ox): 3.4; abrupt and wavy boundary to
C1	70-125	moist; pale yellow (5Y7/3); many (35-70 %) not weathered angular medium gravel (5-20 mm) pumice; common (10-25) very fine (<1 mm) and few (1-10) fine (1-2 mm) roots; carbonate reaction null; abrupt and smooth boundary to
C2	125-142	moist; light yellowish brown (2.5Y6/4); few (<5 %) not weathered angular fine gravel (2-5 mm) pumice; loose single grain structure; carbonate reaction null; abrupt and smooth boundary to
2Ab	142-165	moist; dark brown (10YR3/3); moderate coarse subangular blocky; few (1-10) very fine (<1 mm) roots; carbonate reaction null; Al(ox) + 0.5Fe (ox): 4.65; clear and wavy boundary to
2Bwb1	155-185	moist; dark yellowish brown (10YR3/6); many (15-30 %) coarse (>15 mm) dark brown (10YR3/3) mottles and few (2-5 %) fine (<5 mm) very dark gray (10YR3/1) mottles; few (<5%) not weathered angular fine gravel (2 5 mm) pumice; moderate compound structure medium subangular blocky and coarse subangular blocky; few (1-10) very fine (<1 mm) roots; carbonate reaction null; Al(ox) + 0.5Fe (ox): 3.39; clear and wavy boundary to
2Bwb2	185-230+	very moist; dark yellowish brown (10YR4/4); few (1-10) very fine (<1 mm) roots; carbonate reaction null; Al(ox) + 0.5Fe (ox): 1.95; unknown lower boundary.

Fig. 3 - Description of a representative soil profile (Profile Quindici) located very close to one of the scars which originated the debris flow that reached Quindici and to the edge of the Summit Paleosurface.
Abbr. Al(ox) + 0.5Fe (ox): Aluminium and iron extracted in ammonium oxalate (Schwertmann, 1964).

The initial detachments never exploited the contact surface between cover and limestone, which is generally indented by small karstic trenches and holes. Trees roots do not usually penetrate down to the bedrock, because they find enough nutrients and moisture in the topsoil. Moreover, under the woodlands dominated by wild chestnuts, root penetration has been further decreased by centuries of copse management with a cut every 10 years which solicits the plants to produce new and short roots rather than to reinforce and extend the old ones.

The great number and variety of events that composed the May 1998 crisis (over 150 small slides, tens of which evolved into big debris flows) - surveyed "just post eventum" - enabled us to collect many illuminating examples about the role played by topography (i) in determining the spatial distribution of the points of slide origin and (ii) in controlling the successive spreading of the phenomena.

First of all it was recognised that most of the debris flows had started with summit, initial collapses (planar and rotational slides, generally not larger than few tens of m^3) and soon expanded and bulked downslope (Fig. 4). The initial failures concentrated along abrupt, convex breaks of slope connecting an upper gentle element with abundant pyroclastic cover (thickened by repeated fall-out accumulations and colluviation of volcanic soil materials from upslope) with a steep (> 45°) element able to induce a high gravitative shear stress in the toe

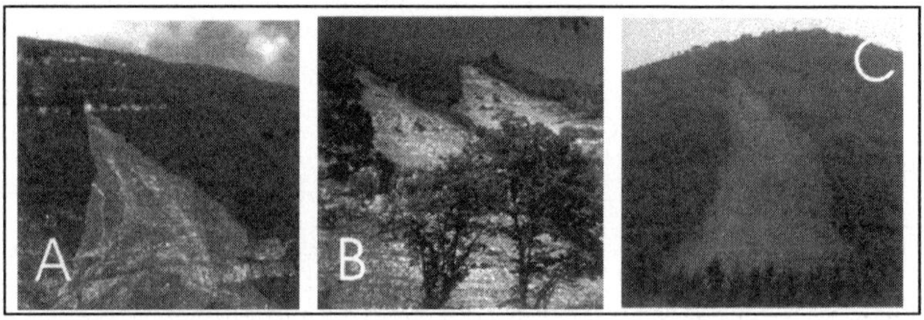

Fig. 4 – Three examples of wide triangular debris flow scars developed from small initial slope failures. The case of photograph (A) also exemplifies the flow enhancement effect of a vertical cliff.

part of the cover resting on the upper element. The higher portions of the MS were preferential places of failure initiation (Fig. 2) because (i) they received more rainfall and because (ii) they are particularly rich of sharp convex breaks of slope (e.g. outer rim of the SP; edges of structure controlled free faces; shoulders of steep flanked incisions). The existence of cliffs (even if only a few metres high) played a decisive role in permitting some of the countless small slides occurred in those days to evolve into long and huge debris flows (Fig. 4A and 7). In fact, clear field evidence were collected that when the sliding or flowing material jumped down a cliff, it was able to shock the pyroclastic cover of the underlying, gentler slope element causing its remoulding, buckling and flowage. A possible major cause of fluidization was, in this case, the sudden increase of pore water pressure brought by the impact (Selby, 1993; pp 307-310). This was a mechanism able to transmit instability even to slope elements much gentler than the threshold for independent failure, which was estimated around 45°. Worthy to note is the fact that also man-made scarplets and convex breaks of slopes (e.g. road-cuts and correlative rubble taluses) favoured the initiation of debris flows in substantially the same way that the above quoted natural topographic features (fig. 5).

Once put in motion on steeper parts of hillslopes, the landslides tended to run over the entire slope length continuing to strip materials away and to increase gradually their width. The latter tendency fully developed only on planar or slightly convex slopes (Fig.4), where the scars took triangular (acute isosceles) plan shapes with rectangular cross sections and high width/depth values (up to 100 and more, toward the triangle base). When, on the contrary, the flows channelised into valleys, ribbon-like scar geometries and roughly triangular cross section developed. It must be noted that virtually none of the observed debris flows was generated along stream beds as a consequence of stream flow erosion and overloading. However, most of the slides which departed from valley-side and valley-head slopes converged quite soon toward a valley floor and followed it as far as the piedmont

Fig. 5 - The road-cut failure which triggered one of the largest debris flows arrived in the town of Sarno.

Fig. 6 – An example of secondary slides in cluster which departed from the steep flanks of a valley probably during the passage of a major debris flow coming from the upper catchment.

area (Fig. 2). Along their way, the flows were also able to incorporate the colluvial/alluvial materials that had previously accumulated - sometimes for millennia - at the bottom of valley

Fig. 7 - An example of the frequent case in which a debris flow passing in a steep valley was reinforced by a drop (central part of the scene) and increased its volume by stripping soil and vegetation from the side slopes hit by skidding waves (foreground).

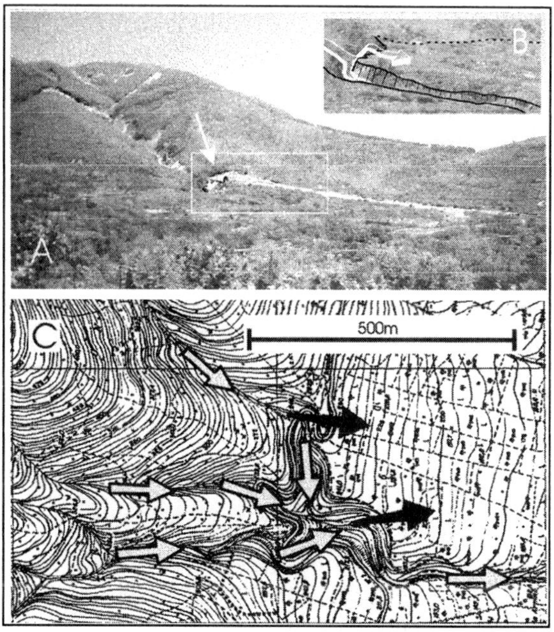

Fig. 8 (after Brancaccio et Al., 1999, modified) - The couple of jump-over event that led a great part of the Tre Valloni debris flow to start running on the surface of a dissected alluvial fan. Due to the perspective of the photograph, the abandoned incision (marked on the inset and on the map) and one of the points of jumping are not visible.

reaches (usually the upper ones) of rare and low water discharge. Moreover, the passage of large debris flows in the valleys triggered many new slides from the valley sides (Fig. 6 and 7) by undermining, by impact of skidding waves of mud (e.g. outside of a bend; at the base of a drop) and, maybe, by vibrations produced by the main flow.

The very high fluidity and velocity of May 1998 debris flows is witnessed by the numerous cases of jump-over of the outer flank of valley bends which have been observed in the upper and middle parts of the PG area. In the representative example of Fig. 8 the jumped valley sides are between 20 and 30 metres high and about 50° inclined. Thanks to phenomena of this kind, or simply for the insufficiency of the pre-existing incisions, the flows finally expanded on the piedmont glacis continuing to erode the vegetation mantle and the topsoil even from surfaces only 5° inclined. Deposition in the form of very thin and broad lobate blankets (Fig. 9) only occurred when the flows reached the sub-horizontal distal part of the piedmont fans, unless flow deceleration was anticipated by the contrast offered by buildings and other structures.

In terms of risk evaluation our study evidenced that the geomorphological situations having the highest potential of landslide initiation are (i) the straight slope elements that exceed 45° of inclination (provided they have a continuous pyroclastic cover at least some 50 cm thick) and (ii) the sharp convex slope-breaks that mark the passage from a gentle upper element and a steep (>45°) lower one. In the second case, the gentler and/or longer the upper element is, the ticker is its cover and, therefore, the risk of failure). It was also noted that length and magnitude of debris flows – little related to the size of the initial, summit failures - have strong relations with the topographic profile downslope of the zone of initial failure. On the concave basal part of the MS, for example, slope elements as gentle as 10° or less can be reached and eroded by a debris flow if the latter arrives there after a hectometrical run along steeper slope elements (>35°), sometimes aided by the energy gained in jumping one or more sub-vertical free-faces. As regards debris flows channelised into deep V-shaped valleys, the external flank of sharp bends and the foot of thalweg drops are the places were the flow may be easier increased by secondary slides. On the other hand, the sharp bends of the lower reach of the valleys (moderately entrenched in the apical part of the PG fans) are places where the flows can come out from the channel and spread on the fan surface.

Fig. 9 - Some aspects of the behaviour of the debris flows in the piedmont area. A) general view of the Sarno's western part, where the flows continued to erode till they reached the built areas; B) One of the flows that reached Quindici. Here too the flow jumped out of the conveying incision and spread on the surface of a dissected fan. The skinned part of the tree in the foreground accounts for the limited thickness of the flow (about 120 cm); C) While spreading on the S. Lucia alluvial fan (Quindici) the flow abandoned the limestone blocks (up to 3 m^3 large) it had taken from the upper valley floor but continued to erode the vegetation and the soil.

REFERENCES

Brancaccio L., Cinque A., Russo F., Sgambati D. (1999) – *Osservazioni geomorfologiche sulle frane del 5-6 Maggio 1998 del Pizzo D'Alvano (Monti di Sarno, Campania).* "Studi Geografici e Geologici in onore di Severino Belloni" a cura di Orombelli G., 81-123, Ed. Glauco Brigati, Genova.

Cinque A., Amato A., Robustelli G., Scarciglia F. (1999) - *Cicli di accumulo e rimozione di depositi piroclastici instabili sui rilievi calcarei prossimi al Vulcano del Somma-Vesuvio (Campania, Appennino meridionale).* Abstract, Simposio MURST, VIII Convegno Glaciologico Italiano, Bormio (SO), 9-12 Settembre 1999.

Guadagno F.M., Celico P., Esposito L., Perriello Zampelli S., Piscopo V., Scarascia Mugnozza G. (1999) - *The Debris Flows of 5-6 May 1998 in Campania*, Southern *Italy.* Landslide News, 12, 5-7.

Schwertmann U. (1964) - *Differenzierung der Eisenoxyde des Bodens durch Extraktion mit Ammoniumoxalat- Losung.* Z. Pflanzenern., Dung, Bodenkunde, 105, 195-202.

Selby M.J. (1993) - *Hillslope materials and processess.* Oxford University Press, 2nd Edition, Oxford, 451 pp.

Soil Survey Staff (1998) - *Keys to Soil Taxonomy.* 8th Edition, USDA, Washington D.C., 325 pp.

Terribile F., Basile A., di Gennaro A., Aronne G., Buonanno M., De Mascellis R., Vingiani S., Malucelli F. (1999) – *The soils of the landslides of Sarno and Quindici.* In: Terribile F. & Adamo P. (Eds) (1999) Cost Action 622, Scientific Program and Guide for the Excursions, Napoli, 27-30 May, 1999.

Varnes D.J. (1978) – *Slope movement types and processes.* In: Schuster R.L.,& Krizek R.J. (Eds) (1978) "Landslides: Analysis and Control", Special Report 176, National Academy of Science, Washington D.C., 235 pp.

The Strategy, Management and Investigation of Coastal Landslides at Lyme Regis, Dorset

A. R. CLARK, D. S. FORT, High-Point Rendel, London, and G. M. DAVIS, West Dorset District Council, UK

INTRODUCTION

Lyme Regis is an ancient town on the west Dorset coast in south-west England (Figure. 1). Records in the early thirteenth century refer to the town and its unique harbour (The Cobb) which was originally constructed as a breakwater on this exposed coastline. In subsequent centuries the coastline and the town have been protected from marine erosion by the construction of a series of seawalls, groynes and coastal defences of various forms. Over the years these defences have been repaired and replaced in a piecemeal fashion in response to storm damage and general decay so that today the town is protected by a variable complex of ageing structures.

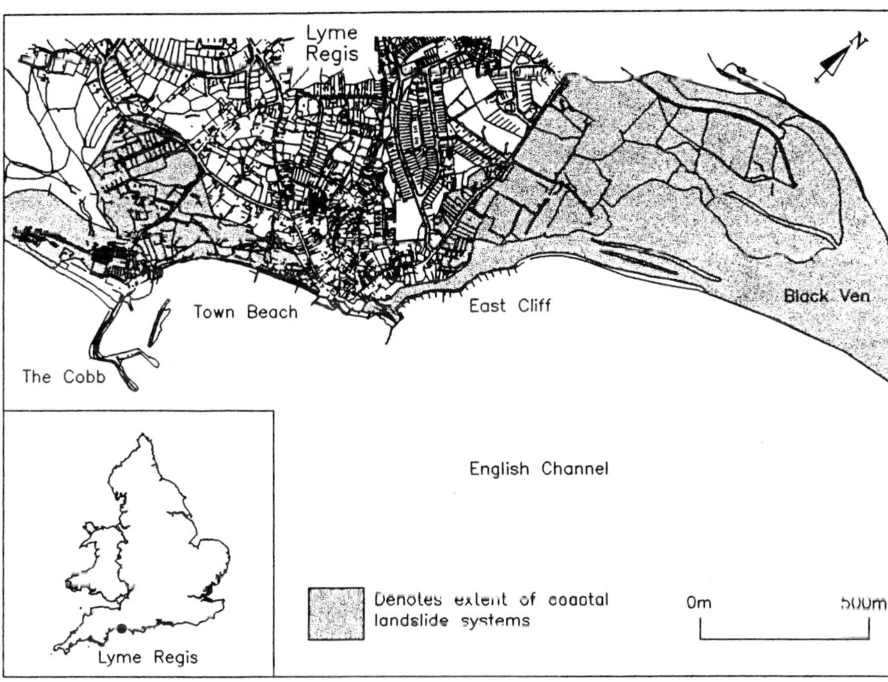

Figure 1. Site Location : Lyme Regis, Dorset, UK

Landslides in research, theory and practice, Thomas Telford, London, 2000.

The town which is a major holiday resort, with a summer population reaching 14000, is developed in an area of post glacial landsliding which is being progressively reactivated as a result of a combination of coastal erosion, climatic factors and anthropogenic triggering. In response to this continuing erosion, vulnerable coastal defences and extending landslide activity in an around the town, an extensive series of multidisciplinary studies and investigations have recently been undertaken as part of the development of a strategy to upgrade possible replacement of the town's coastal defences and to improve ground stability. The long-term objective is to provide adequate levels of coast protection and to improve safety for the town and its residents (Cole and Joy, 1994).

These studies, which have been co-ordinated by the local Coast Protection Authority (West Dorset District Council) have included detailed investigations of geomorphology and geotechnics, structural condition of the existing coastal defences including the Cobb, coastal process and sediment budget studies, offshore geology and geomorphology, the analysis of long term coastal change. Environmental and ecological assessments and a review of potential scheme benefits have also been undertaken.

This paper will focus on the geotechnical studies as a case history to describe the techniques used to investigate the landslides and how a thorough understanding of the mechanisms of instability and coastal erosion processes of the area enabled the development of reliable ground behaviour models. These have been central to the development of the strategy for the upgrading of the coast defences and stabilisation of the slopes.

BACKGROUND TO INSTABILITY

Lyme Regis has been developed on the coastal slopes formed from Lower Jurassic, Lower Liassic mudstones and siltstones with subsidiary limestones which are overlain unconformably in part of the area by Cretaceous sands and clay-with-flints. The whole sequence is covered by variable thicknesses of recent and ancient landslide and periglacial deposits, with the composition varying accordingly to the operating mechanism of instability and the geological source material (Sellwood et al., 2000 for more detail).

The steep slopes within the town undergo periodic movement and disruption to existing development and infrastructure in response to high rainfall events, foreshore erosion and inappropriate excavations (Hutchinson, 1984). This century, 12 properties have been destroyed and many more severely damaged, there have been several seawall breaches and the principal coastal road to the east of the town has been lost.

The cliffs and slopes to the west of the town form the "Landslide Nature Reserve" which is a nationally important conservation area of active coastal instability (Allison, 1992). To the east and more significantly lies the well-documented Black Ven mudslide complex which is one of the largest in Europe (Brunsden, 1969, Brunsden and Goudie, 1997). The average annual cliff top recession rate here caused by the instability is approximately 3m/year although losses of upto 40m/year may be experienced in particularly active years (Chandler and Brunsden, 1995). The area of active movement at Black Ven is gradually extending westwards towards Lyme Regis town as coastal erosion of the unprotected cliffs reactivates the dormant post glacial landslide system up to 600m inland causing the coastal defences on the east side of the town to be progressively outflanked.

GROUND INVESTIGATIONS

Within this context of widespread instability within the town and its surroundings covering a total of 75 hectares (Figure 1), a series of detailed geotechnical investigations were designed. The scope and methods selected were required to provide detailed information to confirm the nature, mechanisms and extent of the instability to develop ground models for each of the failure types; secondly to provide sufficient information to develop conceptual and preliminary designs for new engineering works necessary to ensure the continued integrity and performance of the coastal defences; thirdly to establish a network of instrumentation and monitoring.

Prior to the design of the investigations a desk study and review of published and unpublished information on the area was undertaken and historical research into coastal changes and landsliding. Detailed geomorphological and ground behaviour maps were prepared which enabled preliminary interpretation of ground models for the various forms of instability and identified the full extent of active instability. As it was not feasible to position boreholes on each element of the landslide complex, the mapping provided clear guidance for the most effective positioning of boreholes for the main ground investigation.

The key techniques used within the ground investigation included rotary cored boreholes to provide high quality core, installation of inclinometers and piezometers in boreholes sunk adjacent to the cored boreholes by open hole drilling techniques. The scope of work was supplemented by trial pits and trenches, geological exposure mapping and insitu testing and an extensive programme of disturbed and undisturbed sampling and laboratory testing.
The full scope of the ground investigation is summarised on Table 1.

One of the principal challenges of the investigation and the interpretation of the ground models was distinguishing mudslide and other landslide deposits from material which was weathered and softened insitu but was nevertheless intact. The Liassic sequence consists of a relatively monotonous sequence of massive and fissile mudstones and siltstones with subsidiary limestones, some of which form key marker horizons. To this end, high quality 110mm nominal diameter core was obtained using double tube coring with plastic liners and polymer or air mist drilling fluid. The need for rigorous high quality core logging in such materials with consistent descriptions from all participants was a key issue in the philosophy and management of the investigation. Both Client, Consultants, and Contractor's staff were given a training seminar accompanied by a site specific logging manual prepared prior to the commencement of the works.

Table 1. Scope of the ground investigation

Exploratory Holes		Instrumentation		Testing	
Boreholes	79nr	**Borehole Inclinometers**	34nr	**Borehole natural gamma logging**	873m
Depth range	5-55m	Depth range	10-50m		
Total drilled	1776m	Total length	975m	**Insitu permeability Tests**	50nr
Coring	987m	**Standpipe piezometers**	50nr		
Open hole	744m	Depth range	3-55m	**Laboratory testing**	
Cable percussion	35m	Total length	784m	Full range of classification, strength and chemical tests	
Trial trenches	15nr	Automated loggers	27nr		

The stratigraphic correlation of bedrock across the site using marker horizons was assisted by the use of the characteristic signatures of the natural gamma traces which uniquely distinguish individual horizons of mudstones and limestone (Figure 2). This technique proved very successful in confirming stratigraphic continuity between boreholes and identifying intact bedrock masses. Conversely logs which showed disruption or absence of the distinctive marker traces were valuable in distinguishing areas subject to movement and between intact and disturbed material. This data was then fed into the ground model interpretation which was an integral part of the design process (Figure 3).

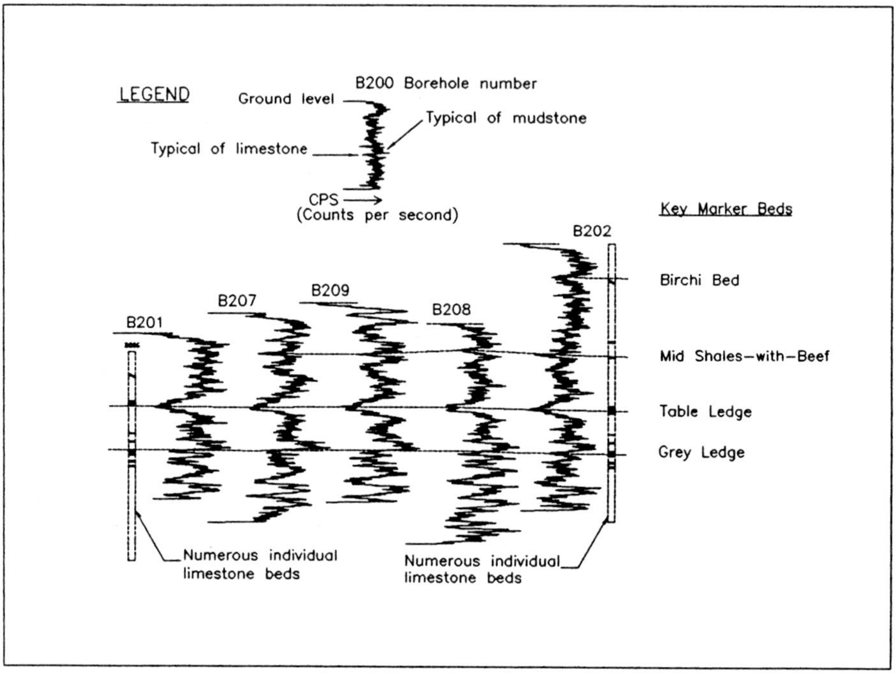

Figure 2. Example of correlation of natural gamma traces

MONITORING AND EARLY WARNING SYSTEMS

The instrumentation installed during the investigation forms part of a network of monitoring positions throughout the town. (Fort et al., 2000). The regular monitoring of groundwater levels and records of movement within the inclinometers is supplemented by GPS (Global Positioning Surveying) of 70 permanent ground surface markers. The results of these data are used to refine the interpretation of the ground models and to adjust projected landslide regression scenarios (Lee et al., 2000). In addition the monitoring has a major role to play as an early warning system for the onset of significant movements which could present a threat to public safety and property.

The early warning function of the monitoring is a key function for coastal sites in the UK where major coastal defence and slope stabilisation works are required (Clark et al., 1996). At Lyme Regis several years may elapse between the ground investigation and instrumentation phase of the work to completion of the construction of new coastal defences

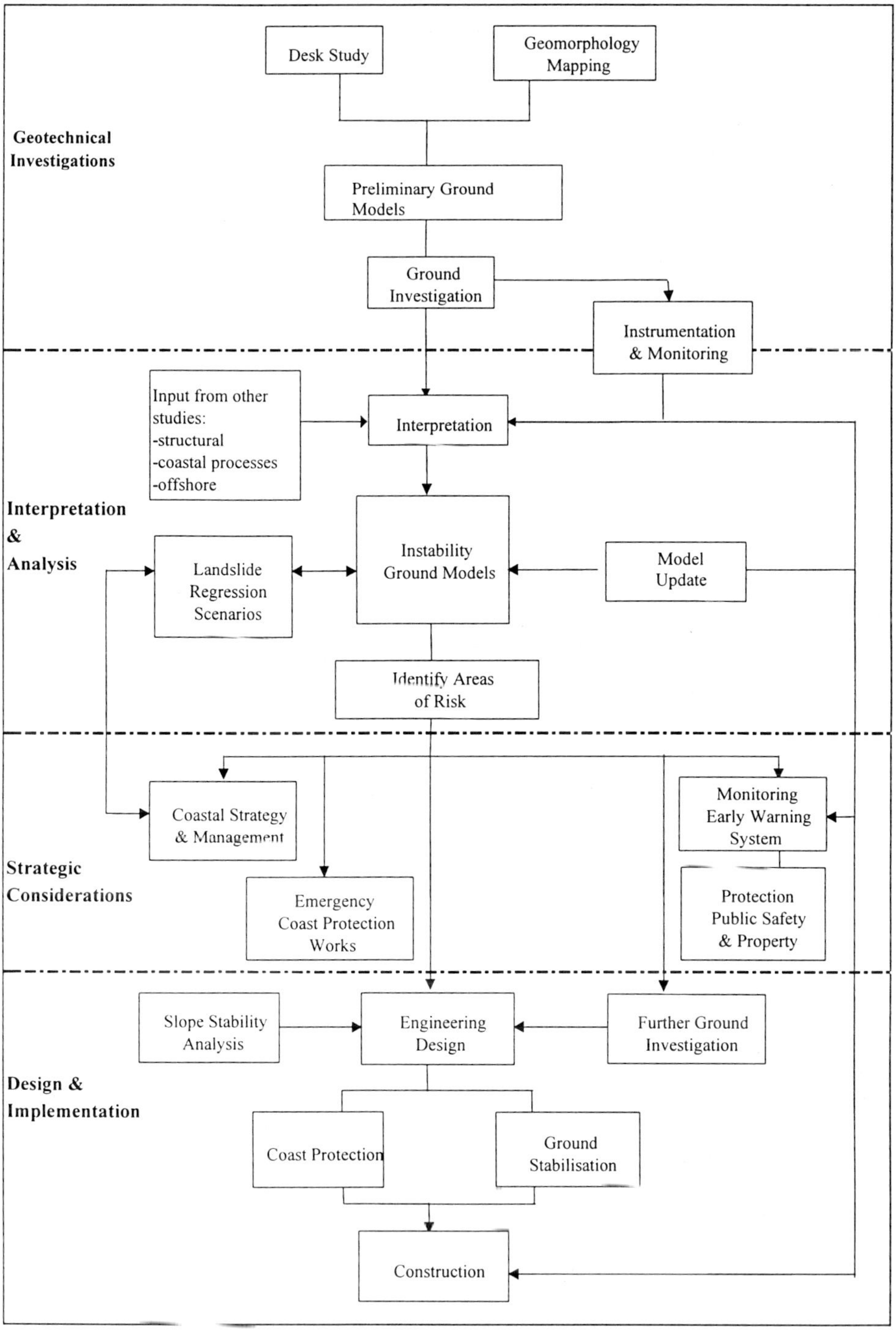

Figure 3. Landslide management strategy framework

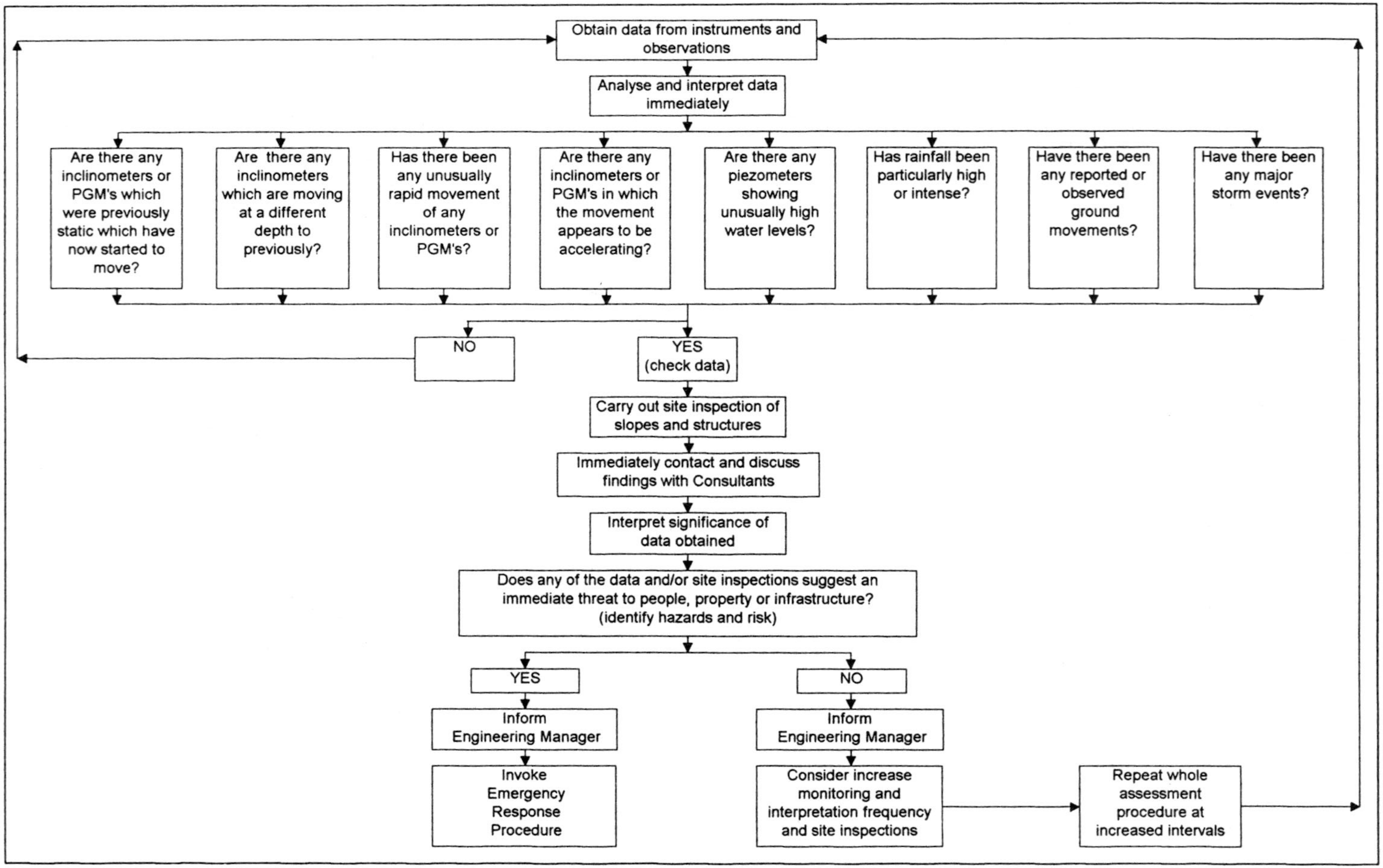

Figure 4. Monitoring Data Analysis and Response Procedure

and slope stabilisation. This is inevitable due to the stages of design, procurement, environmental assessment and funding approvals that have to be completed prior to construction. Thus the monitoring provides interim protection for the area during this period. As part of the strategy for Lyme Regis a detailed procedure has been developed to regularly review the monitoring data and to identify thresholds of significant movement, evaluate the consequences of continued movement, establish communication routes and nominate individual responsibilities and response pathways. A summary of the procedure of actions to be taken in the event of major instability are shown on Figure 4.

DISCUSSION

The investigations undertaken at Lyme Regis have been extensive, systematic and multidisciplinary in their nature, with the geotechnical investigations being by far the largest component. The studies for the future protection of the town have been holistic in approach with the various disciplines contributing to a greater understanding of the problems facing the town in the future.

The detailed results of the geotechnical investigations have enabled the development of ground models for the various forms of instability within the town. These models form the basis of the future control of landsliding enabling a strategic approach to be taken rather than emergency responses to resolve the problems that occur as a result of failures. The holistic approach allows design of the best combination of landslide and coast defence engineering solutions together with measures to deal with other problems such as beach management, seawall maintenance and preservation of ecologically sensitive environments.

ACKNOWLEDGEMENTS

The authors would like to thank West Dorset District Council for granting permission to publish the paper. Thanks are also due to Prof. Denys Brunsden and Prof. Eddie Bromhead for their advice and enthusiasm during the project, and also to past and present colleagues at High-Point Rendel.

The work was funded jointly by the Ministry of Agriculture, Fisheries & Food and West Dorset District Council.

REFERENCES

Allison R.J., (ed) 1992. The Coastal Landforms of West Dorset. Geologists' Assoc. Guide No. 47.

Brunsden D., 1969. The moving cliffs of Black Ven. Geography Magazine, 41,5 372-4.

Brunsden D., and Goudie A., 1997. Classic Landforms of the West Dorset Coast. Geographical Association.

Chandler J.H., and Brunsden D., 1995. Steady state behaviour of the Black Ven mudslide: the application of archival analytical photogrammetry to study of landform change. Earth Surface Process and Landforms, 20. 255-75.

Clark A.R., Moore R., & Palmer J.S., 1996. Slope Monitoring and Early Warning Systems: Application to Coastal Landslides on the South and East Coast of England, UK. Proc 7th International Symposium on Landslides – Trondheim 1531-1538.

Cole K., and Joy D., Lyme Regis Environmental Improvements. Proc. 29th MAFF Conf. River & Coastal Engineers. 7.2.1 – 7.2.13

Fort D.S., Clark A.R., Savage D.T., and Davis G.M., 2000. Instrumentation and Monitoring of the Coastal Landslides at Lyme Regis, Dorset, UK. 8th International Symposium on Landslides, Cardiff.

Hutchinson J.N., 1984. An Influence Line Approach to the Stabilisation of Slopes by Cuts and Fills. Canadian Geotechnical Journal 21(2) 363-370.

Lee E.M., Brunsden D., and Sellwood M., 2000. Quantitative Risk Assessment of Coastal Landslide problems, Lyme Regis, UK. 8th International Symp. Landslide, Cardiff.

Sellwood M., Davis G M., Brunsden D and Moore R., 2000. Ground Models for the Coastal Landslides at Lyme Regis, Dorset, UK. 8th International Symposium on Landslides.

The Influence of Mining on the Bolsover Landslide

A. E. COBB, Geoffrey Walton Practice, Oxford House, Market St, Charlbury, Oxford, D. B. JONES and H. J. SIDDLE, Halcrow UK, 31-33 Newport Road, Cardiff, UK

INTRODUCTION

The old market town of Bolsover with its seventeenth century castle sits at the edge of the north-south trending Magnesian Limestone escarpment in east Derbyshire. In February 1991, appreciable ground cracking was noted at the crest of the escarpment known as the Back Hills. This developed over the following two years into a large compound landslide. Locally severe damage was caused to eighteen properties on the footslope, and gardens and outbuildings of nine properties on High Street at the crest of the escarpment. Ten buildings had to be demolished. Around £1M of losses were incurred by insurers and property owners as a result of which extensive investigations were carried out in order to establish possible liability for the damage. The detailed evaluation of the influence of mining has provided new insight into the circumstances and processes by which mining can affect slope stability.

GEOLOGICAL SETTING

The Magnesian limestone, comprising thinly bedded moderately strong dolomitic limestone, is some 10m thick and rests on about 20m of Lower Permian Marl which consists of weak calcareous mudstone with some thin beds of dolomite (Figure 2). The underlying 0.2m thick basal Permian Breccia unconformably overlies Middle Coal Measures in which argillaceous lithologies predominate. The solid strata are covered with locally derived colluvium which thickens from 1 to 4m away from the escarpment. The regional dip of the strata is about 1° to the east but local cambering created a slight valleyward dip of the limestone at the edge of the escarpment. High angle normal faults, proved in mine workings, affect the Coal Measures and there is evidence that the two larger faults remained active in post Permian times. The most westerly of the faults is responsible for the anomalous north westerly trend of the escarpment at Bolsover. The workable coal seams are in the Lower and Middle Coal Measures, forming part of the East Pennine Coalfield.

DEVELOPMENT OF THE LANDSLIDE

Apart from cambering of assumed Pleistocene age there was no evidence of previous landsliding in the area until 1923. There is some documentary evidence of slope movements from about 1923 that continued into the 1930s. This episode of movement left a prominent scar within the escarpment and ridges of displaced material on the footslope. The slope was regraded after the landslide and the area preserved as an open space.

In the early 1980s slight movements seem to have occurred in some of the gardens on High Street. It was not until early in 1991 that more substantial movements alerted the residents. Monitoring and ground investigations showed steeply inclined displacements at the crest of the landslide and near horizontal displacements of the footslope. The active wedge of the compound failure was delineated by the rear scarp and a counterscarp that extended along almost the entire length of the toe of the escarpment. The passive wedge moved at least 4m along a failure surface located near the base of the Lower Permian Marl.

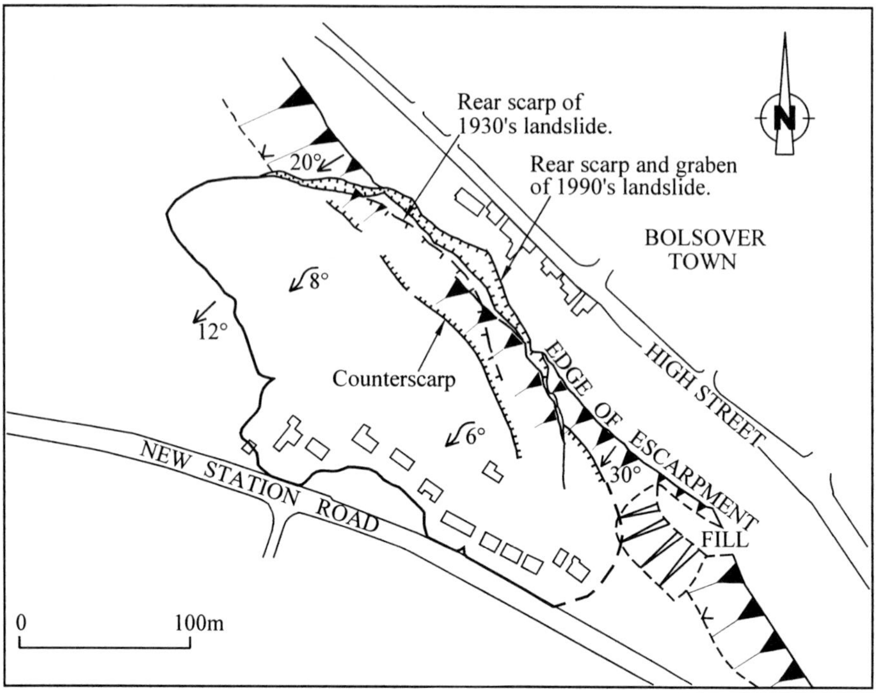

Figure 1. Plan of the landslide

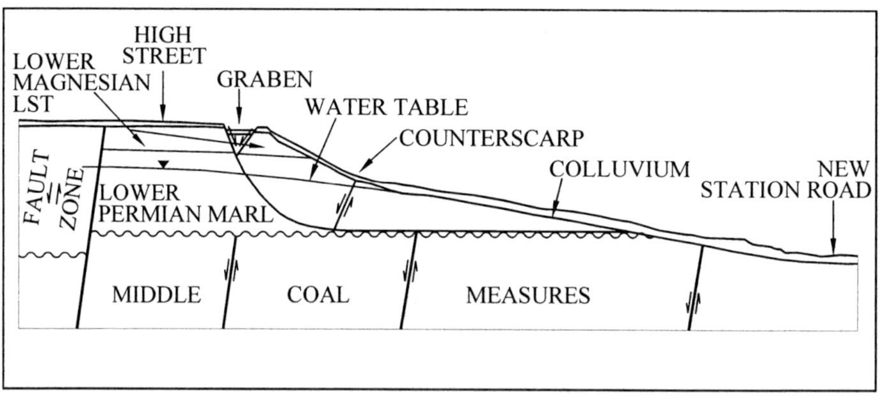

Figure 2. Cross-section through the landslide

The 1990s landslide represented a retrogression and widening of the area of the earlier failure. Although there was heavy rainfall in October 1990, it was only in the order of a 1 in 50 year event, insignificant with respect to the age of the slope. It was concluded that the

1991 landslide event was a progressive failure that represented a further development of the 1920s-1930s landslide. Rainfall may have triggered the 1991 event, but could not have been the main cause of the failure, which had been initiated in the 1980s.

MINING

Mining first approached the Back Hills in 1902, with longwall workings in the Top Hard seam. This seam was about 2m thick, at a depth of about 400m. These workings stopped along a fault running sub-parallel to the escarpment and some 60m south-west of the crest (Figure 3). A pillar of unworked coal was left under Bolsover town and castle. Subsequently, in 1920 to 1925 limited workings took place in this pillar, with narrow, subcritical width extraction panels, most of which were orientated parallel to the faulting.

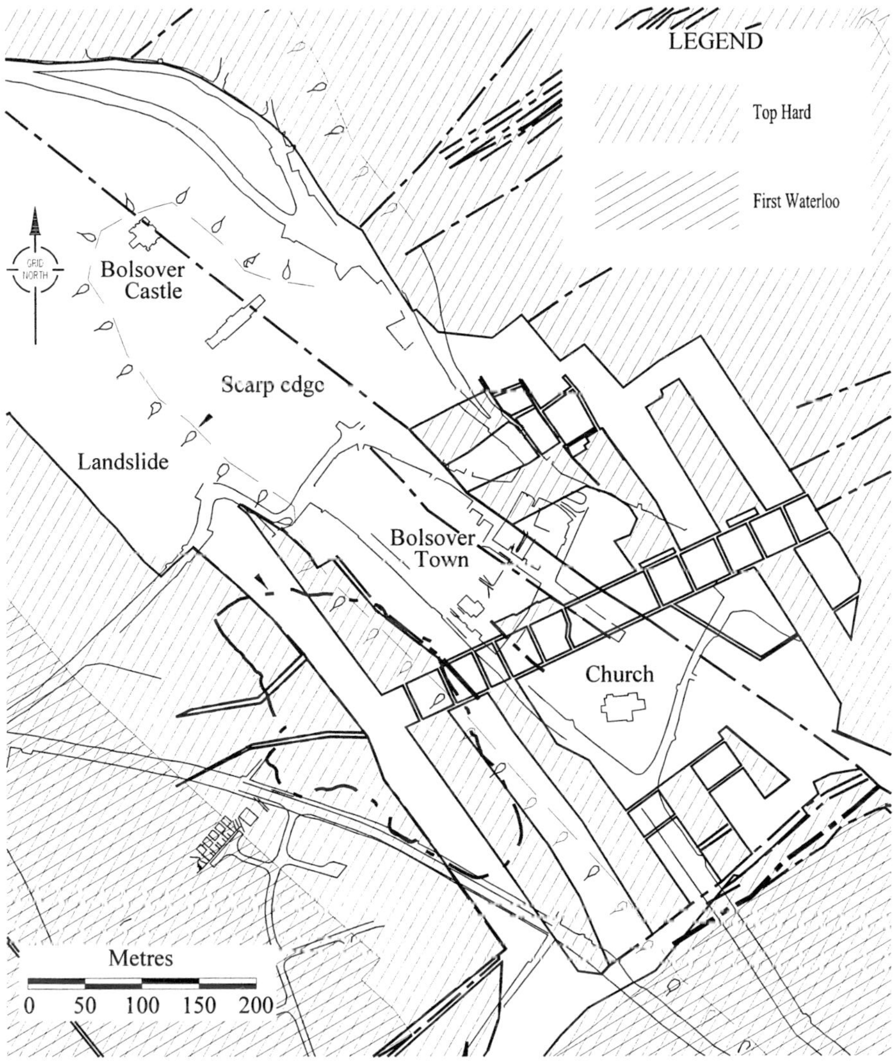

Figure 3. Underground mining in Top Hard and First Waterloo seams

The next seam to be worked (1925-1932) was the First Waterloo, which was about 50m below the Top Hard. It was about 1.0m thick and also worked by longwall methods. This time a larger stand-off was left to Bolsover, with the edge of the workings some 210m from the crest of the scarp (Figure 3). Some six other seams were worked near Bolsover, with three being within the normally defined influence distance of 0.7 times the depth of working from the landslide area. The last workings were in 1984.

SUBSIDENCE

Comparison of the levels of O. S. benchmarks taken in 1875 and 1897, indicated there had been no demonstrable movement in Bolsover town before the advent of mining in the area. Subsequently, O. S. levels indicated subsidence on High Street, behind the landslide, was about 0.27m in 1915, increasing to 0.45m in 1939. (These figures were more than twice what would normally be predicted by the Subsidence Engineers Handbook (SEH) method.) The mining subsidence lead to a steepening of the scarp slope, albeit by only about 0.5°.

BOLSOVER LANDSLIDE ANALYSES

Conventional equilibrium analyses of translational slip surfaces, multi-block wedge-type analyses and finite element and finite difference analyses using the programs 'PLAXIS' and 'FLAC' respectively were performed to assess the stability and cause of the landslide. Each analysis was based on the geological model described previously for a section passing through the central part of the landslide.

Shear strength parameters were determined from a range of laboratory tests carried out as part of the ground investigation and also by using the modified Hoek-Brown failure criterion for jointed rock masses, discussed by Hoek *et al* (1995). A range of parameters representing peak (undisturbed), disturbed and residual shear strengths for the different strata were established. A range of water tables was also analysed to test the effect of changes in water table on stability. Factors of safety were calculated in the equilibrium analyses for a comprehensive set of permutations and combinations of shear strength and water table.

Interpretation of the results of the stability analyses indicated that the area of the landslide should have been stable for all reasonable water tables if peak (undisturbed) shear strength parameters applied and that the area of the landslide was unstable for measured water levels if residual shear strength parameters were assumed. Given that the area of the landslide had been practically stable, at least for deep-seated movements since the 1930s, it was considered likely that the shear strength along the slip surface and potentially along any internal sliding faces had reduced between the 1930s and 1990/91.

There was a dispute about the cause of the landslide. It was alleged that coal mining in the Top Hard and First Waterloo seams had fractured and weakened the strata sufficiently to allow the 1990/91-landslide to develop. The results of the conventional limit equilibrium stability analyses, which relied on assumptions about shear strengths and water tables were subjective and, therefore, inconclusive. The slight downhill tilt of the slope did not appear to give any significant reduction in stability.

To test if there was a pathway between the known mining of 1900-1932, and the 1990/91-landslide, a plane strain finite difference model of the geology was developed using 'FLAC'. The aim of this mathematical model was to simulate the mining and to quantify stress changes in the strata between the coal seams and the ground surface. The finite difference analysis employed stiffness and shear moduli, and shear strength parameters. Various authors

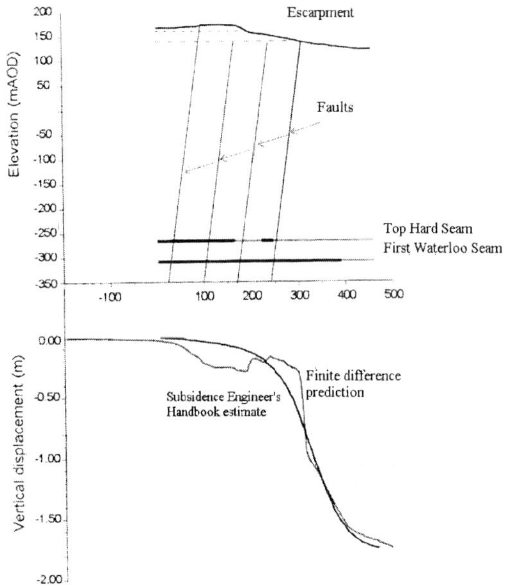

Figure 4. Mining subsidence settlement predictions

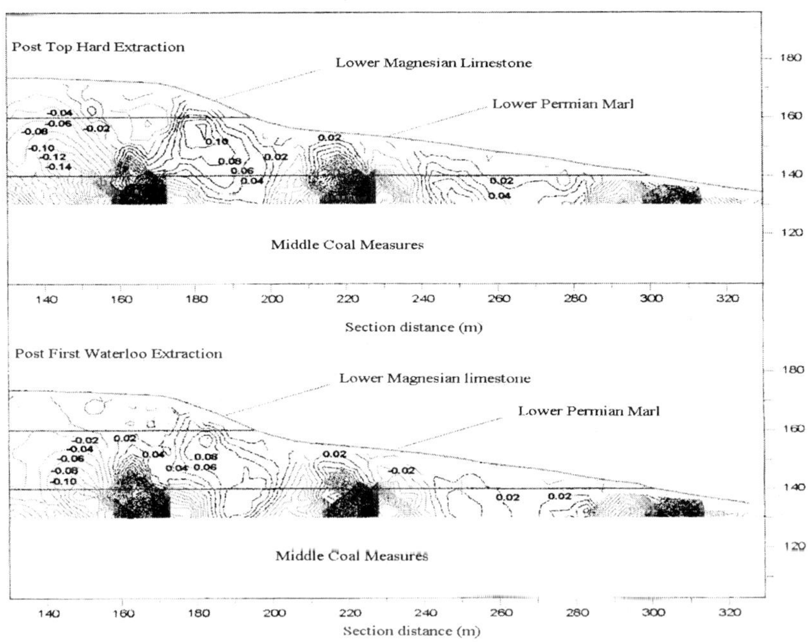

Figure 5. Changes in vertical stress (MPa) caused by mining in the different seams

have commented on the appropriateness and accuracy of finite difference/finite element numerical modelling including Whittaker and Reddish (1989).

To calibrate and test the reliability of the proposed method, a simple geological model assuming a flat ground surface and coal measure strata only was assumed. In this situation, settlements calculated using the empirical SEH method are normally considered accurate. In the first finite different model, the stiffness and shear moduli were obtained from published information, and the previous shear strength parameters were used. To calibrate the FLAC model, finite difference analyses were then performed in which the stiffness parameters were modified until a close approximation to the SEH method was achieved.

The finite difference model was then developed to represent the geological conditions under the escarpment by progressively introducing the known relief, the geological strata, the structure and water table. The findings of each analysis was interpreted in terms of the change in vertical and horizontal stress distributions before and after mining, and the calculated surface settlement vectors. In the final analysis, a non-linear shear strength incorporating strain softening was employed to determine if a failure was predicted that was similar to that observed in 1990/91.

Typical results are shown on Figures 4 and 5. These showed that mining caused:

1. Significant surface settlements and severe differential displacement across faults.
2. Large changes to the historical stress regime and highly stressed zones.
3. A substantial reduction in horizontal stresses at the toe of the landslide instigating strain softening and progressive failure of the strata, which led to the 1990/91-landslide.

CONCLUSIONS

It was concluded that the mining had had a significant effect on the strata. This had probably caused the development of the initial failure and its subsequent extension in 1991, in spite of the latter being many years after the usual cessation of surface movement due to longwall mining.

REFERENCES

Anon. (1975) *Subsidence Engineers Handbook*, 2nd Edition, NCB

Hoek, E, Kaiser, P. K. and Bawden, W. F. (1995) *Support of Underground Excavations in Hard Rock,* Balkema, Rotterdam

Smith E. G, Rhys G. H. and Eden R. A. (1967*) Geology of the Country around Chesterfield, Matlock and Mansfield*, BGS, HMSO, London

Whittaker, B. N. and Reddish, D. J. (1989) *Subsidence, Occurrence, Prediction and Control,* Elsevier, Amsterdam.

Landslide Susceptibility in the Cardoso Slope (Versilia-Italy) Consequences of the Flash Flood of 19th June 1996

L. CODEBO, ENEA CR Casaccia, Rome, G. DELMONACO, C. MARGOTTINI, C. PUGLISI and S. SERAFINI, ENEA CR Casaccia and Ministry of Environment, Rome, Italy

INTRODUCTION

On 19th June 1996 the Versilia area (NW Tuscany, Italy) was affected by an exceptional precipitation event, with a recorded rainfall of 477 mm in 15 hours and a maximum intensity of 158 mm/h, involving an area of about 60 km^2 in the upper part of the Versilia River basin. Such event caused a large flash-flood that buried the village of Cardoso by over 500,000m^3 of mud and debris, killing 14 people and damaging heavily the road network and the economic activities of the whole territory, mainly devoted to the exploitation of marble quarries. The meteorological event triggered over 400 landslides, mostly superficial as debris flows and debris slumps, as direct consequence of critical intensity rainfall. Deeper mass movement occurred, or re-activated, few months later, probably induced by morphological changes and increase of piezometric level due to cumulated precipitations; this lasts constitute today a potential hazardous situation for the safety of the territory also in the light of reconstruction in the damaged area and restoration of shallow phenomena triggered by the crises. Having this in mind, a geological investigation has been undertaken on the slope dominating the area of Cardoso where the re-activation of a dormant rock slide occurred. The stress in the slope was probably induced by the changes in the underneath river morphology, were also the destruction of a bridge, and of some houses, occurred. The present study demonstrated the importance of investigating the slope in all its components and through a multi-disciplinary approach, in order to develop a correct susceptibility zoning, as a basic tool for safety measures and land planning after a disaster.

GEOLOGICAL SETTING OF THE AREA

The study area is situated in the southern portion of Apuan Alps where the Metamorphic Apuan Complex (Palaeozoic) outcrops, tectonically overlain by the Tuscan Complex (Triassic-Oligocene). In particular, the slope of Cardoso is characterised by a thrust of a metamorphic flysch formation, *Pseudomacigno*, over the Palaeozoic bedrock formed by a marble lithotype (*Grezzoni*) and metamorphic sandstones and schists (*Verrucano*) (Fig. 1). This formation is constituted by two distinct units presenting a different bedding: a basal unit, *Pietra di Cardoso,* which is a metamorphic grey compact sandstone, surveyed only in the western side of the slope, overlain by a largely fractured sequence of clayey fillades and sandstones, in alternance; in the eastern portion of the slope the upper unit of the *Pseudomacigno* is tectonically founded over the Palaeozoic bedrock.

The slope of Cardoso is interested by a complex landsliding process (Varnes, 1978), due to the activity of distinct landslide typologies that affect the slope at different depths (Fig. 1); six debris flows were triggered by the meteorological event of June 1996. They can be classified as linear debris flows (Carrara et al., 1985), developed along pre-existing rills, reaching a maximum depth of 2.5 m and generally evolving, in the crown area, as debris slumps. Rockfalls, affecting mainly the sandstone portion of *Pseudomacigno*, are typically punctual events, generated by the selective erosion of fillades due to the infiltration of waters in joints and in the contact sandstone-fillades and by the retrogressive erosion of rills and debris flows. Another rockfall involves the scarp S of Orzale in correspondence of a fault, in the crown area of an ancient rock slide.

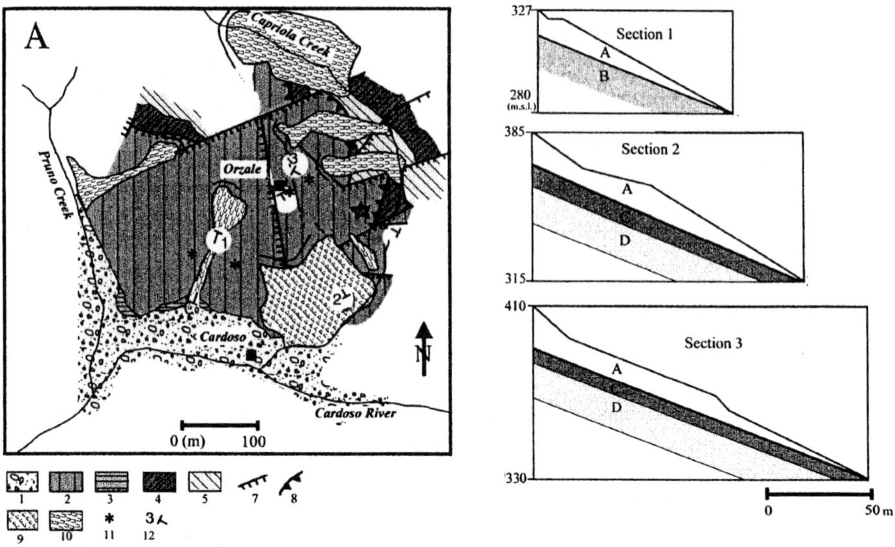

Fig. 1. Geological and morphological map of investigated area. 1)Alluvial; 2)Superoir unit of Pseudomacigno (fillades and sandstone) (**A**); 3) lower unit of Pseudomacigno (Pietra di cardoso)(**B**); 4) Grezzoni (**C**); 5) Verrucano (**D**); 7) normal fault; 8) thrust; 9) ancient rock slide; 10) debris flows; 11) surface cracks; 12) section line.

The latter has been re-activated after the event of June 1996, with a retrogressive strain evidences involving a larger area, downhill and uphill, with respect to the ancient phenomenon. Considering the importance of this slope for the reconstruction of the damaged village of Cardoso a multi-disciplinary approach, to assess the rock slide evolution and the relative slope stability has been performed.

DETAILED FIELD INVESTIGATIONS

The detailed geological survey has permitted to identify a complex setting of the rock slide in two distinct portions of the slope.

In the north-eastern part, the potential failure is represented by the tectonic contact (thrust) between *Pseudomacigno* and *Grezzoni* (Fig 1, sections 2 and 3), with a similar slope bedding and depth varying from 10 m, in the upper part of the slope, to 2 m in the lower portion. The thrust outcrops in a rill and at the top of a karstic cave in the formation of *Grezzoni*. Actual activity evidences are represented by deep tension cracks that locally evolve, due to underground circulating waters, as pseudo-karstic pits and caves. In this portion the Palaeozoic bedrock, that underlies the formation of *Grezzoni*, constitutes the basal piezometric level.

In the western portion, the potential failure is represented by the stratigraphic contact between the superior unit of *Pseudomacigno* (fillades and sandstones) and the lower unit (*Pietra di Cardoso*) (Fig. 1, section 1). This contact, outcropping in an abandoned quarry and in a rill, shows an inclination lower than the slope and involves an average depth of materials of ca 6-7 m; in the middle-lower part of this portion, tension cracks, with a maximum displacement of 10 cm, have been detected, both in the debris and in the sandstone portion of the upper unit, while a stress activity is evidenced in the accumulation zone near Cardoso; in the upper part, only a local debris flow occurred and there are no deeper landsliding evidences.

Geophysical and drilling investigations (seismic refraction, geoelectric profiles and boreholes) seem to confirm that the two portions have a distinct structural setting, being in contact by a normal fault with N-S and E-W directions, that lowers the western one, presenting a substantial different geometry of the potential sliding surface. In the western portion the sliding surface, represented by the contact *Pseudomacigno-Pietra di Cardoso,* dips 30° in SE direction; conversely, in the eastern the sliding surface dips 25° toward E.

Several geotechnical investigations have been carried out, consisting in field and laboratory tests for the assessment of physical and mechanical parameters of the outcropping *Pseudomacigno* formation (compressive strength using the Schmidt hammer, JRC, cohesion and internal friction angle, RQD, RMR). Rock masses present a general medium-high degree of fracture, joints are almost planar smooth to slightly rough, very tight to moderately open and spacing from 0.2-0.5 m to 1.0-1.5 m respectively in the lower and medium sectors of the slope. RQD values have been estimated through the volume frequency index J_v while shear strength parameters have been assessed with laboratory tests and with field tests using the Schmidt hammer. All the data are summarised in the Table 1 with respect to the geological sections of Fig. 1.

Table 1. Mechanical features of *Pseudomacigno* formation

	Section 1	Section 2	Section 3
Cohesion (kPa) of fillades	0-20	0-20	0-30
Cohesion (kPa) of sandstones	185-200	145-155	210-270
Friction angle	28-32°	28-32°	30-35°
Uniaxial strength (MPa) of fillades	25-32	20-25	35-48
Uniaxial strength (MPa) of sandstones	<10	<10	<10
Rock Mass Rating	34-37	29-38	42-54
Class	IV	IV	III
Description	Poor	Poor	Normal
RQD	45-50	45-50	60-70

A dendrogeomorphological analysis (Fantucci, 1998), performed on 14 trees of *Castanea Sativa Mill,* has reconstructed some growth anomalies from 1825 to 1998 that can be attributed to landsliding activity. The comparison between anomalies and precipitation data shows a direct correlation with 5 days cumulated rainfall > 200-220 mm; in addition the re-activation of the slope has occurred in correspondence of severe flooding events (1952, 1959, 1992, 1996). The tree-ring analysis confirms a different response of the slope portions; the eastern side has been frequently re-activated in the period 1820-1950, while the western area shows anomalies occurred mainly in the interval 1970-1996. In all samples, indications of June 1996 event have been recorded.

SLOPE STABILITY ANALYSIS

Slope stability conditions have been performed through a computer program Slope/w (Geo-Slope Int., 1991-1998) analysing the limit equilibrium conditions along the three sections that represent the geological setting in the distinct portions of the slope. The geometry of the potential failure has been designed taking into account the described geological and geomorphological evidences on the field which define the limits of the potential involved materials. The calculation of the safety factor has been performed through the Morgenstern-Price's method (1965), that fit to the non-circular interpreted potential failures geometry of the slope, simulating, in a first iteration, a completely drained situation and values of strength parameters variables from c' = 20-30 kPa and Φ' = 28-35°, as inferred from field tests (Tab. 1). A probabilistic analysis, using the Monte Carlo technique, has been performed on the critical slip surface, taking into consideration the variability of the input parameters, assumed to be normally distributed. In absence of groundwater, the limit equilibrium conditions in the three distinct geological sections envisage a substantial stability and a quasi null failure probability, with lowest values of Fs in the section 3 (1.48), intermediate in section 1 (1.55), highest in section 2 (1.58) (Tab. 2). The same analysis has been performed simulating different levels of the groundwater up to a complete saturation of the slope, in terms of pore-pressure ratio progressively increasing up to r_u=0.40 (complete saturation). A critical situation occurs in sections 2 for values r_u=0.4 with a failure probability of 69% and in section 3 for r_u=0.3 (P=54%); for section 1 a full saturation of materials produces a mean Fs=1.13 with a failure probability of only ca 26%.

Tab. 2. Calculation of mean values of Fs and failure probability P with respect to pore-pressure ratio r_u for each geological section of the slope for c'= 20-30 kPa; Φ' = 28-35°.

	Section 1		Section 2		Section 3	
	Mean Fs	P (%)	Mean Fs	P (%)	Mean Fs	P (%)
r_u=0	1.55	0.69	1.58	0.03	1.48	0.02
r_u=0.1	1.45	3.27	1.42	0.48	1.33	0.26
r_u=0.2	1.34	8.39	1.27	3.41	1.16	6.66
r_u=0.3	1.23	15.24	1.11	19.06	0.99	53.81
r_u=0.4	1.13	26.31	0.95	69.76	0.82	98.87

A second analysis has been carried out considering the residual shear resistance parameters obtained from laboratory tests in the silty-clayey portion of the *Pseudomacigno* formation (c'= 0-5 kPa; Φ' = 25-33°). In the same way, the safety factor has been assessed coupled with an increase of the piezometric level, acting along the failure surface. The factor of safety for all sections drops under the unit for values of r_u in the range 0.15-0.2, which correspond to a piezometric height variable from 3.6 to 4.8 m, for inferred 10 m thickness of involved material. The most critical situation of instability is in the section 3 (Tab. 3); such a water level values are reasonably expectable even considering the presence of discontinuities, pseudo-karstic voids, and the high capacity of drainage of the slope.

Tab. 3. Calculation of mean values of Fs and failure probability P with respect to pore-pressure ratio r_u for each geological section of the slope for c'= 0-5 kPa; Φ' = 25-33°.

	Section 1		Section 2		Section 3	
	Mean Fs	P (%)	Mean Fs	P (%)	Mean Fs	P (%)
r_u=0	1.18	17.74	1.28	0.23	1.15	13.71
r_u=0.1	1.09	34.26	1.13	5.57	1.02	42.32
r_u=0.2	0.96	55.80	0.99	55.32	0.90	83.60

SUSCEPTIBILITY ANALYSIS

The data interpretation of the investigations undertaken in the slope of Cardoso permits to subdivide the slope into three distinct sectors at different landslide susceptibility (Fig. 2).

Sector 1 – Low Susceptibility. The potential failure is represented by the stratigraphic contact *Pseudomacigno-Pietra di Cardoso*, that has been estimated at depth > 18m, progressively increasing at N direction. This sector manifests superficial landsliding involving weathered soils with a general depth < 2m; strain evidences ascribing to the investigated rock slide have not been detected.

Sector 2 – Medium Susceptibility. There are clear evidences of re-activation of pre-existing rock slide with an inferred failure depth of about 6-7m; the potential sliding surface is on the stratigraphic contact *Pseudomacigno-Pietra di Cardoso.* Overstressing in the slope toe is presently active while tensile cracks in the upper part suggest increasing retrogressive activity which may progressively involving the Sector 1 (Fig. 1, section 1).

Sector 3 – High Susceptibility. This sector presents a quiescent rock slide, whose failure is located at a maximum depth of 10m and constitued by the tectonic contact between *Pseudomacigno-Grezzoni* (Fig. 1, section 2 and 3) there are strong evidences of strain, evolving in pseudo-karstic formations (horizontal and vertical hollows, pits, caves) and overstressing at the slope toe. According to numerical modelling a critical situation may occur for fully saturated materials and effective strength parameters (r_u=0.4) and, in terms of residual strength, for pore pressure ratio (r_u) ranging between 0.15-0.2.

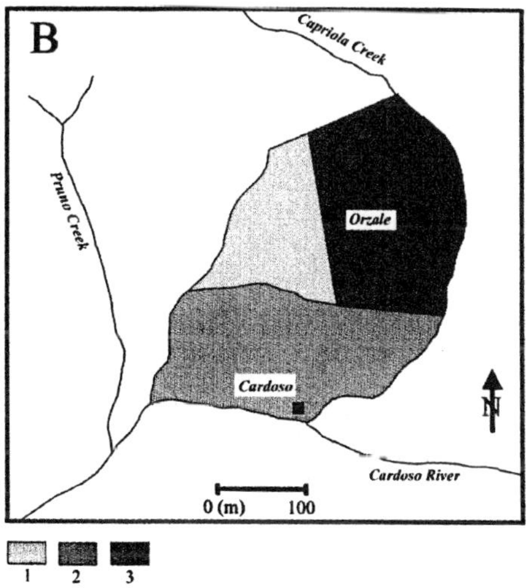

Fig. 2. Susceptibility map of the area. 1) lower susceptibility; 2) medium susceptibility; 3) high susceptibility.

CONCLUSIVE REMARKS

The hydrometorological crises of 19th June 1996 in the Versilia region (Tuscany – Italy), caused more than 400 mainly superficial landslides. In the area of Cardoso, the most damaged village with 14 people killed, the northern slope exhibits six linear debris flows and some unstable blocks. In order to plan the reconstruction activities just below the northern slope a detailed geological and geomorphological investigation, as well as geotechnical, geophysical, and dendrogeomorphological prospectings have been developed, integrated with a local monitoring system.

The results of the present study demonstrate that, critical for future land planning are not the surficial landslides triggered by high precipitation intensity or the isolated unstable blocks but deeper dormant mass movements (rock slide), probably correlated to the shear strength decrease of rock masses due to the fluctuation of the piezometric level.

The analysis of stability, although evidencing a satisfactory stability of rock masses associated to the actually drained conditions, configures at the same time, that a potential re-activation of the ancient rock slide may occur due to pore pressure generation along pre-existing failure surface, identified in this study. The continuos activity of the slope in the past, and than its present residual stage, has been confirmed by:

- tree-ring analysis (decrease of growing up of vegetation due to cutting of roots when laying on surface failure);
- geomorfological analyses;
- preliminary results of a topographic monitoring system set up in the study area, showing stress evidences mainly in the lower part of sectors 2 and 3.

Moreover, open tensile cracks evidences in the sector 2, as well as superficial slumps occurred some months after the flooding event, seem to define a retrogressive uphill activity with a potential extension of involved materials.

In such a way a susceptibility map of the area, mainly constrained by the quiescent rock slide, was developed. This map identify three areas at different degree of susceptibility but, mainly, recognise that even after a dramatic hydrometeorological cries, like in 1996, there is a need of an exhaustive investigation on the conditions of slopes, before any reconstruction planning. In fact, the crises of 1996 triggered mainly surficial phenomena induced by high hourly intensity of precipitation without affecting deeper mass movement that probably needed more infiltration (less intensity and longer duration), to increase pore pressure till collapse. On the other side the permanent modification of morphology caused by the flash flood (fig. 3), probably it weakened the northern slope of Cardoso that today seems to be closer to collapse than before.

Figure 3 – The village of Cardoso in 1994 and after the flash flood of 19th June 1996 (N on left).

ACKNOWLEDGEMENTS

This study is part of the European Commission project TEMRAP (The European Multi-hazard Risk Assessment Project); the authors wish to thank the scientific responsible Dr. Denis Peter. The authors are also grateful to Comunità Montana 'Alta Versilia' for funding the 1998 spring investigations and C.G.A. (Bologna) for field geoelectric survey and data interpretation.

REFERENCES

Bieniawsky, Z.T. (1989). Engineering rock mass classifications. John Wiley & Son, New York, 251 pp.

Carrara A., D'Elia B. & Semenza E., (1985). Classificazione e nomenclatura dei fenomeni franosi. Geol. Appl. ed Idrogeol., Vol. XX, Part. II, 223-243.

Fantucci, R. (1998). Relazione dendrogeomorfologica sulla frana nei pressi dell'abitato di Cardoso (LU) in Alta Versilia. Documento ENEA, Roma, 28 pp.

Morgenstern N.R. and Price V.E. (1965). The analysis of the stability of general slip surfaces. Geotechnique, V15,, pp. 79-93.

Varnes D.J. (1978). Slope Movement Types and Processes. In *Landslides: Analysis and Control* (Eds R.L. Schuster and R.J. Krizek). Spec. Report 176, Transp. Res. Board, National Academy of Sciences, Washington, DC, 11-33.

VV.AA. (1998). Valutazione della stabilità geomorfologica del versante di Orzale (Stazzema-LU). Documento ENEA, Roma, 50 pp.

Characterisation of a Structurally Controlled Flowslide in a Kaolinised Granitic Slope

J. S. COGGAN, D. STEAD, Camborne School of Mines, University of Exeter, Cornwall, and
J. H. HOWE, ECC International, Cornwall, UK

ABSTRACT

Slope failures in mixed soil/rock slopes frequently involve complex interaction of primary and secondary mechanisms. Examples of this type of environment occur within the china clay open pit mines of south-west England where kaolinised granite is exploited for kaolin. The paper provides an evaluation of a structurally controlled flowslide that occurred in late October 1998 in one of eighteen pits operated by ECC International near St. Austell, Cornwall, UK. Instability followed a period of prolonged rainfall; the initial failure appears to have been structurally controlled with a transition to a flow mechanism in response to the saturated ground conditions. The paper also presents a novel technique to quantify and visualise slope regression during instability by slope profiling using a remote laser-based system.

INTRODUCTION

The potential instability of one particular slope within a china clay pit operated by ECC International had been identified during production planning resulting in a preliminary investigation, including piezometer installation. Within the slope the kaolinised granite rock mass has resulted in a mixture of, primarily, china clay matrix and relatively unkaolinised rock. The structural geology of the rock mass was characterised through discontinuity surveys and shown to be dominated by steeply dipping persistent discontinuities that also control the regional kaolinisation.

The paper describes the site characterisation and results of the preliminary back-analyses of the failed slope. Back-analysis of the failure included both a limit equilibrium and a numerical modelling approach. The emphasis of the numerical modelling was to improve the understanding of the mechanism of failure. The paper highlights the problems associated with analysis of complex slope failure mechanisms.

In order to monitor the slope a Quarryman ALS system was used to undertake periodic slope profiling. Initial results suggest that this represents a novel technique to monitor and quantify slope regression, which will assist slope engineering and landslide management.

Landslides in research, theory and practice, Thomas Telford, London, 2000.

SLOPE CHARACTERISATION

Several researchers, including Irfan and Dearman (1978), Baynes and Dearman (1978) and Hencher *et al* (1990), have highlighted the varying engineering properties associated with the hydrothermally altered granites of the south west of England. Characterisation of weathered or altered rock usually involves categorisation into classes, zones or grades according to readily recognised or simply measured changes in their characteristics which are perceived to indicate significant changes in engineering behaviour.

Characterisation of the kaolinised rock mass used the description and classification of weathered rocks suggested by the Geological Society Engineering Group Working Party (Anon, 1995) and suggestions of Hencher and Martin (1982). In upper regions of the slope, in relatively unkaolinised granite, structural mapping was also performed. Figure 1 provides a schematic representation of weathering grades determined for the slope together with the failure profile. The figure highlights variations from grade I to grade V material. Characterisation of the slope is particularly complex as it contains regions of primarily china clay matrix and relatively unkaolinised rock. The kaolinisation varies both vertically and laterally along the slope profile. Hencher *et al* (1990) noted that altered zones were associated with abundant quartz veins which were much rarer in the unaltered zones. Alteration was also noted to decrease laterally away from such veins.

Discontinuity survey results confirm dominant joint sets trending NW-SE and ENE-WSW. Exley (1959) recognised that the kaolinised zones within the granite are typically orientated parallel to these directions. During mapping 'relict' discontinuities were observed dipping at 30 and 60 degrees, striking sub-parallel with the slope face, within moderately kaolinised regions.

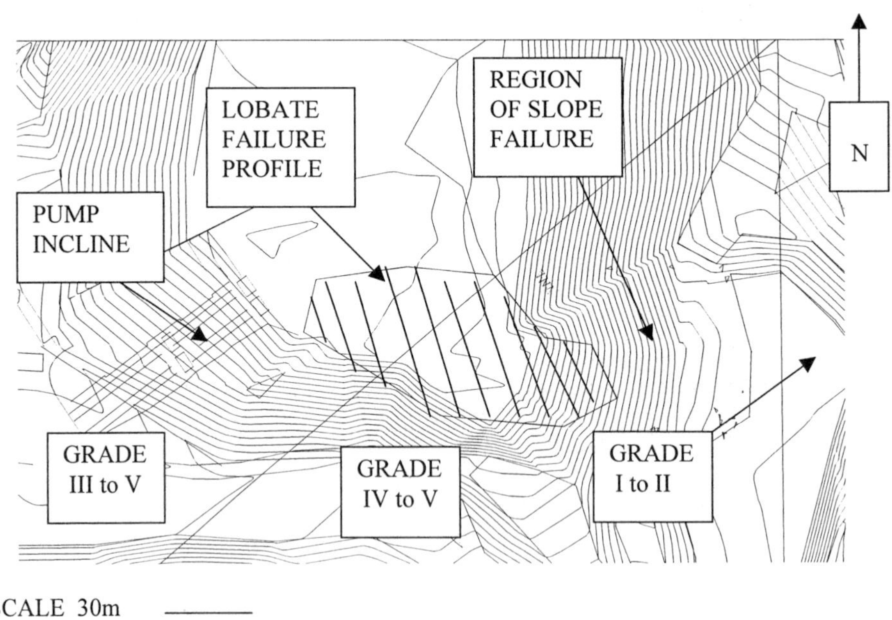

Figure 1. Schematic representation of weathering grades together with plan view of failure

SLOPE FAILURE

A slope failure occurred on 31/10/98 in one of the eighteen operating china clay pits within the south west of England. The failure followed a period of prolonged rainfall and was associated with a 50m high, 40 degree section of slope that resulted in a failure debris volume of approximately 91000m^3 of intensely kaolinised granite. 140mm of rainfall was recorded for October 1998 (compared with an October average of 105mm over the previous five years) with 82mm occurring in the preceding 6 days. Immediately prior to the failure groundwater was observed emanating from the lower section of the slope face, and piezometer records show an increase in groundwater elevation immediately behind the failure location.

Figure 1 shows a plan view of the failure highlighting the lobate form of the failure profile. Figure 2 shows a photograph of the failed mass taken from the top of the failure. Figure 3 shows a photograph taken of the upper section of the failure that appears to be controlled by steeply dipping discontinuities.

Figure 2. View of lobate nature of failed mass taken from slope crest

Figure 3. View of upper region of failure

BACK ANALYSIS OF SLOPE FAILURE

Hencher *et al* (1984) highlight the values of field studies in providing information on the mechanisms which control slope failure in weathered granites. Coggan *et al* (1998) provide a review of the application of various investigative techniques to analysis of quarry slope behaviour. Examples of the use of numerical modelling are provided to highlight their potential application to more complex scenarios that would be difficult to analyse with conventional limit equilibrium techniques. Back analysis of the failure in this study utilised both a limit equilibrium and a numerical modelling approach. Samples were taken for laboratory shear strength testing and compared with published shear strength data on weathered granite.

Limit Equilibrium Approach

Back analysis of the failure, and sensitivity analyses on the effects of varying groundwater levels, using both Hoek and Bray circular failure charts (Figure 4) and the SLOPE/W software (Geoslope International Ltd 1998) (Figure 5), suggest that the failure material has a mass cohesion of approximately 15 kPa for a friction angle of 42 degrees. This is comparable to shear strength magnitudes determined for grade IV/V material. Figure 4 emphasises the influence of cohesion on the factor of safety of the slope. Work by Ebuk *et al* (1993) noted that the relict crystalline structure is the main contributor of cohesion/bond strength in grade IV material.

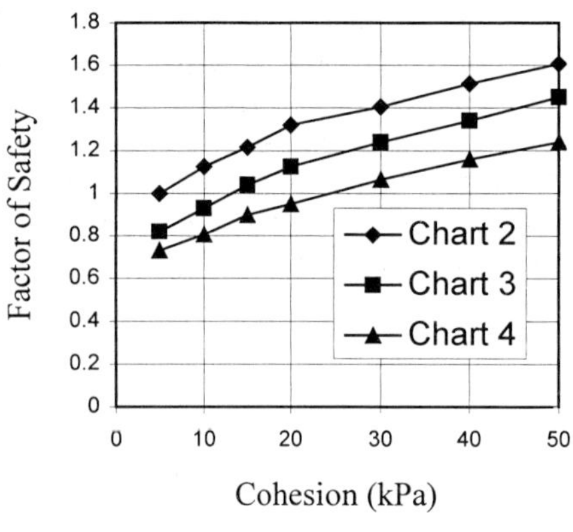

Figure 4. Effect of cohesion on Factor of Safety

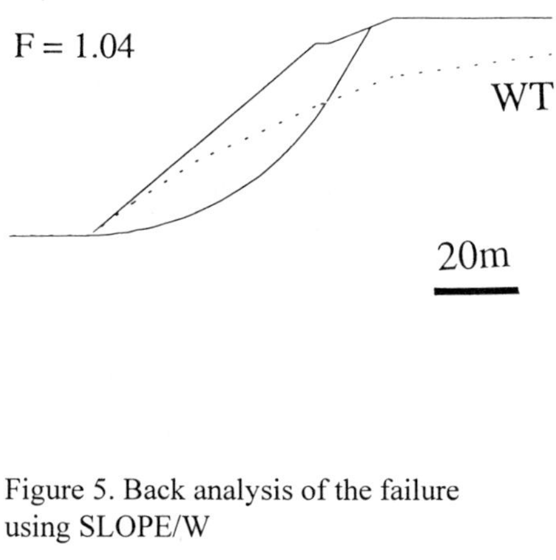

Figure 5. Back analysis of the failure using SLOPE/W

Numerical Modelling Approach

Simplified models were developed within FLAC (Itasca Consulting Group Inc. (1995) in order to investigate the influence of not only groundwater but relict discontinuities on the failure mechanism. Modelling utilised the approach suggested by Starfield and Cundall (1988). Both a Mohr-Coulomb plasticity and ubiquitous joint model were used to perform analyses incorporating different groundwater tables. The ubiquitous model corresponds to a Mohr-Coulomb material that exhibits a well-defined strength anisotropy. Figures 6 and 7 show typical computed contours of maximum shear strain increments for the two constitutive criteria adopted. The figures can be used to provide an indication of the likely failure surface geometry, with Figure 7 highlighting the potential influence of any steeply dipping relict discontinuities on the modelled failure profile. In this case the ubiquitous joint was set at 60 degrees to simulate the steeply dipping relict features.

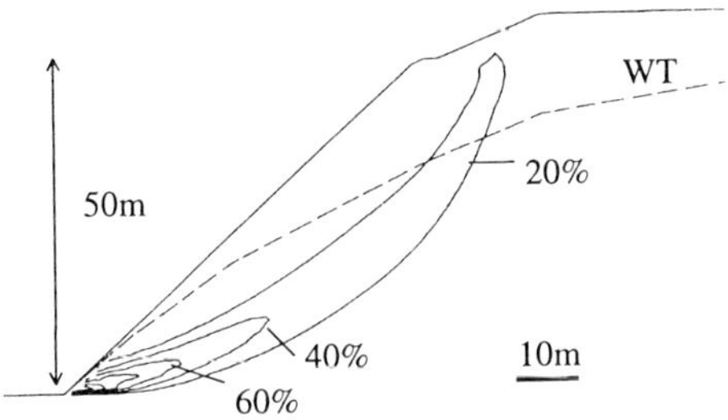

Figure 6. Shear strain increment contours for Mohr-Coulomb model

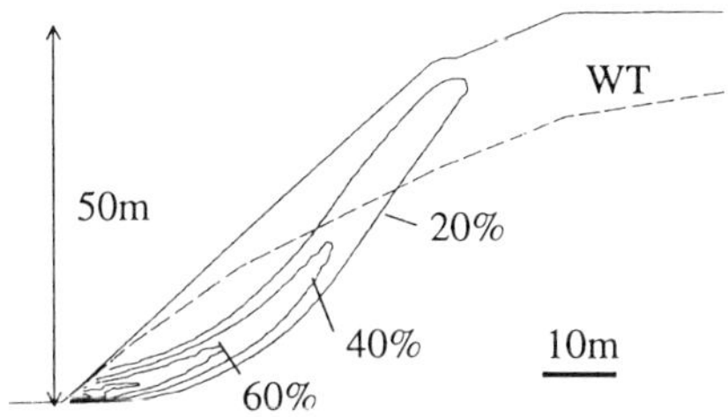

Figure 7. Shear strain increment contours for ubiquitous joint model

VISUALISATION OF SLOPE REGRESSION

Figure 8 shows an image of the upper region of the slope failure generated by slope profiling using a Quarryman ALS system (Measurement Devices Ltd 1998). Importing the recorded slope x-y-z data into the DATAMINE mine design software package (DATAMINE International 1998) allows generation of a three dimensional image of the upper slope profile. Image manipulation in three dimensions can then be performed using visualisation software. This is particularly useful for enhanced analysis of the failure scar. Using a similar methodology periodic profiling can be utilised to both visualise and quantify slope regression Figure 9 shows a combined image of profiles carried out during January (lighter profile) and August 1999 (darker profile). The figure provides a visual indication of slope regression in the vicinity of the slope crest for the seven-month period.

Figure 8. Profiled image of upper region of failure scar

Figure 9. Combined profile image for 7 month period January-August 1999

CONCLUSIONS

The slope failure involved both a translational and flow mechanism. FLAC modelling has been used to simulate the effects of relict discontinuities and the adverse effects of a high groundwater table that complements conventional limit equilibrium analysis. Numerical modelling provides additional information on the failure mechanism that would be difficult to obtain with conventional limit equilibrium techniques. Further investigation is required to fully evaluate the potential detrimental influence of piping and preferential groundwater flow within the slope. The case study highlights the need for use of a coupled groundwater-mechanical model to realistically model liquefaction and flowslide behaviour. Slope profiling provides a useful method for visualisation and quantification of periodic slope regression. This technique could have more widespread application in slope/landslide monitoring and management.

ACKNOWLEDGEMENTS

The authors would like to acknowledge permission from ECC International to publish this paper.

REFERENCES

Anon (1995). The description and classification of weathered rocks for engineering purposes. Geological Society Engineering Group Working Part Report. *Quarterly Journal of Engineering Geology*, 28, 207-242.

Baynes F.J. and Dearman W.R. (1978). The relationship between the microfabric and the engineering properties of weathered granite. *Bulletin of International Association of Engineering Geology*, 18, 191-197.

Coggan J.S., Stead D. and Eyre J.M. (1998). Evaluation of techniques for quarry slope stability assessment. *Trans Inst. Min. & Metall.*, (*Sect. B: Appl. earth sci.*), 107, B139-B147.

DATAMINE International (1998). DATAMINE 5 Users Guide, Wells, UK.

Ebuk E.J., Hencher S.R. and Lumsden A.C. (1993). The influence of structure on the shearing mechanisms of weakly bonded soils. *The Engineering Geology of weak Rock*, Cripps *et al* (eds). Balkema, Rotterdam, 207-215.

Geoslope International Ltd (1998). SLOPE/W. Version 4, CD Rom users manual.

Hencher S.R., Ebuk E.J., Abrams J.H. and Lumsden A.C. (1990). Field description and classification of hydrothermally altered granite from SW England. *10th Southeast Asian geotechnical Conference*, 16-20 April, Taipei, 303-308.

Hencher S.R. and Martin R.P. (1982). The description and classification of weathered rocks in Hong Kong for engineering purposes. *7th Southeast Asian Geotechnical Conference*, 22-26 November, Hong Kong, 125-142.

Hencher S.R., Massey J.B. and Brand E.W. (1984). Application of back analysis to Hong Kong landslides. *Proceeding of the 4th International Symposium on Landslides*, Toronto, 631-638.

Irfan T.Y. and Dearman W.R. (1978). Engineering classification and index properties of a weathered granite. *Bull. of International Association of Engineering Geology*, 17, 79-90.

Itasca Consulting Group Inc. (1995). FLAC (Fast Langrangian Analysis of Continua). Version 3.3 User's manual, Minneapolis, Minnesota, USA.

Measurement Devices Ltd (1998). Quarryman ALS system, Malton, Yorkshire.

Starfield J.L. and Cundall P.A. (1988). Towards a methodology for rock mechanics modelling. *Int. J. Rock Mech. Min. Sci. & Geomech. Abstr.*, 25, 99-106.

Influence of Recent Morphological Evolution on the Stability of Two Slopes in the Euganean Hills

S. COLA, University of Padova, Italy, and P. SIMONINI, University of Padova, Italy

INTRODUCTION

The Euganean Hills were formed during eruptive activities in the Upper Eocene and Inferior Oligocene periods. They cover a restricted area of about 120km^2 in the Venetian alluvial plain and reach a maximum altitude of 600m a.s.l. Magmatic rocks - basalts, ryolithes, trachytes and latites – form the upper and steeper parts with sedimentary rocks, limestone and marls with marine origin, usually present in the lower and more gentle part of the slopes: layers of weathered clay materials normally overlie these rock formations.

A recent census of unstable areas (Dal Prà et al., 1995) singled out that about 4% of the total surface of the hills are affected by landslides due to antrophic as well as natural causes; in 1997 alone, about 60 movements of various sizes and importance were recorded. The majority of slides are superficial and involve the colluvial stratum. For 2000 years the quarrying of marls and trachytes among the human activities has had the greatest impact on the morphology of the Euganean Hills. Its influence on the stability conditions of two landslides in the south-east sector of the hills will be presented and analyzed in the paper using both the limit equilibrium method and finite element analysis.

LANDSLIDE MORPHOLOGY AND HISTORY

The first landslide examined here lies at the base of the Rusta Hill, close to the village of Cinto. Figures 1 and 2 show a plan view of the area and a cross-section of the slope. A marl quarry was opened on this site in the early 60's. In 1976 some small sliding movements involving the colluvial sheet were observed behind the top of the excavation front. Again, in 1985 a larger failure with an extension of about 1000 m^2 occurred at the eastern border of the quarry: at that time the scarps reached a height of about 60 m with a slope of 20°. Thereafter, the Mine and Quarry Regional Department decided to suspend the excavation works and the scarps were re-profiled, as shown in the cross-section of Figure 2. At the same time, the presence of tension cracks on the ground and damages to private buildings such as the ancient S. Lucia chapel and the water system indicated that a larger area of approximately 20,000 m^2 would be involved in translational movements of the detrital cover. The movements took place along gently sloping surfaces (8°-10°) and continued at a variable rate. Note that the prevailing direction of sliding is NNE-SSW (Section A1 of Figure 1), which does not coincide with the section crossing the quarry (Section B1).

The second landslide is located at the base of Mount Murale, near the city of Este and is an area of about 60.000 m^2 (Figure 3). The slope is characterized by a natural dip angle of about 10°. As in the previous case, quarrying was carried out in the 1970's to provide materials

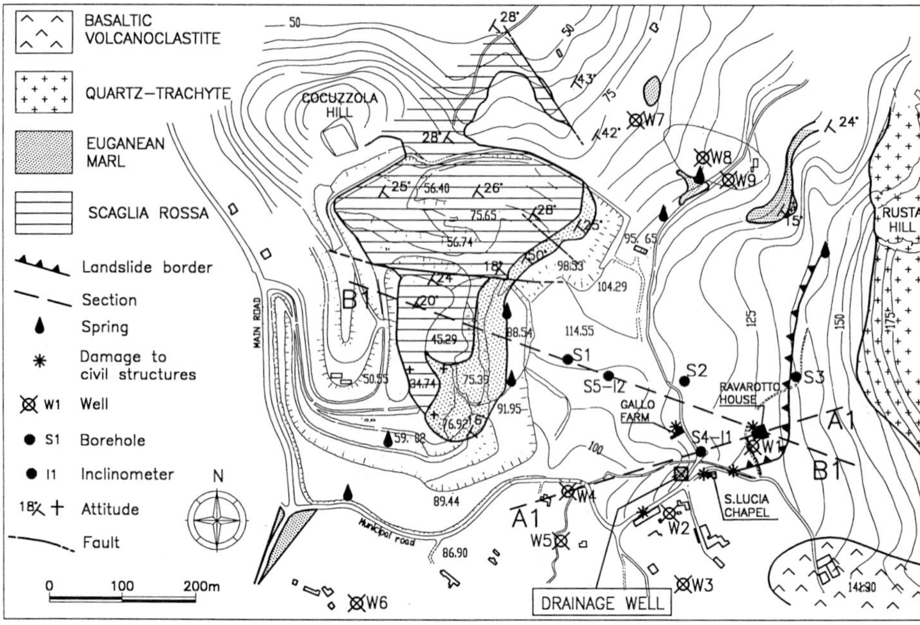

Figure 1. General view of Rusta Hill landslide area

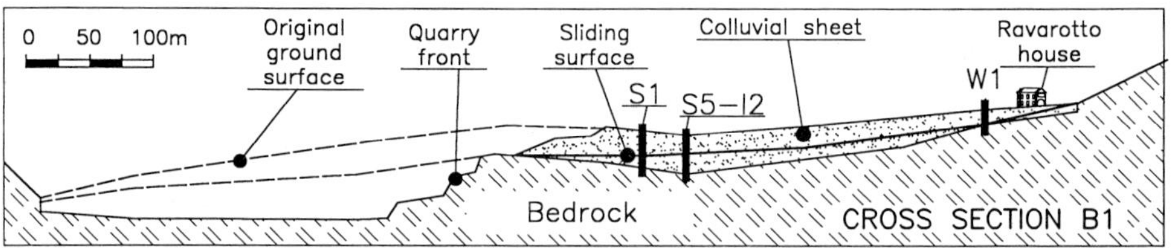

Figure 2. Cross section of Rusta Hill slope intersecting the quarrying area

for cement production. In 1995, a trench crossing transversely from East to West inside the slope was excavated by the National Gas and Oil Public Company in order to install pipelines for methane transport. Although the census of 1995 classified the whole area as unstable, no movements were noted until 1997. The sliding phenomenon occurred during early 1997, after a period of persistent rainfall. A large fissure up with to 1 m of relative sliding was formed in the upper part of the slope. Some damages and fissures in the masonry of old buildings located in the area were claimed by the habitants. Due to a sharp decrease of gas pressure in the pipeline, the Gas and Oil Company interrupted the methane flow and substituted the original pipe with a more flexible one, able to allow for large horizontal displacements without any damage. The cross-section A2 of Figure 4 shows the slope profile together with the maximum extension of quarry when the activity was suspended.

To understand the nature and extension of the landslides of Rusta Hill and Mount Murale and to design stabilization intervention, geotechnical investigations were carried out both in situ and in the laboratory (Aquater, 1986, Favaretti *et al.*, 1991, Trevisan, 1998, Bianchi, 1999). Monitoring systems, consisting in piezometers and inclinometers, were also installed. The location of in-situ investigations is reported in Figure 1 and 3 for the two landslides.

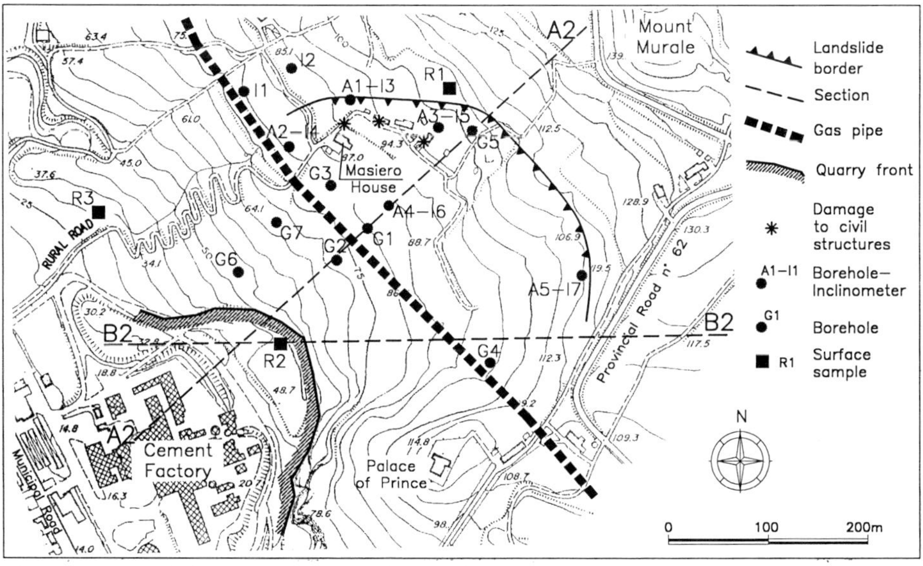

Figure 3. General view of Mount Murale landslide area

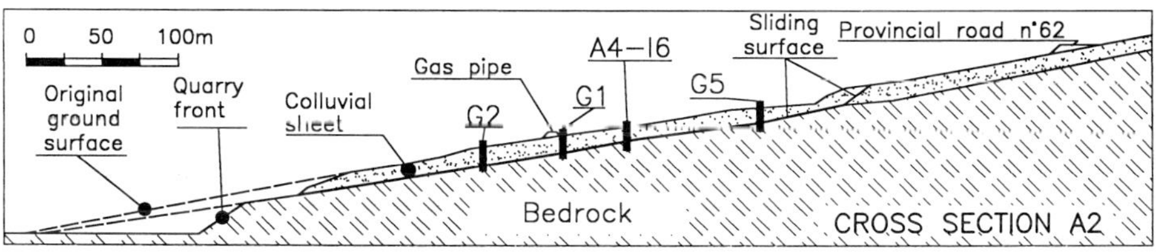

Figure 4. Cross section of Mount Murale slope with the reconstruction of original profile

GEOTECNICAL CHARACTERIZATION OF MATERIALS

Two types of sedimentary soft rock formations emerge in the investigated areas, namely the "Scaglia Rossa", a marly limestone, and the Euganean Marl. In the first area, rock strata are of about 50 m thickness and are sloping in a E-SE direction with a dip angle of 25°-30°. The same formations emerge with sub-horizontal layering in the quarry scarps at Mount Murale. The bedrock lies at depths ranging between 10 m and 30 m in the first area and between 10 m and 15 m in the second one. The sedimentary rocks appear to be medium-highly fractured and moderately weathered at the surface.

The colluvial sheets are in both cases constituted by overconsolidated detrital materials, mostly composed of weathered trachytic elements in a cohesive matrix. Due to this heterogeneity undisturbed sampling was possible only in some cases, therefore, samples subjected to geotechnical laboratory tests were prevalently in remoulded conditions.

The cohesive matrix presents a plasticity index *PI* of 10%÷40% and a liquid limit *LI* of 30%÷66%, the higher values determined on the Mount Murale formations were due to the presence of more clayey components forming the colluvium. The altered marl has a *PI* equal to 20%÷29% and *LI* equal to 37%÷53% without showing any difference between the two sites.

The residual friction angle ϕ_r of the clayey matrix and of the altered marl was determined using the Bromhead apparatus. Both ranges of residual strength (20°÷31° for colluvium and 16°÷21° for altered marl) are very similar in the two sites.

HYDROLOGY AND CLIMATIC CONDITIONS

In the area of the Rusta Hill landslide the water circulation is prevalently confined to the detrital materials resting above the impermeable marly bedrock. An underground valley with orientation E-W was discovered using geophysical investigation. Its presence was confirmed by some springs at the interface detrital cover/marly bedrock in the southern part of the quarry.

The ground water level was measured over the period December 1985 - May 1986. Despite the limited period of observation, strong oscillations of the phreatic surface, up to 7 m, were measured in all the wells and piezometers (except S3) according to climatic conditions. In the S-W zone of the landslide, i.e. piezometer S3 and wells W1 and W2, the maximum water level was measured at depths approximately between 1.0 and 3.0 m below the ground surface. No pore pressure was recorded in piezometer S1 and the water level did not rise over 10.0 m below ground level (g.l.) in S2. It seems that the underground valley, in which the piezometer S1 and S2 are installed, acts as a deep drainage system.

In Mount Murale area no efficient surface drainage systems are present, so in rainy periods water cannot be rapidly discharged and, therefore, easily infiltrates into the ground. In addition, the presence of the trench where the methane pipe was installed might have increased the amount of water infiltration.

As observed in the first case, the impermeable marly bedrock represents a lower boundary for seepage flow. Some measurements of ground water level were collected during Summer 1997 and Winter 1998 by means of piezometers installed in boreholes G1-G7. No significant oscillations of water table were noted; unfortunately the available data do not provide a sufficient database to directly relate the effect of raining on the pore pressure variations in the ground. The maximum water level was measured at a depth between 2.0 m and 7.0 m in the boreholes G1-G4 and at 10.0 m in G5. No water was found in G6 and G7.

The most significant movements of the two landslides occurred in different seasons: the Rusta Hill landslide displaced mostly in Summer 1985, which was an unusually dry period, whereas the largest sliding of Mount Murale was observed in Winter 1997, after three particularly wet months (total rain amount = 420 mm). So, while the decrease of effective stress in the ground might be the main cause inducing the Mount Murale slide, no direct correlation may be found between rain and sliding movements in the first case.

LIMIT EQUILIBRIUM AND FINITE ELEMENT ANALYSES

In order to evaluate the stability condition of the two areas, limit equilibrium (LE) analyses were performed with the Bishop simplified method assuming a residual strength ϕ_r=20° along slip surfaces. Moreover, numerical analyses were carried out using combined finite and boundary element method (BEFE - Beer, 1986) with the aim of studying the effect of excavation on slope movements. The material model for detrital sheet was elastic-perfectly plastic with Young's modulus E=20 MPa, Poisson's ratio ν=0.30 and friction angle ϕ'=20°, as result of laboratory tests on cohesive matrix samples: in this context, it was assumed that the relatively small percentage of boulder elements in the matrix did not significantly influence the deformability and strength of the detrital cover. The material model for the sedimentary bedrock was linear elastic with E=200 MPa and ν=0.30 as suggested by Meigh (1976).

LE analyses were carried out along the sections A1 (maximum slope) and B1 (intersection with the quarry) of the Rusta Hill landslide taking into account two positions for the phreatic surface, i.e coinciding with the g.l. or corresponding to the maximum level measured in piezometers and wells. In both cases, the safety factor resulted greater than unity, approaching the instability in section A1 when the water table reaches the g.l.

BEFE analyses showed that stress relief as a result of the excavation propagated backwards up to the upper border of the landslide. Swelling of the marly bedrock induced a general heave of the whole area, especially in the zone close to the quarry. Small vertical settlements, ranging from 0.02 to 0.04 m, were calculated at the g.l. in that zone where some tension cracks were observed in the field. In the same zone, horizontal displacements between 0.01 and 0.02 m were determined in the direction of the quarry (Figure 5).

As regards the Mount Murale slide, LE analyses examined the effect of pore pressure conditions on slope stability with the present configuration (i.e. excavation at the toe). Two cross-sections A2 and B2 were studied, considering a possible excursion of the water table from bedrock to the ground surface. With the phreatic level coinciding with the ground surface, the safety factor was less than unity. Assuming the maximum water levels measured in the piezometers it turned out to be equal to 1.5 and 1.7 for the two sections respectively. LE analyses were repeated considering the original profile of the slope before quarrying. In this case, the slope reached the failure condition only with the soil fully saturated. On the basis of the comparison between the LE analysis before and after the excavation, a reduction of the factor of safety to between 5% and 20% was observed, depending on the section and hydraulic condition under consideration.

Numerical simulations of the excavation were performed with BEFE software in order to show the effect of excavation also on the displacements behind the quarry (Figure 6). The analyses showed a trend of induced displacements similar to that observed for the Rusta Hill

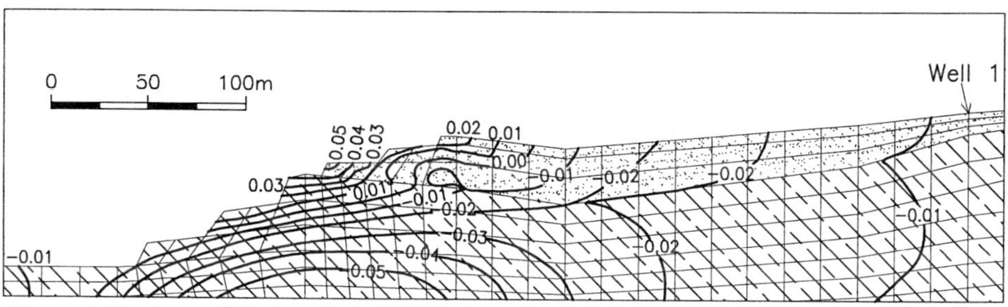

Figure 5. Horizontal displacement distribution due to quarrying in Rusta Hill slope

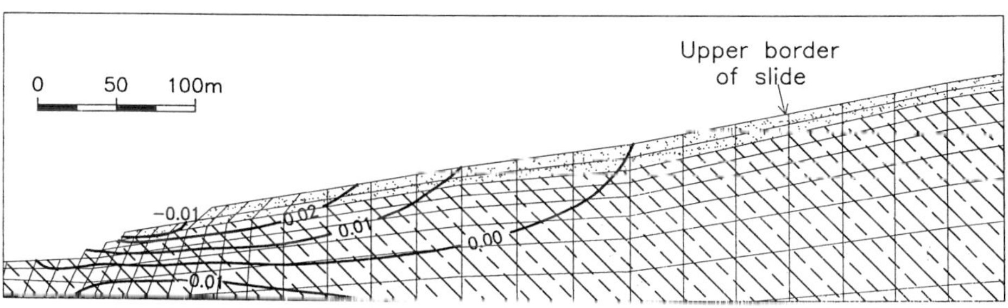

Figure 6. Horizontal displacement distribution due to quarrying in Mount Murale slope

landslide but, in this case, the influence of the stress relief on the cover did not cause any appreciable displacement within the detrital material. Calculated displacements of the ground surface at the upper boundary of the landslide did not exceed 10^{-3} m and $4 \cdot 10^{-3}$ m in the horizontal and vertical direction respectively. The difference between the results of two cases may be explained considering the different volumes excavated: at Rusta Hill it was 10000-12000 m^3 / m with a quarry front of 60 m in height whereas at Mount Murale the volume was only 2000-2500 m^3 / m with a front not over 30 m.

FINAL REMARKS

From the analyses and comparisons between the two cases the following comments and observations can be drawn.

In the case of the Rusta Hill landslide, due to the uncoupling between slide and excavation directions it is very difficult to provide a satisfactory understanding of the sliding phenomenon in the detrital cover by means of the LE method. Additional difficulties in its explanation arise also from the absence of any evident relation between pore pressure variation and landslide activation. An attempt to analyze possible coupling effects between excavation and induced slope movements was carried out with combined finite and boundary element method. In this way it was possible to consider the influence of the very large excavation not only on the overall stability but also on slope displacements. It was shown that the influence of excavation back-propagates, even though with small movements, up to the upper border of the landslide, about 400 m far from the crest of the quarry. On the basis of these results, it may be presumed that the displacements induced by the excavation could represent a decisive factor of sliding when shear stress on possible pre-existing sliding surfaces are very close to failure conditions.

On the contrary, in the case of Mount Murale, the sliding movements did not seem a straightforward consequence of excavation at the base, even though the directions of sliding and quarrying coincide. In fact, the largest movements were observed many years after the end of excavation and after a long period of a heavy rainfall over Northern Italy. Consequently, it may be explained as a consequence of the relevant pore pressure increase in the slope. However, the effect of quarrying cannot be totally excluded: a reduction of the overall stability factor of about 5-20% was calculated with the limit equilibrium method showing that the slope was significantly closer to instability after excavation. With regards to the movements induced by the quarrying on the slope, finite element analyses showed that swelling movements due to the excavation are very small compared with those of the previous case.

REFERENCES

Beer G. (1984). BEFE - A combined Boundary-Finite Element Computer Program. *Advances in Engineering Software*, 6, No. 2.

Bianchi D. (1998). Effetti di un intervento di consolidamento sulla stabilità di un versante nei Colli Euganei. MSc Thesis, University of Padova.

Favaretti M., Previatello P., Soranzo M. (1991). Stability analysis of landslides occurred close to a marl and limestone. *Proc. of the Sixth Int. Symp. on Landslides*, Vol. 1, pp. 397-402.

Meigh A.C. (1976). The Triassic rocks, with particular reference to predicted and observed performance of some major foundations. Rankine Lecture. *Geotechnique*, Vol. 26, No. 3, pp. 391-452.

Dal Prà A., Di Lallo E., Passuto A., Sedea R., Silvano S. (1995a). Le frane nei colli euganei. University of Padova, *Mem. Sci. Geol.* Vol. 47.

Trevisan A. (1998). Analisi di un movimento franoso nei colli Euganei Sud-occidentali. MSc Thesis, University of Padova.

Presentation of Monitoring Data for Unstable Slopes

M. R. COOPER, Gifford & Partners, Southampton, UK

ABSTRACT

This paper considers the key relationships between the nature of data obtained from slope instrumentation and the intended function of the instrumentation. Common practice in presenting instrumentation data is reviewed through an analysis of the contents of the 7th International Symposium on Landslides. As a result a more considered selection of data presentation methods is recommended, and different presentation styles are suggested for different purposes. Specific examples are given of particularly useful but not widely employed presentation styles, with emphasis on the presentation of inclinometer data.

INTRODUCTION

Considerable time, money and thought are generally applied to the design of instrumentation systems for the investigation of landslides and other geotechnical applications. Even more expense is then incurred in the purchase and installation of the chosen system. All too frequently however the results that are obtained from this exercise are presented at best unimaginatively, at worst in a barely decipherable "standard" output. Too often the manufacturer's general plotting software, deliberately designed to be wholly neutral for the widest possible applicability, is used to prepare output intended to illustrate a very specific result or observation.

This paper offers no new field data, no new analysis methods and no great technical breakthrough. It simply seeks to demonstrate how the value of much of what justifiably enters the landslide literature under these headings could be increased. All that is required is the application of as much thought to the final process of data use, that is its presentation, as is routinely applied to all the earlier stages.

INCLINOMETER DATA

Probably the most commonly used instrument to determine and present movement patterns in landslides is the inclinometer. In the 7th International Symposium on Landslides, Trondheim 1996 only a disappointing 15% of papers reported any new field measurement data, but almost two thirds of these gave inclinometer results. Of these papers 70% gave the data in the standard displacement/depth form, for a single face, as shown in Figure 1, from Leroueil et al 1996. But in appropriate situations the data obtained from inclinometer installations has far greater potential than this, and can be used to make much clearer representations of quite complex movement patterns. An inclinometer installation read at reasonably regular time intervals has the potential to yield the following outputs: the true shape and inclination of the tube at any reading date; the displacement at any point, either incremental or cumulative, relative to any earlier reading; the time related movement pattern at any depth and variations

Landslides in research, theory and practice, Thomas Telford, London, 2000.

in the rate of movement. Moreover each of these outputs can easily be presented in true vector format.

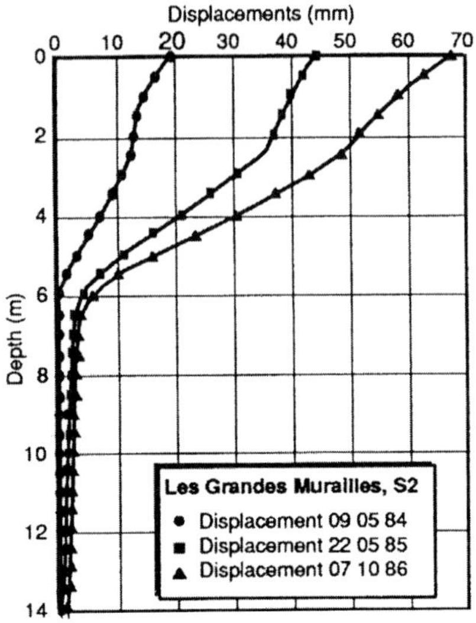

Figure 1. Good example of a standard format inclinometer plot

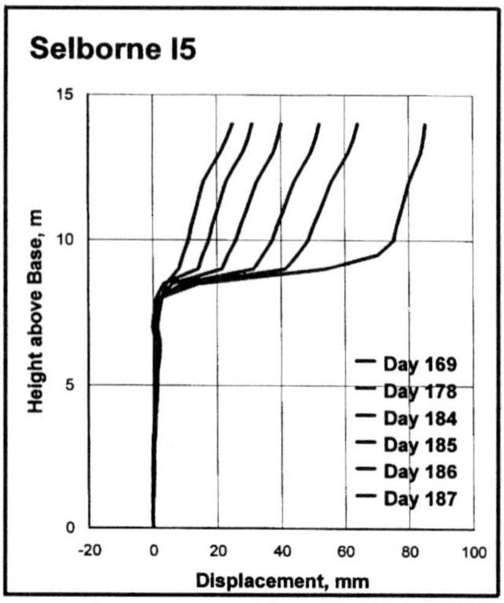

Figure 2. Obscured time relationship

Compare this with what is offered by the standard single face plots, which show only either deflected shape or displacement profiles in arbitrary directions at often haphazard reading dates. This immediately generates the potential for illusory effects.

In Figure 2 data from the Selborne slope study (Cooper et al, 1998) is presented as a group of displacement/depth profiles from six different dates. Without careful appreciation of the legend the impression is of a fairly uniform development of movement at around 5 metres below ground level, but closer inspection of the reading dates reveals that the movements shown in the last four sets of readings in fact took place over a period of only 3 days after over 5 months of very little activity.

The plots in Figure 2 are all from the face indicating movement in the directly downslope direction, which might, without further information, be taken as the true direction of movement in a failing slope. In fact 8 mm of movement is also recorded at Day 187 in the conjugate face giving a true movement direction of 6° off the direct downslope direction, which in this case allowed important conclusions to be drawn on the effects of the geological structure. (Barton and Cooper, 1998).

The following sections show how this data might better be presented to make its meaning and value more accessible to its readership in respect of the two most often overlooked aspects of the data; time and vector information.

Time relationships

As inclinometers are frequently used to monitor the development of an instability, data can often be made more meaningful by plotting the displacement at carefully chosen depths against time. In Figure 3 the complete data set from which the Figure 2 plots were developed is presented as three displacement/time plots: one at depth, one just above the developing slip surface, and one near ground level. The time pattern of the failure's development is immediately more apparent.

Vector relationships

The full value of displacement data can only be realised if the direction is also shown. This can be achieved in a number of ways, two of the most useful are discussed here.

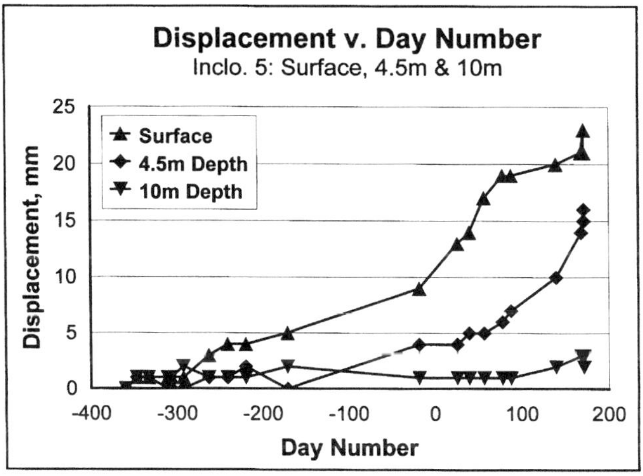

Figure 3. Time based inclinometer plot

In Figure 4 a dual plot shows both the true displacement magnitude and the direction of movement against a common depth axis. In preparing such a plot some care is needed in placing the depth axis on the displacement bearing scale so as to produce a single continuous line. If the wrong value is chosen the line will "jump" from one side of the plot to the other since the two extremities of the direction scale represent the same value. The effect of this is illustrated in Figure 5 where an inappropriate position for the depth axis has resulted in a fragmented presentation of the data from Figure 4.

From this type of plot presenting directional information it is immediately evident whether the movement of the inclinometer is reasonably uniform or whether different influences at different depths may be causing variations in the directions of movement down the depth of the installation. This can be an important consideration in some complex environments where the movement is not just a single slip mass moving in a single direction.

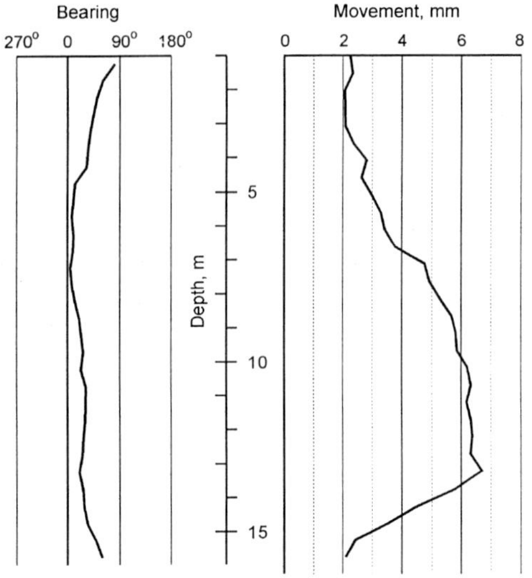

Figure 4. Dual magnitude and bearing plot

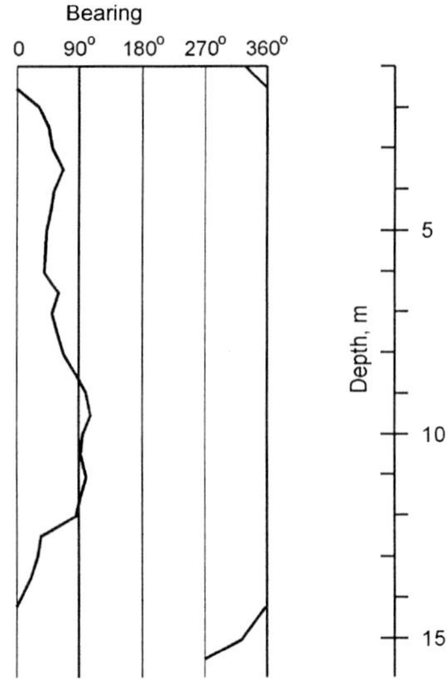

Figure 5. Poor bearing origin

One such case occurs at a cliff stability study on the Isle of Wight, U.K., where a high sea cliff is incised by both a valley (or "chine") and a cliff recession feature, as shown in Figure 6. The effects of these different influences are apparent at different depths and to represent this most clearly a plot is required which gives emphasis to the direction of movement.

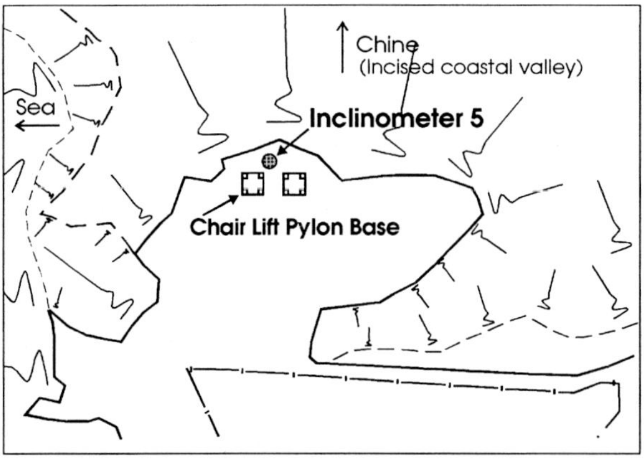

Figure 6. Plan of inclinometer location

Such a presentation is shown in Figure 7, where the direction of movement is shown directly with a depth scale now constructed as the radius of the circle on which the direction is plotted. Thus the centre the circle in Figure 7 represents the top of the tube, whilst the outside represents the bottom. The variations in the direction of movement are immediately obvious. The purely directional plot in Figure 7 should not be used in isolation since it gives no indication of the magnitude of the movement, but it is particularly useful where movement direction requires emphasis. The dual plot of Figure 4 probably represents the best combination of usefulness and versatility for comprehensive presentation of inclinometer data.

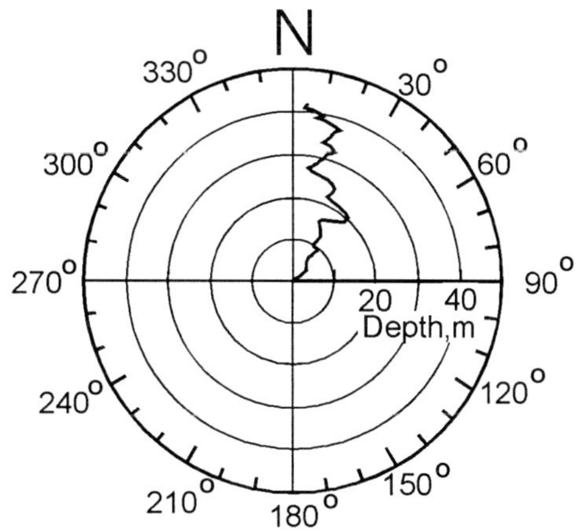

Figure 7. Compass direction-depth plot

PIEZOMETER DATA

Whilst the main aim of this paper has been to encourage more considered presentation of inclinometer data in published work it should also be stressed that similar improvements are achievable in most areas of geotechnics. For instance in the slope engineering problem of rapid embankment construction a key factor considered in many papers is the dissipation of excess pore pressure in relation to the stability condition. In many such papers however authors do their readership a disservice by merely showing the pore-pressure data. The key information required here must be the difference between the measured pore pressure and the presumed long term equilibrium value, which will not always be the measured value pre-construction.

Figures 8 and 9 both show excess pore-pressure data from embankments on soft ground at St Philips Causeway, Bristol presented first in Figure 8 as raw pore-pressure data, where the degree of equalization is not at all apparent, and then in Figure 9 as excess pressure for each piezometer.

CONCLUSION

The purpose of this paper has been, through the presentation of examples, to encourage authors to offer the maximum possible clarity and value to their readership by considering carefully the message they intend to be appreciated from their presented data. The truth of the maxim that "a picture paints a thousand words" is seldom more true than in the presentation of technical data. A well planned and constructed diagram capturing the true essence of a body of data can even breach language barriers and will always repay the modest extra effort involved.

ACKNOWLEDGEMENTS

The data used in this paper has already appeared in other publications with the permission of the Engineering and Physical Sciences Research Council, the Selborne Brick and Tile Company, and Balfour Beatty Civil Engineering Ltd. Data for Figures 4, 5, 6 and 7 is used with the permission of The Needles Pleasure Park, and Figure 7 was developed in conjunction with Claire Jarvis whose study of data presentation methods in geotechnical engineering was undertaken at the University of Southampton, Department of Civil and Structural Engineering.

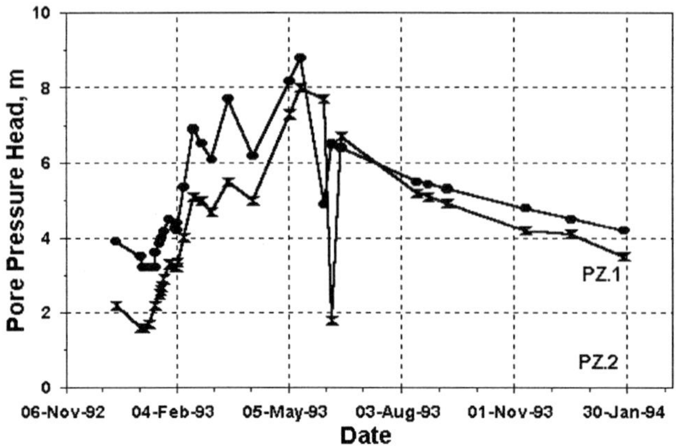

Figure 8. Raw pore pressure data

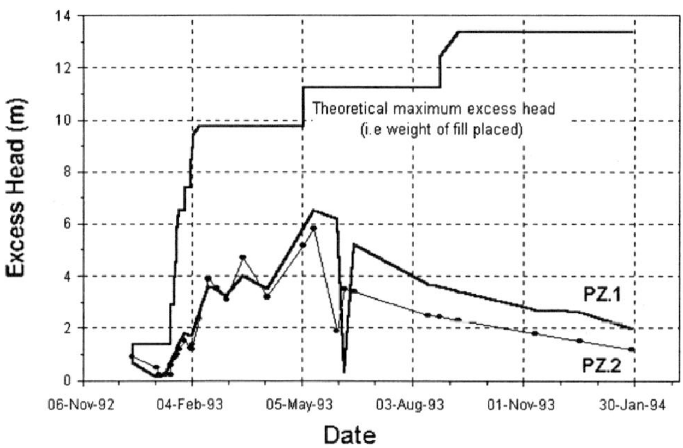

Figure 9. Excess pore pressure clarified

REFERENCES

Barton, M. E. and Cooper, M. R. (1998). Major joint trends in the Gault Clay at Selborne. *Geotechnique* 48, 123-128.

Cooper, M. R., Bromhead, E. N., Petley, D. J. & Grant, D. I. (1998) The Selborne cutting stability experiment. *Geotechnique* 48, 83-101.

Leroueil, S., Locat, J., Vaunat, J., Picarelli, L., Lee, H. & Faure, R. (1996) Geotechnical characterization of slope movements. *Landslides (proc. 7th Int. Symp. Landslides), ed. Senneset.*

Seismic Reactivation of Pelton Park Landslide

D. H. CORNFORTH, Landslide Technology, Portland, Oregon, USA

INTRODUCTION

Pelton Park is on a bench about 2 miles (3.2 km) upstream from Pelton Dam in central Oregon. The bench is an erosional feature left behind when the Deschutes River cut its channel through beds of basalt and tuff. The entire bench is a marginally-stable landslide which tapers from narrow ledges at the ends to a maximum length of about 550 feet (170 m) near the center (Fig. 1). The width of the landslide is approximately 2,300 feet (700 m); the volume is about 2,000,000 cu.yd. (1,530,000 cu. m).

FILL LOAD REACTIVATION

The slide reactivated in 1990 when an earthmoving contractor pushed a large quantity of soil over the edge of a cliff above the park (see Plan, Fig. 1). This work began on about May 1, 1990. Trucks hauled debris away from the base of the spoil heap to an off-site disposal area; however, the heap grew faster than it was being removed. The spoil heap reached its maximum size around June 5 when the earthmoving contract was halted. It remained in place for two months and was removed between July 25 and August 2, 1990.

Two inclinometers, SI-1 and SI-2, had been installed on the bench in early 1990. As the spoil heap piled up, the inclinometers showed that the ground was shearing horizontally at elev. 1,540 (470 m), approximately 75 to 85 feet (23 to 26 m) below the ground (Fig. 2). Shear was occurring within a stiff, tuffaceous silt (white volcanic ash) immediately above basalt bedrock, as shown on one of the boring logs reproduced in Figure 3. The toe of the landslide is below the reservoir surface. Subsequently, two piezometers were installed on the park slide, each with the tip sealed near the slide plane. Groundwater levels were 1 to 2 feet (0.3 to 0.6 m) above the reservoir level in Lake Simtustus (elev. 1,577).

The tuffaceous silts in this part of Oregon are on essentially horizontal bedding planes with some crossbedding. Slide failures generally occur as a double wedge with the upper wedge providing the push to the lower wedge. The lower wedge, typically much larger and having zero driving force, provides resistance along a very weak, slickensided shear zone. Residual strengths in these tuffaceous silts are around $\phi'_r = 8$ degrees.

The headscarp is exposed on the thin cover of talus above the bench. Once it reaches the tuff beds, the shear plane of the upper wedge probably dips at about 60 degrees to the horizontal. The downslope edge of the upper wedge is defined by a reverse scarp, as shown on the plan and section. This interpretation is shown on cross-section A-A (Fig. 2). It can be seen that the large spoil heap sits atop the upper wedge and increases the driving force of the landslide. Although limited in width compared to the entire slide, it was sufficient to reactivate the entire landslide.

When ground cracks first appeared, Portland General Electric (PGE) installed several survey

Landslides in research, theory and practice, Thomas Telford, London, 2000.

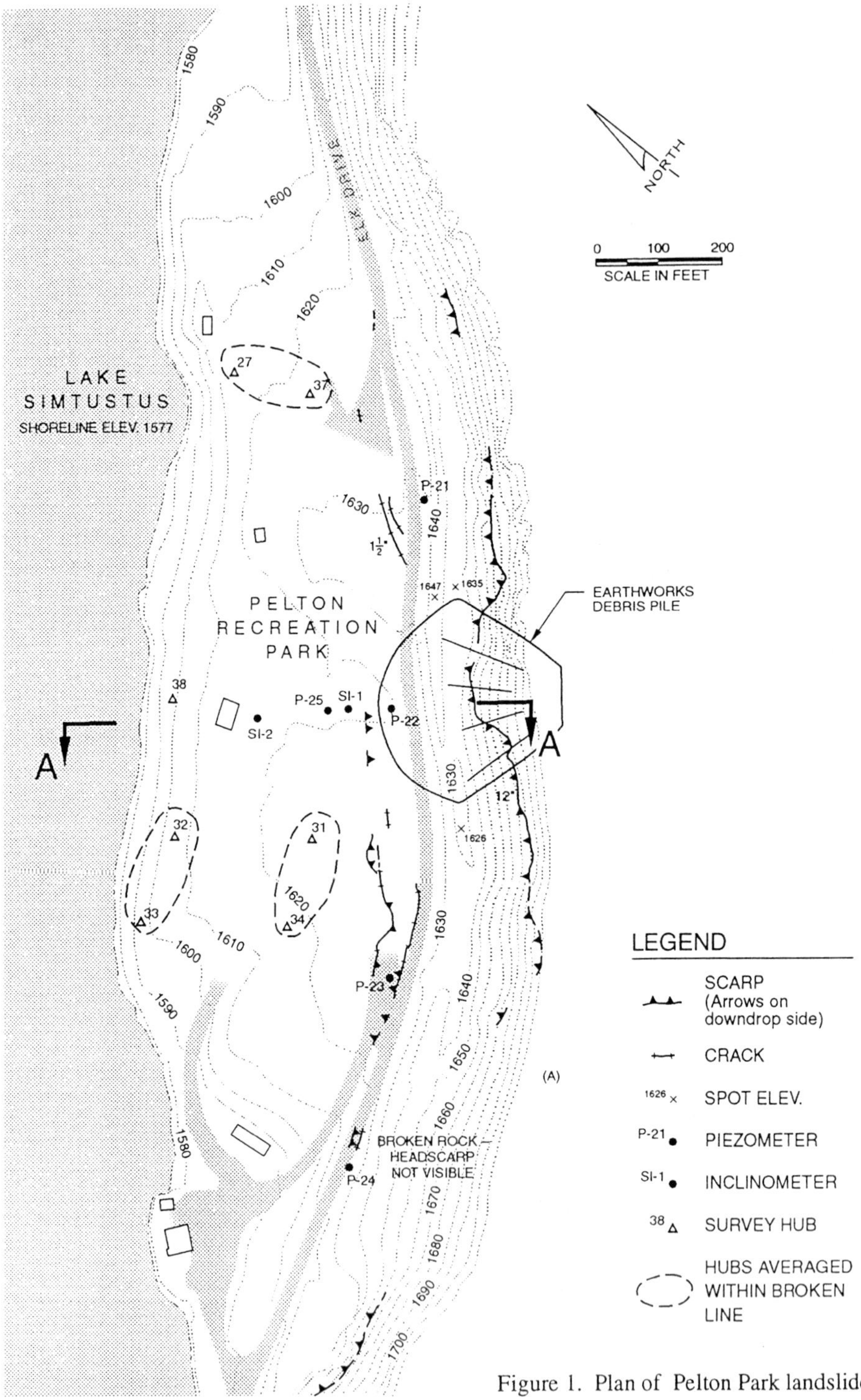

Figure 1. Plan of Pelton Park landslide.

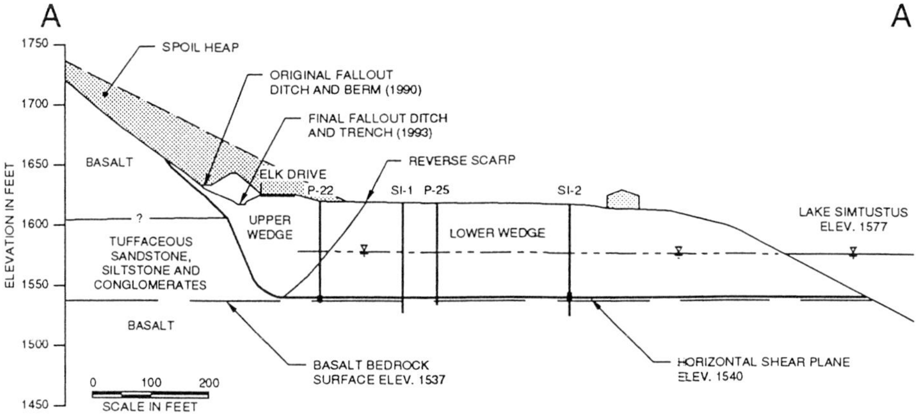

Figure 2. Cross-section A-A through the center of the landslide.

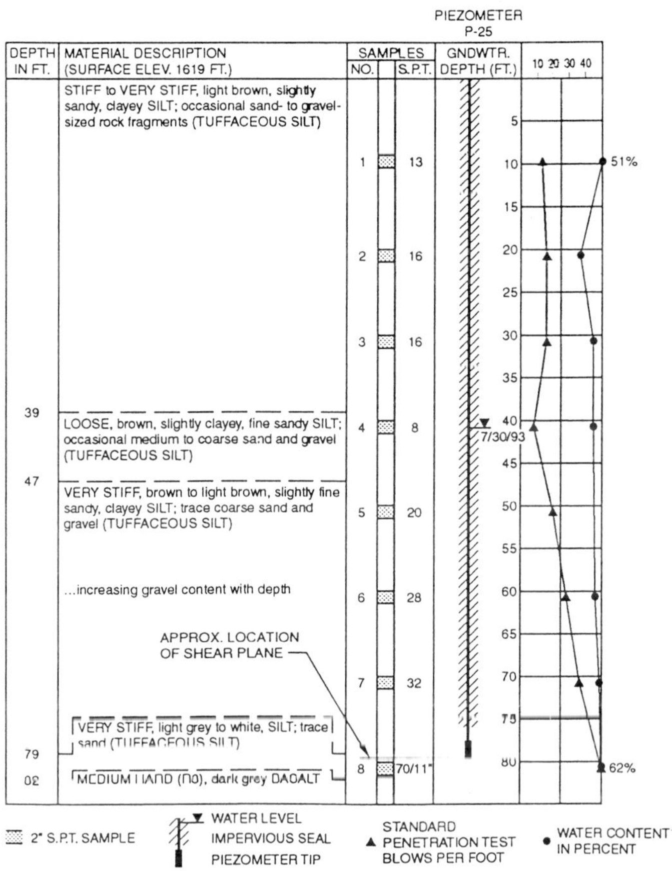

Figure 3. Summary boring log for inclinometer SI-1, and piezometer P-25 details.

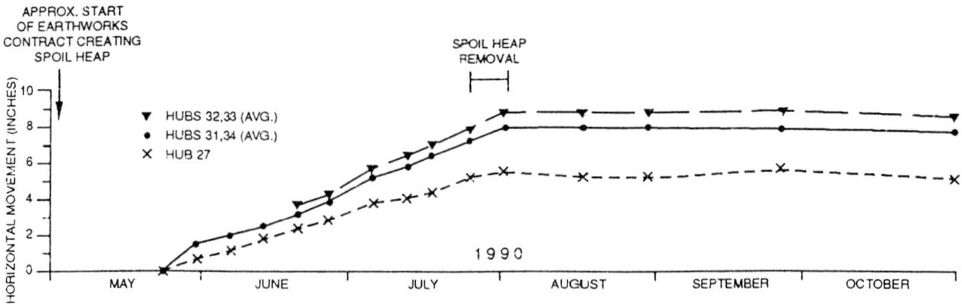

Figure 4. Movement vs. time for landslide reactivated by fill heap.

hubs to monitor the amount and rate of movement in different areas of the park. Three areas have been selected for analysis; the survey hubs in each area are surrounded by a broken line on the plan (Fig. 1). Each hub within these selected areas had essentially the same movements and have been averaged. The movement-time plots (Fig. 4) show a near-linear rate of movement from May 23 (when they were set) to August 2 (when the heap was removed). The rates of movement were as high as 1 inch (25 mm) per week. Movements stopped when the spoil heap was removed.

After the heap removal, PGE reshaped the upper wedge area to provide a fallout ditch and berm (Fig. 2). These catch any rocks falling off the slope above and thus protect the perimeter road (Elk Drive) which runs along the base of the basalt cliff. Numerous readings of the survey hubs showed no movements over the next two years.

SEISMIC REACTIVATION

On March 25, 1993, almost three years after the fill load reactivation had ceased, a crustal earthquake of magnitude 5.6 occurred with its epicenter near Scotts Mills, about 72 miles

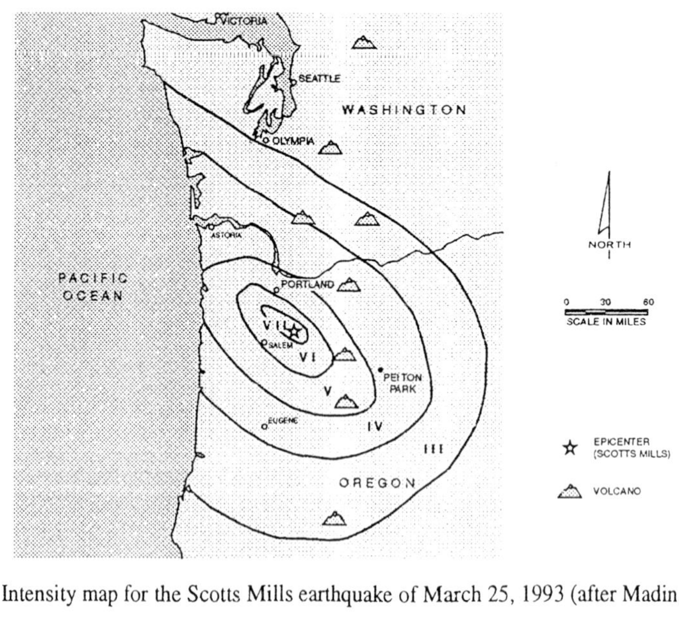

Figure 5. Intensity map for the Scotts Mills earthquake of March 25, 1993 (after Madin, et al., 1993).

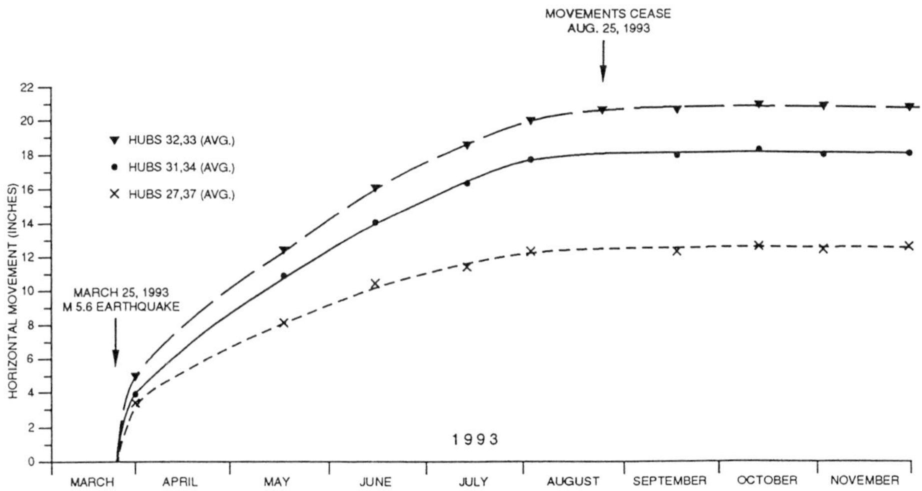

Figure 6. Movement vs. time for landslide reactivated by the Scotts Mills earthquake.

(112 km) from the site. The focal depth was 10 miles (16 km). Although there are no instruments on Pelton Dam, the maximum ground acceleration at this distance is estimated to be around 0.01g (Madin, et al., 1993).

The intensity maps produced after the earthquake (Madin, et al., 1993) indicate that the park landslide would have experienced Intensity level IV bordering on V (Fig. 5). The Modified Mercalli scale Intensity IV is equivalent to "Vibration like passing of heavy trucks; or sensation of a jolt like a heavy ball striking the walls". Intensity V: "Felt outdoors. Small unstable objects displaced or upset".

PGE surveyors were on the site within six days of the earthquake and measured renewed movements at the survey hubs. Subsequently, readings were taken at about monthly intervals and showed substantial continuing movements (Fig. 6). Total movements were as high as 21 inches (530 mm).

The horizontal movements can be interpreted as follows:

(i) an initial rapid movementof $3\frac{1}{2}$ to 5 inches (90 to 130 mm) at these six survey points, followed by

(ii) a subsequent velocity of about 1 inch (25 mm) per week which slowly decreases over a three-month period, and

(iii) cessation of movements around August 25, five months after the earthquake.

A surprising aspect of this case history is that such a modest earthquake shaking could cause this prolonged and significant amount of movement. The question arises: could there have been some other cause?

Pelton Dam has year-round maintenance staff at the hydroelectric project. There was no change in dam operations which could account for the landslide reactivation.

Another possible cause is exceptional rainfall. The dam is located in one of the more arid climates of Oregon, with average annual precipitation of only 9 inches (230 mm). March

1993 was wetter than normal, but at least two of the previous ten March precipitations were higher. Since no movements occurred during these earlier high rainfall periods, it is highly unlikely that March rainfall was the causal factor, and rainfall does not account for the prolonged movements into summer.
Based on the earlier movements caused by the fill, it is clear that Pelton Park slide is an example of a landslide at the very limit of stability. Possible explanations for the mechanics of the earthquake-reactivated movements are:

(A) The lower wedge slip plane, being a pre-existing ancient slide, is at residual strength. The shaking may have caused small excess pore water pressures to develop on the horizontal residual shear plane, thereby allowing movements to occur which lasted until the excess pore pressures dissipated. (e.g. Seed & Chan, 1966)

(B) The horizontal shaking from the earthquake may have increased the width of tension cracks at the headscarp. An increased separation would reduce the shearing resistance on the steeply-inclined upper wedge and increase the near-horizontal force (push) of the upper wedge on the lower wedge.

With respect to Explanation (A), the two piezometers referred to earlier were installed between June 29 and July 8, 1993, several weeks after the earthquake when movements were almost finished. Five sets of readings from July 9 to August 5 show no significant changes in groundwater levels over the four-week period.

MARGINAL REMEDIATION
PGE was interested in improving the stability of the Pelton Recreation Park to provide some small reserve of stability. Given the prevailing conditions at this site: deep slip plane, toe below water, infrastructure in place, there were few choices.

It was decided to unload the upper wedge to the maximum practicable extent to decrease the driving force. This solution required the fallout ditch to be excavated deeper, removing the protective berm and replacing it with a wire mesh fence at the roadside (Fig. 2). These measures reduce the load on the upper wedge.

After regulatory approval, reconstruction work occurred between early November 1993 and late January 1994. The owner excavated about 30,000 cu. yd. (23,000 cu. m) of soil. Calculations indicate that a 5 to 6 percent improvement of the factor of safety is achieved by the remedial work. This marginal level of remediation should help to prevent or reduce movements which might occur in the future from rapid reservoir drawdown, exceptional precipitation, or another minor earthquake.

Continued measurements of the hubs over the past five years have shown additional movements of less than 0.2 inch per year. Crack healing could account for these small movements.

REFERENCES
Madin, I., Priest, G. R., Mabey, M. A., Malone, S., Yelin, T. S. and Meier, D. (1993), "March 25, 1993 Scotts Mills earthquake–western Oregon's wake-up call," Oregon Geology, Vol. 55, No. 3, May, 51-57. Also published as EERI Special Earthquake Report, May 1993, 1-7.

Seed, H. B. and Chan, C. K. (1966), "Clay strength under earthquake loading conditions," Proc. ASCE, Journal of Soil Mechanics Div., Vol. 92, SM2 March, 53-78.

Continuous Monitoring of Groundwater Response to Winter Rainfall

D. H. CORNFORTH, Landslide Technology, Portland, Oregon and P.E. MIKKELSEN, Geoengineering & Instrumentation Consultant, Bellevue, Washington, USA

INTRODUCTION

Instrumentation for continuous monitoring of groundwater is becoming increasingly easy to install and is affordable for most major landslide studies. Rainfall can be monitored concurrently at the site so that the direct relationship between rainfall and groundwater response can be analyzed without relying on distant rain gauging stations. This paper presents the continuous record of rainfall and piezometer response at Frog Lake landslide, Oregon during the winter of 1998-99.

BACKGROUND INFORMATION

The Pacific Northwest region of the United States has a wet, temperate climate on the west side of the Cascade mountains. The onset of winter rainfall typically occurs in mid-October and continues until March. Rainfall then tapers off until the end of June; dry summer weather occurs from July to mid-October.

Large areas of western Oregon are covered by ancient landslides which are believed to have originated during the late Pleistocene. Because of the wet climate, most, if not all, of these ancient landslides are marginally stable and manmade changes to the topography can trigger more active movements on portions of the ancient landslide terrain. One such case is at Frog Lake reservoir, situated at about elev. 1,900 feet (580 m) on the western flanks of the Cascades.

In 1953, a 70-foot (21 m) high dam was built to retain water for a Portland General Electric (PGE) hydroelectric project. It reactivated a portion of ancient landslide terrain approximately 3,000 feet (910 m) wide and 5,000 feet (1,520 m) long above the Clackamas River with an average slope of 8 degrees to the horizontal. The shear plane is at least 250 feet (76 m) deep near the head of the active lobe where the Frog Lake reservoir adds weight to the landslide mass. Further details of the landslide are omitted as irrelevant to the scope of this technical paper.

PIEZOMETER INSTALLATION

Piezometer LT-21 was drilled in July 1995 about 65 feet (20 m) north of the embankment toe. The log is reproduced on Figure 1. All materials encountered in the drillhole are landslide debris. The upper 71 feet (22 m) of hole was drilled through sandy silt and sand by the ODEX technique using air feed. The lower 79 feet (24 m) was drilled with HQ coring and passed through highly broken rock.

On removing the drilling rods, the lower part of the hole collapsed. A plastic standpipe piezometer, nominal 1 inch (25 mm) inside diameter, was installed to a depth of 90 feet (27 m) in the remaining open hole. The lower 20 feet (6 m) of the plastic pipe has 0.010 inch (0.25 mm)

wide slots and is surrounded by No. 6 to No. 9 clean sand (approx. 2.2 to 3.4 mm grains). The upper 70 feet (21 m)of the hole is grouted with a cement-bentonite (8:1 mix by weight) slurry.

Drilling took nine days due to the broken rock within the hole. The total cost of drilling, including a full-time geotechnical engineer on site, was around $22,000 (in 1995).

MONITORING EQUIPMENT INSTALLATION

Piezometer

The standpipe piezometer was automated in the spring of 1997. A ¾-inch (19 mm) diameter vibrating-wire pressure sensor was lowered into the standpipe by a cable and suspended at the 90-foot (27 m) depth below the top of the riser. The measured water head is added to the stationary tip elevation to obtain the piezometric elevation.

The PVC riser is enclosed in a 6-inch (150 mm) diameter steel protective casing at the surface. Here, a small junction box with a surge protector for the sensor was installed which serves to terminate the sensor cable, provide lightning protection, and connect to the trenched-in telephone cable. The new trench extends a distance of about 600 feet to another instrument location at the toe of the embankment. From there, wires run another 400 feet up the slope of the embankment to the control house/emergency generator building. All wires were buried in 2-foot (0.6 m) deep trenches. The wire pair was connected to a Campbell Scientific CR10X data logger inside the building.

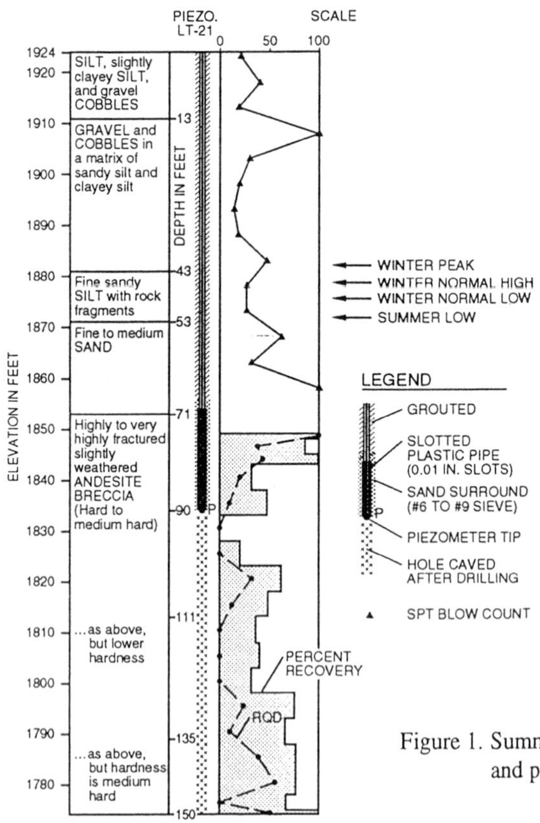

Figure 1. Summary log of drillhole and piezometer LT-21.

The data logger was programmed to store sampled data every hour on the hour. Vibrating-wire frequency readings, heads in feet, and elevations were stored along with other data.

Rain Gauge

Rainfall in inches and intensity in inches per hour were measured with a gauge placed on a steel pole next to the control house. A pair of telephone wires were used to carry the pulse signal from the rain gauge to the CR10X data logger some 200 feet away.

A tipping bucket style rain gauge (Met-One) was installed. Since power was available at the site, a heater for the gauge funnel was included. The thermostatically controlled heater activates below 40 degrees Fahrenheit, melting any snow or ice.

The bucket in the gauge tips for every 0.01 inch (0.25 mm) of accumulated rainfall. Each tip activates a magnet/reed switch and sends a pulse signal to the data logger. The logger keeps track of the number of pulses and the program calculates the cumulative rainfall and intensity. The results are stored each hour.

The approximate additional costs to provide the automated data acquisition system (ADAS) were: instruments: $8,000; design, install: $6,500; total $14,500.

The readings can be downloaded from the site by the owner (PGE) or consultant via telephone lines and graphed in the manner shown in this paper. The installation costs at this site were reasonable, and future monitoring is less expensive than the cost of sending personnel to the site for manual readings.

GROUNDWATER RESPONSE ANALYSIS

In western Oregon, groundwater levels fall steadily through the drier summer months. As winter rainfall begins, groundwater rises and reaches a seasonal high in mid-winter. During the winter, periodic heavy storms provide high intensity rainfall which can temporarily raise groundwater to short-term highs. These spikes usually cannot be caught by weekly or monthly site visits, yet are frequently the triggering event for movements on marginally stable slopes such as ancient landslide terrain. It is of benefit to the geotechnical engineer performing back-analysis slope stability calculations to know the height reached by groundwater during storms because it provides the groundwater model for using factor of safety F = 1.00 in the analysis. In this regard, an in-place inclinometer (Cornforth & Mikkelsen, 1996) can also be included in the ADAS to accurately relate groundwater levels to the onset of shear movements.

At the Frog Lake site, the groundwater levels fall 600 feet (183 m) over a distance of 4,500 feet (1,370 m) from piezometer LT-21 to the Clackamas River, a hydraulic gradient of 13 percent. Groundwater levels within the slope depend upon the rate of infiltration and the groundwater flow regimes.

The groundwater response to winter rainfall has been subdivided into three 13-week periods on Figures 2, 3 and 4. Although one must be cautious in reading too much significance into a single site record, Frog Lake piezometer LT-21 does offer some insights into groundwater response which probably apply to many other sites. For simplicity, an elevation such as elev. 1,875.0 is stated as elev. 75.0; these units are in feet and the conversion of elevation readings to metric units is omitted.

Late Summer, 1998

No rainfall occurred at the site from July 30 to September 18. The end of this period is shown by A on Figure 2 and groundwater levels fell throughout the period at a rate of about ¼ inch/day (6 mm/day).

During the seven-week period from September 18 to November 10, there was sporadic rainfall but no major storms. A total of 7 inches (178 mm) of rainfall was recorded, but groundwater continued to fall, reaching the annual low of elev. 71.5 near the end of September (B, Fig. 2). Therefore, the rainfall infiltration was essentially rewetting the upper 53 feet (16 m) of soil above the groundwater following the dry summer months, and had no influence on groundwater levels.

Winter, 1998-99

The first major storm produced 7.4 inches (188 mm) of rainfall in 6.6 days; the groundwater level rose by 5.1 feet (1.55 m), as shown by CD on Figures 2 and 3. Thereafter, winter storms kept groundwater levels generally within a band fluctuating between about elev. 75.4 and elev. 78.5, a range of 3.1 feet (0.95 m) for about 16 weeks. These limits (Fig. 3) can be termed the 'normal' highs and lows of groundwater during winter and are well above the late summer low of elev. 71.6.

In late December, a groundwater spike reached a peak of elev. 82.0 (E, Figs. 3 and 5). The peak reading was caused by a severe storm with rainfall of 6.0 inches (152 mm) over 1.8 days which raised the groundwater level by 6.1 feet (1.86 m). This peak value is 3.5 feet (1.07 m) above the 'normal' winter high. It is of interest that the peak lasted for only two to three hours (9:00 p.m. to midnight on December 28, see Fig. 5) and the groundwater level declined at almost 2 feet per day (0.6 m per day) for the next two days (F, Fig. 5).

The delay time from the peak rainfall intensity to peak piezometric level was about 25 hours (Fig. 5). In other storms, the delay time was generally about 19 hours.

Spring 1999

Groundwater fell below the 'normal' winter range in late March. However, groundwater fluctuations in the spring appear to be more sensitive to rainfall during this period (G, Fig. 4) than they were in the late summer of 1998 (B, Fig. 2). The spring fluctuations may have been affected by runoff of snowmelt from the high slope above the site.

Continuous vs. Periodic Groundwater Readings

Suppose that winter groundwater observations were being taken at monthly intervals occurring on the first day of each month. The readings for December through March are shown by open

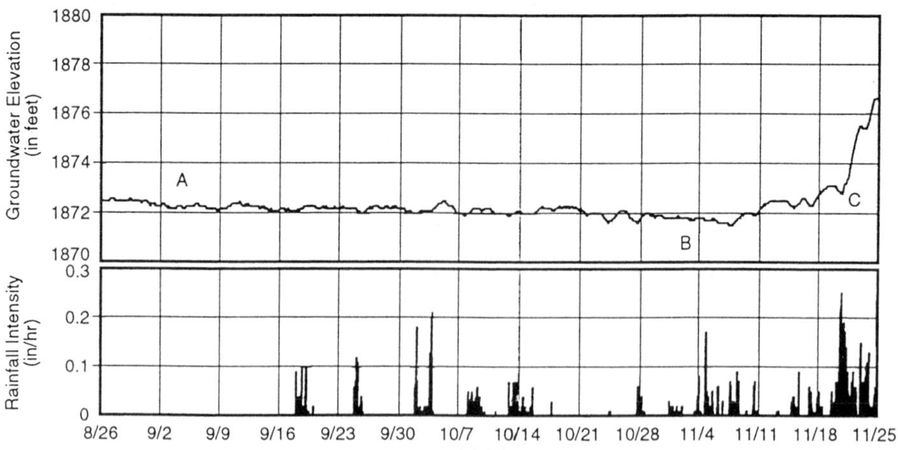

Figure 2. Piezometer data and rainfall intensity 8/26/98 – 11/25/98

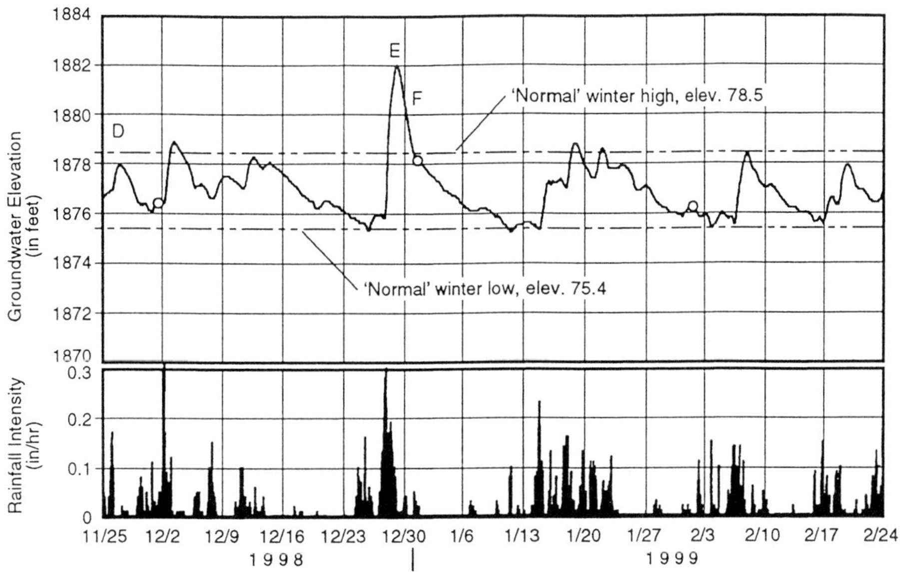

Figure 3. Piezometer data and rainfall intensity 11/25/98 – 2/24/99

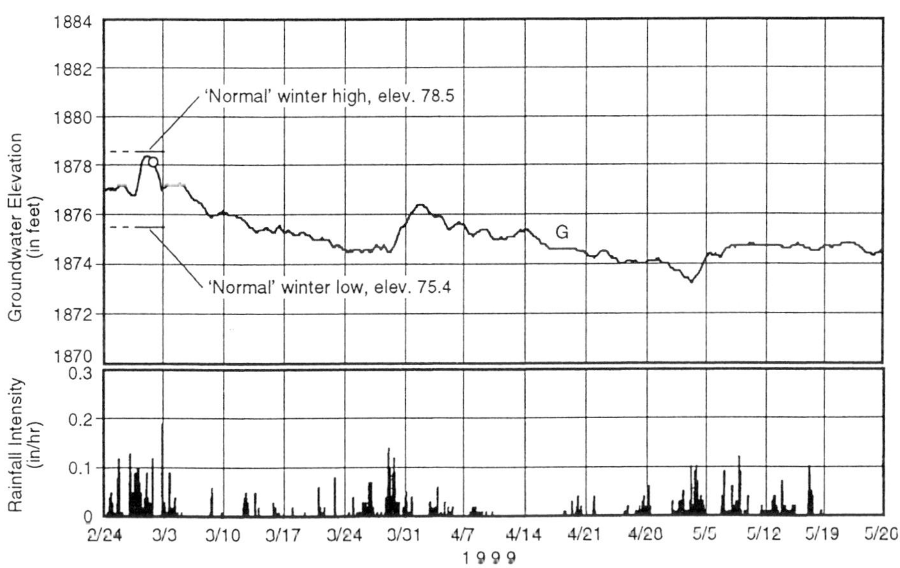

Figure 4. Piezometer data and rainfall intensity 2/24/99 – 5/26/99

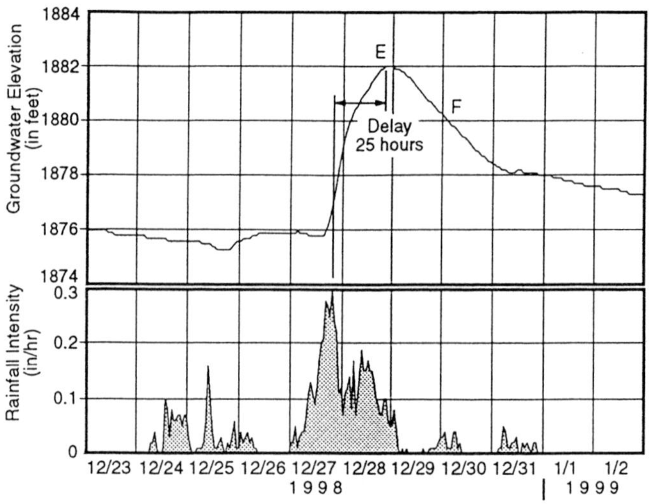

Figure 5. Spike of groundwater in December 1998

circles on Figures 3 and 4. The data shows that the observations would approximately cover the 'normal' range of groundwater levels for the winter, but they would have missed the peak groundwater level (E, Fig. 3).

CONCLUSIONS

1. Continuous groundwater monitoring can measure the precise maximum head occurring during a major storm event. In this case history, the winter peak occurred at around midnight on December 28, 1998 when manual readings would not have been taken.

2. Major spikes of groundwater develop and subside quickly and have a low probability of being recorded by groundwater readings taken on a set schedule, such as weekly or monthly intervals.

3. (i) The groundwater cycle for the winter of 1998-99 at this site was: Groundwater slowly declined through the dry summer months to a base elevation of 71.5 feet. The late summer rate of groundwater decline in dry weather is only ¼ inch (6 mm) per week (A, Fig. 2).
   (ii) Rain in the early fall had essentially no effect on groundwater level, and probably is infiltrating the voids and cracks above groundwater (B, Fig. 2).
   (iii) The first major storm caused a sharp rise in groundwater, moving the water level into the winter range (C, Fig. 2 to D, Fig. 3).
   (iv) Winter storms kept the groundwater generally within a range fluctuating from a low of elev. 75.4 to a high of elev. 78.5, for a period of 16 weeks (Figs. 3, 4).
   (v) A severe mid-winter storm of 6.0 inches (152 mm) in 1.8 days caused a spike of groundwater to elev. 82.0, which is 10.5 feet (3.2 m) above the summer low (E, Fig. 3).
   (vi) Groundwater levels generally dropped below the winter range in mid-March and began the seasonal decline to the summer base level (G, Fig. 4).

REFERENCE

Cornforth, D. H. and Mikkelsen, P.E. (1996), "Continuous monitoring of the slope above an excavation within a marginally stable landslide," Proceedings of the 7th International Symposium on Landslides, Trondheim, Norway, Vol. 3, pp. 1539-1544.

Landslides and Climate Change in the Alps Since the Late-Glacial: Evidence of Case Studies in the Dolomites (Italy)

A. CORSINI, Department of Earth Sciences, University of Modena and Reggio Emilia, Modena, Italy, A. PASUTO, C.N.R.-I.R.P.I., Research Institute for Hydrological and Geological Hazard Prevention, Padova, Italy, M. SOLDATI, Department of Earth Sciences, University of Modena and Reggio Emilia, Modena, Italy

INTRODUCTION

Interest in studying the relationships between landslides and climate from the Late-glacial to the present has recently increased in Europe. Indeed, it is widely acknowledged that climate changes have influenced the onset of landslides in this period, either directly or indirectly. On the other hand, the correlation between series of dated slope movements and climate changes is not so immediate, nor is it certain.

Numerous investigations have been carried out on the recurrence in time of landslides on mountain chains in Europe. Slope movements of various dimensions and types have been dated by means of different methods: stratigraphic, radiometric, dendrochronological and historical (e.g. Schoeneich, 1991; Corominas et al., 1994). In particular, a concentration of events during certain periods of the most recent geological history has been recorded; this has led to speculation on the existence of correlation between the distribution of landslides in time and climate changes (e.g. Starkel, 1966, 1985; González Díez et al., 1996; Berrisford and Matthews, 1997; Lateltin et al., 1997; Moya et al., 1997; Panizza et al., 1997; Margielewski, 1998). Moreover, several authors have proposed comparison methods and concise correlation for the various areas investigated in Europe and have pinpointed the complexity of this topic both for the variability in time and space of climate changes in Europe and for the amount and reliability of available data (cf. Matthews et al., 1997).

Since 1991, investigation on the relationships between landslide occurrence and climatic change from the Late-glacial have been carried out in the areas of Cortina d'Ampezzo and Alta Badia (Italy) within the realm of EU-funded projects. Such research, which also dealt with geomorphological and landslide hazard mapping, brought to a better understanding of the geomorphological evolution of these areas and took large benefit from radiocarbon dating of several mass movements. In particular, the study of the past distribution of landslides was fundamental for the definition of landslide hazard and for planning monitoring systems to be installed on the most active slope movements as well as for making speculations on the relationships between landslide occurrence and climate changes (Panizza et al., 1996, 1997; Soldati, 1999; Corsini et al., 1999). The results achieved are presented and discussed below.

CASE STUDIES IN THE DOLOMITES

The study areas of Cortina d'Ampezzo and Alta Badia, located in north-eastern Italy in the central sector of the Dolomites (Figure 1), show similar geological and geomorphological features. The geological structure of both areas is characterised by a repeated succession of dolomitic and pelitic rocks, belonging to the Triassic sedimentary sequence, which are largely covered by Quaternary deposits (Pasuto et al., 1997; Corsini et al., 1999). This results in softly degrading morphologies in the medium and lower part of the slopes, where bedrock is made up of incompetent rocks, and on steep rock walls in the peripheral parts of the basins, where dolomitic rocks make up massifs which reach altitudes of over 3000m.

Landslides in research, theory and practice, Thomas Telford, London, 2000.

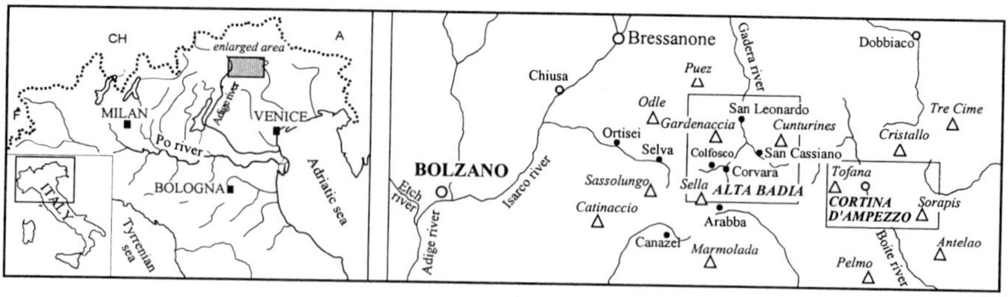

Figure 1. Geographical setting of the study areas.

As regards the area of Cortina d'Ampezzo, more than thirty landslides of different type, size, age and degree of activity have been identified (Panizza, 1990; Panizza et al., 1996; Pasuto et al., 1997), several of which have been dated by means of radiocarbon method (Table 1). The oldest dates obtained, ranging between 13,000 and 10,000 cal. yr BP, refer, on the one hand, to huge rock slides, now inactive, which affected dolomitic rocks after the retreat of the Würmian glaciers and, on the other hand, to slides and earth flows (mudslides according to Brunsden, 1984) mainly involving pelitic rocks (Figure 2). The rock slides (Col Drusciè, Zuel, Pierosà), which occurred when slopes were no longer sustained by ice masses, markedly modified the morphology of the Cortina valley with their hill-shaped accumulations which, later on, have been partly remobilised by successive slides and flows. An example of this is given by the Pierosà rock slide body which was partly reactivated by a series of earth flows (Cortina landslide) since 10,000 cal. yr BP. Early Holocene dates have been collected also for other earth flows, such as that of Lacedel, probably triggered in the Late-glacial. The Lacedel landslide had a recurrent activity during the Holocene and also at present shows a high rate of movement. For other earth flows (La Riva, Pezzié, Chiamulera, Chiave) early Holocene movements are not documented, but a series of dates which show a cluster of events between 5500 and 2500 cal. yr BP have been achieved (Figure 2). However, it was not possible to define whether the dated events were first time failures or reactivations of older movements.

In Alta Badia a series of radiocarbon dates of landslides and related lacustrine deposits have been collected (Corsini et al., 1999). Many of the dates available refer to movements of the Corvara landslide (Table 2), a complex phenomenon (slide-flow) located just south of Corvara in Badia. The oldest events dated range from about 10,000 to 9000 cal. yr BP (Figure 2) and refer to massive earth flows subsequent to large-scale rotational slides, which affected the bedrock down to 40-50m in depth. More recent movements, mainly spanning from about 5000 to 2000 cal. yr BP, refer to earth flows which episodically moved along more shallow shear surfaces. Apart from a reactivation of the Arlara landslide, dated to ca. 7700 cal. yr BP, other landslides in Alta Badia fit in a similar time span. In fact, in the northern part of the Badia valley landslide events have been dated at about 5600 cal. yr BP (San Leonardo earth flow) and about 5000 cal. yr BP (Colfosco rock fall). More recent dates (2800 to 2500 cal. yr BP) refer to two superimposed earth flows, separated by a 20cm thick soil, which detached from the accumulation of a much larger rotational slide near San Cassiano. The research in Alta Badia has also identified the recurrent formation, in the past, of a dam-lake downstream of Corvara in Badia. Valley damming has been related to the Col Maladat rotational slide. Charcoals from cores of the dam-lake deposits, together with sedimentary evidence, date damming conditions in two distinct periods: from about 10,000 to 7000 cal. yr BP and from about 4000 to 1000 cal. yr BP. It is likely that the onset of the oldest damming event closely post-dates an early sliding event. On the other hand, the most recent lacustrine phase, which appears to have taken place after the oldest lake was extinguished, should be related to a partial reactivation of the landslide due to debris slides and flows.

Table 1. Results of radiocarbon dating in Cortina d'Ampezzo (Calibrated data from Radiocarbon Calib. Program 4.1 by Stuiver and Reimer, 1993; Dataset: Stuiver et al., 1998; Cal. method: Intercept).

	Landslide (name)	**Sample** (lab. code)	**^{14}C age** (yr BP)	**Cal. age 2σ** (cal. yr BP)	**Landslide event type** (Cruden and Varnes, 1996)
1	Col Drusciè	-	9000 +/- 150	10,500 to 9605	rock slide
2	Cadin	GX-16117	12,150 +/- 435	15,744 to 13,168	earth flow
3	Chiave	B-748002	4520 +/- 60	5441 to 4970	earth flow
4	Lacedel	GX-17696	10,035 +/- 110	12,277 to 11,201	earth flow
4	Lacedel	GX-17697	9270 +/- 105	10,734 to 10,218	earth flow
5	Pierosà	B-63341	10,850 +/- 80	13,132 to 12,639	rock slide
6	Cortina	B-83624	8710 +/- 70	10,108 to 9535	earth flow
6	Cortina	B-63338	4350 +/- 60	5209 to 4830	earth flow
6	Cortina	Gx-17699	3315 +/- 140	3891 to 3214	earth flow
6	Cortina / Alverà	B-63340	2810 +/- 60	3137 to 2778	earth flow
6	Cortina / Alverà	B-83623	2560 +/- 80	2839 to 2357	earth flow
6	Cortina / Alverà	R-1753	1460 +/- 30	1410 to 1295	earth flow
6	Cortina / Staulin	GX-17698	2465 +/- 125	2835 to 2161	earth flow
7	Chiamulera	B-63339	4700 +/- 60	5590 to 5307	earth flow
8	Pezzié	B-83622	6570 +/- 70	7578 to 7324	earth flow
8	Pezzié	B-83621	6190 +/- 50	7248 to 6912	earth flow
8	Pezzié	B-74803	5730 +/- 70	6721 to 6323	earth flow
9	La Riva	GX-17689	8280 +/- 100	9521 to 9011	earth flow
9	La Riva	B-63342	4220 +/- 60	4868 to 4551	earth flow
10	Zuel	GX-17687	9440 +/- 105	11,156 to 10,292	rock slide
10	Zuel	GX-17688	9215 +/- 105	10,674 to 10,190	rock slide
	Dam-lake deposit (name)	**Sample** (lab. code)	**^{14}C age** (yr BP)	**Cal. age 2σ** (cal. yr BP)	**Geomorphic event type**
11	Campo-Zuel	GX-17690	7180 +/- 200	8387 to 7613	Deposition in dam-lake

Table 2. Results of radiocarbon dating in Alta Badia (Calibrated data from Radiocarbon Calib Program 4.1 by Stuiver and Reimer, 1993; Dataset: Stuiver et al., 1998; Cal. method: Intercept).

	Landslide (name)	**Sample** (lab. code)	**^{14}C age** (yr BP)	**Cal. age 2σ** (cal. yr BP)	**Landslide event type** (Cruden and Varnes, 1996)
12	Corvara	B-112032	8820 +/- 50	10,152 to 9632	rotational slide - earth flow
12	Corvara	B-112033	8560 +/- 90	9709 to 9334	rotational slide - earth flow
12	Corvara	B-112031	7920 +/- 70	9009 to 8543	rotational slide - earth flow
12	Corvara	B-112030	4260 +/- 70	5025 to 4575	earth flow
12	Corvara	B-105976	3830 +/- 60	4417 to 3999	earth flow
12	Corvara	B-105977	2860 +/- 60	3207 to 2792	earth flow
12	Corvara	B-93975	2490 +/- 60	2750 to 2352	earth flow
12	Corvara	B-112034	2260 +/- 50	2351 to 2129	earth flow
13	Arlara	B-105975	6870 +/- 50	7789 to 7592	rotational slide - earth flow
14	Col Alto	B-93976	2350 +/- 60	2708 to 2183	earth flow
15	Pocol	B-128367	950 +/- 50	954 to 736	earth flow
16	Colfosco	B-112024	4420 +/- 70	5030 to 4978	rock fall
17	San Cassiano	B-128365	2260 +/- 60	2869 to 2715	earth flow
17	San Cassiano	B-128366	2610 +/- 60	2845 to 2496	earth flow
18	San Leonardo	B-128368	4890 +/- 70	5839 to 5475	earth flow
	Dam-lake deposit (name)	**Sample** (lab. code)	**^{14}C age** (yr BP)	**Cal. age 2σ** (cal. yr BP)	**Geomorphic event type**
19	Corvara-Col Maladat	B-112026	9080 +/- 70	10,401 to 10,154	deposition in dam-lake
19	Corvara-Col Maladat	B-112028	8810 +/- 70	10,173 to 9557	deposition in dam-lake
19	Corvara-Col Maladat	B-112025	7740 +/- 80	8696 to 8386	deposition in dam-lake
19	Corvara-Col Maladat	B-112027	6460 +/- 90	7563 to 7213	deposition in dam-lake
19	Corvara-Col Maladat	B-112023	3550 +/- 70	4075 to 3640	deposition in dam-lake
19	Corvara-Col Maladat	B-112022	3210 +/- 60	3626 to 3272	deposition in dam-lake
19	Corvara-Col Maladat	HD-19408	1249 +/- 22	1263 to 1090	deposition in dam-lake

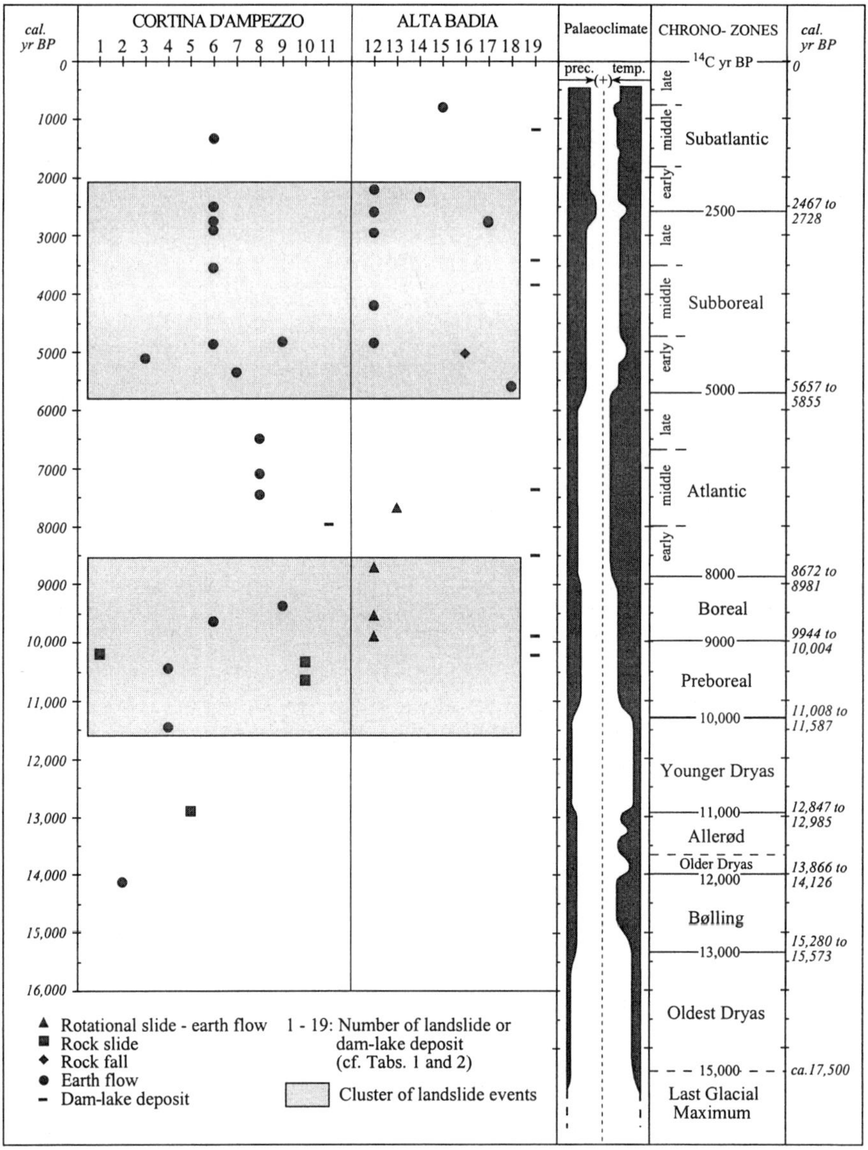

Figure 2. Temporal distribution of landslide events since the Late-Glacial. Palaeoclimate curves are modified after Orombelli and Ravazzi (1996) and Goudie (1992); chrono-zones boundaries are after Orombelli and Ravazzi (1996).

DISCUSSION

On a large time-scale, the dates gathered so far in the area of Cortina d'Ampezzo and in Alta Badia show clusters of landslide events in two periods characterised by marked climatic changes: the early Post-glacial (Preboreal and Boreal) and the Subboreal (Figure 2). The causes and types of mass movements of these two clusters are, however, substantially different.

The first cluster includes landslides of various types which occurred in the early Post-glacial (Figure 2) when, after the retreat of the Würmian glaciers, slopes became prone to mass movements. On the one hand, these movements consist of rock slides of considerable size (e.g. Zuel and Col Drusciè landslides) detached from the steep dolomitic rock walls no longer sustained by ice masses, as a consequence of glaciopressure implications (cf. Panizza, 1973); on the other hand, they consist of complex movements (Corvara), rotational slides (Col Maladat) and earth flows (Lacedel, Cortina, La Riva) which affected the pelitic formations, outcropping in the middle and lower part of the slopes, as a consequence of higher availability of groundwater; the latter is likely to be connected to increased precipitation (cf. Figure 2) and permafrost melting (cf. Soldati, 1999). In a European perspective, the period during which these landslides took place coincides with the oldest period of frequent post-glacial slope movements identified in Europe by various authors (e.g. Starkel, 1966; González Díez et al., 1996; Berrisford and Matthews, 1997; Lateltin et al., 1997; Margielewski, 1998).

The second cluster of landslides, which coincides with the Subboreal period, includes, in both study areas, mostly earth flows involving pelitic rocks. The dated movements are likely to be reactivations of older landslides rather than first time failures, since the same slopes appear to have been mobilised more than once in the past. The period of occurrence corresponds with an increase of precipitation reported in literature (cf. Figure 2). However, the hypothesis of a direct influence of this climate change on landslides, partially confirmed by other European case studies (e.g. González Díez et al., 1996; Berrisford and Matthews, 1997; Lateltin et al., 1997), still needs further verification especially through the comparison with other alpine sites.

However, the Cortina d'Ampezzo and Alta Badia areas have been affected by landslides also during other periods of the Holocene: the recurrent activity has been favoured by non-climatic factors such as seismic events and litho-structural characteristics of the rock types involved which are either affected by dense networks of joints (dolomites) or markedly ductile (pelitic rocks). Also extreme meteorological events, not specifically related to climate changes, may have triggered mass movements.

ACKNOWLEDGMENTS

The research was funded by the projects MURST 1997 ("Response of geomorphological processes to environmental changes", National Resp. Prof. A. Biancotti; Local Resp. Prof. M. Panizza) and CNR 05 1997 ("Application of new physical dating techniques to the Holocene reconstruction in Dolomitic areas", Resp. Prof. M. Panizza). The authors are grateful to Prof. M. Panizza for the critical review of the manuscript.

REFERENCES

Berrisford M.S. and Matthews J.A., 1997. *Phases of enhanced rapid mass movement and climatic variation during the Holocene: a synthesis.* Paläoklimaforschung - Palaeoclimate Research, 19, 409-440.

Brunsden D., 1984. *Mudslide.* In: D. Brunsden and D.B. Prior (eds.), Slope Instability. Wiley & Sons, Chichester, 363-418.

Corominas J., Weiss E.E.J., Van Steijn H. and Moya J., 1994. *Use of dating techniques to assess landslide frequency, exemplified by case studies from European countries.* In: R. Casale, R. Fantechi and J.C. Flageollet (eds.), Temporal occurrence and forecasting of landslides in the European Community - Final Report. European Commission, Bruxelles, 1, 71-93.

Corsini A., Pasuto A. and Soldati M., 1999. *Geomorphological investigation and management of the Corvara landslide (Dolomites, Italy).* JGU Transactions, 20(3), 169-186.

Cruden D.M. and Varnes D.J., 1996. *Landslides Types and Processes*. In: A.K. Turner and R.L. Schuster (eds.), Landslides: Investigation and Mitigation. Transportation Research Board, National Academy of Sciences, Washington D.C., Special Report 247, 36-75.

González Díez A., Salas L., Díaz de Terán J.R. and Cendrero A., 1996. *Late Quaternary climate changes and mass movement frequency and magnitude in the Cantabrian region, Spain*. Geomorphology, 15, 291-309.

Goudie A., 1992. *Environmental change*. Third Edition, Oxford Univ. Press, Oxford, 329 pp.

Lateltin O., Beer C., Raetzo H. and Caron C., 1997. *Landslides in Flysch terranes of Switzerland: Causal factors and climate change*. Eclogae geol. Helv., 90, 401-406.

Margielewski W., 1998. *Landslide phases in the Polish outer Carpathians and their relation to climatic changes in the Late Glacial and the Holocene*. Quat. Studies in Poland, 15, 37-53.

Matthews J.A., Brunsden D., Frenzel B., Gläser B. and Weiß M.M. (eds.), 1997. *Rapid mass movement as a source of climatic evidence for the Holocene*. Paläoklimaforschung - Palaeoclimate Research, 19, 444 pp.

Moya J., Vilaplana J.M. and Corominas J., 1997. *Late Quaternary and historical landslides in the south-eastern Pyrenees*. Paläoklimaforschung - Palaeoclimate Research, 19, 55-73.

Orombelli G. and Ravazzi C., 1996. *The Late Glacial and Early Holocene chronology and paleoclimate*. Il Quaternario, 9(2), 439-444.

Panizza M., 1973. *Glacio Pressure Implications in the Production of Landslides in the Dolomitic Area*. Geol. Appl. e Idrogeol., 8(1), 289-297.

Panizza M., 1990. *The landslides in Cortina d'Ampezzo (Dolomites, Italy)*. In: A. Cancelli (ed.), ALPS 90-6th ICFL, Switzerland-Austria-Italy, Aug. 31st - Sept. 12th, Conference Proceedings. Università degli Studi di Milano, 55-63.

Panizza M., Pasuto A., Silvano S. and Soldati M., 1996. *Temporal occurrence and activity of landslides in the area of Cortina d'Ampezzo (Dolomites, Italy)*. Geomorphology, 15, 311-326.

Panizza M., Pasuto A., Silvano S. and Soldati M., 1997. *Landsliding during the Holocene in the Cortina d'Ampezzo region, Italian Dolomites*. Paläoklimaforschung - Palaeoclimate Research, 19, 17-31.

Pasuto A., Siorpaes C. and Soldati M., 1997. *I fenomeni franosi nel quadro geologico e geomorfologico della conca di Cortina d'Ampezzo (Dolomiti, Italia)*. Il Quaternario, 10(1), 75-92.

Schoenheich P., 1991. *Les datation des glissements de terrain*. In: D.H Bell (ed.), Landslides. Proc. Sixth ISL, Christchurch, 10-14 February 1992, Balkema, Rotterdam, 205-212.

Soldati M., 1999. *Landslide Hazard Investigations in the Dolomites (Italy): The Case Study of Cortina d'Ampezzo*. In: R. Casale and C. Margottini (eds.), Floods and Landslides: Integrated Risk Assessment. Springer-Verlag, Berlin, 281-294.

Starkel L., 1966. *Post-glacial climate and the moulding of European relief*. Proc. Int. Symp. on World Climate from 8000 to 0 B.C., Royal Meteorological Society, London, 15-33.

Starkel L., 1985. *The reflection of the Holocene climatic variations in the slope and fluvial deposits and forms in the European mountains*. Ecologia Mediterranea, 11(1), 91-97.

Stuiver M. and Reimer P.J., 1993. *Extended 14C database and revised CALIB radiocarbon calibration program*. Radiocarbon, 35, 215-230.

Stuiver M., Reimer P.J., Bard E., Beck J.W., Burr G.S., Hughen K.A., Kromer B., McCormac F.G., v.d. Plicht J. and Spurk M., 1998. *INTCAL98 Radiocarbon age calibration 24,000 - 0 cal BP*. Radiocarbon, 40, 1127-1083.

Design Strength Parameter of a Natural Slope on Gneissic Residual Soil

R. Q. COUTINHO, J. B. SOUZA NETO and F. Q. COSTA , Federal University of Pernambuco, Recife, Brazil

ABSTRACT
This paper presents basic engineering properties of a gneissic residual soil from a slope in which a greater landslide occurred. Emphasis is giving to the design strength parameter obtained through laboratory tests. A stability analysis is done taking into consideration unsaturated condition. The landslide causal factors are also presented and discussed. The study results have shown that the strength parameters mobilized in the first time failure are characterized by the laboratory results (maximum or 10mm horizontal deformation) considering the low value of the initial soil matric suction.

INTRODUCTION

The State of Pernambuco presents some roads with slope instability problems that periodically require repair and a great cost to the state government. In this context, the Geotechnical Group of the Federal University of Pernambuco established a joint research project with the Pernambuco State Road Department. This project aimed at making geological-geotechnical study of the materials (residual soil/rock), slope stability evaluation, and a proper understanding of the failure mechanism. The site studied is a natural non-homogeneous and unsaturated slope located in the road PE-89, Machados, Pernambuco, Brazil, where a significant landslide occurred years ago (Figure 1). This paper will present and discuss the basic engineering properties stressing the design strength parameters of a gneissic residual soil together with the stability analysis results considering the unsaturated condition and the landslide causal factors.

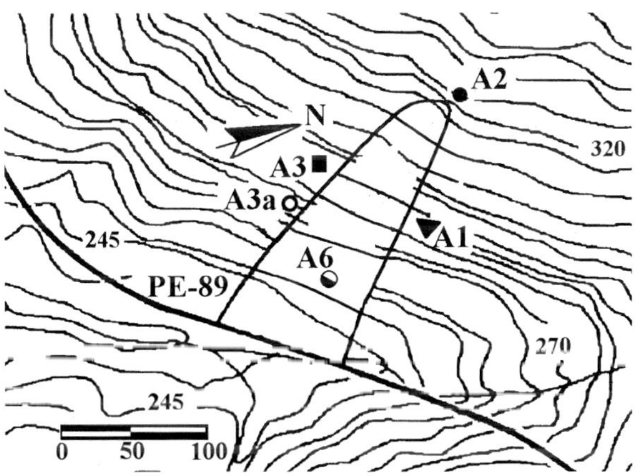

Figure 1. Landslide 1 – Investigated areas (A1 to A6)

Landslides in research, theory and practice, Thomas Telford, London, 2000.

SITE CHARACTERIZATION OF THE INVESTIGATED AREA

The natural slope investigated is formed of a gneissic residual soil/weathered rock presenting an elevation of 350m above sea level and an average declivity of 35°. The landslide presents a ruptured surface having a length of 170m and an average width of approximately 80m (Figure 1). The region has a hot and humid climate with rainfall averaging 1200 to 1600mm per year laving a higher intensity from april to august period (wet season). In the year of the landslide the rainfall average per month during the wet period was in the order of 200mm.

The field investigation performed in the landslide was carried out by using topographic profile of cross sections, engineering-geologic mapping, subsurface exploration - SPT/RQD, vertical trenches, soil moisture content profiles, field permeability, instrumentation-piezometer, and sampling in soils and in rocks. Figure 1 shows the locations of the activities in the field exploration program. The information presented in this paper is from Areas 1 and 3. More details can be found in Coutinho et al. (1997) and Coutinho et al. (1998).

Figure 2 shows the typical results of a complete borehole performed together with some physical and mineralogical properties of the materials. In general, the mature residual soil (horizon B) is composed of porous sandy clay having an average thickness of 2 meters. The young residual soil (horizon C/Cr) is a silty sandy with mica, but still preserving some characteristics of parent rock (foliation of around 30° with horizontal), and thickness that can reach 15 meters. The rock (biotite gneiss) presents a high degree of fracturing with low RQD values in some depths. The results suggest that the failure surface correspond to depth limit of the standard penetration test (the approximated definition of the contact soil-rock).

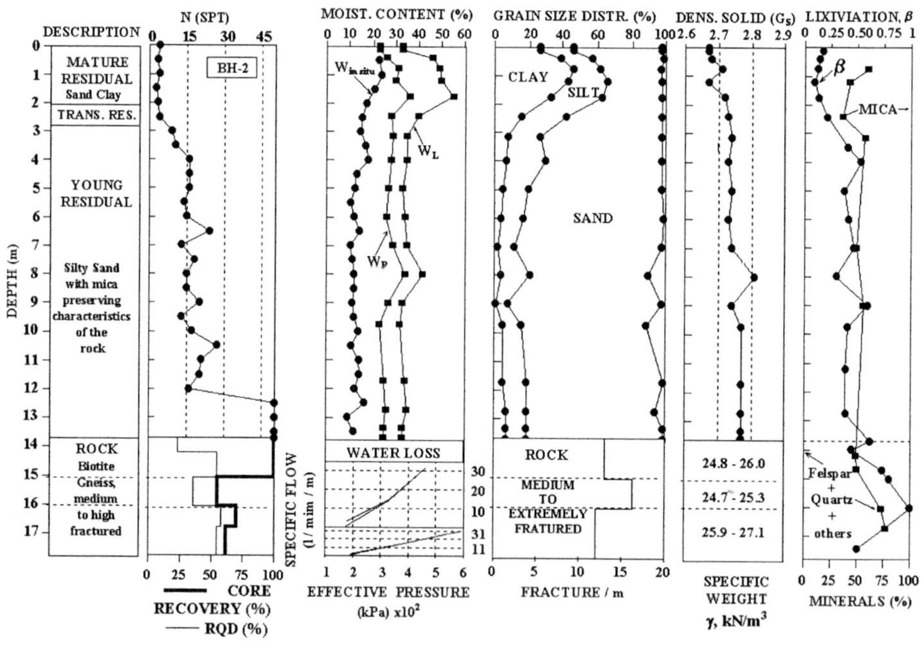

Figure 2. Typical result of a borehole (SPT / RQD) and some characteristic properties

The moisture content profiles were periodically determined in Areas 1, 2 and 3a, to observe the moisture content variation during the dry and wet seasons. The results show the moisture content values between 10% and 24% in general depth decreasing. In the profiles located on the slope the most significant variations occurred at a depth of approximately two meters (mature residual soil), and approximately four meters in the profile located on the top of the landslide area.

The hydraulic conductivity of the soils was determined through a field test using the Guelph permeameter. The results of the tests (Figure 3) show that the mature residual soil, a sandy clay presents a higher saturated hydraulic conductivity (K_{fs}=1.1 ±0.48 x 10^{-5}m/s) than the young residual soil, a silty sand (K_{fs} = 0,41 ± 0,15 x 10^{-5} m/s). This behavior is due to the granular and porous structure of the mature residual soil. The variation with suction of the relative hydraulic conductivity to the young residual soil (K_r=K/K_{fs}, where K is unsaturated hydraulic conductivity), indicate a strong reduction in the Kr values to small values of suction in this soil influencing the flow behavior of the slope.

GEOTECHNICAL PROPERTIES

An intense laboratory investigation was done to evaluate geotechnical properties of the materials. A summary is presented in this paper with emphasis on the strength parameters. More details can be found in Coutinho et al. (1998) and Souza Neto (1998).

The results of the characterization tests in the mature residual soil showed clay fraction around 40 % and a plasticity index in the interval of 12 to 20 %, being classified as CL/MH in the Unificated Classification. The young residual soil presented around 70 % of sand fraction, and plasticity index in the range of 6 to 16 % has being classified as SM/SC with a significant presence of mica. The plasticity index in the young residual soil was high considering the clay fraction in the order of 4% and the kaulinite as the predominant clay mineral. This fact may be related to the high mica content (43 to 51 %) in this material (Figure 2).

The matric suction - moisture content relationship was obtained by pressure plate test using four or five specimens to each curve. Figure 4 presents the suction-moisture content curves obtained from the different soils investigated. The observed behavior is in accordance with the texture of the soils. The values of the suction in the field were determined during June-August/95 period (wet season) by the filter paper method from the young residual soil. The results ranged from 10 to 80 kPa. The initial suction of the specimens used in the pressure plate tests in laboratory was estimated in the range of 20-70 kPa similar to the field values.

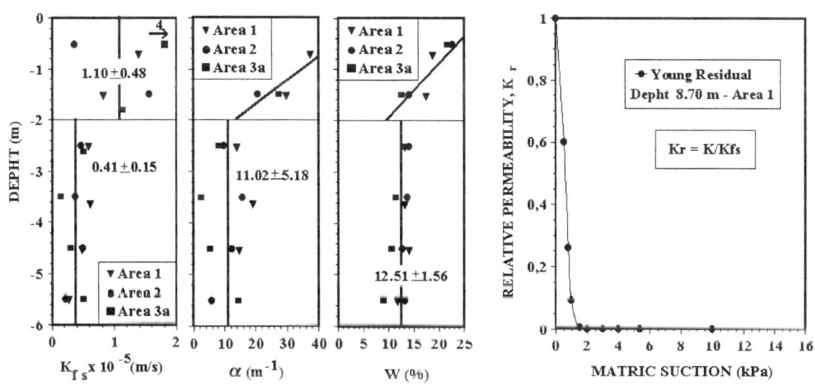

Figure 3. Results of permeability test (Guelph permeameter)

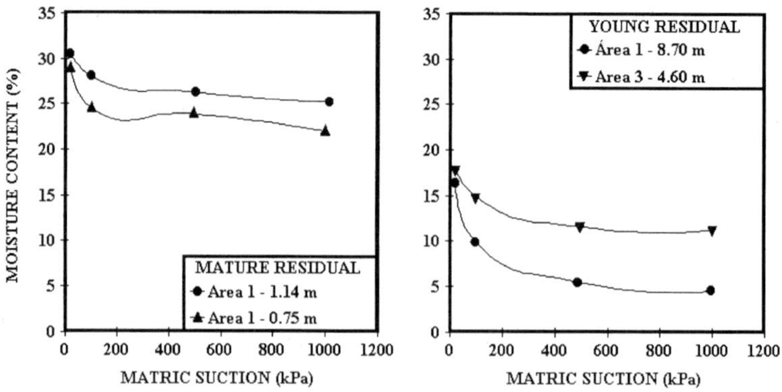

Figure 4. Suction - moisture content relationship

The conventional direct shear tests were performed by using square specimens of 101.6 mm side and 25 mm of height to obtain drained shear strength parameters. The tests were done on soaked specimens prepared perpendicular and parallel directions to the foliation plane, and on natural moisture content specimens prepared parallel direction to the foliation plane. Figure 5 presents a summary of the results obtained in the young residual soils from the two investigated areas.

The young residual soils showed the effective friction angle (ϕ') values in the range of 29.3° to 31.3° with an average of 30° while the effective cohesion parameter (c') values were in the following ranges: c'=2.9 - 7.4kPa to the young residual soil from Area 1: and c'=2.2 - 40kPa to the young residual soil from Area 3. The higher c' values from the tests with specimens in the natural moisture content, and the lower values to the tests with soaked specimens prepared parallel to foliation. The strength parameters of soaked specimens prepared parallel and perpendicular to the foliation plane are similar to both areas investigated. The results from the natural moisture content specimens pointed out some difference in the initial matric suction. Considering 10 mm of horizontal displacement to the soaked specimens parallel to foliation, ϕ' was 25° (Area 1) and 30° (Area 3), with c'=0 kPa to the two areas.

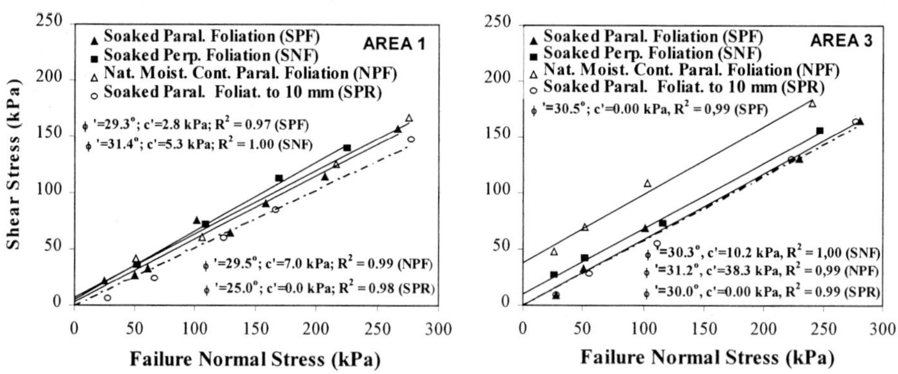

Figure 5. Results of conventional direct shear tests in the young residual soil

Sandroni (1977) (from Lacerda & Almeida 1995) studied the gneissic residual soils. It was found that the mineralogy of the sand fraction, including the mica and feldspar proportions, correlates well the soil strength which tends to be higher as the feldspar content increases and the mica content decreases. The results (c', ϕ') obtained in this study are in agreement with Sandroni (1977) investigation. The specimens prepared perpendicular to foliation showed higher values of c', indicating some influence of the anisotropy in this soils. Considering the in situ total stress, the increase in the shear strength is 11% in Area 1 and 17% in Area 3. The foliation plane parallel direction to the slope direction (adversely oriented mass discontinuities) plus the high mica content are potentially predisposition factors.

The study of the collapsibility (Coutinho et al., 1997) established that the soils can be classified as a non-collapsible soil if they meet natural condition and $\sigma_v=\sigma_{vo}$ indicating that this phenomena cannot be considered as a predisposition factor to the landslide occurrence. In the same way laboratory tests showed that the residual soils investigated can be classified as a non expansible soil.

STABILITY ANALYSIS

The slope stability evaluation in this research project was done as a two-dimensional unsaturated stability analysis. It was divided into two steps: 1) slope flow analysis for the rainfall history just before the failure; 2) slope stability evaluation using a limited equilibrium analysis (Sarma, 1979), considering the pore pressure distribution obtained from the flow analysis. More information about this analysis is presented in Costa (1996).

Two conditions were considered in the analysis: the cross section before the road construction (slope natural condition) and the cross section after the construction (failure condition). For the last condition, two hypotheses were considered for the landslide mechanism. In the first, the landslide occurred in only one step, which means, one complete cross section for all slopes (cross section 2 - Figure 6a). In the second, the landslide occurred in two steps: one small landslide close to the foot of the slope, and the other one giving continuity to the top of the slope (cross sections 10-8 - Figure 6a). Figure 6b presents the typical soil profile / cross section for the case of landslide in only one step (section 2).

Flow Analysis

In the flow analysis due to the unknown initial suction in the slope during the failure period, a parametric study considering a constant initial suction in all slopes was done in the range of 4 to 60kPa values. The geotechnical parameters (hydraulic relationship between moisture content and suction, etc) used in the analysis were those described before, taking into consideration the saturated hydraulic conductivity of the saprolite (Ks) higher or equal to the young residual soil (Ks > 5 x 10^{-5} m/s or Ks = 5 x 10^{-5} m/s, respectively).

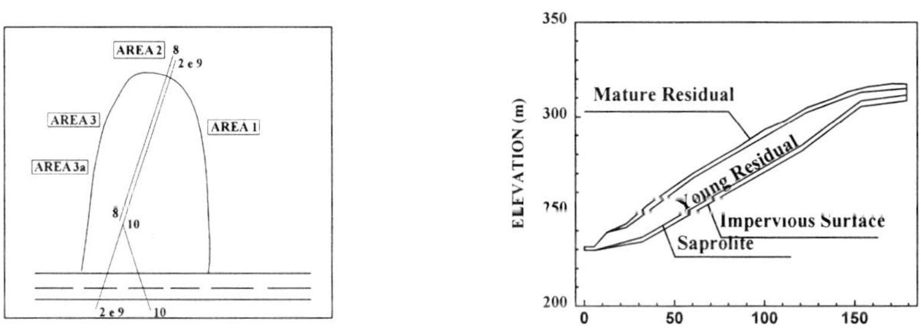

(a) (b)

Figure 6. a) cross-sections-stability analysis; b) typical soil profile - cross section

The results of the analysis showed basically two types of behavior: i) for initial suction less or equal to 10kPa, the rain infiltration reached the top of the rock (adopted as the impervious surface), saturating this part of the slope and causing positive pore pressure along this impervious surface; ii) for initial suction higher than 10kPa, the rain infiltration did not saturate the region of the impervious surface, keeping all slope in an unsaturated condition with negative pore pressure.

Stability analysis

The drained shear strength parameters adopted in the stability analysis were in function of the results obtained from direct shear tests of the mature and young residual soils. Figure 7a presents the hypotheses considered in the study to the drained shear strength in the saturated condition for all soils adopting c'=20 kPa and ϕ'=45° to the saprolite. The "ϕ^b parameter related to the matric suction (Fredlund et al., 1978) was adopted equal to ϕ', in function of the small values of suction observed in the field during the wet season. The research of the critical sliding surface was done considering a non-circular surface varying inside a region that included the observed failure surface.

Figure 7b shows the critical factor of safety results for the landslide case in only one step (cross section 2) for the different hypotheses included in the cohesion and suction study considering Ks > Kyr. The behavior of the FS shows a continuous reduction with the reduction of the initial suction considered, which presents a higher velocity from the initial suction equal to 10 kPa, due to the positive pore pressure occurrence in the slope during the rain infiltration. For the case presented here, it can be also observed that in the conditions of c'=0 to 5kPa (hypotheses a and b), the FS is very close to 1.0 for initial suction in the order of 7 to 10kPa. In the same conditions as above the FS results for the two-step landslide cases (cross sections 10-8) were very similar to the one presented, pointing out the possibility of occurrence in both landslide mechanisms.

The critical sliding surface obtained in all conditions considered in the stability analysis presented a similar behavior. All of them were closed to the failure surface observed in the field. The slope stability evaluation for the cross section 9 before the road (Figure 6a) showed that for initial suction ≤ 10kPa, the natural condition FS was in the order of 15% higher than the FS after the construction of the road (Figure 7b). Such result characterizes the possibility of low natural FS occurrence and a significant landslide hazard in the area. This result represented an indicative that the road cut was an aggravating triggering factor in the landslide.

Hypotheses for cohesion	Cohesion -c' (kPa)			
	MR	YR	SP	
(a)	0,00	2.88	20.00	ϕ' = 30°
(b)	5.00	5.00	20.00	ϕ' = 30°
(c)	10.00	10.00	20.00	ϕ' = 30°
(d)	15.00	15.00	20.00	ϕ' = 30°

MR = Mature Residual
YR = Young Residual
SP = Saprolite

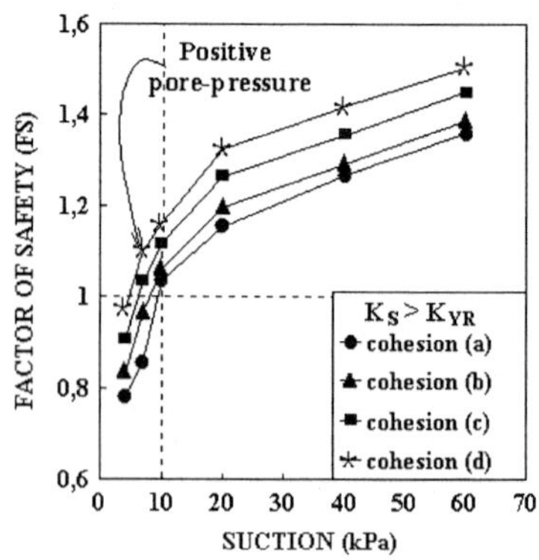

a) b)

Figure 7. a) strength parameters considered; b) FS-Suction Relationship / section 2

Back analysis

A back-analysis was done considering apparent cohesion $c^* = c' + (u_a - u_w) \tan\phi^b$ mainly to find which strength parameters can explain the failure, and to evaluate the strength parameters from laboratory tests. Table 1 presents a summary of the results obtained to the section 2 (Figure 6a). The results show that strength parameters obtained from back-analysis were well resembling to those determined through laboratory tests, indicating that results from laboratory tests can explain the failure of the slope considering low value of initial suction.

Table 1. Summary of the results of back-analysis to the section 2

Friction angle, ϕ' (^{o})							
28.5		29.0		29.5		30.0	
c* (kPa)	FS	c* (kPa)	FS	c* (kPa)	FS	c* (kPa)	FS
5	1.019	3	1.007	1	1.003	1	1.024
7	1.046	5	1.039	3	1.024	3	1.044
--------	---------	--------	--------	5	1.053	5	1.073

Comment

Leroueil et al. (1996) proposed a geotechnical characterization of slope movements considering four distinct stages: a) the pre-failure stage; b) the onset of failure; 3) the post-failure; and 4) the reactivation stage. Each of the four stages involves mechanical phenomena, controlling laws and parameters, which are very different. In a first time failure, even in homogeneous soils, the general failure of a slope is the result of a combination of several processes. First, the local failure develops associated with this process and so, progressive failure takes place in the soil mass. It can remain limited ("confined failure") or generate a general failure. The overage strength parameters at failure can be somewhere between the peak and residual strength values. For intact clays, the shear strength parameters can be characterized by a friction angle corresponding to critical state, and by an empirical cohesion. In the case of ductile materials, the strength mobilized in situ is close to that obtained in the laboratory tests on specimens of appropriated size so as to include typical features of the materials. In unsaturated soils their strength increase with the matric suction in the soil. Thus, an important point is the permeability or the rainwater infiltration of the unsaturated soil. The authors consider that the failure is the result of a complex process, and the selection of representative in situ parameters is generally difficult.

Popescu (1994) and Leroueil et al (1996) presented a method for reporting and investigating landslide causal factors considering ground conditions, geomorphological processes, physical processes, and man-made processes. In this context causal factors can be classified as: 1) preparatory causal or in situ predisposition factors which make the slope susceptible to movement; 2) triggering or aggravating causal factors which initiate movement; and 3) revealing factors which give an evident evolution of slope.

In the present study, the stability analysis using the peak strength parameters obtained in laboratory tests can explain the failure of slope. The initial results also show that in order to obtain FS $\cong$ 1.0 it is necessary a low or very low value of suction. This hypothesis is expected to be possible from in situ suction measurement. Considering the strength parameters for 10 mm of horizontal displacement ($c' = 0$ and $\phi' = 25^o - 30^o$; see Figure 5), the necessary value of the initial suction to obtain FS $\cong$ 1 is a little higher, which it is also possible to occur. The probable predisposition factors are the direction of the foliation plane parallel to the direction of the slope (adversely oriented mass discontinuities), and the high mica content in the young residual soil. The excavation of the slope to construct the road is an aggravating triggering factor. The intense rainfall or the prolonged high precipitation is the probable triggering factor through the reduction in the suction and possibly the positive pore pressure.

CONCLUSIONS

The basic engineering properties and the strength parameters results from laboratory tests has been in accordance with what is described in literature. It is also observed a small anisotropic behavior in the shear strength. The slope stability analysis shows that it is possible to have a low factor of safety (1.15-1.20) for the natural condition of the slope during the wet season, characterizing a significant landslide hazard in the area. The results of stability analysis indicate that the laboratory strength parameters (maximum or 10 mm horizontal deformation) may explain the failure with a low or very low value of suction. The probable predisposition factors are the high mica content in the young residual soil and the direction of foliation plane parallel to the direction of the slope (adversely oriented mass discontinuities). For the natural condition, the soils are not collapsible or expansible suggesting that these phenomena cannot be considered a predisposition factor. The aggravating and triggering factors are the excavation of the slope to construct the road and the intense rainfall or prolonged high precipitation.

ACKNOWLEDGEMENTS

The authors are grateful for all support received from the MCT/FINEP, CNPq-Brazilian Research Council and the Road Department of the State of Pernambuco. Professors Willy A. Lacerda (COPPE / UFRJ), Tácio M.P. de Campos (DEC/PUC-RJ) and Edmilson Lima (DEGEO/UFPE) for their helpful contributions to the Project as member consultors.

REFERENCES

Costa, F. Q. (1996). Problemas de Estabilidade de Taludes em Solo Residual de Gnaisse – Análise do Escorregamento do Espinhaço da Gata, PE-89 / Pernambuco. *Tese de MSc.*, Eng. Civil/UFPE, 113p.

Coutinho, R. Q., Costa, F. Q. & Souza Neto, J. B. (1997). Geotechnical characterization and Slope Stability Evaluation of a Slope in Residual Soil from Pernambuco, Brazil. *2nd Pan-American Symposium on Landslide.* Rio de Janeiro/Brazil. Vol. 2, p.287-298.

Coutinho, R. Q., Souza Neto, J. B., Barros, M. L. S., Lima & E. S., Carvalho, H. A. (1998). Geotechnical characterization of a young residual soil/gneissic rock of a slope in Pernambuco, Brazil. *In* 2nd International Symposium on The Geotechnics of Hard Soils and Soft Rocks, Naples, Italy. Vol.1, p.115-126.

Lacerda, W. A.& Almeida, M. S. S. (1995). State-of-the-Art-Lecture: Engineering Properties of Regional Soils: Residual Soils and Soft Clays. *IX PCSFE.* Guadalajara / Mexico.

Leroueil, S., Vaunat, J., Picarelli, L., Locat, J., Lee, H. & Faure, R. (1996). Geotechnical Characterization of Slope Movements. *In: Proc. of the Seventh International Symposium on Landslides.* Trondheim, p.53-74.

Popescu, M. E. (1994). A suggested method for reporting landslide causes. *BULLETIN of the International Association of Engineering Geology.* LCPC, Paris. n° 50, p. 71-74.

Sarma, S.K. (1979). Stability analysis of embankments and slopes. Journal of the Soil mechanics and Foundation Division. ASCE, v.105. n. GT12, p. 1511-1524.

Souza Neto, J. B. (1998). Características geotécnicas do solo residual de gnaisse da encosta Espinhaço da Gata, PE-89, Machados - PE. *Tese de MSc.*, Eng. Civil/UFPE, 250p.

Risk Scenarios for Complex Debris Slump-Debris Flow Phenomena

G. CROSTA, Dip. Scienze Geologiche e Geotecnologie, Università degli Studi di Milano – Bicocca, Italy

INTRODUCTION

Large landslides in detritic materials often show their complex nature being capable of sudden changes in behaviour, passing from slumping to flowing; and of relatively high mobility. These features are the cause of the extremely high severity and hazard typical of such phenomena. The study and modelling of these complex slope failures are often influenced by the difficulty of direct observations. This is the result of multiple factors such as: the rapidity of the phenomena or part of them, the continuous changing of material properties, e.g. due to water feeding from converging channels, and the resultant need of different rheological models to simulate different phases of the event, and, eventually, the morphometric characteristics along the flowing path.

A case history, including all the evolutionary stages from sliding to flooding, is here presented. This case history is a good example of the evolution this kind of instabilities are sometimes subjected to and it allows to discuss how risk scenarios for such events could be developed. Historical data, pre- and post-event observations are fundamental for such a kind of study. As a consequence the use of such data is presented and discussed in this paper together with some forecasting approaches that have been adopted in this case and are adoptable to reconstruct event risk scenarios.

SETTINGS AND HISTORY OF THE SLIDE

The study area is located in the Grigna valley (104.8 km^2 catchment area), a left hand tributary of the Val Camonica (valley bottom at 256 m a.s.l., in the southern Alps, Lombardy, Italy, Fig. 1). This tributary valley is characterised by the presence of a narrow and deep rocky gorge about 6 km in length. This site, and especially the upper catchment area, is quite peculiar for the thick glacial deposits covering the hillslopes. The here described phenomenon, known as Sesa landslide (about 1450 m a.s.l.), has been described since the early Thirties when large slumps induced debris flow accumulation at the hillslope foot and along the upper creek sectors. Reforestation

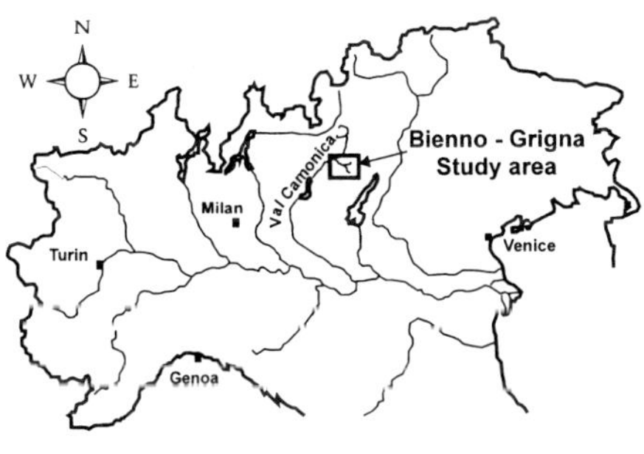

Fig. 1 - Location map of the study area (Grigna catchment), N. Italy.

Landslides in research, theory and practice, Thomas Telford, London, 2000.

and early biotechnical stabilisation works were realised after this first historic event.

In 1987, a slope failure manifested itself by the development of large tension cracks (500 m in length) bounding a huge mass (about 500 m high) of glacial deposits in the same slope area. Topographic surveyings were started and continued with monthly frequency for almost two years while some minor superficial water drainages were installed. On October 9th 1993, after a 15 days long rainy period, we assisted to the restart of the movement and debris flow activity and to the total collapse of the slope. Rainfall (Tab. I) was relatively low in intensity but quite conspicuous when considering the total antecedent rainfall for which a 100 yr recurrence time resulted. Rainfall analyses have been performed on data available from 5 rain gauges located in proximity of the site.

A 1.6 Mm^3 mass of glacial deposits (Fig. 2a, b), with a thickness ranging between 60 and 70 m, was involved together with about 2000-3000 m^3 of wood trunks. This thick glacial deposits, deeply cut just before the event, outcrops along the major slide scarp (up to 80 m high), are still visible on the scar after few years since the failure. Below this deposit, the bedrock (sandstone and siltstone) was dipping at 30° in the same direction of the slope. A slump characterised the initial movement along the hillslope but some indications point out for a flow to occur already along the intermediate slope sector. No run-up took place on the opposite valley flank and no sign of impoundment was observable after the event. This is quite interesting because shows that no landslide dam was generated and as a consequence neither was its collapse. After that, the material flowed within the main creek for about 7 km up to the outlet of the valley. Clear indications of a debris flow and finally of a hyper-concentrated flood (Pierson & Costa, 1987) where observed along the rocky gorge. Flow tilting and changes in thickness (3.5 to 30 m) , alluvial debris and slope scouring, and deposition of coarse and fine materials were observed and described along the entire path. The duration of the phenomenon was estimated in about 17-20 minutes (average velocity = 7.5 m s^{-1}) according to witness reports and the time of occurrence of damages at different structures (bridges, hydroelectric power plant located along the path at known elevations from the valley bottom).

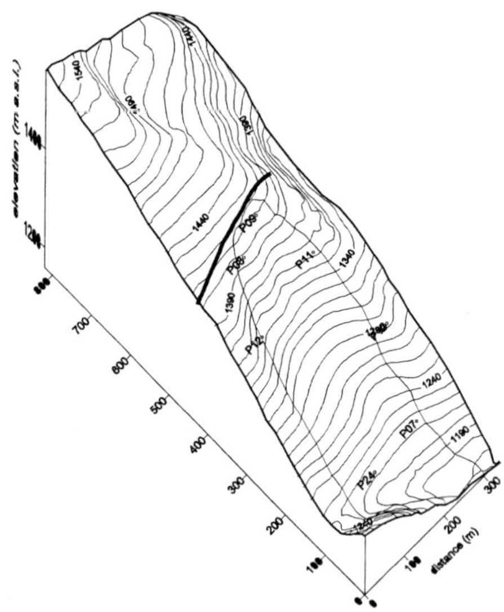

Fig. 2a - Pre-failure topography for the Sesa Landslide area. On the pre-failure topography are represented 7 from the 12 topographic reference points positioned after 1987 and used for forecasting the time of failure. through the Voight & Fukuzono methods. The landslide boundary and a superficial drain-collector (thick line), realised after the 1987 reactivation, are also drawn.

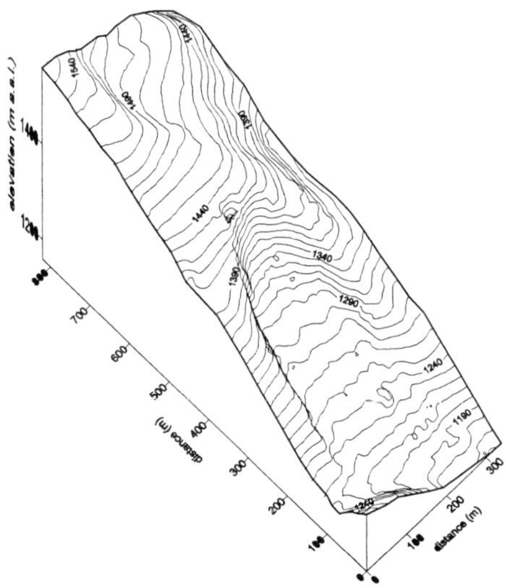

Fig. 2b - Post-failure topography for the Sesa Landslide area. The depleted volume is about 1.6 Mm^3 and less than 0.4 Mm^3 were deposited along the slope and at its foot where the Grigna Creek tributary (Travagnolo creek) flows. No upgrading of the Travagnolo creek bed was observable at the slope foot.

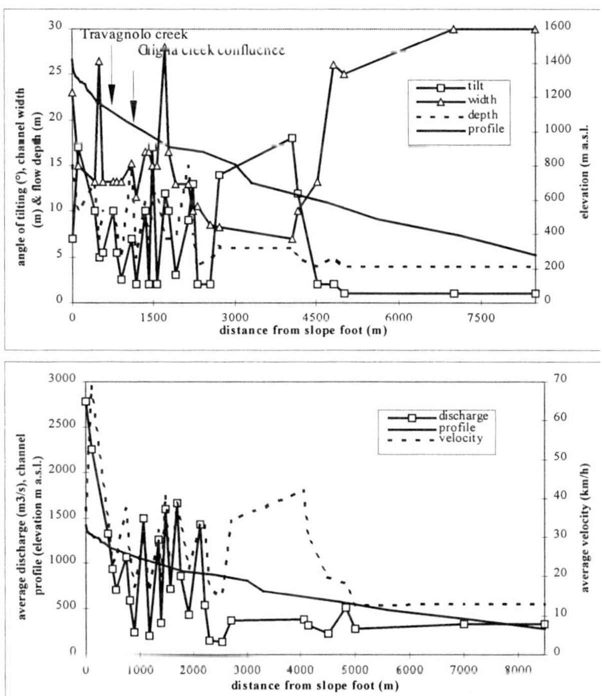

Fig. 3 -a) - Main geometrical parameters measured at 28 cross sections along the flow path. b) - Velocity and discharge along the flow path as obtained according to Johnson & Rodine (1984)

FLOW DESCRIPTION

The analysis of flow features along the landslide path is vital for the understanding of the phenomenon. Indirect (or post flow) measurements of flow characteristics have been performed along the entire path (about 7km).

In particular, flow thickness and tilting, thickness and type of deposits, together with their volumes, have been investigated during field surveying.

We must point out that difficulties have been encountered during field work both because of logistics (a narrow rocky gorge with frequent steps) and the lack of information about alluvial debris deposits antecedent to the event. The recognition of layering or its absence within the deposits has been the most useful feature, beyond their freshness, in distinguishing between old alluvial deposits and recent debris flow deposits along the main gorge.

Of the initial mass, about 1.2 Mm^3 flowed along the gorge: 10% of it being formed by silt and clay, as supported by samples of slope material collected along the main scarp. This mass of finer particles was transported farther down of the valley outlet under hyper-concentrated flow conditions, as suggested by large deposits occurring downward from the valley outlet.

About 0.8 Mm^3 of coarse debris have been deposited along the 6 km long gorge (713 m of relief). All along this sector the deposits are mainly formed by boulders (up to almost 100 m^3 in size) and pebbles (55%) with abundant gravel and sand (45%).

They are characterised by inverse grading, with undefined stratification, high density and compaction, gradual downstream decrease of the average grain size and with average larger size in correspondence of path bending. Their thickness ranges between 0.5 and 3.5 m. On the contrary, flow thickness reached much higher values, up to 30 m, with a gradual downstream decrease.

Evidences of tilting of the upper flow surface have been along the frequently recognised in the field entire path within the gorge. 28 cross sections have been located and surveyed along the gorge describing for each of them the main sedimentological and geometrical characteristics of the flow scar (Figure 3a, 4a, b, c). These observations have been used to assess the average velocity and discharge along the path, following Johnson & Rodine (1984) (Fig. 3b). Looking at the results it is possible to say that the mass moved rapidly (between 40 and 70 km/h) at the real beginning (from 0 m to 400 m) slowing down up to the confluence of the Travagnolo Creek into the Grigna Creek (about 1100 m from the start).

From this point up to the hydroelectric power plant (located about 4000 m from the slide area), the debris flowed at average speed (15-40 km/h) with local increase of velocity and discharge (Figure 3a, b) due to morphological control and eventual water inflows from minor secondary valleys.

It is evident the narrowness of the channel in the upper sector (Fig. 3a) with almost constant width and with frequent tight curves that might have caused pulsing of the flow. The second half of the path is more rectilinear with minor bending and a more continuous flow. Nevertheless, damages to bridges and small crossings downstream of the valley outlet still occurred with an average deposit thickness of 1 meter.

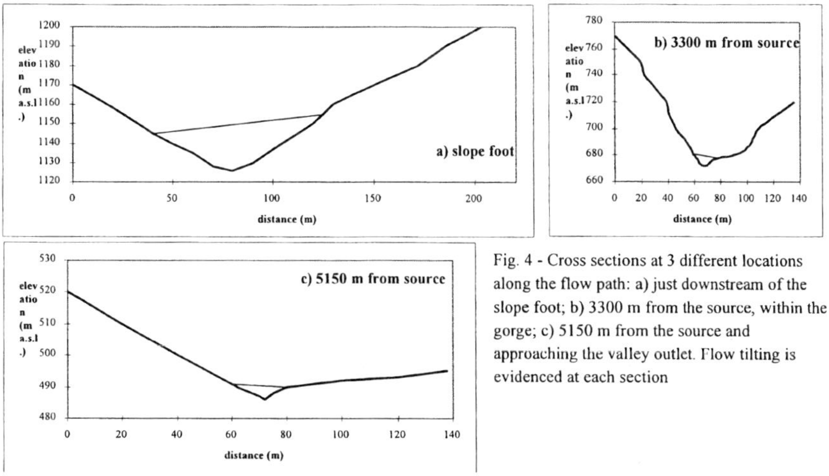

Fig. 4 - Cross sections at 3 different locations along the flow path: a) just downstream of the slope foot; b) 3300 m from the source, within the gorge; c) 5150 m from the source and approaching the valley outlet. Flow tilting is evidenced at each section

Table 1. Rainfall data and their relative recurrence times, for the September - October 1993 period as measured at the Grigna valley outlet (Bienno rain gauge)

Time Period	Total rainfall	Rainfall Duration	Average rainfall	Average Recurr. Time
	mm	*h*	*mm/h*	*yr*
23-30 sept.	122.3	83.5	1.46	10
5-9 oct.	123.0	65.0	1.89	25
7-8 oct.	91.8	29.0	3.16	14
23 sept. -8 oct.	337.1	177.5	1.90	100

LANDSLIDE ANALYSIS AND MODELLING

Three main subjects have been analysed in this study: stability of the slope, forecasting of the slope failure time, modelling of the flowing mass.

Slope stability has been analysed both for pre- and post-failure conditions. According to limit equilibrium analysis, introducing material properties evaluated by means of geotechnical laboratory tests (GP, GW USCS class, bulk density = 1.74 (looser) - 2.05 (denser) g/cm^3, angle of internal friction = 38°, cohesion = 0 - 100 kN/m^2), in situ description supported by image processing techniques and groundwater conditions, as suggested by post-failure evidences (seepage front), the slope was close to critical conditions (factor of safety ranging between < 1 and 1.5).

Monitoring data, from 12 topographically surveyed points, for the 1988-1989 period and successive to the 1987 slide reactivation, have been analysed to evaluate the possibility of their use for the *forecasting of slope failure time*. The analysis was based on the Fukuzono's (1985) and Voight's (1988) approach and showed interesting results for 6 surveyed points (Fig. 3a) because of their relatively large displacements. Voight's method is based on the general formula:

$$\Omega^{-1} = [A(1-\alpha)(t-t_o)+\dot{\Omega}^{1-\alpha}]^{1/(\alpha-1)}$$

with A (usually between 0 and 0.5) and α (usually between 1 and 2) empirical constants, t and t_o are any time and the time of initiation of measurements, Ω is any measurement of deformation and $\dot{\Omega}$ its velocity or change in unit time.

Two sets of parameters from data fitting are summarised in Table II by considering the failure time (t_f) directly resulting by the application of the method and an imposed time of 5 years (since onset of topographic measurements, 1988, to slope failure, 1993). The difference among the times of failure could be related to: major activity on the left hand flank of the landslide (P11) contrasting with that of the right flank (P12), involvement of the maximum debris thickness in the upper area (crown area, P10, P08, P09), different mobility at the slide toe because of the presence of containing structures (P07, P24). All the values obtained for the controlling parameters are within the valid range of values according to Fukuzono and Voight and fitting resulted in extremely high correlation coefficients (between 0.952 and 0.999 using t_f, and between 0.51 and 0.952 for the 5 years time interval).

The study of the flow evolution after the initial slumping was the aim of the performed *dynamic modelling*. The dynamic model proposed by Hungr (1995) has been implemented in a numerical code introducing changes about: mass transfer between adjacent mass slices, capability to simulate scouring-amalgamation and deposition of material along the path, numerical solution. The main advantage of this approach consists in its ability to simulate material flow according to different rheological models corresponding to different flow regimes. The main disadvantage consists in the difficulty to model the flow behaviour in consequence of channel narrowing. This problem has been overcome by minimising channel width changes. The developed code allowed to model the frictional regime, typical of the initial movement, and the successive changes to a turbulent regime, as witnessed by field surveying evidences, and the gradual changes due to the increase of water content (tributary valleys, deposition along the path). Among the results we can list the forecasting of velocity, thickness of flow and deposit, total and partial times of flow, mass distribution during flow and after deposition. The results have been compared to and the model has been calibrated by means of field observations (Fig. 5).

DISCUSSION AND CONCLUSIONS

Complex landslides are generally the most difficult slope instability processes to be analysed. This is often the result of combination of different causes, mechanisms and effects. As a consequence the evaluation of properties, time of failure, types of movements and regimes is a quite complex process but vital to outline and understand risk scenarios. Then, site characterisation, monitoring data and sound modelling are necessary.

This case history put in evidence the importance of field observations and the applicability of Voight's method for time of failure forecasting. The need for monitoring data becomes evident when considering all the surveyed points on the slide area where a good model fitting were obtained and those where relatively small displacements were observed after the installation of the monitoring network. Unfortunately, as in this case, the costs to maintain

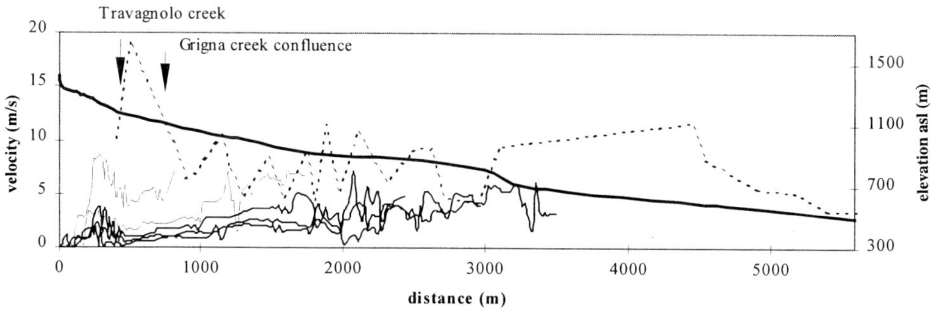

Fig. 5 - Flow velocity vs. distance as from dynamic modelling (continuous lines for different flow times) and field survey data (dashed line, as Fig. 4a, b). Field data analysis result in an envelope of the maximum velocities reached along the path and is clearly sensitive to channel bending and/or narrowing.

functional a monitoring network are often considered too high when failure is not evident. This fact can hinder data analysis because of the superposition of movement seasonality to a long term failure tendency that is usually more intelligible by means of methods like the Voight's one.

Finally, risk scenarios at the valley outlet are usually the most important problem especially in alpine areas, where secondary catchments are often characterised by the presence of large landslides. Numerical models showed their usefulness in this case history. They have been applied to the studied slide and to some other areas within the same catchment, where similar phenomena could occur.

Table 2. Total measured displacements and parameters for the Voight's method

Point	**Total displacement**	**t_f, time of failure**	α_f	**A_f**	Ω_f	α_{5yr}	**A_{5yr}**	Ω_{5yr}
	cm	*h*	-	-	-	-	-	-
P07	31.7	12500	1.80	0.0182	0.153	1.41	0.00061	0.3336
P08	110.6	12500	1.90	0.0075	1.039	1.56	0.00049	5.1166
P09	461.3	12500	1.35	0.0012	7.251	1.47	0.00026	4.6217
P11	106.7	8300	1.94	0.0562	496.2	1.32	0.00037	4.7017
P12	38.1	15000	2.02	0.0026	0.377	1.55	0.00087	0.8680
P24	99	10000	1.20	0.0039	505.2	-	-	-

The analysis allowed us to develop a risk scenario for this and similar instabilities. Risk zonation in the distal sector has been performed considering: natural damming by deposition and obstruction of the river course, river diversion and invasion of morphologically prone sectors, hazard for bridge occlusion or destruction on the basis of deposit thickness (as obtained by numerical modelling); damages to building and structures; cumulated effects from confluence of various valleys. Risk zonation has been carried out both parallel and transversal to flow direction distinguishing among high, medium, low risk areas and according to different types of damages (fine or coarse debris accumulation, partial or total destruction, flooding, etc.).

REFERENCES

Fukuzono T. (1985) A new method for predicting the failure time of a slope. Proc. 4th int. Conf. and Field Workshop on Landslides, Tokyo

Hungr O. (1995) A model for the runout analysis of rapid flow slides, debris flow and avalanches. Can. Geotech. J., 32, 610-623.

Johnson A.M. & Rodine J.R. (1984) Debris flow. In Slope Instability, Anderson ed., 257-357

Pierson T.C., Costa J.E. (1987) A rheologic classification of subaerial sediment-water flows. In Debris Flows/Avalanches: process, recognition and mitigation, Rev. Eng. Geol., 7, Costa J.E. Wieczorek G.F., eds., Geol. Soc. Am., Boulder, Co., 1-12

Voight B. (1988) A method for prediction of volcanic eruptions. March 1988, Nature, 332

Deep Seated Slope Deformations: Huge, Extraordinary, Enigmatic Phenomena

G. CROSTA, Dip. Scienze Geologiche e Geotecnologie, Università degli Studi di Milano – Bicocca, A. ZANCHI, Dipartimento di Scenze dell'Ambiente e Territorio, Università degli Studi di Milano – Bicocca, Italy

INTRODUCTION

Deep Seated Slope Gravitational Deformations (DSSGD): a definition which reflects the complexity of peculiar slope stability problems. DSSGD have been described worldwide starting almost 60 years ago but their study can still be considered in its youth. They have been found initially in the Alps where it is still possible to discover them, larger then ever and with a large variety of evidences. The authors present and discuss four examples, never recognized before but so interesting for their size, characterizing features and tectonic settings. These examples are taken from Italian Central Alps and Prealps. Different lithologies (phyllites, mica-schists, shales and limestones) control the evolution of the phenomena and the formation of morpho-structural features. Furthermore, each case is characterized by specific tectonic and neotectonic settings. These settings have been studied by field surveys, stereo-photo interpretation and by applying paleostress analysis techniques. Structural and tectonic lineaments have been surveyed and their influence evaluated, and introduced in the numerical modelling (not here presented) performed to support a conceptual model for each one of the cases.

DEEP SEATED SLOPE GRAVITATIONAL DEFORMATIONS

Deep seated slope gravitational deformation is a definition that has become more and more important during the last decades as suggested by the abundant technical papers related to such a problem. As a consequence, a large number of examples has been produced from different locations, classified with different definitions and explained with different mechanisms and evolutionary models (Crosta, 1996). At least one example of deep-seated gravitational deformation has been presented for each important mountain belt in the world, linking these phenomena to any high relief environment with different rock mass lithology and geomechanical characteristics.

A large number of terms have been suggested to describe these phenomena since their first description. In particular we can remember the following terms: deep-seated large scale rock slides, sackung (Zischinsky, 1966), deep seated gravitational deformations, deep seated distortion (Varnes et al., 1989), sagging (Hutchinson, 1988), mass rock creep (Chigira, 1992), rock flow, etc. Even if so many terms have been suggested it is still difficult to find a complete and correct definition for these phenomena. For the aim of this paper, we suggest to consider among these phenomena those characterised by: strong morphological evidences (scarps, up-hill facing scarps, trenches, etc.) extended to large part of the extension of high relief slopes (> 500 m) starting from its foot and frequently up to or beyond the main upper ridge; large volumes and mass thickness; relatively slow and small actual displacements. We don't think that the possibility for future catastrophic evolution or the presence or absence of a sliding or failure surface can be considered really discriminating features. At the same time the authors support the idea that recent geological history (neotectonic or recent tectonic events, including actual or recent tectonic stress fields and seismicity, deglaciation or valley deepening) can play a fundamental role in large slope deformations, as suggested in part of the literature, together with the presence of main tectonic and structural features (fault, thrust, etc and bedding or schistosity dip) active or not.

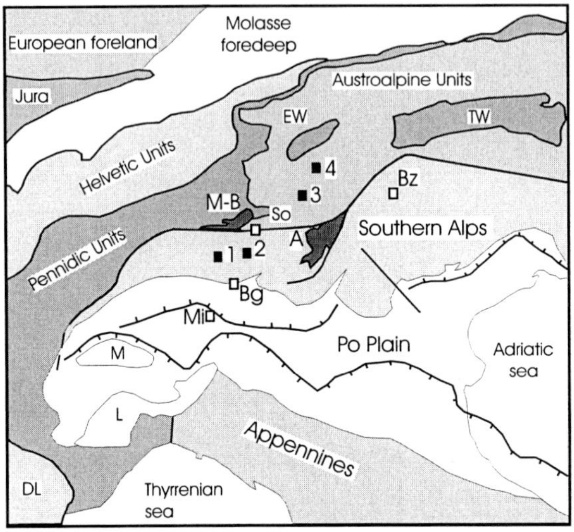

Fig.1 Tectonic sketch map of the Alps. Black squares refer to the studied areas: 1)Vedeseta, Taleggio valley 2) Gandellino, Mt Calvera - Vigna Soliva, Seriana valley, 3)Valfurva, Ruinon, Bormio, 4) M. Vatles, Venosta valley. A: Adamello Pluton, M-B: Masino-Bergel Pluton, M: Monferrato Arc, L: Ligurian-Piedmont Basin, DL: Delphinese units, EW and TW: Engadine and Tauern windows. Empty squares: Mi: Milano, Bg: Bergamo, So: Sondrio, Bz: Bolzano.

Finally, a specific mention could be done to those processes that are usually defined as lateral spreading. These spreadings generally involve sites where hard rock tabular elements overlay soft or weak rocks or overconsolidated clays and argillites. In these cases the mechanisms involved seem more evident whereas only in relatively few cases very deep deformations are associated with block lateral spreading.

CASE STUDIES

The Italian Alps and Prealps are quite well known from a geological and geomorphological point of view. Nevertheless, it is still possible to find sites where huge deep seated slope gravitational deformations have been incompletely described or absolutely neglected. This is the case for the four case studies presented in this paper (Figure 1): 1) the Vedeseta sackung, 2) the Mt.Calvera - Mt. Vigna Soliva sackung, 3) the Ruinon sackung and 4) the Upper Val Venosta area and in particular the Mt. Vatles sackung. These sites are characterised by quite different geological and geomorphological settings making more intriguing their study.

1) The Vedeseta sackung

The Vedeseta sackung is the first deep seated slope gravitational deformation described in the Taleggio Valley (Bergamo, Italy) within the Middle-Late Triassic successions of the Southern Alps. This sackung, together with that of the Corno del Bruco, occurs where rigid carbonatic massifs (Corno Zuccone: Dolomia Principale Fm.; Corno del Bruco: Esino Limestone Fm.)

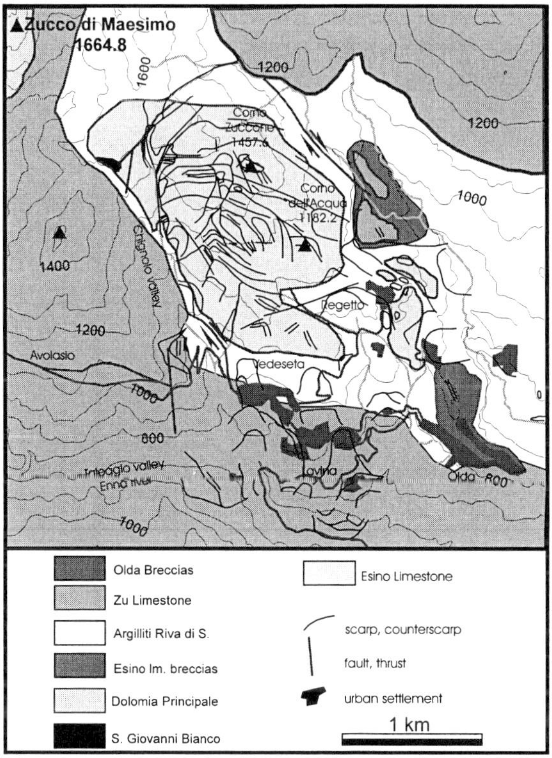

Fig. 2- Map of the Vedeseta sackung area.

have been thrust upon the more ductile shales (Late Triassic Argillite di Riva di Solto Fm.). In fact, the Taleggio Valley is dominated by the occurrence of a widespread basal unit (*Parautoctono Bergamasco,* GAETANI & JADOUL, 1979) consisting of a 1000 m thick carbonatic and shaly succession (Argillite di Riva di Solto Fm., 400 m thick, and Calcare di Zu, 500 m, Dolomia a Conchodon Fm., Norian to Liassic in age), forming a regular locally folded SW- to W-dipping monocline. Three main thrust sheets, including Middle to Late Triassic successions, are stacked above the basal unit. The Corno Zuccone klippe (stratified dolostones belonging to the Dolomia Principale Fm., JADOUL, 1985). is limited at its base by a gently southward dipping thrust. Several Quaternary units occur in the area. In particular, poorly sorted rounded conglomerates with breccias (Breccia di Olda Unit, Early Quaternary fluvial fan) form a isolated ridge which testifies a fossil talweg developed before the excavation of the present E-W canyon. As a consequence their activation could possibly be post-Early Quaternary. The main morpho-structures of the Corno Zuccone sackung consist of a spectacular set of subvertical E-W and ENE-WSW trending fractures developed within the upper carbonatic massif and suggest a dominant S-SSW slope movement (Fig. 2). The main scarp (1.5 km in length), just below the upper ridge, suggests a vertical throw of at least 100 m, whereas relatively small throws are recognizable along secondary fracture sets. Despite their continuity, these scarps and fractures have been successively cut by NNE-SSW trending fractures. This assemblage splits the sackung into different sectors with different kinematic behaviours. The NW part of Corno Zuccone moved toward the Valle del Chignolo, and the SW portion toward Reggetto. Two main E-W concentric counterscarps, along the northern slope of Corno Zuccone, together with three counterscarps controlling large NE-SW trending saddles, suggest a northward propagation of S- to SE-dipping sliding surfaces north of the most deformed area of the sackung. The analysis of these systems suggests that also the upper member of the ARS, present below the Corno Zuccone klippe, is affected by the DSGD. This conceptual model is confirmed by the 200 m southward displacement of the valley axis, the presence (Fig. 2) of large slide scarps and instabilities, presently active, within this formation in the lower bulging half of the slope.

2) The Mt. Calvera-Vigna Soliva Sackung

The Mt. Calvera sackung is located in the western part of the Southern Alps in the upper Seriana Valley, together with some other deep seated slope deformations (Mt. Toazzo). The slope gravitational structure occurs close to the eastern closure of the Trabuchello-Cabianca Anticline belonging to the ENE-WSW en échelon Orobic Anticlines. These open folds, interpreted as out-of-sequence ramp folds (Laubscher, 1985), include the crystalline basement (mica schists and phyllites), the Early Permian continental volcano-sedimentary Collio Fm, the Verrucano Lombardo (fluvial conglomerates) and at the top the Early Triassic Servino Fm. (transgressive marine marls and oolitic dolostones). The sackung developed in the lower part of the Sedornia valley (left hand tributary of the Seriana Valley, Fig. 3), along the very steep southern slope of Mt. Calvera-Vigna Soliva, close to the tectonic boundary between the anticline and the Vigna Vaga Unit (Early to Middle Triassic sedimentary successions). This fault, interpreted as a steep south dipping wedging structure, rotates in this area from an E-W regional trend to a NE-SW one, following some Permo-Triassic paleofaults. Extensional shear zones at the boundary between the crystalline basement and the sedimentary cover have been recognized in the area as well as other minor faults bewteen the Collio and Verrucano formations. In the Vigna Soliva massif the thickness of the Collio Fm. is exceptionally small and may be related to the same paleogeographic setting (permian transtension). The main structural control on the sackung seems to be related to NNW-SSE fractures which are perpendicular to the Trabuchello-Cabianca fold. N-S joint sets have been also identified at the top of Cima Calvera and play a major role in the activation of the Seriana valley facing structures. The slope structure, recognized by De Jong (1979) with no description and suggestion about the mechanism, have never been mentioned in other studies. A throw of at least 30-40 m at the top of the sackung is suggested by lowered layers of Servino within a graben dissecting the upper main ridge (Fig. 3). Large open fractures split the upper ridge in huge blocks which moved downward along a convex failure plane. Impressive up-hill facing scarps and large landslide scarps, both caracterised by high continuity, are spreaded along the entire upper slope sector.

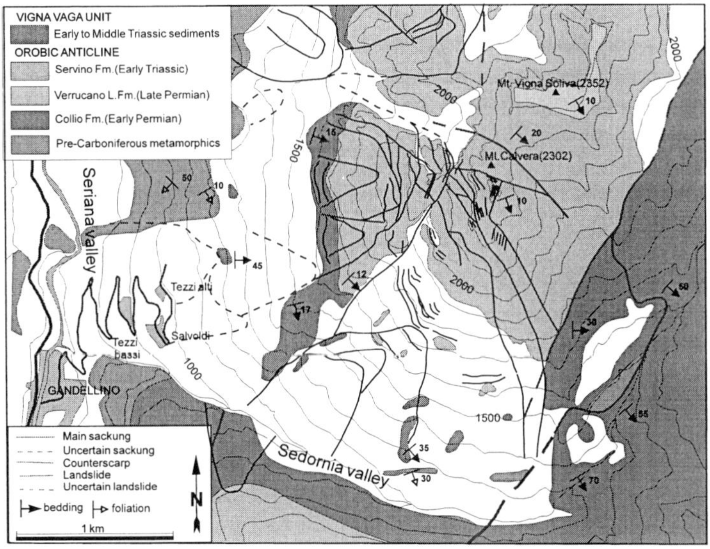

Fig. 3 - Map of the Mt. Calvera - Mt. Vigna Soliva sackung

3) Ruinon sackung

The Ruinon sackung is located east of Bormio along the middle part of Valfurva. The sackung occurs within the metamorphic rocks of the Campo nappe which form a large part of the Austroalpine crystalline complexes laying below the Ortles and Silvretta sedimentary units (Fig. 1). Here the rock mass consists of the Bormio Phyllite metapelitic unit including minor transposed and isoclinally folded marble and prasinitic intercalations. The main foliation (Pre-alpine in time), has been refolded by open alpine structures. The sackung developed along WNW-ESE and N-S fracture systems. The former has been related by

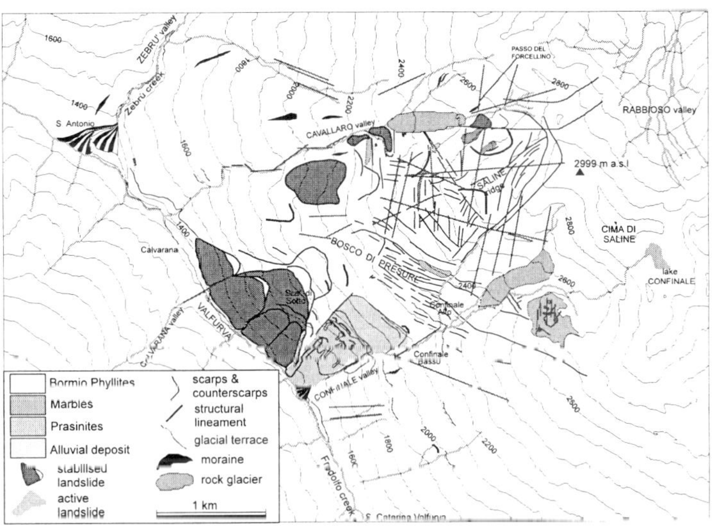

Fig. 4 - Map of the Ruinon sackung (Valfurva)

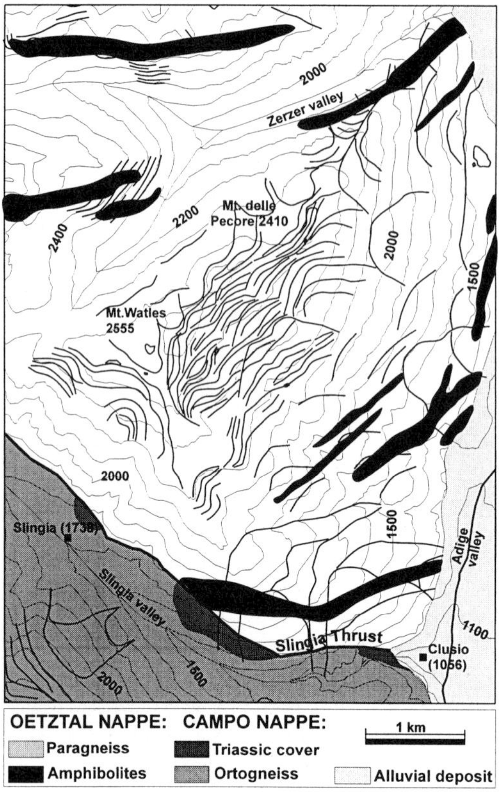

Fig. 5 - Map of the Mt. Vatles sackung within the Upper Venosta Valley

Forcella & Orombelli (1984) to neotectonic activity. Evidence of recent reactivation of these structures is suggested by gravitational faults which cross the middle sector of the slope and the Holocene rock glacier of the Cavallaro valley (active between 10 and 6 ka) (Fig. 4). The throw along the main surface in the upper part of the sackung, Saline Ridge, amounts at about 150 m on the basis of the displacement shown by marble layers and by the upper scarp length close to the ridge. Presently, a great interest is directed towards this area because of the accelerated development and movements of a 30 Mm^3 complex landslide at the slope toe (between 1400 m and 2150 m a.s.l). This phenomenon interests the lower "bulging" slope sector where other major past instabilities took place (Fig. 4).

4) M.Vatles sackung

The Upper Venosta Valley (eastern Alps) is an interesting site both for the geological and the incredible geomorphological characteristics (steep and very large debris fans and other indicators of a very rapid geomorphological evolution). Here, deep-seated deformations develop in a large region occupying entire valley flanks with relief up to 1700 m and various kilometers in length. It is interesting to note that some adjacent DSSGDs of the area are so much different among each other. This difference is evident in the type and location of morpho-structural features (scarps and counterscarps up to some tens of meters in height, trenches, lake filled depressions, and bulging, etc.), the indicators of toe slope activity, the location of main tectonic features, the subdivision of the main DSSGD body in secondary elements. The M. Vatles massif is located just above the village of Burgusio, a few kilometres south from the Resia Pass, in the metamorphics of the Austroalpine Oeztal Unit (Fig. 1). The Oeztal nappe overrides the Campo nappe along the Slingia Line (Schmidt & Haas, 1989), an important Cretaceous shear zone crossing the westernmost flank of M. Vatles, where non metamorphic Triassic and Jurassic sedimentary slices occur. The sackung entirely develops within the Oeztal basement rocks, including isoclinally folded paragneiss with subordinate lenses of ortogneisses and amphibolites (Fig. 5). Isoclinal folds with steeply dipping axes characterize the structure of the metamorphics.

Whereas no direct control of the Slingia overthrust-related tectonic structures can be observed at the surface in the sackung area, a relatively deep influence could be suggested because of the locally eastward dipping of the fault plane. A more evident link exists with the Late Tertiary-Quaternary brittle structures of the area. The entire region is still characterised by an intensive microseismicity (Slejiko et al., 1989) related to very superficial strike-slip faults. Fault plane solutions obtained by these authors around M.Vatles suggest the activity of NE-SW left lateral strike-slip faults or NW-SE right-lateral faults. By the analysis of mesoscopic structures, performed during our study, the Upper Venosta valley develops along a N-S left-lateral strike-slip fault, successively reactivated as a normal fault. NW-SE and NE-SW conjugated strike-slips related to a N-S compression have been detected at several sites in the study area enhancing the importance of the recent tectonic ddeformation. As a matter of fact the M. Vatles sackung clearly develops along NE-SW structural trends which may be the most active ones as well as along N-S structures possibly related to the main fault crossing the Venosta valley north of Glorenza (Fig. 5). It is possible that secondary shear zones related with the Slingia thrust may also be present in depth and then influence the main sliding surface of the structure. At the same time the differences in the rock rheology, below and above the Slingia Fault surface, could have contributed to localise deformations in the upper weathered paragneiss. This model is also suggested by the results of performed numerical modelling.

DISCUSSION AND CONCLUSIONS

Even if so many examples of deep seated gravitational deformations have been studied in the past, very few reliable data and descriptions exist about the actual influence of structural and geological features on the triggering and development of such huge slope instabilities (Crosta, 1996). This paper presents four cases from four different areas each one characterised by peculiar geologic and geomorphologic conditions and by the presence of multiple deep-seated slope deformations. Here, the structural and geological site conditions are presented to discuss their influence as preparing and controlling factors, but both structural and engineering geology studies have been performed together with numerical modelling (by finite difference and distinct element models). Furthermore, to improve the analysis of these slope movements and the understanding of the involved mechanisms, GIS techniques have been adopted. In fact, GIS allows a better perception of the 3D geometry, as well as the description and representation of morpho-structural features. As a general comment to the part of the study, here presented, it is possible to derive some conclusions or underline some ideas. Deglaciation is only one of the possible triggering causes and could be subdivided in different factors (drop of contrasting forces, groundwater level fluctuations, ice melting within fractures, climatic changes, etc.) and was not always active at the studied sites. Post-glacial triggering is often demonstrated by scarps and counter scarps cutting through glacial deposits (moraine, rock glacier, etc.). At the same time it must be considered the role played by stress relief under conditions of rapid valley erosion also in absence of deglaciation. Structural control is often involved as natural for so large phenomena and its influence can change with the structural settings of the area. Bedding and foliation have different importance according to their dip direction and inclination and frequently originate deep seated slope deformation if present with high dip angles. Involved lithologies can be quite different, both for rock masses and fault rocks (cataclasite, mylonite, breccia, etc.).

Seismological conditions can be quite important and in particular the stress states sequence occurred during recent geological hustory can control te evolution of such phenomena. Valley flanks with similar phenomena have in general a high topographical relief and are characterised by recent or actual slope instabilities at their foot. Catastrophic or rapid evolution is typical only to these phenomena along the lower slope sectors and here technical staff and planners or risk assessors must concentrate their attention.

REFERENCES

Chighira M. (1992) Long-term gravitational deformation of rock by mass rock creep. Eng. Geol. 32, 3, 157-184.

Crosta G. (1996) Landslide, spreading, deep seated gravitational deformation: analysis, examples, problems and proposals. Geogr. Fis. Dinam. Quat., 19, 2, 297-313

De Jong K.A. (1979) Overthrusts in the central Bergamasc Alps, Italy. Geol en Mijnbouw, 58, 2, 277-288.

Forcella F.and Orombelli G. (1984) Holocene slope deformations in Valfurva, Central Alps, Italy. Geogr. Fis. Dinam. Quat., 7, 41-48.

Gaetani M., Jadoul F. (1979) The structure of the Bergamasc Alps. Accad. Naz. Lincei, 8-66/5: 411-416.

Hutchinson J.N. (1988) Morphological and geotechnical parameters of landslides in relation to geology and hydrogeology. Proc. 5th Int. Symp. on Landslides, Lausanne, CH, A.A. Balkema, 1, 3-35.

Jadoul F., (1985) Stratigrafia e Paleogeografia del Norico nelle Prealpi Bergamasche occidentali. Rivista It. Paleont. e Stratigrafia, 91, 479-512.

Laubscher H.P. (1985) Large-scale thin skinned thrusting in the Southern Alps: Kinematic models. Geol. Soc. Am. Bull., 96: 191-211.

Schmid S. M., Haas R., (1989) Transition from near surface thrusting to intrabasement decollement, Schlinig thrust, Eastern Alps. Tectonics, 8, 697-718.

Slejko D., Carulli G.B., Nicolich R., Rebez A., Zanferrari A., Cavallin A., Doglioni C., Carraro F.,

Varnes D.J., Radbruch-Hall D., Savage W.Z. (1989)Topographic and structural conditions in area of gravitational spreadig of ridges in the western United States. U.S.Geol.Surv. Prof. Pap. 1496, 1-28.

Zischinsky U. (1966) Movement of unstable valley sides. Gesellschaft der Geologic und Bergbaustudenten, Mitteil ungen, v. 17, June, 127-168 (in german).

Landslide Modes in the Canadian Rockies

D. M. CRUDEN, Department of Civil and Environmental Engineering, University of Alberta, Edmonton and X.-Q. HU, Department of Earth and Atmospheric Sciences, University of Alberta, Edmonton, Canada

INTRODUCTION

The Canadian Rockies, over 100km wide, stretch 1400km northwestwards from the U.S. border, 49° N, at 114°. They are formed by Proterozic clastics and Paleozoic carbonates and clastics which were moved northeastwards up listric thrusts and over Mesozoic clastics (Gabrielse et al., 1992). Landslides from the steep slopes of the Rockies can travel large distances and have catastrophic impacts on structures in their paths. Site reconnaissances in the mountains seek to identify slopes which may move and to locate facilities away from the consequences of these movements. Such reconnaissances require identifying potentially unstable slopes and their modes of movement.

When a rock mass contains penetrative discontinuities which are parallel to each other, such as bedding or foliation, their orientations can be used to classify rock slopes formed in the rock mass. In cataclinal slopes, the penetrative discontinuity dips in the same direction as the slope; in anaclinal slopes, the penetrative discontinuity dips in the direction opposite to the slope (Figure 1). Cataclinal slopes may be further divided into overdip slopes which are steeper than the dip of the penetrative discontinuities and underdip slopes which are less steep than the dip of the penetrative discontinuities. Dip slopes follow the dip of penetrative discontinuities. We propose to term anaclinal slopes which are perpendicular to the dip of the discontinuity "normal escarpments", slopes steeper than normal are "steepened escarpments" and slopes less steep than normal are "subdued escarpments". (Cruden, Eaton, Hu, 1988).

Based on the kinematics of rock mass movements in stratified rocks, a process diagram can indicate modes of slope movements on cataclinal and anaclinal slopes. The diagram (Figure 2) includes movements from two dominant sets of discontinuities, the penetrative discontinuities, bedding planes, and the non penetrative kathetal joints which are perpendicular to bedding but have the same strike as bedding. In the diagram, the friction angle is assumed as 30 degrees, cohesion is zero and the rock mass is without pore pressure. These values are reasonable first approximations in the Canadian Rockies (Cruden and Hu, 1996). We continue with field examples of slope movements on each type of cataclinal and anaclinal slope.

CATACLINAL SLOPES – OVERDIP SLOPES

The behavior of overdip slopes can be modeled by the behavior of a sliding block of the rock mass. In the simplest model, the block slides when the dip, ψ, of the discontinuity down the slope is greater than the friction angle, ϕ, of the rock block on the slope. Cruden (1985) has demonstrated that the basic friction angle, ϕ_b, the rock is a useful estimate of the lower bound of rock avalanche activity in aseismic areas in the Canadian Cordillera. It seems reasonable

Landslides in research, theory and practice, Thomas Telford, London, 2000.

then to take ϕ_b as an estimate of the friction angle on bedding surfaces except when the rock has softened and altered or the discontinuity has been polished by displacement. The discontinuity then may be at its ultimate friction angle (Krahn and Morgenstern, 1979). When an over-dip slope is less steep than this, the rupture surface remains covered with debris and a rapid movement does not take place (Simmons and Cruden, 1980).

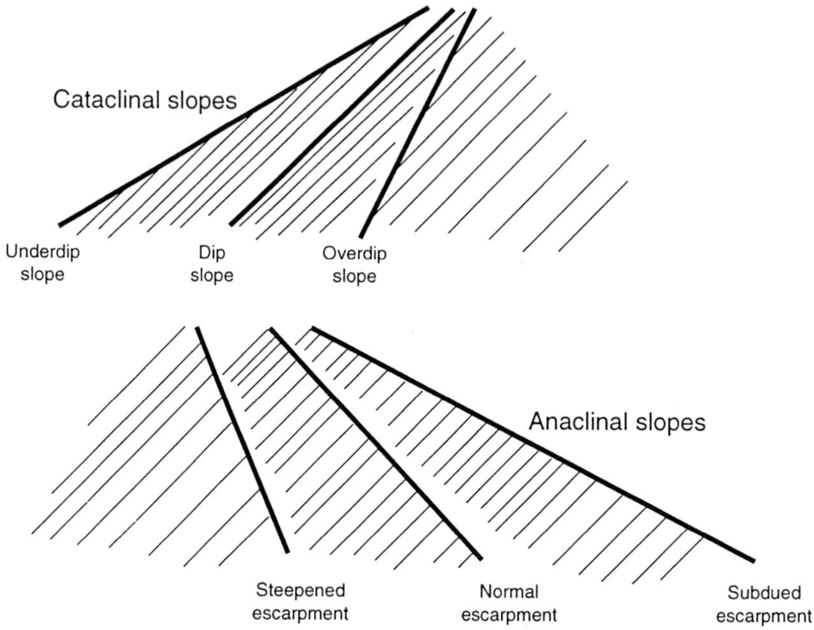

Figure 1. Types of slopes in rock masses containing a single penetrative discontinuity such as bedding. Thin lines represent bedding, thick lines represent slopes. The 6 types of slopes diagrammed are plotted as triangles on Figure 2

A major rock slide can be expected on a cataclinal overdip slope in a cohesive rock mass along discontinuities dipping at relatively low angles which slightly exceed their basic angle of friction. An example is a prehistoric rockslide at Jonas Ridge (Cruden, 1976; Bruce and Cruden, 1980). Bedding at ϕ_b, thus forms a threshold for the occurrence of catastrophic instabilities in the Rockies. Steep overdip slopes give way, with the loss of cohesion over time, to dip slopes which, as characteristic landforms (Brunsden and Thornes, 1979), contribute to the fabric relief.

Lower bounds of apparent cohesion were calculated for 14 overdip slopes in Kananaskis Country where the dip angles of the bedding surfaces exceed the basic friction angles. For 12 of the overdip slopes, the lower bounds of cohesion are between 28kPa and 47kPa, equivalent to less than 0.1% of rock bridges along the bedding surfaces. Generally, then, the present topography of the Rockies seems to require little cohesion to support it (Cruden and Hu, 1992).

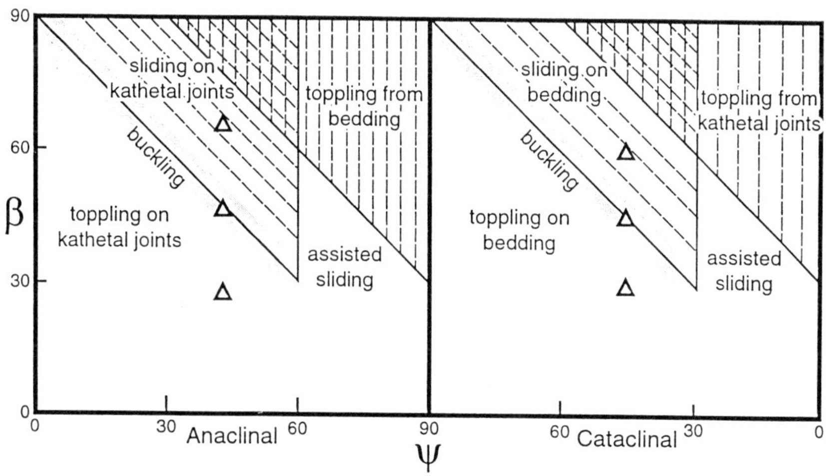

Figure 2. Processes on anaclinal and cataclinal slopes. The bedding dip is ψ and the slope angle is β. The 6 types of slopes in Figure 1 are plotted as triangles on Figure 2

DIP SLOPES

Detailed descriptions of historic catastrophic instability in the Rockies all documented movements on cataclinal slopes (Cruden, 1985). The study by Cruden and Krahn (1973) has contradicted McConnell and Brock's claim (1904, p. 12) that "The Frank Slide was a Bergsturz", a breaking away of the mountain mass across the bedding planes. The Frank Slide demonstrated that catastrophic movements were possible on dip slopes if other low angle discontinuities daylighted on the slope. The Frank Slide now appears to have taken place on a rupture surface following bedding and a thrust fault at its toe, joints control the scarp of the slide (Cruden and Krahn, 1973).

When discontinuities suitably oriented for sliding do not exist, movements can occur in steeply dipping rock masses by buckling. Classical buckling theory suggested thinly-bedded, cohesionless, low-modulus rock masses would deform under low loads (Hu and Cruden, 1993). Once rupture surfaces have been formed, the displaced masses behave in the same way as slides on overdip slopes, sliding on bedding and the rupture surfaces of the buckles. Buckling occurs where the bedding layers are steep. Buckling can be associated with toppling and may involve many layers. The displaced rock masses are deposited at the toe of the rock wall.

UNDERDIP SLOPES

Cruden (1989) has demonstrated that toppling from bedding is possible on all underdip cataclinal slopes steeper than the angle of friction, ϕ. We have found topples developed on underdip slopes in the Highwood Pass, Alberta (Cruden and Hu, 1994). Beds in the sandstones and the shales of the Triassic Spray River Formation dipping at 240/75-80 (dip direction/dip) were rotated to 060/20-50. Depending on the geological structure, topography and strength of the rock masses, the topples in the Highwood Pass are either block flexure topples or block topples (McAffee and Cruden, 1996).

Block flexure topples occur where the spacing of strike joints is less than twice the bedding thickness. Where the beds are thicker than 50 cm, topples can be as large as 30m. The toppled beds form rounded folds.

Block toppling occurs where the spacing of strike joints is more than twice the bedding thickness. Bedding dips can change over 10 degrees between adjacent blocks separated by strike joints along bedding layers. Small-scale toppling is found where the bedding thickness is less than 5cm and the present slope is less than 35°. Rupture surfaces form oblique to both bedding surfaces and strike joints. Rupture surfaces extend less than 10m and dip at around 35° downslope.

ANACLINAL SLOPES

Goodman and Bray's (1976) condition for common toppling can be conveniently shown on the process diagram given estimates of ϕ and assuming zero cohesion (Cruden, 1988). The development of toppling requires discontinuities perpendicular to the penetrative discontinuities.

Common toppling on anaclinal slopes is limited to substantially steepened escarpments. These might develop by glacial erosion of the toes of the rock walls of strike valleys. The landslide on Elk Ridge, west of the Highwood Pass, provides a typical example. The lower part of the rupture surface is along sheeting joints; the upper part is oblique to the bedding surfaces and the strike joints. Sliding on the sheeting joints was probably followed immediately by toppling from the bedding (Hu and Cruden, 1992). Identical rock sequences in the debris and in the rock wall indicate that the movement was a single event. Sheeting joints develop parallel to slope surfaces and are spaced from 1 to 2 metres apart but penetrate less than 5 metres into the slope. Sheeting joints have been found on anaclinal slopes only in thick, massive limestone sequences.

Sliding along joints, which may be followed by falls has formed talus. Most of these accumulations are not from single events, but rather fragmental falls occurring over a period of time.

CONCLUSION

The process diagram (Figure 2) indicates that normal escarpments and dip slopes are characteristic land forms in the Rockies. They evolve from glacially steepened escarpments, and overdip slopes. The evolution may be sudden and movements can be catastrophic.

The process diagram also shows some of the uncertainties remaining. Lower slope boundaries of flexural toppling and buckling of bedding are arbitrary. Presumably, these boundaries are also controlled by block geometries and material properties of rock masses besides discontinuity and slope orientations. Boundaries for assisted sliding and toppling will depend on the processes assisting slope movement. No falls are shown in the diagram, as generally, the rock mass moves first as a slide or a topple. A local steepening of the slope then projects the displaced mass into the air and it falls.

The process diagram assumes friction angles along discontinuities are 30 degrees. When the friction angle is different, the boundary of different modes of movements should be adjusted accordingly. Processes have been analyzed assuming cohesion and pore pressures are absent from the rock masses. Such assumptions may not be valid above the tree line, in the presence of permafrost or of glaciers feeding meltwater into rockslopes. Two fabric elements have been considered, a penetrative discontinuity and kathetal joints. Where other fabric elements such as faults or sheeting joints occur, kinematic possibilities become more complex and the simple generalizations proposed break down. The simple rock mechanics proposed here explains the broad outlines of the structurally controlled relief that characterizes the Rockies. Individual sites still require careful investigation and may offer challenging complexities.

In summary, modes of rock slope movement in the Rockies are usually determined by the relationship between the orientation of the slope and the orientation of the bedding in the sedimentary rocks forming the slope. Major rock slides occur on overdip slopes and move down bedding surfaces whose dips slightly exceed their basic angles of friction. Overdip slopes become dip slopes after the slides. Buckling occurs on dip slopes. Block flexure topples and block topples may develop on underdip slopes. In the block topples, rupture surfaces can form which allow sliding. On steepened escarpments, rock beds may topple from bedding and slide along kathetal or sheeting joints. These modes of movement can be placed on a process diagram, allowing knowledge of the dip of the beds, their basic angle of friction and the slope angle to predict slope performance.

ACKNOWLEDGMENTS

The work reported here has been supported by grants and contracts from Alberta Environment, Alberta Research Council and Alberta Transportation, the Geological Survey of Canada and the Natural Sciences and Engineering Research Council.

REFERENCES

Bruce, I, Cruden, D.M., 1980. Simple rockslides at Jonas Ridge, Alberta, Canada. Proceedings, 3rd International Symposium on Landslides, New Delhi, Vol. 1, pp. 185-190.

Brunsden, D., Thornes, J.B., 1979. Landscape sensitivity and change. Transactions, Institute of British Geographers. 4: 463-484.

Cruden, D.M., 1976. Major rock slides in the Rockies. Canadian Geotechnical Journal, 13:8-20.

Cruden, D.M., 1985. Rockslope movements in the Canadian Cordillera. Canadian Geotechnical Journal. 22: 528-540.

Cruden, D.M., 1989, The limits to common toppling. Canadian Geotechnical Journal. 26: 737-742.

Cruden D.M., Eaton, T.M., Hu, X-Q., 1988. Rockslide hazard in Kananaskis Country, Alberta, Canada. Proceedings, 5th International Symposium on Landslides, Volume 2, 1147-1152.

Cruden, D.M., Hu, X-Q, 1992. Exhaustion and steady state models for predicting rockslide hazards in the Canadian Rocky Mountains. Geomorphology, 8:279-285.

Cruden, D.M., Hu, X-Q., 1994. Topples on underdip slopes in the Highwood Pass, Alberta, Canada. Quarterly Journal of Engineering Geology, 27:57-68.

Cruden, D.M., Hu, X-Q.,1996. Hazardous modes of rock slope movement in the Canadian Rockies. Environmental and Engineering Geosciences, 2: 507-516.

Cruden, D.M., Krahn, J., 1973. A reexamination of the geology of the Frank Slide, Canadian Geotechnical Journal, 10:581-591.

Gabrielse, H., Monger, J.W.H., Wheeler, J.O., Yorath, C.J., 1992. Tectonic framework. In Geology of the Cordilleran Orogen in Canada, edited by H. Gabrielse and C.J., Yorath, Chapter 2, pp. 15-60, Geological Survey of Canada, Ottawa.

Goodman, R.E., Bray J.W., 1976. Toppling of rock slopes, Proceedings of the Specialty Conference on Rock Engineering for Foundations and Slopes, American Society of Civil

Engineers, Boulder, Colorado 2:201-234.

Hu, X-Q. Cruden D.M., 1992. Rock mass movements across bedding in Kananaskis Country, Alberta, Canadian Geotechnical Journal. 29:675-685.

Hu, X-Q., Cruden, D.M., 1993. Buckling deformation in the Highwood Pass, Alberta, Canadian Geotechnical Journal, 30:276-286.

Krahn, J., Morgenstern, N.R., 1979. The ultimate frictional resistance of rock discontinuities, International Journal of Rock Mechanics and Mining Science, 16, 127-133.

McAffee, R., Cruden, D.M., 1996. Landslides at Rock Glacier Site, Highwood Pass, Alberta, Canadian Geotechnical Journal, 33: 685-695.

McConnell, R.G., Brock, R.W., 1904. Report on the great landslide at Frank, Alberta, Department of the Interior, Annual Report for 1903, Ottawa, Part 8, 11 p

Example: a Cluster of Slope slides in Tibet, China

CUI JIANHENG, First Highway Survey & Design Institute, Communications Ministry, Xian, P. R. China

ABSTRACT
In the rainy season of 1991, the mountain slope, along a river, suddenly slid down into the river, formed a barrage and blocked the stream. Scarcely had the barrage been burst when the torrent scored the riverside and caused a series of slope slides. More than 1kilometer highway vanished with the slope. Since the slope slides had formed, the soil on the slide slope was washed away as debris flew, and the makeshift road was destroyed times and again. The section of highway lies in a narrow valley with precipitous and disastrous slope, there is no way to make a detour. It is necessary to control the cluster of slides so as to reinstate the highway.

NATURAL SURROUNDED
During the recent geological period, the crust has a speed rising, and formed high mountain peaks and deep valleys in the plateau. As the river rapidly cut down, remnant of loose sediment was hanged on the steep slope bedrock.

The disaster site lies in the nearby of the Yearlong Zangbu narrow valley. Between mountain peaks, which are about 6,000 meters high above sea level, there is the branch of the Yarlung Zangbu River. The valley floor is 2000 meters above sea level.

The dam and hot air from the Indian Ocean flows north along the narrow valley and meets cold current, causes heavy rainfall (about 1100mm/annul) in the area. There are rainy season and dry season in a year in this area. The rainfall in rainy season is about 70% of annul rainfall.

The river has a drainage area of 28,600 square kilometer, and a flow capacity of 1277cubin meter per second. The rapid stream has a velocity of 7.5 meters per second, and a powerful scoring.

The site lies at the seism belt between Indian Slab and Tibetan Slab, dozens of earthquakes have occurred during past years within the near region. The strongest shock was of 8.4 magnitude. Frequent earthquakes caused the stability of slope weaker and weaker.

THE DISASTER SITE
The site lies at a branch of the Yarlung Zangbu River. In this section, the river passes through a narrow valley, and the side slope is precipitous. A highway crosses the middle of the steep slope. In the rainy season of 1991, part of the slope suddenly slid down into the river, formed a barrage and stopped the flow up. Scarcely had the barrage burst, the torrent washed the riverside and caused a series of slope slide. The caused slides were disintegrated at once and left no remains. The section of the highway vanished with the slope.

Landslides in research, theory and practice, Thomas Telford, London, 2000.

The slips lie at the front edge of a platform, which is formed of glacial debris and about 350 meters higher than the riverbed. And the front slope has gradient of about 40 degree. At the foot of the slope, along the river, there exist base rock outcrops. The platform seems hanging on to the base rock slope.

Behind the slide slope, there are some marshes on the platform, and some springs on the exposed main scarp of the landslide. These springs are turned from the marshes. Before the slope slid down, these springs existed as plenty groundwater in the slope, and weaken the strength of the soil, reduced the stability of the slope.

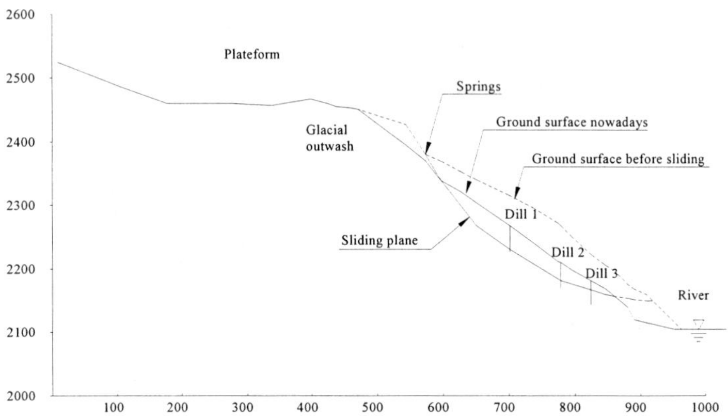

Figure 1 The vertical section of the largest slope slid

The present situation
Since the slope slid in 1991, no measure has been taken for controlling. A makeshift road was built for traffic. At present, more than half component of sliding slope have been washed away as debris, gullies have formed on the surface and change day after day. The high and precipitous main scarp behind the slips slides frequently. The makeshift road was destroyed every year, and the traffic was hold up during every rainy season.

The information from dynamic detecting shows that the first slope, which caused others, is still in creeping situation (see Figure 2). At the back part of the slid slope, the displacement is at the depth of 29 meters, and consistent with the surface of rupture based on drilling.

It is shown as a gravitational slide, and still unstable.

CHARACTERISTICS
The slid slopes are formed of glacial debris, which hang on the surface of base rock slope as platform. The thickness of glacial deposit is more than 120 meters in vertical. Expect the first one, which caused others, the slid slopes disintegrate at once, and no remains exist.

The first slip, having a largest scope, remained a volume of about 5,100,000 cubic meters. Its limits are 420 meters in width from left flank to right flank and 450 meters long from the crown to the foot. And the top is 220 meters higher than the foot. The surface of slope

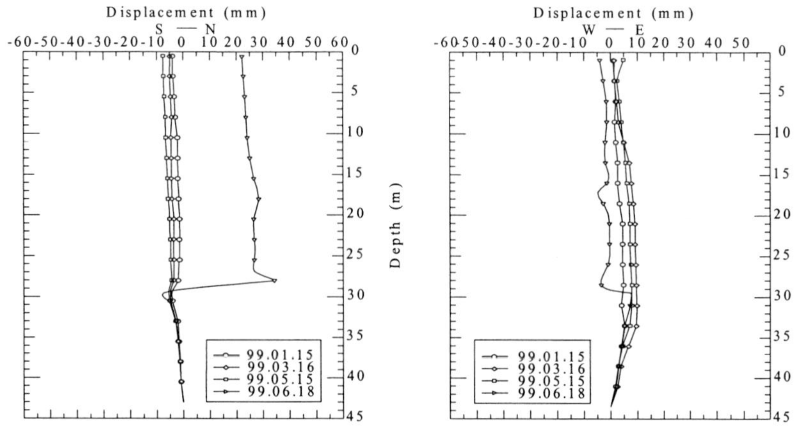

(a) Curves of displacement VS depth for Drill.3

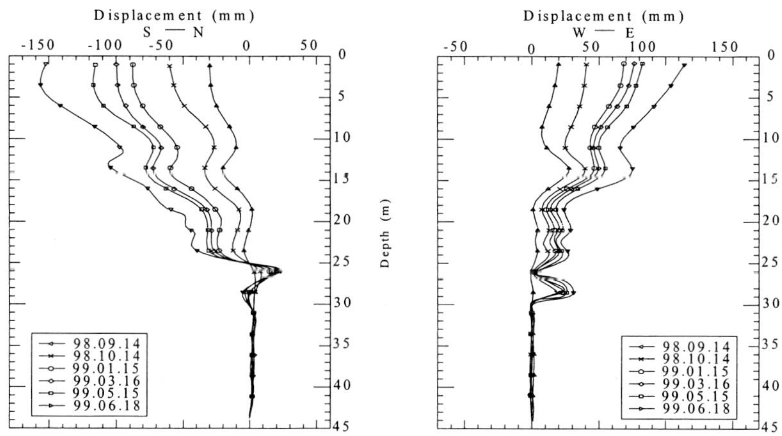

(b) Curves of displacement VS depth for Drill.3

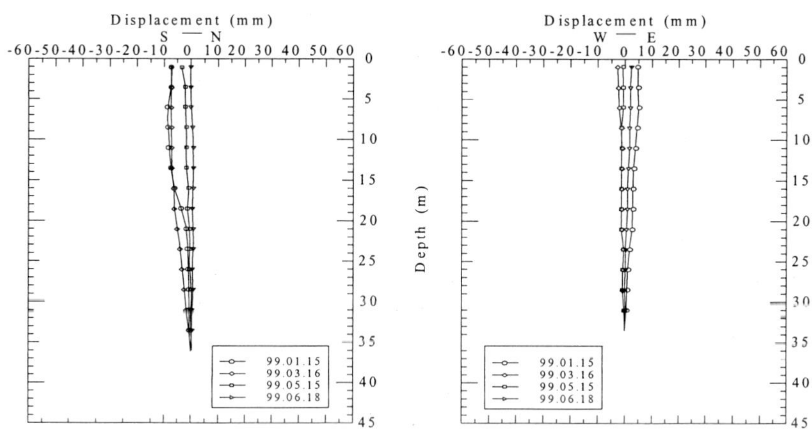

(c) Curves of displacement VS depth for Drill.3

changed from 43 degrees before sliding to 37 degrees nowadays. Based on drilling, it is known that the surface of rupture was formed at depth of 39 meters in glacial deposit.

Actually, before long after the highway has been constructed, the slope has been found displacing. And in 1960's, it displaced distinctly.

CONCLUSIONS

1. As a platform and hanged on base rock slope, the glacial deposit is unstable;
2. The plenty groundwater, turned from rainfall, is the inducement for slope sliding;
3. The largest slide, which caused others, is unstable and still displaces.

Soliflucted Deposits in the Vicinity of Belfast

S. J. CUMMINGS, I. G. DORAN and V. SIVAKUMAR, School of Civil Engineering, Queens University Belfast, UK

ABSTRACT
The paper considers the geological history of the Belfast area, and describes in detail the glacial processes, which formed the soliflucted material overlying a significant area of north and west Belfast. An upper and a lower soliflucted layer have been identified and the paper describes the two layers in detail. The difficulties in identifying each layer and its distinct properties are also discussed. The paper identifies the zone of instability in the area based on a large number of site investigations conducted in the region.

INTRODUCTION
The most striking scenic feature of the Belfast area is the line of steep hills which overlooks the city from the north and west. This line of hills is an escarpment forming the south eastern edge of the Antrim Plateau. Below the bottom of the scarp the ground surface slopes gently towards the city.

For many years slips have taken place in the north west region of Belfast. The effects of the slips are quite varied ranging from fractures in pipelines and maintenance problems in roads to more serious structural damage. The slips occur in a mantle of soliflucted material which extends below the base of the escarpment. Although the problem is long-standing the basic causes are not well understood and difficulty is experienced in identifying the conditions under which instability is likely to occur.

The soliflucted materials are more widespread than is generally recognised. Figure 1 shows a map of north west Belfast and the approximate extent of the soliflucted material is indicated. Clearly this significant covering of soliflucted material has a major influence on construction in the area.

GEOLOGY
The geological sequence of the solid rock formations is as follows:

Tertiary	Lower Basalt
Cretaceous	Upper Chalk
	Hibernian Greensand
Jurassic	Lower Lias
Triassic	Rhaetic
	Mercia Mudstone

A cross section through the escarpment is sketched in Figure 2.

Landslides in research, theory and practice, Thomas Telford, London, 2000.

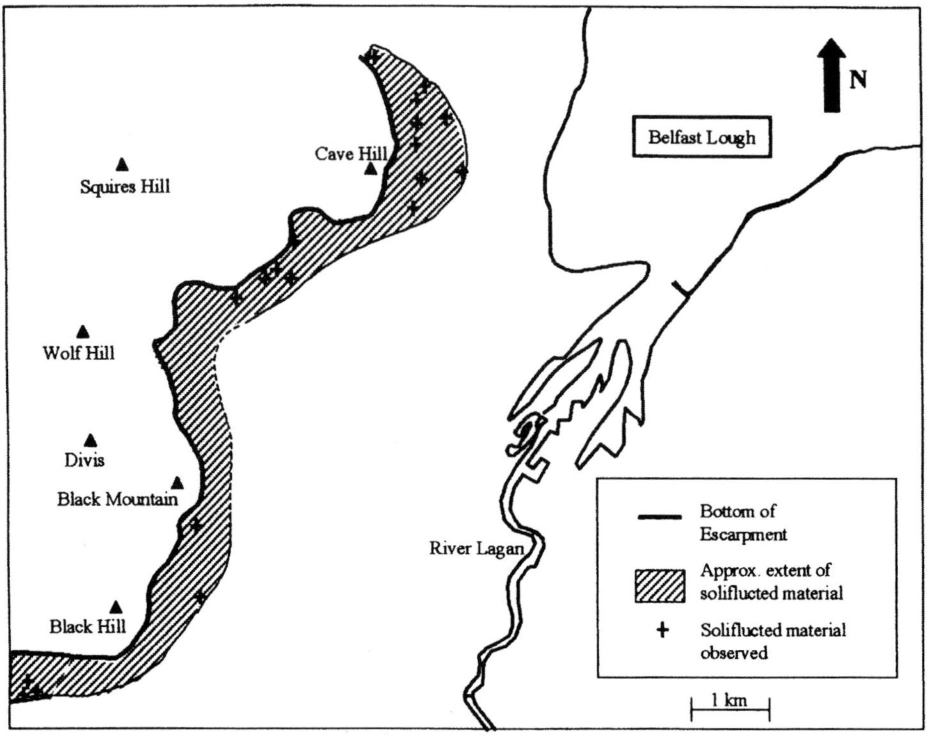

Figure 1. Map showing extent of soliflucted material in north-west Belfast

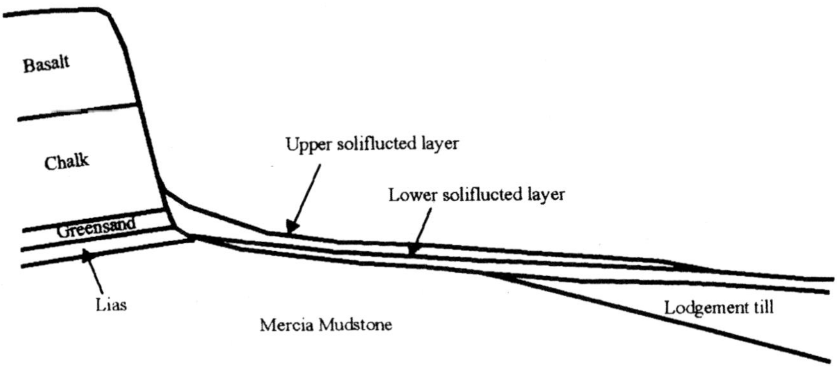

Figure 2. Sketch of cross-section through north-west Belfast

It should be noted that the Upper Chalk (White Limestone) in Northern Ireland differs from the contemporaneous English Upper Chalk in that it is exceptionally hard Wilson (1972). The other solid formations below the Hibernian Greensand, not all of which are necessarily present at any given locality, underlie a gentler slope which continues to flatten as the distance from the escarpment increases. In the lower parts of the slope a layer of glacial deposits, usually lodgement till, overlies the solid formations.

The outcrop of relatively pervious Hibernian Greensand is the source of many springs and the run off from the bottom of the escarpment is significant. Below the scarp, the solid formations are covered by extensive soliflucted deposits, extending up to 1000 m beyond the escarpment. In Figure 1 the crosses indicate where the soliflucted deposits have been encountered. They are often although not exclusively the sites of slips.

FORMATION OF THE SOLIFLUCTED MATERIALS

In a study of the stability of natural slopes in south east England Weeks (1969) identified two distinct soliflucted layers which he described as follows: "The upper layer of structureless sand, silt and clay with coarse fragments of resistant rock was clearly not related to the clay comprising the slope. The lower layer consisted entirely of material from the weathered zone of the underlying stiff fissured clay". In this paper Weeks's terminology has been adopted.

Ice contact deposits, that is subsoils laid down in direct contact with ice, can be considered under three headings, namely, sub-glacial, pro-glacial and supra-glacial deposits, depending on the source of the materials from which they have been derived.

Sub-glacial deposits consist of glacial debris transported near the base of a glacier or ice sheet, which is deposited beneath the ice as lodgement till. Lodgement till is a stiff heavily over-consolidated clay usually containing stones. It has a high undrained shear strength.

Pro-glacial deposits, often referred to as flow tills, melt-out tills and ablation tills, are composed of materials transported within the ice which are deposited when the ice melts during the retreat phase of the glacier. These materials were laid down in the presence of large quantities of water and were therefore originally in a near liquid state. Because of the manner of their formation they are either normally or lightly over-consolidated and they therefore have lower undrained shear strengths than lodgement tills. Because of the high stone content of tills generally and the consequent sampling and testing difficulties, and more particularly because the actual constituents are similar, it is often a matter of considerable difficulty to distinguish between ablation and lodgement tills with certainty.

Supra-glacial deposits are formed from debris composed of angular fragments, often of very great size, shattered from the rock forming the valley sides and accumulating on top of and at the sides of the glacier. They are deposited during glacier retreat as lateral moraines.

The last ice sheet to affect the north of Ireland was Scottish Ice (approximately 13000BP). The Scottish re-advance extended as far as the mouths of the River Foyle at Londonderry in the west and the River Lagan at Belfast in the south. This ice as it finally retreated in a north easterly direction along the south eastern edge of the Antrim Plateau deposited pro-glacial and supra-glacial materials at the base of the escarpment. It seems probable that the upper soliflucted layer is derived largely from these materials. It is characterised by a high content of sharp edged angular rock fragments of chalk, basalt and occasionally greensand.

Since the end of the ice age the continuing improvement in climate has resulted in a gradual drying out of the upper soliflucted layer with the result that it is often found in a reasonably firm condition. However it softens readily in the presence of water and under conditions of poor drainage instability can develop and slip failures are fairly common. Many fossil slip surfaces have been identified throughout the material.

As the Pleistocene Period came to an end the land surface beyond the margins of the ice sheets was subject to severe peri-glacial climatic conditions. Where material such as Mercia Mudstone was exposed at or near the surface it was evidently subject to severe weathering to considerable depths. Mercia Mudstone in its intact condition is a very stiff fissured heavily overconsolidated indurated clay with a high undrained shear strength. The effect of weathering commences as an opening up of the fissures followed by softening along the fissure planes. This softening process results in a material which consists of angular pieces of intact mudstone (lithorelicts) embedded in a softer clay matrix. Chandler (1969) has proposed a weathering scheme for Mercia Mudstone which is based mainly on the size of the lithorelicts. The weathering process will ultimately result in complete disappearance of the lithorelicts and the degradation of the mudstone to a plastic clay.

Where the mudstone is exposed on a sloping surface subject to alternate freezing and thawing a thin surface zone known as the "active layer" is formed. The underlying permanently frozen subsoil forms an impermeable barrier and as a result soft saturated material, the lower soliflucted layer, develops at the surface and spreads downwards as a soliflucted mantle covering the slope {Skempton & Weeks (1976), Morgenstern (1981) and Ballantyne & Harris (1984)}.

As is the case with the upper soliflucted layer, the lower soliflucted layer has also dried out and under present day conditions is generally quite firm. It contains fossil slip surfaces which are often particularly well developed at its base.

GEOTECHNICAL PROPERTIES OF THE SOLIFLUCTED LAYERS

Upper Soliflucted Layer

The upper soliflucted layer is extremely variable and ranges from large boulders of basalt, chalk or Hibernian Greensand to a dark grey silt consisting of very fine particles of chalk and basalt. Large boulders are however uncommon and by far the greater part of the material is dark brown sandy, silty clay with stones. The proportions of the various constituents vary widely. Occasional continuous thin layers of sand have been encountered in shallow trench excavations. The shear strength of the material depends largely on moisture content and the undrained shear strength characteristics of the material vary from largely frictional to entirely cohesive.

The results of all available unconsolidated undrained triaxial tests of the upper soliflucted layer have been plotted on an s, t diagram. All the test results lie within the shaded area shown in Figure 3. It is important to note that because of the variable nature of the material the strength measured in any particular test is to a large extent fortuitous since sampling is effectively random. In addition the strength actually measured is a function of the drainage conditions existing at the time of boring and these may well deteriorate. In other words, when testing is carried out on a selection of samples, particularly those taken from borings it is not necessarily the case that any of the test results obtained will give the actual operative strength on a failure plane.

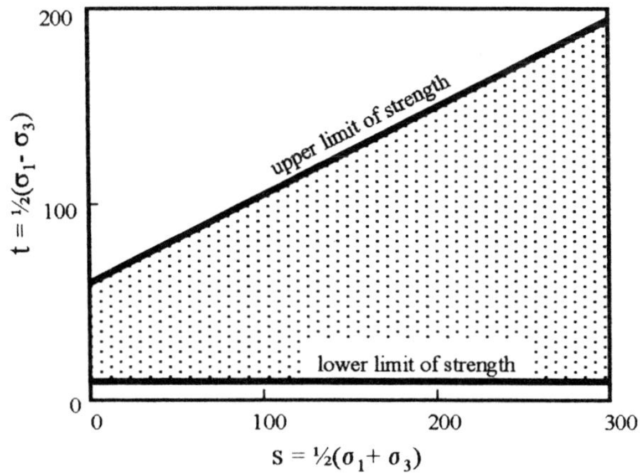

Figure 3. s,t plot for the upper soliflucted material

Where the existence of a failure plane is recognised, usually in a trial pit, the strength of samples from the failure plane can be determined. The data referred to above includes occasional tests on failure planes and the results, as would be expected, are in all cases near the lower limit of strength. The absolute minimum undrained shear strength of the upper soliflucted layer measured in all the investigations was 10kPa.

Lower Soliflucted layer

As already discussed the lower soliflucted layer in the Belfast area is the top most zone of a severely weathered heavily overconsolidated clay, usually Mercia Mudstone but occasionally a weathered lodgement till largely derived from Mercia Mudstone. It corresponds to Chandler's weathering zone IV, Chandler (1969). At the time of deposition it was very soft but under present conditions the material has dried out and is now a firm plastic clay with undrained shear strengths in the range of 50-100kPa. There is no likelihood of the material returning to its original near liquid condition but fossil slip surfaces occur particularly at its junction with the underlying, less weathered subsoil. In several major investigations continuous slip surfaces extending over considerable areas have been identified at depths up to 2 metres below the surface. Shear strengths estimated from back analysis of actual slips and measured in laboratory tests lie in the range 10-20kPa.

SLOPE STABILITY IN THE AREA

In the upper part of the slope adjoining the base of the escarpment the upper soliflucted layer occurs immediately below the topsoil. The undrained shear strength of the material is very low, the minimum recorded value being 10kPa. The slope in its present condition is only marginally stable and minor slips are common. These minor slips give rise to high maintenance costs for roads and occasional breaches of water mains. Lower down the slope stability is generally increased but severe disruption and structural damage occur occasionally due to changes in drainage and more particularly because of injudicious excavations.

In many parts of the area where the Mercia Mudstone is close to the surface fossil slip planes occur with undrained shear strengths corresponding to the residual strength of the material. Serious damage has been caused by the re-activation of these slip surfaces due to excavation and large scale dumping.

CONCLUSIONS

It is clear that considerable areas adjoining the escarpment to the north and west of Belfast are relatively unstable and the zone of instability has been indicated in Figure 2. The basic cause of the many slips in the area is softening of the soliflucted subsoils in the presence of water. Ingress of water is often caused by lack of drainage, ponding due to earth-moving and dumping, prolonged heavy rainfall and seepage from the junction between the Hibernian Greensand and the underlying impervious clays. It is important to note that large scale earth moving and dumping of fill anywhere on the soliflucted material, even on quite gentle slopes, may be an extremely hazardous operation.

The most effective practical step which can be taken to reduce the high maintenance costs of roads and pipelines in this soliflucted material is to introduce proper drainage measures which will prevent, as far as this is possible, access of water to the soliflucted layers. Large scale excavation and dumping should only be carried out under the strictest possible control.

ACKNOWLEDGEMENTS

Most of the data available to the authors and referred to in this paper was obtained from investigations carried out for various Authorities when the senior author was a partner with or a consultant to Dr. I.G. Doran & Partners, consulting engineers. The authors wish to thank the Queen's University of Belfast School of Civil Engineering for the opportunity to carry out this work.

REFERENCES

Ballantyne, C.K. & Harris, C. (1994) "The Periglaciation of Great Britain". Cambridge University Press.

Chandler, R.J. (1969)"The effect of weathering on the shear strength properties of Keuper Marl". Geotechnique 19.

Morgenstern, N.R. (1981) "Geotechnical engineering and frontier resource development". Geotechnique, 31, 305-365.

Skempton, A.W. & Weeks, A.G. (1976) "The Quaternary History of the Lower Greensand Escarpment and Weald Clay Vale near Sevenoaks, Kent" Philosophical Transactions of The Royal Society, A283.

Weeks, A.G. (1969) "The stability of natural slopes in north-east England as affected by peri-glacial activity". QJ Eng, G 2.

Wilson, H.E. (1972)"Regional Geology of Northern Ireland". HMSO.

Case Study of a Landslip in Glacial Sands and Gravels in N. Ireland

S. J. CUMMINGS, I. G. DORAN and V. SIVAKUMAR, School of Civil Engineering, Queen's University Belfast, UK

ABSTRACT

This paper will describe the geological history of the area in which the landslip occurred, study the contributing factors, consider the effect on the surrounding community, discuss the lessons learned and recommend measures which should prevent a similar landslip occurring again.

INTRODUCTION

Ballycastle is located on the north coast of Ireland as seen in Figure 1. It is situated in a rural area with very little industrial development and the land is currently used for hill farming.

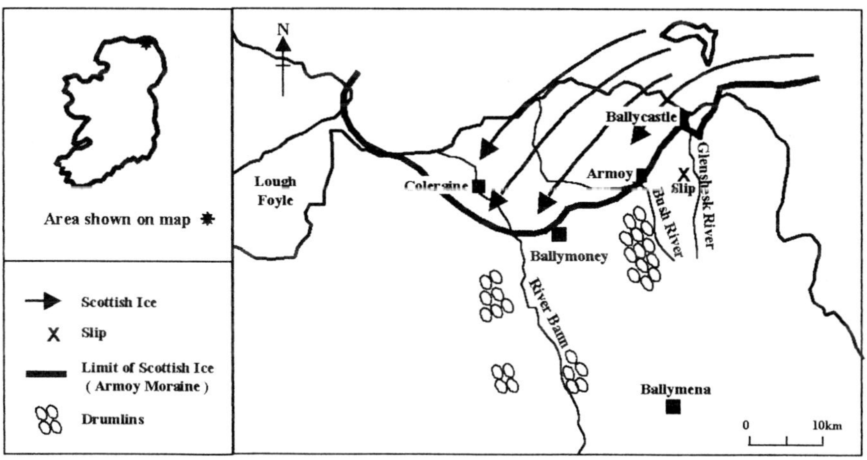

Figure 1. Map showing the location of the Slip

At a late stage during the last glaciation a re-advance of Scottish ice moved along the north and east coasts of Ireland but did not override the whole island. Much of the Ballycastle area is covered with fluvio-glacial sands and gravels deposited during this re-advance of Scottish ice.

Landslides in research, theory and practice, Thomas Telford, London, 2000.

Greenan Road is a cul de sac 2km long cut into the Glenshesk River Valley in the Ballycastle region. The road serves approximately 15 families engaged in hill farming.

On the 2nd November 1998 after a sustained period of heavy rainfall a 30m length of the Greenan Road collapsed and slid into the Glenshesk Valley. In Plate 1 the extent of the slip and the damage to the road can be seen. At the site of the landslip the river valley rises at an angle of approximately 45 degrees.

Plate 1. Extent of the slip on the Greenan Road

No one was injured by the landslip but the small farming community served by the Greenan Road was cut off. Safe access for the families and emergency services was the first priority of the road engineers in the area. The road was reopened to the public on the 7th November 1998 after sufficient remedial work had been carried out. Over 13000 tonnes of rock armour was placed from the river bed to road level.

GEOLOGY

Although the quaternary glacial period lasted for about two million years only the last events can be determined as the most recent ice sheet movements destroyed any evidence of previous ice sheet movements. In Northern Ireland four stages have been identified:

1. an ice-sheet flowing east or north from a centre probably located in Donegal
2. an ice-sheet from Scotland streaming south or west across N. Ireland
3. an Irish ice-sheet moving east and north
4. a re-advance of Scottish Ice which moved into north Antrim and north-east Londonderry

The evidence for a late Scottish Ice re-advance is provided by the Armoy moraine (Figure 1). The Scottish ice re-advanced on to coastal areas of Co. Antrim along the line Armoy, Ballymoney, and Coleraine, the ice edge crossing the river Bann to the south west of Ballymoney and then up to the mouth of Lough Foyle. Lough Foyle was blocked off by the ice and formed a glacial lake in which at least 37 m of stratified clay was deposited over a large area Cummings (1996). This material provided stratigraphic evidence for the re-advance. The morainic deposits and surface boulder clays contain abundant material from Scotland such as Ailsa Craig microgranite. Other evidence is the absence of drumlins north of Coleraine, Ballymoney and Armoy. As can be seen in Figure 1 the belt of drumlins streamlined by ice flowing from the south is cut off sharply by the limit of the Scottish ice re-advance.

Mtichell (1972) noted that the Antrim coast re-advance was a distinct late glacial event which occurred approximately 13000 B.P. and originated from Scotland.

Bedded fluvio-glacial sands and gravels occupy a considerable area in the valleys around Ballycastle and they usually overly boulder clay {Charlesworth (1939), Wilson (1966)}. It is in these sands and gravels in which the slip at Greenan Road occurred.

CONDITIONS IN THE AREA OF SLIP

Greenan Road is a 2 km cul de sac along the side of the Glenshesk River valley. This is an asymmetric valley due to continual lateral erosion of one side by the river. The landslip occurred on the steep side of the valley which rises at an angle of approximately 45° with the road cut into the side of the valley. The slope of the river valley is presented in Figure 2a.

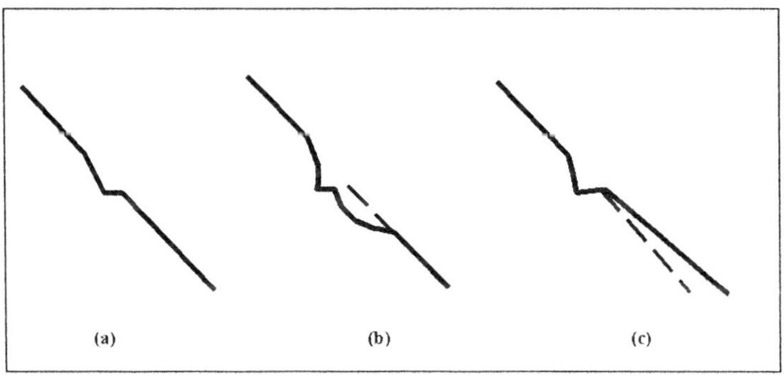

Figure 2. Sketch of the valley side

CONTRIBUTING FACTORS

A combination of factors contributed to the slip on the Greenan Road. On the inside of the road an old wall existed with what appeared to be a drainage channel behind it. The drainage channel had become blocked over the years and the water from the upper slope was discharging through the sand and gravel underneath the road way. The slopes of the valley side above and below the road were at an angle of approximately 45° which is above the angle of internal friction for normal sands and gravels. Therefore the safety factor of this slope was at best 1.

During the period before this slip occurred there had been very heavy rainfall for a prolonged period of time with considerable flooding in the lower lying areas.

Another major factor was the erosion of the base of the slope by the Glenshesk river. The meandering nature of the Glenshesk river coupled with the topography of the area have combined to ensure that the river was continually eroding away the side of the river valley on which the Greenan Road had been built. (This is identified by the asymmetric nature of the valley). With the high water level of the river due to the heavy rainfall the erosion of the base of the slope by the river was accelerated.

The effects of the erosion by the river were clearly seen further up the river valley were large slips have already occurred. As this is a rural area they have not caused any damage to roads or buildings and therefore remain unnoticed by most people.

EFFECT ON THE COMMUNITY

The cost to local communities of landslides around the world is unquantifiable. Information collected in the USA suggests that annual costs run into billions of dollars. In addition to financial costs, there may also be the loss of life which is a disaster for friends and family and the community at large (Bromhead, 1992). In the case of the Greenan Road landslip there were no fatalities although there was a significant impact on the small farming community.

A thirty metre section of the Greenan Road was washed in to the Glenshesk river following several days of extremely heavy rainfall. The Greenan Road is the only access residents in the area have to their homes. An alternative route for transport into the area was not possible but repair work on the site started immediately to ensure the road was in a safe condition as soon as possible. It was very unusual to have families stranded inland. Despite rationing the hill farmers in the area ran out of meal for their livestock within a few days. They requested help to get food through to their livestock. After a short time the Agriculture Minister asked the army to take immediate steps to transport supplies of animal feed to those farmers in the area affected by the flash flood. These supplies were delivered as quickly as possible to ensure the farmers involved could continue feeding their livestock until the necessary access road repairs were carried out.

The landslip occurred on 2nd November and on the 6th November two army helicopters carried out an airlift of meal for the livestock. On Saturday 7th November after the initial remedial works were complete the Greenan Road was reopened to the public.

Although the hill farmers were satisfied with the airlift of meal for their livestock some residents in the area expressed their concern regarding the unnecessary waste of public money incurred through employing an expensive airlift of animal feed.

Often it is the engineers on site who are confronted by the general public with their grievances. The latter highlights the difficult job the engineer has in not only resolving the engineering problem but also having an adequate public relations strategy with the media.

REMEDIAL ACTION

The road was closed immediately to reduce the risk of damage or injury to the public. The toe of the 30 m long slip was approximately 20 m below road level. The effect of the slip on the valley side can be seen in the section presented in Figure 2b. 2500 tonnes of rock armour was placed from a depth of approximately 25 m to road level to reduce the slope to approximately 40 °. A drainage pipe to channel the water from the upper slope was also installed. This process took three days after which the road was reopened to the public.

Further long term measures were required to ensure that the slope remained reasonably stable in the future. After a site inspection a further quantity (10500 tonnes) of rock armour was placed at the bottom of the slope at the river level up to the existing rock armour. This required the placement of large boulders at the base to prevent them being washed away at high water levels in the river. Figure 2c shows a section through the repaired slope. The outfall of the drainage pipe was located well down the valley side close to the river level to prevent erosion of the present slope. Dead trees and branches were removed as they would potentially weaken the stability of the slope. The rock armour has been covered in soil in preparation for planting trees and vegetation to improve stability. A safety barrier rather than a large bank will be placed at the side of the road on the lower slope. In Plate 2 the repaired slope can be seen.

Plate 2 The repaired slope

DISCUSSION & CONCLUSIONS

By improving the drainage from the upper slope and disposing of the water directly into the river this should reduce the erosion of the slope.

Stabilisation of the slope by placing rock armour from the river level to the road at a slope of 40° should significantly reduce erosion by the river of the toe of the slope.

Greenan Road is an example of a slip, which can occur at any time especially in rural areas. The problem was dealt with in a satisfactory manner with the safety of the public not put at risk.

A risk assessment was carried out considering the size and scale of the problem and produced a safe but suitable solution. The correct decisions were made quickly to ensure the safety of the community but without being influenced by others with a hidden agenda. The effect from this landslip on the small community was significant and highlights the non engineering problems which engineers may encounter.

The slip raises the question at what angle is a slope safe? Since the material was deposited approximately 13000 years ago it has had sufficient time to become stable. It is apparent that the erosion of the base of the slope by the river was the most influential factor in causing the slip. In an asymmetric valley the side on which the slip occurs will influence the remedial action required. The process of erosion by the river is particularly strong when water levels are high. The risk of a similar slip occurring and damaging the road in the region is low. As this is a rural area the risk of injury to the public if a slip did occur is also minimal. The basic measures of reducing the angle of the slope, redirecting the run off water from the upper slope and strengthening the base by using rock armour in this case were sufficient.

A general survey of the slopes of the valley and effects of the erosion process of the river at regular intervals should alert the engineer to any further possible slips in the area.

ACKNOWLEDGEMENT

The authors would like to thank the Department of Environment Roads Service N.I. for their support and co-operation with writing this paper and the Queens University Belfast School of Civil Engineering for providing the opportunity to carry out this work.

REFERENCES

Bromhead, E.N. 1992 "The Stability of Slopes". Second Edition, Published by Blackie Academic & Professional.

Charlesworth, J.K. 1939 "Observations on the Glaciation of North-East Ireland." Proc. Roy. Irish Acad., 45, Sect. B, pp. 255-295.

Cummings, S.J. 1996 "An investigation of the properties and behaviour of a glacial lake deposit". PhD Thesis, Queens University Belfast.

Mitchell, G.F. 1972 "The Pleistocene history of th Irish Sea:second approximation". Scient. Proc. R. Dub. Soc., Series A, 4, pp181-199.

Wilson, H.E. and Robbie J.A. 1966 "Geology of the Country around Ballycastle". Memoirs of the Geological Survey. HMSO

Wilson, H.E. 1972 "Regional Geology of Northern Ireland." HMSO

Stephens, N., Creighton, J.R. & Hannon, M.A. 1975 "The Late-Pleistocene period in north-eastern Ireland: An assessment 1975". Ir. Geog., 8, pp1-23.

Geologic and Geomorphic Factors of the Landslides Triggered in the Cardoso T. basin (Tuscany, Italy) by the 19th June, 1996 Intense Rainstorm

G. D'AMATO AVANZI, R. GIANNECCHINI and A. PUCCINELLI, Dipartimento di Scienze della Terra, Università di Pisa, Italy

INTRODUCTION

On June 19, 1996 an exceptionally intense rainfall event fell in the southern Apuan Alps (Tuscany, Italy): ca. 474 mm rainfall was recorded within 12 hours (21% of the annual rainfall) and the maximum rainfall intensity recorded was 158 mm/hour. The rainstorm hit a territory ca. 150 km^2 wide in the area of the main watershed between the Versilia and the Serchio River basin. It caused hundreds of landslides, flooding of wide areas, 14 fatalities and very heavy damage (Caredio et al., 1998; D'Amato Avanzi, 1999).
Here, the results of the studies on the landslides occurred in the Cardoso Torrent basin are summarized. This basin (Fig. 1), 13 km^2 wide, was the most severely damaged area, where almost 400 landslides were triggered. Its morphology is typically mountainous, with deeply cut valleys and steep slopes, mostly underlain by mainly metamorphic, impervious rocks; among which the Pseudomacigno Fm. (metamorphic sandstone interbedded by phyllite) prevails. These slopes are largely covered by thin colluvial deposits, with a 30°-40° slope gradient and covered with woods. Permeable calcareous and dolomitic rocks form the highest peaks of the area, where the slope steepness is often greater then 60°-70°.

LANDSLIDE CHARACTERISTICS

The rainstorm triggered at least 382 mappable landslides (ca. 30 landslides/km^2): 357 were first time landslides, while the others partially mobilized dormant landslide bodies. According to some witness, many landslides were activated in a short time, when ca. 250 mm rainfall within 8 hours had been recorded (ca. 31 mm/hour). This rather high value may represent the landslide-triggering rainfall threshold in the study area.

Type of movement

According to Cruden and Varnes's (1996) classification, the types of movement of the landslides in the Cardoso T. basin were sub-divided into: a) fall, b) translational slide, c) flow, d) complex, translational slide-flow, e) complex, rotational slide-flow and f) complex, rotational-translational slide-flow. The percentage of each type of movement is represented in Fig. 2.

The typical landslide was a shallow landslide (0.5-2.5 m thick) with a high length to breadth ratio ($3<L/B<20$). This feature is probably due to loss of structure and fluidification of materials. In fact, because of the slope steepness, many debris slides attained high velocities (some m/sec) and turned in debris flows. The sliding and flowing masses often poured into the riverbeds, mixed with the runoff and overloaded streams with sediments and wood. Thus, the most common landslide movements were complex, very or extremely rapid, debris slides-debris flows (Cruden and Varnes, 1996); according to Campbell (1975), they may be named soil slips-debris flows. These phenomena represent typical effects of intense rainstorm: the rapid infiltration of rainfall, causing soil saturation and rise in pore-water pressure is probably the mechanism by which most shallow landslides were triggered (Wieczorek, 1987, 1996).

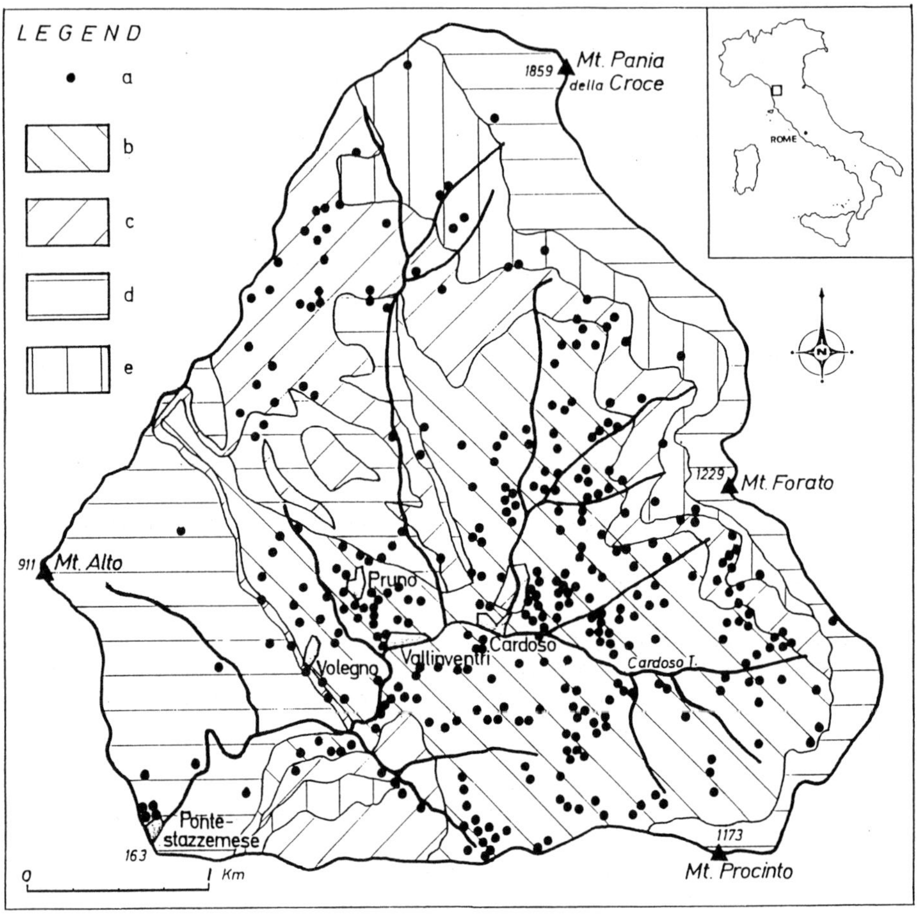

Fig. 1. Sketch map of the study area: a) landslide of June 19, 1996; b) metamorphic sandstone and siltstone (Pseudomacigno Fm.); c) phillyte and schist; d) limestone and dolomite; e) cavernous limestone.

Material involved

The landslides generally involved only the regolith cover of slopes (0.5-2.5 m thick). Among them, 96.1% involved colluvium (Fig. 3), consisting of a poorly sorted, matrix-supported, mixture of angular rock fragments and fine-grained materials; 3.3% involved talus deposits, mainly rock-supported, formed of calcareous and dolomitic angular rock fragments and forming screes and talus cones at the base of cliffs or rocky chutes. The failure surface usually

developed at the contact between the regolith cover and the bedrock, while sometimes the superficial portion of bedrock was also involved. Very few landslides occurred in man-made deposits (quarry dumps or filling materials). The scarcity of landslides in talus and man-made deposits was mainly due to their high permeability and also to their limited spread.
The most significant parameters of the landslides were analysed in order to identify the factors which most influenced their activation and effects. The studies in the landslide sites highlighted many geologically and geomorphically recurrent factors, explained below.

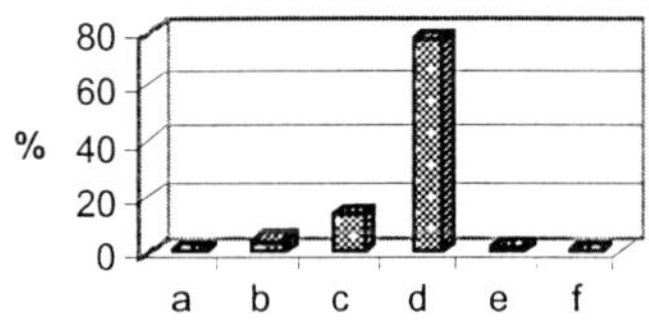

Fig. 2. Landslide distribution related to type of movement: a) fall; b) translational slide; c) flow; d) complex, translational slide-flow; e) complex, rotational slide-flow; f) complex, rotational-translational slide-flow.

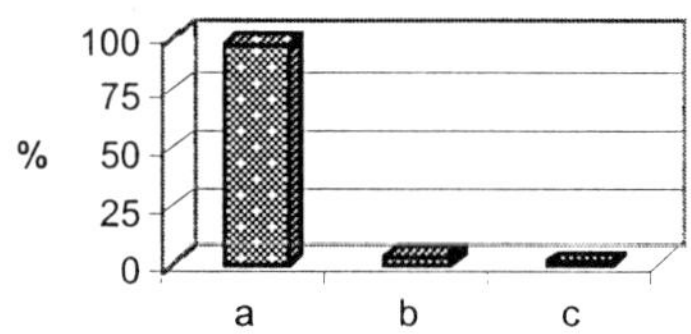

Fig. 3. Landslide distribution related to material involved: a) colluvium; b) talus; c) man-made deposits.

Bedrock characteristics
As shown above, the landslides of June 19, 1996 mainly involved the regolith cover of the slopes. Some characteristics of the bedrock (lithology, permeability, discontinuity) in landslide sites were investigated. The formations underlying the basin were grouped according to lithology: a) metamorphic sandstone and siltstone, b) phillyte and schist, c) limestone and dolomite, d) cavernous limestone. Percentage of basin area, landslide density (surface of landslides divided by total surface), percentage of total landslides and surface involved in sliding were calculated for each group. The main results are summarized by Fig. 4: in landslide sites, the bedrock was generally made up of impervious or scarcely pervious rocks; among them, the Pseudomacigno Formation (metamorphic sandstone with interbedded phillyte) was the most common one (8.4% of landslide density).

74.9% of landslides occurred on slopes underlain by metamorphic sandstone with interbedded phillyte (Pseudomacigno Fm.), while 17.8% occurred on slopes underlain by phillyte and schist; these types of rock respectively occupy 36.2% and 20.8% of the whole surface of the Cardoso T. basin. Then, we deduce 92.7% of landslides occurred in the covers of impervious or scarcely pervious rocks, occupying 57% of the area of the basin, while 7.3% only of landslides occurred in the covers of pervious rocks, occupying 43% of the basin. This is confirmed by the landslide density of the Pseudomacigno Fm. (8.4%), very higher than others.

The attitude of main discontinuities (bedding or schistosity) of the bedrock with respect to the slope immersion was also analized for 205 landslides, involving the cover of the Pseudomacigno Fm. in a sample area 3.25 km^2 wide (ca. 3/4 of its outcropping surface). This area was divided in three domains, according to the attitude of the discontinuity (downslope, upslope or oblique); area of domain, percentage of landslides and landslide density were calculated for each domain. Most of landslides (45.9%) occurred in the "downslope" domain

(Fig. 5), occupying only 33.5% of the sample area. This domain also favoured a large extent of landslides, which there reached 52.6% of the total landslide area.

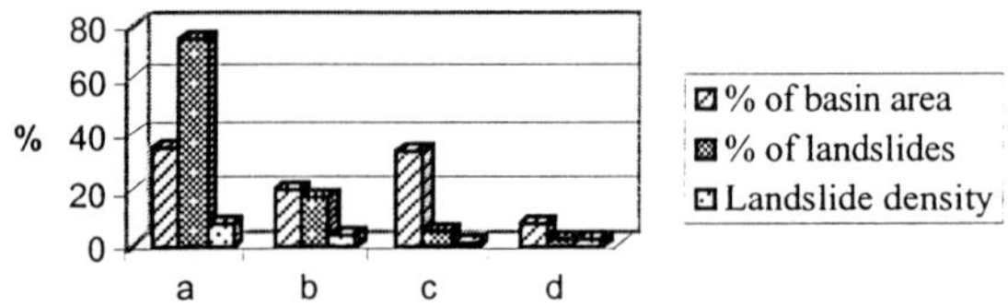

Fig. 4. Landslide distribution related to bedrock lithology: a) metamorphic sandstone and siltstone; b) phillyte and schist; c) limestone and dolomite; d) cavernous limestone.

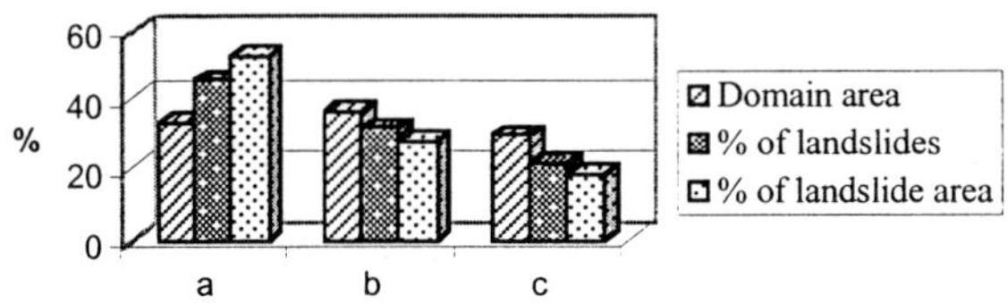

Fig. 5. Landslide distribution related to the attitude of main discontinuities: a) downslope; b) upslope; c) oblique.

Slope shape and gradient

Landslides were related to profile, surface (Panizza, 1992) and gradient of the slope in landslide sites.

Slope profile

According to Panizza (1992), the possible profiles for slopes are: rectilinear (the gradient is uniform from top to bottom), concave (the gradient decreases from top to bottom), convex (the gradient increases from top to bottom), concave-convex, convex-concave and complex. In the Cardoso landslide sites, the rectilinear profile was the most frequent one. Nevertheless, it should emphasized that this particular profile is very common in the studied basin and is probably not a significant parameter.

Slope surface

According to Panizza (1992), the possible shapes for slope surface are: planar (rectilinear counturs), truncated cone convex shape (counturs convex towards the slope toe), truncated cone concave shape (counturs concave towards the slope toe). In 55.8% of the examined landslide sites, the shape of the slope was truncated cone concave (Fig. 6), corrisponding to a small basin or hollow. In these situations the concave geometry of the regolith/bedrock interface favoured the convergence of groundwater flow and the build-up of pore pressure, leading to failure.

Slope gradient

Most of landslides occurred where the slope gradient was from 36° to 40° (Fig. 7); 89.5% of landslides occurred on rather steep slopes (31°-45°). On steeper slopes (less frequent), landslides were uncommon, probably because on these slopes the regolith cover is little developed or absent; moreover, these slopes are often underlien by pervious rocks.

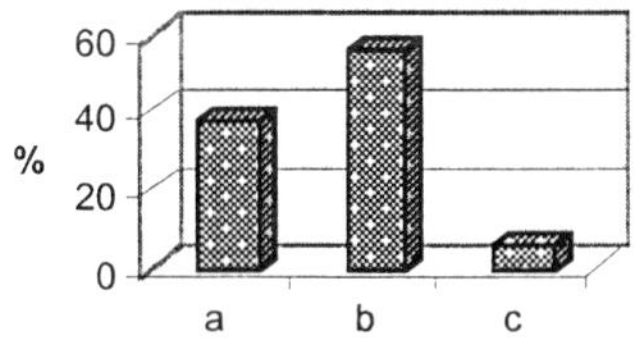

Fig. 6. Landslide distribution related to slope surface: a) planar; b) truncated cone concave shape; c) truncated cone convex shape.

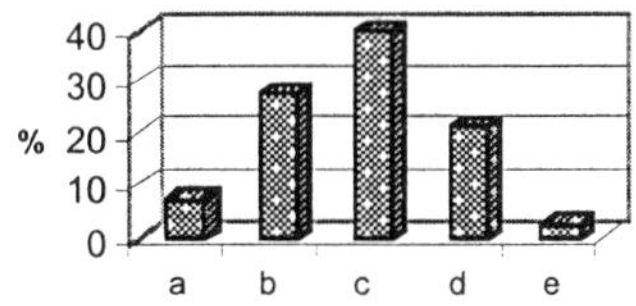

Fig. 7. Landslide distribution related to slope gradient: a) 25°-30°; b) 31°-35°; c) 36°-40°; d) 41°-45°; e) > 45°.

Land use and vegetation

By means of on-site investigation and using a vegetation map (Amorfini et al., 1997), the vegetation cover was examined and subdivided into four classes: chestnut wood, mixed wood (oak, hornbeam, beech, maple, ash), grassland (Brachypodium genuense) and cultivated or ex-cultivated, terraced areas. We can note that the chestnut wood was the area most involved in sliding (76.7% of landslides). On the other hand, chestnut woodland represents the most widespread coenosis throughout the basin and generally grows on impervious or scarcely pervious rocks (such as Pseudomacigno Fm., where most of the landslides occurred).

Surface involved and volume of mobilized material

The total surface involved in landslides of June 19, 1996 was estimated at ca. 0.55 Km^2, 4.2% of the whole basin surface. 81.4% of this surface was covered by chestnut trees: thus, more than 4,000 chestnut trees was uprooted by the landslides and fell into the riverbeds. It significantly contributed to the widespread destruction and blockage of bridge spans (with flooding), also downstream of the study area. We estimate the total volume of material mobilized by landslides at ca. 850,000 m^3: most of this volume (ca. 755,000 m^3) poured into the riverbeds, while the rest (ca. 95,000 m^3) remained on the slopes involved. The total volume of the material mobilized on June 19, 1996 by both slope and stream processes was estimated at ca. 1,400,000 m^3 in the whole Cardoso T. basin; thus, ca. 645,000 m^3 were mobilized by stream erosion in the riverbeds.

Conclusions

The results of the studies on almost 400 landslides triggered in the Cardoso T. basin by the June 19, 1996 rainstorm enable us to make some remarks. These studies identified some factors which most influenced location and characteristics of landslides.

The most common landslides were complex, very or extremely rapid debris slides-debris flows and involved the colluvium cover of the slopes. The characteristics of the bedrock (lithology, permeability and attitude of discontinuities) were of primary importance in determining landslide locations: most of landslides involved slope covers underlain by the Pseudomacigno Fm. (impervious, metamorphic sandstone with interbedded phillyte), which often had a significant discontinuity dipping downslope. This may explain the particularly high landslide density in some tributary basins, such as the Colombetta T. basin, where the landslide density reached 66.7 landslides/km^2. In fact, this particular basin is almost totally underlain by Pseudomacigno Fm. The distribution of landslides related to the slope morphology evidenced a rather typical slope configuration, characterized by rectilinear profile, truncated cone concave, hollow type shape, rather high gradient (31°-45°) and densely covered by chestnut wood.

On the other hand, some slopes showing these characteristics were not involved in sliding. This is probably due to the great heterogeneity of the rainstorm, so that a different rainfall intensity and/or amount poured on neighbouring slopes. Moreover, the present study considered only the most important parameters of landslides. The research is still in progress and involves particularly three areas of interest: determination of landslide-triggering rainfall thresholds, response of pore pressure to rainfall and geotechnical parameters of hillslope materials. These data would help us to evaluate better the conditions of slope stability and the critical landslide-triggering rainfall thresholds. However, the present results (through the knowledge of the recurrent factors in landslide activation) already enable us to perform risk scenarios and plan works of prevention, also in neighbouring territories, where risk conditions are similar.

REFERENCES

Amorfini A., Bartelletti A. & Zocco Pisana L. (1996) - Dissesto idrogeologico e soprassuoli boschivi: il caso di Cardoso e Fornovolasco, nelle Alpi Apuane, durante gli eventi del 19 giugno 1996. Proc. Meet. "Piano di bacino dell'Arno e dissesto idrogeologico", Pisa, 7/3/97, 10 pp.

Campbell R.H. (1975) - Soil slips, debris flows and rainstorms in the Santa Monica Mountains and Vicinity, Southern California. U.S. Geological Survey Professional Paper 851, 51 pp.

Caredio F., D'Amato Avanzi G., Puccinelli A., Trivellini M., Venutelli M. & Verani M. (1998) - La catastrofe idrogeologica del 19/6/1996 in Versilia e Garfagnana (Toscana, Italia): aspetti geomorfologici e valutazioni idrauliche. Proc. meet. "Prevention of hidrogeological hazard: the role of scientific research", Alba (CN), 5-7/11/1996, 75-88.

Cruden D.M. & Varnes D.J. (1996) - Landslide types and processes. In: "Landslide: Investigation and Mitigation", Spc. Rept. 247, Transp. Res. Board, Nat. Acad. of Sciences, Washington, 36-75.

D'Amato Avanzi G. (1999) - Landslides triggered by the intense rainstorm of June 19, 1996 in the southern Apuan Alps (Tuscany, Italy). Trans. Jap. Geomorph. Un., 20-3, 203-218.

Panizza M. (1992) - Geomorfologia. Pitagora Editrice, Bologna, 396 pp.

Wieczorek G.F. (1987) - Effect of rainfall intensity and duration on debris flows in central Santa Cruz Mountains, California. In: Costa J.E. & Wieczorek G.F. (eds) "Debris flows/avalanches: process, recognition and mitigation". Geol. Soc. Am., Rev. Eng. Geol., 7, 93-104.

Wieczorek G.F. (1996) - Landslide triggering mechanisms. In: "Landslide: Investigation and Mitigation", Spc. Rept. 247, Transp. Res. Board, Nat. Acad. of Sciences, Washington, 76-90.

Anthropogenic Triggering of Weathering: a Case Study

DEBASHIS DAS, Visva-Bharati University, Santiniketan Birbhbum, West Bengal. India

INTRODUCT ION

The vastly accelerated process of soil removal brought about by human interference with the normal disequilibrium between soil building and soil removal is designated as weathering due to anthropogenic triggering.

OBJECTIVES

1. To study the anthropogenic causes of weathering
2. To evaluate the effects of anthropogenic weathering on the agricultural economy.
3. To uncover the problems of anthropogenic weathering and to *suggest* policy measures.

ANALYTICAL PROCEDURE

The data are collected from fifteen randomly selected villages and from three hundred and sixty randomly selected agricultural farms in Birbhum district of West Bengal. The absolute and causal relationships are computed between the causes and effects of anthropogenic weathering.

ANTHROPOGENIC CAUSES OF WEATHERING

It is often found that man and its activities cause weathering. Due to meeting of the growing demand of large population, man often tries to cultivate the land intensively but unscientifically, which results in soil erosion. Man tries to adopt some preventive and protective measures to preserve soil which are not accepted by all, rather gives rise to social conflict among the inhabitants of the villages.

Birbhum district has a small proportion of forest covered area (3.55%) and due to deforestation the forest covered lands gradually decline. Deforestation takes place due to two reasons (i) stealing of logs (ii) meeting of fuels to the poor population. Due to this deforestation weathering takes place during heavy rainfall. Canal irrigation in the district takes place without the alignment of field channels. This flood irrigation causes weathering to a significant extent. The amount of eroded soil increases during the last phases of each monsoon season. Cultivation does not take place on the uncultivable and other than current fallow land. Therefore. these are liable to weathering. The permanent pasture and other grazing land represent some of the worst weathered areas. These are characterised by unchecked misuse. The grazing is by and large unrestricted and widely practised in the forest areas, the livestock feeding on undergrowth and trees, pass into the regenerated areas. This type of indiscriminated grazing has caused great damage to forestry and ultimately encourages erosive agents to work continuously on the unprotected soil to erode it away.

Some social constraints come in the way of the management of soil resources. The ignorance of farmers, illiteracy, lack of agricultural education, traditional mentality, sticking to old and

Landslides in research, theory and practice, Thomas Telford, London, 2000.

obsolete practices, poverty, lack of essential inputs and infrastructures etc. are the main causes that deteriorate productivity value of soil. Although some social and administrative institutions demonstrate the measures to control weathering and to retain productivity value of soil, yet the adverse influence of this 'social vicious circle' often go against the practice of judicious soil and moisture conservation measures. The study area suffers from weathering due to excessive overland flow of irrigation water from the barrages. During the last phase of the rainy season the barrages cannot store surplus rainfall, which is allowed to flow down the rivers, rivulets and canals during heavy rainfall. This causes erosion when heavy overland flow takes place down the slope of terrain.

These are some of the many causes of anthropogenic weathering. Man through some socio-economic process and activities are responsible for it.

EFFECTS OF ANTHROPOGENIC WEATHERING

Anthropogenic weathering has some aggravated consequences on the regional socio-economy. The volume of overland flow, soil characteristics and nature of slope together determine the magnitude of weathering in the region. The lands of the western part of the area are relatively less affected by erosion. Although the slope of this part is steeper (10 feet per km.) than eastern part (2 or 3 feet per km) but the limited flow of drained water and rainfall do not cause acute weathering there. On the other hand, although the slope of the land of central and eastern parts are comparatively gentler, but the surplus water of canal irrigation, rainfall, and drained water combine together and flow, causing weathering intensively in those parts. Flood erosion causes low fertility and productivity of soil. Therefore, these affected villages have smaller amount of net value of crops per acre of gross cropped area, but higher intensification of crops, because the farmers have to depend upon the production of crops grown at non kharif seasons for their survival and struggle for existence. Following table illustrates it.

Table 1 Relationship between weathered land and other variables (N=15 sample villages)

Variable X	Variable Y	r value	level of significance
% of weathered land	% flood affected land	0.75	0.01
ditto	Nett value of crops per acre of GCA	(-)0.29	0.50
Ditto	Cropping intensity	0.60	0.50

Table 2 Regression analysis: weathered land & nett value of crops per acre of G.C.A.

Variable X	Variable Y	b value	a value	R^2 value	level of significance
% of weathered land	Nett value of crops per acre of GCA	(-)1.65	998.16	0.08	0.05

Present study reveals that the major proportion of land holdings of the weaker sections of the sample villages are located near the agents of weathering which are weathered seriously. These farmers have to invest a large amount of money to control it. These poor farmers do not have an adequate amount of money and infrastructures to tackle this weathering problem. They cannot upkeep their damaged land in time because poverty compels them to work on the land of rich farmers of the villages either as share croppers or as agricultural labourers.

The big farmers often encroach upon or divert the alignment of drainage channels and make them more serpentine in appearance. These channels cannot clear away the water, rather surplus water overflows these channels and causes weathering. In this way fertility and productivity value considerably, the level land becomes uneven, then large sized plots are required to be reduced into many small pieces to avoid the problem of unevenness. This not only causes a decline in the productivity of crops but also increases the cost of production. Cropping Pattern is more diversified and productivity values of crops are higher irrespective of farm size on non-weathered land than on weathered land .

Table - 3 Size of agricultural farm and weathered land

Size of farm in acres	Total No of farms	% of weathered farms	% of weathered land	Value of damage per acre weathered land in Rs
More than 10 (Big)	42	83.64	15.56	77.62
7.5-10 (Medium)	38	83.33	16.30	74.69
5-7.5 (Low medium)	49	75.32	22.87	77.91
2.5-5 (Small)	60	68.76	20.54	75.69
Less than 2.5	50	73.00	26.66	76.19
Non recorded share cropper	28	79.98	17.94	75.59
Recorded share cropper	39	80.0	21.06	79.99

SUGGESTIONS

Some policy measures are suggested to combat above problems.

1. Controlled irrigation should be provided.
2. Proper drainage facilities should be augmented.
3. Indiscriminate grazing should be controlled.
4. Uneven land should be levelled and proper bunding should be made.
5. Preventive and protective measures to be taken to control flood and waterlogging.
6.

Incentives to be provided specially to the poor farmers and healthy functional relationship should be established between land lord and tenants.

CONCLUSIONS

Weathering distorts the economy of the rural area and particularly the poor are the victims of it. In order to combat this unwanted situation and integrated policy decisions at the level of government calls for which should be implemented by involving farmers of the affected villages.

Table 4 Cropping pattern and productivity of crops in kg/acre in weathered and non-weathered land

Size of farm in acres	Weathered land	Non-weathered land
>10	Paddy (900), Potato (2400), Mustard (300)	Paddy (1200), Wheat (600), Potato (5400), Mustard (375), Onion (1800)
7.5-10	Paddy (900), Potato (2400), Mustard (300)	Paddy (1200), Wheat (750), Potato (6000), Mustard (375)
5-7.5	Mustard (300), Potato (600)	Paddy (975), Wheat (750), Potato (3300), Mustard (330), Onion (1500)
2.5-5	Paddy (675) Potato (2400)	Paddy (1275), Wheat (510), Potato (4500), Mustard (255), Onion (1500)
Less than 2.5	Paddy (660), Potato (2300), Mustard (90), Onion (1200)	Paddy (870), Wheat (825), Potato (2400), Mustard (195), Onion (1500)

Field Monitoring of Cryogenic Mass Movement

M. C. R. DAVIES, Department of Civil Engineering, University of Dundee, Dundee DD1 4HN C. HARRIS, F. W. SMITH Department of Earth Sciences, Cardiff University, P.O. Box 914, Cardiff CF1 3YE and I. T. BERTHLING, Geografisk Institutt, University of Oslo, Postboks 1042 - Blindern, 0316 Oslo

INTRODUCTION

The annual thawing of ice-rich slopes in periglacial environments is generally associated with slow mass movements caused by gelifluction. The ability to assess the susceptibility of a slope to such processes, and to quantify rates and mechanisms of movement, is important where negative economic or safety impacts may result. Even slow annual soil displacements may have a significant effect on structures (**Figure 1**), communication routes and services. Thickening of the depth of seasonal thawing due to climate change may increase these effects in near future.

Figure 1 Foundation damaged by soil movement

Although surface rates of gelifluction have been measured in a wide range of periglacial environments by Harris[1] and Lewkowicz[2], quantitative field studies of processes and their controlling factors are rare. Field studies by Washburn[3], Benedict[4], Harris[5,6], Smith[7] and Matsuoka[8] emphasised the significance of frost heave and soil moisture content on gelifluction rates. Because such field studies require monitoring over many years in remote and climatically hostile environments only limited and often qualitative date have been collected.

Taking advantage of the controlled conditions of laboratory experiments, Harris et al[9,10] simulated gelifluction in a large scale slope model in which frost heave, thaw settlement, downslope soil displacements and pore water pressure were monitored. In these experiments annual freeze thaw cycles were replicated in approximately one month. Despite these advantages the logistical difficulties in conducting such experiments precluded the repetition necessary for parametric studies. The geotechnical centrifuge modelling technique allows this problem to be overcome and is currently being used to investigate cryogenic mass movement processes[11]. However, to provide confidence in the results of laboratory simulation experiments, it is still important that comparable "benchmark" quantitative data from field monitoring be collected. This paper reports on the design and installation of a highly instrumented field experiment devised to provide such data.

FIELD LOCATION

The site selected for monitoring at Finse, in southern Norway (**Figure 2**), comprises a south facing slope of gradient 17° at an altitude of 1370m. The bedrock in the lower part of the valley consists of Finse granite on the northern side, and gneiss on the central and southern part of the valley (Sigmond et al.[12]). Both belong to the Precambrian basement rocks, which were eroded down to a peneplain at the end of Precambrian time. Above the peneplain, which can be seen as a bench on both sides of the valley at about 1350m a.s.l., the basement rocks are overlain by phyllite. Mean annual air temperature measured at an automatic weather station located close to the site in 1998 was -1°C and precipitation was 885mm. Snow cover is normally thickest in March, with a mean of 220 cm but thickness varies considerably across the slope due to drifting. Snow accumulation usually starts in October, and the snow cover lasts until June or early July . The monitoring station is located on a large gelifluction lobe some 50 m in length (**Figure 3**).

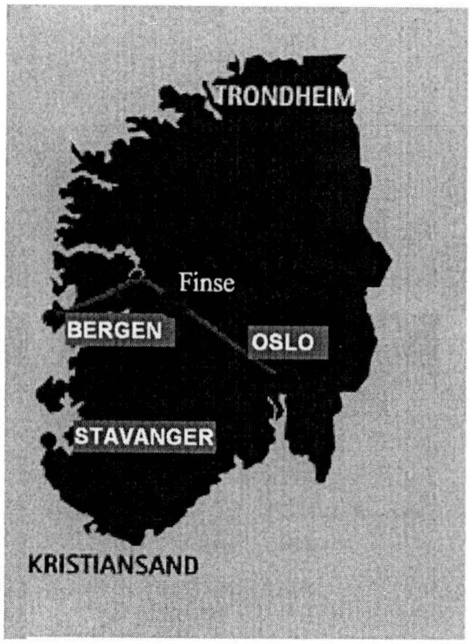

Figure 2 Location of Finse, Norway

MONITORING STATION

In September 1998 a monitoring station was established on a gelifluction slope at two locations (A and B in **Figure 3**) at each of which are installed six thermistors, 3 pore water pressure transducers and pairs of LVDTs (supported by tubular aluminium frames) arranged to monitor frost heave, thaw settlement and down slope soil surface displacements. The data acquisition consists of a Campbell CR23X data logger. Readings are taken at one minute intervals and averaged at intervals of 4 hours. A GSM telemetry system allows interrogation of the data logger from the UK. Typically, data is downloaded at weekly intervals.

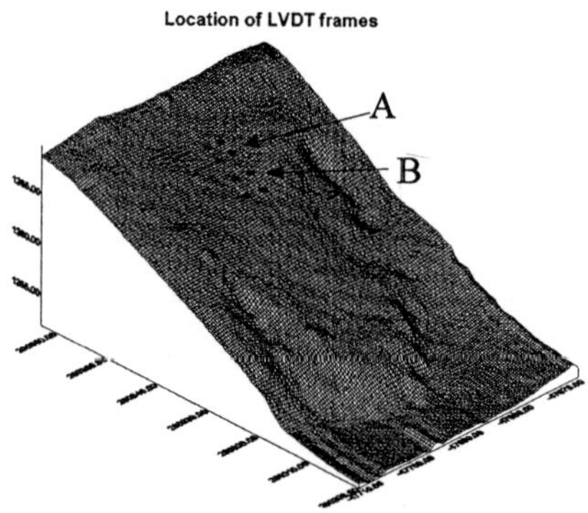

Figure 3 Digital terrain model of site showing location of frames (A & B)

Thermistors

At each location a string of thermistors was installed at depths of 5, 10, 25, 50, 75 and 100 cm. To minimise soil disturbance these were inserted into the side of a 150 mm diameter bore hole, which was carefully backfilled. The thermistors are accurate to within 0.1C.

Pore water pressure transducers

Three type PDCR 81 Druck miniature pore water pressure transducers were also installed in the same bore in a similar manner to the thermistors at depths of 10, 25 and 50 cm. Prior to insertion in the ground the transducers were de-aired under vacuum in a polyethelyne solution to protect them from damage during soil freezing (this technique has been shown to be most successful in large scale laboratory tests, Harris et al[9].

LVDTs and supporting frame

At locations A and B a frame was fabricated from 48 mm diameter tubular aluminium (**Figure 4**) The vertical members are buried to a depth of 1.5 m. They were installed in a 150 mm diameter bore hole and the lower 0.5 m of each vertical member was embedded in concrete. The upper 0.5 m of the buried length was first greased and then sheathed in lengths of plastic tube to isolate it from the surrounding soil. In this way the effects of frost heaving on the frame is minimised. To monitor frame distortion due to snow loading, strain gauges were fixed to a number of the horizontal members. These were calibrated to measure deflection and have been arranged to detect horizontal and vertical displacements.

Figure 4 View of support frame

A pair of LVDTs are mounted on an aluminium track and form a fixed base triangle with the apex connected to an aluminium footplate which has four 2 cm deep anchor points (**Figure 5**). The footplate is embedded in the ground surface and progressive surface displacements due to frost heave and gelifluction are detected by changes in the geometry of the LVDT triangles.

Figure 5 Arrangement of LVDT's

Care was taken during the installation of the instrumentation and the support frame to cause minimal disturbance to the ground, as this would modify the soil fabric and, thus, change the established soil response to cryogenic processes.

PRELIMINARY RESULTS

Soil temperatures

The first snow fell on the 15th October (day 16 on graphs); 13mm precipitation followed by many days with significant precipitation (**Figure 6**). The soil temperature cooled very slowly following arrival of snow. The lowest air temperature was recorded between days 100-125 but the freezing front did not reach its maximum depth until day c225 (**Figure 8**). It is estimated that the freezing front reached 50-75 cm depth. Above zero air temperatures were recorded on 4 occasions prior to day c240 but the soil temperature did not respond to above zero temperatures until snow cover cleared on day c270, when all thermistors went positive together. This would imply that the ice content is low because had it been higher there would have been a delay as the ice absorbed latent heat to cause a phase change; as observed in laboratory model tests [9,10,11].

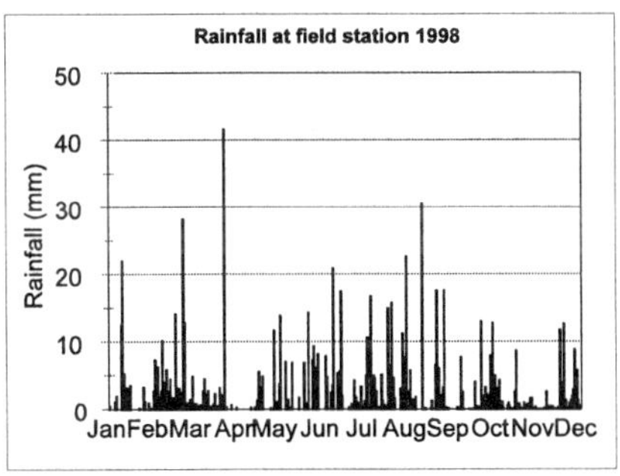

Figure 6 Precipitation at field station 1998

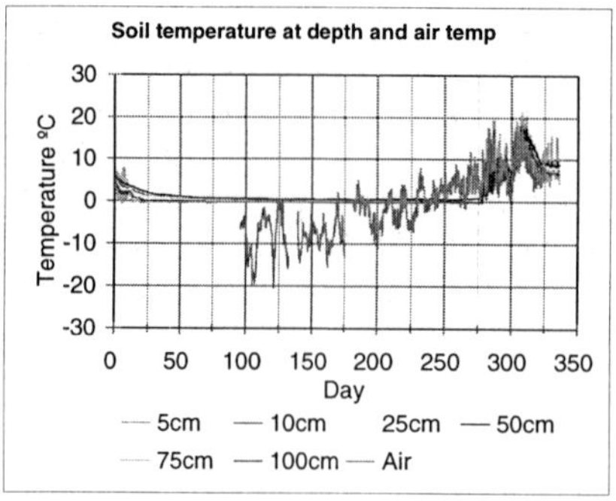

Figure 7 Soil and air temperatures at site A (day 94=1 Jan, day 274=1 Jul)

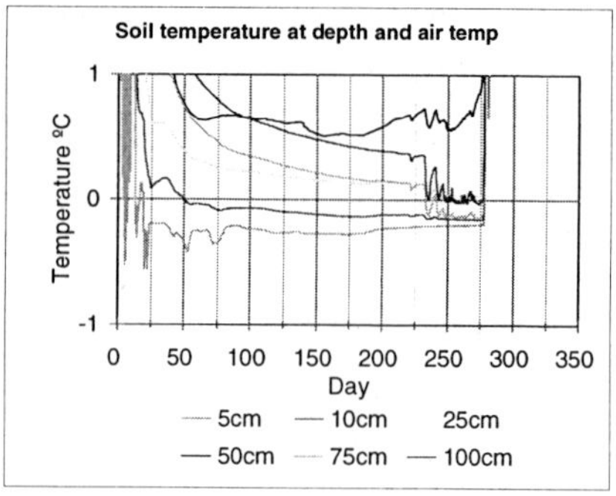

Figure 8 Expanded temperature chart at site A

Porewater pressures

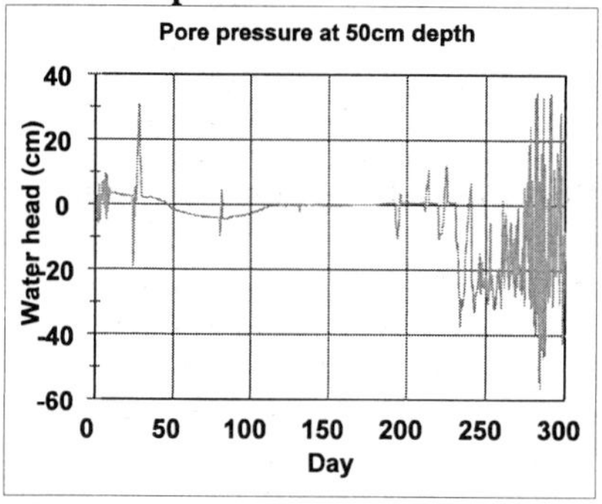

Figure 9 Pore pressure at 50cm depth

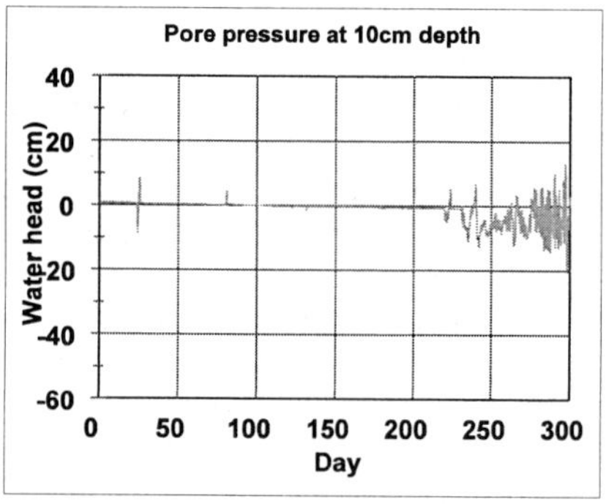

Figure 10 Pore pressure at 10cm depth

When the instrumentation was installed it was noted that the ground was moist but not saturated. Water did not rise in a 1m deep bore hole left open overnight. At the beginning of the period the pore water pressure at 10cm indicated little or no fluctuation because the soil was partially saturated and the porous stone in the transducer was not high air entry (i.e. measuring atmospheric pressure). The three above zero air temp fluctuations between days 200 and 240 are reflected in variations in pore water pressure recorded at 10cm and 50cm depth, probably reflecting influx of melt water from upslope. Since there is a response to snow melt when the soil temperature is still below zero ($t>225$), it suggests that the soil was partially frozen and therefore retains some permeability.

Performance of frame

To allow for easier transportation to the site and to make allowances for possible construction inaccuracies the frame was assembled from aluminium scaffold tubing and pressed steel scaffold clips. As noted above some of the horizontal members were fitted with strain gauges to allow the effects of snow loading to be quantified and eliminated from the displacement data. During the winter of 98/99 the snowfall was quite significant with avalanche activity in the vicinity. Examination of the data showed that the frame was not loaded for an initial period but increasing strains after about day 50 for the top frame and day 75 for the lower frame indicated that the frames had become buried in the snow. The strains increased until about day 250 when a rapid reduction indicated that thawing was occurring. When steady state conditions were achieved again it was apparent from the strains that some permanent deformation of the members had occurred. It had not been anticipated that downslope movement of the snow would be significant so no instrumentation was installed to measure this. However it is likely, given the magnitude of the strain in one of the horizontal members, that there was considerable force exerted on the frame parallel to the slope. It is therefore only safe to make assessments of the soil movements during the periods when the frame was not buried.

LVDT heave and displacement

Figure 11 shows the calculated displacement of the soil in the period before the frame was buried by the snow. Heave is apparent but the downslope movement is quite irregular. It is likely that this is due to the downslope force of the snow on the vertical legs of the support frame as the snow depth increased.

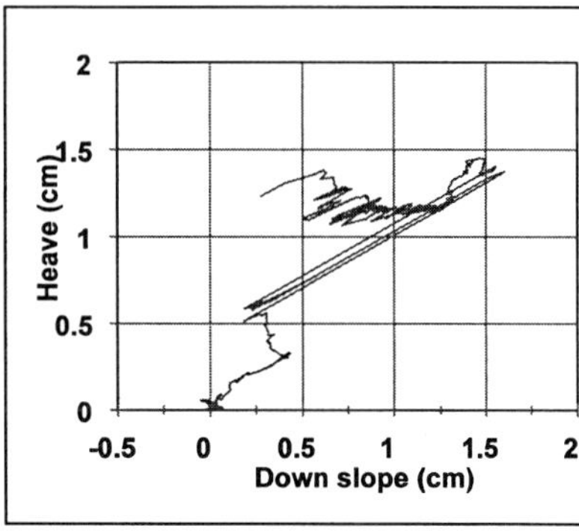

Figure 11 Monitored soil movement

DISCUSSION

Performance of station - snow distribution

The amount of force exerted by the snow on the frame was greater than expected. It had been assumed that as the snow settled around the frame tubes that it would flow slowly around the tubes and thus relieve them of much of their load. This does not seem to have been the case. The tube deflections calculated from the strain gauge readings and observed as the frame emerged from the snow make it difficult to calculate the heave and down slope movement while the frame was buried. In addition the snow load on the LVDTs caused the rod on the down hill gauges to bend thus affecting the results. The LVDTs had been used successfully in a climatic chamber but in the exposed conditions of this site two gauges failed within a month and were replaced. An additional frame is to be constructed at the site and a number of modifications have been incorporated that should improve the quality of the data received. The frame will be constructed from steel rather than aluminium tube and with a shorter span. Sleeves have been manufactured for the LVDTs to greatly increase their resistance to sideways loading and the side bays of the frame will be cross braced with steel wires to prevent sway of the frame due to down slope movement of the snow. The balance between strengthening the frame by increased section size and the increased loading it will attract is difficult to reconcile especially when all materials have to be carried across difficult terrain to reach the site so the modified frame will still be a compromise.

It was fortunate that the site at Finse was located only 80m from a mobile telephone ariel. This allowed the Campbell data logger to be interrogated remotely from the UK and data downloaded at intervals. Intermittent problems prevented calling at some times but the memory capacity of the logger and the frequency of recording were such that it was able to hold at least one year's data so no data was lost. The telephone was able to communicate even during the period of deepest snowfall although the closeness of the phone mast would have greatly assisted in maintaining communication.

Field Measurements

The temperature data collected has allowed the extent of the active layer to be determined although the readings from the 50cm deep sensor appear to have drifted relative to the others. The porewater pressure sensors have not replicated the results from the laboratory simulations, possibly because the soil does not appear to have been saturated. A suction gradient showing increasing suction with depth appears to be due to a water front advancing downwards from the surface. The pulsing would appear to be a result of intermittent thawing (cf **Figure 8**)

The results from the second year should provide better information with respect to movement and should also help to make a better interpretation of the porewater pressure response.

REFERENCES

1. Harris,C. 1987. 'Mechanisms of mass movement in periglacial environments' in Anderson, M. G. and Richards, K. S. (Eds), Slope Stability, Wiley, Chichester, 531-559.
2. Lewkowicz, A. G. 1988. 'Slope processes' in Clark, M. J. (Ed), Advances in Periglacial Geomorphology, Wiley, Chichester, 325-370.
3. Washburn, A.L. 1967. Instrumental observations of mass-wasting in the Mesters Vig district, Northeast Greenland. Meddelelser om Grønland 166, 318pp.
4. Benedict, J.B. 1970. Downslope soil movement in a Colorado Alpine region: rates, processes and climatic significance. Arctic and Alpine Research, **2**, 165-226.
5. Harris, C. 1972. 'Processes of soil movement in turf-banked solifluction lobes, Okstindan, northern Norway', in Price R. J. and Sugden, D. E. (eds), Polar Geomorphology, Institute of British Geographers Special Publication 4, 155-174
6. Harris, C. 1977. Engineering properties, groundwater conditions, and the nature of soil movement on a solifluction slope in North Norway. Quarterly Journal of Engineering Geology, **10**, 27-43
7. Smith, D. J. 1988. 'Rates and controls of soil movement on a solifluction slope in the Mount Rae area, southern Canadian Rocky Mountains', Zeitschrift fur Geomorphologie, Supplementband, **71**, 25-44
8. Matsuoka, N. 1994. Continuous recording of frost heave and creep on a Japanese Alpine slope. Arctic and Alpine Research, **26**, 245-254
9. Harris, C., Davies, M.C.R. and Coutard, J.-P. 1995. Laboratory simulation of periglacial solifluction: significance of porewater pressure, moisture contents and undrained shear strength during thawing, Permafrost and Periglacial Process, 1995, **Vol. 6**, 293-311.
10. Harris, C., Davies, M.C.R. and Coutard, J.-P. 1997. Rates and processes of periglacial solifluction: an experimental approach, Journal of Earth Surfaces Processes and Landforms, **22**, 1997, pp 849-868.
11. Harris, C. and Davies, M.C.R. 2000. Gelifluction processes: some observations from large scale laboratory simulations, Accepted for publication in Arctic and Alpine Research, May, **Vol. 2**, 2000.
12. Sigmond, E., Gustavson, M., Roberts, D. 1984. Berggrunnskart over Norge - M. 1:1000000 (Bedrock Geology of Norway), Geological Survey of Norway

The Avalanche Lake Rock Avalanche: an Alternative Emplacement Mechanism

T. R. DAVIES, Natural Resources Engineering, Lincoln University PO Box 84, Canterbury, New Zealand

INTRODUCTION

The extraordinary deposit of the Avalanche Lake rock avalanche, Mackenzie Mountains, Northwest Territories, Canada, has been the subject of considerable description and debate (Eisbacher, 1977, 1979; Evans, 1989; Kaiser and Simmons, 1990; Evans, Hungr and Enegren, 1994). This discussion examines the emplacement mechanism postulated by Evans, Hungr and Enegren (1994; hereafter referred to as [1]) in the light of some anomalous consequences it implies, and suggests an alternative mechanism that removes these anomalies.

EMPLACEMENT PROCESS (EVANS, HUNGR AND ENEGREN, 1994)

The sequence of events proposed in [1] is, briefly, as follows (see Fig.1, after Evans, Hungr and Enegren (1994), and Fig.2a):

1. About 600 years ago, $200x10^6$ m^3 of rock, in the form of a slab about 2 km long and 2 km wide by 50 m thick, detached on a 30° plane and slid into the valley now occupied by Avalanche Lake.

2a. The slab reached a very high velocity (greater than 80 m/s) in moving across the narrow valley bottom and began to move up the opposite slope towards the "Shelf".

3a. As the (by now shattered) slab moved up the steep valley side (the "Wall"; C, Fig.1) below the Shelf, it experienced increased rigidity as a result of active pressure conditions; this allowed the front part of the debris mass (about $5x10^6$ m^3) to be pushed up onto the Shelf, about 640 m above the valley bottom, forming the Shelf Lobe. A considerable volume of wet till excavated from the valley bottom was present in this blocky mass.

4a. The rest of the debris mass fell back from the Wall below the Shelf into the valley bottom forming "fallback" and "reverse runup" ridges (respectively at C and J, Fig.1).

2b. Meanwhile, the southern part of the detached, shattered slab was deflected to its left by the south-eastern end of the Wall and moved along the valley now occupied by South Lake to form the South Lobe comprising about $40x10^6$ m^3 of material.

3b. A small part of this mass moved due west up Shelf Creek (S, Fig.1) onto the Shelf, and subsequently a part of this flowed back down Shelf Creek again forming conspicuous levees.

5. Some of the debris on the Shelf, following emplacement from a north-easterly direction, turned through more than a right angle and flowed about 1 km down the gently-sloping Shelf to the north-west (as shown by flow-lines in the debris at E, Fig.1), eventually leaving the abrupt NW edge of the Shelf with sufficient velocity to emplace an isolated deposit of material on the valley side 500 m farther on and about 200 m lower down (A, Fig 1)

Landslides in research, theory and practice, Thomas Telford, London, 2000.

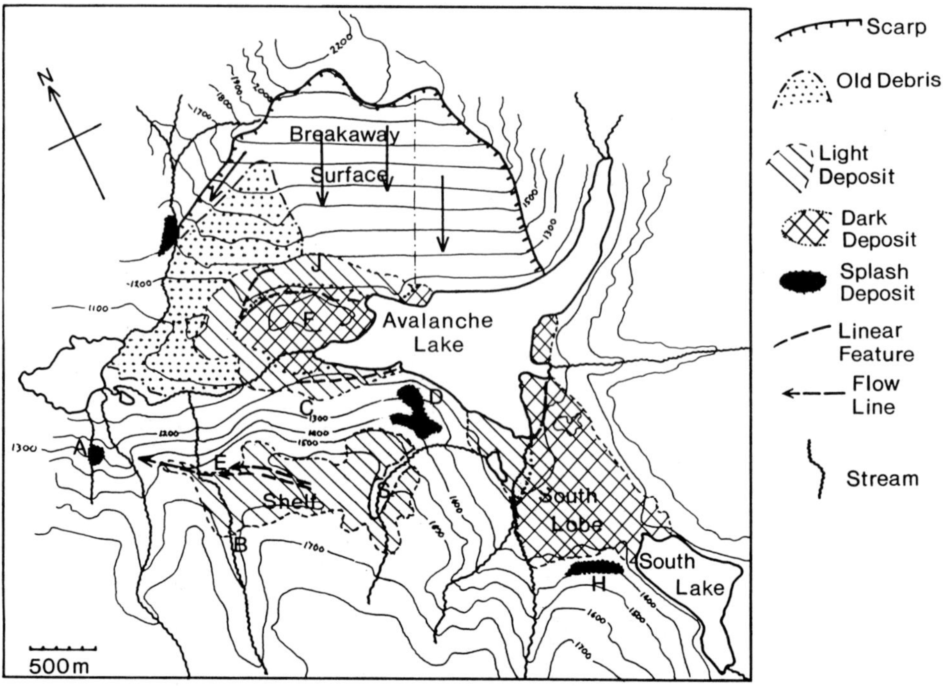

Figure 1. Map of the avalanche lake rock avalanche (after [1])

ANOMALIES OF THE EMPLACEMENT PROCESS

1. The extraordinary mobility of the Shelf deposit required for 5 above is difficult to reconcile with the material properties necessary for its proposed emplacement. If, as suggested by Hungr (1995), the shattered material of the original slab became more rigid as a result of compressional stresses as it ascended the Wall, and the leading material was pushed up onto the Shelf, then these same material properties would prevent the material becoming sufficiently mobile to accelerate in a direction perpendicular to the emplacement, to the velocity needed to explain the isolated deposit beyond the NW end of the Shelf.
2. The runup onto the Shelf is anomalously high. Data from 24 other rock avalanches show an upper envelope corresponding to $h_2 = 0.3h_1$, where h_2 is runup and h_1 is initial fall height: whereas the Avalanche Lake event has $h_2 = 0.52h_1$, [1]. By contrast, the same material in the South Lobe does not exhibit characteristics suggesting such unusual mobility.
3. If the Shelf deposit were indeed emplaced from the north-east, the runup in this direction into the high valley at the NW end of the Shelf (B, Fig.1) is surprisingly short.
4. The Wall below the Shelf (C, Fig.1) shows little or no sign that tens of millions of cubic metres of rock material have passed up and down it; the screes below the Shelf look very similar in character to undisturbed screes farther down the valley ([1], Figs. 2 and 3).
5. By contrast, the buttress comprising the SE end of the wall (D, Fig.1) looks "sand-blasted", as one might expect if it had suffered such an event ([1], Figs 2 and 3; Evans (1989) Fig.1; Kaiser and Simmons (1990)).
6. If the South Lobe deposit came from the eastern part of the initial breakaway slab, it should be about 40% of the total material. In fact it is about 20% [1].

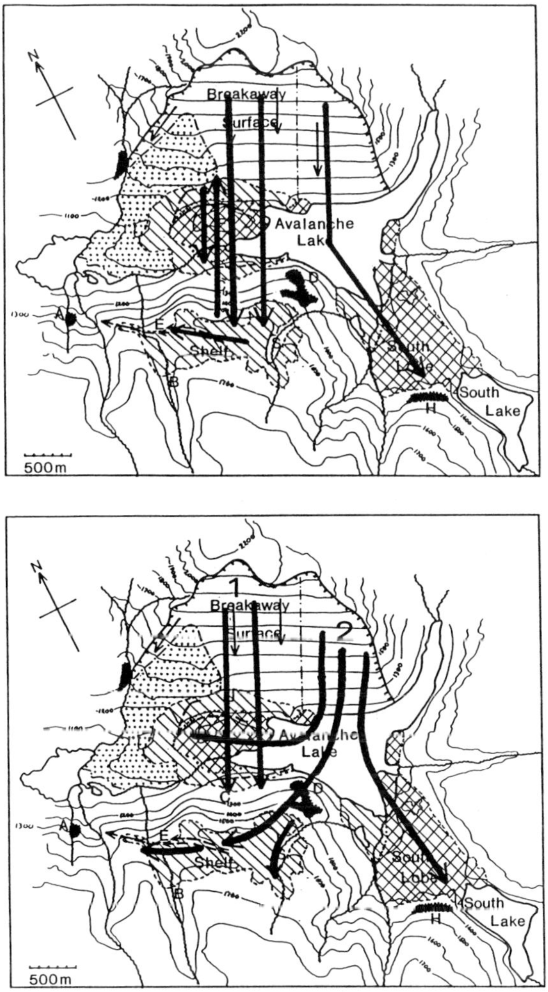

(a)

(b)

Fig.2 Emplacement event sequences (a) according to [1] and (b) according to the present hypothesis.

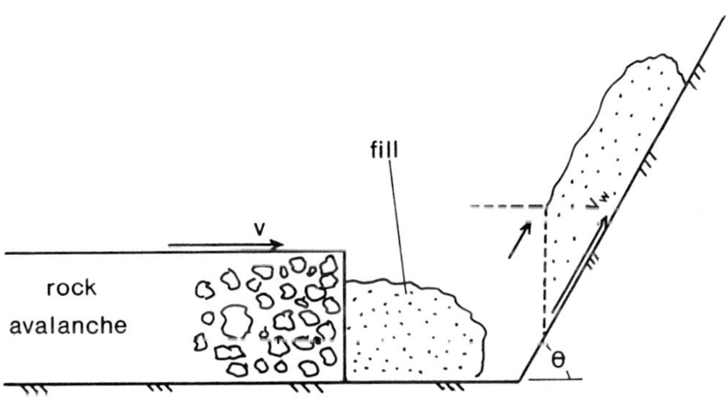

Fig.3 Diagram illustrating mechanism of high-velocity upward ejection of valley fill from base of buttress east of the Wall.

7. The deposits seem to be made up of two different types of material (Figs 2 and 3 of [1], Fig.1 herein); most of the South Lobe and the central hump of the main deposit (F, Fig.1) have a dark, blocky appearance (cross-hatched on Fig.1), while the remainder of the deposits have a lighter coloured smoother appearance (diagonally shaded in Fig.1).

DISCUSSION

Only the first of these anomalies has been addressed in [1], in which it was suggested that the "spectacular" mobility of the Shelf deposit could have resulted from the presence of snow or of saturated ground resulting from melted permafrost. It is undeniable that much of the material that reached the Shelf was very wet indeed, but whether even debris-flow material could accelerate sufficiently rapidly in a direction perpendicular to that of its emplacement (hence from essentially zero velocity) is doubtful. As shown in Fig.1, some of the material arriving on the Shelf from the NE flowed NW on a slope of about 10° for about 1 km before becoming airborne and landing about 500 m farther on and 200 m lower down (A, Fig.1). Assuming a ballistic trajectory for this flight, the velocity required on leaving the Shelf is about 100 m/s - an order of magnitude faster than the largest recorded terrestrial debris flows on such slopes (Davies et al, 1992). There must be doubt as to whether such an acceleration is realistic; if not, the only possible explanation is that the material that undeniably made this flight possessed a substantial initial velocity in a NW direction when it arrived on the Shelf, ie *it was emplaced from the SE* rather than from the NE, and simply continued on across the Shelf towards the NW.

The amazingly high runup of the Avalanche Lake event (Evans, 1989) is admitted to be beyond present explanation ([1], p 766). It is there calculated that the maximum "normal" runup for a fall height of 1220 m would be 366 m rather than 640 m, giving a maximum runup elevation of 1246 m. *This is very close to the maximum elevation of the "fall-back" ridge below the Wall* (at C, Fig.1). An obvious suggestion to make is that the main deposit, with its complex of ridges, is in fact the limit of travel of the material of the slab that fell from the NW part of the initial breakaway, and that *no* material moved up the Wall; this would also resolve anomalies 3 and 4. Recent experiments (Davies, McSaveney and Hodgson, 1999) also cast doubt on the ability of the "fall-back" process proposed in [1] to yield the complex of ridges present in the Avalanche Lake deposit; stationary granular material does not remobilise easily as a result of a subsequent impact by similar material.

If these hypotheses are accepted for present purposes, four matters still remain to be explained:
1. how was the Shelf deposit emplaced?
2. why is the South Lobe so small?
3. why does the east buttress of the Wall look "sandblasted"?
4. why do there appear to be two types of material in the deposits?

ALTERNATIVE EMPLACEMENT HYPOTHESIS

The emplacement of the Avalanche Lake rock avalanche could have been quite different from that proposed in [1]. As Fig.3 of [1] shows, there are signs that more than one breakaway surface exists on the detachment slab, and thus there could have been more than one detachment event. For example, the northwest part of the rockfall could have detached slightly before the rest. The sequence of events could then have been as follows (Fig.2b):

1. The NW part of the breakaway slab moved across the valley and came to rest up against the Wall, having shattered and shortened by about 25%, to form the lower part of the main

deposit; J and C, Fig.1. The large boulders at the base of the detachment surface ([1], Fig. 15B), rather than marking the limit of reverse runup, could well have fallen from the detachment surface later and come to rest against the rear edge of the fallen mass.

2. The SE part of the slab detached later (how much later does not matter). It impinged almost directly against the buttress at the east end of the Wall, and this split the moving mass into two parts. One part moved south to form the South Lobe, while the other part moved north-west to over-ride the deposit from the first rock avalanche and form the central mound of the main deposit (F, Fig.1), hence the similar dark, blocky appearance of these two later deposits. Thus the material from the later rockfall, which fell from the SE part of the detachment surface, is divided between the South Lobe and the central part of the main deposit, and the South Lobe is then not anomalously small.
3. This later event, in moving south-west across the valley floor towards the eastern buttress of the Wall, moved before it a mass of till and other saturated valley fill deposits; this material was propelled ahead of the main rock mass onto the deposit of the first rock avalanche, causing the lighter colour of the "fall-back" and "reverse runup" ridges of the first rock avalanche and also the "splash" deposit beyond the limit of the South Lobe.
4. Some of the fill material became trapped between the advancing rock mass of the second detachment and the eastern buttress of the Wall (D, Fig.1); the geometry of the lower part of the Wall is appropriate for this fill material to be forced up the Wall at very high velocity (see below) and to attain the Shelf, carrying with it a small part of the rock mass. Thus the Shelf Lobe was emplaced at high velocity from the south-east. It comprised a mixture of wet valley fill and rocks from the original breakaway slab, much of which continued to flow across the Shelf and down into the valley, while some reversed its direction and flowed back down Shelf Creek. This process is elaborated in the following section. Six hundred years of rainfall erosion could have removed much of the fine-grained material originally present.

EMPLACEMENT OF THE SHELF LOBE

A crucial component in this alternative emplacement scenario is the presence in the valley bottom of wet valley fill. As the SE part of the rock slab fell from its original position it accelerated to about 100 m/s (centre of mass falling about 600m, $v = (2gh)^{1/2}$). This mass of rock moved across the valley bottom, pushing wet fill ahead of it. When the fill encountered the steeply sloping base of the east buttress of the Wall, the fill became squeezed between the rapidly advancing deformable rock mass and the rigid rock wall (Fig.3); with the result that the fill was forced rapidly upwards at a velocity related to the velocity of the rock mass. If the front of the rock mass is vertical, then after it encounters the inclined wall the material between the front of the mass and the wall moves upwards along the wall at $v/\cos\theta$, where v is the horizontal velocity of the rock mass and θ is the slope of the wall (Fig. 3). If $v = 100$ m/s and $\theta = 60°$ then this upward velocity is

$$v_w = 100/\cos 60° = 200 \text{m/s}.$$

The vertical component of the initial upward velocity is

$$v_w \sin 60° = 200 \times 0.866 = 170 \text{ m/s},$$

which would convert into an increase in elevation of

$$\Delta h = v^2/2g = 170^2/20 = 1450 \text{ m},$$

and is ample to project the material up the 600 m or so to the Shelf.

The initial horizontal velocity component is

$$v_w \cos 60° = 200 \times 0.5 = 100 \text{ m/s},$$

which will provide a substantial initial velocity towards the north-west when the fill and rocks arrive on the Shelf. This velocity will in practice be reduced by a number of factors and because the assumptions herein will not be completely fulfilled; however, it appears that in principle the mechanism outlined could emplace the Shelf Lobe as required, and that the order of magnitude of velocities required is not unreasonable in the circumstances of the Avalanche Lake event.

CONCLUSIONS

1. The sequence of events proposed in [1] as the emplacement of the Avalanche Lake deposit does not satisfactorily explain the highly anomalous runup of the Shelf Lobe; the behaviour of the Shelf Lobe material; the small volume of the South Lobe; the apparent presence of two distinct deposit textures; the lack of erosion evident on the Wall; the extreme erosion apparent on the east buttress of the Wall; or the lack of runup into a valley that runs north-east into the Shelf.
2. An alternative sequence of events is proposed that is capable of explaining these anomalies. It requires that the deposits represent two distinct sequences of events, separated in time but perhaps only by a very short time.
3. The alternative explanation also requires that saturated valley deposit material was ejected upwards at high velocity as it became compressed between the rapidly-advancing front of the rock avalanche and the steep lower part of the east buttress of the Wall.

REFERENCES

Davies, T.R.H. et al 1992. Debris flow behaviour - an integrated overview. *In:* Erosion, Debris Flows and Environment in Mountain Regions (Proceedings of the Chengdu Symposium), IAHS Publication **209:** 217-225.

Davies, T.R.H. , McSaveney, M.J. and Hodgson, K.A. 1999. A fragmentation-spreading model for rock avalanche runout. *In press*, Canadian Geotechnical Journal.

Eisbacher, G.H. 1977. Rockslides in the Mackenzie Mountains, District of Mackenzie. *In* Report of Activities, part A. Geological Survey of Canada, **Paper 77-1A**, pp 235-241.

Eisbacher, G.H. 1979.Cliff collapse and rock avalanches (sturzstroms) in the Mackenzie Mountains, northwestern Canada. Canadian Geotechnical Journal, **16**: 309-334.

Evans, S.G. 1989. Rock avalanche runup record. Nature (London), **340**: 271.

Evans, S.G., Hungr, O. and Enegren, E.G. 1994. The Avalanche Lake rock avalanche, Mackenzie Mountains, Northwest Territories, Canada: description, dating and dynamics. Canadian Geotechnical Journal, **31**, 749-768.

Hungr, O. 1995. A model for the runout analysis of rapid flow slides, debris flows and avalanches. Canadian Geotechnical Journal, **32**, 610-623.

Kaiser, P.K. and Simmons, J.V. 1990. A reassessment of transport mechanisms of some rock avalanches in the Mackenzie Mountains, Northwest Territories, Canada. Canadian Geotechnical Journal, **27**, 129-144.

Three Dimensional Visualisation of Landslides in SE Spain using a Digital Elevation Model (DEM) and Orthophotography

A. M. DAVIS, P. J. MASON, HME Partnership, P. O. Box 3036, Romford and J. McM MOORE, Earth Sciences, T.H. Huxley School of the Environment, Earth Science & Engineering, Imperial College, London, UK

ABSTRACT
This paper demonstrates how digital elevation models (DEM) can be used with orthophotographs to produce three dimensional views of landslides near Sorbas, Almeria Province, S.E. Spain. ERDAS OrthoMAX software was used to generate DEMs from conjugate relationships of 60% overlap stereo aerial photography. The DEMs were then used to orthorectify the photographs to produce terrain corrected image maps.

Three dimensional viewing of the orthophotographs using the DEM allows improved interpretation of the landslides and shows the relationship between land-cover (rock types, soil and vegetation) and terrain. Edge enhancement techniques were applied to the DEM to accentuate the landslide morphology, using textural information. The results show that a DEM provides additional information about geomorphology beyond that available from two-dimensional imagery alone.

Virtual mapping of the landslides can be used to identify specific localities for detailed field measurements. This enhanced geo-morphological information integrated with imagery and geotechnical data in a GIS, can contribute significantly to the landslide hazard (and other geohazard) analysis.

INTRODUCTION
Automatically generated digital elevation models (DEMs) provide three-dimensional information about geomorphology and can be used to create virtual views of the terrain. These views together with DEM image processing facilitate identification of geohazards e.g. the geomorphology and areal extent of slope instabilities.

This paper reports an experimental study based on air-photos of the Rio de Aguas valley, east of Sorbas town, Almeria Province, S.E. Spain (lat. 37:5:30N, long. 2:3:44W). Three landslides i.e. Carrasco, Los Perales and El Tesoro escarpment, previously identified in satellite imagery (Mason *et al.*, 1998, Eyers *et al.*, 1998), were used as test sites for this study.

The main objectives are to (i) appraise the accuracy of air photo derived DEMs, (ii) generate selected 3D virtual views and (iii) to critically assess the use of DEMs for landslide hazard assessment.

Landslides in research, theory and practice, Thomas Telford, London, 2000.

DEM GENERATION

ERDAS OrthoMAX software package was used to generate four DEMs from 60% overlapping stereo 1:13,000 scale aerial photographic diapositives. Height values are calculated from the conjugate relationships of stereo aerial photography.

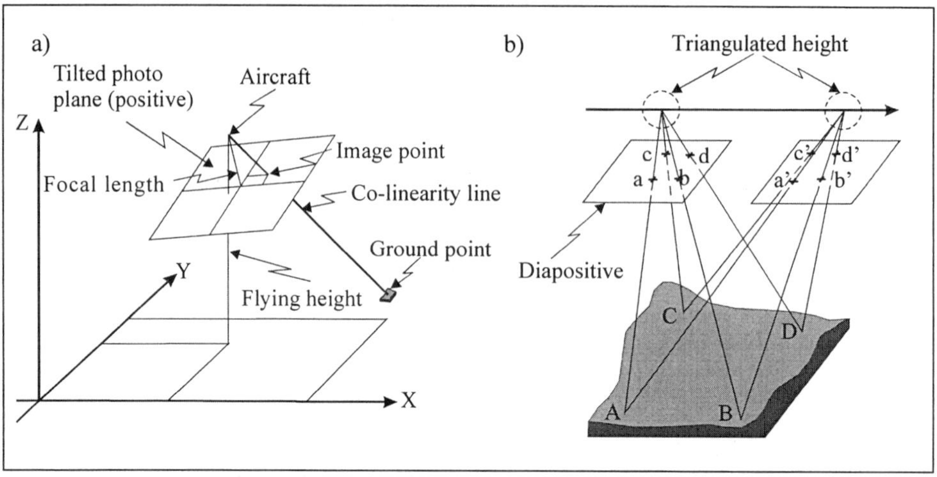

Figure 1. Schematic representation of (a) co-linearity condition (after Wolf, 1983);
(b) conjugate relationships of four ground control points used to triangulate exterior orientation of the photographs

Figure 1(a) illustrates the relationship between a ground point (GP), image point and the aircraft flying height. These points lie on a straight line and this condition is known as co-linearity (Wolf, 1983). ERDAS OrthoMAX uses this condition firstly to triangulate the flying height and secondly to generate the DEM.

Using the condition of co-linearity, conjugate relationships and several ground control points or GCPs (i.e. absolute points measured from a map) together with their corresponding image points it is possible to triangulate the flying height of the aircraft (i.e. exterior orientation). Figure 1(b) shows how four GCPs are used to triangulate the flying height for each photograph. Once the exterior orientation is determined, the DEM is automatically generated using the conjugate relationships of the flying height and the two photographs.

Figure 2 outlines the methodology sequence involved in generating the DEMs in ERDAS OrthoMAX. Six diapositive photographs representing the area of interest were photogrammetrically scanned at 1200 dpi (22.5 microns). The block file stage involves the input of the raw data and calculation of the exterior orientation of the aircraft. The camera calibration details i.e. focal length, principal point offsets, fiducial point positions, aerial photographs (tiff images) and the GCPs are input to the block file. In this study GCPs were selected by comparing photo images with maps. Ideal GCPs are well defined fixed points e.g. road junctions. Four GCPs were selected and located using 1:25000 Spanish topographic maps (ED50 datum and projected in the UTM 30 north grid.).

Interior orientation of the images involves the manual measurement of the fiducial points on the photographic images. Each fiducial point measurement is recorded and compared to its corresponding calibrated fiducial point in the camera file. The interior orientation indicates any distortion due to diapositive shrinkage. GCP measurement involves the manual positioning of the GCP on screen. The GCPs were selected to be evenly distributed within the onlap of the photographs.

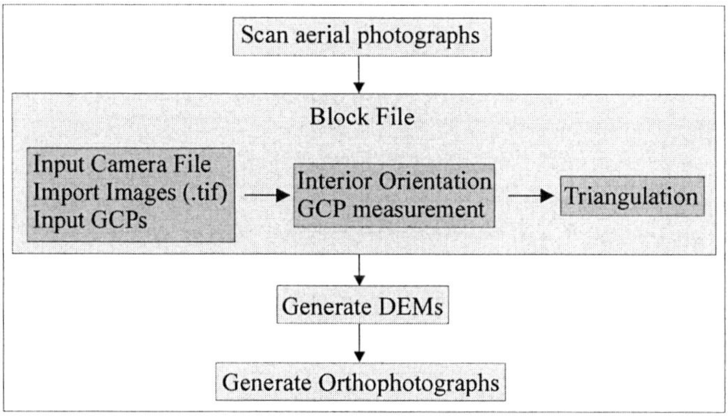

Figure 2. Flow diagram of DEM process (after ERDAS, 1994)

Triangulation involves the planimetric resection of the two photographs and, through an iterative process, calculation of the exterior orientation of the photographs (Wolf, 1983). If there are large errors in triangulation results at this stage, it is possible to re-measure the GCPs and re-compute the triangulation. The computed block file now provides the information for the DEM generation i.e. it uses conjugate relationships of the photograph images and their corresponding triangulated exterior orientations to create a regular grid of height values. In this case the DEMs were generated at a grid spacing of two metres.

Visual inspection of the DEMs indicated that errors had occurred in the DEM generation, e.g. spurious height values occurred where the motorway crossed the Rio de Aguas. ERDAS OrthoMAX has fourteen strategy parameters that allow the user to customise the algorithm that automatically generates the DEM. The DEMs were generated using the default parameters in the Orthomax package. By changing some of the parameters, the accuracy of the DEM can be improved (Gooch *et al.*, 1999). In this study four parameters: minimum threshold; maximum parallax; minimum and maximum templates were changed and a second DEM generated. The DEMs were exported from ERDAS into ERMapper software.

The two DEMs were subtracted from each other and the result indicated areas of potential error in the automatic DEM generation (Gooch & Chandler, 1999). Figure 3a shows the subtracted DEM result, where medium grey tones indicate no change in height while white and black tones indicate relative differences in height suggesting potential errors. Figure 3d indicates potential errors where the motorway crosses the Rio de Aguas (indicated at E).

Orthorectification removes geometric and height distortions from aerial photographs. The results from the block file and the DEM are used together to correct these distortions to produce an orthophotograph which can be treated as a image map. The orthophotographs were generated at a pixel size of 0.4 m.

IMAGE PROCESSING

The DEMs and orthophotos were integrated using ERMapper image processing software. The four DEMs and orthophotos were mosaiced to produce two data sets covering the landslide area.

An effective representation of the Rio Aguas landslide area is given by the greyscale drape on a pseudo-illuminated 'sun angle' intensity layer (both derived from the DEM) (Figure 4a). Textural enhancement of the DEM was carried out using neighbourhood convolution filters (Mason *et al.*, 1998). Figure 4b represents slope angle, calculated using a convolution filter (black tones = slopes $< 10°$; white = slopes $> 45°$). The georectification of Landsat Thematic Mapper (TM), and Airborne Thematic Mapper (ATM) imagery to the orthophotographs can provide significant additional information on vegetation, rock and soil types and moisture distribution when used as drapes on a DEM.

SLOPE STABILITY APPLICATIONS

Three of the products from this preliminary study have practical uses for the geotechnical engineer and engineering geologist concerned with slope stability geohazards.

The virtual perspective terrain model combining DEM and photographic or scanned image information (Figure 3b), provides a method to view terrain from a variety of viewpoints including some which could be inaccessible or dangerous for the field geologist. Interactive study of a perspective DEM can improve the speed and efficiency with which the investigator can appreciate terrain. It can also be used in conjunction with 2D maps and images to 'visualise' terrain and represent specific terrain features e.g. scarps and landslide debris accumulations to complement or replace field photographs as report illustrations.

Greyscale or colour composite images, draped on DEM-derived, pseudo-illuminated, 'sun-angle' intensity images can be used by skilled interpreters to detect and map a variety of terrain stability related geomorphological features. Figure 4a illustrates the form of unstable escarpments (locality 1) and landslide back-scarps or crown walls (localities 2 & 3). Fissuring in competent rocks (gypsum and limestone) related to block sliding can be seen, as can the 'terrace' surface of detached slip blocks. A major advantage of the 'sun-angle' illumination technique for DEMs is the ability to manipulate the pseudo-illumination azimuth and inclination. To photo-interpreters familiar with the value of 'low sun angle' air photographs for geotechnical applications, this increased flexibility in 'illumination' conditions will be immediately appreciated.

The value of the classified slope-angle filter image (Figure 4b) will be most useful to those concerned with terrain slope analysis. It effectively classifies scarps (45°+), potentially unstable slopes (10°- 45°) and relatively flat terrain (0°- 10°). In practical applications, slope class boundaries would of course be adjusted to match the engineering stability criteria for the area and project concerned.

Photographic and digital imagery incorporating infra-red information (VNIR, SWIR or thermal) reveals significant amounts of information about soil moisture and vegetation, beyond that provided by panchromatic or 'true colour' photography (Eyers *et al.*, 1998). Combined use of false colour infra-red imagery with DEMs can significantly increase the information available to the competent engineer or geotechnical interpreter.

Digital Elevation Models, combined with photographic or scanned imagery represent a significant improvement on the value of air photography alone as tools for landslide geohazard terrain investigation.

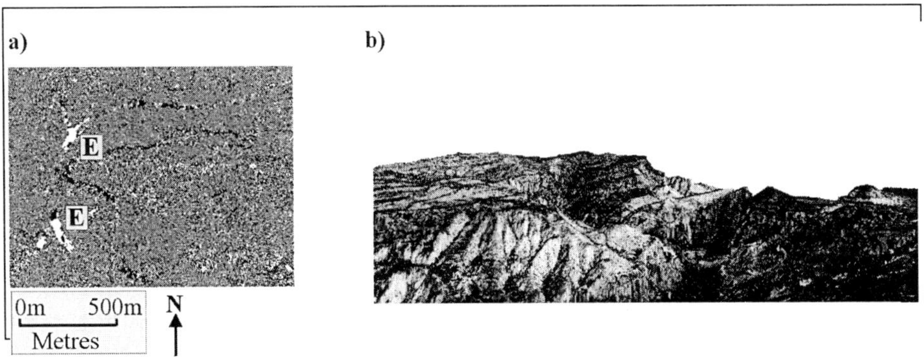

Figure 3. a) DEM accuracy assessment generated by subtracting the default generated DEM from the customised generated DEM (grey tones indicate no change in the DEM, while white and black tones indicate potential errors); b) three dimensional view using the orthophotograph and the DEM

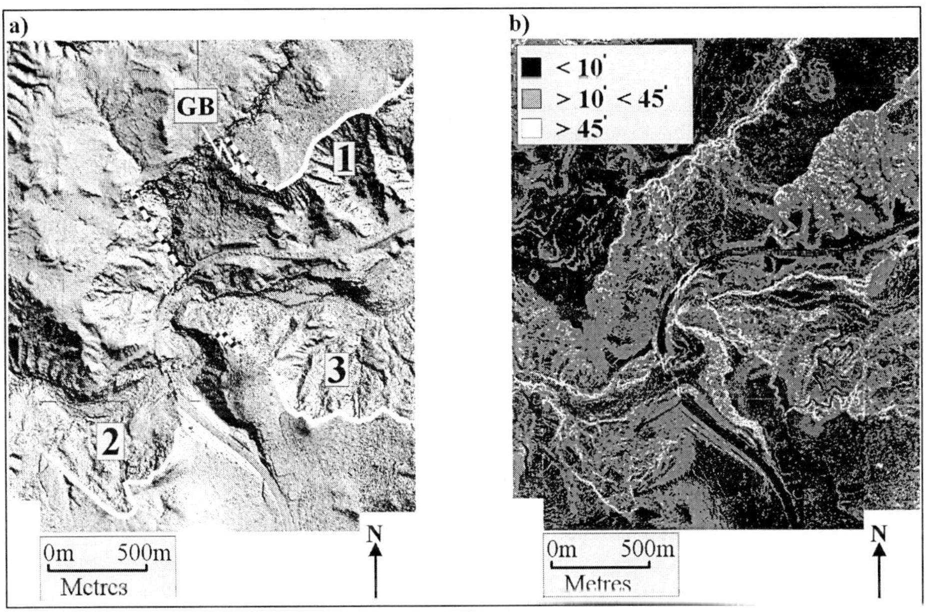

Figure 4. Rio de Aguas 2m DEM: a) greyscale drape over a sun-angle intensity layer showing the El Tesoro escarpment (1), the Carrasco landslide (2) and the Los Perales (3) scarp. White lines indicate landslide scarps, dashed lines indicate fissures and GB indicates a slipped gypsum block; b) slope angle filter image (black tones indicate slopes < 10°, grey tones indicate slopes between 10° and 45°, and white tones indicate slopes > 45°).

DISCUSSION & CONCLUSIONS

This study shows that the accuracy of automatically generated DEMs can be improved by changing the default strategy parameters in the ERDAS OrthoMAX. The triangulation in the block file can also be improved by selecting GCPs from a smaller scale map e.g. 1:10000, or by using differential global positioning system (GPS) GCPs collected in the field.

Processing of the 2m DEMs has significantly improved the interpretation of the landslide morphology in the Rio de Aguas valley. The need to compile the dataset into a consistent datum and projection has lead to a base image layer stack that can now be used in a GIS to manipulate the image data for geohazard assessment and monitoring. This work continues the geohazard monitoring in S.E. Spain (Moore *et al.*, 1999), using integrated DEMs and remotely sensed imagery.

Rapid developments in computer technology have lead to a number of specialist centres that create "virtual reality", three dimensional immersive environments. Three dimensional viewing of remotely sensed imagery draped over DEMs displayed in such an environment can allow the geoscientist to interactively roam around the imagery. This virtual mapping can be used to not only identify geohazards but to plan logistics and site localities for future fieldwork.

ACKNOWLEDGEMENTS

The authors would like to acknowledge the University of London Intercollegiate Research Services (ULIRS); ERMapper; ERDAS Imagine/ Orthomax; Prof Ian Dowman for the use of Geomatic Engineering facilities at UCL; Andy Bussell at Greenwich University; NERC for the aerial photographic positives.

REFERENCES

ERDAS, 1994. Imagine OrthoMAX Users Guide, Vision International.

Eyers, R., Moore, J. McM., Hervas, J. & Liu, J.G. 1998. Integrated use of Landsat TM and SPOT panchromatic imagery for landslide mapping: case histories from southeast Spain. Geohazards in Engineering Geology, *Engineering Geology Special Publications*, 15, Geological Society, London, 133-140.

Gooch, M.J., Stojic, M.J. & Chandler, J.H., 1999. "Accuracy assessment of digital elevation models generated using the ERDAS Imagine OrthoMAX digital photogrammetric system", *Photogrammetric Record*, 16(93): 519-531 (April 1999).

Gooch, M.J. & Chandler, J.H., 1999. Failure prediction in automatically generated digital elevation models. Web article found at http://www-staff.lboro.ac.uk/~cvjhc/gc050.htm.

Mason, P.J., Rosenbaum, M.S. & Moore, J. McM. 1998. Digital image texture analysis for landslide hazard mapping. Geohazards in Engineering Geology, *Engineering Geology Special Publications*, 15, Geological Society, London, 297-305.

Moore, J. McM., Mason, P.J. & Davis, A.M., 1999. Geohazard Monitoring in Southeast Spain using Integrated Imagery and Digital Elevation Model, *Proceedings of the Thirteenth Thematic Conference on Applied Geologic Remote Sensing*, Vancouver, Canada, 1-3 March 1999.

Sorbas and Polopos 1995. Mapa Topografico Nacional de Espana, Instituto Geografico Nacional 1:25000 map sheets, Madrid.

Wolf, P.R., 1983. *Elements of Photogrammetry*. McGraw Hill, Singapore.562p

Rock Mechanics Studies to Analyse Toppling Failure

C. DEANGELI and A. M. FERRERO, Dipartimento di Georisorse e Territorio, Politecnico di Torino, Italy

INTRODUCTION

The study is devoted to the analysis of toppling failure of a rock slope. The rock wall is part of a quarry of ornamental stone, situated in Ossola valley (Italy). The rock mass is constituted by ortogneiss and is interested by a main set of discontinuities, the foliation. Three minor sets of joints are also present, isolating slabs and blocks, leading to flexural toppling and block flexural toppling kinematism.

A face of the quarry was interested in the past by an instability phenomenon (figure 1), due to water seepage along joints and due to the height of the overhanging rock wall. This phenomenon was back analysed by a 2D numerical modelling based on the DEM (Deangeli et al., 1996). On the basis of this analysis a new stable configuration of the rock face has been verified, and the exploitation activity in this zone is exhausted. The excavation is presently moved to a zone of the quarry with the same geostructural scenery but with a lower weathering degree of the rock mass (figure 2). The exploitation planning of the future development of the quarry has been carried out by means of analytical and numerical procedures. Parametric analyses of the main structural and mechanical features of the rock mass have been set up taking into account different values of the joint spacing and dip, tensile strength and deformability of the rock matrix in different geometrical configuration of the rock face.

The results of the geomechanical studies together with some insights for the interpretation of toppling phenomena allowed the optimisation of the quarry layout in a safe stability condition

TOPPLING FAILURE

Toppling failures of slopes have been observed in a wide range of rock masses and can occur in both natural and excavated slopes. The rock masses susceptible to this mode of failure are subdivided into a series of slabs or columns formed by a set of fractures that strike approximately parallel to the slope face and dip steeply into the face . In a toppling failure the rock column or slab rotates about an essentially fixed point at/or near the base of the slope and slippage between layers occurs at the same time. As described by Hoek & Bray (1981) there are four principal types of toppling failure, namely: block, flexural, block flexural and secondary toppling.

Research into toppling failures has historically concentrated on block toppling, which is a failure mode associated with sliding and toppling of rock columns along pre-existing basal failures plane formed by a discontinuity dipping into the excavation. The method developed by Goodman and Bray (1976) for block toppling is based on the Limit Equilibrium Method and considers each column in turn proceeding from the crest of the slope to the toe and determines one of the three stability conditions: stable, plane sliding or toppling.

Landslides in research, theory and practice, Thomas Telford, London, 2000.

Figure 1. Zone of the quarry where the instability phenomenon occurred

Figure 2. Actual zone of exploitation

Flexural toppling is a mode of failure involving the bending of interacting rock columns formed by a single set of steeply dipping discontinuities (such as regular bedding planes, foliation). The columns can be visualised as an array of interacting cantilever beams fixed at a certain depth, and free to bend into the excavation. In such cases, the rock column bend forward under their own weight and transfer load to the underlying columns, thus giving rise to tensile and compressive bending stresses. Failure is initiated when the tensile (bending) stress in the toe column exceeds the tensile strength of the rock. Aydan & Kawamoto (1992) developed a limit equilibrium method for flexural toppling, taking into account the rock tensile strength. On the basis of the experimental results of a series of centrifuge tests on small scale manufactured models, Adhikary et al. (1997) performed the calibration of the theoretical model proposed by Aydan & Kawamoto (1992).

The numerical methods that are currently used to model joint stiffness, rock block deformability, cracking propagation and progressive slope failure, (which can play an important role in primary and secondary toppling) are: the finite element methods (FEM), the boundary element method (BEM), the finite difference method (FDM), the distinct element method (DEM). The FEM can be applied to study the stability of a system that can be actually considered as a continuum, with a very few joints isolating blocks subjected to very small displacements. The FDM was utilised by Orr et al. (1991) to model rock mass with an high degree of fracturing and weathering. The BEM can result an useful tool to study crack propagation problems. The DEM can be applied to study problems of block toppling, with rigid blocks, or flexural toppling, by taking into consideration the movements between the interfaces and the deformability of the rock matrix.

ROCK MASS CHARACTERISATION

The quarry area is part of the Domodossola (Verbania, Italy) quarrying basin mainly composed by a wide range of gneiss which differ due to original petrography and structural variation (Del Greco et al., 1991). These gneiss rocks are formed mainly of porphyroblastic gneiss amphibolitic gneiss and micaschist. This area is part of the Pennidic domain and the geologic structure of the area id dominated by the geometry of the Antigorio nappe. The direction of thrusting in this area is toward the North and the north-west i.e. the major principal stress varies from N-S to NW-SE during the deformation. This controlled the orientation of the major discontinuity, the schistosity, giving a general NE-SW strike to this structure. This schistosity is almost vertical in the root zone of the nappe, which occurs in the region of Crevoladossola valley where the quarry studied in the paper is located. In situ measurements and laboratory tests have been carried out for the rock mass characterisation in the investigated quarry. Table 1 reports rock matrix physical and mechanical properties while Table 2 reports rock discontinuity structural and mechanical properties. Figure 3 shows equal-area projections of the discontinuities surveyed in the examined site. The rock mass is interested by a principal set of discontinuity, P, represented by the foliation, with an average spacing of 2.5m in the actual zone of exploitation, and of 0.5m in the weathered zone, where the instability phenomenon occurred.

Table 1. Physical and mechanical characteristics of rock materials (c: cohesion; ϕ: friction angle; E: Young modulus; ν: Poisson ratio; σ_t: tensile strength; γ: unit weight)

c (MPa)	ϕ(°)	E(GPa)	ν	σ_t(MPa)	γ (kN/m³)
30	30	40	0.25	10	26.5

Table 2. Structural and mechanical characteristics of rock discontinuities

Joint set	Dip direction (°)	Dip (°)	Spacing (m)	Persistence (%)	Base friction angle (°)
P	150	70	0.5-2.5	100	30
J_1	70	72	7	100	-
J_3	317	15	4	10	30
J_2	240	40	20	30	-

Three other sets of discontinuitics arc present, showing an higher spacing and/or a lower persistence: the J_1 family is represented by a set of small faults; the J_3 family is a sub horizontal set with a great spatial variability in the excavated face of the rock slope; the J_2 family is characterised by a wide spacing (about 20 m) and a low persistence. These sets of joints can isolate large rock volumes as the one involved in the occurred landslide, consisting in flexural toppling. The rock wall is about 40 m long and because of different distributions of joints can be subjected to different types of toppling: flexural toppling can occur where the system J_3 is widely spaced and the spacing of foliation P reaches the minimum value of 0.5m; block flexural toppling can occur as a principal mode of failure in the central zone of the rock slope where the exploitation activity is in progress.

ANALYSIS OF THE STABILITY CONDITIONS OF THE SLOPE

Both analytical and numerical approaches has been applied to analyse the stability of present and future configuration of the slope. Flexural toppling has been analysed by applying the limit equilibrium method proposed by Aydan and Kawamoto (1992). The assumptions of the model are:

- each column or layer can be treated as a cantilever beam subjected to gravity forces and side forces;
- a state of limiting equilibrium exists simultaneously across all the columns along the future fracture; the maximum tensile stress acting in each column is equal to the tensile strength of the material σ_t.

A column subjected to such loads and bending moments in plane deformation yields the following distribution of axial stress σ_x at distance y from the centroid of the section:

$$\sigma_x = -\frac{N}{A} + \frac{M}{I} y \tag{1}$$

where N is the axial force in the column, M is the bending moment, A is the cross-sectional area and I the second moment of inertia.
By rearranging equation 1 and substituting for the forces indicated in figure 4, Aydan and Kawamoto (1992) derived the following expression:

$$P_{j-1} = \frac{\max(P_j,0)\left(\chi h_j - \frac{1}{2}\mu b_j\right) + \frac{1}{2}T_j h_j - \frac{2I_j}{b_j}\left(\sigma_{tj} + \frac{N_j}{b_j}\right)}{\left(\chi h_{j-1} + \frac{\mu b_j}{2}\right)} \qquad (2)$$

where $I_j=b^3/12$, $W_j=\gamma b_j h_j$, $N_j=W_j \sin\alpha$, $T_j=W_j \cos\alpha$ $h_j=0.5(h_j+h_{j-1})$, $\mu=\tan\phi_j$, ϕ_j is the joint friction angle α is the column inclination angle, b_j is the column thickness and γ is the unit weight of the material.
Equation (2) can be used by commencing at the upper most column that may be unstable under its own weight and progressing down to the toe of the slope. The value of P_{j-1} must be evaluated at each step and then substituted as P_i for the next step, until the resultant side force at the toe of the slope P_o is found. P_o is an indicator of the stability of the slope: if $P_o>0$ the slope is unstable; if $P_o=0$ the slope in at limiting equilibrium; if $P_o<0$ the slope is stable.

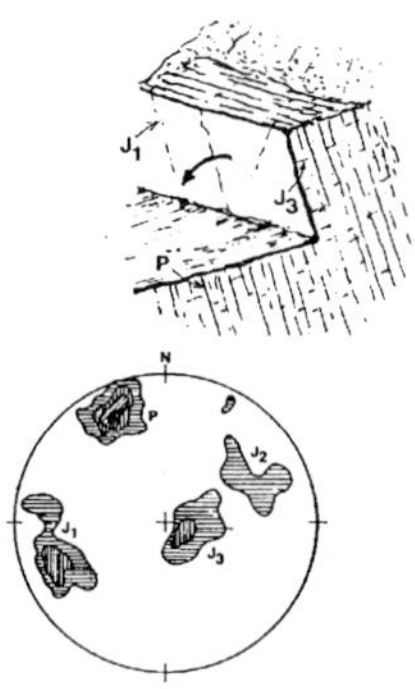

Figure 3 Equal area projection of the discontinuities poles of the rock mass

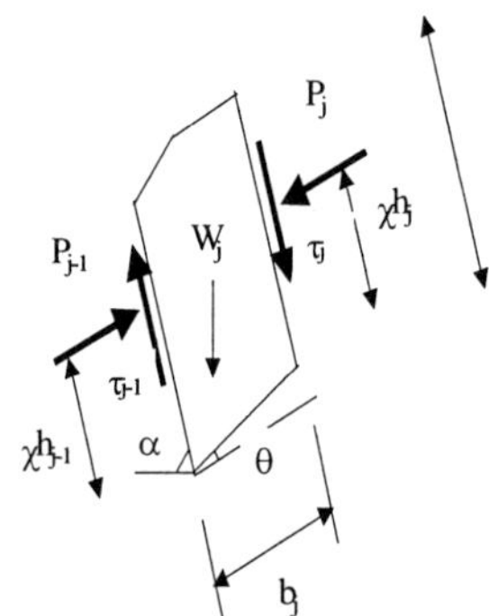

Figure 4. Forces acting on a single column (by Adhikary et al., 1997)

The analysis carried out with this procedure (by assuming the following values of the parameters: b=0.5÷2.5m; θ=0°; χ=1; ϕ=30°; α=70°; γ=26.5 kN/m³) showed as an overhanging slope geometry (figure 5) would not guarantee stability of the face whilst the new proposed configuration (figure 6) would be stable with a large safety margin. The factor of safety has been computed in this second case by varying the rock tensile strength down to the limit equilibrium condition. The ratio between the available tensile strength (10 MPa) and the tensile strength required to reach limit equilibrium condition (0.5 MPa) gives an indication on the actual factor of safety (20) when 2.5m spacing is considered. A parametric analysis has been performed taking the foliation spacing into account. The average joint spacing, in fact, showed wide variation within the rock domain showing a minimum value of 0.5m in the quoted weathered area of the quarry. The results of this study agree with the results of the DEM model set up by Deangeli et al. (1996). Anyhow, stability conditions would have a large margin of safety (2.5) being the limit equilibrium condition reached at 4MPa of tensile strength.

The block flexural toppling has been analysed by numerical modelling by using UDEC, which is a code based on Distinct Element Method. The Distinct Element Method simulates the behaviour of jointed media subjected to quasi static or dynamic conditions. The medium is simulated as an assemblage of blocks which interact through corner and edge contacts. Discontinuities are regarded as boundary interactions between these blocks. The method utilises an explicit timestepping (dynamic) algorithm which allows large displacements and rotations and general non-linear constitutive behaviour for both matrix and discontinuities. Blocks are fully deformable with an ideal elasto-plastic behaviour. The strength criterion adopted is the Mohr-Coulomb.

Two series of analysis have been carried out: a parametric analysis has been performed on a simplified scheme of the slabs (figure 7) by varying the slab width from 2.5m down to 0.5m; the second model reproduce the real geometry of the quarry by taking the quarrying sequence into consideration. The maximum excavation depth is reached by two subsequential descending slices of 3+3m each (down to 12m). The results of the first model showed as the maximum induced tensile stress in the slab is 0.269MPa and the water pressure increase it up to 0.704MPa. The same trend has been found in all analysed configurations where the maximum induced stress is induced in the thinnest slab in presence of water. Anyhow, even in the worst condition the maximum induced stress does not determine rock plasticisation. These results have been confirmed on the overall quarry model where no plasticisation areas have been observed during the various excavation phases. Since joint opening has been observed, possibly due by water pressures, rock bolting is suggested to avoid increasing water presence. Since blocks of small volume ($<1m^3$) can move from the slope crest during the excavation they have to be detached at the end of each phase.

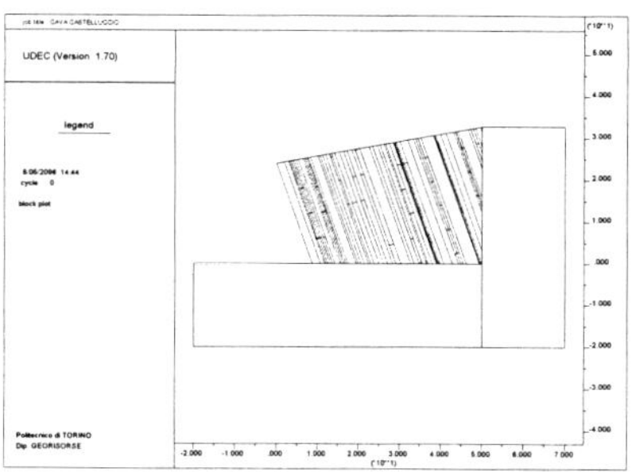

Figure 5. Cross section of the failed zone utilised for both the analytical and DEM models.

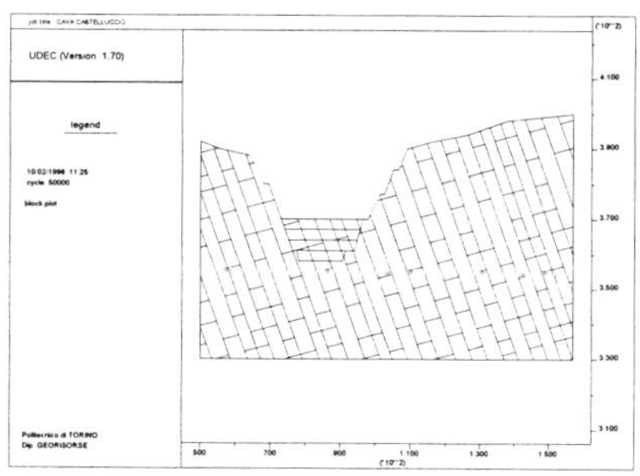

Figure 6. Cross section of the actual exploitation zone with the proposed configuration.

CONCLUDING REMARKS

Different methods of analysis has been applied to study the stability of a rock wall prone to toppling failure. The use of different approaches was necessary for the high spatial variability of distribution of joints in the rock mass. The studies carried out lead to an improvement of the safety conditions of the overall quarry. Analytical and numerical results are in good agreement showing as the new proposed configuration can guarantee a good margin of safety. However the DEM analyses have also allowed to outline slab opening movements occurring during the excavation. Since water can induce major weathering processes as the ones observed in the collapsed zone, an appropriate bolting mesh is suggested to avoid any possible evolution of the phenomena. This consideration showed as a more sophisticated numerical approach (in comparison with the analytical one) can give further useful suggestions for an optimal quarry design.

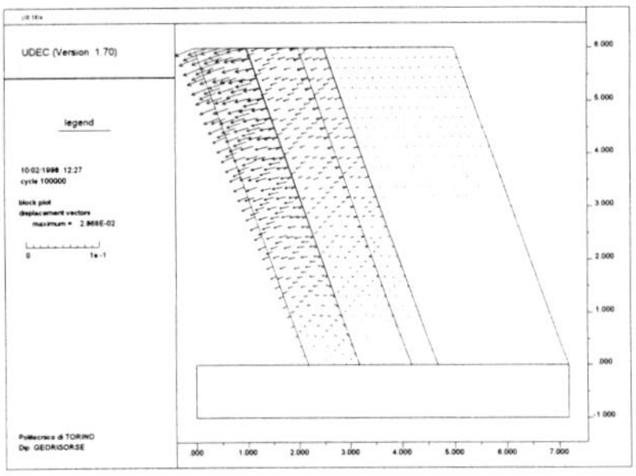

Figure 7. Simplified scheme of slabs, corresponding to a phase of exploitation.

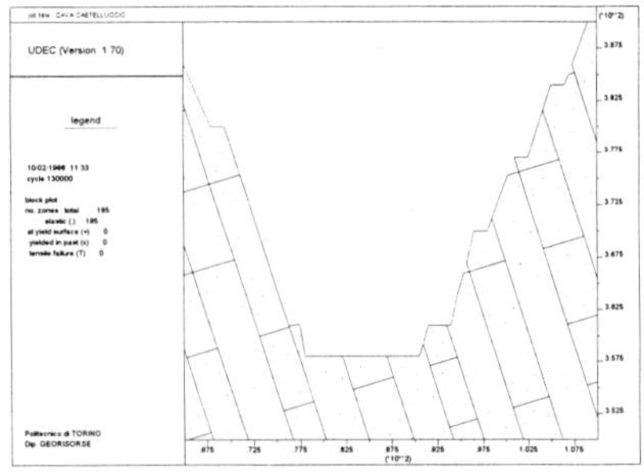

Figure 8. Configuration of the slope at the end of the planned exploitation activity.

REFERENCES

Adhykary D.P., Dyskin A.V., Jewell R.J. & Stewart D.P., 1997 *A study of the mechanism of flexural toppling Failure of rock slopes*. Rock Mech. Rock Engng.30 (2), pp. 75-93.

Aydan O. & Kawamoto T., 1992. *Stability of slopes and underground openings against flexural toppling*. Rock Mech. Rock Engng.25 (3), pp. 143-165

Deangeli C., Del Greco O., Ferrero A. M., Pancotti G. & Giani G.P., 1996. *Rock mechanics studies to improve intact rock block exploitation and slope stability conditions in a quarry basin*. EUROCK 96 Torino, Balkema, pp. 561-568.

Del Greco O., Fornaro M. and C. Oggeri, 1992. *Structural conditions and quarry development for dimension stone in the Western Alps, Italy*. ISRM Sypm. on Rock Slopes. New Delhi, pp 429-437.

Goodman R.E. & Bray J.W., 1976. *Toppling of rock slopes*. ASCE Spec. Conf. on Rock Engng for Foundations and Slopes, Boulder Colorado, vol2, pp. 201-234.

Hoek & Bray J.W., 1981, *Rock slope engineering* 3rd ed. Institution of Mining and Metallurgy, London, pp. 402.

Turner K. & Schuster R.L., 1996. *Landslides, Investigation and mitigation*. Special Report 247 Transportation Research Board, National Research Council, National Academy Press, Washington, D.C., pp. 673.

A Historical Landslide in the Province of Belluno (Eastern Dolomites, Italy): Study of the Past Events and Provision for the Future

E. DE BIASIO, Via A. Manzoni, 13 – 32036 Sedico (Bl), Italy, M. C. TURRINI and W. BUSIN Dipartimento di Scienze Geologiche e Paleontologiche, Corso E. I D'Este, 32 – 44100 Ferrara, Italy

INTRODUCTION

An understanding of the geologic model of an area in which a large landslide has occurred is of paramount importance for the reconstruction of the predisposing factors of mass movement. If the landslide is historical, in order to understand the triggering factors, instead, bibliographical research in the many municipal and canonical archives, available throughout Italy, are of great help.

This study, which is an example of the above mentioned principles, deals with a large landslide that dammed the valley of the Cordevole river in the province of Belluno in 1771, and gave rise to Lake Alleghe, a popular tourist spot in the Eastern Dolomites. The landslide was composed of two translational slides (Varnes, 1978): the first, greater in size, (accumulation of about 20 million cubic metres), took place in the month of January and dammed the valley, creating extensive damage and causing many deaths; the second one, with an estimated accumulation of about 4 million cubic metres, occurred four months later and fell into the lake, which had formed following the first incident, causing further damage and loss of life. At the present time great concern has been aroused by a rock wedge, still undisplaced, whose volume has been estimated to be about 5 million cubic metres, and which represents the last remnant of the ancient slope.

GEOLOGICAL MODEL: PREDISPOSING CAUSES

The formations involved in the two movements are the Morbiac Limestone and the Contrin Formation, both of which are Upper Anisian. The Morbiac Limestone consists of layered marly limestone, characterised by the presence of numerous marly interlayers which are some centimetres thick; these interlayers easily alter turning soft soil with low mechanical parameters. The Contrin Formation, at its bottom is made up of layered dolomites, for about 30 metres of thickness, where there are rare marly interlayers; the upper part of the formation, on the other hand, is constituted by massive dolomite for a thickness of around 80 - 90 metres. The slope is affected by an anticline fold which has caused the layers to dip in the same direction as the slope, with a minor inclination than the ancient one in the upper middle part (about 30°), up to 50°-60° in the lower part. This orientation of the layers, together with the low mechanical parameters of the aforesaid marly interlayers, has been the most important predisposing causes of landslides.

Landslides in research, theory and practice, Thomas Telford, London, 2000.

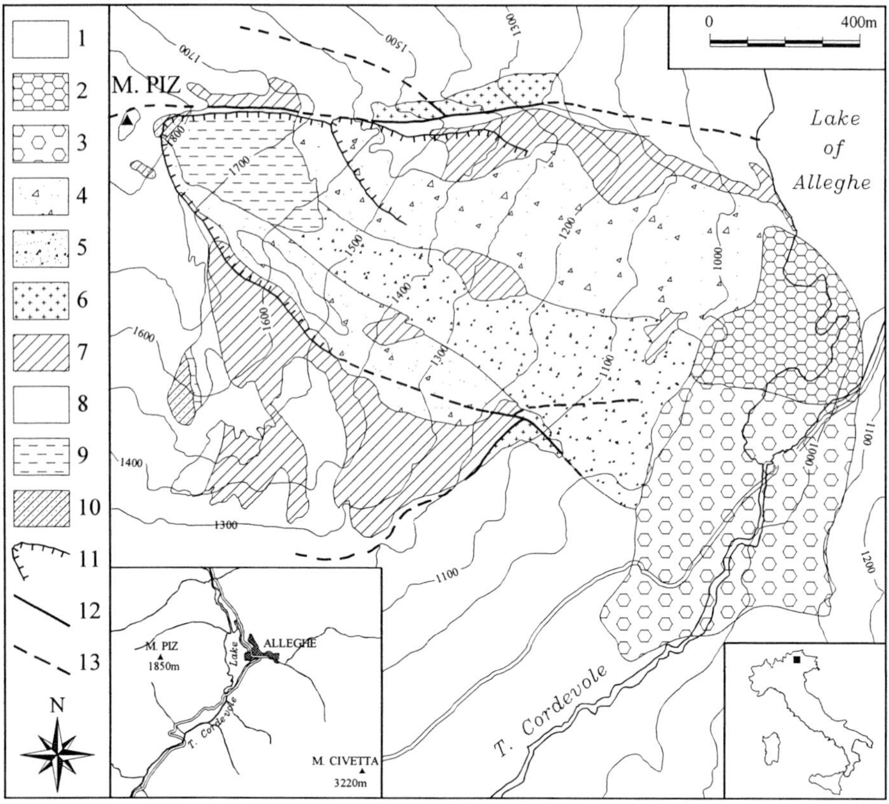

Figure 1. Geologic – geomorphologic map of the study area. 1) Eluvial/colluvial deposit; 2) Accumulation of the second landslide; 3) Accumulation of the first landslide; 4) Dolomitic talus; 5) Calcareous talus; 6) Igneous Formations; 7) Contrin Formation, upper part; 8) Contrin Formation, lower part; 9) Morbiac Limestone; 10) Werfen Formation; 11) Main scarp; 12) Fault; 13) Inferred fault

HISTORICAL RECONSTRUCTION OF EVENTS: PREPARATORY AND TRIGGERING CAUSES

The first mention of the Alleghe landslide which has come to our notice dates back to 1771 and may be found in the rectory of Alleghe in the Book of Harvests which indicates the dates on which the movements occurred and those on which religious services were held in memory of the deceased.

The first scientific description, instead, dates back to a few years after the formation of the lake (Gualandris, 1780), and the Author gives a detailed account of the slope before the landslide had occurred; from this we can deduce that this slope had already been strongly damaged by minor landslides. The presence of a small spring at the foot of the landslide mass induces the Author to state that the preparatory cause must have been erosion by water of an interposed layer between the displaced mass and the stable slope. This would drastically have reduced the strength along the slip surface, making the mass move under its own weight.

Even later reports give analogous descriptions of the event, emphasising the influence of the autumnal rainfall.

No detailed descriptions of the second landslide are available. However, since it took place in spring, it may be deduced that once again the abundant rainfall and the melting of the snow must have been the causes of the movement. The precipitation distribution of this area does, in fact, present two peaks, one in autumn and the other in spring.

FIELD AND LABORATORY STUDIES

Field studies consisted of a detailed geologic - geomorphologic survey to the scale of 1:5,000 (Fig. 1); a census of the springs; a grainsize distribution analysis using sieves and metric tape of 10 samples of accumulation deposits; a geomechanical survey of the undisplaced rock mass and the waviness of the slip surfaces. The field survey was then completed with photo-interpretation of aerial photographs to the scale of 1:25,000 which allowed us to connect the scarps of the two landslides with important discontinuities (Fig. 2).

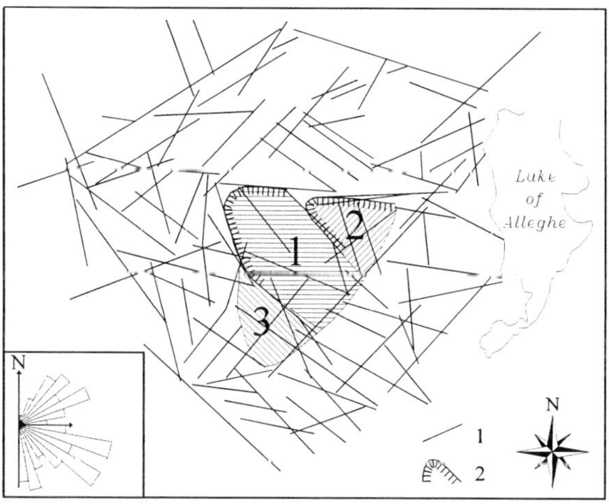

Figure 2. Lineation map: 1) Lineation; 2) Landslide scarp. The hatched areas are the slip surfaces of the first (1) and of the second (2) landslides and the undisplaced rock wedge (3). The rose diagram shows the frequency of lineaments. The longest radius represents a frequency of 10%

Identification of the springs demonstrated that the landslide slope did not contain any springs, in that the orientation of the layers and the debris cover were such that no groundwater could reach the surface. During long spells of rainfall and the thaws, however, a groundwater is likely to form within the Contrin Formation, having Morbiac Limestone as its base. Furthermore, the numerous surveys carried out in different periods have shown that runoff water does not exist in any season owing to the high secondary permeability of the dolomite.

Analysing the grainsize distribution of the accumulations, it was possible to ascertain that the fraction lower than 75μm is always below 7%, whereas the boulders of greater volume (up to 1,000 cubic metres) are concentrated in the first accumulation; the second accumulation, instead, is characterised by the presence of boulders whose maximum volume is of 20 cubic metres. The mechanical parameters of the remoulded soil obtained from direct shear tests in accordance with ASTM standards are: c = 0kPa, ϕ = 29.5° - 31.8°.

Three samples taken from the marly interlayers of the Morbiac Limestone were then analysed. The percentage of the fraction lower than 75μm exceeded 95%, whereas that lower than 2μm reached 30%. The $CaCO_3$ content was almost 50%. According to the Casagrande plasticity chart, the soils were classified as inorganic silt of medium plasticity. The residual values of the mechanic parameters, obtained on remoulded samples, are c = 0kPa and ϕ varying from a minimum of 23.2° to a maximum of 27.5°.

Finally, as indicated above, the waviness of the slip surfaces were measured using the compass and disc clinometer method (ISRM, 1978). This analysis gave a modal value of the orientation of the first surface of 134°/32 ° (dip direction/dip) with limits of 115°-178°/15°-50°; for the second landslide the values are, instead, 132°/42° (120°-155°/27°-56°).

The geomechanical survey of the rock mass was performed on 12 stations (ISRM, 1978). The most interesting results were found to be the presence of a set of discontinuities, revealed in 4 stations, sub-parallel to the layering, which may have facilitated the translational slide in the massive dolomite. Another set exists, evident in 3 stations, which dips inside the slope, practically perpendicular to the layers. These joints may have favoured the movement, both by breaking up the Contrin Formation into large boulders, and by favouring the penetration of rainfall as far as the marly interlayers. Finally, a third set of joints exists, well represented in 6 stations, which are practically vertical and oriented similarly to the large discontinuities on which the main scarps of the landslides have settled.

STABILITY ANALYSIS

The stability analysis was carried out using both the limiting equilibrium method (Hoek and Bray, 1981) and finite elements software (Z-Soil, 1993). Since the results obtained are comparable only those obtained with finite elements software will be reported. The analysis made for the two accumulations was carried out bearing in mind that the residual friction angle measured in the laboratory refers to the finest fractions of the deposits, thus the friction angle of the whole material is certainly greater. However, safety conditions exist for all the situations analysed, which is clearly evident also in the field. The only hazard, confirmed during past floods, may derive from the slope undercutting due to the Cordevole river.

The first landslide

The study of the first landslide was carried out by analysing the safety factor for a section drawn along the direction of the movement, almost parallelly to the dip direction of the layers. The analysed surface coincided by 85% with the Morbiac Limestone, whereas the remaining 15% with a joint of the first set mentioned above, present in the Contrin Formation. The back analysis was made considering the movement as a simple translational slide. In reality it must have been a wedge slide (Fontaive, 1981), in fact, while 80% of the slide surface coincides, as previously mentioned, with an interlayer of the Morbiac Limestone, the southern part, near the still undisplaced rock wedge, is set on another plane in the Contrin Formation; the latter plane forms an angle of about 150° with the former. Given the width of this angle, the wedge factor has little effect on the safety factor (Hoeck and Bray, 1981), and so it was not taken into account. Neither were the frictions between the body in movement (area 1 of fig. 2), the still undisplaced body of the second landslide (area 2 of fig. 2), and the most southerly rock wedge (area 3 of fig. 2), in that the stretches of the scarp of the first landslide, where it was in contact with those bodies, coincide with the faults oriented about N 130°-140° (Fig. 2). The field studies show a horizontal movement for this fault, and, therefore, the contact walls between the different bodies must have been very smooth, as can be seen from the northern wall of the still undisplaced wedge. Since historical documents report movements of the mass prior to the final one, the friction angles used are the residual ones which emerged from the laboratory tests for the marly- interlayers of the Morbiac Limestone (23° and 27°), whereas for the part of slip surface imposed on the joints sub-parallel to the layers which cut the Contrin Formation, the values used were those obtained from the geomechanic survey (ϕ_p = 53° and ϕ_r = 40°). In order to obtain the cohesion value along the joint, i.e. to assess the extension of the rock bridges, a parametric analysis was carried out in three different saturation conditions: the first without groundwater, the second with groundwater height of 10 metres above the level of the Morbiac Limestone, and the third with a groundwater 20 metres height. This height was considered to be the maximum possible given the conditions of fracturing and release of the rock mass. The analyses allowed us to evaluate the extension of the rock bridges at the failure, which varied from a minimum of 6% in dry conditions and with the highest value of interlayers friction, to a maximum of around 14% in highest groundwater conditions and with the lowest value of interlayers friction.

Second landslide

A back analysis was made for the second landslide on a section traced along the mean dip direction of the layers, considering the movement as translational. Along the section analysed, the slip surface was located in a higher stratigraphic level than the first landslide; it coincides for 19% with the stratified bottom of the Contrin Formation, whereas the remaining 81% cuts through the massive part of the same formation. Hypothesising that the sliding of the body partly occurred along a marly-clayey interlayer of the layered dolomite, the rock bridges of the joint, at the failure, must have extended from a minimum of 0.5% to a maximum of 2%: certainly less, therefore, than that of the first landslide. This finding enabled us to deduce that in the time that elapsed between the first and second landslide there was an acceleration in the progressive failure of the rock bridges. If, on the other hand, the movement of the stratified dolomite is hypothesised to have taken place along a bed plane and not along an interlayer, then in order to have the slide, the friction angle along the two parts of the slip surface must have overcome the peak value. The calculations made give a value around 49°. No modelling was done to take an earthquake into account since no mention is made in any of the descriptions of a seismic event.

Present slope

The study on stability of the present slope was carried out on a section traced along the mean dip direction of the layers and considering that the slip surface has a 13% coincidence with the marly interlayers of the Morbiac Limestone, 50% with the layered dolomite of the Contrin Formation, and a remaining 37% with a joint sub-parallel to the layer in the massive dolomite of the same formation. In this case, too, the same three conditions of saturation were considered. The results obtained conform with those for the preceding landslides, i.e., sliding occurs only when the movement is considered to take place along a marly interlayer of the layered dolomite. In this case, the rock bridges of the joint, at the failure, must extend from a minimum of 0.5% to a maximum of 5%. If, instead, the movement in the layered dolomite is considered to occur along a layer, no movement would occur even when the friction values reaches residual values: the safety factor obtained is, in fact, higher than the unit (Fs = 1.06). In this case, an earthquake acceleration, even if very small (0.01g), would have to be introduced into the model in order to have a slide, whereas in conditions of peak friction a horizontal acceleration of 0.09g is required. The minor values of horizontal acceleration refer to the groundwater at 20 metres, whereas the higher ones to conditions of the slope devoid of water.

Finally, since the field studies revealed a persistent joint, parallel to the layer, which practically splits the massive dolomite into two for most of its outcropping, an analysis on the stability of the rock wedge was carried out also along this discontinuity. The safety factor was again found to be above the unit. In this model, too, a slide occurs only in the presence of a horizontal acceleration (0.01g for residual friction values, 0.13g for peak values).

CONCLUSIONS

The whole study aimed at the acquisition of knowledge for the evaluation of rock wedge stability still undisplaced and, therefore, the risk factors for the inhabited areas and infrastructures located in the valley bottom. Considering the lack of financing, the approach used was of a classical nature and the results obtained allowed us to identify the conditions that must be verified in the case of a translational slide along a surface which is similar to those of the previous two landslides. On the basis of the Authors' experience and judgement these conditions cannot be far off, since the field surveys revealed the presence, at the summit of the wedge, of active tension cracks of 50 - 60 cms in width, perpendicular to the direction of the likely movement, testifying to the progressive failure of this part of the slope, similarly to what took place for the other landslides. More precise data would, undoubtedly, be achieved if the latter part of the eastern slope of the M. Piz still undisplaced were to be monitored with the appropriate equipment.

ACKNOWLEDGMENTS

Authors are grateful for the invaluable suggestions of Prof. E. Semenza and Prof. R. Genevois.

REFERENCES

ASTM (1994) – Annual Book of ASTM Standards, Philadelphia.

Fontanive F. (1981) – *Analisi della frana del Monte Piz (Dolomiti Bellunesi); primi risultati.* Quaderni dell'IMGA, 2-22.

Gualandris A. (1780) – Lettere Odeoporiche. Venezia.

ISRM, Commission on the Standardisation of laboratory and Field Tests (1978) – *Suggested method for the quantitative description of discontinuities in rock masses.* Int. Journ. Rock Mech. Min. Sc. Geomech. Abstr., 15(6), 319-368.

Hoek E. & Bray J.W. (1981) – *Rock Slope Engineering.* Revised 3rd Edition, Institution of Maining and Metallurgy, London.

Varnes D.J. (1978) – *Slope Movements Types and Processes.* In: Schuster R.L. & Krizek R.J. (eds), Landslides: Analysis and Control. Transp. Res. Board, Special Report 176, 11-25.

Z-Soil (1993) – User Manual. Zace Service ltd. Lausanne.

Influence on Landsliding and Slope Development of the Particular Environment of the Dead Sea Region: A Case Study for the Wadi Mujib Canyon (Jordan)

C. DE JAEGER, University of Ghent, Belgium

INTRODUCTION

One of the earth's most impressive tectonic belts is the East African Rift which has an overall length of about 6000 km. The northern tail end of this major rift is a transform fault running through the Near East up to Turkey. Starting at the Red Sea in Aqaba this sinistral strike-slip fault (currently the most widely accepted view) passes the Wadi Arava and the Dead Sea graben, to continue along the Jordan river towards the northern border of Jordan (Ginat *et al.*, 1998; Freund *et al.*, 1970; Garfunkel, 1970; Ben-Avraham & Ten Brink, 1989). The Dead Sea graben is the deepest subaerial point on earth and contains a 100 km long pull-apart basin separated into two bodies by the Lisan peninsula. The graben is bordered by normal faults, divided into several smaller units by transverse faults and characterised by locally occurring salt diapirs and sinks, for example the Arnon (= biblical Wadi Mujib) sink (Ben-Avraham & Ten Brink, 1989 ; Rabinowitz *et al.*, 1996). The Dead Sea graben is still subsiding; the water body is shrinking due to high evaporation and human consumption of the Jordan River water (level presently at –408m b.s.l.) (Neev & Emery, 1995, p.40). The eastern 'shoulder' of the graben is the (Trans)Jordanian plateau which is deeply incised by wadis draining into the Dead Sea. The most spectacular incision is that of the Wadi Mujib Canyon (Figure 1). Apart from the Grand Canyon in the U.S.A. and the Fish River Canyon in Namibia, it forms one of the biggest canyon systems in the world.

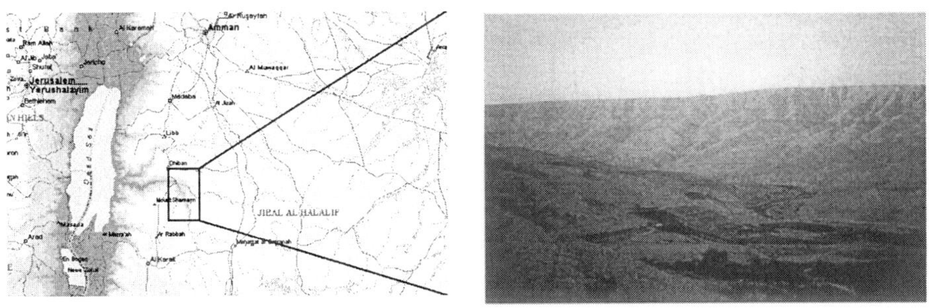

Figure 1. Localisation of the study area in the Wadi Mujib Canyon

However, the slopes in the Mujib Canyon are subject to a fast rate of degradation and landsliding due to several natural but even antropogenic factors.

Landslides in research, theory and practice, Thomas Telford, London, 2000.

BASIN EVOLUTION DURING THE QUATERNARY

The geology of the Dead Sea Region has been the subject of much debate and many theories have been advanced over the past decades in order to explain the structural development in the area since the Miocene. However, except for some very local case studies, research on the effects of this quite young tectonic history on the geomorphology of the surrounding areas, especially of the eastern or Jordanian side, is still limited.

Effects of the Dead Sea graben on the early drainage of the Wadi Mujib Basin

Since the Late Oligocene / Early Miocene the Dead Sea graben should have started to subside, so that the uplifted Jordanian plateau became drained to the west : a proto-Wadi Mujib basin developed but at that moment most probably the drainage lines were merely incised in the marine sediments of the plateau. Due to Early Miocene-Pliocene subsiding movements basaltic flows, pouring from several fissure eruptions, covered the plateau and partly flowed into the proto-Wadi Mujib (Bender, 1975 ; Khalil, 1992 ; Neev & Emery, 1995). Outcrops of these basalts can be observed on the plateau edges in the uppermost parts of the canyon walls. From this observations it may be concluded that the incision of the canyon proper only started after the eruptions; if not basalt should have flown into the canyon and this is not the case. A canyon with a mean depth of 600 m and with the young age of about 2.5-2 million years testifies to the important role of a lowering base level. Even more, since the Plio-Pleistocene the base level of the east-west draining proto-wadis shifted from the Mediterranean Sea to the Dead Sea and due to the sinistral fault movement the Jordanian plateau has moved from the Miocene period on about 100-110 km to the north within two important phases. During the second movement the heads of the Late Miocene wadis have been displaced approximately 43 km and were disconnected from their lower reaches (Freund *et al.*, 1968).

Effects of the lithology on the development of the Wadi Mujib Canyon

The Wadi Mujib has cut through the nearly horizontally bedded marine sediments of limestone (hard and soft layers), marl, chalk and chert, with alternations of gypsum, tripoli, clay and shale (Bender, 1975; Khalil, 1992). This means that the Wadi Mujib cuts layers with different erodibility. As a result a series of knick points is developed along the river profile due to regressive erosion of the canyon heads. Close to the heads steep slopes are created and are prone to mass movements.

SLOPE PROCESSES RELATED TO THE EXTREME ENVIRONMENT

The geomorphology and her processes in the study area can be seen as a true reflection of causing and triggering factors : topography, climate, structure, lithology, tectonics, hydrology and Man. The very complex slopes of the canyon walls are actually prone to several slope processes moving rock masses downslope (Selby, 1993): falls (rock-fall, slab failure) and slides (rotational slumps, translational slides). Some are the result of previous slope processes, others are directly caused by one or a combination of causal and triggering factors: such mass movements sometimes happen suddenly and may be catastrophic!

Mass movements indirectly caused by previous slope processes

The complexity of the slopes is not only the result of different types of mass movements but rill erosion on the upper slopes, turning into gully erosion on the middle and lower parts of the slopes is very abundant and affects both the original slopes and earlier downslope moved rock masses. This fluvial slope degradation is facilitated by the solution of the $CaCO_3$-rich sediments producing subterranean holes and pipes which can suddenly collapse favouring rill and gully erosion, or the headcutting erosion of already existing gullies. The rock masses situated between two or more strongly developed gullies confluencing downslope then become unstable by systematic undercutting and will be driven downslope due to the gravity force.

Directly caused mass movements

Topography. The basic factor responsible for the mass movements in the canyon is the topography: slope degree usually is a very critical causal factor for slope instability. The canyon slopes aim at a physical equilibrium. Selby (1993) and Kirby et al. (1993) have given a value of 8° below which landslides normally do not occur, and a degree (30-35°) above which it can no longer be considered as a landslide, but rather as a rock-fall. Statistical analysis and GIS applications on the basis of remote sensing (Russian stereoscopic and high resolution TK-350 satellite imagery) of the study area showed that 21/44 detected past landslide events occurred on a slope degree of 10-15° and 10/44 on 15-20°. The other 13 detected landslides occurred on 5-10° and 20-25°. On the steepest parts of the upper canyon walls rock-fall and slab failures are abundant and cause the retreat of the escarpment.

Climate and microclimate. The study area is situated on the transitional zone of a dry mediterranean to a semi-arid and arid climate. Main characteristics are the high variability and the high intensity of the scarce (winter) rainfall. A short cloudburst can end in a flood (when the infiltration capacity < the intensity of the rainfall and so runoff starts) with strong fluvial erosion, transport of weathered material, fallen rocks or sediment and finally with lateral erosion undercutting unstable slopes in concave meander curves by which suddenly giant landslides occur changing the landscape in a few moments. During summer or the dry season the day temperatures are high (mean temperature 30-33° in July or August) on the plateau and very high in the canyon (>43°C !). The night temperatures are much cooler, so that there is a constant shift between high and low temperatures during summer favouring the physical weathering of the rock. Locally, microclimatic effects on the slope processes are obvious and are the result of the slope

Structure. The lithological layers are nearly horizontally bedded or only slightly disturbed by tectonic influence. Only at the mouth of the basin near the Dead Sea graben the sediment layers dip down to 90°. This horizontal structure favours the slope stability but the fact that soft and hard layers alternate means that the hard layers are undercut by the easy weathering and erosion of the soft layers below ensuing slides and rock-falls. The joints in the limestone rocks (Awad & Khatib, 1991) often are expoited by the runoff water and are easily weathered zones forming slide planes depending on their orientation in respect to the slope direction and gradient.

Lithology. Apart from the topography and the structure the lithological composition is crucial in making the slopes more easily affected by degradation and landsliding, especially in the middle and upper reaches of the basin where marine limestone, marl, chalk and chert dominate. All these $CaCO_3$-rich sediments are sensitive to mass movements, except for some very hard limestone banks and chalks. This is especially the case for the very soft and permeable marls; in case of wetting the pore pressure becomes positive quite rapidly and translational slides are common. Shale and clay lenses occur locally and a gypsum layer of >2m thickness and of large extent over the basin (Khalil, 1992) forms a very important 'zone' along which the slopes are very affected by landsliding due to temporal swell and shrink.

Tectonics. Lineaments and faults occur in a dense 'network' due to the neotectonic activity of the Dead Sea fault system. Earthquakes occur regularly (Kovach *et al.*, 1990). Due to the vibrations and fault movements, landslides (translational slides, rotational slumps) are a logical consequence. Fault zones are areas of likely weakness where the lesser cohesion of the rock due to fractures would move more easily in an event of seismic activity or further readjustment. According to the statistical analysis of the detected past landslides 31/44 occurred in a distance <400m from a fault or lineament. However this analysis may be biased by the high density of faults in the area!

Surface and subsurface hydrology. In analogy to the distance from faults or lineaments, the distance to the nearest thalweg is of importance for landsliding too: 26/44 detected landslides occurred within a distance <400m to the nearest important thalweg due to higher pore pressures and undercutting of the slopes. Unless the seasonality of surface runoff the subsurface zone of the thalwegs stays permanently humid or the water table there is high. Further, there is an important aquifer (B2/A7 Upper Cretaceous limestone) with a saturation thickness of 250-300m. The groundwater flux depends not only on the topography but even on the presence of major faults (e.g. the Siwaqa fault (Awad & Al-Khatib, 1991)): the water-level contours change abruptly their direction so that a high pore pressure gradient may trigger landsliding along this zone again.

Man. Last but not least Man plays an additional role in triggering the slope degradation. Since the Neolithic Revolution, some 10,000 years ago, Man tried to get hold of nature and started deforestation (De Dapper *et al.*, 1997). Because the protecting vegetation disappeared the thin soil layer washed away, so that the canyon walls now appear to be bare. Today, there still live a lot of shepherds in the area feeding their cattle by overgrazing the slopes. Looking for the scarce vegetation the cattle travels contourwise up to very steep slope degrees and by trampling they create small vibrations and compaction of the surface leading to the formation of a dense network of terracettes. As the area is sparsely populated only a few roads are constructed but, unfortunately, with low perception of landslide risk : winding roads are constructed to narrow the canyon gap as shortly and fastly as possible. Some parts of roads are constructed between two rotational slump masses on the headscarp of the lowest rotational slump : compaction and vibration caused by traffic trigger the two rotational slump masses to reactivate ! The same can be said about the practice of irrigation of fields upon the headscarps of past mass movements augmenting the water infiltration and reactivation of the movement. In the near future a dam will be constructed in the canyon to gain fresh water for irrigation and to fill up the groundwater table (Loy, 1994). The storage lake is endangered to loose capacity by the fill up of landslide sediment, but the slope instability itself will be enhanced too by higher infiltration of water into the marly slopes (Figure 2) (De Jaeger et al., 1999; De Jaeger et al., 1999; De Jaeger, 1999)!

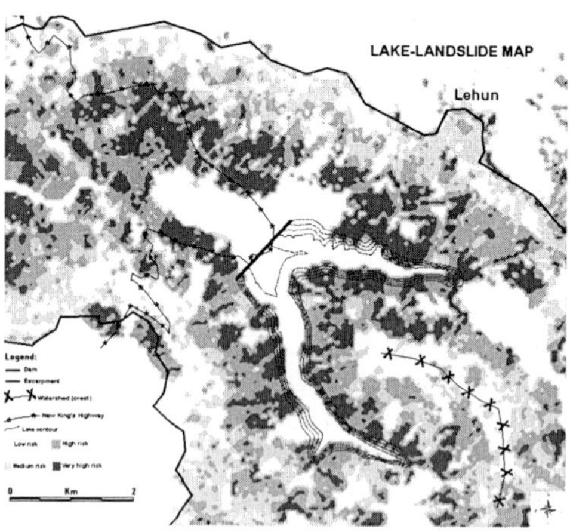

Figure 2. Landslide risk map (based on TK-350 satellite imagery) with dam simulation (storage lake contours corresponding to dam heights of : 10m, 20m, 30m, 40m, and 50m)

CONCLUSIONS

The geomorphology of the Dead Sea Region is susceptible to extreme environmental conditions. Wadi Mujib is the most impressive fluvial incision draining into the Dead Sea: a canyon with a mean depth of 600m mainly developed during the Quaternary due to the lowering base level of the subsiding Dead Sea graben. The slopes of the canyon are degrading, since the Neolithic deforestation, at a higher rate than the sediment evacuation by the Wadi Mujib. Bare rocky slopes are strewn with huge masses moved downslope as slides, slumps and falls. Rocky masses can become unstable by previous slope processes such as gully erosion or systematic weathering of a joint plane, but a landslide may also occur suddenly, triggered by an earthquake and accompanying fault movement, a flood... The influence of Man in the study area, however sparsely populated, is considerable by overgrazing, irrigation on unstable zones, construction of winding roads... Constuction of a planned dam through the Wadi Mujib will result in more slope instability due to water infiltration into the rocks (marls !) and the storage lake is at risk of filling by landslides.

REFERENCES

Awad, M. & Khatib, L. (1991) 'Geology and hydrogeology of B2/A7 aquifer system in the Mujib Basin, Jordan', Natuurwetenschappelijk Tijdschrift, Gent, Belgium, 73: 113-128.

Ben-Avraham, Z. & Ten Brink, U. (1989) 'Transverse faults and segmentation of basins within the Dead Sea Rift', Journal of African Earth Sciences, Pergamon Press plc., Great Britain, 8 (2/3/4): 603-616.

Bender, F. (1975) 'Geology of the Arabian Peninsula : Jordan', Geological Survey Professional Paper, US Govt Printing Office, Washington, U.S.A., 560 (1): 36pp.

De Dapper, M., De Jaeger, C., De Vliegher, B.-M. & Mertens, J. (1997) 'La recherche géo-archéologique / Geo-archeologisch onderzoek', in: Homès-Frédéricq, D., Lehun et la voie royale – Les fouilles belges en Jordanie / Lehun en de Koningsweg – De Belgische opgravingen in Jordanië, Belgisch Comité voor Opgravingen in Jordanië, Brussel, Drukkerij Cultura, Wetteren, Belgium: 129-135.

De Jaeger, Ch., De Dapper, M., De Vliegher, B.-M., Goossens, R., Risack, S., Closson, D., Cornet, Y. & Ozer, A. (1999) 'The use of very high resolution and stereoscopic satellite imagery (Russian TK-350) for landslide risk mapping around a future dam site in the Wadi Mujib canyon (Jordan)', in: Geoinformatics beyond 2000, Proceedings of the International Symposium, 9-11 March 1999, IIRS Dehra Dun, India (in print).

De Jaeger, Ch. (1999) 'Geomorphological hazard mapping in the Wadi Mujib canyon (Jordan) based on Russian stereoscopic TK-350 satellite images', in: Proceedings of the Advanced Study Course on Desertification in Europe: Mitigation strategies, Land use and Planning, 1-9 June June 1999, Alghero, Italy (in print).

De Jaeger, Ch., De Dapper, M., De Vliegher, B.-M. & Goossens, R. (1999) 'The use of Russian TK-350 satellite images for the geomorphological mapping of the semi-arid Wadi Mujib canyon (Jordan)', in: Proceedings of the Second International Symposium on Operationalization of Remote Sensing, ITC, Enschede, The Netherlands (in print).

Freund, R., Garfunkel, Z., Zak, I., Goldberg, M., Weissbrod, T. & Derin, B. (1970) 'The shear along the Dead Sea Rift', Philos. Trans. R. Soc. London, London, Great Britain, A267: 107-130

Freund, R., Zak, I. & Garfunkel, Z. (1968) 'Age and Rate of the Sinistral Movement along the Dead Sea Rift', Nature, 220: 253-255.

Garfunkel, Z. (1970) 'Tectonics of the Western Margins of the Southern Arava', PhD. Dissertation, Hebrew University of Jerusalem : 204pp.

Ginat, H., Enzel, Y. & Avni Y. (1998) 'Translocated Plio-Pleistocene drainage systems along the Arava fault of the Dead Sea transform', Tectonophysics, Elsevier Science B.V., Amsterdam, 284: 151-160.

Khalil, B. (1992) 'The geology of the Ar Rabba area – Map sheet NO.3152 IV', Geology Directorate, Geological Mapping Division, Amman, Jordan, 22: 106pp.

Kirby, M.J., Naden, P.S., Burt, T.P. & Butcher, D.P. (1993) 'Computer Simulation in Physical Geography', 2nd Edition, John Wiley & Sons, Chichester, UK: 180pp.

Kovach, R.L., Andreasen, G.E., Gettings, M.E. & El-Kaysi, K. (1990) 'Geophysical investigations in Jordan', Tectonophysics, Elsevier Science Publishers B.V., Amsterdam, The Netherlands, 180: 61-69.

Loy, W. (1994) 'Waterconflicten', Meded. Zitt. K.Acad. Overzeese Wet., Brussel, Belgium, 39: 281-290.

Neev, D. & Emery, K.O. (1995) 'The Destruction of Sodom, Gomorrah, and Jericho : Geological, Climatological, and Archaeological Background', Oxford University Press, Inc., New York, U.S.A. : 155pp.

Rabinowitz , N., Steinberg, J. & Mart, Y. (1996) 'New evidence of magmatic diapirs in the intermediate crust under the Dead Sea', Tectonics, 15 (2): 237-243.

Selby, M. J. (1993) 'Hillslope materials and processes', 2nd Ed., Oxford U. Press: 451pp.

Seismic Rockfall Hazard in the Sele River Valley, Southern Italy

V. DEL GAUDIO, Dipartimento di Geologia e Geofisica - Osservatorio di Geofisica e Fisica Cosmica, Università di Bari, J. WASOWSKI, CNR - CERIST c/o Istituto di Geologia Applicata e Geotecnica, Politecnico di Bari, Italy

INTRODUCTION

Rockfalls are among the most common earthquake-induced slope failures (e.g. Keefer, 1984). Although rockfalls frequently occur in the south central Apennine Mountains, so far they have received little attention as compared to other types of landslides.

Like in many other Apennine intermontane basins, the rockfall hazard in the Sele valley (Fig. 1) is closely linked with the fault-bounded mountainous ranges characterised by sharp elevation contrasts and presence of tectonically sheared and fractured carbonate rocks (Fig. 2). We focus on the Sele valley for the following reasons: presence of towns and roads close to the steep carbonate outcrops; high seismicity of the area (see Table 1, Fig. 3); availability of a large amount of data on rockfalls triggered by the southern Italy 6.9 magnitude earthquake of 1980.

The distribution of the source areas of recent rockfalls in the Sele valley has already been mapped by several workers (Agnesi et al.,1983; Carrara et al., 1986; Esposito et al., 1998). For some specific sites (towns, infrastructures) these data need to be integrated with seismic rockfall susceptibility estimates that rely on rock structure and on fracture characteristics. An adequate hazard evaluation, however, requires that the issue of the recurrence time is also addressed. In first approximation the temporal rockfall hazard can be assessed by defining the probability of seismic triggering (e.g. Wasowski and Del Gaudio, in press).

HISTORICAL DATA AND PROBABILISTIC ANALYSIS

The review of historical data on earthquake-induced slope failures shows that events of intensity VI-VII (Mercalli, Cancani, Sieberg - MCS - scale) are capable of triggering rockfalls in the Sele valley. On the basis of regional seismic activity evaluation, we calculated hazard parameters for seven towns situated in the valley: Castelnuovo di Conza, Caposele, Calabritto, Laviano, Senerchia, Colliano e Oliveto Citra. For each of them, we estimated the number of earthquakes felt with intensity not less than a given threshold value, occurred in a time interval for which the available earthquake catalogue is complete with respect to the threshold selected. Assuming a Poissonian time distribution of the seismic events, the number of their occurrence in a fixed interval allows to calculate parameters (e.g. the mean return time T) useful to estimate occurrence probability of earthquakes whose intensity is equal or higher than the threshold considered. The calculations were carried out assuming threshold values (VI and VII M.C.S.) that are relevant for rockfall triggering.

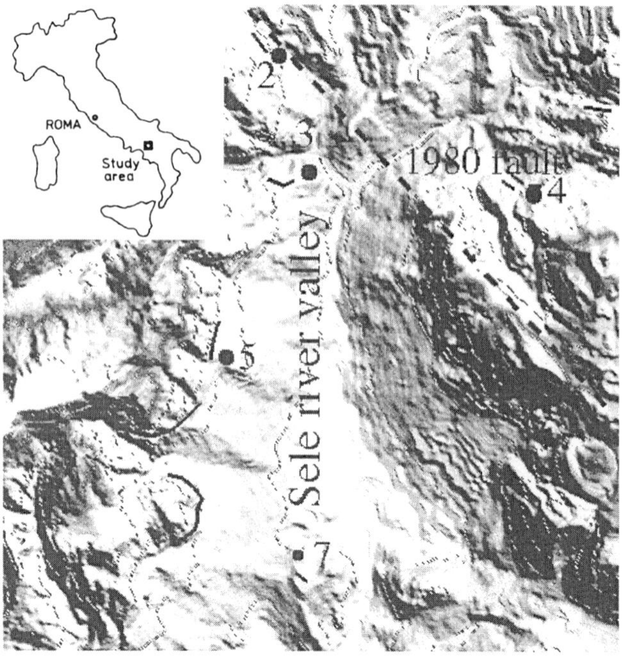

Fig. 1. DEM of the upper Sele river valley (area shown ~ 300 km^2). The topographically uppermost zones correspond to Mesozoic age fault-bounded carbonate massifs (main source of rockfalls). Dots mark seven towns: 1) Castelnuovo di Conza, 2) Caposele, 3) Calabritto, 4) Laviano, 5) Senerchia, 6) Colliano, 7) Oliveto Citra; dotted and continuous black lines mark, respectively the seismogenic fault and the rockfalls areas of the 1980 earthquake.

Fig. 2. Photo showing carbonate units making up the fault-bounded massifs; these rocks are characterised by high seismic susceptibility due to the presence of numerous discontinuities.

Table 1. Main seismic events occurred in the southern Apennine area. The events which caused documented rockfalls in the upper Sele valley are marked in grey (data from NT4.1 catalogue, Camassi and Stucchi, 1998).

DATE	TIME	LAT. N	LONG. E	EPICENTRAL AREA	INTENSITY
1125 10 11	-	41.133	14.783	BENEVENTO	VIII
1273 -- --	-	40.633	15.800	POTENZA	VIII-IX
1293 9 --	-	41.483	14.467	BOIANO	VIII-IX
1361 7 7	-	41.200	15.600	ASCOLI SATRIANO	IX-X
1456 12 5	-	41.150	14.867	BENEVENTANO	X
1550 8 25	-	41.150	15.017	ARIANO IRPINO	VIII-IX
1561 8 19	-	40.540	15.490	VALLO DI DIANO	IX-X
1688 6 5	16.15	41.317	14.567	MATESE	XI
1694 9 8	11.30	40.900	15.433	CALITRI	X-XI
1702 3 14	5.15	41.117	14.950	BARONIA	IX-X
1732 11 29	12.30	41.083	15.117	IRPINIA	IX-X
1805 7 26	21.01	41.500	14.533	MATESE	X
1831 1 2	14.07	40.067	15.800	LAGONEGRO	VIII
1836 11 20	7.27	40.150	15.800	LAGONEGRO	VIII
1851 8 14	13.20	40.950	15.650	VULTURE	IX X
1853 4 9	12.45	40.850	15.250	VALLE DEL SELE	IX
1857 12 16	21.15	40.350	15.833	BASILICATA	X-XI
1910 6 7	02.04	40.979	15.320	CALITRI	VIII-IX
1930 7 23	00.08	41.050	15.300	IRPINIA	X
1962 8 21	18.19	41.167	14.967	SANNIO	IX
1980 11 23	18.34	40.800	15.267	IRPINIA-LUCANIA	IX-X

A specific catalogue of Italian earthquakes aimed at hazard evaluation (NT4.1; Camassi and Stucchi, 1998) was employed. The local intensities of historical earthquakes reported in the NT4.1 catalogue were estimated by applying four different attenuation law:
logarithmic (Cornell, 1968),

$$Io - I = a + b \cdot D + c \cdot \log(D + 10.) \quad (1)$$

exponential ("Bottari law" in: Berardi et al., 1994)

$$Io - I = Io \cdot [1 - \exp(-\alpha \cdot D)] \quad (2)$$

cubic root (Berardi et al., 1994)

$$Io - I = \beta \cdot \sqrt[3]{D} + \gamma \quad (3)$$

Grandori et al. (1987)

$$Io - I = \frac{1}{\ln \Psi} \ln\left[1 + \frac{\Psi - 1}{\Psi_0}\left(\frac{D}{D_0} - 1\right)\right] \quad (4)$$

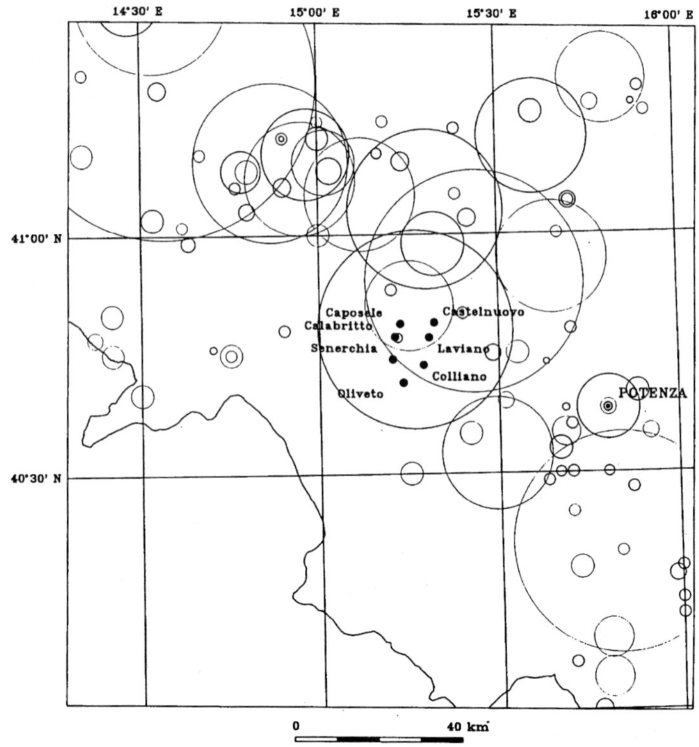

Fig. 3. Location of the damaging earthquakes occurred in the southern Apennine area (data from NT4.1 catalogue, Camassi and Stucchi, 1998). Circles represent the earthquake focal volumes calculated according to Bath and Duda (1964).

where *Io* is the epicentral intensity and *I* the intensity felt at distance D (expressed in km); a, b, c, α, β, γ, Ψ, Ψ_0, D_0 are coefficients calculated for specific areas. The coefficients for the first three relations were calibrated on the basis of the 98 macroseismic observations referring to the upper Sele valley reported in an Italian data base (Monachesi and Stucchi, 1998). Through a least square criterion the coefficients reported in Table 2 were obtained. The equation (4) employs different coefficients for seismic sources belonging to different seismogenic zones (Peruzza, 1996).

Table 2. Coefficients of the local attenuation laws

Law type	Logarithmic (Cornell, 1964)				Exponential ("Bottari law")		Cubic root (Berardi et al., 1994)		
	a	b	c	r.m.s.	α	r.m.s.	β	γ	r.m.s.
	-3.186	0.017	3.358	0.89	0.0102	1.04	1.69	-2.44	0.91

The macroseismic observations were also employed to measure the scattering of real intensity values around the theoretical attenuation laws employed. This allows to estimate the probability that the intensity values obtained from the attenuation laws differ from the real intensity; this probability was taken into account to correct the estimated number of events which reached or exceeded the selected threshold (Berardi et al., 1994).

RESULTS AND CONCLUSIONS

Overall, the different estimates of the mean return periods (T) for the intensity thresholds of VI and VII range from 20 to50 years and from 50 to 90 years, respectively (see Table 3). The probability of seismic triggering decreases down valley (from north to south): the T values for northernmost town are 20-30 and 50-60 (respectively for VI and VII MCS), whereas the T values for southernmost town are 30-50 and 70-90 (respectively for VI and VII MCS). The explanation for this trend is the following: even though some of the major earthquakes that affected the Sele valley had their source within the valley (e.g. the 1853 and 1980 events: see Tab. 1) or south of it (as the 1857 event), the most frequently activated seismogenic structures are located north of the valley.

Seismically induced rockfalls are highly unpredictable instantaneous movements and represent a considerable hazard to people and private/public property. The expansion of towns and road networks in the Sele valley implies that the levels of risk will increase in the future. The example of this work shows that the rate of occurrence of seismic events capable of mass movement triggering can be inferred on probabilistic grounds. This information should be helpful to the regional administration and civil protection department involved in landslide risk assessments and in setting the priorities for the allocation of funds for preventive and mitigation efforts.

Table 3. Seismic hazard parameters: n° = number of events felt with an intensity not less than the specified threshold during the time interval for which the earthquake catalogue is complete with respect to the same intensity threshold; T = mean return period for events of intensity at least equal to the specified intensity threshold. The completeness interval is from 1860 to 1992 for I ≥ VI and from 1780 to 1992 for I ≥ VII.

		Cornell		Bottari		Cubic root		Grandori	
		I ≥ VI	I ≥ VII	I ≥ VI	I ≥ VII	I ≥ VI	I ≥ VII	I ≥ VI	I ≥ VII
Castelnuovo	n°	3.8	3.8	6.2	4.3	3.7	3.7	4.0	3.9
di Conza	T	34.7	55.7	21.1	49.2	35.9	56.8	33.2	54.7
Caposele	n°	4.0	3.8	6.0	4.0	4.2	3.7	3.8	3.7
	T	32.6	55.6	21.9	53.5	31.8	58.1	34.8	58.1
Calabritto	n°	3.8	3.6	5.8	3.7	4.2	3.1	3.7	3.4
	T	34.8	59.7	22.7	57.2	31.6	67.5	36.0	61.5
Laviano	n°	3.8	4.1	6.0	4.0	3.4	3.3	3.9	3.7
	T	34.7	52.3	22.9	53.3	38.9	64.0	33.6	57.8
Senerchia	n°	3.2	3.0	5.2	3.2	3.3	2.9	3.4	3.1
	T	41.0	70.9	25.6	65.4	40.3	72.2	39.0	67.5
Colliano	n°	3.2	3.4	5.4	3.5	3.3	3.0	3.5	2.9
	T	41.5	61.9	24.6	60.3	39.7	71.7	38.1	72.4
Oliveto Citra	n°	2.8	2.8	4.6	2.7	2.7	2.4	3.2	2.6
	T	46.4	76.1	28.4	77.7	48.9	87.4	41.4	82.1

ACKNOWLEDGEMENTS
This work was supported in part by the European Community (EV5V-CT94-0451grant) and the Italian Ministry of University and Scientific and Technological Research

REFERENCES

Agnesi V., Carrara A., Macaluso T., Monteleone S., Pipitone G. and Sorriso-Valvo M., 1983: Elementi tipologici e morfologici dei fenomeni di instabilità dei versanti indotti dal sisma del 1980 (Alta valle del Sele). Geol. Appl. Idrogeol., 18, n.1, 309-341.

Båth M. and Duda S.J., 1964: Earthquake volume, fault plane area, seismic energy strain, deformation and related quantities. Ann. Geof., 17, 3, 353-368.

Berardi R., Petrungaro C., Zonetti L., Magri L. and Mucciarelli M., 1994. Mappe di sismicità per l'area italiana. ISMES-ENEL.

Camassi R. and Stucchi M., 1998. Un catalogo parametrico di terremoti di area italiana al di sopra della soglia del danno. GNDT, (technical report), Versione NT4.1.1 con aggiornamenti 1981-1992, marzo 1998, http://www.emidius.itim.mi.cnr.it/NT/home.html.

Carrara A., Agnesi V., Macaluso T., Monteleone S. and Pipitone G., 1986: Slope movements induced by the southern Italy earthquake of November 1980. Geol. Appl. Idrogeol., 21, n.2, 237-250

Cornell C.A., 1968: Engineering seismic risk analysis. Bull. Seism. Soc. Am., 58: 1583-1606.

Esposito E., Gargiulo A., Iaccarino G. and Porfido S., 1998: Distribuzione dei fenomeni franosi riattivati dai terremoti dell'Appennino Meridionale. Censimento delle frane del terremoto del 1980. Proceedings of International Conference "Prevention of hydrogeological hazards: the role of scientific research", Alba (Italia), 5-7 November 1996, 409-427.

Grandori G., Perotti, F. and Tagliani A. On the attenuation of macroseismic intensity with epicentral distance; in: Ground Motion and Engineering Seismology (ed A.S. Cakmak), pp. 581 to 594, 3rd Int. Conf. on Soil Dynamics and Earthquake Eng., Princeton, USA, 1986, Elsevier, Amsterdam, 1987.

Keefer D.K., 1984: Landslides caused by earthquakes. Geol. Soc. Am. Bull., 95, 406-421.

Monachesi G. and Stucchi M., 1998. DOM4.1, un database di osservazioni macrosismiche di terremoti di area italiana al di sopra della soglia del danno. Gruppo Nazionale Difesa dai Terremoti, http://www.emidius.itim.mi.cnr.it/DOM/home.html.

Peruzza L., 1996: Modalità di attenuazione dell'intensità macrosismica. Rapporto sintetico del gruppo di lavoro "Rischio Sismico" del Gruppo Nazionale per la Difesa dai Terremoti. http://emidius.itim.mi.cnr.it/GNDT/alle_att_sism.html.

Wasowski J. and Del Gaudio V. in press. Probabilistic evaluation of mass movements and seismic hazards in Caramanico Terme, Italy. Engineering Geology.

Overburden Stability of Rock Slopes in Quarries

O. DEL GRECO, M. FORNARO, C. OGGERI, Dept. of Georisorse e Territorio, Politecnico di Torino, C.so Duca degli Abruzzi 24, 10129 Torino, Italy

GENERAL ASPECTS

Many ornamental and industrial stone quarries are located along steep slopes or within particular geostructural domains in the Italian Alpine Range. The stability of a quarry face should be considered within the exploitation method and can be solved by means of rock mechanics studies and by adopting appropriate excavation techniques. A detailed geostructural survey and a characterization to determine the strength parameters are necessary to plan an exploitation in order to perform a stability analysis, from which the various failure mechanisms can be highlighted. At the same time, the choice of the excavation tecnique is related to the type of rock and to the structure of the mass, and it influences, for example, the heigth of the benches, their length and the general exploitation plan in terms of spatial development of the excavations. The stability of the overburden formations, such as detritic or cohesionless materials, blocky and altered elements or morainic formations, can be compromised by exploitation activitics. In other cases there is a natural evolution of the slope, due to singular but diffused phenomena such as block falls, or due to global movements caused by gravity and water interaction. All these facts involve safety aspects not only for the quarrying activities but also because of the possible interference with other nearby structures such as roads, electric power lines and railroads.

In order to ensure correct safety margins it is therefore necessary to apply a comprehensive approach: knowledge of the geostructural context of the quarry slopes, study of the possible evolution of the phenomena and the related causes, a numerical modelling of the phenomenon, monitoring of the area and the choice of adequate passive and active protection and stabilisation works. Some points should however be mentioned: the economic costs for preventive and remedial works must be consistent with the budget of the quarrying activities; monitoring is often involved with fast and sudden phenomena, and it should therefore be suitable to particular requirements, in order to be effectively useful; the final configurations of the residual slopes should allow a reclamation of the area, in particular if landfills are foreseen. On the basis of the following examples, some suggestions are given in order to obtain an operative scheme, that is also easily adaptable to monitoring and protection of civil excavations of slopes.

STRUCTURAL FEATURES OF QUARRIES

Italian quarries in mountainous areas (Alpine Range, Apuane Range) are exploited both for ornamental stones (gneiss, diorite, ophicalcites, marbles, slate) and for aggregates (andesite, olivine, limestone). In some regions open pit mines also exist (quartzite, feldspar, albite).

Landslides in research, theory and practice, Thomas Telford, London, 2000.

Overburden is usually intended as the geological formation that covers the useful part of a rock mass or of an orebody. In the studied cases three main situations have been encountered. The first one is relative to morainic strata on a bedrock surface. The thickness of this overburden formation varies, but the common range is between less than 1 m to about 15 m, at least where the exploitations are carried out. These formations are characterized by a very heterogeneous material, with a very wide grain size distribution; a cohesion is sometimes present. Water is present in concomitance with meteoric events, because of the relative high permeability of the morain. In some cases lenses of fine materials can act as local reservoirs of water inside the formation.

The geotechnical parameters that should be determined in order to perform the relative stability analyses are, in particular, cohesion and friction angle. Due to the different scales of the laboratory tests and of the actual formation it has not been possible to carry out a direct determination of the parameters. Generally, it was only possible to carry out a back analysis of limit equilibrium situations (safety factor $\approx$ 1.0), and this was integrated with a detailed description of the material and the applied loads. Some information can also be obtained from a grain size distribution, provided it can be performed and that a relationship between the size distribution and friction angle can be applied. In some cases an overall dip angle of 35° for residual slopes has been assumed for long term stability and a value of 40° has also been accepted for local or short term conditions.

The second case is relative to detritic and colluvial sediments (materials that have been transported) or soils (in site origined). The geometry of these deposits is very irregular. A direct determination of the grain size distribution is possible, and back analyses in critical situations can give appropriate values of the friction angle and cohesion for the mass.

The detritic masses are usually made up of angular elements, as they are broken rock fragments with no reshaping due to transportation. The talus cone can belong to this category of overburden, as can the waste muck piles constructed along the slopes. In this case the parameter angle of repose is sometimes adopted in the analyses, and this corresponds to the angle of the slope that can be held by a loose pile of rock fragments. Talus cone slopes

Fig.1. Overall view of an albite open pit mine with morainic cover during exploiation.

usually exhibit angles of between 34° and 38°, even though single benches could probably be excavated with steeper angles, but in conditions of temporary and limit equilibrium (42° to 45°). Some laboratory tests, such as direct shear tests are also possible. It should be recommended to perform both tests and consequently analyses, with different water contents and different shearing velocities, as slides can occurr under both dry and wet site conditions. New computational codes (such as PFC from Itasca, based on the equations of distinct element method) are used both for back analyses and design.The third case is relative to altered and weathered rock masses, with the possible filling of soils and clay inside the voids among the blocky elements.

The geomechanical parameters can be determined by means of compression, tensile and shear tests on the rock matrix, and by means of shear tests on discontinuities and on the filling materials. It would be very useful to have tilt tests at large scale. Once again a back analyses can be performed by using, for example, distinct element codes (e.g. UDEC by Itasca). In this case, the geometry of the overburden is usually more regular and can be similar to the structure of the exploitable rock mass. The debris elements are, in general, angular and the mass is quite permeable.

Fig. 2. Ornamental stone quarry with detritic and colluvial overburden.

A common element for the three situations is the need to determine the geometry of the contact between the overburden and the bedrock, as this morphology, together with the geomechanical parameters are the elements which affect the type of surface slides, the mechanisms and the evolutions of the phenomena. It is important to try to recognize, from the large variety of cases of quarries and overburden morphologies, the common features that allow one to classify possible slide types and, in particular, the operative suggestions for retaining and mitigation structures.

In addition, the thickness of an overburden of rock masses for ornamental stones and also for various industrial minerals or aggregates, when it is not negligeable, influences both the rentability and also the method of exploitation today adopted for the quarries.

Apart from the case of an overburden ratio that is so unfavourable that an underground exploitation has to be considered (1:1 the ratio between the thickness of the exploitable strata to the overburden, for the case of ornamental stones, with a recovery factor of at least 50%), the general rule for open pit quarries is the adoption of descending horizontal slices, both when the overburden is cohesive or not, and the thickness is limited. The environmental reclamation of the uphill slope is therefore carried out progressively at the same time as the advancing of the downhill exploitation. Fot intermediate situations with medium thickness of overburden the old traditional method of ascending vertical slices still seems to be the most common adopted, as it allows production in a relatively short time without creating removal of that large surface layers. A compromise between the previously described methods could be that of the adoption of sequential and superimposed stripes in which the vertical slices are exploited progressively in descending order. This would allow the reduction of the removal of the overburden during the initial productive cycle of the quarry, and would easily carried with the environmental rules. Another important decision factor for the exploiting method is the selectivity of the mineral or rock blocks, in order to improve rentability. For example, when it is possible, the selection is operate at different benches in a open pit exploitation of limestone for the cement industry, in order to separate waste dolomitic layers from in the rock mass, while also taking the dip of the benches imposed by stability features into account.

FAILURE MECHANISMS AND EVOLUTION

The here described cases can be classified, in terms of both the failure causes and the evolution of the slide events. These two phases are obviously not independant, as each type of rock structure or material may preferrably induce a particular slide mode. In this note the slope movements along overburden formations focuse on their correlation with the exploitation activity, especially when this activity together with structural features and meteoric events are the causes of a slide. In other cases a quarry can be subjected to damage due to consequences of slides that are not directly linked to the quarry itself. In both cases the study of the landslide can help in the choice of the appropiate remedial and protection works.

The basic approach of analyses methods, such as the limit equilibrium method, and the common characterization of the parameters are suitable, at least for simple geometries. When also considering the evolution of the movements, other sometimes new and innovative approaches are more suitable. In this sense it is of great interest to classify the type of landslide, following, for example, one of the recently available engineering classifications (Varnes, Takahashi, Hutchinson). As different slide types depend on the overburden, the following remarks can summarize the main situations.

Rock falls. These are present in altered rock layers, in particular in ornamental stone quarries. They are often associated with toppling failures of both the productive strata and of the surface blocks and are very dangerous for the safety of the workers.

Soil slides. These are more common in colluvial deposits. The movements can be progressive or assume high velocities and are often associated with meteoric events.

Debris flows. These phenomena are generally observed in quarries at the beginning, when considering colluvial overburden but also morainic covers.

Complex failures. These are present in irregular contact bedrock geometries. They are usually progressive and result in the accumulation of pile cones in quarries.

Plane failures. These are relative to altered blocky strata and are dangerous because of their quick evolution and the large volumes involved.

Landslides involving thick colluvium are usually relatively slow moving rotational slides that rarely exhibit subsequent disaggregation, while when the overburden is thin the initial translational movement can easily transform into a debris flow. This combination of soil slide and debris flow seems the most common failure mode involving colluvial slopes.

In slope talus cones of detritic elements there is a continuous feeding of fragments, and rock falls and movements are frequent. Talus deposits are mainly made of coarse grained sub

Fig. 3 . Overall view and detail of a marble quarry: lanscape and protection from rock chutes.

angular shaped fragments with a variable percentage of fine fragments and large rock elements. When the percentage of coarse material is high the talus can be more unstable, because the overall deposit may have a configuration in which the loads are transmitted with sudden realignements of the fragments, whilst when the fine matrix is compacted a more cohesive behaviour can be observed and larger elements float in the finer matrix. Each of these talus categories behaves differently in a kinematic sense.

The slopes of these talus cones often reflect the repose angle of the material. As a consequence, when steeper angles are obtained with artificial cuts thanks to temporary cohesion, rock fragments can ravel from the artificial faces resulting in a continuous failure. This situation requires a stable passive protection of the quarry area, and active works for hydraulic stabilization of the deposits. The energy of the mass and the morphology of the slope can be described by means of the travel angle α = arctan H/L, where H is the difference in elevation between the crown and the tip of the landslide, while L is the length of the landslide.

Particular attention must be paid to rock chutes or rock avalanches above detritic deposits or piles because of the possibility of starting off debris flows, even in the case of dry deposits, which are similar to "sturzstrom" phenomena, even when small volumes of rock are involved.

REMEDIAL WORKS AND FINAL REMARKS

The removal of a portion of an overburden at the crown of a quarry determines the exposure of a face in which the material is free to fall and slide down.

It is necessary to carry outa series of stabilisation (preventive) and mitigation (protective) wotks. Within the economic balance of a quarry the final reclamation and also the provvisional operation for safety and stability must be paid by the production of the stones. In the case of two or more quarries with common or dependant for structures a coordination should also be ensured by the official mining bureau. A classical example is that of the debris waste landfill of a quarry that may influence the safety of another quarry at a lower elevation (e.g. the Apuane basin in Tuscany), or the contemporary exploitation of old landfills for secondary materials while the quarry is still active for raw material extraction.

Among the stabilisation works, the following can be mentioned: a) resloping of the overburden face in order to reach less steep slopes; b) scaling, in order to remove the loose rock elements; c) stabilisation of debris piles and cones by means of ground sills and check dams in order to reduce the erosion effect of surface water; d) adoption of an appropriate sequence of exploitation in the quarry in order to allow the gradual exposure of the overburden; e) perimetral collection and removal of surface water. Among the protective works the following can be mentioned: a) benching of the quarry face in order to alterate the free fall path of rock elements; b) artificial barriers and fences at the crown of the quarry in order to catch the blocks that fall from the overburden; c) creation of areas inside the quarry, like yards, for the accumulation of any possible collapsed material; d) construction of embankments and ditches at the base of the quarry.

The contemporary revegetation of surface zones at the crown of a quarry and eventual monitoring of surface movements of singular elements or discontinuities or of the whole area above a quarry are complementary but indispensable tools that guarantee a higher level of safety during work operations and land reclamation.

Fig. 4. Protection at the top of a geiss quarry with fences.

REFERENCES

Cravero M., Fornaro M., Iabichino G., Rinaldi L. (1998) "Stone quarries in difficult conditions: considerations on design and static safety appraisal limits". Proc. 3rd NARMS'98 Rock Mechanics in Mining, Petroleum and Civil Works, Cancun.

Del Greco O., Fornaro M., Oggeri C. (1994) "Modification of a quarry face: stabilisation criteria and environmental reclamation" Proc. 7th Int. IAEG Congr., Lisboa, Balkema Publ.

Hutchinson J.N., (1988) "Morphological and geotechnical parameters of landslides in relation to geology and hydrogeology" Proc. 5th Int. Symp. on Landslides, Lausanne, Balkema Publ.

Turner A.K., Schuster R.L. (1996) "Landslides: Investigation and Mitigation" Transp. Research Board, National Research Council, National Academy Press, Washington D.C.

Slow Movements of Earthflow Accumulations along the Ionic Coast (South-Eastern Italy)

B. D'ELIA, G. LANZO, Dept. of Structural and Geotechnical Engineering, "La Sapienza" University of Rome, M. ROSSI-DORIA, Dept. of Hydraulics, Transportation and Roads, "La Sapienza" University of Rome, Italy

INTRODUCTION

Slow and deep movements in old landslides formed of clayey materials are normally a consequence of pore water pressure changes within existing shear zones and are therefore determined by rainfall distribution in time. In most cases, these movements can be classified as "extremely" slow as they occur at rates smaller than 1.6 cm/y (WP/WLI, 1995). A peculiar feature of deep-seated landslides is that the sliding process is a consequence of the rainfalls that cumulate over long periods of time (Lanzo and D'Elia, 1997; Sangrey et al., 1984).

Slow slope movements are widespread in several areas in Italy. Nowadays there is an increasing need to consider the impact of engineering structures on the territory and of the use of new areas for civil and industrial purposes. Because of its intrinsic uncertain and elusive character, the evaluation of the landslide risk connected with those movements is a difficult task. The only means for attempting to make predictions is to use accurate monitoring data covering long periods of time.

In the case under study, the slow movements involve earthflow accumulations located at the toe of a slope on a coast, along which a main road runs parallel to a railway track (D'Elia et al., 1991a and 1991b). The data collected over 15 years of inclinometric monitoring (1984-1999) are examined as well as the relationships between movements and rainfall and between movements and sea erosion.

SLOPE STABILITY

The slope is modelled in sequences consisting mainly of intensely fissured and/or sheared clayshales, with some limestone intercalations. The slope extends for 1 km along the coast and 1 km perpendicularly to the coast line up to about el. 225 m a.s.l. The distribution of the accumulations, their extent and the age of their last emplacement are illustrated in Fig. 1. The accumulations have differing thicknesses, the maximum measured being about 16 m (accumulation A). The emplacement of the accumulations has occurred as a result of a paroxysmal activity of slump-earthflows that developed in the upper portion of the slope. This activity recurrently occurs at intervals of some tens of years, namely when cumulative effective rainfall reaches critical values (D'Elia et al., 1991b).

Landslides in research, theory and practice, Thomas Telford, London, 2000.

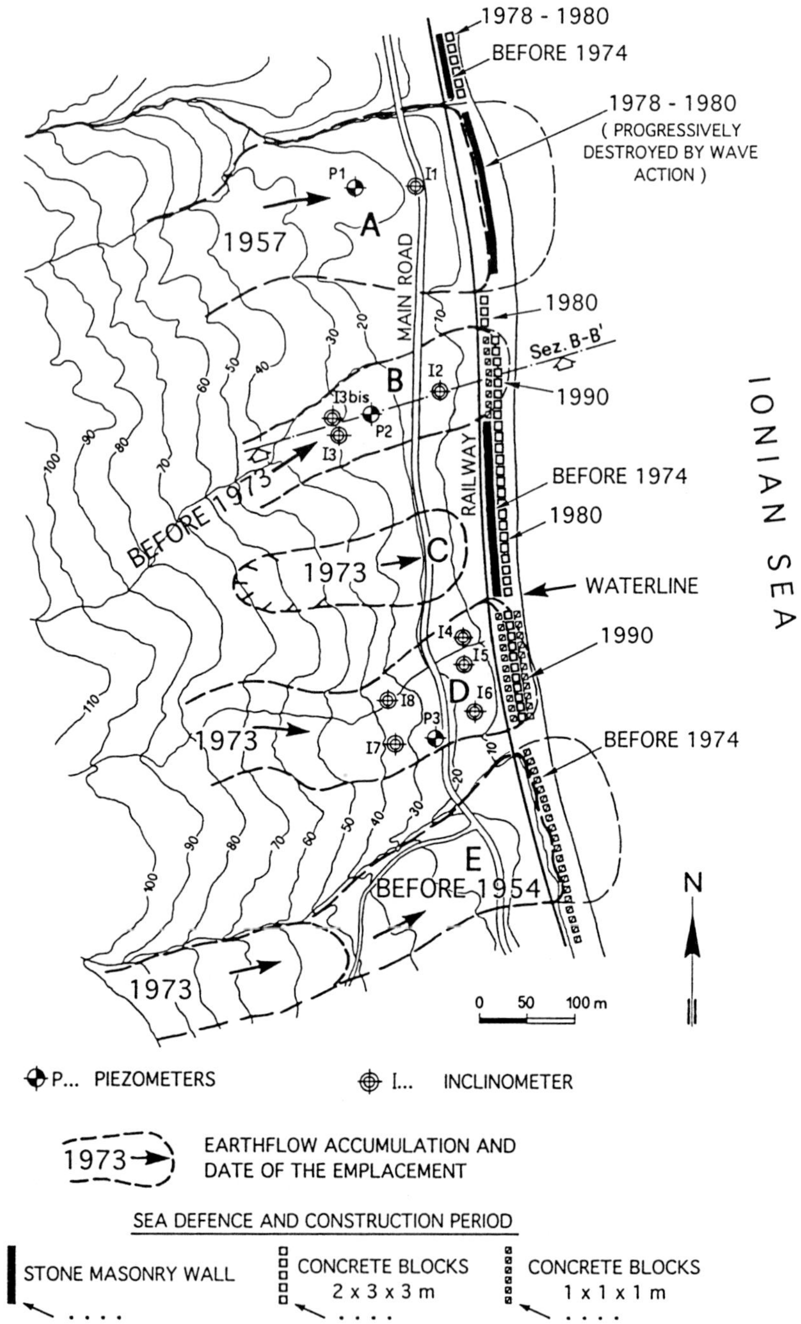

Fig. 1 - Site plan (modified from D'Elia et al., 1991b)

Movements in the earthflow accumulation zones depend essentially on feed and toe erosion. When feeding stops and there is no erosion at the toe, the movements gradually die down and the accumulations tend to assume a stable configuration. In this phase the evolution of movements depends not only on geometrical factors, on the initial stress field and on mechanical soil parameters, but also on the pore water pressure regime which in turn is a function of climatic conditions.

During the observation period, there was virtually no feed from above, while instead erosion at the toe was active at Accumulations B and D up to 1990 when protection works against wave action were built (in the 1980-1989 the toe of accumulation D had withdrawn by 20 m). The sea defences built during the 1978-1980 period at Accumulation A were gradually destroyed by the action of the waves (Fig. 1).

MONITORING

In November 1984, 3 piezometric stations and 8 inclinometers were installed (Fig. 1). Only during the first period of observation (November 1984-February 1987) was it possible to make piezometric measurements at regular intervals (D'Elia et al., 1991b).

During the first period of observation 8 inclinometer readings were taken at one-metre depth step spacings; during the second period (August 1987-June 1999) 26 readings were taken, with a new reference reading and depth step spacing of 0.61 m. To make the displacements-depth profile of inclinometers I4 and I5 easier to understand only the readings taken at the end of the winter-spring season are plotted.

The inclinometers have indicated that, apart from superficial deformations, the movements of the accumulations are due essentially to sliding within a shear zone at the landslide material-bedrock contact, and where residual shear strength is mobilized (D'Elia et al., 1991b). Hence the displacement values considered in the following are those recorded along such shear zone.

CLIMATIC CONDITIONS

The climate in this area is of the semiarid type with a slight surplus in winter (Thornthwaite, 1948). Average monthly temperature ranges between a minimum of 9°C in January and a peak of 26°C in July/August. Rainfall data were measured at the Montegiordano Marina station (elevation 7 m a.s.l.) at 4 km from the area under study. Rainfall trends were analysed with reference to monthly values and to effective monthly rainfall (Hutchinson, 1971), cumulated in the October-March semester which is the time when the groundwater system is recharged (Table 1).

Table 1. Cumulative monthly rainfall (CR) and cumulative effective monthly rainfall (CER) in the October-March semester

Semester	84-85	85-86	86-87	87-88	88-89	89-90	90-91	91-92
CR (mm)	412	206	213	306	140	146	519	143
CER (mm)	251	92	93	145	14	28	397	0
Semester	92-93	93-94	94-95	95-96	96-97	97-98	98-99	
CR (mm)	148	412	249	256	299	507	286	
CER (mm)	33	300	79	148	159	327	160	

In the 1984-1999 period, the average values of cumulative rainfall (279 mm) and of effective cumulative rainfall (151 mm) were sharply lower than the values measured for the 1951-1984 period (412 and 223 mm respectively).

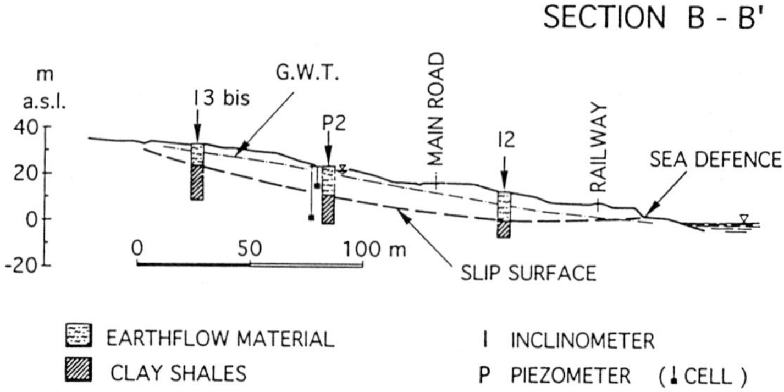

Figure 2. Longitudinal cross section of Accumulation B

MOVEMENTS AND RAINFALL

To ascertain the degree of correlation between rainfall regime and movements, a comparison was made between displacement increases, indicated by the inclinometer readings made at the end of two consecutive winter-spring seasons, and rainfall and effective rainfall, both being cumulated in the October-March semester (Table 1).

1) *Accumulation A* - Inclinometer I1 showed up an overall displacement of the order of 10 mm. However the annual displacements are very small and scattered, so that the movements are difficult to define and therefore cannot be correlated with rainfall.

2) *Accumulation B* (Fig. 2) - Both inclinometric casings installed (I2 and I3bis) sheared off during the first period of observation.

Inclinometer I2 pointed out the presence of a basal slip surface at some 12 m depth. In February 1985, at 3 months after installation, the reading showed a displacement of 33 mm in conjunction with a period of high rainfall after which the inclinometer sheared off.

Inclinometer I3bis pointed out the presence of a basal sliding surface at a depth of about 9 m (Fig. 3). In February 1987 it gave a reading of a 115 mm displacement after which it sheared off. The movements match the cumulative rainfall for that period (Table 1). Accumulation B was subject to rather high average displacement rates (>50 mm/year), being highest at the toe of the accumulation.

3) *Accumulation D* - The movements involved only the zone of inclinometers I4 and I5; no displacements were indicated by inclinometers I6, I7 and I8.

Inclinometer I4 highlighted a deformation band at a depth between 3.5 and 5.5 m, and a shear zone at 15 m depth (Fig. 4). The overall displacement was 33 mm (average displacement rate: about 2 mm/y). Annual displacements increased in time but they do not correlate with cumulative rainfall, nor with effective cumulative rainfall (Fig. 6).

Inclinometer I5 showed up a sharp basal shear zone at a depth of 14 m (Fig. 5). In June 1994 the sliding reached 50 mm (average displacement rate: about 5 mm/y); subsequently the inclinometric casing sheared off. The annual displacements correlate with both cumulative rainfall (r^2=0.580) and cumulative effective rainfall (r^2=0.659) (Fig. 6).

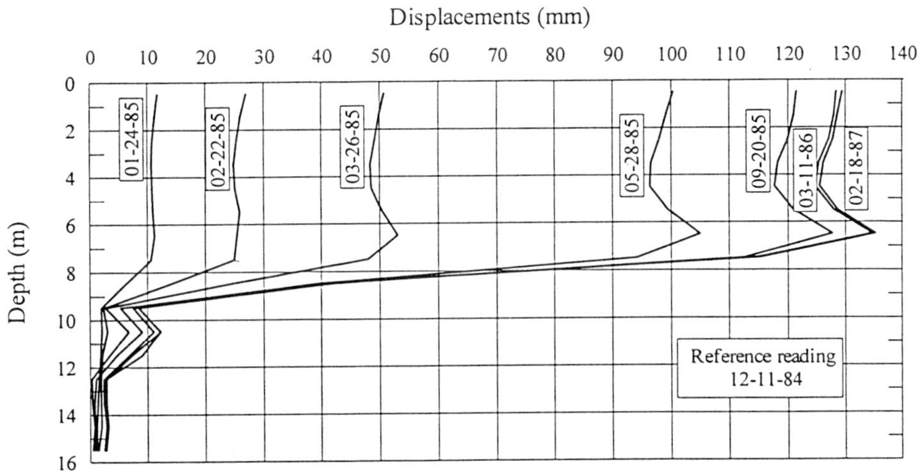

Fig. 3 - I3bis inclinometer (Accumulation B)

As far as the relationships between accumulation movements and cumulative rainfall is concerned, the inclinometric data indicate that where the rate of movement is very low (I4) the annual displacements do not correlate with the rainfall regime. On the contrary, where the rate of movement is relatively high (I3bis and I5) such a correlation does exist.

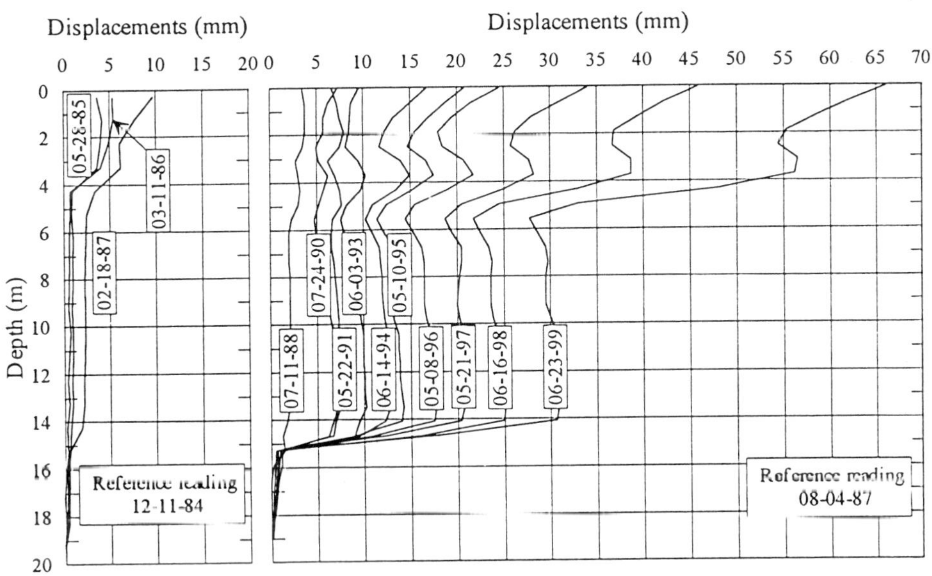

Fig. 4 - I4 inclinometer (Accumulation D)

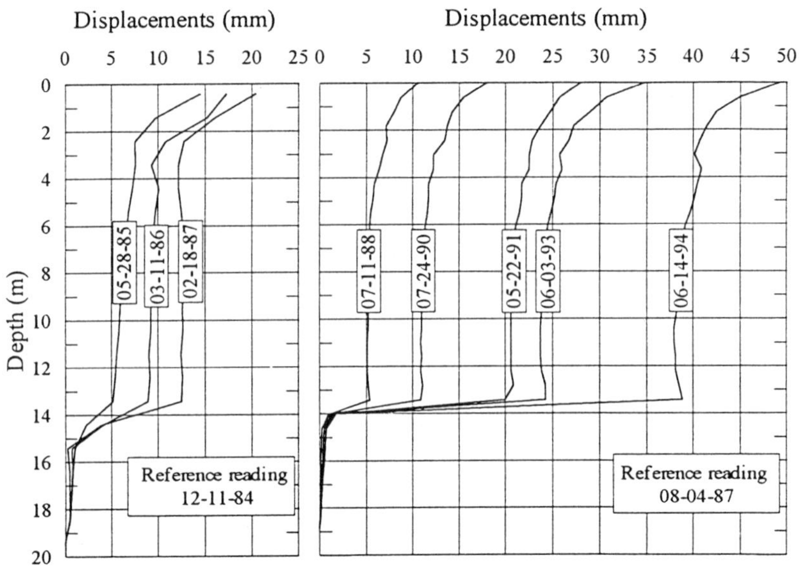

Fig. 5 - I5 inclinometer (Accumulation D)

This result can be interpreted in a sense that accumulations are subjected to permanent creep-like movements (very low rate of movement) when the factor of safety is somewhat greater than 1.0 and to slidings (relatively high rate of movement) when the factor of safety drops to 1.0 due to pore pressure increase (Patton, 1984).

MOVEMENTS AND SEA EROSION

1) *Accumulation A* - During the monitoring period, sea defence, consisting in a light stone masonry wall, was progressively destroyed by the action of the waves but there was only a limited retreat of the accumulation toe. Furthermore, taking into account also the near total lack of movements, sea erosion may be assumed to have had a negligible influence on the stability conditions of the accumulation, which appears to be involved only in small residual movements during the decay phase. The slowness of the movement may be related to the time that had elapsed from the emplacement of the accumulation (1957).
2) *Accumulation B* - The accumulation, which in 1990 was protected with defence works, has continued to move also during the period of observation, as pointed out by the relative displacement of the upper part of the I2 inclinometric casing which was still accessible, and by the disarrangement of the concrete blocks at the accumulation toe. The high displacement rate, which, besides, was greater in the zone of the toe, suggests that the sea erosion, which was quite intense prior to the construction of the defence works, acted in such a way as to produce conditions conducive to the movement.
3) *Accumulation D* - Even though defence works were built in 1990, the increase in displacement rate in time measured by inclinometers I4 and I5, may be due to the conditions resulting from erosion which persist for some time after the erosion action has ceased.

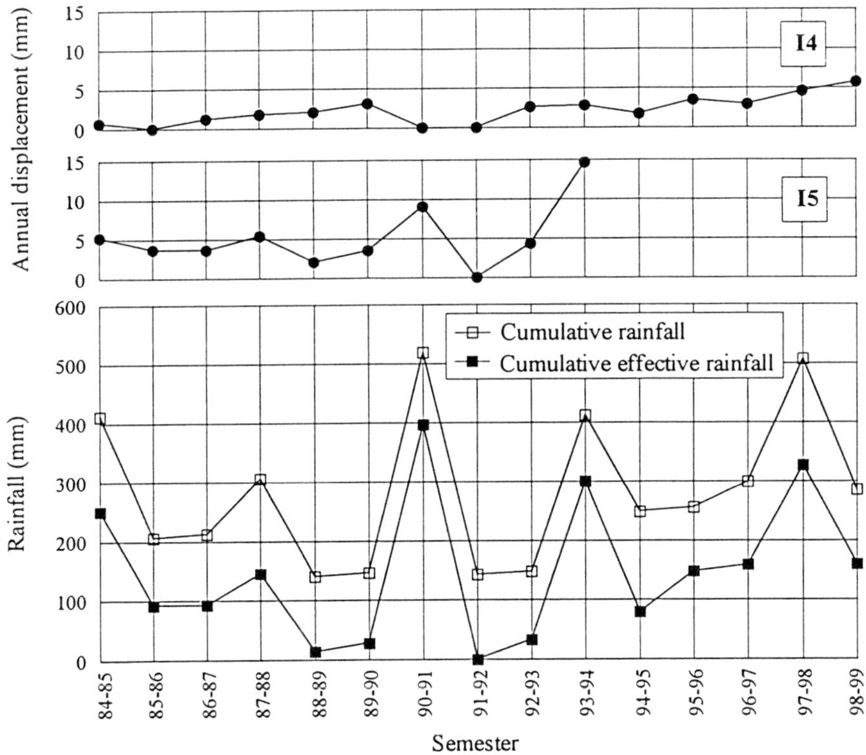

Fig. 6 - I4 and I5 inclinometer annual displacements, cumulative rainfall and cumulative effective rainfall versus time

CONCLUSIONS

1) The study has highlighted that in a relatively restricted area there are slump-earthflow accumulations which are subject to different movement characteristics, even within the same accumulation.
2) It was possible to identify the characteristics of the extremely slow movements thanks to the unusually long monitoring period and to the reliability in time of the inclinometric measurements.
3) The collected data show that the characteristics of the movements can be defined differently according to the duration of the observations and to the period when they are made.
4) The inclinometer data indicate that, apart from the superficial deformations, the accumulations move as a result of the sliding within a shear zone at the contact between the landslide material and the bedrock.
5) During the observation period (1984-1999), characterized by a markedly lower rainfall than that of the preceding 30 years, the average displacement rate of the accumulations was found to be between some 0.5 mm/y and over 50 mm/y.

6) In the cases in which the average rate of displacement was greater than 4-5 mm/y, it was possible to recognize a significant correlation between annual displacements and both rainfall and effective rainfall, cumulated during the October-March semester.
7) Especially in the zone at the toe of the accumulations, sea erosion was one of the controlling processes of the movements even though it was not possible to identify a direct relationship between erosion and the evolution in time of the movements.
8) The time elapsed between the emplacement of the accumulations, the dying effects of sea erosion and the slowness and even absence of movement suggest that overall the accumulations have approached a stable configuration.

ACKNOWLEDGEMENTS

This research was carried out with the financial support of GNDCI (Publ. n° 2044) (U.O. 2.40) and of *Centro di Studio per la Geologia Tecnica* - IRTR (CNR). Since 1987 the inclinometric and piezometric measurements have been carried out by A. Cittadini and R. D'Inverno, the latter having also processed the inclinometer data.

REFERENCES

D'Elia B., Lanzo G., Rossi-Doria M. (1991a) - *Landsliding in the coastal slopes of Capo Spulico area in the Gulf of Taranto (SE Italy).* Proc. Int. Conference on "Slope Stability Engineering - Developments and Applications", Isle of Wight (U.K.), Ed. Thomas Telford, 297-302

D'Elia B., Lanzo G., Rossi-Doria M. (1991b). *Condizioni di Stabilità dei versanti costieri tra Rocca Imperiale e Capo Spulico (Alto Ionio - Calabria).* In Frane Costiere a cura di F. Esu, Esagrafica Roma, 85-131

Hutchinson J.N. (1971). *Field and laboratory studies of a fall in Upper Chalk cliffs at Joss Bay, Isle of Thanet.* Proc. of the Roscoe Mem. Symp. "Stress-strain behaviour of soils", Cambridge, Ed. R.H.G. Parry, 692-706

Lanzo G., D'Elia B. (1997) - *Experiences on slope movements in clayey soils.* Rivista Italiana di Geotecnica, XXXI, 15-27

Patton F.D. (1984). *Climate, Groundwater pressures and Stability analyses of landslides.* Proc. ISL, Toronto, 3, 43-59

Rat M. (1984). *Meteorologie, hydrogeologie et glissements de terrain.* Proc. ISL, Toronto, 3, 33-41

Sangrey D.A., Harrop-Williams K.O., Klaiber J.A. (1984). *Predicting ground-water response to precipitations.* Journal of Geotech. Eng., ASCE, 110, 957-975

Thornthwaite C.W. (1948). *An approach toward a rational classification of climate.* The Geogr. Rev., 38, 55-94.

WP/WLI (1995). *A suggested method for describing the rate of movement of a landslide.* Bull. IAEG, n. 52, 75-78

The St. Boniface Landslide of April 1996: a Large Retrogressive Landslide in Sensitive Clay with Little Flow Component

D. DEMERS, D. ROBITAILLE, Géotechnique et géologie, Ministère des Transports du Québec, Québec, Canada, and D. PERRET, Commission géologique du Canada, Centre géoscientifique de Québec, Québec, Canada

ABSTRACT
On April 21st, 1996, a massive landslide occurred along the Rivière Machiche, about 5 km south of Saint-Boniface-de-Shawinigan, in the sensitive clays of Eastern Canada. The movement is one of the largest to have occurred in the St. Lawrence Lowlands over the last century. The slide is characterized by 10 unbroken ridges, some of which run parallel to the river for a distance of 700 m. It produced a horst and graben lateral spread with very little debris flow. Piezocone tests, vane tests and soil samples confirm the type of landslide and show that the failure surface was not uniform and controlled by contrasting mechanical properties of the soil.

INTRODUCTION

The Saint-Boniface-de-Shawinigan landslide that occurred in the afternoon of April 21st, 1996 along the banks of the Rivière Machiche (Figure 1) is a major retrogressive slide. Its size (approximately 14 ha.) makes it one of the largest to have occurred in Eastern Canada this century. The slide cannot be considered as an earthflow due to its little flow component. Successive enormous and narrow strips of terrain characterise it (Figure 2). Some of these strips are over 700 m long, which is exceptional.

Even if similar landslides occur every 2.5 years in this part of Canada (Lebuis *et al.*, 1983), few of them have been investigated. A detailed terrain investigation was undertaken for the Saint-Boniface-de-Shawinigan landslide in order to understand more fully the failure mechanisms that take place in such phenomena. The slide was previously described in two summary publications (Bégin *et al.*, 1996, Evans *et al.*, 1997). The current article will focus on presenting the principal results obtained from soundings and boreholes in and around the landslide. As the piezocone has already proven itself capable of detecting the location of failure plans for landslides occurring in sensitive clays (Demers *et al.*, 1999), this field instrument has been extensively used in preparing this study.

GEOLOGICAL AND GEOMORPHOLOGICAL CONTEXT

The landslide occurred in post-glacial sea sediments, dating from approximately 10 000 years ago. A mainly sand pro-deltaic and deltaic sequence of roughly 10 m in thickness is underlain by clay sediments which are at least 40 m thick. The entire sedimentary body is contained in a deep pre-glacial valley, with the contretorts of the Canadian Shield to the north and the St. Narcisse moraine to the south.

Landslides in research, theory and practice, Thomas Telford, London, 2000.

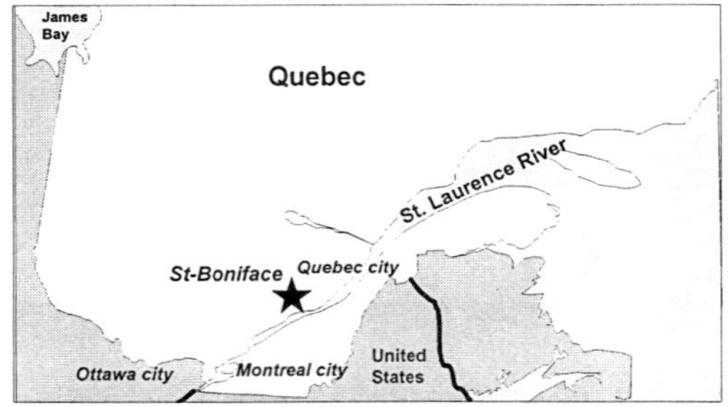

Figure 1 : Location map of the site.

Figure 2 : Photograph taken in the hours following the landslide. View toward the south showing the three lobes and the ridges.

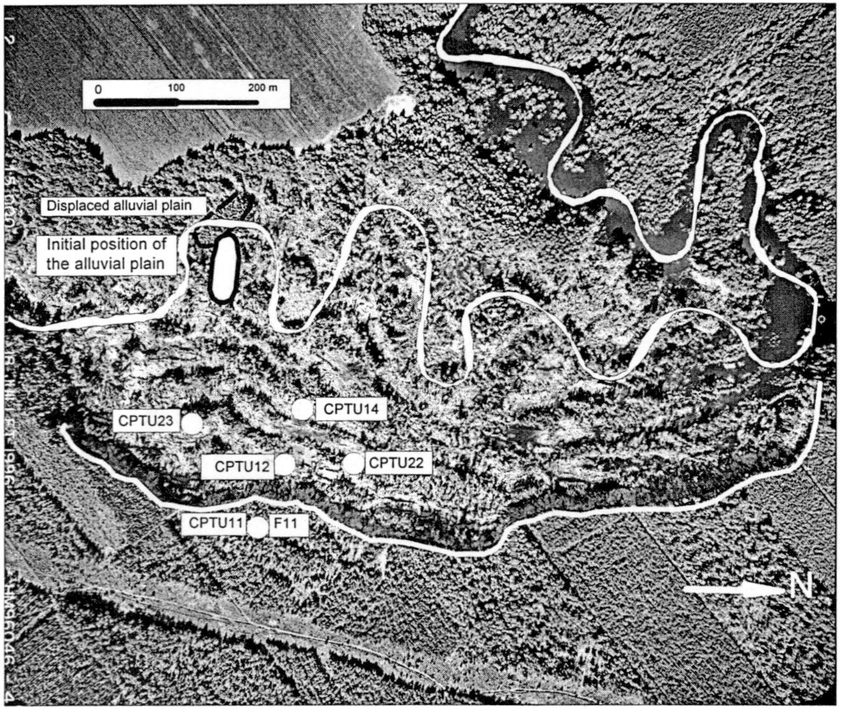

Figure 3 : Aerial photograph taken in July 1996 showing the scar, the location of the principal borings and the displaced alluvial plain.The river's initial course is also shown.

The Rivière Machiche, situated in the St. Lawrence plain, follows a sinuous course in a gully that is between 30 to 35 m deep. Alluvial plains form plateaus within the meanders that can be up to 7 m above the river bed. The lower few meters of the densely forested river banks are eroded. At the landslide site, no less than 4 major erosion points were active (Figure 3). The river banks are precariously stable, since deep or superficial rotational landslides had previously occurred. Downstream from the site, two scars left by major retrogressive landslides are evident.

SITE DESCRIPTION

The Saint-Boniface slide runs parallel to the river over a distance of 950 m (Figure 2 and 3). It is formed by a succession of roughly ten ridges that have remained more or less intact. The ridges are heavily forested and can attain a length of over 700 m. The trunks of trees on the ridges indicate both forward and backward tilting, although many are still upright. Strips of very remoulded material, forming rugged surfaces in which some pinacles of intact clay are visible, separate the forested ridges.

The average retrogression at the top of the cliff is only 150 m, meaning the scar differs greatly in appearance from other major landslides that have occurred in Eastern Canadian clays (Chagnon, 1968; Tavenas, 1984). When viewed from above, the crest of the landslide amphitheatre resembles a stretched arc that is comprised of three distinct lobes (Figure 3). Each lobe corresponds to an increase in the number of ridges involved in the movement and whose size is related to the valley's initial width. The larger portion of the colluvions came to rest against the opposite side of the valley, completely obstructing it. Very little debris was carried downstream. The debris accumulated in a compact mass, whose relatively uniform surface had a maximum relief difference of 2.5 m. An alluvial plateau on the same side of the river as the landslide was transported intact by the slide, coming to rest on the opposite side of the river. This plateau can be readily identified in aerial photos by the coniferous forest it supports (Figure 3).

STRATIGRAPHY AND GEOTECHNICAL PROPERTIES

Stratigraphic and geotechnical characteristics of the site are very uniform (Bégin *et al.*, 1996). Data from borehole F11 (Figure 4) and piezocone CPTU11 (Figure 5A and B), located just behind the scar (Figure 3), provide information about the initial soil conditions. The profile begins with a fine to medium sand with loose to dense compaction to a depth of 6 m. The sandy deposit then becomes very compact, containing some sandy silt interbeds, for another 4 m. Between 10 to 14 m in depth, the deposit changes to a clayey silt with sandy stratifications that become thinner with depth. Piezocone measurements of the tip resistance (q_t) vary because of the numerous stratifications, though the results are always greater than 1000 kPa.

Between 14 to 28 m down, the clayey deposit is more homogeneous and thin sandy strata only occasionally appear. The deposit is comprised of approximately 70% silt and 30% clay. However, small grain size variations cause rhythmic fluctuations of soil properties. Thus, more silty layers have a plasticity index (I_p) ranging from 3 to 6, remoulded shear strength (τ_{fur}) ranging from 0.1 to 0.2kPa, water content (w%) ranging from 28 to 34% and liquidity index (I_L) of up to 5. Inversely, the more clayey layers have a I_p ranging from 8 to 11, w% between 37 and 55% and τ_{fur} of 0.4kPa.

Between 28 and 42 m down, (the maximum depth measured), the soil comprises up to 52% clay. Rhythmic fluctuations in the physical and mechanical properties of the soil are still present. This lithographic change is clearly registered by the piezocone due to an increase in lateral friction (f_s), as well as by the ratio between the pore pressure (u_b) and the tip resistance (q_t) (Figure 5A and B). Between 16 and 42 m down, the q_t profile shows regular peaks, though it generally increases gradually with depth, ranging from 1200 to 3000kPa. The shear strength of intact clay varies between 70 and 150 kPa, making it a stiff to very stiff clay. The sensitivity recorded at borehole F11 is even higher (S_t up to 1500) than the first regional results presented by Bégin *et al.* (1996). The overconsolidation ratio drops from 1.6 to 1.3.

Seven piezocone borings were done inside the scar. CPTU22 (Figure 5C), similar to CPTU12, first reveals the presence of stratified sandy material to a depth of 5.5m. Then the borings encounter clayey material with a q_t profile ranging from 250 to 1000kPa, indicating that significant remoulding has occurred. The failure plan, situated at an elevation of 75.0 m, is easily identifiable due to a slight reduction in u_b and q_t, followed by a rapid increase of the values until they reach the intact profile values as measured by CPTU11.

CPTU14 (Figure 5D), similar to CPTU23, gives results that differ from those of the two previous tests. Clayey material is found at the surface. This material has low q_t measurements until a depth of 4m is reached, then the profile increases past 1000kPa. A simultaneous reduction of the u_b and q_t at an elevation of 75.32m indicates the location of the failure plan. Deeper into the ground, the profile is identical to the intact soil profile (CPTU11). The difference between the u_b and q_t profiles on either side of the failure plan reveal that approximately 4 to 5m of soil is missing. This soil was probably remoulded and extruded upwards during the slide.

ANALYSIS OF FAILURE MODE

The position of the failure plan varies between 66,3 m and 77,0 m. It seem to be at the same elevation along the same ridge but may differs from a ridge to an another. It is not controlled through a lithologic unit but it may correspond to weaker resistance layers. According to the piezocone measurements, the undrained shear strength of successive layers can varies, for example, from 70 to 130 kPa on a distance of 0,3m only.

All the field observations and test results, incorporated into the diagram accompanying Figure 5, allow us to conclude that the main failure mechanism is linked to the development of horsts and grabens, as described by Odenstad (1951), Seed and Wilson (1967) and Carson (1977, 1979). After the floodwaters receded, triangular ridges representing horizontal stratifications were observed, which meant that the pinnacles had not been displaced through rotation. They are often covered by a large quantity of remoulded material extruded from deep within the failure zone. As detected in the CPTU14 boring, the portions of intact clayey soil between 4 and 11 m deep reveal the presence of these pinnacles. The remoulded material covering the pinnacles comes from the missing layer located immediately above the slide plan. Furthermore, the inclination of the tree trunks and the presence of numerous horizontal soil plates suggest that the slide occurred by a subside and lateral spread. These intact plates lay on a large quantity of remoulded clayey soil, as indicated by CPTU22. The last ridge involved in the slide is a circular slip associated with the backscarp readjusting itself.

CONCLUSIONS

The information gathered indicates that the slide surface is not controlled by change in the lithology but may be associated with strong contrasts in the shear strength resistance. The elevation of the failure plan varies from one ridge to another. The principal failure mode consists of a horst and graben lateral spread. Extrasensitive clayey soil along the failure plan was very remoulded and was extruded to the surface between the more or less intact ridges.

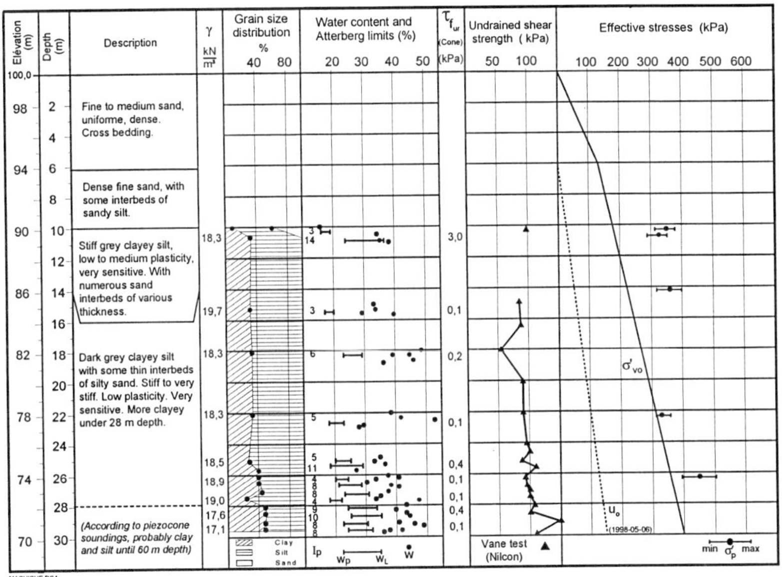

Figure 4: Geotechnical profil near the backscarp (F11).

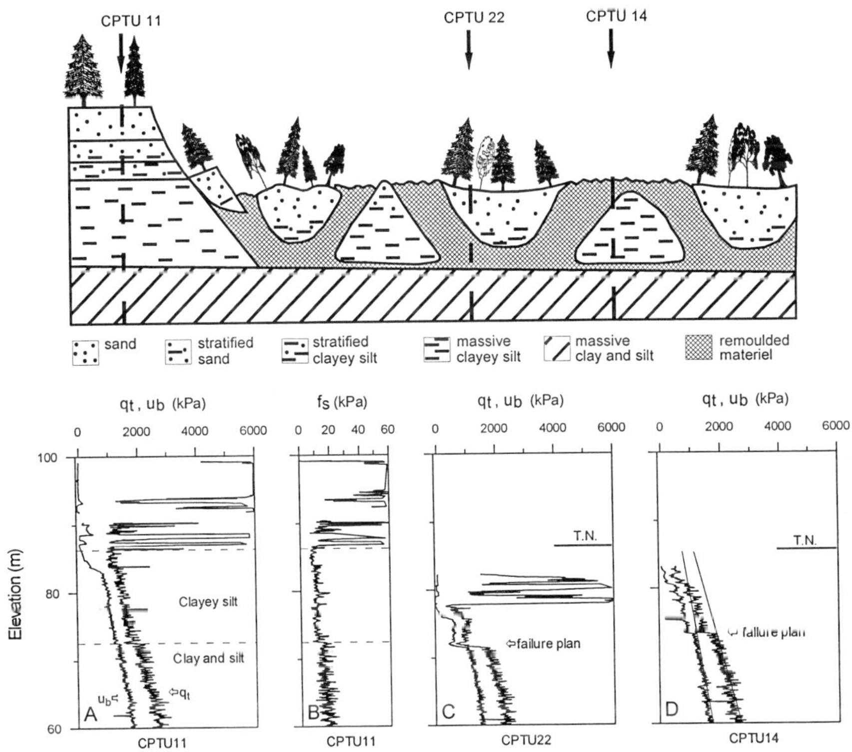

Figure 5 : Piezocone profiles and the diagram illustrating the failure mode.

Despite the sensitive nature of the deeper material, there is little flow component for this movement. This is probably explained by the thickness of the non-sensitive sandy material near the surface, the rapidly blocked narrow valley and the horst and graben failure mode, which could not provide the necessary energy to remould the material. During classic circular slips, sensitive material is heavily remoulded after it falls which obviously is not the case here. The retrogression was controlled by the width of the valley, which limited the number of ridges involved in the slide, as had been previously observed by Carson (1981). For the time being, investigation results have not allowed us to understand what triggered the failure. However, the displacement of an alluvial plain indicates that the horst and graben mechanism was present at the onset of the slide. This failure mode by ridges occurs very frequently during major retrogressive landslides in Eastern Canadian clays, according to recently observed slides (Grondin and Demers, 1996, Evans and Brooks, 1994, Carson, 1979). The hypothesis normally proposed regarding retrogression by successive circular slips can no longer be considered as the main failure mode leading to large movements in the clayey soils of Eastern Canada. We must learn more about these phenomena in order to improve prevention and mitigation methods. Others investigations are going on to verify the present hypothesis.

BIBLIOGRAPHY

Bégin C., Evans S.G., Parent M., Demers D., Grondin G., Lawrence D.E., Aylsworth J.M., Michaud Y., Brooks G.R., Couture R., 1996. *Le glissement de terrain d'avril 1996 à Saint-Boniface-de-Shawinigan, Québec : observations et données préliminaires.* In Recherches en cours 1996-E, Commission géologique du Canada, p. 215-223.

Carson M.A., 1977. *On the retrogression of landslides in sensitive muddy sediments.* Can. Geotech. J., 14, p. 582-602.

Carson M.A., 1979. *On the retrogression of landslides in sensitive muddy sediments : Reply.* Can. Geotech. J., 16, p. 431-444.

Carson, M.A., Lajoie, G., 1981. *Some constraints on the severity of landslide penetretation in sensitive deposits.* Géographie physiqe et Quaternaire, 35, p. 301-316.

Chagnon J.Y., 1968. *Les coulées d'argile dans la province de Québec.* Naturaliste canadien, 95, p. 1327-1343.

Demers D., Leroueil S., d'Astous J. 1999. *Investigation of a Landslide in Maskinongé, Québec.* Can. Geotech. J., 36, p. 1-14.

Evans S.G., Brooks G.R., 1994. *An earthflow in sensitive Champlain Sea sediments at Lemieux, Ontario, June 20 1993, and its impact on the South Nation River.* Can. Geotech. J., 31, p. 384-394.

Evans S.G., Bégin C., Lawrence D.E., Demers D., Grondin G., Aylsworth J.M., Parent M., Michaud Y., Brooks G.R., 1997. *A Massive Retrogressive Landslide in Sensitive Champlain Sea Sediments near Saint-Boniface-de-Shawinigan, Québec, Canada, April 1996.* Landslide News, no 10, p. 17-19.

Grondin G., Demers D., 1996. *The 1989 Saint-Ligouri flakeslide : characterization and remedial works.* Proc. of the 7th Int. Symp. On Landslides, Balkema, p. 743-748.

Lebuis J., Robert J.M., Rissmann P., 1983. *Regional mapping of landslide hazard in Québec.* Symp. on Slopes on Soft Clays, Swedish Geotech. Institute, Report no17, p. 205-262.

Odenstad S., 1951. *The landslide at Sköttorp on the Lidan River.* Proceedings of the Royal Swedish Geotech. Institute, 4, p. 1-38.

Seed H.B., Wilson S.D., 1967. *The Turnagain Heights Landslide, Anchorage, Alaska.* J. Soil Mech. and Found. Div., 93, p. 325-353.

Tavenas F.A., 1984. *Landslides in Canadian Sensitive Clays – A State-of-the-Art.* Proc. 4th Intern. Symp. On Landslides, p. 141-153.

Basis for Criterion for Warning Systems on Expected Landslides

A. M. DEMINE, Comprehensive Institute for Mineral Resources Exploitation, Russian Academy of Sciences, Russia

GENERAL PROPOSITONS

Owing to opencast operations, landslides occur at open pit mines periodically. Landslides deteriorate the terms of open cast operations, decreasing the effectiveness of extraction. Table 1 gives the examples of largest landslides, which took place at modern open pit mines for the last 17 years.

Table 1.

Year	Country	Landslide conditions	
		Location	Volume, mln m^3
1981	former GDR	Pit wall of «Berzdorff»	90
1985	former CSR	Outward dumps of «Merkur» open pit mine	120 -140
1987	former USSR	Pit wall number 3 of Prikaspiyskiy GOK	25
1990	former CSR	Outward dumps of «Iirzhi» open pit mine	50 -70
1992	Russia	Outward dumps of open pit mine of Norilsk GMK	60
1997	Russia	Outward dumps of open pit mine of Michailovskiy GOK KMA	20

Ordinary landslides occur more widely and have volume of tens and hundreds of million m^3, they involve very high hazard levels. Mining engineers need to predict the moment of failure at dangerous part of open pit mines, especially when cracks are present. This evaluation gives an opportunity - to elaborate warning system on expected landslides.

As many factors influence on landslides, the method of analysis of data of observations for moving of slopes at open pit mines has been chosen. For collection, generalization and analysis of data we've elaborated 2 methods: classification of landslides at open pit mines and methods for systematization of descriptions of large landslides at open pit mines. Using this method, we've collected and generalized data of dimensions of landslides, which took place in Russia and primary mining countries for last 70 years. Database contains about 400 cases.

Landslides in research, theory and practice, Thomas Telford, London, 2000.

METHOD FOR APPROXIMATE ESTIMATION FOR MOMENT OF DISASTROUS PHASE OF LANDSLIDE.

The determination of the approach of a landslide's disastrous phase is the primary task of the whole complex of problems on the prognosis of landslide danger in open pit mines. There are 5 indexes for determination of this moment at present:

1. The velocity of vertical displacement of the top ground of the surface mine-working or dump. Judged by the voluminous literature, which is devoted to this index, it is of world-wide use in practice. However, its main defect lies in the fact, that the velocity is indirect, but not directly measurable value and, so, it depends on the measurable interval time. Besides that, numerical significance of critical velocities are changed at wide scope, differing from each other by 40 orders . So vcr. changed at 8-300 mm/day in a rock mass and 1 -30 mm/day in a clay mass.
2. The number of noises per unit time at the sloping zone of the surface mine working or dump. In this case boring of deep holes with transducers placed in them and long duration continuous observations is necessary. However, field research, executed in this way, has not given real results.
3. Components of Earth's magnetic field. This needs the use of bulky apparatus, and arrangement of special mine workings for its placement. Use of this method is impossible at iron ore deposits.
4. Stresses in the electrical field at the sloping rock mass. The electrical sound was tested at the open-cast Malo-Chkalovsky . This method needs expensive apparatus, and measuring may be conducted during "technological windows" only, through hindrances of working mining equipment.
5. Changes of microinclinations (in angular seconds). This method needs using highly exact apparatus, which must be placed in special building at underground mining workings.

The observational practice for landslides in quarries shows that except in the method N. 1, by which the indirect value is measured, which hardly depends on real (accepted) time interval of measurings, the remainder of methods have not become widespread. Using this fact 2 main requests for new informative criterion have been determined influenced by the moment of disastrous phase of landslides: a) opportunity for measurement it by marksheider survey of open pit mine; b) variation of critical values of this criterion.

Based on generalization and analysis of results of field instrumental observation for landslides in Russia, and main mining countries of the world, the basic criterion selected are - the value of relative vertical deformation of the upper part of benches - h_{ver}/ H, where h_{ver} - vertical settlement of the upper part of bench; H - the height of bench. The critical value of relative vertical deformation is corresponds to moment of disastrous phase of landslides. Used the data observation for landslides, we've determined the critical value h_{ver}/ H, presented at the Table 2. Variations of critical deformation values are so small and they are equal for benches of dumps and natural slopes. This method is acceptable for landslides with brittle rupture of rock mass (landslide failures). For viscous - plastic destruction it's necessary to use another criteria.

So, the most informative criterion for monitoring for landslides dangerous at open pit mines and natural slopes is the value of relative vertical deformation of the upper part of bench part of slope which is variated at the diapazon 0.01 - 0.05 . Used this criterion increase the truth of prognosis and decrease expenses for monitoring.

Table 2. Critical values of relative vertical deformation

N	Landslides location	Year	Country	$(h_{ver}/H)*10^{-3}$	Information
	Landslides of dumps and bench of open pit mines				
1.	Inside dump of open pit mines Otto	1936	Germany	12	Heineman, 1939
2.	Bench of of Leonhardt open pit mines	1937	Germany	14	Lang, 1939
3.	Railway embankment .	1962	Japan	22	Rasrushenie, 1964
4.	Outward dump of open pit mines of Turov.	1966	Poland	12	Rzepisko, 1969
5	Outward dump of open pit mines at Apatit group of mines	1967	USSR	25	Permjakov, 1969
6	Bench of Chuquicamata open pit mines	1969	Chile	18	Kennedy, 1971
7	South bench of Korkinskiy open pit mines	1970	USSR	13	Sapozhkov, 1986
8	South bench of Korkinskiy open pit mines	1971	USSR	42	Sapozhkov, 1986
9	East bench of «South» PO «Vahrushevugol» open pit mines	1976	USSR	50	Sapozhkov, 1986
10	South bench of Korkinskiy open pit	1977	USSR	38	Sapozhkov, 1986
11	West bench of Volchanskiy open pit	1978	USSR	30	Sapozhkov, 1986
12	Bench of Berkely open pit mines	1977	USA	15	Goldberg & Frizzel, 1989
13	Bench of Berkely open pit mines	1982	USA	22	Goldberg & Frizzel, 1989
14	Bench of trench in open pit mines of Yaroslavskiy	1986	USSR	11	Glozman, 1988
15	Outward dump of open pit mines in British	1993	Canada	25	Hungr, 1995
16	Bench of trench in open pit mines of Yaroslavskiy	1994	Russia	15-18	Bakhaeva, 1995
17	Bench of open pit mines «Damian»	1970	Yugoslavia	12	Metodika, 1980
	Landslides of natural slopes				
18.	Agojama	1972	Japan	10	Saito, 1982
19.	Tama	1976	Japan	12	Saito,1982
20.	Jukovka	1986	USSR	15	Erish, 1989

REFERENCES

Bakhaeva, S.P., Bakushkin, B.P. Analyz marksheiderskykh nabludeniy za ustoichivosti bortov razreza - Kemerovo , 1995 , № 9 - s. 78-83

Demine, A.M, Erseev, V.N. Upravlyaemoe sdvizhenie otvalov -M, 1982, 27 s.

Demine, A.M. Problema preduprezhdeniya deformatsiy obkritikh gornikh virabotok i otvalov // Gorniy zhurnal. 1984, - № 7 - s. 20-23

Demine,A.M. Zakonomernosty proyavleniy deformatsiy otkosov v karierach: M:navka, 1981.

Glozman, G.R. i dr. Opit otstroiki borta kariera v usloviyakh ego deformirovaniya - L., Gipronikel , 1988 , s 74-83

Heinemann . Rutschungen im Tagebau Otto -Tanneberg // Braunkohle, 1939 , H 9 S. 233-237

Hungr. ,O Kent, A. Coal mine waste dump failure in British Columbia , Canada // Landslide News , 1995 , № 9 - p.26 -28

Iskusstvennaya aktivizatsiya opolzney / Erish , I.F. , Postoev , G.P. Solomatin ,V.N. i dr. - M. Nedra ,1989 - 134 s.

Kennedy , V.A. and others . A case study of slope stability at the Chuqui - camata mine, Chile // Trans. Soc. Mining Eng. AIME , 1971 , 250 , № 1. - P.55-61.

F.Lang .Uber eine Kohlenrutschung im Tagebau Leonardt // Braunkohle,1939, H 9 S., 129-132

Metodika inzherno- geologicheskikh issledovaniy visokikh obvalnikh i opolznevikh sklonov - M. : Izd - MGU , 1980 - 184 s.

Nikolashin , V.M. , Dushutin , A.K. Kontrol razrusheniy v porodakh borta kariera metodom seismoakustiki // Geol.- geophiz . obespechenie gornorudnikh predpriyatiy ratsionalnoe ispolzovanie nedr - Belgorod , 1982 - s. 73-75 .

Parikovskiy, V.V. i dr. Otsenka sostoyaniya otvalov skalnikh smeshannikh porod s ispolzovaniem akusticheskoy emissii // Gorniy zhurnal, 1995 - №9 - s.13-14.

Permyakov R.S. Osobennosti razrabotki nagornikh mestorozhdeniy Zapolyarjya - L Navka 1969 - 234 s.

Pit slope monitoring and back analysis of the Berkeley Pit / Goldberg , W.C. , Frizzel , E.M. // Mining Eng . / USA / . - 1989 - 41 № 12. - P. 1191 -1195 .

Polovov, B.D., Lyadskiy, V.L. Klassifikatsia gorno-tekhicheskikh objectov i otsenka vozmozhosti effectivnogo upravleniya ikh sostoyaniem // Ekologicheskie problemi gornogo proizvodstva, pererabotka u razmeshenie otkhodov - M., MGGY, 1995 - s. 553-557.

Rasrushenie osnovaniya nasipi // Soil and Foundat, 1964, № 2 - p.24-33

Rzepisko, R. Obsunience skarpy zwali wewnetrznego w jednej z koplan odkrywkowych welga brunatnego . // Przegl . geologiczny , 1969 ,17 , № 8 . 393 -397

Saito, M Fakticheskiy analyz sluchaev predskazaniya obrusheniy sklonov // Problemi geomekhaniki (Erevan) , 1983 , № 8 - s 45-74

Sapozhkov , V.T. , Kim , D.N. , Afanasjev , B.G. Harakter deformirovaniya bortov uralskikh razrezov v okolopredelnom sostoyanii // Sposobi upravleniya deformatsiyami gornogo massibva - L . Vnimi , 1986 - s.64-71

Shirokov ,I.A. , Nikolashin , V.M., Anokhin, K.M.. Opit izuzhenia deformatsionnogo regima borta glubokogo kariera s pomoshiu naklonomerov // Prilivi i vrasheniya Zemli - Kiev , 1985 - s.1 39-47

Sobolev , E.G. Asarbaev , K.A. , Danilin N.A. Razrabotka metodiki postroenia razvivaysheisya v otkose poverkhnosti skolzheniya .

Sobolev , E.G. Prognoz i kontrol opolznevikh protsessov na mestorozhdeniyakh // Gorniy zhurnal , 1986- s. 43-57.

Unson , V.K. , Montrel , M.F. Prognozirovanie ustrichivosti gornikh porod na karierakh Nikolaevskogo margantsevorudnogo besseina s pomoshju electrometricheskoy apparaturi.

Databases and Geographic Information Systems for Medium Scale Landslide Hazard Evaluation: An Example from Typical Mountain Watershed in Nepal

S. DHAKAL, Graduate School of Agricultural Sciences, University of Tsukuba, Japan,
T. AMADA, Institute of Agricultural and Forest Engineering, University of Tsukuba and
M. ANIYA, Institute of Geoscience, University of Tsukuba

ABSTRACT

The landslide distribution data extracted from aerial photos, and thematic databases derived from maps (e.g., slope gradient, geology, land use/cover) were used for modeling landslide hazard, at a medium scale (1:25,000 - 1:50,000), using GIS. The site characteristics of landsliding were evaluated using multivariate statistical (Quantification Scaling Type II) and univariate statistical (Failure Rate) methods. Two failure rate analyses were carried out. Three hazard maps were produced from the results of the analyses. The spatial agreements between three hazard maps were evaluated. The hazard maps produced from different methods agreed up to 60 percent to 80 percent depending on whether the study area is classified into four hazard classes or two hazard categories, respectively.

INTRODUCTION

Assessment, prediction and mapping of areas associated with landsliding are necessary at medium scale (1:25,000 – 1:50,000) for different uses. The hazard maps produced at this scale are useful for sediment modeling, prioritization of the areas for the management, delineation of debris flows sites, feasibility studies of roads and dams etc. In Nepal; 83 percent of which is comprised of mountains, about 20,000 landslides have been estimated to occur in just one season (Ives and Messerli, 1981). Hence, hazard mapping is critical for mitigation activities and for watershed management

Data handling for assessment of landslide hazard is often difficult, time consuming and costly. Geographic Information Systems (GIS; Burrough, 986) has overcome the many working difficulties associated with data handling (Carara 1983; Van Westen 1993; Dikau *et al.*, 1996). On larger areas, the statistical approaches of landslide hazard analysis are usually more appropriate (Mulder, 1991). The relative contributions of the slope instability causing factors are assessed on the assumption that the landslides will be more likely to occur under conditions similar to those of previous landslides (Varnes, 1984; Brabb, 1984). Hazard maps can be produced from the results of the analyses. Using GIS, the classified grid-cells at the same location on different hazard maps (hereafter termed "spatial agreement"), can be evaluated to compare the results of different methods. The purposes of this study are (1) to present methods of landslide hazard mapping utilizing GIS and statistical methods, and (2) to compare the spatial agreement between the hazard maps produced from different methods.
The study area is Kulekhani watershed, located in the Lesser Himalayan region of the Himalayan belt in Nepal (Figure 1), with the elevation ranging from 1,500 m to 2,600 m. The average annual rainfall is about 1600mm. The area is drained into the Kulekhani reservoir, which supports one third of the total electric power generation of Nepal.

Landslides in research, theory and practice, Thomas Telford, London, 2000.

ACQUISITION OF DATA BASES AND GIS DATA LAYERS

The stereopairs of black and white vertical aerial photographs (1: 20,000) were interpreted using a stereo zoom-transferscope to plot landslides on a topographic map at a scale of 1:12,500 (enlarged from 1:25,000), and was digitized after field verifications.

From digitized contours (contour interval 20m), a Digital Elevation Model (DEM) was generated from a Triangulated Irregular Network (TIN) model. Slope gradient, slope aspect and elevation layers were derived from the DEM. Ridges and valleys were also defined from the DEM. Employing the Strahler (1957) method, the drainage basin order layer was derived from the topographic map. A geological layer was derived from digitizing the 1 : 50,000 scale geological map. Land use/cover layer was created from land use/cover map (1 : 25,000) of 1991. These factors are classified into different classes (Table 1). The sample landslides surveyed in the field guided the selection and classifications of these factors.

METHODS OF ANALYSIS

The hazard assessments were done by the Quantification Scaling Type II (Q-S II; Hayashi, 1952, 1954), and the Failure Rate (FR; Aniya, 1985; Dhakal *et al*, 1999) analyses. Two FR analyses were employed using either area or number of landslides in a class of a factor. Results of the analyses were used to produce three hazard maps. The spatial agreements in the hazard maps were evaluated. Figure 2 is the flow diagram of the study.

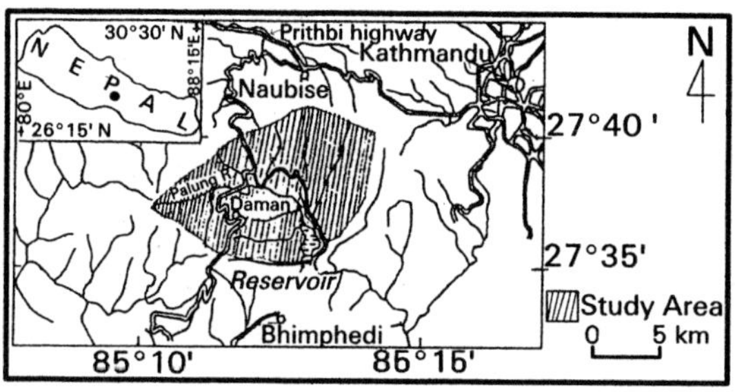

Figure 1 Location of the study area

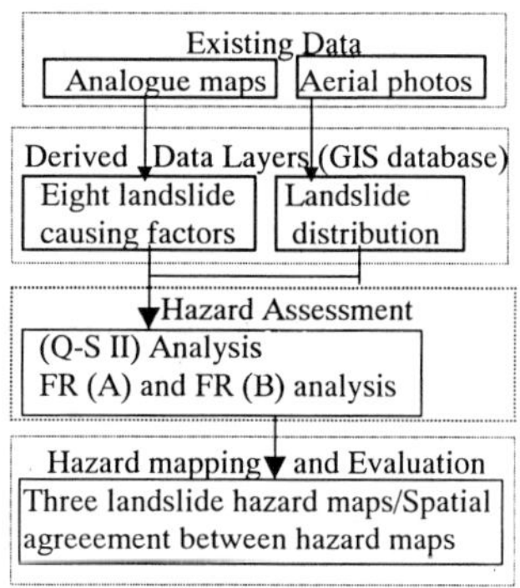

Figure 2 Flow diagram of the study

Q-S II Analysis

The Q-S II is a discriminant analysis, which incorporates nominal data. The method is suitable tothe landslide hazard assessment, because nominal variables (factors) such as geology and land use are often most important to discriminate between landslide and non- landslide groups. The linear Q-S II function (score; αa_q) for a sample belonging to a group q in n factors and m classes in a factor can be written as;

$$\alpha a_q = \sum_{j=1}^{n} \sum_{i=1}^{m} \delta a(ji)\, Xji$$

where, $\delta a(ji) = 1$, if sample *a* belongs to the i-*th* class of factor *j*, otherwise 0; X*ji* = score of the i-*th* class of factor *j*

Table 1 Derived data bases for GIS analysis

Factors	Class code						
	1	2	3	4	5	6	7
Slope gradient	< 15°	15° - 25°	25° - 35°	35° – 45°	> 45°	-	-
Slope aspect	North	East	South	West			
Elevation	< 1800 m	1800 m – 2000 m	2000 m – 2200 m	> 2200 m	-	-	-
Drainage basin	First	Second	Third	-	-	-	-
Distance from ridge	< 50 m	50 m - 100 m	> 100 m	-	-	-	-
Distance from valley	< 50 m	50 m - 100 m	> 100 m	-	-	-	-
Geology	Slates with quartzites or limestones	Limestones	Slates with meta-sandstones and phyllites	Schists with quartzites and marbles	Biotite schists and micaceous quartzites	Granite	Alluvium
Land use/cover	Crops	Coniferous forest	Broad leaf forest	Mixed forest	Shrub land	-	-

The separation between groups (landslide and non-landslide in this case) is maximized by the frequencies of factor-classes of the sampled data of the two known groups. As a large class score in a factor contributes more than a small one, class score and the range of scores (difference in maximum and minimum scores between the classes) can be interpreted to determine their importance. For the data, which are not sampled, the factor-classes are measured, and the group to which they belong is predicted from the score of the classes. The Q-S II analysis is available in the Japanese version of SPSS statistical package.

For sampling, the study area was tiled into grid-cells of 25 m x 25 m, a set of grid-cells representing the landslides was randomly chosen. Subsequently, the area was stratified into rectangular blocks, and a set of non-landslide group was obtained from the unaligned stratified random sampling method. The classes of eight terrain factors (see Table 1) were assigned to each grid-cell of landslides and non-landslides for the Q-S II analysis.

FR Analysis

The FR analysis is a univariate statistical approach in which the importance of each class in a factor is individually analyzed. A numerical score is computed using either number of the landslides, or area of the landslides falling on the class. It was calculated as, FR Score = R (L) / R (A), where, R (L) = a ratio of the landslide grid-cells in a particular class of a factor to the total landslide grid-cells, or a ratio of number of landslides to the total landslide area; R (A) = a ratio of the grid-cells belonging to that particular class to the total study area. In FR (A) analysis, the score for the classes were computed from area of landslides falling in a class, whereas in FR (B) analysis, number (frequency) of landslides were used. The score greater than 1 suggests that the class contributes to landsliding.

RESULTS

Q-S II Analysis

As higher range of scores indicate higher importance, geology is found to be the most important factor for landsliding (Table 2). Elevation, land use/cover and slope aspect are ranked in the second group, while slope gradient is ranked in the third group of importance. Drainage basin order, distance from ridge and valley have the least importance.

As for classes for geology, the score is high for "granite", followed by "biotite schists with micaceous quartzites". In elevation, a zone of "2,000 m - 2,200 m" is most susceptible, followed by " 1800 m -

2000 m". In land-use/cover, "coniferous forest" shows the highest importance followed by "shrub land". "East" followed by "South" facing slopes are critical. For slope gradient, the score is highest at class "35° - 45°". The first order drainage basin shows higher importance. The greater the distance from the ridge (>100 m) and the shorter the distance from the valley (<50 m) the susceptibility increases.

Table 2 The results of the Q-S II analysis, a large range of scores and a large partial correlation coefficient indicate more discriminating potential of the factor, and a negative class score implies the contribution to landsliding. The results of FR (A) and FR (B) analyses are also listed, in which a large class scores indicate the critical classes to landsliding (first five important factors are listed).

Factor	Class	Q-S II analysis			FR (A)		FR (B)	
		CS	RS	PCC	CS	RS	CS	RS
Geology	Slates with quartzites or limestones	1.607	2.333	0.301	0.17	1.70	0.13	1.45
	Limestones	0.818			1.69		0.88	
	Slates with meta-sandstones &	0.112			1.17		1.09	
	Schists with quartzites & marbles	0.110			0.97		0.79	
	Biotite schists & micaceous	-0.210			1.32		1.27	
	Granite	-0.726			1.16		1.45	
	Alluvium	0.888			0		0	
Land Use/cov	Crops	0.348	0.789	0.153	0.62	1.05	0.71	0.766
	Coniferous forest	-0.441			1.51		1.48	
	Broad leaf forest	0.179			0.93		0.72	
	Mixed forest	-0.193			0.89		1.23	
	Shrub land	-0.395			1.66		1.32	
Elevation	< 1800 m	0.318	0.856	0.145	0.40	1.4	0.49	1.172
	1800 m - 2000 m	-0.134			1.0		1.12	
	2000 m - 2200 m	-0.485			1.80		1.66	
	> 2200 m	0.371			1.06		0.94	
Slope aspect	North	0.484	0.834	0.151	0.80	0.77	1.04	0.780
	East	-0.350			1.32		1.27	
	South	-0.286			1.32		1.18	
	West	0.149			0.55		0.49	
Slope gradient	< 15°	0.109	0.438	0.057	0.51	0.94	0.65	1.31
	15° - 25°	-0.088			1.20		1.39	
	25° - 35°	0.038			1.44		1.50	
	35° - 45°	-0.288			1.45		0.83	
	> 45°	0.150			1.13		0.19	

CS = Class score, RS = Range of scores, PCC = Partial correlation coefficient

FR Analysis

Table 2 also lists the scores obtained for FR (A) and FR (B) analyses. Since FR analysis considers the importance of classes of a factor independently, the interpretation of ranges of scores is difficult. The results of FR (B) analysis are very much close to the results of Q-S II analysis. It might be because the Q-S II analysis also tends to depend on the number of landslides in a class, because each sample of landslide grid-cell represents a single landslide. Though the scores obtained for FR (A) also resemble with Q-S II analysis the class "shrub" replaces "coniferous forest" in land use/cover, and "limestones", replaces "granite" in geology, though the score for granite and coniferous forest are also higher. These differences are met because the class "limestones" (geology) where a few large landslides are located received the highest score in the FR (A) analysis, whereas "granite", where the large numbers of small landslides are located, received the highest score in the Q-S II analysis. In summary, the association of classes "granite" (Limestones in FR (A)), "2000 m - 2200 m", "coniferous forest" ("shrub land" in FR (A)), "East" "35° - 45° (25° -35° in FR(B))", "first order" <50 m (distance from valley) and >100 m (distance from ridge) are most critical to landsliding.

LANDSLIDE HAZARD MAPPING

The factor layers with scores for the classes were superimposed to obtain a cumulative score for each grid-cell. For Q-S II analysis, the point of intersection of two cumulative frequency curves plotted by the score for landslide and non-landslide groups obtained from the analysis was taken as a critical value separating the unstable and stable hazard category (Figure 3). In order to differentiate between very unstable and marginally unstable and between very stable and marginally stable, scores at which the accuracy of decision would be about 80 percent were selected to divide each hazard category into two classes, creating four hazard classes: high, moderate, low and least. The term hazard category is used for unstable and stable, and the term hazard class is used for high, moderate, less and least. For FR analysis, the scores that separate the different hazard classes into the same number of grid-cells as that of the Q-S II analysis were obtained and hazard maps were produced. Figure 4 shows hazard maps produced from the results of Q-S II, FR(A), and FR(B) analyses.

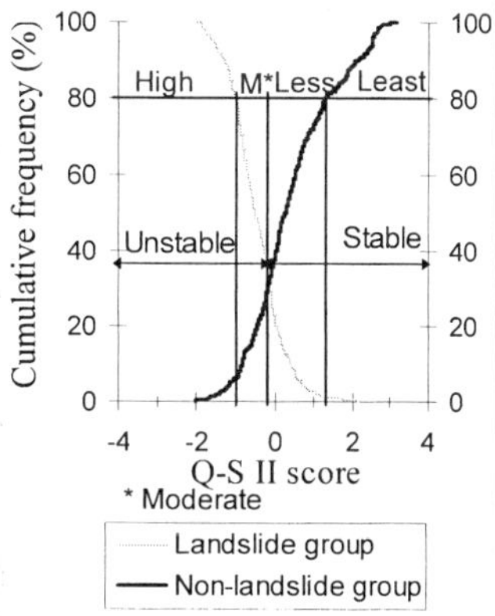

Figure 3 Cumulative frequencies and Q-S II scores

SPATIAL AGREEMENT BETWEEN THE HAZARD MAPS

Spatial agreements between the hazard maps were evaluated to quantitatively examine the difference in the hazard maps produced. Two of the three hazard maps were overlaid in turn and grid-cells classified into the same hazard class (agreed grid-cells) were counted. The "overall spatial agreement" was then calculated by taking the proportion of agreed grid-cells to the total number of grid-cells, in the similar manner to the evaluation of overall accuracy from the error matrix. Figure 5 depicts the agreed and disagreed hazard classifications for each map crossed with another. It depicts two third of the area into same hazard classes when hazard maps are classified into four classes. However, if study area is classified into only two hazard categories, unstable and stable, the overall spatial agreement will be about 80 percent.

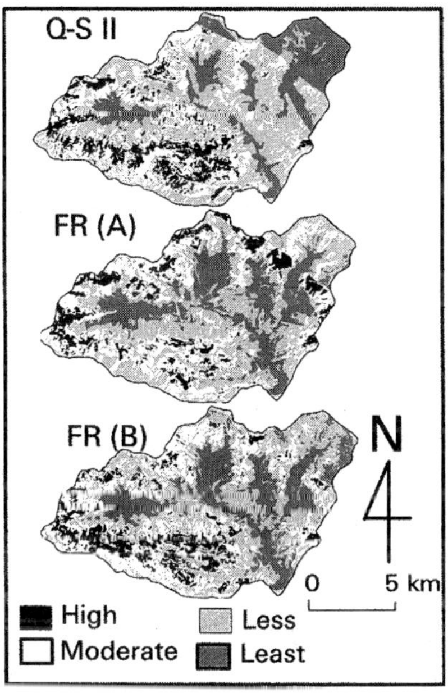

Figure 4 Hazard maps produced from the results of three analyses

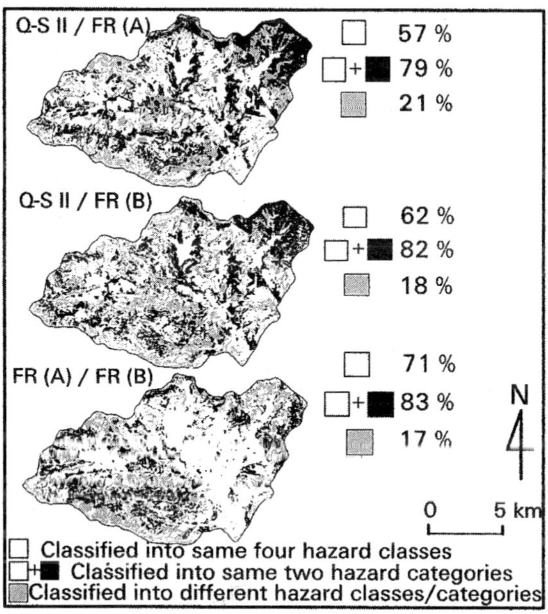

Figure 5 Spatial agreement between the three hazard maps (indicated upper left in each map)

DISCUSSIONS AND CONCLUSIONS

The study demonstrated the method of medium scale landslide hazard assessment and mapping using GIS and statistical methods. The use of a GIS in hazard assessment and mapping is invaluable in reducing the complexity associated with the hazard assessment. The methods used exhibit comparative advantages and limitations. The FR analysis did not consider the correlation between the factors because a variable was considered independently when calculating the score. However it's operation is simple and easy. The Q-S II analysis takes into account the interrelationships between the factors; however, the outcome of the analysis may depend to some extent on the sampled data. Irrespective of the method employed or condition introduced (in FR analysis), two third of the areas were classified into same hazard class, when hazard map was produced with four hazard classes. If the purpose of the hazard map would be to delineate only two categories unstable, and stable, 4/5 of the area will be classified into same category. This suggests that depending on method, the sites classified into marginal hazard classes may differ in different maps. Hence, areas that fall into marginal hazard classes may need attention during the use of the hazard map.

Acknowledgments

Nepal Geotechnical Society is thanked for the logistic support provided during the field work. The research grant awarded by Japan Society of Erosion Control Engineering in the year 1998 is acknowledged. Stephen Bird is thanked for the editorial comments.

References

Aniya, M., 1985. Landslide-susceptibility mapping in the Amahata river basin, Japan, *Annals of the Association of American Geographers*,75(1): 102-114.

Brabb, E. E., 1984. Innovative approach to landslide hazard and risk mapping. *Proceed. of the 4th Int. Symposium on Landslides*, Toronto, Canada, Volume 1, pp. 307-324.

Burrough, P.A., 1986. *Principles of Geographical Information Systems for Land Resources Assessment*, Clarendon Press, Oxford, 194 p.

Carrara, A., 1983. Multivariate models for landslide hazard evaluation, *Mathematical Geology*, 15(3):403-427

Dhakal, A. S., T. Amada, and M. Aniya, 1999. Landslide hazard mapping and the application of GIS in the Kulekhani watershed, Nepal, *Mountain Research and Dev.*, 19 (1): 3-17.

Dikau, R., A Cavallin, and S. Jager, 1996. Databases and GIS for landslide research in Europe, Geomorphology, 15:227-239.

Hayashi, C., 1952. On the prediction of phenomena from qualitative data and the quantification of qualitative data from the mathematico-statistical point of view, *Annals of the Inst. of Statistical Mathematics*, 3:69-98.

Hayashi, C., 1954. Multidimensional quantification - with the applications to analysis of social phenomena, *Annals of the Institute of Statistical Mathematics*, 5:121-143.

Ives, J.D. and Messerli, B. 1981. Mountain hazard mapping in Nepal- introduction to an applied mountain research project. Mountain Research and Development, 1 (3-4): 223-230

Mulder, H.F.H.M., 1991. *Assessment of Landslide Hazard*, Faculty of Geographical Sciences, Univ. of Utrecht, The Netherlands,150p.

Strahler, A. N. 1957. Quantitative analysis of watershed geomorphology, *Transactions of the American Geophysical Union*, 38:913-20.

Van Westen, C. J., 1993. *Application of Geographic Information System to Landslide Hazard Zonation*, ITC Publication number 15, ITC, Part 1, Enschede, 245 p.

Varnes, D. J., 1984. *Landslide Hazard Zonation: A Review of Principles and Practice,* IAEG, UNESCO, Natural Hazard No. 3, 61 p.

Influence of Pore Liquid Composition on the Shear Strength of an Active Clay

C. DI MAIO and R. ONORATI, Università della Basilicata, Potenza, Italy

SUMMARY

This paper describes the results of triaxial tests and direct shear tests on an active clay of the Italian Apennine. The tests were carried out on specimens reconstituted with distilled water and on specimens reconstituted with a 1 M NaCl solution. A set of tests was performed by exposing the specimens to a liquid equal to their pore liquid, and another set on specimens prepared with the solution and exposed to distilled water. The results show that the pore liquid composition strongly influences the peak strength as well as the residual shear strength. Furthermore, exposure of solution-saturated material to distilled water causes noticeable effects on the residual shear strength.

INTRODUCTION

It is known that the behaviour of clays is influenced by pore fluid composition (among others: Mesri and Olson, 1970; 1971; Sridharan, 1991; Mitchell, 1993). Recently, Anson and Hawkins (1998) published some experimental results showing that, in the case of montmorillonite, a pore solution of 80 mg/l Ca^{2+} is sufficient to reach a residual shear angle much higher than that obtained for samples tested with deionized water. The validity of this finding is related not only to pure montmorillonite. In fact, the behaviour of soils is greatly influenced by their clay minerals; in particular, soils with c.f. > 50% exhibit values of the residual shear strength equal to those of their clay part (Lupini et al., 1981). Hence their behaviour is expected to be strongly influenced by pore liquid composition. Di Maio (1996a) published data relative to some natural clays reconstituted with distilled water and with a saturated NaCl solution, showing that the clays exhibit almost the same values of residual shear strength as the Ponza bentonite - which is a practically pure montmorillonite - both in distilled water and in the salt solution. Natural state pore liquid is often a compound solution, hence the use of distilled water for laboratory tests would be not appropriate. Anson and Hawkins briefly discussed the implication for slope stability analysis of using deionized water in laboratory tests, suggesting that the difference between results calculated by back analysis and those obtained in laboratory could be due, at least in part, to the use of deionized water. In the case of the Bisaccia clay - a smectitic soil of marine origin - previous results (Di Maio, 1996a) show that the residual shear strength is heavily influenced by pore liquid composition. Furthermore, exposure to distilled water of specimens reconstituted with a saturated NaCl solution causes slow processes of volume increase and of residual shear strength decrease. The purpose of this paper is to analyse the influence of pore liquid composition on the peak strength of the clay. To this aim, triaxial and direct shear tests have been carried out on the material reconstituted with and exposed to distilled water, as well as on the material reconstituted with a 1 M NaCl solution. The latter has partly been exposed to the same solution and partly to distilled water.

EXPERIMENTAL RESULTS

The investigated soil is the Bisaccia clay. Some of its features are expounded in a paper presented at this Symposium by the same authors. The experiments spoken about in the present paper were carried out on soil taken from an open pit. Its c.f. is about 65% and the liquid limit evaluated with distilled water is 110%. The liquid limit evaluated with a 1 M NaCl solution is equal to 51%. As expected, the intrinsic compression line is strongly dependent on the pore liquid composition, as shown by Fig.1 which reports the one-dimensional curves of void ratio against effective axial stress for the material reconstituted with and exposed to distilled water and for the material reconstituted with and exposed to a 1 M NaCl solution. Points A^I, A^{II} and A^{III} represent the equilibrium conditions after isotropic normal consolidation for three water-saturated specimens, and points B^I, B^{II}, B^{III} and B^{IV} for four solution-saturated specimens. All the specimens (height=72 mm and diameter=33 mm) underwent common triaxial CU tests which were carried out with constant cell pressure and with the axial stress increased in a strain controlled manner, at a rate of 0.03 mm/min. The results are reported by Fig.2a in terms of $q' = \sigma'_a - \sigma'_r$ against $p' = (\sigma'_a + 2\sigma'_r)/3$, and by

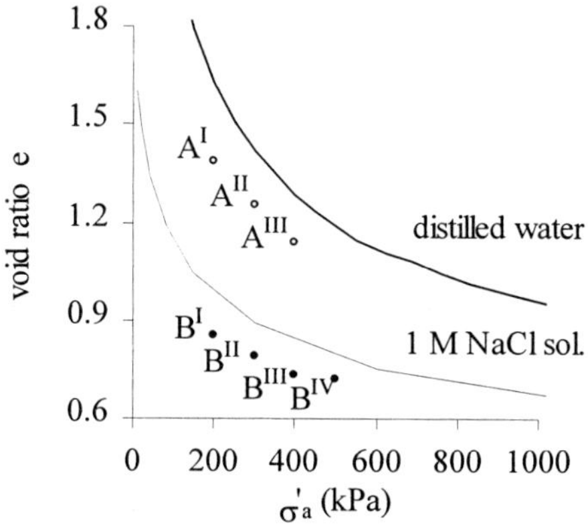

Fig.1. Intrinsic one-dimensional compression lines of the Bisaccia clay reconstituted with water and with a 1 M NaCl solution, and void ratio against axial stress after isotropic consolidation of specimens which subsequently underwent triaxial tests

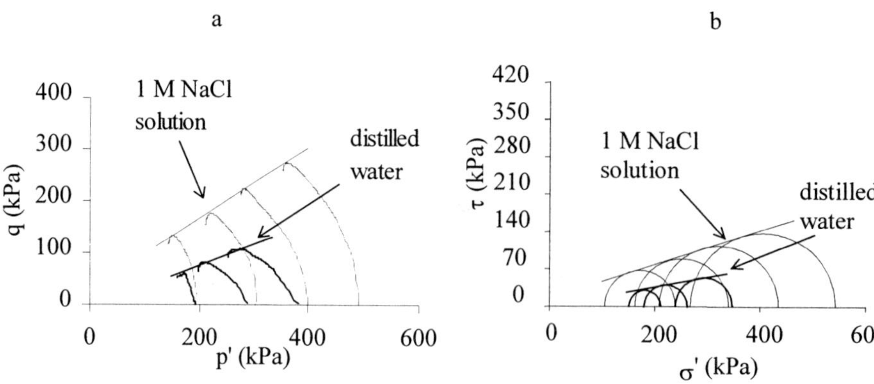

Fig.2. CU triaxial tests on normally consolidated specimens of the Bisaccia clay in terms of Mohr's circles of effective stress at failure

The figures show that the material reconstituted with the salt solution is characterised by much higher values of shear strength than the material reconstituted with distilled water. Other tests were carried out in order to evaluate whether the results could have been influenced by possible pore pressure gradients within the specimens. The material was reconstituted with a 1 M NaCl solution, and also the hydraulic circuits were filled with the same solution. Some specimens were consolidated to 800 kPa, and afterwards each of them was unloaded to a given value of *p'*. At equilibrium, some specimens were sheared in drained conditions, with a rate of axial displacement of 0.0015 mm/min, and some others in undrained conditions at a rate of 0.03 mm/min.

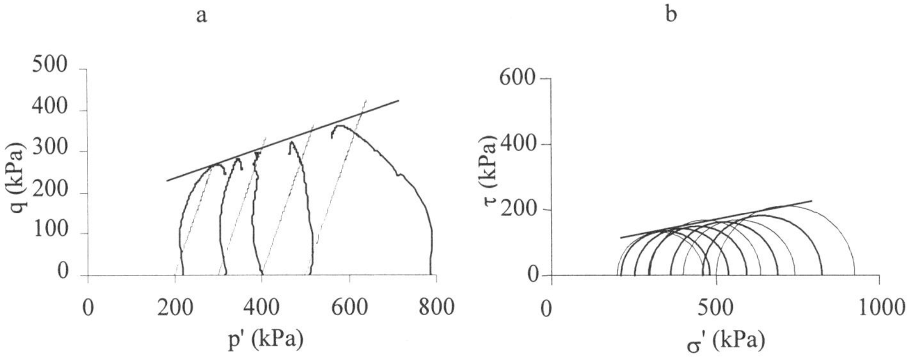

Fig.3. Triaxial tests on overconsolidated specimens of the Bisaccia clay.

Fig.3a reports the results in terms of *q:p'* and Fig.3b in terms of Mohr's circles. It can be seen that the differences between the results obtained with the two types of tests are negligible. Large differences between the behaviour of the material saturated with distilled water and that of the material saturated with the salt solution were also furnished by direct shear tests. Fig.4 reports the peak values of shear strength for specimens reconstituted with and immersed in distilled water, and for specimens reconstituted with and immersed in a 1 M NaCl solution. The specimens (area = 36 cm^2, thickness = 2 cm), normally consolidated, were sheared with a rate of displacement of 0.005 mm/min.

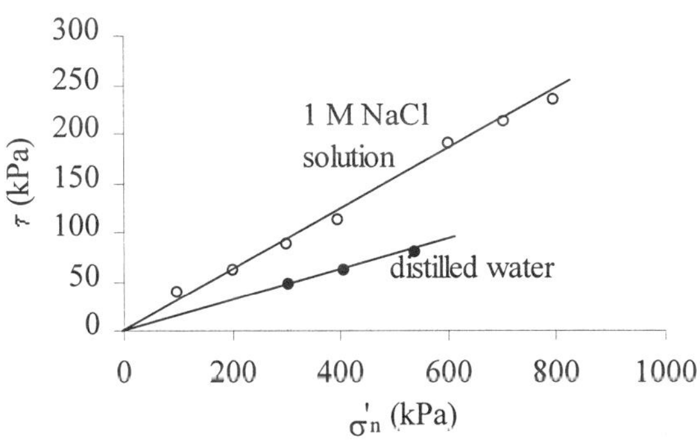

Fig.4. Peak shear strength evaluated by means of direct shear tests on normally consolidated specimens.

The friction angle φ' was found to be about 15° for the material in the salt solution, and 8.4°

for the material in distilled water. Once the peak strength had been reached, the specimens were sheared back and forth over a distance of 6 mm until the minimum strength was obtained. A time interval of about 24 h was allowed to elapse between the end of one cycle and the beginning of the next. Fig.5 reports the residual shear strength against normal stress for the two types of specimens. The residual friction angle φ'_r is 8.6° for the material in the salt solution, whereas it is less than 4° in distilled water.

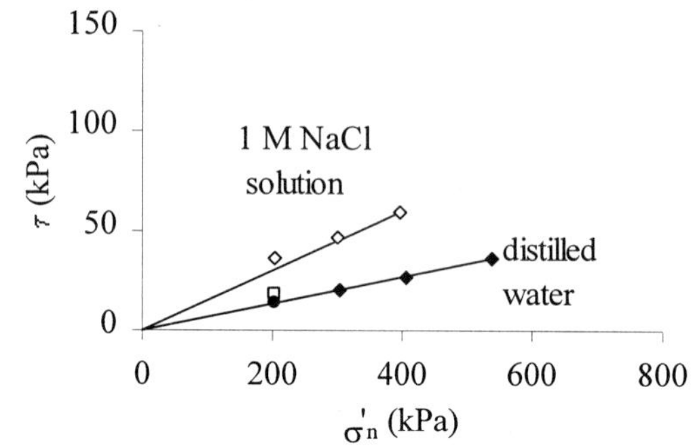

Fig.5. Residual shear strength.

All the above tests, both triaxial and direct shear tests, were carried out by exposing the specimens to liquids equal to their pore liquids, thus avoiding chemically induced water or ion flows. The only exception was constituted by the CU tests on normally consolidated material reconstituted with the salt solution. However it is reasonable to suppose that the undrained conditions, on one hand, and, on the other, the very small volume of water in contact with the specimens prevented such flows. The following results refer to direct shear tests in which a chemical exchange between the pore liquid and the cell liquid could occur.

Fig.6 compares the curves of shear strength against shear displacement for two specimens which - both reconstituted with a 1 M NaCl solution – were overconsolidated while in a bath of the same solution. As soon as the shear test began, the cell solution of one of the specimens was substituted with distilled water. It can be seen that the substitution did not cause differences in the peak values, whereas it caused differences in the post peak behaviour. In fact, the residual angle of the specimen exposed to distilled water slowly decreased to a value equal to that obtained for the water-saturated specimens of Fig.5.

Fig.7 reports axial displacements against time recorded during the shear tests. The two arrows point at the peak strength conditions. It can be seen that the two specimens reached the peak strength with almost the same increase in height. For further displacements, they exhibited different behaviour. In particular, the specimen exposed to distilled water began to swell. Since, unavoidably, during the shear test a loss of material continuously occurred, the axial displacements are not related to the real increase in void ratio. The process of volume increase caused by exposure to distilled water can be evaluated better through the results of an oedometer test. A specimen of the Bisaccia clay was reconstituted with a 1 M NaCl solution, it was consolidated to 5000 kPa and afterwards unloaded to 10 kPa while in a bath of the salt solution. At equilibrium, the cell solution was substituted with distilled water. This caused the osmotic swelling reported by Fig.8 whose time trend is similar to that reported by Fig.7. Hence it is reasonable to suppose that, during the shear test, the specimen exposed to distilled water underwent an osmotic process of swelling and softening. The process is very slow, hence it could not influence the peak strength which was reached in about 7 hours. On the other hand, in the case of the residual strength, the duration of the test was long enough to allow for the development of a large volume increase which, in turn, caused a large decrease in strength.

Similar experimental results were found also for other smectitic clays and for the Ponza bentonite (Di Maio, 1996a; 1996b), and were interpreted in terms of ion diffusion and osmotic water flows which influence the thickness of the diffuse double layer.

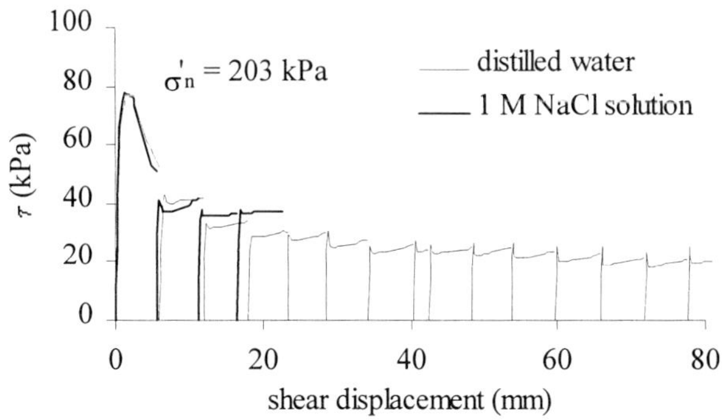

Fig.6. Shear strength against shear displacements for two specimen reconstituted with a 1 M NaCl solution, one exposed to distilled water and the other to the solution.

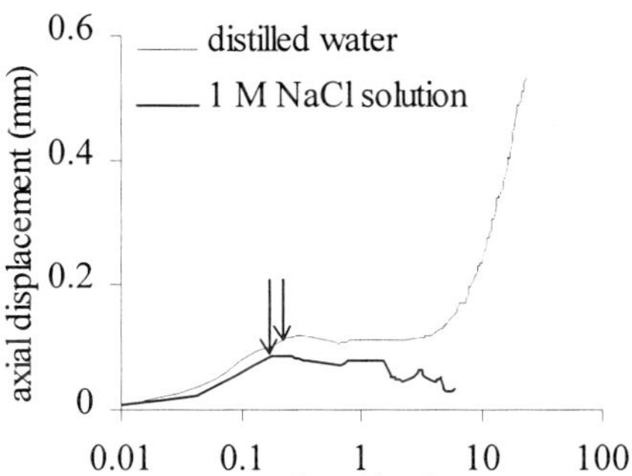

Fig.7. Axial displacement against time during the shear tests (h_0 = 15 mm).

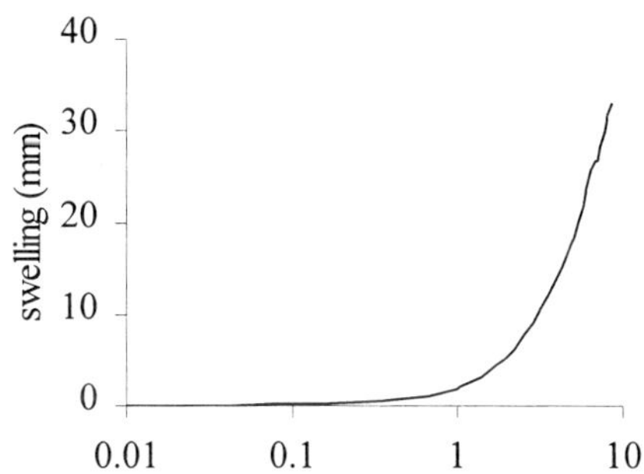

Fig.8. Osmotic swelling caused by exposure to distilled water (h_0 = 20 mm).

CONCLUSIONS

The experimental results show that pore liquid composition influences noticeably both the peak and the residual shear strength of the reconstituted Bisaccia clay. Furthermore exposure to distilled water of specimens reconstituted with a NaCl solution causes noticeable volume increase and residual shear strength decrease.

The results have direct practical consequences on laboratory testing; in fact they show that, in the case of active clays, the evaluation of intrinsic properties and of residual shear strength requires the use of a liquid similar to the undisturbed pore liquid. Furthermore, in the authors' opinion, the comparison between results obtained by using distilled water and those obtained by using a concentrated NaCl solution gives an idea of the variations which would have occurred – and that can still occur - in marine origin clays as an effect of exposure to rainwater.

ACKNOWLEDGEMENTS

The authors would like to thank Dr. Francesco Gambacorta who helped in carrying out the laboratory tests.

REFERENCES

Anson, R. W. W. e Hawkins A. B. (1998). The effect of calcium ions in pore water on the residual shear strength of kaolinite and sodium montmorillonite. *Géotechnique* 48, No. 6, 787-800.

Di Maio C. (1996a). The influence of pore fluid composition on the residual shear strength of some natural clayey soils. *Proceedings of the VII international Symposium on landslides*, Trondheim , Balkema, Vol. 2, 1189 - 1194.

Di Maio C. (1996b). Exposure of bentonite to salt solution: osmotic and mechanical effects. *Géotechnique* 46, No. 4, 695 -707.

Lupini J. F., Skinner A. E. , Vaughan P. R. (1981). The drained residual strength of cohesive soils. *Géotechnique* 31, No. 2, 181 - 213.

Mesri,G., Olson R. E. (1970). Shear strength of montmorillonite. *Géotechnique* 20, No. 3, 261 - 270.

Mesri,G., Olson R. E. (1971). Consolidation characteristics of montmorillonite. *Géotechnique* 21, No. 4, 341 - 352.

Mitchell, J. K. (1993). *Fundamentals of soil behaviour*. 2nd ed. John Wiley & Sons, Inc. New York.

Sridharan A. (1991). Engineering behaviour of fine grained soils - a fundamental approach. *Indian Geotechnical Journal*, 21, No. 1, 1 - 136.

Swelling Behaviour of Active Clays: the Case of an Overconsolidated, Marine Origin Clay

C. DI MAIO and R. ONORATI, Università della Basilicata, Potenza, Italy

INTRODUCTION

Clay formations outcropping along the Italian Apennine are often affected by severe problems of slope instability. In the case of the Bisaccia clay shale, earthflows are particularly widespread. The Bisaccia clay shale has been studied and described by many authors (among others: Cicolella and Picarelli, 1990; Picarelli and Urciuoli, 1993; Picarelli et al., 1998). It is a structurally complex material which formed in a marine environment some tens millions years ago. The clay is heavily overconsolidated because of an intense erosion which probably began about 300,000 years ago (Fenelli and Picarelli, 1990). One of its characteristics is a large difference between the physical properties of the upper layer, a few meters thick, and those of the lower material. The former is in fact characterised by much higher water contents and compressibility, as well as by a quite different fabric (Picarelli et al.).

Some experimental results show that exposure to distilled water of the Bisaccia clay reconstituted with a concentrated salt solution causes a slow process of osmotic swelling and of shear strength decrease (Di Maio, 1996a). It seems reasonable to suppose that similar processes could occur also *in situ*, as an effect of interaction of the marine origin clay with rainwater. In order to control this hypothesis, a series of one-dimensional compression and swelling tests were carried out both on undisturbed and on reconstituted specimens. In order to evaluate the influence of rainwater and seawater, the tests were carried out by exposing the material to, respectively, distilled water and a 1 M NaCl solution.

EXPERIMENTAL RESULTS

The investigated material was taken from a borehole drilled west of the Bisaccia hill, in the Ferrelli valley. Its c.f. varies between 36% and 65%, the plasticity index between 55% and 100%, the activity index between 85% and 160%. Fig.1, which reports some characteristics against depth from the ground level, gives an idea of the type of material and of its variability. The values of the liquid limit reported in Fig.1 were obtained by mixing the powdered clay with distilled water. By mixing the clay with a NaCl solution, the liquid limit decreases as the solution concentration increases, as shown by Fig.2. Hence the initial void ratio of specimens reconstituted at about the liquid limit varies with the composition of the employed pore liquid. Consequently, also the "intrinsic" characteristics (Burland, 1990) vary, as shown by Fig.3 which reports the one-dimensional compression and swelling curves of two specimens, one of which was reconstituted at the liquid limit with distilled water and the other with a 1 M NaCl solution. The specimens were exposed to liquids equal to their pore liquids, thus avoiding chemically induced water flows. It can be seen that, at any value of stress in the investigated range, the void ratio of the water-saturated material, and the slopes of its compression and swelling curves are higher than those of the solution- saturated material. This kind of results is typical of clays characterised by high specific surface (Bolt,

1956; Sridharan, 1991; Mitchell, 1993).

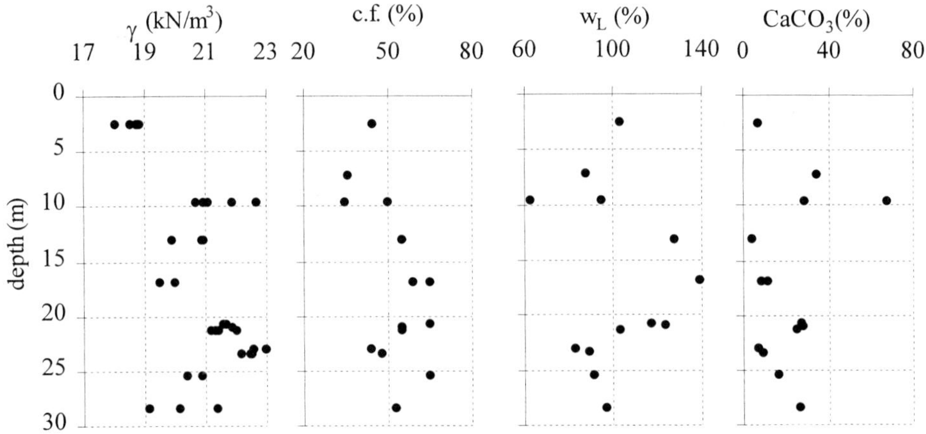

Fig.1. Unit weight, clay fraction, liquid limit and water content against depth from the ground level.

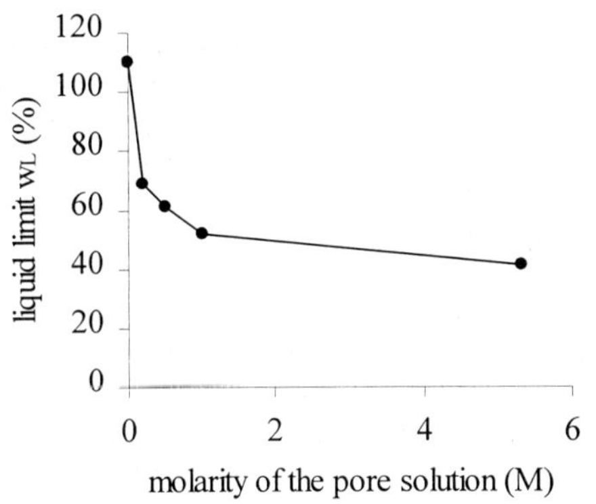

Fig.2. Liquid limit against pore solution molarity

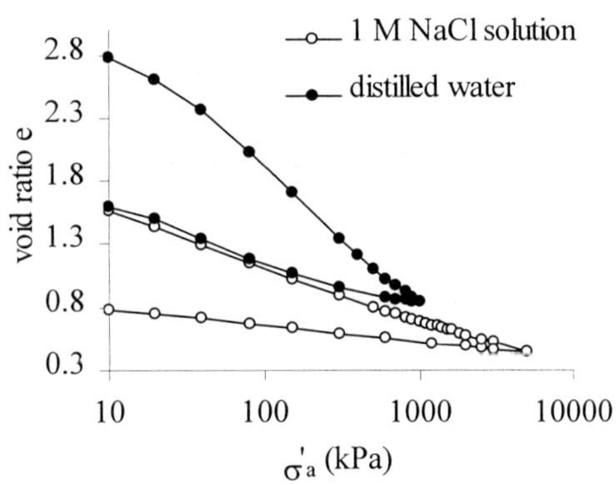

Fig.3. Oedometer curves of two reconstituted specimens of sample C10 (depth = 25 m).

Volume change behaviour is also influenced by the composition of the liquid the clay is exposed to. Fig.4 reports the oedometer curves of two couples of undisturbed specimens, coming from two different samples: C1, taken at about 2.5 m from the ground level , and C5bis, taken at about 21 m. A specimen of each couple underwent oedometer test in a bath of 1 M NaCl solution, and the other in a bath of distilled water. It can be seen that the specimens exposed to water experienced much higher swelling than those exposed to the NaCl solution. It is worth noting that water was never renewed, hence it probably became a ionic solution during the test.

Once swelling at 10 kPa ended, the specimens which had been exposed to the salt solution were exposed to distilled water which was continuously renewed. This caused noticeable swelling, under constant external load, represented by segments normal to the abscissa in Fig.5. A noticeable extrusion required the test on specimen C5bis to be interrupted.

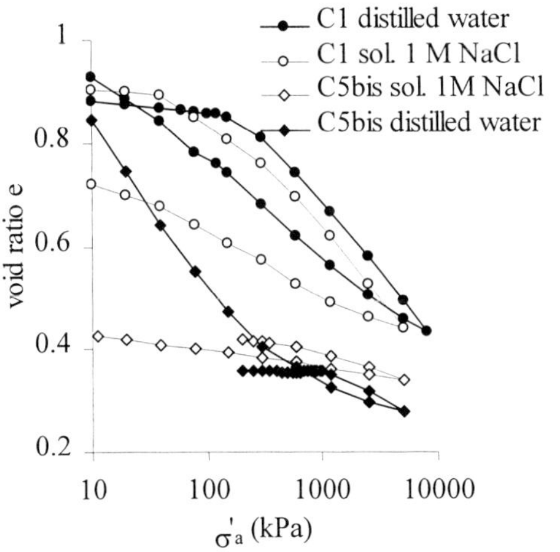

Fig.4. Oedometer curves for specimens exposed to distilled water and to a salt solution.

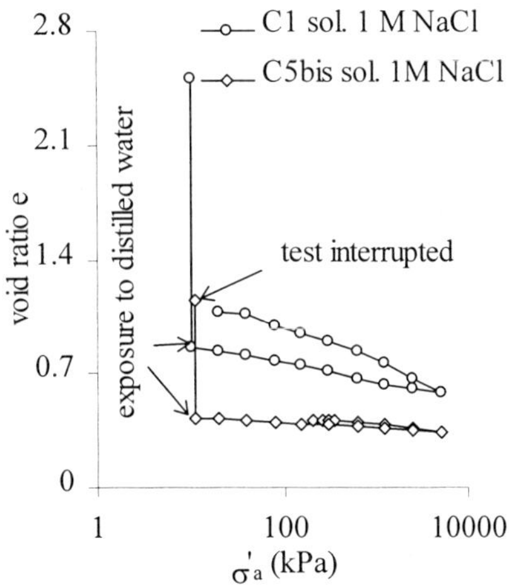

Fig.5. Omotic swelling caused by exposure to distilled water.

On the contrary, specimen C1 could swell until equilibrium and its final void ratio was much higher than that reached by the companion specimen (Fig.4) which had been exposed to distilled water for the whole duration of the test. This was probably due to ion exchange which could have occurred during previous exposure to the salt solution, as well as to the continuous water renewal. However, it seems as though a necessary condition for a large osmotic swelling is a low stress level, as shown by the following results.

Fig.6 reports the swelling curves obtained for two specimens which were reconstituted with a 1 M NaCl solution at about the liquid limit. They were consolidated until 5000 kPa and then unloaded. Specimen 1 was exposed to distilled water at 10 kPa, and specimen 2 at 1200 kPa. As in the case of specimen C5bis, a noticeable extrusion required the test on specimen 1 to be interrupted. Even so, the recorded swelling was much higher than that of specimen 2. The strong dependence of osmotic swelling on the stress level has been found also in the case of the Ponza bentonite (Di Maio, 1996b) and it can be qualitatively interpreted by considering the interaction between the diffuse double layers (Bolt, 1956).

The time trend of osmotic swelling underwent by specimen C1 is reported by Fig.7. It can be seen that although its initial height was as low as 2 cm, the specimen swelled noticeably for over forty days! The coefficient of swelling c_v is of the order of $10^{-7}cm^2/s$. The composition of the liquid to which the clay is exposed influences also the swelling pressure. Table 1 reports the swelling pressure evaluated in the course of oedometer tests, by adding weights to prevent swelling caused by soaking of the undisturbed specimens. Such a pressure has been evaluated both for specimens in a bath of distilled water and for those in a bath of 1 M NaCl solution. The data show that the swelling pressure depends on the cell liquid composition and on the activity of the clay more than on depth of sampling. The osmotic pressure model qualitatively explain the observed behaviour (Mitchell, 1993).

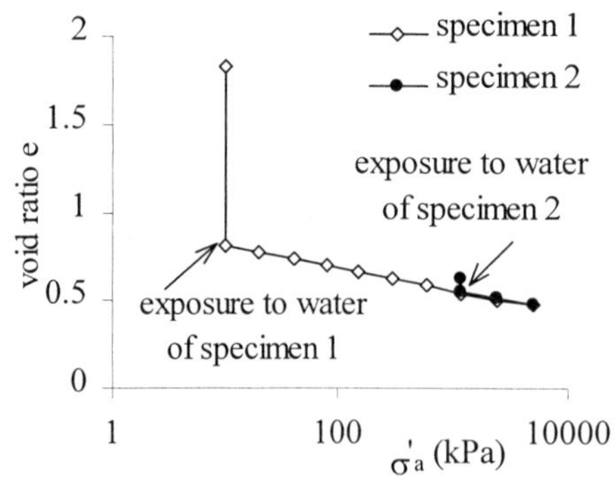

Fig.6. Dependence of osmotic swelling on the stress level.

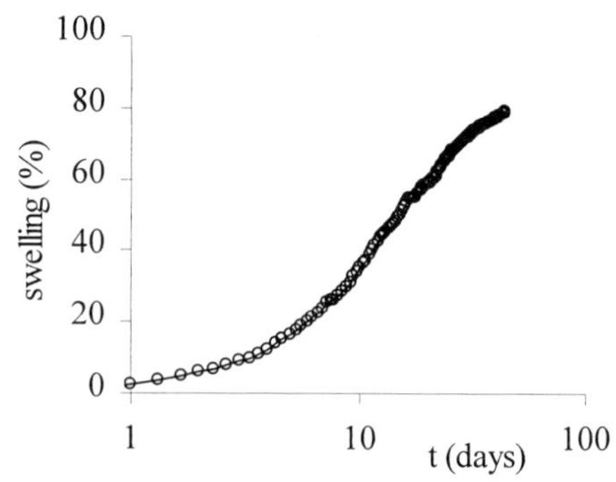

Fig.7. Swelling caused by exposure to distilled water (sample C1) under constant external load.

Table 1. Swelling pressure at various depth from the ground level.

Sample	Depth (m)	c.f. (%)	A (%)	Swelling pressure (kPa)	
				Distilled water	1 M NaCl solution
C1	2.5 – 2.9	45	1.26	150 – 300	40 – 80
C3	9.6 – 9.9	50	1.28	190 – 490	80 – 150
C4	13.1 – 13.5	55	1.65	1000 – 1200	150 – 300
C5bis	20.8 – 21.0	65	1.31	1000 – 1200	200 – 250
C7	21.3 – 21.7	55	1.30	1100 – 1200	
C8	23.0 – 23.4	44	1.19	850 – 900	
C9	23.4 – 23.8	48	1.20	600 – 1200	300 - 350

An indirect evidence of the chemical nature of the interaction between solid and liquid is also furnished by the results obtained by using a continuous oedometer and swelling test equipment in strain controlled manner. Fig.8, which reports swelling pressure against time for an undisturbed specimen (C7) exposed to distilled water, shows that the specimen achieved equilibrium in four days, whereas, in the absence of physicochemical processes, it should have reached equilibrium much faster (degree of saturation was S=1), in theory instantaneously.

It would be interesting to compare swelling pressure with the *in situ* stresses. However the latter were not measured, nor can they easily be calculated because of the extremely complex stress history of the considered soil. In order to test material having a known stress history, the strain controlled tests were performed on two specimens which were reconstituted with a 1 M NaCl solution and were anisotropically consolidated in a triaxial cell (back pressure = 200 kPa, cell pressure = 400 kPa, deviator stress = 90 kPa). Afterwards they were unloaded in undrained conditions and the swelling pressure was measured in the one-dimensional equipment. One specimen was exposed to distilled water and the other to the very same solution as its pore solution. Notwithstanding the similarity of the initial void ratio and stress conditions, the specimens exhibited very different behaviour. In particular, the value of swelling pressure of the material exposed to distilled water, even though probably not yet an equilibrium value, is higher than the mean effective stress imposed during anisotropic consolidation, whereas the value relative to the material exposed to the salt solution is much lower.

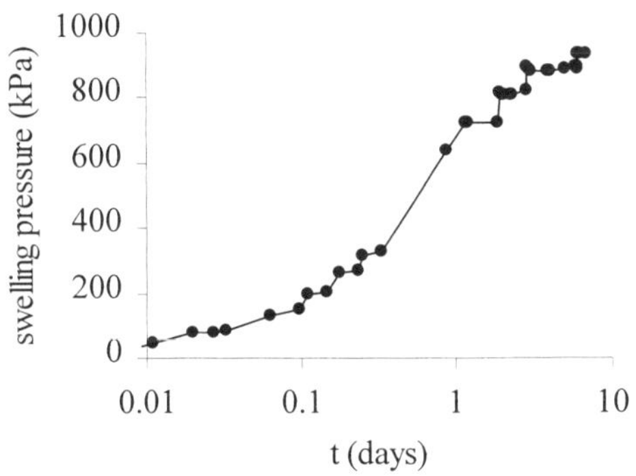

Fig.8. Swelling pressure against time for an undisturbed specimen exposed to distilled water.

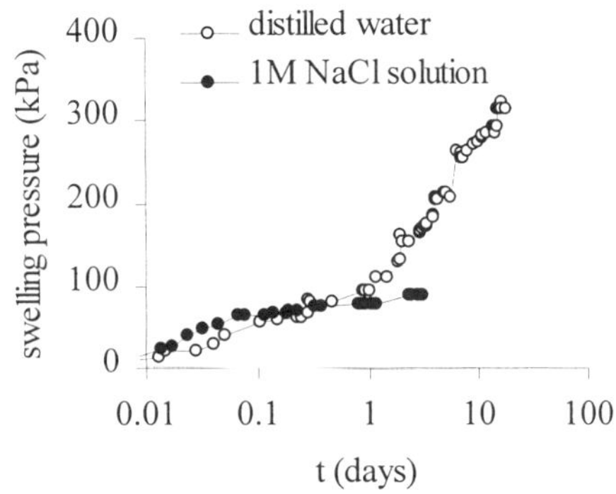

Fig.9. Swelling pressure against time for specimens exposed to different liquids.

All the above results show that the behaviour of the Bisaccia clay is strongly influenced by the composition of its pore liquid and of the liquid to which it is exposed. Hence it seems reasonable to try and interpret the *in situ* conditions in physicochemical terms. Fig.10 reports void ratio against depth as determined in four boreholes. Since the material is strongly overconsolidated by erosion, the experimental points can be considered as the field geological rebound points.

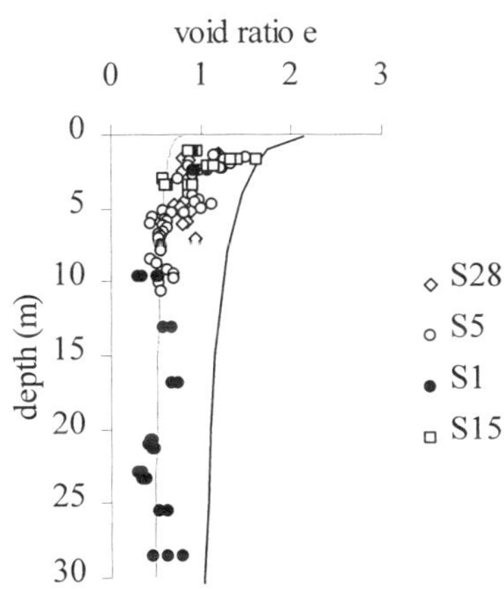

Fig.10. Theoretical and experimental void ratio against depth.

On the basis of the results obtained by oedometer tests on the reconstituted material, two theoretical *in situ* rebound curves were constructed, both in the simplified hypothesis of one-dimensional swelling and in the hypothesis of parallelism of the swelling curves of all the points in the vertical line. The thin line represents the rebound curve of the Bisaccia clay in a 1 M NaCl solution. It has been calculated taking into account secondary compression. The thick curve is the theoretical curve for the material in distilled water, in the absence of secondary compression, calculated in the reasonable hypothesis that erosion corresponds to the removal of an overburden of 1000 kPa.

It can be observed that the experimental data practically overlap the first theoretical curve for depths larger than about 6 m, whereas, for smaller depths, they gradually approach the second theoretical curve. One of the possible explanations of this trend is that exposure to rain water caused in the upper soil an osmotic swelling of the type shown in Figs. 6 and 7 for specimens exposed to distilled water. The long time intervals required for osmotic swelling, as well as its dependence on the stress level, probably explain why it could have occurred only in the upper clay.

CONCLUSIONS

Exposure of the Bisaccia clay to distilled water causes large osmotic swelling. A first simplified analysis of the *in situ* conditions seems to show that the high water content of the upper soil was probably achieved by a similar type of swelling caused by exposure to rainwater.

ACKNOWLEDGEMENTS

The authors would like to thank Mrs Lucia Santoli and Mrs Monica Larocca who helped in carrying out the laboratory tests.

REFERENCES

Bolt G. H. (1956). Physico-chemical analysis of the compressibility of pure clays. *Géotechnique*, 6, No.2, 86 – 93.

Burland J. B. (1990). On the compressibility and shear strength of natural clays. *Géotechnique*, 40, No.3, 329 – 378.

Cicolella A., Picarelli L. (1990). Decadimento meccanico di una tipica argilla a scaglie di elevata plasticità. *Rivista Italiana di Geotecnica, anno XXIV,* No.1, 5-23.

Di Maio C. (1996a). The influence of pore fluid composition on the residual shear strength of some natural clayey soils. *Proc. of the VII Int. Symp. on landslides*, Trondheim , Balkema, Vol. 2, 1189 - 1194.

Di Maio C. (1996b). Exposure of bentonite to salt solution: osmotic and mechanical effects. *Géotechnique* 46, No. 4, 695 -707.

Fenelli G.B., Picarelli L. (1990). The pore pressure field built up in a rapidly eroded soil mass. *Canadian Geotechnical Journal*, 27, No. 3, 387-392

Mitchell, J. K. (1993). *Fundamentals of soil behaviour*. 2nd ed. John Wiley & Sons, Inc. New York.

Picarelli L., Urciuoli G. (1993). Effetti dell'erosione in argilliti di alta plasticità. *Rivista Italiana di Geotecnica*, anno XXVII – No.1, 29-47.

Picarelli L., Di Maio C., Urciuoli G., Olivares L. (1998). Properties and behaviour of tectonised clay shales in Italy. *II Int. Symp. on hard Soils and Soft Rocks*, Napoli. In press.

Sridharan A. (1991). Engineering behaviour of fine grained soils - a fundamental approach. *Indian Geotechnical Journal*, 21, No. 1, 1 - 136.

Behaviour of Woody Debris Carried by a Debris Flow of Impermeable-Type "Sabo" Dam already Filled with Sediment

Y. DOI, Doctoral Program in Agricultural Sciences, Institute of Agricultural Engineering, University of Tsukuba, T. YAMADA, N. MINAMI, Erosion and Sediment Control division, Public Works Research Institute, Ministry of Construction and T. AMADA, Institute of Agricultural Engineering, University of Tsukuba, Japan

INTRODUCTION

There are many sediment and woody debris related disasters that occur during the rainy season in Japan. Most mountains are covered with trees. Once a slope collapses, the trees turn to woody debris. When the woody debris reaches a river-crossing structure such as bridges, culverts and etc, it can disrupt water flow and cause flood disasters. When it reaches the reservoirs, the stored water surface can be covered with woody debris and damage the intake function. When it reaches the seashore, boats, ports and fishing activities are disturbed. Trapping woody debris before it reaches human habitat is necessary in order to avoid these kinds of disasters. Therefore, woody debris trapping facilities are required in such disaster prone streams. To design the capacity of the facility, it is important to estimate the volume of woody debris.

On the other hand, many impermeable type "Sabo" dams (sediment trapping dams) are already installed along the debris flow prone streams in Japan. This type of dam is designed to trap sediment and it is believed that less woody debris can be trapped when the dam is filled with sediment. However, it has been reported that these types of dams trapped remarkable volumes of woody debris. (Yamada, et al, 1999) If these "Sabo" dams can trap a large amount of woody debris, the size of woody debris trapping facility can be minimized or it may not be necessary to install them at all. The purpose of this study was to evaluate the woody debris trapping ability of sediment filled "Sabo" dams. Authors conducted flume experiments of woody debris flow to analyze the function of the trapping patterns.

FLUME EXPERIMENTS

Procedure

The assumed model scale was 1/50. The experiments were conducted with a 5 m long, 20 cm wide and 40 cm high flume. A sediment trapping zone model with a 90 cm long, 60 cm wide and 40 cm high (Fig.1(c)) was attached to the end of the flume. A 12 cm wide and 4 cm high notch was placed at the end of the model of the sediment trapping zone as a spillway for the dam (Fig.1(b)). The flume was installed with a 17^0 gradient and the sediment trapping zone model was filled with sediment with half of the flume gradient being 8.5^0. The specific gravity of the sediment was 2.62, particle sizes of 95 % and 50 % were 14.5 mm, 6.1 mm, respectively. Sediment volume used for the experiments was determined by past

Landslides in research, theory and practice, Thomas Telford, London, 2000.

data of debris flows which occurred in Japan (Ishikawa, 1990) and which was converted to Froude similarity and calculated as 30,000 cm^3. The volume was fixed during all the experiments. Regarding the woody debris, the total volume of the wood models was fixed at 600 cm^3. Three different sizes of wood models (6 cm, 10 cm and 20 cm) were prepared. The mixture rate of the longest wood model (20cm) of the total volume of the wood models was defined as "wood mixture rate". The numbers of each wood model were adjusted in accordance with mixture rate of the 20 cm wood model (Table 1). The wood mixture rates between 0 to 100 % were prepared to confirm the effects on the woody debris trapping rate and the sediment trapping rate. The water discharge in order to generate debris flows were determined 2.3 l/s, 1.7 l/s, 1.4 l/s and 1.1 /s. The flow time of each discharge was settled by debris flow tests and fixed at 120 sec., 120 sec, 300 sec, 300 sec, respectively.

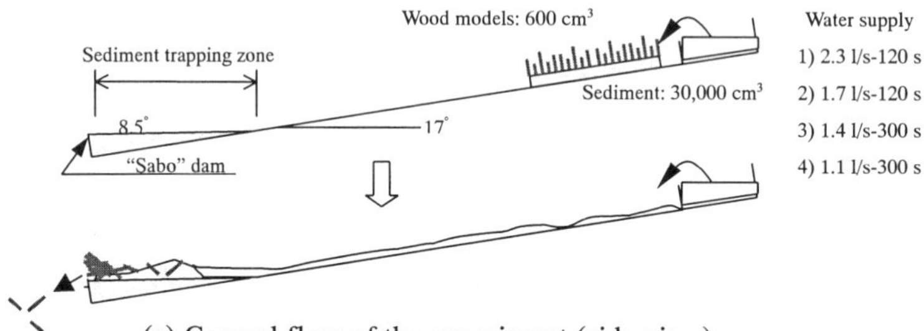

(a) General flow of the experiment (side view)

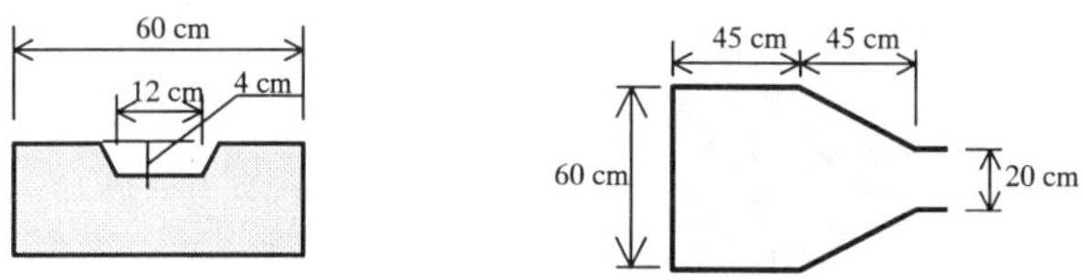

(b) Front view of "Sabo" dam (c) Sediment trapping zone

Fig.1 Schematic illustrations of experimental flume

Table 1 Numbers of wood models used in the experiments

Wood mixture rate (20 cm) (%)	Numbers of wood models		
	L 20 cm ϕ 0.8 cm	L 10 cm ϕ 0.6 cm	L 6 cm ϕ 0.4 cm
0	0	168	168
10	6	151	151
20	12	134	134
30	18	117	117
40	24	101	101
50	30	84	84
60	36	67	67
70	42	50	50
80	48	34	34
90	54	17	17
100	60	0	0

* Total volume of woody debris: 600 cm^3

The debris flow sediment of with a length 1.5 m was prepared at the upper end of the flume and the wood models were placed randomly in the sediment (Fig.1(a)). Water was supplied to generate a debris flow using 4 discharges. Woody debris trapping rate was defined as the ratio of the trapped wood models in the sediment control zone and the prepared sediment in volume. Not only woody debris trapping rates, but also sediment trapping rates were obtained. The movements of the woody debris flows were observed live and by video during the experiments.

Woody debris moving patterns on the Sediment control zone

Woody debris carried by debris flow became concentrated at the front of the debris flow as it flowed down (Ozaki, et al. 1998). When the woody debris reached the sediment control zone, the woody debris stopped flowing, because the gradient became gentle and the channel was widened. Immediately, the sediment covered over the woody debris. The next stages could be categorized in the following 4 patterns.

A: Most of the woody debris was trapped and remained in the sediment trapping zone. (Fig.2(a))

B: Woody debris, carried by the debris flow, clogged the spillway. However, woody debris was finally flushed out through the spillway. (Fig.2(b))

C: Woody debris, carried by debris flow, clogged the spillway and was trapped. Sediment was also trapped behind the clogged woody debris. (Fig.2(c))

D: Woody debris was flushed down through the spillway. (Fig.2(d))

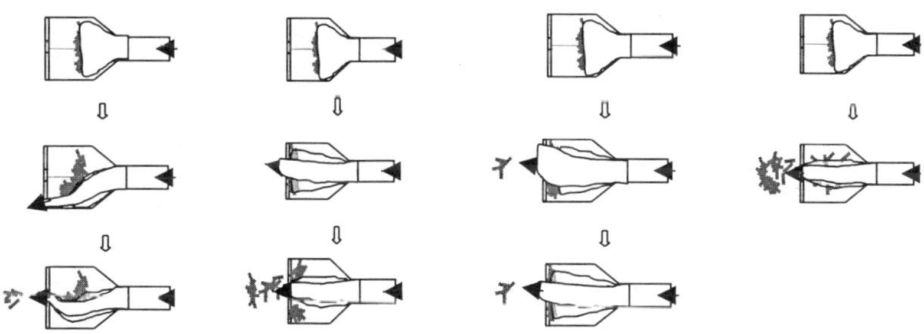

(a) Pattern A (b) Pattern B (c) Pattern C (d) Pattern D

Fig. 2 Woody debris moving patterns in Sediment trapping zone

Wood mixture rate and Woody debris trapping rate separated by woody debris moving patterns

Fig. 3 shows the relation between wood mixture rate and woody debris trapping rate. Each event was categorized by the patterns introduced in the former chapter.

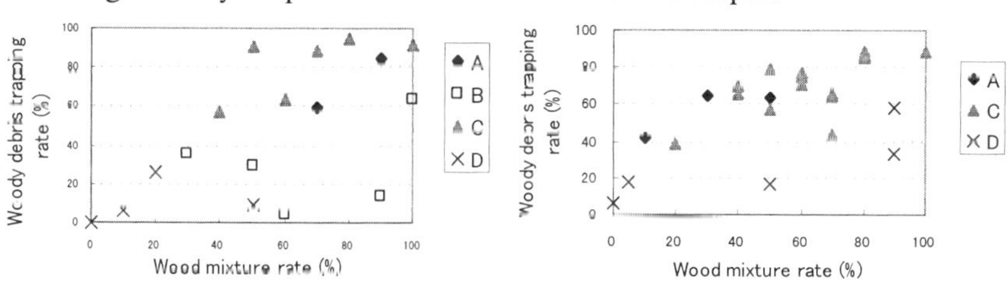

(a) 2.3 l/s – 120 sec (b) 1.7 l/s – 120 sec

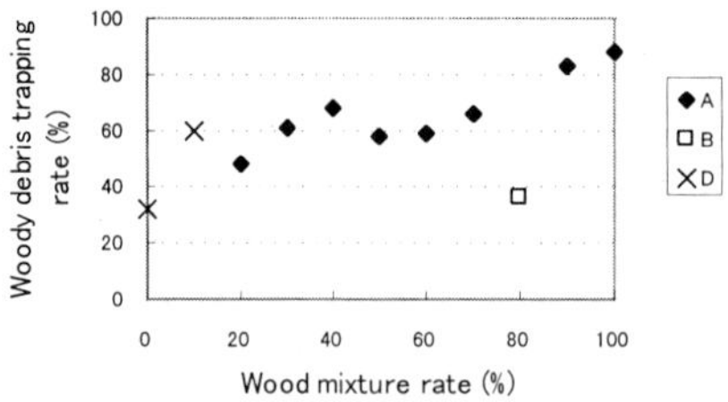

(c) 1.4 l/s – 300 sec

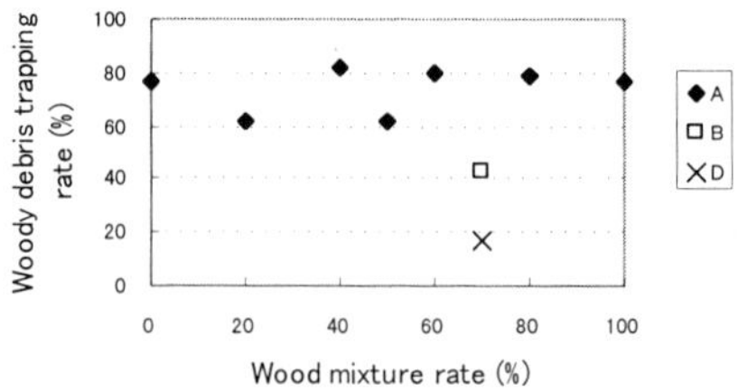

(d) 1.1 l/s – 300 sec

Fig. 3 Relation between Wood mixture rate and Woody debris trapping rate separated by woody debris moving patterns

Fig.3(a) shows the results with 2.3 l/s discharge. Woody debris with low trapping rates showed patterns B or D and high trapping rates showed patterns C or A. Trapping rates increased in proportion to the wood mixture rates till it reached 50 %, then stabilized. Low trapping rates were observed in spite of high wood mixture rate, which are shown in pattern B. In the case of 1.7 l/s (Fig.3(b)), the tendency was similar to the 2.3 l/s case except pattern B was replaced with pattern D. When the discharge was 1.4 l/s (Fig.3(c)), trapping rates increased with wood mixture rates. However, pattern C was not observed. The difference was that the trapping pattern with high trapping rates was only in pattern A. In the case of 1.1 l/s (Fig.3(d)), there seemed to be no relation between these two factors, and high trapping rates with pattern A were obtained.

Discharge and Trapping rate

Fig.4 shows the relation between the sediment trapping rate and the woody debris trapping rate of patterns A and C separated with discharge. These two patterns can be expected to obtain high woody debris trapping rate.

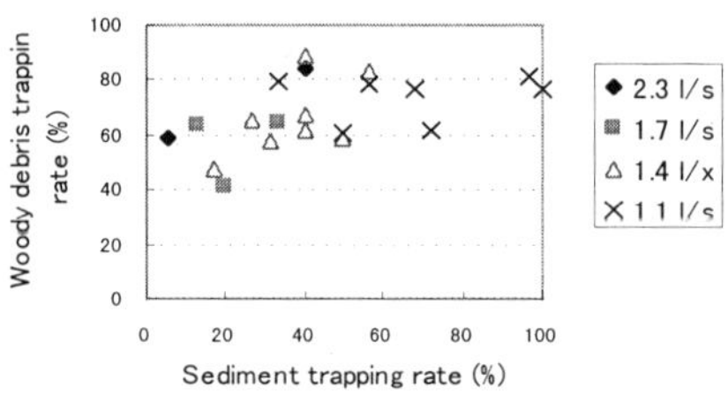

(a) pattern A

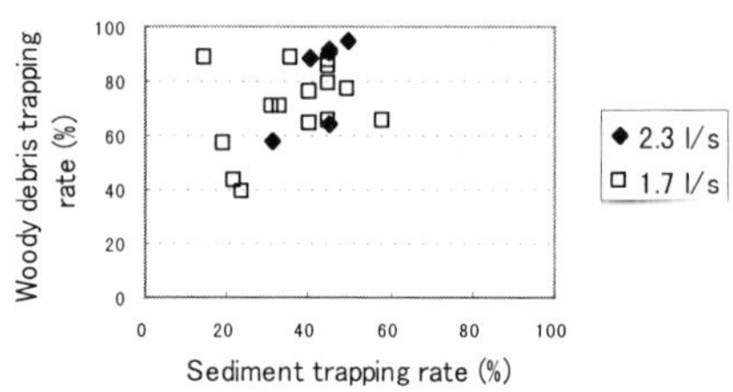

(b) pattern C

Fig.4 Relation between the sediment trapping rate and the woody debris trapping rate separated with discharges (patterns A and C)

According to the results of pattern A, the woody debris trapping rates were in the range of 40 – 90%. However, the sediment trapping rates were scattered from 5 – 100%. It is clear that lower discharges showed high sediment trapping rates from Fig.4(a). On the other hand, the sediment trapping rates of the pattern C were in the range of 20 – 60%, more stable trapping rate compare with pattern A case, and no pattern C could be observed when the discharges were 1.4 and 1.1 l/s.

DISCUSSION

Woody debris moving patterns in the sediment control zone were classified and it was confirmed that patterns A and C showed high woody debris trapping rates, while patterns B and D showed low trapping rates.

The pattern A, trapped in the sediment control zone, occurred when the resistance force became stronger than the drag force. The longer wood had a large attaching area to the streambed, which then produced a stronger resistance force caused by the friction. Short wood had a small attaching area, therefore a smaller frictional resistance force. In the case of high wood mixture rate, the strong frictional resistance force worked between the woody debris and the streambed. When the water level increased, short and thin wood could float down easier than the long and thick wood. Woody debris with high wood mixture rate contained greater numbers of long and thick wood. Consequently, this produced high woody debris trapping rates.

Pattern C could be observed when the discharges were high at 2.3 l/s and 1.7 l/s. Pattern C could be observed when the woody debris received enough force to be carried to the spillway. However, if the long wood unstably clogged the spillway, the clogged woody debris collapsed and flushed out. This was pattern B and it showed a low trapping rate. The ratio of the longest wood in the woody debris and the width of the spillway was an important factor to show pattern C. In the case of high wood mixture rate, there was a high possibility that long wood would be clogged and show a high trapping rate. This means the trapping rate is determined by the possibility of that if long wood can clog the spillway firmly.

In regard to patterns A and C, whose trapping rates were controlled by the scales of the discharge (Fig.4). Woody debris in the stream trap sediment (Kasai, et al. 1994), however, trapped sediment was eroded by water when discharge was high, while the sediment remained stable when discharge was small. On the other hand, the sediment could be trapped behind the woody debris in pattern C. Difference against pattern A was that even though the discharge was high, the trapped sediment could not be eroded and therefore this pattern obtained stable trapping rates.

Following the discussion above and the experiment results, relations between the woody debris and sediment trapping rates and the wood mixture rate of pattern C can be expressed with following differential equation.

$$\frac{dT}{dM} = kM\frac{T_{max} - T}{T_{max}} \quad \cdots (1)$$

Where; T: Trapping rate of woody debris or sediment of pattern C
M: Wood mixture rate
T_{max}: Maximum trapping rate
k: Experimentally determined coefficient

Increase rate (dT/dM) of the woody debris and sediment trapping rates (T) increase the proportion to the wood mixture rate (M), though the trapping rates reach peak (T_{max}) after the wood mixture rate increases. The equation (1) can be solved by giving the conditions when the wood mixture rate M=0 and the trapping rate T=T_0 , then obtained as follows,

$$T = T_{max} - \left(T_{max} - T_0\right)\cdot e^{-\frac{k}{2T_{max}}M^2} \quad \cdots (2)$$

Where T_0: Trapping rate when the wood mixture rate is 0 %

"T_{max}" and "T_0" of woody debris could be read as 90 %, 0 % and that of sediment could be read as 50 %, 0 % from Fig.5. Coefficient "*k*" was determined by observing the Fig.5 with trial and error, which resulted as 15 for woody debris, 10 for sediment. Curves on Fig.5 show the calculated results by the equation (2) of the maximum trapping rate of woody debris and sediment.

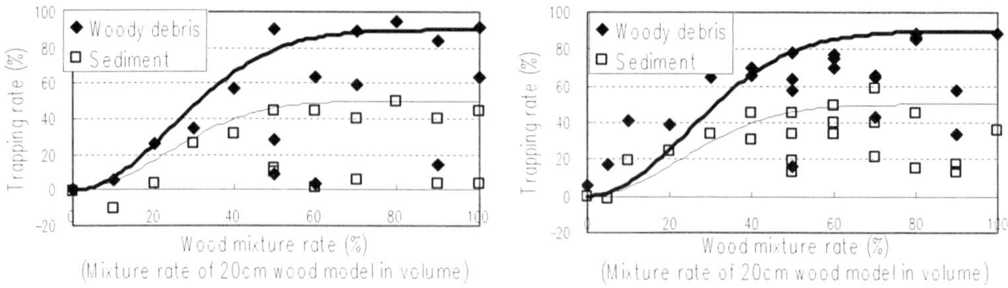

(a) 2.3 l/s – 120 sec (b) 1.7 l/s – 120 sec

Figure 5. The highest limit of the trapping rates of woody debris and sediment computed by the equation (2)

The thick curves are show the highest limit of woody debris trapping rate and the thin curves show the highest limit of sediment trapping rate. Highest limit curve of woody debris in Fig.5 (b) shown here is lower than the obtained data. However, other results nearly corresponded to the experimental data and it proved that the trapping rate at the highest limits could be estimated by the equation (2).

CONCLUSIONS

Woody debris moving patterns in the sediment trapping zone were classified, then A and C patterns produced high woody debris trapping rates. Sediment can be trapped behind the woody debris, however the trapping rate of pattern A is controlled by its discharge. The equation to estimate the woody debris and sediment trapping rates of pattern C was introduced. This equation is helpful in estimating woody debris and sediment trapping rates and in planning woody debris trapping facilities in woody debris prone streams.

It is important to note that, this experiment was conducted with limited conditions, such as scale of debris flow, volume of woody debris, particle size of sediment, etc.. Further analyses are required to minimize these kinds of disasters.

REFERENCES

Ishikawa, Y., (1990), Studies on Disasters caused by Debris Flows carrying Floating Logs down Mountain Streams, Ph-D thesis, presented to Kyoto Univ., 121 pp.

Kasai, M. and T. Marutani (1994), Sediment storage by Woody debris on a streambed of a mountainous river channel, J. Jpn. For. Soc. 76, pp.560-568 (in Japanese)

Ozaki, Y., Y. Kamogawa, T. Mizuyama, S. Kasai and J. Shima (1998), A debris flow with Woody debris trapped by a Steel-pipe gridded Sabo dam, JSECE Vol. 51. No.2, pp.39-44 (in Japanese)

Yamada, T., Y. Doi, N. Minami and T. Amada (1999), Experimental Analysis of Woody debris trapping by Impermeable type Sabo dam, filled with sediment, -Woody debris carried by Debris flow-, JSECE Vol.3, pp.18-23 (in Japanese)

Fault Reactivation in the Vicinity of Landslides in South Wales, UK

L. J. DONNELLY, IMC Consulting Engineers Ltd, PO Box 18, Huthwaite, Sutton-in-Ashfield, Nottinghamshire, K. J. NORTHMORE, British Geological Survey, Keyworth, Nottingham, UK and C. A. JERMY, University of Natal, Department of Geology & Applied Geology, Durban 4041, Republic of South Africa

ABSTRACT
During mining subsidence geological faults may undergo reactivation, resulting in the appearance of fault scarps on the ground surface. In the South Wales Coalfield numerous fault scarps and ground fissures occur on the moorland plateaux. These features differ markedly from other cases of mining-induced fault reactivation that have been documented throughout the coalfields of the UK, due to their magnitude and extent. Many large landslides have developed in association with these features and it appears that they may play an important role in the initiation of first-time slope movements and the reactivation of existing landslides. Mining has been practiced in the South Wales valleys over a period of at least 150 years and available evidence suggests that it is the effects of mining subsidence along pre-existing faults that has significantly contributed to their dramatic present-day development. However, the fault scarps probably represent several phases of reactivation and the possibility that these features, albeit in a subdued form, existed prior to mining cannot be ruled out, although this is difficult to prove.

INTRODUCTION
Geological faults, when located within an area undergoing mining subsidence, are susceptible to reactivation. Associated ground movements along the fault planes can result in the formation of fault scarps (or 'steps') and ground fissures, causing severe disruption to the ground surface. In the valleys of the South Wales Coalfield, numerous fault scarps and fissures exist on elevated moorland plateaux capped by hard and well-jointed Pennant Measures Sandstones. Glaciated U-shaped valleys are incised into the plateaux surface and the underlying beds are dominated by fissile mudstones. Landslides, developed on the upper and middle valley slopes, are numerous and present a significant hazard to the well-populated lower flanks of the valley sides. The objectives of this paper are to highlight the presence of the distinctive fault scarps and fissures in the region, their association with landsliding and to put forward possible mechanisms for their generation, whether entirely mining-induced or due to other processes.

CHARACTERISTICS OF THE FAULT SCARPS
The magnitude and extent of the South Wales Coalfield fault scarps make them distinct from similar mining-induced features observed in other UK coalfields (Donnelly, 1994). The fault scarps appear remarkably fresh and unweathered and, where best developed, occur as steeply-dipping features up to 4m high and several kilometres in length. They can influence the geomorphology of entire moorland slopes but are restricted in distribution mainly to the upland plateaux capped by thick, jointed sandstones. The scarps extend to the slope crests but tend to die out towards the middle valley slopes underlain by mudstones and shales, although the faults continue across the valley floor. The fault scarps vary from single isolated features, with or without sub-scarps, to parallel major sets with intervening grabens. (Figure 1).

Landslides in research, theory and practice, Thomas Telford, London, 2000.

Figure 1. The Tableland fault scarp and the deep-seated, slump-earthflow, Darren Goch landslide (after Donnelly 1994).

Often, they occur as a series of antislope (uphill-facing) scarps, or in an en-echelon array with inter-connecting fault ramps and splays. Along strike, the scarps may be linear, gently sinusoidal or saw-toothed in plan. In places, the scarps act as barriers to surface run-off and give rise to springs and seeps along the fault lines. Waterlogged ground and small ponds are frequently observed adjacent to the fault walls on the downthrow side. The general trend of the fault scarps is NW to SE or NNW to SSE, slightly oblique to the NW-SE oriented valleys. This is consistent with the main structural elements of the region and reflects a general south to north compression associated with the Hercynian orogeny in late/post- Carboniferous times (Woodland and Evans 1964).

CHARACTERISTICS OF THE GROUND FISSURES

The ground fissures are not always associated with faults. They are developed oblique or sub-parallel to the valley sides, in en-echelon arrays, as conjugate sets, or in box-work and saw-toothed networks. In places, they form interconnecting, sub-surface void systems that can be followed for considerable distances with individual fissures generally ranging from 1-6m, and occasionally exceeding 10m, in width. They are frequently bridged with rock fallen from their sidewalls. In many areas, a bridging cover of grassy turf masks their presence on the ground making access across large tracts of moorland particularly hazardous (Figure 2).

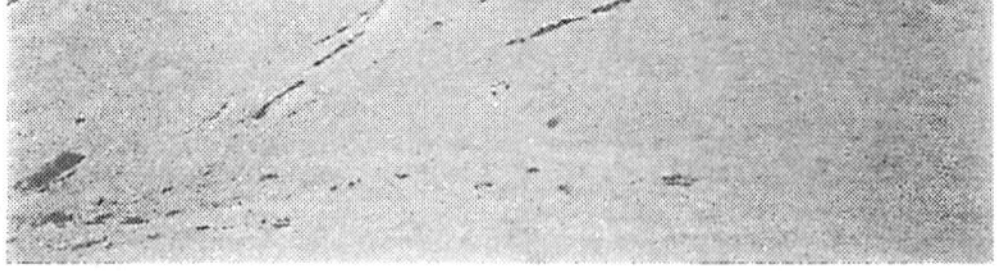

Figure 2. The Tableland fault and extensive development of ground fissures on the plateau, east of the Darren Goch landslide (after Donnelly 1994).

Many fissures are developed sub-parallel to the slope crests near the rim of the plateaux. Close to the slope crests, these fissures represent the dilation of existing joints in the sandstone caprock in response to cambering, removal of slope support due to landsliding and general mass wasting of the valley sides. However, extensive fissures are also developed more than 0.5km from the slope crests in the central plateaux surfaces between adjacent valleys. These may be related to more significant movements involving deep-seated lateral spreading of the jointed sandstone caprock.

ASSOCIATION OF FAULTING AND LANDSLIDES

Landslides often occur where geological fault scarps and graben structures intersect or pass close to the slope crests. In addition to weakening the rock mass, the faults also influence groundwater flows and can provide pathways for the rapid transmission of groundwater to the valley sides following intense or prolonged rainfall. This has a significant influence on the stability of the valley slopes and the reactivation of pre-existing landslides. Three typical examples of the many localities where landslides and major faults occur in close association are given below:

Darren Goch landslide, Garw Valley (NGR SS907940)
The Darren Goch landslide is a moderately deep-seated, complex landslide of the slump-earthflow type, with extensive rockfall debris from a 20m high sandstone backscarp being incorporated into and covering the slip mass. The Tableland Fault scarp, over 3.5km long and 3m high, passes within 30m of the backscarp of the landslide. The downthrow of the fault is to the east ('upslope'), resulting in an extensive up-hill facing, slickensided scarp wall. The fault has offset several small, dry stream valleys, both vertically and laterally, by about 3m. The Aber Fault (Mant-y-Moel Fault on earlier maps) runs parallel to and approximately 400m east of the Tableland Fault, forming a similar east-facing scarp wall up to 4.5km long and 3m high (Figure 3). The ages of the landslide and fault scarps are not known exactly. However, the Tableland Fault scarp pre-dates the Ordnance Survey map of 1876 which indicates its approximate position with the annotation "*major fault scarp*" (Halcrow 1989). The early geological maps of 1896 and 1900 record a small Darren Goch landslide and the inferred position of the faults, with no mention of scarp features. Following re-survey in 1949, the geological map indicates that the landslide had tripled in size since 1896 (Woodland and Evans 1964). Furthermore, the surface position of the Tableland fault is decisively marked, with the annotation "*subsidence crack*" recorded along its surface position for a length of approximately 600m. Immediately east of the Darren Goch landslide a note on the geological field slip reads "*well marked subsidence step 10 feet high*".

East Pentwyn and Bournville landslides, Ebbw Fach valley (NGR SO 207075; SO 206068)
The adjacent East Pentwyn and Bournville landslides are deep-seated, complex rotational slump-earthflows. Extensive rockfall debris, from 50-60m high backscarps, partly covers the upper and middle parts of the large slip masses. The main landslide movement at East Pentwyn occurred dramatically in 1954 (Halcrow, 1983, 1985). The precise date of the Bournville landslide is not known, but on the 1896 geological field sheet the date of *"1893"* is marked for the slip and may correspond to the date of initial movement. Movements at both sites have been recorded more or less continuously from these initiation dates. The Brynmawr Fault scarp, 350m long and 4m high, follows the N-S valley trend some 370m upslope of the landslides in the central part of the plateau between the Ebbw Fach and Cwmtillery valleys. Three distinct fault bounded grabens trend NW-SE, oblique to the Brynmawr Fault. Two 'crop out' at the Bournville landslide backscarp (Figure 3). The other graben feature and a single fault scarp are most prominent in the central part of the plateau. The single fault scarp crosses and appears to laterally displace the Brynmawr fault scarp. When projected to the valley crests, both the central graben and this fault align with the backscarp of the East Pentwyn landslide. Although clearly annotated on the current geological field sheets, there appears to be no reference to fault scarps on the early 1896 and 1902 maps.

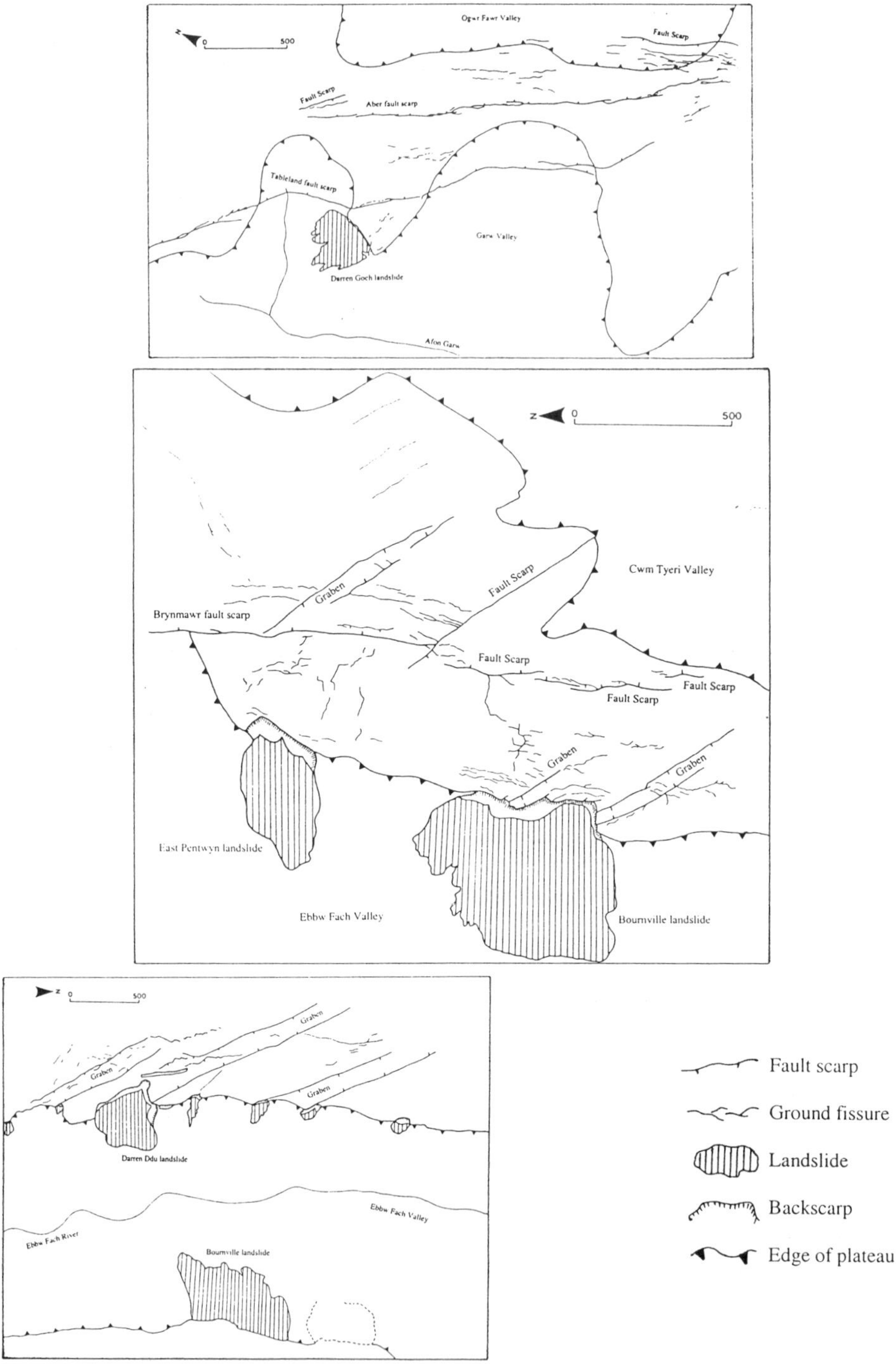

Figure 3. Aerial photogeological interpretation of geological fault scarps and ground fissures: The Darren Goch landslide (NCB Opencast)(upper); the East Pentwyn and Bournville landslides (Meridian Airmaps Ltd and HSL, UK) (middle); and the Darren Ddu landslide (RAF) (lower).

Darren Ddu landslide, Ebbw Fach valley (NGR SO198060).
The Darren Ddu landslide is a rockfall failure composed of large blocks of Pennant Sandstone. Figure 3 shows the position of this landslide in relation to a series of NW-SE trending fault scarps, 2km long, which intersect the valley side. The fractures have created three graben features with maximum vertical displacements of 5-6 m near the slope crest, decreasing to about 1m towards the centre of the plateau surface. The combination of vertical and horizontal movements has clearly been sufficient to cause the rockfall, which continues to extend back into the slope crest. The date of the initial failure is not known, but a smaller landslide is indicated on topographic maps surveyed between 1876-1879. Sporadic movements were recorded in 1911 (Knox 1927) and an extensive rockfall occurred in 1975 (Halcrow 1989). The 1894 geological field sheets indicates a fissure at the present day scarp position, approximately 300m long and annotated "*rift 1foot wide, great depth in sandstone*." The 1901, second edition geological map, and the subsequent 1925 revision, do not show any indication of this scarp, although the landslide is indicated.

LATERAL SPREADING OF THE MOORLAND PLATEAU

The abundance of extensive open fissures and graben structures, particularly away from the valley crests near the centres of the moorland plateaux, indicate that lateral extension (spreading) of the plateaux caprock has taken place. Mining has been practised in the area since at least the mid-1800s to 1985 (Donnelly 1994). However, it was not until the mid 1900s that extensive, modern longwall extraction, and hence major ground subsidence movements, took place below the faults. There is no doubt that mining subsidence has created or exacerbated many of these features, but the possibility exists that many of these features may have developed, albeit less dramatically, by other mass movement processes. For example, during the cold stages of the Pleistocene the Coalfield valleys would have remained essentially stable, supported by the mass of valley ice and deep permafrost. On retreat of the ice, the valley sides would have lost lateral support and been generally weakened by melting of permafrost and elevation of pore pressures. The weaker mudstones forming the valley sides would have responded differently from the competent sandstone caprock. Cambering, valley bulging, landsliding and other forms of periglacial mass wasting would have been facilitated. Under these conditions the jointed and faulted caprock may have experienced some lateral spreading and fault reactivation, later exacerbated by mining-induced ground movements. South Wales has also suffered earthquakes in historical time and this may also have contributed to fault reactivation in the area. Fault reactivation on hill slopes and palaeoseismicity have also been observed in recently de-glaciated terrains throughout Europe (e.g. Davenport et. al 1989, Kujansuu 1964, Lagerback 1979, Morner 1978 and Ringrose 1987) and in Canada (Bovis 1982).

CONCLUSIONS

The purpose of this paper is to draw attention to the dramatic fault scarps and ground fissures on the sandstone plateaux of South Wales Coalfield, the possible mechanisms for their formation and association with major landslides. They differ significantly in magnitude and extent from the similar mining-induced features elsewhere in the UK. Where the fault scarps and associated grabens intersect the valley sides, landslides are developed. It is likely that these faults and fissures have undergone several phases of mining-induced reactivation, with each phase of movement exacerbating the magnitude of the fault scarps. It is not possible to determine the age of the fault scarps with accuracy. There were few detailed recordings of the fault scarps on British Geological Survey and Ordnance Survey in the late 1800s and early 1900s. However, the absence of fault scarps on maps cannot be taken as evidence that the fault scarps did not exist.

It is known, from some early maps, that fault scarps were present in the late 1800's (e.g. the

Tableland fault is shown on an 1876 map). However, modern longwall mining was not practised at this time and it is uncertain whether earlier mining methods could have generated these features.

The extensive graben structures are indicative of lateral spreading of the plateaux surface between glacially incised valleys. It is difficult to envisage how a mining subsidence mechanism alone may be responsible for the extension of the ground to this magnitude, although large lateral ground movements in the vicinity of cliff faces have been well documented. Fault reactivation has been observed in recently de-glaciated terrains throughout Europe and Canada. The possibility that similar movements along the South Wales faults were caused by glacial stress relief mechanisms prior to large-scale mining cannot be ruled out, although they have clearly been exacerbated by subsequent mining subsidence.

ACKNOWLEDGEMENTS
This paper has been published with the permission of the Director, British Geological Survey, IMC Consulting Engineers Ltd, UK and the University of Natal, Republic of South Africa.

REFERENCES

Bovis, M. J. (1982). Uphill-facing (antislope) scarps in the Coast Mountains Southwest British Columbia. Geological Society of America, **93,** 804-812.

Davenport, C. A., Ringrose, P. S., Becker, A., Hancock, P. and Fenton, C. (1989). Geological investigations of late and post glacial earthquakes activity in Scotland. In: Gregersen., S. and Basham, P. W. (eds). Earthquakes at North-Atlantic Passive Margins: Neotectonics and Postglacial Rebound, 174-194.

Donnelly, L. J. (1994). Predicting the Reactivation of Geological Faults and Rock Mass Discontinuities during Mineral Exploitation, Mining Subsidence and Geotechnical Engineering. Ph. D. Thesis, University of Nottingham, British Coal Corp. Industrial Sponsor.

Halcrow, Sir William and Partners (1983). East Pentwyn Landslip. First report on geotechnical investigations. Unpublished report for Gwent County Council.

Halcrow, Sir William and Partners (1985). East Pentwyn and Bournville Landslip Research Project. Report for the Department of the Environment and the Welsh Office.

Halcrow, Sir William and Partners (1989). Landslides and Undermining, Research Project Landslide Inventory. Joint project undertaken by Sir William Halcrow, Consultant Engineers and Nottingham University, Department of Mining Engineering on behalf of the Department of the Environment, (unpublished).

Knox, G (1927). Landslides in South Wales Valleys. Proceedings South Wales Institution of Engineers, **43**, 203-205.

Kujansuu, R. (1964). Recent faults in Finnish Lapland. Geologi, **16,** 30.

Lagerback, R. (1979). Neotectonic structure in northern Sweden. Geologi Foren, Stockholm Fourhand., **100(3),** 263-668.

Lundqvist, J and Lagerback, R. (1976). The Parve Fault: A late glacial fault in the Precambrian of Swedish Lapland. Geol. Foren. Forhand., **98,** 45-51.

Morner, N. A. (1978). Faulting, fracturing and seismicity as functions of glacio-isostasy in Fennoscandia. Geology, **6(1),** 41-45.

Woodland and Evans (1964). The Geology of the South Wales Coalfield IV. The Country around Pntypridd and Maesteg. Memoir Geological Survey of Great Britain, 3rd ed, HMSO, London.

Ringrose, P. R. (1987). Fault Activity and Palaeoseismicity During Quaternary Times in Scotland. Ph.D. Thesis, University of Strathclyde, Glasgow.

Management of Steep Rock Slopes after Studies on Reservoir Slopes by Electricité de France

J-Y. DUBIÉ, Division Géologie et Géotechnique, Electricité de France, Aix en Provence, P. DUFFAUT, Consultant, Paris, France

INTRODUCTION

In 1248 a huge vertical slice of Mont Granier, close to Chambéry, French Alps of Savoy, fell down and buried a lot of villages, with 3 to 5 thousands of casualties (Goguel & Pachoud, 1972). The limestone mountain had, and has yet, a vertical face about 300 m high, towering 1300 m above a flat valley floor. The volume displaced was between 300 and 500 hm^3 ($1\ hm^3 = 10^6\ m^3$), the surface covered 23 km^2, and the maximum reach 7 km.

Every year smaller events occur along steep slopes, either in mountain valleys or along sea cliffs. So the chalk cliffs along the French Channel are known to recede a few metres in a few decades, keeping their faces sub vertical. Erosion along mountain valleys is by far slower, leaving enough time for weathering to occur. But where glacier erosion has worked, it has taken out all soft material, leaving very steep slopes, among the highest and steepest in the world (Duffaut, 1993).

As a matter of fact, many dammed reservoirs occur along mountain valleys, and their safety problems may be extended to any other facilities; including cities and industrial plants. Though by far rarer than snow avalanches and icefalls, massive rock falls also rank among natural hazards and their power of destruction must not be understated. Usually steep rock slopes only show "skin" instabilities, leading to scree curtains, but rare massive failures have to be considered, though they go beyond the field of classical stability theories.

Since the fall of Monte Toc into the Vajont dam reservoir in 1963 (Italian Alps), owners and operators of reservoirs are fully aware of this kind of hazard. Norway fjords have given many examples of rock falls generating huge waves travelling far along the shores (Bjerrum & Jörstad, 1968). It is unclear that water always impairs the slope stability, as the contrary is sometimes observed (Saint Pons slide below). Whatever the answer, reservoirs increase the hazard of slope failure, and they also increase the potential of damage, as showed by the Vajont case.

In France EDF (the national electricity company) has reviewed most of its reservoirs, and at the world scale ICOLD (the International Commission on large dams) has set a working group on landslides and reservoirs, the report of which is to be published soon. If operation of reservoirs may affect the stability of steep slopes, the other factors are the same as for slopes without reservoirs. So the paper has a wider scope than only reservoir slopes). Many cities lie close to the mountain toes, as did the villages beneath Mont Granier. Too few attention is paid to those

Landslides in research, theory and practice, Thomas Telford, London, 2000.

slopes supposed to keep stable for long times, as they did not yet show any sign of fatigue or illness!

LANDSLIDES INTERESTING FRENCH RESERVOIRS

Sainte Croix Reservoir, Saint Pons slide (Baldy & Thérond 1976)
This reservoir on the Verdon, second in France by volume, asked for long studies because of a karstic spring in the reservoir. A 500 m length scar was known long before along the right bank of the reservoir, some kilometers upstream of the dam, over a place named Saint Pons. The reservoir side is a straight 40% slope, from the top of a flat poudingue plateau, el. 680 m, and the valley floor, el. 400 m (the planned reservoir level being 477 m). This scar appeared about el. 510, as the upper limit of a significant slide, and the lower part of the slope exhibited the classical features of a sub cylindrical slide : sloping trees, and a bulging of the alluvial plain, pushing the river away. The moving volume is about 15 hm^3.

Geodetic measurements showed that the lower part was moving outwards horizontally at a rate 250 mm/yr, confirmed along the last 25 years by a series of aerial photographs ; the higher part moved downward vertically about half this rate. Given the total displacement and the rate measured, initiation of the slide was ascribed to an event about 70 years earlier : at this period two historical events may have triggered the slide, either a flood in 1906 or an earthquake in 1909. Maps from before this time showed the river was flowing along the toe of the slope.

The measurements were carried since 1966 : the average yearly displacement was reduced to 80 mm (somewhat more with lower water levels and less when the reservoir is full up). So the presence of water had slowed the movement. In the latest years, due to a new reservoir management with an average level lower than before, the rate has increased to 150 mm/yr. In spite of a rather huge volume, about 15 hm^3, such rates have been thought of as harmless (but measurements twice a year are kept mandatory).

Table 1. Main data on dams and reservoirs quoted

Dam name	river	near city	year	dam type	height	reservoir volume
Sainte Croix	Verdon	Marseille	1974	arch	95 m	767 hm^3
Monteynard	Drac	Grenoble	1963	arch	155 m	275 hm^3
Grandmaison	Eau d'Olle	Grenoble	1984	rock fill	160 m	137 hm^3
Soulcem	Mounicou	Toulouse	1983	rock fill	78 m	29 hm^3

Monteynard Reservoir
The river Drac valley proved very convenient for hydropower as it alternates wide basins with narrow gorges, thanks to cycles of glacial and epigenetic erosion. The gorge upstream of Monteynard dam has a steep right bank, sloping about 50° along the dip of Lias lime stones, on a height up to 400 m over reservoir level ; over this height and for 1000 m more the slope is gentler. At a particular spot a protruding spur, la Pelloud, has an over dip volume 100000 to 150000 m^3, laying on a stratification joint, with a sub vertical face, height 200 m; such a volume could initiate a huge wave in case of falling down into the reservoir.

Though not any rock fall or movement has ever been noticed, this spur was considered menacing, and a thorough visit of the place has been made in 1998, thanks to mountain guides and climbing ropes. The visitors were directed to salient features by observers from the opposite bank. No evidence of any continuous weak joint or fracture has been discovered and the spur is no longer object of fear (but it is kept under yearly observation).

Grandmaison dam and reservoir

This large dam recently built has provided three examples of slope instabilities :

i) The first project was an arch dam more than 180 m high, between high granite abutments, with scree fans at their toe. 400 m higher, the exposures of rock polished by the glacier appeared crisscrossed by open fractures, as if the whole slope had been twisted by a tangential thrust from the flow of ice. This location has not been retained and a fill dam was preferred 800 m upstream.

ii) A large quarry has been opened downstream of the new dam site, along the slope of the left bank to provide rock fill. High over its level, some crevices had been discovered many years before, up to 1 m wide and 20 m deep. Monitoring had showed that the slope was stationary. But after two years of operation of the quarry, those crevices were reactivated, opening a few centimeters more after each snowmelt ; the quarry had to be extended laterally instead of vertically (as has first been planned).

iii) The main problem, the *Billan slide* (Dubié et al, 1988), aroused in 1986 during the first filling of the reservoir, 1 km upstream of the dam. A road formerly sited in the valley bottom had been rerouted over the future reservoir level ; in between, moraine material had been taken from this slope for inclusion in the dam shoulders. Leveling of the road showed movements about 1 cm/yr, increasing to 2-3 cm/yr during the removal of material. When the lake level was close to the maximum, a continuous crack appeared 50 m high over the new road on a length about 250 m, indicating a deep slope movement. Immediately bore holes were bored in order to document the cross section inside the slope : two formations were known, gneiss above, and calcschists below, but instead of dipping inward they dipped outwards due to *tilting downhill,* a well known phenomenon where the strike of bedding or schistosity is parallel to the slope (defined by Swiss geologist Lugeon as *"balancement des têtes de couches"*, and shortly named in French *fauchage*, after the scythe tool, *la faux*). The tilted volume is about 20-30 hm^3 on 100 m depth.

Again, 500 m higher on the slope, open fractures were found. Geodetic measurements as well as extensometers and tiltmeters in bore holes provided data on the volume which had moved, about 1 hm^3. This volume is but a small part of the tilting rock mass, and similar slow movements occur at other locations. A drainage gallery supplemented with bore holes proved efficient and the landslide came to rest (though the tilted zone goes on moving as slowly as before, about 5 mm/yr. It has been noticed that the drainage attenuates the seasonal changes of groundwater level, chiefly during snowmelt time).

The slide may be ascribed to the addition of two causes : first a lot of material has been taken from the scree along the slope toe ; and the exceptionally fast snowmelt of spring 1986 has raised the water table higher than usual. Other slopes around have shown movements up to a few centimeters in ten days, even though they keep perfectly fixed during the rest of the year.

Soulcem Dam and Reservoir
The reservoir lies in a glacial valley of central Pyrénées mountains, where an alluvial basin is followed by a narrow gorge, both with steep rock slopes in fractured gneiss. A few rock fall events were known over and close to the dam site (left bank), and farther upstream over the reservoir site, the latest in winter 1944-45. Part of the material of an ancient rock fall has been buried inside the alluvium. After a detailed analysis of the rock mass structure and the valley morphology (Lagasquié, 1978), the slope over the left abutment was explored by an adit bored along the basal discontinuity surface (a minor fault rather than a regular joint). As inclinometers from this adit showed no movement, no measures were taken for increasing the stability. Instead the dam structure has been adapted, close to this left bank, to make it able to receive a substantial rock fall without significant damage.

Table 2. Main characters of problems quoted

Reservoir	gradient	slope height	slide height	rock type	movement type	volume
Sainte Croix	22°	280	110	poudingue	cylindrical slide	15 hm^3
Monteynard	> 50°	800	200	limestones	potential rockfall	100000 m^3
Grandmaison	40°	1000	500	tilted schists & gneiss	slide	1 hm^3
Soulcem	> 45°	700	> 100	gneiss	potential rockfall	1000 m^3

La Clapière and Séchilienne
Two cases of rock slides along glacial valleys in the French Alps (without reservoir, except in case of the fallen rock dams the valley) were identified about twenty years ago (when too many rolling stones menaced the safety of the roads below) ; they are monitored since (Collective, 1998). The latter came to a rest after a tentative forecast of the time of failure (Vibert et al, 1988).

CLASSIFICATION OF SLOPES AND MAIN FACTORS OF INSTABILITIES
Slopes may be classified along *height*, up to about 4000 m, along *gradient*, up to 90° plus some overhangs, along *shape*, convex or concave or composite, as well horizontally as vertically, and finally along *material* and *structure*, including weathering and jointing. Gradient alone, as used in slope maps, does not have any general significance ("the steepest slope a soil can stand decreases with increasing height", Terzaghi, 1942). The *age* of slopes is another key for classification, as younger ones always are steeper than old ones.

It looks even more important to classify slopes along their genesis and history, which does control the stress field inside (Duffaut, 1994). Scree, sand dunes, moraines, and volcanoes are *built slopes*, similar to man made fills, heaps, walls and dams ; marine cliffs, erosion valleys, karstic and solution shafts are *carved slopes*, similar to man made cuts and open pit mines. One may notice that the height of man made slopes is far behind that of natural ones.

Only no cohesion slopes may have a fixed angle irrespective of their height, either scree slopes or rock slopes along a natural joint (Claps de Luc, table 3) ; conversely, cohesion always introduce a space scale : there are stable vertical slopes and even overhangs in any material (except dry sand), from a few meters height in soft clays to hundreds of meters in massive limestones and granites. A graph on high and steep granite slopes (Duffaut, 1993) allowed Habib (1993) to assign to such rock masses a large scale cohesion about 2 MPa.

As some mountains look to stand stable for long times, people think they shall stand for ever, in spite of evidence of some slope failures here and there, whatever height, gradient, slope material and structure. Most of slope stability studies apply to *failed or failing slopes*, the features and movements of which called for attention; in such cases a back analysis can explain why the slope did fail or is failing. Up to now, not any method is likely to help understand how to recognize the slopes which *will* fail next time, and the time and extent of their failure. *No method can explain why actual steep slopes do stand, and for how much time.*

Main factors of instability are well known: *gravity*, the main law of terrestrial mechanics, *water pressure* (Duffaut & Louis, 1972, Duffaut, 1978, water may be seen as the "libido" of the rock mass), and far behind *internal stresses*, depending on depth and tectonics, but also on genesis and evolution of the slope. Some mining or civil works have triggered slope failures, as did some powerful earthquakes. In short, slope stability mainly relies on *geology*, through materials and structures, and on *groundwater surges*. Weathering, mechanical fatigue and other damaging processes are at work to decrease the stability along time.

So slope failures are *rare events*, just as are extreme floods and earthquakes, their distribution along space and time belongs to extreme value statistics. Most of slope failures are related to floods (through toe erosion), to quakes (through shaking) and/or to groundwater surges from extreme rains. Nobody can assign any upper bound to rain intensity and to transient bursts of water pressure. If alone, neither a geologist nor an engineer can fully understand slope failures : any slope may be compared to a lock, the mechanism of which is hidden behind an opaque steel plate ; engineers underrate the rock mass structure they do not see, geologists underrate the forces from water ; well, the water pressure may turn the key to open the lock.

Table 3. Some rare events in the Alps

year	name	close to	volume, hm^3	details
1248	Mont Granier	Chambéry, France	300 to 500	7 km reach on flat land
1442	Claps de Luc	Die, Drôme valley, France	1,1	huge limestone blocks
1963	Monte Toc	Vajont dam, Italy	300	very high speed
1985	Rocher Blanc	Belledonne, Grenoble, France	1	spread on scree fan
1994	Randa	Zermatt, Switzerland	20 in two events	spread on scree fan

Along Cruden and Varnes classification (1996), Granier failure began as a rock topple and ended as a rock flow ; Claps de Luc and Monte Toc were rock slides ; Rocher Blanc & Randa rock falls.

As far as a safety factor may make sense, it is close to unity for most of them. If friction keeps constant along time in most of cases, cohesion irreparably decreases. When Terzaghi (1952) wrote that "all slopes have experienced the worst (climatic) conditions", he was wrong, as the worst always is to come; we are, all of us, born since the last exceptional rain.

CONCLUSIONS

Four slopes of reservoirs have been presented, as follows:

i) Saint Pons, a slow active slide, identified before the construction of the dam, along an ill-cemented conglomerate slope with moderate height and angle, slower after filling the reservoir.

ii) La Pelloud, a small over dip rock volume along the very steep rock slope of a very young epigenetic gorge, in bedded and fractured limestone, without any signs of pending instability.
iii) Le Billan, a slide inside a tilted rock mass along a high slope; triggered by excavating a moraine cover and an exceptional snowmelt ; stopped by a drainage gallery.
iv) Soulcem, steep rock slopes (gneiss in a glacial valley) with many previous rock falls, small to medium size, the fill dam has been designed to withstand such falls.

Mass movements downhill are a normal way of the evolution of slopes, which flatten and recede along millions of years. Work of daily infinitesimal erosion and deformation is surpassed by rare crises, just as along rivers the work of major floods surpass that of the average discharge along centuries. Except rejuvenation, either by tectonics or toe erosion, any slope is condemned to disappear along time. Ageing and gravity are everlasting laws of Nature. Slope stability relies on *geology* through materials and structures, and on *groundwater surges*. As any slope is a dam for water inside, dam engineers are better prepared to fully understand this power of water.

As has been stated long since, *"Contra impetus naturae, nihil est stabile"* (Lucretius, *De natura rerum*), and even before, *"The mountain falls and crumbles away"* (Holy Bible, Job 14-18), any slopes are to fall down, just as every living beings are to die; but "neither the day, nor the hour is known" (or year, or millenium …).

REFERENCES

Collective, Regione Piemonte & LIRIGM, 1998 ; Risques générés par les grands mouvements de versant ; Grenoble Univers.
Baldy P. & Thérond R., 1976 ; Le remplissage de la retenue de Sainte Croix sur le Verdon : *Travaux*, 243, mars.
Bjerrum L. & Jorstad F., 1968 ; Stability of rock slopes in Norway, NGO paper 79.
Cruden D.M. & Varnes D.J., 1996 ; Landslides types and processes, in *Landslides, investigation and mitigation*, US Transportation research Board special report 247.
Dubié J.Y., Bénéfice P. et Guitton C.; 1988 ; Télétransmission de l'auscultation d'un glissement ; Retenue de Grandmaison, glissement du Billan ; *Proc.* 5th *ISL*, Lausanne, Ch. Bonnard ed.
Duffaut P. & Louis C., 1972 ; L'eau souterraine et l'équilibre des pentes naturelles ; *Proc.* 24th *Int. Geological Conf.*, Section 13.
Duffaut P., 1978 ; Le rôle de l'eau dans le terrain, c'est celui de la libido dans le comportement humain ; Mém CERGH, Univers. Montpellier.
Duffaut P.,1993 ; Sur la stabilité des pentes et des pics ; *Rev. Franç. de Géotechnique*, 63.
Duffaut P., 1994, Stabilité des barrages et des pentes naturelles ; *Proceedings* 18th *ICOLD Conf.*, Durban, vol 4.
Goguel J. & Pachoud 1972 ; Géologie et dynamique de l'écroulement du Mont Granier ; *Bull. BRGM*, III.
Habib P., 1994 ; Discussion de Duffaut P. ; *Rev. Franç. de Géotechnique*, 64.
Lagasquié J.J., 1978 ; Analyse morphologique de la vallée du Pla de Soulcem ; thesis Toulouse Univ.
Terzaghi K., 1942 ; Theoretical Soil Mechanics, Wiley, New York,
Terzaghi K., 1952 ; Landslides, in *US Geological Society Berkey Volume*,
Vibert C., Arnould M., Cojean R. & Le Cleac'h J.M., 1988 ; Essai de prévision de rupture d'un versant montagneux à Saint Etienne de Tinée, France ; *Proc. Symp. on Landslides*, Balkema.

Conformities of Creeping Development in the East of the Russian Plain

A. P. DVINSKIH and V.A. YELKIN, Kazan University, Department of Geography and Geoecology, Russia

INTRODUCTION

The slow movements of the soil-ground masses on grassed slopes mostly result from soil-creep. V.M. Devis the first used this term in 1899. The fluctuations of the volume of the ground resulting from the change of its temperature, moisture content, and by freezing and thawing underlie this process. Besides, there arise in the ground forces directed in all directions. These forces are reinforced by the gravitational downslope component. Just this fact determines the slow displacement of the ground downwards. The investigations in the East of the Russian Plain began in the 1950's and 60's by A.P.Dedkov and V.A.Duglav (1967). In the East of the Russian Plain in the basins of Middle Volga and Low Kama there were built the whole net of stations for creep studies. This includes seventy pits, to give adequately reliable data on the rate of this process.

Pits 1.5m deep were dug at each station. Metal (bench - mark) rods - 0.5-0.6 m long were driven vertically in the base of the pit side walls. Lines were drawn on the end of each rod. In the wall of a pit strictly perpendicular to the slope's inclination there were placed metal plates (4-6 cm^2) with the interval of 10-20 cm oriented vertically and horizontally relatively to an immovable bench-mark in the pit unit. Then the pit was filled and a precise topographical bearing was taken. After the period of 2-5 years the pits were opened and the horizontal displacement of the plates with respect to the vertical line were measured. The soil-creep intensity is shown by the following points: (1). The linear rate of the displacement separated metal plates in a direction of a slope's fall (mm. yr^{-1}). The summary graph of vertical velocity profile of creep is built on this linear rate. (2). The average speed of ground movement occurs at a different depth. (3). The volume of the ground displaced in cm^3 per cm of slope length per year. ($cm^3 cm^{-1} yr^{-1}$)

THE MAIN RESULTS OF THE INVESTIGATIONS:

Creeping stops at the depth of 100-120 cm. Maximum displacement velocities make up to 3mm a year with mean values of 1-1.5 mm. The amount of displacing material reaches 1-20 cm^3 cm^{-1} yr^{-1} . Most of the pits have their maximum displacement at the depth of 20-40 cm. The speeds of movement on the slopes are depended on geomorphologic, geologic, landscape - climatic and others factors. The geomorphological significance of this creep is relatively small: it gives rise only to a weak smoothing down the edges and feet of the slopes. The mechanical composition of the soil-ground masses has influence on the speed of movement. Creeping is the exposed best of all in clay sediments, but in sand sediments it decreases. The movements are more intensive by a factor of 1.5-2 on the clay loams, than on the light loams.

Table 1. Intensity of soil movements on the slopes of different aspect and mechanical composition (a-- $cm^3 cm^{-1} yr^{-1}$)

Mechanical Composition	Exposition			
	South-west		North-east	
	Slope angle		Slope angle	
	<10	>10	<10	>10
Light loam	a) ------	a) 6.79	a) ------	a) 10.8
	b) ------	b) 2.23 mm yr^{-1}	b) ------	b) 2.9 mm yr^{-1}
	c) ------	c) 3 pits	c) ------	c) 3 pits
Sandy loam	a) 6.70	a) 15.0	a) ------	a) 13.9
	b) 1.65 mm yr^{-1}	b) 2.9 mm yr^{-1}	b) ------	b) 3.15 mm yr^{-1}
	c) 10 pits	c) 6 pits	------	c) 2 pits
Clay loam	a) 9.2	a) 15.1	a) ------	a) 27.8
	b) 1.7 mm yr^{-1}	b) 4.32 mm yr^{-1}	b) ------	b) 4.80 mm yr^{-1}
	c) 4 pits	c) 11 pits	c) ------	c) 5 pits

The clays are more changed of volume with fluctuations of the moisture content. In the sand the water leaks down the layers. The speed of the movement is connected with fluctuations of the mechanical composition. The variants of summary graphs of displacement and the change of the mechanical composition: **Case 1.** The mechanical composition doesn't change with increasing depth. The graph of vertical velocity profile of creep completely corresponds. The maximum speeds are observed at the surface and decrease with the depth; **Case 2**. The sand fractions increase with the depth. The graph corresponds to case 1, but speeds decrease more quickly; **Case 3** The clay fractions increase with the depth. The maximum speeds are observed not at the surface but at the depth of 20-40 cm.

It has been established that the speeds of movement depend upon the aspect of slopes. All the other conditions being equal, the ground speeds of movement and the volumes displaced on the slopes exposed to the north, north-east are 1.5 times higher than on other aspects. The slopes are situated in the sun's shadow is more smooth due to their higher moisture content. They are covered by snowfields until May. The slopes of south, south-west aspect are steeper and drier. Creeping intensity is directly proportional to a slope's steepness. The speeds of movement increase with increasing slope's angle. Gravitational downslope component determines the slow displacement. This dominates the influence of other factors connected with a slope's steepness. Creep is observed on the slopes until –30 to 45° whereupon the mode of failure changes to slumps and earthflows. The movements are observed on the slopes covered with forest, and in the non -forest areas. It impossible to distinguish any deposits created by creep but only rather considerable thickening of the soil horizon at the foot of the slope. Thick vegetation cover does not lead to the complete cessation of slope denudation.

REFERENCES

1.Дэвис.В.М. Геоморфологический очерк. М.; Изд-во иностранной литературы,1962
2.Дедков.А.П, Дуглав В.А. Медленные движения почвенно-грунтовых масс на задернованных склонах.- Известия А.Н. СССР, серия геогр,1967, № 4

Morphology and Interpretation for a Recent Multiple Peat Slide Event on Cuilcagh Mountain, Northern Ireland

A. P. DYKES and K. J. KIRK, Geographical Sciences, University of Huddersfield, Huddersfield, UK

INTRODUCTION

Mass movements of peat in Ireland have been recorded, and their descriptions published, since at least 1697 (Sollas et al., 1897). These early accounts described the bursting of raised and blanket bogs of great area and depth, but more recently reported peat failures have tended to be rather smaller and of greater variety. The most common are 'bogflows', in which liquid basal peat escapes from beneath the less humified surface peat. The source areas of these failures typically leave scars up to 200 m wide (e.g. Alexander et al., 1986), and the flow-paths may be extensive (e.g. Tomlinson, 1981; Coxon et al., 1989). Bogflows are rare in mainland Britain, but shallow landslides involving peat have been reported from the Pennines in particular. Most of these involved failure of the substrate material beneath the peat (Crisp et al., 1964; Carling, 1986), and have been referred to as 'peat slides'.

Almost all published accounts of peat mass movements in Ireland involve failure of the peat itself, either by flowing of liquid basal peat (e.g. Alexander et al., 1986) or sliding along a distinct structural surface within the peat (Wilson and Hegarty, 1993). The notable exception is the report of seven separate 'peat slides' in Co. Antrim in which a distinct failure surface within a clay substrate is identified (Tomlinson and Gardiner, 1982). The mass movement discussed in this paper appears very similar in nature to these Antrim slides.

SITE DETAILS AND METHODS

Cuilcagh Mountain is situated approximately 20 km southwest of Enniskillen, Co. Fermanagh, Northern Ireland (Fig. 1a). On the north and east sides of the mountain lies an extensive area of blanket bog which covers the middle slopes to a depth of up to three metres in places. The bog-covered slopes descend from around 500 to 200 m O.D. over a distance of 4 km, but are dissected by small valleys cut by surface streams (Fig. 1b). The peat has formed on a variable thickness of till thought to date from the Midlandian glaciation (Gunn, 1995), which overlies Namurian sandstones and shales (Leitrim Series). Much of the land has been used as rough grazing for sheep in recent times. The blanket bog on Cuilcagh mountain is scarred by many mass movements and other forms of erosion and disturbance including burning (Mr R. Watson, Marble Arch Caves, pers. comm.) and cutting (Walker, 1998).

The head of the mass movement examined in this paper is located near the northeastern margin of the summit massif of Cuilcagh Mountain at an altitude of around 455 m (Fig. 1b). The vegetation around the south and west of the failure is dominated by *Calluna vulgaris*, *Polytrichum commune* and *Eriophorum angustifolium* with some *E. vaginatum* and *Luzula* spp. on the western side, although locally along the centre of the upper scar edge and along linear depressions away from the southwest corner *Polytrichum commune*, *Spagnum cuspidatum* and *S. fuscum*(?) dominate, thought to be associated with tension cracks and subsurface peat pipes. *Juncus effusus* covers most of the ground north and east of the upper scar.

Landslides in research, theory and practice, Thomas Telford, London, 2000.

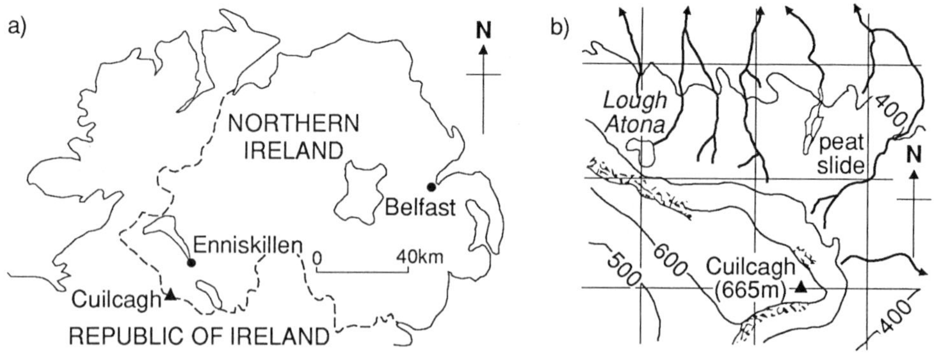

Figure 1. (a) Location of the study site at Cuilcagh; (b) location of the peat slide on Cuilcagh mountain (1 km grid squares; contours in metres).

According to local residents, the mass movement occurred on 23 or 24 October 1998. It was photographed on 26 October, and by 17 November it had been surveyed using a Total Station system. Intact core and bulk samples of peat and clay were collected from the margins of the main scars during the same period for the determination of physical properties using standard methods.

ANTECEDENT AND EVENT RAINFALL

According to annual average isohyets (1941–70) for the Cuilcagh area (source: N.I. Met. Office, Belfast), the failure site would expect to receive 1900–2000 mm of rain p.a.. Rainfall data recorded in an instrumented subcatchment 2.5 km north of the failure site for the three water years 1993–96 (Walker, 1998) showed an annual mean of 1890 mm, a monthly mean of 152 mm, and an average of 5–6 days per month with at least 10 mm of rainfall. Hourly rainfall intensity was usually very low, with a mean maximum during storm events of 5.3 mm h^{-1}, but on one occasion 27 mm of rain fell in one hour. The highest recorded daily rainfall was 65 mm.

Readings from a manual raingauge at Marble Arch Caves, 5 km north of the failure site, show that between 10.00 am on 23 October and 10.00 am on 26 October 1998 (Table 1) a large amount of rain fell in the area. Continuous monitoring of discharge through two flumes in the instrumented subcatchment, one of which included flow from a tributary into which the peat slide moved, resulted in a hydrograph showing a storm runoff event commencing sometime on 24 October and lasting 29 hours, with a peak discharge of 48.4 l s^{-1} ha^{-1}. Using a statistical model developed and tested previously (Walker, 1998), analysis of this runoff event yielded a predicted storm rainfall of between 62.9 and 99.7 mm, with a maximum 60-minute intensity of 19.1–20.1 mm h^{-1}. It is thus clear that heavy rainfall occurred on 24 October 1998, which was probably the trigger for the peat slide following a wet few days.

Table 1. Rainfall readings at 10.00am every weekday at Marble Arch Caves: October 1998

Day	Date	Rainfall (mm)	Day	Date	Rainfall (mm)	Day	Date	Rainfall (mm)
Sat.	10		Fri.	16	0.0	Thu.	22	7.5
Sun.	11		Sat.	17		Fri.	23	8.2
Mon.	12	12.1	Sun.	18		Sat.	24	
Tue.	13	3.8	Mon.	19	22.9	Sun.	25	
Wed.	14	0.0	Tue.	20	5.5	Mon.	26	54.2
Thu.	15	0.0	Wed.	21	37.7	Tue.	27	24.2

DESCRIPTION OF THE MASS MOVEMENT

Upper failure

The event comprised two distinct slope failures: an initial small upper failure and a subsequent, larger failure of a lower part of the slope (Fig. 2). Letters in the following text refer to the map and caption in Figure 2. The initial failure occurred on a 7° slope, above a degraded drainage ditch (A) cut across the slope just above a steeper slope segment. Immediately east of the failure is a depressed area containing only a thin peat cover, thought to be a much older mass movement. Both ditches (A,B) were cut between 1989 and 1991 (as determined from air photographs). The upper ditch had originally been about 15–20 cm wide and cut down through the peat and into the underlying clay. On excavation during heavy rain, an estimated discharge of around 0.5 l/min occurred from the edge of the ditch into the scar.

Around the scar a clearly visible boundary between the peat and the underlying clay occurred at a fairly consistent depth of around 70 cm, with local variations from this of up to ±20 cm. The depth of the failure surface below the ground surface was around 110–120 cm. The lower 40–50 cm of the edge of the scar comprised a layer of pale grey-brown coloured clay, the water content of which exceeded its Liquid Limit when sampled three weeks after the event.

The bare scar area comprised a dark blue-grey sandy clay, covered with a thin and discontinuous veneer of small (< 5 cm) angular stones. Despite the stones, the failure surface had a smooth appearance. Only one small raft and a number of small pieces of peat, and some patches of the pale clay, were left behind on the upper part of the failure surface when the slide occurred: it appears that the sliding peat remained largely intact and that there was no liquid peat as observed in nearby bogflows. Two distinct striations (C and another) were left in the blue clay surface by the sliding mass, caused by stones within the base of the sliding mass.

No peat pipes were visible in the headwall of the scar although there was indirect evidence of pipes leading to or from the scar area (D). However, five small pipes were found within the pale clay layer (E). Four were circular, 2–4 cm in diameter, located 15–20 cm below the base of the peat, and with estimated discharges of up to 0.5 l/min on 17 November (after several hours of rain which followed at least three dry days). The fifth was 35 cm below the base of the peat, 10 cm wide and 1–4.5 cm high, and discharging around 2 l/min. Two further flat, horizontal pipes around 10 cm wide and 1 cm high were observed discharging water from the base of the peat. It is likely that there were other undetected pipes in both the clay and the peat.

At the western side of the scar, a large intact raft of peat (F) slid a short distance, rotated a little, and stopped moving when it began to over-ride the *in situ* peat downslope of the ditch. The upper edge of this raft appears to have had a large (*c*.30 cm) peat pipe running into it at the base of the peat, although this was partly submerged and could not be investigated further. Below the lower ditch the slope steepens to 12.5° and a track (H) was eroded to the depth of the dark clay. To the side of this the sliding peat rode up over the *in situ* peat and spread across the slope, leaving the grass flattened, pointing downslope (I) and covered with a thin veneer of clay, small stones and fragments of peat from a few centimetres to over a metre across. Only one notable peat raft (J) slid over the *in situ* peat on the western side. The erosion track ends where the remainder of the sliding peat also rode up over the *in situ* peat.

Lower failure

The lower failure scar (K) is similar to the upper scar in most respects, although there were no visible signs of pipes at any depth and no ditches. Some rafts remained in the scar area, and two large tension cracks (L) had developed. A deep erosion track was excavated (M) roughly in line with the upper erosion track, which contained disrupted peat debris along its eastern side. East of this, the extent of damaged vegetation can be related to the sliding of individual peat rafts (N). One raft that stopped within the affected area (O) derived from the lower failure.

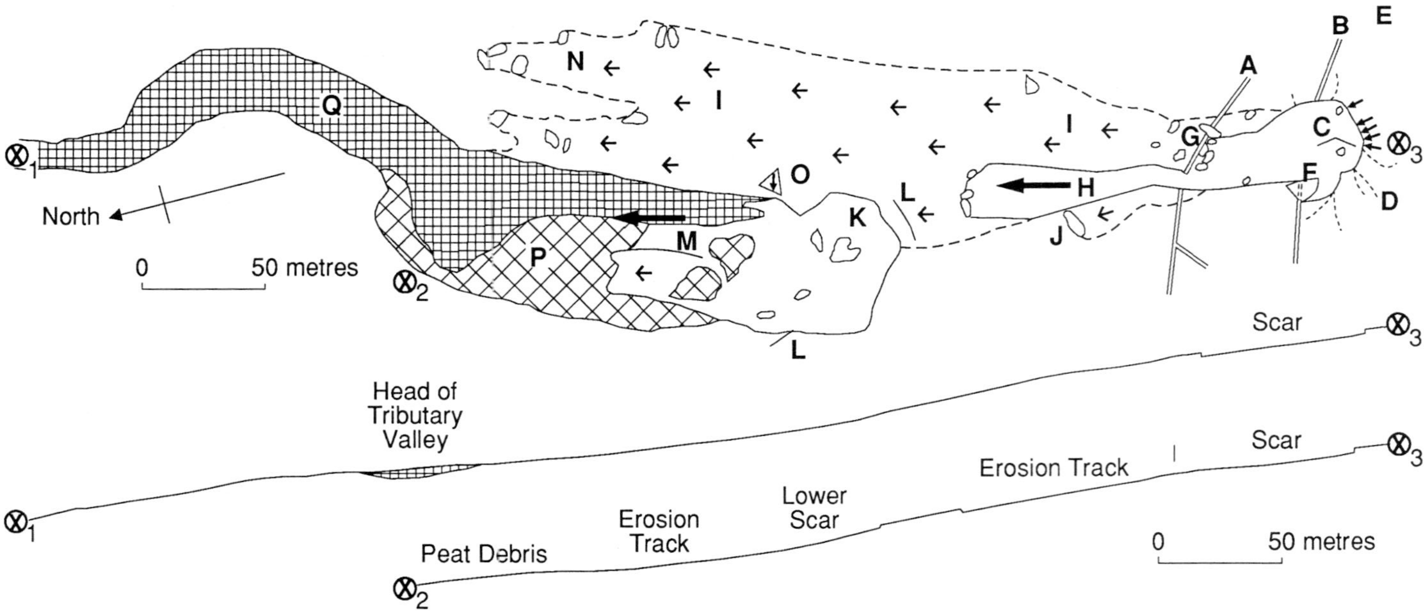

Figure 2. Surveyed map and profiles (no vertical exaggeration) of the Cuilcagh peat slide. Letters identify features and zones as follows:

A	Lower drainage ditch	J	Large peat raft
B	Upper drainage ditch	K	Head of the lower failure scar
C	Striation in the exposed clay surface	L	Large tension crack in the peat
D	Approximate courses of subsurface peat pipes	M	Erosion track below the lower failure scar
E	Approximate locations of pipes in the clay	N	Limits of travel of large peat rafts defining affected area
F	Intact block of displaced peat	O	Rotated peat raft from the head of the lower failure scar
G	Cluster of intact peat rafts	P	Peat debris mostly comprising large intact rafts of peat
H	Erosion track resulting from the upper failure	Q	Disrupted peat fragments and wet peat slurry
I	Mud-covered vegetation flattened by sliding peat rafts		

An area of mostly intact but displaced peat defines the extent of the lower failure. Large cracks filled with water separate intact blocks and rafts in this area (P). From the lower erosion track (M) to the upper reaches of a pre-existing tributary stream (Q), the peat debris occupying the failure area and the stream valley is highly dispersed, broken up into very small intact fragments or partially liquified. However, only very small quantities of peat were carried further downstream than the northern extent of the surveyed area (Fig. 2).

PEAT AND CLAY CHARACTERISTICS

Unlike other parts of Cuilcagh mountain, where highly humified basal peat has been associated with a number of flow failures, the peat in this case appears more strongly fibrous throughout its depth. This explains why most of the peat moved as large intact rafts rather than as a semi-liquid slurry. The properties of the pale clay are summarised in Table 2; in the lower scar it appeared to contain more peat fibres than in the upper scar (loss on ignition = 10.0%, n = 3).

Table 2. Properties of the pale clay at the upper failure scar

	Saturated moisture content (% dry weight)	Saturated unit weight (kN m^{-3})	Saturated hydraulic conductivity (m s^{-1})	Loss on ignition (%)	Plastic Limit (% dry weight)	Liquid Limit (% dry weight)	ϕ' (peak) (°)	c' (peak) (kPa)
Clay	64.0	16.48	4.2 x 10^{-8}	3.9	34–38	55–61	26.5	2.75
n =	6	6	4	3	4	4	3	3
s.d.=	4.2	0.52	2.7 x 10^{-8}	0.0	-	-	-	-

INTERPRETATION

The irregular margins of the pale clay extending out beyond the peat around the edges of the scar areas, and the layer of pale clay attached to the base of most of the peat rafts, initially suggested that it is this material which failed. There was no evidence of the clay having flowed despite its near-liquid state. However, the peak shear strength of the pale clay was very high (Table 2): for an infinite slope analysis to give a factor of safety of 1.0 then either artesian water pressures or a lower shear strength would be required (the latter possibly indicating failure at the interface between the two types of clay, c.f. Tomlinson and Gardiner, 1982, or that the pale clay was shearing at its residual strength).

Detailed field examination and morphological interpretation of this mass movement suggest a distinct sequence of events. The movement and final position of raft F, and the over-riding of *in situ* peat just a few metres below ditch B on the eastern side of the top scar, suggest that failure of the pale clay upslope of ditch B started the whole event. The change of direction in striation E corresponds with the rotation of raft F, and the number of pipes at depth within the clay could be expected to increase pore water pressures very quickly, possibly giving rise to artesian conditions in response to rapid inputs of rainwater. Whether or not water from ditch B contributed to the development of high water pressures can not be determined, although degraded ditches have been implicated in previous peat failures (Wilson and Hegarty, 1993).

It appears that the peat between the two ditches was pushed downslope; it is unlikely that it could have been pulled up and over the marginal scars intact. To some extent the process was aided by the lower ditch providing a structural discontinuity between moving peat from upslope and *in situ* peat. A cluster of small rafts with insufficient momentum failed to ascend the northern side of the ditch (G) The force exerted by the moving peat was sufficient to dislodge a narrow strip of peat below the ditch, although whether erosion track H was formed by erosion or basal shear is not clear. In any case, *in situ* peat resisted disturbance on the steepest part of the slope and was over-ridden. East of the erosion track, the steeper slope and wet grass with a very low friction coefficient enabled large rafts to slide over 300 metres.

The next event was probably undrained shear failure of clay beneath the peat to form the lower erosion track (M) caused by rapid loading by sliding peat from the upper failure and erosion track. This unloaded the toe of the steeper slope at K which in turn immediately failed. This sequence provides the simplest explanation for raft O, which was over-ridden by upper failure peat before sliding from the head of scar K and rotating to the alignment shown in Fig. 2 (determined from the flattened grass on top) before coming to rest on intact ground affected by sliding of upper failure peat rafts. The peat from the main lower failure slid almost a distance of only 50–100 metres, although at the lower end the head of a first order stream valley provided a concentrated route for peat which had broken up and combined with excess rainwater to be transported beyond the main mass.

CONCLUSIONS

An initial mass movement involving about 770 m^3 of peat and clay occurred as a result of high rainfall probably combined with the generation of high water pressures in the clay due to pipeflow inputs to the failure area, although this, and the influence of the ditches, is not yet certain. The initial movement probably caused shear failure of the adjacent downslope block of peat by pushing on it, and an additional 2650 m^3 of material was lost from the upper scar and erosion track. Some of this peat triggered a failure lower down the slope by undrained loading, and this lower failure in turn unloaded the steeper slope above which then also failed. Thus a further 5700 m^3 of peat and clay moved downslope. The total land area affected by sliding peat was 30,500 m^2. Whilst the role of surface and subsurface drainage in peat failures remains open to debate, the main outcome of this study is the identification of a causal mechanism for a larger failure which resulted from a very small initial event.

ACKNOWLEDGEMENTS

Our thanks go to Prof. J. Gunn for logistical support and for alerting us to the event, D. Russell and R. Harding (fieldwork), staff at Marble Arch Caves (field assistance and aerial survey), Dr J. Labadz (ground survey data processing), Dr C. Walker (streamflow analysis), G. Rushworth (sample preparation), and S. Pratt for drawing the maps.

REFERENCES

Alexander, R. W., Coxon, P. and Thorn, R. H. 1986. 'A bog flow at Straduff townland, County Sligo'. *Proceedings of the Royal Irish Academy*, **86B**, 107-119.

Carling, P. A. 1986. 'Peat slides in Teesdale and Weardale, Northern Pennines, July 1983: Description and failure mechanisms'. *Earth Surface Processes and Landforms*, **11**, 193-206.

Coxon, P., Coxon, C. E. and Thorn, R. H. 1989. 'The Yellow River (County Leitrim Ireland) flash flood of June 1986'. In Beven, K. and Carling, P. (Eds.), *Floods: Hydrological, Sedimentological and Geomorphological Implications*. Wiley, Chichester, 199-217.

Crisp, D. T., Rawes, M. and Welch, D. 1964. 'A Pennine Peat Slide'. *Geographical Journal*, **130**, 519-524.

Gunn, J. 1995. 'Environmental change and land management in the Cuilcagh karst, Northern Ireland'. In McGregor, D. F. M. and Thompson, D. A. (Eds.), *Geomorphology and Land Management in a Changing Environment*. Wiley, Chichester, 195-209.

Sollas, W. J., Praeger, R. L., Dixon, A. F. and Delap, A. 1897. 'Report of the Committee consisting of . . . appointed by the Royal Dublin Society to investigate the recent Bog-flow in Kerry'. *Scientific Proceedings, Royal Dublin Society*, **8**, 475-508.

Tomlinson, R. W. 1981. 'A preliminary note on the bog burst at Carrowmaculla, Co. Fermanagh, November 1979'. *Irish Naturalists' Journal*, **20**, 313-316.

Tomlinson, R. W. and Gardiner, T. 1982. 'Seven bog slides in the Slieve-an-Orra hills, County Antrim'. *Journal of Earth Sciences Royal Dublin Society*, **5**, 1-9.

Walker, C. 1998. *Modelling the hydrological impacts of mechanised peat extraction on an upland blanket bog*. Unpublished PhD thesis, University of Huddersfield, UK.

Wilson, P. and Hegarty, C. 1993. 'Morphology and causes of recent peat slides on Skerry Hill, Co. Antrim, Northern Ireland'. *Earth Surface Processes and Landforms*, **18**, 593-601.

Slope Movements and Active Tectonics in the Izbor River Basin (Granada, Spain)

R. EL HAMDOUNI, C. IRIGARAY, Departamento de Ingeniería Civil, Universidad de Granada, Facultad de Ciencias, Granada, España, T. FERNÁNDEZ, Departamento de Ingeniería Cartográfica, Geodésica y Fotogrametría, Universidad de Jaén, España, C. SANZ DE GALDEANO, Instituto Andaluz de Ciencias de la Tierra, CSIC, Universidad de Granada, Facultad de Ciencias, Granada, España and J. CHACÓN, Departamento de Ingeniería Civil, Universidad de Granada, Facultad de Ciencias, Granada, España

ABSTRACT

In this paper an inventory of landslides and its relationships to the Quaternary active tectonics are analysed taking in account geomorphic indexes and field observations. The terrain affected by landslides represents a 7.4% of the total surface of the study area and most of the movements are incipient slides on Nevado-Filábride and Alpujárride metapelites, or advanced earth flows on Neogene silts and sand deposits. Also, there are debris flows and rock falls developed on Alpujárride marbles and calcareous schists.

In general, the observed unstable zones are related with very steep slopes in the margins of rapidly incised river channels, resulting from environmental processes and the recent active tectonics. A comparison of the landslide inventory and the distribution of geomorphic indexes, related to the river channels network, is made. From this, different areas of increasing active tectonics are distinguished: the Southern margin of Sierra Nevada and Northern slopes of Sierra Los Guájares, where a relatively high degree of tectonic activity is established and the landslides are abundant, in one side, and the Northern slopes of Sierra de Albuñuelas and the SW of the Padul basin, where the tectonic activity is relatively lower and the slope instability is not frequent.

INTRODUCTION AND SETTING OF THE STUDY AREA

The extraordinary high incision of the rivers excavating the Western margin of Sierra Nevada and the abundant scarps affecting young deposits (Pleistocene to Holocene alluvial fans) are expressing a recent tectonic activity of continental uplift. Nevertheless, in order to quantify its incidence, the existence of Quaternary active tectonic in the observed faults needs to be evaluated. This is made following methods based on different geomorphic indexes (Keller and Pinter, 1996) on the Izbor river and subsidiary channels network, and Lecrin valley faults at the SW margin of Sierra Nevada. The study area is a part of the Southern Granada Province (Andalucía, Spain) settled at the most South-eastern limit of the former Granada basin (Figure 1). Geologically, the area is composed by outcrops of metamorphic rocks of the Internal Zone of the Betic Cordillera, and also Neogene and Quaternary deposits filling up the Padul and Lecrín basins, derived from the former Granada basin. The effect of uplift of Sierra Nevada is most apparent in this Southern margin than in the Northern one (Keller et al., 1996), what is a particular feature since Miocene times.

Landslides in research, theory and practice, Thomas Telford, London, 2000.

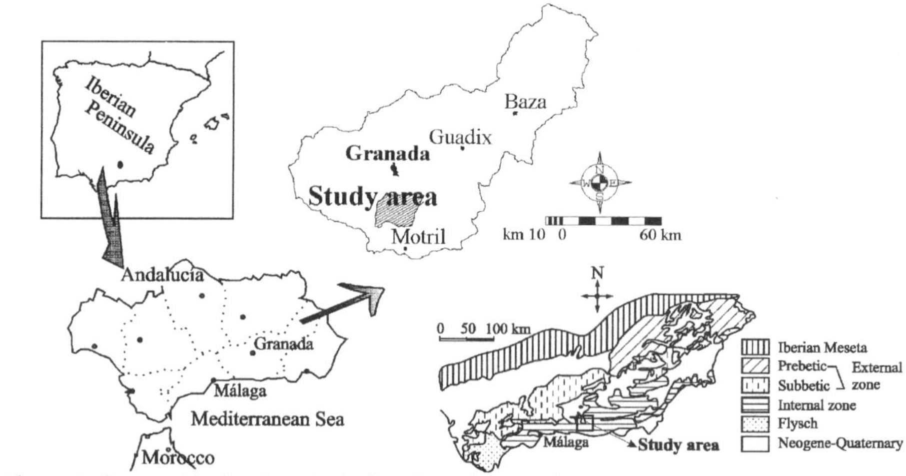

Figure 1. Geographical and geological setting of the study area

GEOMORPHIC INDEXES AND FIELD EVIDENCES OF ACTIVE TECTONICS

Many definitions of active faults include the use of geomorphic evidences as main indicators of the degree of tectonic activity and also as reference features for the classification of the fault activity. This is usually based on different geomorphic indexes (Keller and Pinter, 1996). Given homogeneous lithological and structural conditions, these indexes help to the identification of current or recent tectonic activity along faults, folds and mountain fronts. Seven of these indexes have been here applied to the analysis of tectonic activity in 27 sub-basins in the study area (Figure 2). The necessary data were obtained mainly from topographical maps, aerial photographs and a digital elevation model (DEM) performed in GIS.

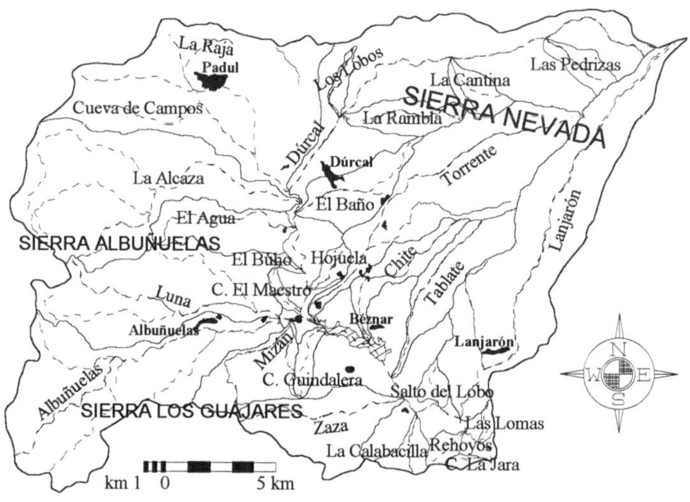

Figure 2. Streams and sub-basins analysed in the study area.

Stream profiling of the rivers and the SL index.

The stream profile of a river is very sensible to the tectonic uplift process and may be useful for the recognition of currently active structures crossing that river (Seeber y Gornitz, 1983).

In the NE of Sierra Los Guájares and SW margin of Sierra Nevada, most of the streams show convex upward profiles and/or segments with variable slope angle. This convexity is less apparent in the SW of Padul and N of Sierra Albuñuelas. The SL stream-length gradient index (Hack, 1973) is defined as: $\mathbf{SL = (\Delta H/\Delta L).L}$, where $\Delta H/\Delta L$ is the channel slope of the considered reach, and L is the total channel length from the point of interest to the highest point on the channel. In this area, the SL index shows values from 5 to more than 5000. Typically a sudden appearance of relatively high values along a channel may be related to tectonic activity along faults or folds crossing that stream channel. In the SW border of Sierra Nevada, a concentration of high SL index values was observed along the intersections of border faults and stream channels (Keller et al., 1996) with no relation with lithological changes, what is considered as indicative of a Quaternary tectonic activity. Along the margins of Sierra Los Guájares and Southern border of Sierra Albuñuelas also some high values have been measured in boundaries between different lithologies and, therefore, related to changes in the strength of the materials, although some few SL index high values appear in homogeneous lithologies. In the Northern border of Sierra Albuñuelas and SW Padul basin, the SL index shows relatively low values and no anomalous values are observed.

The asymmetry factor index AF

It is defined as $\mathbf{AF = 100\ (A_r/A_t)}$, where A_r is the area of the basin to the right of the trunk steam, and A_t is the total area of the drainage basin (Hare and Gardner, 1985). The AF index is used to detect tectonic tilting of a basin. In a normal basin, with regular distribution of the slopes along the main channel and subsidiaries, an AF index of 50 is obtained. Lower and higher values may be related to a tectonic tilting or to structural or sedimentary features of the lithological units. The obtained values in the study area (Table 1) were expressed as AF-50, in order to show anomalies. Values "(AF-50) > 10" were plotted on the lithological map in order to disregard those anomalies associated to lithological changes.

Hypsometric curve and Hypsometric integral of the basin: IH index

The hypsometric curve describes the distribution of the elevations of a particular region and gives a general overview of the land forms (Strahler, 1952). A method for the featuring of the hypsometric curve is based in the calculation of its integral (Mayer, 1990) which is defined as the area included under the curve, interpreted as the non eroded volume of the basin. A high hypsometric integral indicates young land forms (high mountains and incised rivers), until intermediary values and sygmoidal shape of the hypsometric curve indicate an advanced stage in the land form development. Low values are related to old land forms. In this paper values above 0.5 are considered as derived from a recent tectonic uplift. Table 1 shows the values observed in the study area.

Ratio of valley-floor width to valley heigth (V_f index)

The V_f index (Bull and McFadden, 1977; Keller and Pinter, 1996) is defined as $\mathbf{V_f=2V_{fw}/[(E_{ld}-E_{sc})+(E_{rd}-E_{sc})]}$ where V_{fw} is the with of the valley floor, E_{ld} and E_{rd} are elevations of the left and right valley divides, respectively, and E_{sc} is the elevations of the valley floor. The V_f values show a reverse relation to the tectonic activity. In table 1 the obtained values are listed for the main valleys of the study area, showing a range between 0.039 (Dúrcal river) and 11.43 (Cuevas de Campo creek). Values below 1 are considered as evidence of tectonic activity.

Basin form index: B_s

A basin affected by tectonic uplift shows typically a tendency toward a elongated form which derives to a circle along time when the process of uplift ends. The index is expressed as:

$B_s = B_l/B_w$, where B_l is the length of the basin, measured from its mouth to the most distant drainage divide, and B_w is the width of the basin measured across the short axis (Cannon, 1976, Ramírez, 1998). High values in the border of Sierra Nevada and lower in all the rest are obtained. Values above 4 are considered as indicators of tectonic activity.

From all these different indexes the sub-basins are classified in three classes of tectonic activity (TA classes): 1 (low); 2 (moderate) and 3 (high). Table 1 shows the obtained results.

Table 1. Geomorphic indexes and classification of the sub-basins

Nº	Sub-basin	Area (Km^2)	Stream Profiles	SL	AF - 50	IH	Vf	Bs	TA Class
1	Bº Cueva de Campos	25.36	-	-		0.28	1.25	1.35	1
2	Bº Alcaza	46.72	-	-	-5.60	0.46	1.90	2.22	1
3	Bº El Agua	11.51	-	-	16.60	**0.51**	1.11	3.14	1
4	Bº El Búho	0.70	-	-	-2.90	**0.65**		3.26	1
5	Bº Cortijo del Maestro	1.08	+		**-10.19**	0.40	**0.23**	**4.38**	3
6	Bº Luna	33.90	+	+	-5.49	**0.50**	**0.25**	2.76	3
7	Albuñuelas river	44.63	+	+	8.35	**0.66**	**0.10**	1.86	3
8	Bº Mizán	10.97	+	+	**15.09**	**0.50**	**0.68**	1.70	3
9	Bº Cortijo Guindalera	2.29	+	+	-1.09	0.38	**0.38**	**4.62**	3
10	Bº Zaza	16.42	+	+	19.00	**0.55**	**0.09**	2.04	3
11	Bº La Calabacilla	2.96	-	-	4.39	**0.59**		1.42	1
12	Bº Rehoyos	1.73			**-12.43**	**0.57**		1.56	3
13	Bº Cortijo de la Jara	0.71			**10.56**	**0.62**			3
14	Bº Las Lomas	0.81			**-19.14**	**0.56**		3.51	3
15	Lanjarón river	50.56	+	+	5.86	0.48	**0.15**	**5.72**	3
16	Bº Salto del Lobo	1.85			-5.14	0.46		1.46	1
17	Bº Tablate	11.52	+	+	**-29.25**	0.48	**0.05**	**5.81**	3
18	Bº Chite	7.55	+	+	5.23	0.42	**0.22**	**5.11**	3
19	Torrente river	35.79	+	+	**-15.00**	0.46	**0.08**	**4.35**	3
20	Bº Hojuela	1.98	+	-	-2.53	0.46		2.36	1
21	Bº El Baño	9.82	+	+	-7.76	0.30	**0.23**	2.94	2
22	Bº La Rambla	6.18	+	+	-4.50	**0.58**		3.29	2
23	Bº La Cantina	3.95	+	+	**-14.18**	**0.50**		1.96	3
24	Bº Las Pedrizas	3.57	+	+	3.22	**0.60**		2.66	2
25	Dúrcal River	50.34	+	+	**-15.35**	0.43	**0.04**	**4.20**	3
26	Bº Los Lobos	1.72	+	+	**-17.44**	**0.56**		**4.78**	3
27	Bº La Raja	28.01	+	+			**0.23**		3

Stream profiles: a sign "-" refers a concave-linear profile, "+" for a convex profile. Bº : Creek
SL: "+" indicates presence of high values anomalies. "-" no anomalies are observed

Field evidences of active tectonics

In the sector Padul–Dúrcal sets of faults associated to the main Padul-Nigüelas fault are observed deforming alluvial fans of Middle to Upper Pleistocene age, and Holocene peat deposits of Padul. The main fault scarp is tilted in some segments due to the effect of the latest displacements (Keller et al., 1996; Sanz de Galdeano et al., 1998). There is also a current sismicity associated to this fault showing, at least in the historical record, shallow epicentres of moderate to low magnitude at depths around 10-12 kms (Chacón, 1999).

INVENTORY OF LANDSLIDES

A number of 983 landslides were inventoried from field and aerial photography surveys (Figure 3). From these, 71 are complex movements (affecting, in surface area, a 27.9% of the disturbed lands), 157 earth flows (20.1%), 198 slides (25.1%), 234 rockfalls (10.7%) y 323 debris flows (16.2%). The areal extension of the lands disturbed by the landslides is a 7.4% of the total surface of the study area. Most of the slides are in micaschist and calcareous schists (49.46% of all the landslide events). The earth flows are in silts, sands and marls (55.27%). A 81.63% of the debris flows affect to Alpujárride crushed marbles and the rest to colluvials developed on these materials. The rock falls are widespread on calcareous schists (35.76%) and marbles (35.28%). The complex movements are mostly on schists (30.54%), in the case of the combination slide-flow-fall, on marbles (25.03%), in the case rockfall-slide and in phyllites in the case slide-flow (17.47%).

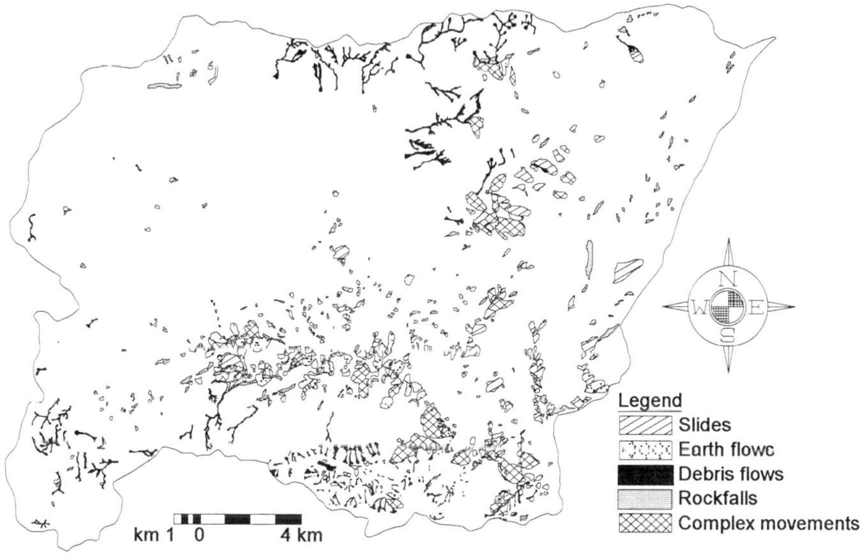

Figure 3. Inventory of slope movements

RELATIONSHIPS BETWEEN SLOPE MOVEMENTS AND ACTIVE TECTONIC

In the regional framework, the boundary of the former Granada basin shows a very high concentration of unstable areas, large landslides, current seismic activity including the epicenter of a historical regionally catastrophic earthquake (Arenas del Rey-Alhama, 1884) (Chacón, 1999). These landslides are usually affecting Betic materials (Nevado-Filábride and Alpujárride units). Some are in the outside part of the limit, as the landslides of Izbor and Olivares, until other are in the outside part of the limit, as the large landslides of Albuñuelas, Monachil and Güevéjar, this last one triggered by the catastrophic earthquake of 1884. Particularly, those of Albuñuelas and Izbor are in the study region in areas showing a high degree of tectonic activity (class 3). As it can be seen in Table 2, most of the landslides occur in class 3 of high tectonic activity.

Table 2. Distribution of landslide frequency in relation to the degree of tectonic activity. E: Events; AL: affected land.

TA class	Area	All the landslides		Flows		Slides & Complex		Debris flows		Rockfalls	
	(Km^2)	E (%)	AL (%)	E (%)	AL (%)	E (%)	AL (%)	E (%)	AL (%)	E (%)	AL (%)
1.Low	91.08	3.72	1.09	6.56	0.32	3.37	0.51	1.48	0.09	4.93	0.18
2.Moderate	19.57	3.16	4.32	0.00	0.00	3.28	2.30	7.48	2.01	0.01	0.00
3.High	301.94	93.12	8.25	93.44	1.395	93.35	4.25	91.04	1.59	95.06	1.02

CONCLUSIONS

The Southern border of the Granada basin is tectonically active as shown not only by anomalies in geomorphic indexes, but also by field observation of the fault system and the current seismic record. The slope instability is conditioned not only by unfavourable factors such as lithology, slope angle, water pressure, etc., but also by the tectonic activity at the border of the former Granada basin and South-western margin of Sierra Nevada, as it is shown by the coincidence of areas with higher frequency of landslides and also high tectonic activity. Therefore, for the analysis of the susceptibility to landsliding of these areas, and any other with similar features, the tectonic activity during the Quaternary, the active tectonics, is one factor to be included and also one of the main geological factor to be taken in account in terrain evaluation for engineering purposes, land use planning and natural hazard assessment.

Acknowledgements: This research has been partially funded by Group RNM 121 of the Andalusian Research Plan, Regional Government of Andalucía, Spain.

REFERENCES

Bull, W.B. and McFadden, L.D. 1977. Tectonic Geomorphology north and south of the Garlock Fault, California. In Doehring, D.O. (Ed.). Geomorphology in Arid Regions, Proceeding of Eighth Annual Geomorphology Symposium, State University of New York, Binghamton, p. 115-138.

Chacón, J. (1999). Riesgos naturales en el borde suroriental de la depresión de Granada. En: P.Rivas & R.Gómez (Coords). *Ciclos Naturales y Desarrollo Sostenible*. Grupo Editorial Universitario. FETE-UGT.

Cannon, P.J. (1976). Generation of explicit parameters for a quantitative geomorphic study of Mill Creek drainage basin. Oklahoma Geology Notes, 36(1), 3-16

Hack, J.T. (1973). Stream-profiles analysis and stream-gradient index. *Journal of Research. U.S. Geological Survey*, 1 (4): 421-429.

Hare, P.W. and Gardner, T.W. (1985). Geomorphic indicators of vertical neotectonism along converging plate margins, Nicoya Peninsula, Costa Rica, In M. Morisawa and J.T. Hack (eds). *Tectonic Geomorphology*: Proceedings of the 15th Annual Binghamton Geomorphology Symposium, September 1984. Allen & Unwin: Boston.

Keller, E.A. and Pinter, N. (1996). Active tectonics: Earthquakes, Uplift, and Landscape. Prentice Hall. New Jersey.

Keller, E.A. Sanz de Galdeano, C., and Chacón, J. (1996). Tectonic geomorphology and earthquake hazard of Sierra Nevada, Southern Spain. In J.Chacón & J.L.Rosúa editors. 1ª Conferencia Internacional Sierra Nevada. Volumen I. pp. 201-218.

Mayer, L. (1990). Introduction to quatitative geomorfology. Prentice Hall: Englewood, Cliffs, NJ.

Ramírez-Herrera, MT. (1998). Geomorphic assessment of active tectonics in the Acambay Graben, Mexican volcanic belt. Earth Surface Processes and Landforms, Vol 23, 317-332.

Sanz de Galdeano, C. El Hamdouni, R. y Chacón, J. (1998). Neotectónica de la fosa del Padul y del Valle de Lecrín. En: Gómez Ortiz, A.; Salvador Franch, F.; Schulte, L. y García Navarro, A. (editores). *Itinerarios Geomorfológicos por Andalucía Oriental*. Publicacions de la Universitat de Barcelona. Barcelona, pp. 65-81.

Seeber, L and Gornitz, V. (1983). River profiles along the Himalayan arc as indicators of active tectonics. Tectonophysics, 92: 335-367.

Strahler, A.N. (1952). Hypsometric (area-altitude) analysis of erosional topography. *Geological Society of America Bulletin*, 63: 1117-1142.

Landslides due to the Propagation of Cracks and Plastic Zones

S. A. ELSOUFIEV, Odessa State University, Ukraine

INTRODUCTION

An estimation of strength and rigidity of many structures is made as a rule by the scheme of an elasticity theory although many materials (soils first of all) have non-linear strain-stress diagram with residual deformations almost from the loading beginning. In this situation the study of non-linear problems is of high interest (Elsoufiev 1997; Elsoufiev, Nesterova 1997). Here the fracture of different media near the ends of cracks and stamps at antiplane deformation and in tension-compression is investigated. In order to have a complete picture of the problem and to use the analogies between its different parts some known results are also included.

MAIN EQUATIONS

In a longitudinal shear (antiplane deformation) the displacement u, strains γ_r, γ_θ (in polar co-ordinates r, θ) and stresses τ_r, τ_θ are linked by expressions

$$\gamma_r = \partial u/\partial r,\ r\gamma_\theta = \partial u/\partial \theta,\ \partial(r\tau_r)/\partial r + \partial\tau_\theta/\partial\theta = 0,\ \gamma_i = \Omega(t)(\tau_e)^{m-1}\tau_i \quad (i = r, \theta;\ \tau_e = \sqrt{(\tau_r)^2+(\tau_\theta)^2}) \quad (1)$$

where Ω, m are experimentally determined creep function of time t, an exponent of the hardening law, and τ_e – an equivalent (maximum shearing) stress.

In the conditions of the plane deformation strains ε_r, ε_θ, $\gamma_{r\theta}\equiv\gamma$ are linked with displacements u_r, u_θ by expressions

$$\varepsilon_r=\partial u_r/\partial r,\ r\varepsilon_\theta=u_r+\partial u_\theta/\partial\theta,\ r\gamma = r^2\partial(u_\theta/r)/\partial r+\partial u_r/\partial\theta,\ r\partial/(\partial(r^2\varepsilon_\theta)/r\partial r)\partial r-\partial^2\varepsilon_\theta/\partial\theta^2 = \partial(r\partial\gamma/\partial\theta)/\partial r \quad (2)$$

The last formula here corresponds to the compatibility condition. The stresses in a weightless material subdue to equilibrium equations

$$r\partial\sigma_r/\partial r + \sigma_r - \sigma_\theta + \partial\tau/\partial\theta = 0,\ \partial(r^2\tau)/\partial r + r\partial\sigma_\theta/\partial\theta = 0 \quad (3)$$

and are linked with strains by expressions similar to the 4th one in (1) as follows

$$\varepsilon_r = \Omega(t)(\tau_e)^{m-1}(\sigma_r - \sigma_\theta)/4,\ \gamma = \Omega(t)(\tau_e)^{m-1}\tau.\ (\tau_e = 0.5\sqrt{(\sigma_r - \sigma_\theta)^2 + 4\tau^2}). \quad (4)$$

Landslides in research, theory and practice, Thomas Telford, London, 2000.

The experiments (Elsoufiev 1982) show that the rheological law of (4) and the 4th of (1) type may be used at a monotonous loading while an equivalent strain does not diminish, and at stepwise or interrupted regimes for the parts where the equivalent stress is not less than on the portions before it if the time is calculated from the moment of a new loading. Buryshkin and Elsoufiev (1994) proved its applicability to the cycling stress change too. The correlation of the equations with test data is better for unsteadier creep. When Ω = const they turn to be the plasticity basic expressions of the deformation type generalized here.

SOLUTIONS IN ELASTIC RANGE

While solving elastic problems the functions of complex variable $z = x + iy$ as well as potential functions $\omega(z)$ for the longitudinal shear and $F(z)$ for the plane problem with loads perpendicular to axis x (dashed arrows in Figure1) are used. In these terms we have

$$u=\mathrm{Re}\,\omega(z)/G,\ \tau_r-i\tau_\theta=\omega' e^{i\theta};\ \sigma_r+\sigma_\theta=4\mathrm{Re}F',\ \sigma_\theta-\sigma_r+2i\tau=e^{2i\theta}4iy\overline{F''},2G(u_r+iu_\theta)=e^{-i\theta}(kF-\bar{F}-2iy\overline{F'}) \quad (5)$$

where ', " denote the derivatives by z, Re –the real part of a complex variable, $k=3-4\nu$ for the plane deformation and $(3-\nu)/(1+\nu)$ for the plane stress state (ν-the Poisson's ratio), G-the shear modulus.

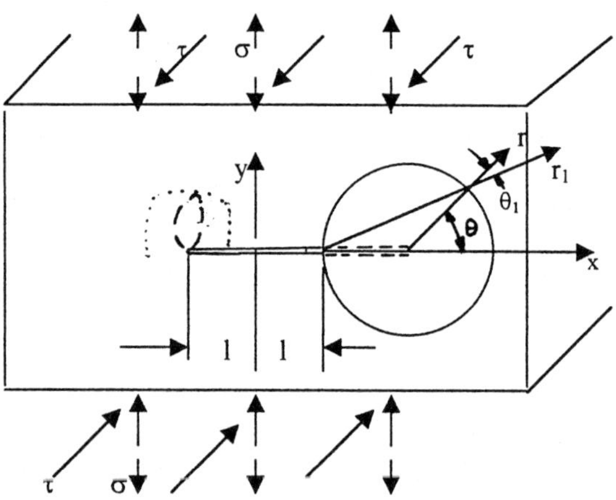

Figure 1. Crack in antiplane or plane deformation

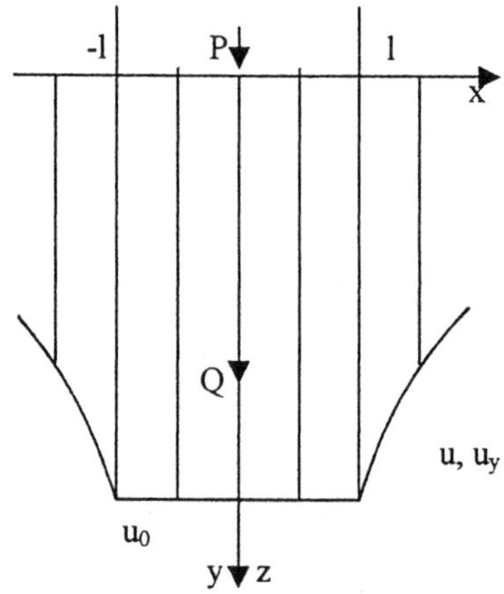

Figure 2. Constant displacement

Using these variables we can represent the solution for the shear problem of the material with a crack by stresses τ in infinity (solid arrows in Figure 1) as $\omega(z) = -i\tau(z^2-l^2)^{-0,5}$. In order to appreciate the stress-strain distribution near the ends of the crack we apply to the last formula the asymptotic decomposition $z-l = r(\exp i\theta)$ that gives

$$Gu=\tau\sqrt{2rl}\sin\theta/2,\ \tau_r = \tau\sqrt{l/2r}\sin\theta/2,\ \tau_\theta = \tau\sqrt{l/2r}\cos\theta/2,\ \tau_e = \tau\sqrt{l/2r}. \quad (6)$$

In a similar way the solution of the problem for constant shear displacement u_0 in the length–l, l of axis x (Figure 2) can be represented as $\omega(z) = Gu_0 - (Q/\pi)\ln(z/l + ((z/l)^2-1)^{0.5})$ where Q is the resultant of the shearing stresses in the length –l, l. It is interesting to notice the analogy between the last two solutions as the functions $\omega'(z)$ for them differ by a constant factor and one of them may be received from another by the replacement of z and l by $1/z$ and $1/l$ respectively. In the asymptotic representation the last solution is

$$u = u_0 - (Q/G\pi)\sqrt{2r/l}\cos\theta/2,\ \tau_r = (Q/\pi\sqrt{2rl})\cos\theta/2,\ \tau_\theta = (Q/\pi\sqrt{2rl})\sin\theta/2,\ \tau_e = Q/\pi\sqrt{2rl}. \quad (7)$$

From the 4th expressions (6), (7) we can see that in the both cases condition τ_e = const gives a circumference with the centre at x = /l/ (a circle in solid line at the right end of the initial crack in Figure 1) where a fracture or the first plastic zones appear.

There is another analogy – between antiplane deformation and the considered type of plane one. Their solutions differ by a constant multiplier which can be found from a static condition. So from the solution for a crack in longitudinal shear we find results for a crack in tension by stresses σ in infinity (Fig.1) as $F(z) = 0.5\sigma(z^2-l^2)^{0.5}$, $F' = 0.5\sigma z(z^2-l^2)^{-0.5}$, $F'' = -0.5\sigma l^2(z^2-l^2)^{-3/2}$ where the answer is somewhat simplified without a loss of truth in asymptotic decomposition in which we compute (the condition τ_e=const is shown by dashed line in the left part of Fig.1)

$$\sigma_r = \sigma\sqrt{l/2r}(1+\sin^2\theta/2)\cos\theta/2,\ \sigma_\theta = \sigma\sqrt{l/2r}\cos^3\theta/2,\ \tau = 0.5\sigma\sqrt{l/2r}\sin\theta\cos\theta/2,\ \tau_e = 0.5\sigma\sqrt{l/2r}\sin\theta,$$

$$u_r = (\sigma/G)\sqrt{rl/2}((1-k)/2 - \sin^2\theta/2)\cos\theta/2,\ u_\theta = (\sigma/G)\sqrt{rl/2}((1+k)/2 - \cos^2\theta/2)\sin\theta/2. \quad (8)$$

In a similar way the case of the stamp pressure can be considered. Sadowski (1928) used the mentioned above analogy between the antiplane and plane deformations and gave the solution in a form $F(z) = Gu_l-(Pi/\pi)(z+(z^2-l^2)^{0.5})$, $F' = -(Pi/2\pi)(z^2-l^2)^{-0.5}$ where u_l is a constant, P-the resultant of the forces moving the stamp (Figure 2). According to (5) we receive for a plane deformation of incompressible body in the asymptotic approach

$$\sigma_r = (P/2\pi\sqrt{2rl})(2+\sin\theta)\sin\theta/2,\ \sigma_\theta = (P/2\pi\sqrt{2rl})(2-\sin\theta)\sin\theta/2,\ \tau = (P/2\pi\sqrt{2rl})(\cos^2\theta/2)\sin\theta/2,$$

$$\tau_e = (P/2\pi\sqrt{2rl})\sin\theta,\ u_r = u_2 - (P/\pi G)\sqrt{r/2l}(\cos^2\theta/2)\sin\theta/2,\ u_\theta = u_3 - (P/\pi G)\sqrt{r/2l}\cos^3\theta/2 \quad (9)$$

Here the case of the plane deformation is supposed; u_2, u_3 are constants, and the condition τ_e=const gives again the dashed line in the laft part of Figure 1.

PERFECT PLASTIC BODY

Rice (1968) gave the solution for the problem of the crack in antiplane deformation for the ideal elastic-plastic body. He supposed that the plastic zone is a circle with a centre in the end of the propagated crack (the dashed line in the right part of Figure 1). A radius r_0 of the boundary circumference must be found from the uninterruption condition on it of the stresses τ_i and the displacement *u*. According to the vector's transformation formulae (see Fig.1) $\tau_{r1} = \tau_r\cos\theta_1 - -\tau_\theta\sin\theta_1, \tau_{\theta 1} = \tau_r\sin\theta_1+\tau_\theta\cos\theta_1$ we find with the help of (6) $\tau_{r1} = 0$, $\tau_{\theta 1} = \tau\,(l/2r)^{0,5}$. Taking $\tau_{\theta 1} = \tau_{yi}$ (yielding point at shear) we find the radius r_0 as $r_0 = \tau^2 l/2(\tau_{yi})^2$. Now we consider the displacement and from the elastic part of the border we have according to (6) $u = (\tau/G)(2lr_0)^{0.5} \times\sin\theta$. Then starting from the plastic zone we derive from (1) on the circumference

$$u = (\tau_{yi}/G)2r_0\int_0^{\theta_1}\cos\xi d\xi = 2(\tau_{yi}/G)r_0\sin\theta_1. \quad (10)$$

Taking into account the equality $\theta_1 = \theta/2$ and comparing (10) to the u-value found before we derive r_0-value received earlier. Lastly we determine the displacement in the tip of the crack in the form $\delta = 2u(r=r_0,\theta=\pi) = 2\tau^2 l/G\tau_{yi}$.

In the same manner the task of the strip's longitudinal move can be considered. Using (7) and the vector's transformation formulae above we receive on the circumference $r = r_0$: $\tau_{\theta 1}=0$, $\tau_{r1} = (Q/\pi)(2rl)^{0,5}$ Supposing $\tau_{r1} = \tau_{yi}$ we find the radius r_0 of the plastic zone as $r_0 = Q^2/2\pi^2 l\tau_{yi}$. If we replace in the first formula (7) *r* by r_0 the expression for *u* in the elastic range is valid.So, from the plastic zone of the border we have according to (1) $u=u_0-\tau_{yi}r_1/G = u_0-2r_0(\tau_{yi}/G)\cos\theta/2$. Comparing this expression with the first formula (7) we derive the r_0-value as before. The position of the centroid will be corrected later.

An elastic-plastic solution of the considered problems in tension-compression is unknown. However the results for an ideal rigid plastic body is obtained. The first solution of this kind was got in 1920 by L.Prandtl for a more common type of soil.We give here a more real construction proposed later by R:Hill (see Figure 3 where the slip lines on which $\tau_e = \tau_{yi}$ are shown) that can be used for both problems considered.The ultimate stamp pressure (left part in Figure 3) is $p^*=\sigma_{yi}(1+\pi/2)$. The same p*-value is valid for the crack in tension (right part of Figure 3) where in square ODCD' $\sigma_x = \sigma_{yi}\pi/2$, $\tau_{xy} = 0$, $\sigma_y = \sigma_{yi}(1+\pi/2)$ and hence $\sigma_r = \sigma_{yi}(\pi/2+\sin^2\theta)$, $\sigma_\theta = \sigma_{yi}(\pi/2+\cos^2\theta)$, $\tau = \tau_{yi}\sin 2\theta$. In the same manner we find in triangle OAB $\sigma_y = \tau_{xy} = 0$, $\sigma_r = \sigma_{yi}\cos^2\theta$, $\sigma_\theta = \sigma_{yi}\sin^2\theta$, $\tau = \tau_{yi}\sin 2\theta$. In sector OBD $\tau = \tau_{yi}$ and $\sigma_r = \sigma_\theta$ change as linear function of θ. Diagrams of $\sigma_\theta/\sigma_{yi}$, σ_r/σ_{yi}, τ/σ_{yi} are given in Figure 4 by solid, dashed and interrupted by points lines respectively.

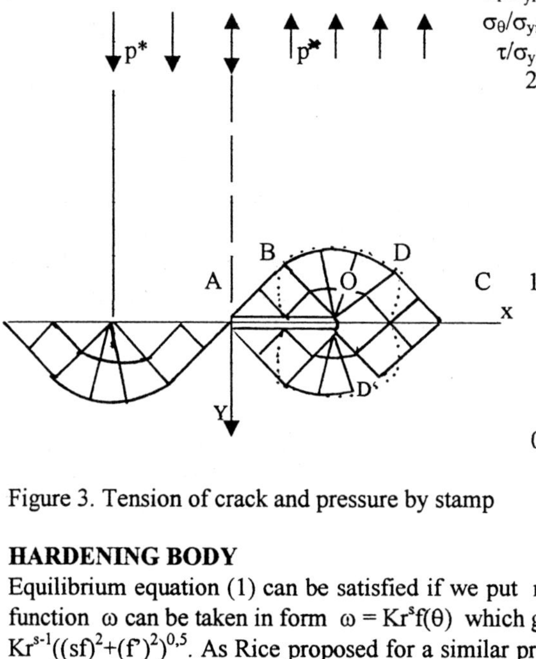

Figure 3. Tension of crack and pressure by stamp

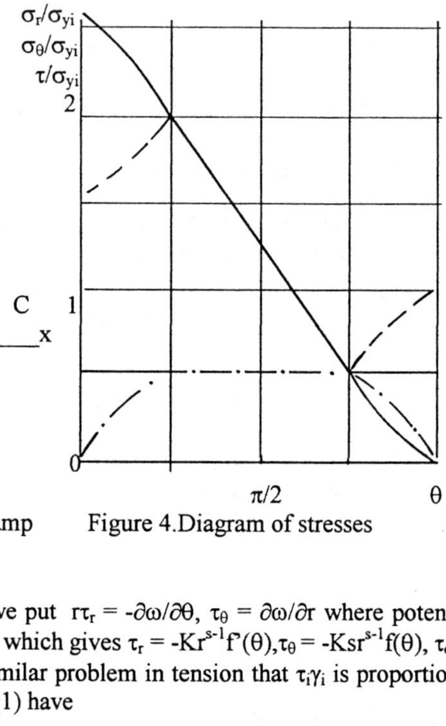

Figure 4.Diagram of stresses

HARDENING BODY

Equilibrium equation (1) can be satisfied if we put $r\tau_r = -\partial\omega/\partial\theta$, $\tau_\theta = \partial\omega/\partial r$ where potential function ω can be taken in form $\omega = Kr^s f(\theta)$ which gives $\tau_r = -Kr^{s-1}f'(\theta)$, $\tau_\theta = -Ksr^{s-1}f(\theta)$, $\tau_e = Kr^{s-1}((sf)^2+(f')^2)^{0,5}$. As Rice proposed for a similar problem in tension that $\tau_i\gamma_i$ is proportional to r^{-1} we find $s = m/(m+1)$ and according to (1) have

$$u = \Omega(t)(m+1)K^m r^{1/(1+m)}((sf)^2+(f')^2)^{(m-1)/2}f'(\theta),\ \tau_\theta = Kmr^{-1/(m+1)}f(\theta)/(m+1). \quad (11)$$

From the compatibility equation for strains $\partial\gamma_r/\partial\theta - \partial(r\gamma_\theta)/\partial r = 0$ and (1), (11) we derive the differential law for f(θ) as

$$f'' + (1+ m(m-1)(((m+1)f')^2+f^2)/(((1+m)f')^2+mf^2))f/(m+1)^2 = 0. \quad (12)$$

In order to get the rigorous solution we put in (12) $f = e^v$, $f' = e^v U = fU$, $f'' = f(U^2+U')$ that gives (C is a constant)

$$U=m(\sqrt{(1+m)^2+\tan^2(C-\theta)} - (1+m))/2(1+m)\tan(C-\theta). \qquad (13)$$

On the other hand, $d(\ln(f/D)=Ud\theta$ (D is a const), and with consideration of (13) we have

$$f=D((U^2+m^2/(m+1)^2)^m(U^2+1/(m+1)^2)^{-1/(m+1)} \qquad (14)$$

Pütting (13) in (14) we find

$$f=2^{(m+2)/2(m+1)}(1+m)m^{-m/2(m+1)}\tan(C-\theta)/(((1+m)^2+4\tan^2(C-\theta)\pm(m+1)\sqrt{(1+m)^2+4\tan^2(C-\theta)})^m\times$$

$$\times((m^2+1)((1+m)^2+4\tan(C-\theta))\pm2m^2(m+1)\sqrt{(1+m)^2+4\tan^2(C-\theta)})^{1/2(m+1)}. \qquad (15)$$

For the case of crack $f(0)=1$, $f(\pi)=0$ which give $C=0$ and $D = m^{m/(1+m)}/(m+1)$. From Fig.5 where according to expression (15) and those for stresses the diagrams $\tau_r/\tau_e=\phi_1(\theta)$, $\tau_\theta/\tau e=\phi_2(\theta)$ for m = 1, 3, 15 are constructed by solid, dashed and interrupted by points lines we can see that with the increase of *m* the role of one component of τ_e in a certain part of the plane is growing and in other part – is diminishing. In order to find the K-value we compute integral $J=d\Pi/dl$ in which $d\Pi = \int_0^{dl} \tau_\theta(,r,0)u(dl-r,\pi)dr$ is a free energy of crack propagation. Taking into account u, τ_θ -formulae (11) we have $J = \Omega(t)K^{m+1}m(f'(\pi))^m(\text{sign}f'(\pi))\int_0^1(\xi/(1-\xi))^{1/(1+m)}d\xi$.

According to Rice (1968) the J-value in a similar problem of tension does not depend on the properties of the material. Taking this assumption for our task too we can equal J to its meaning at m=1 that gives $K^{m+1}= -\tau^2\pi l\,\text{sign}f'(\pi)/2G\Omega(t)/f'/^m I(m)$ where $I(m)=(m+1)\Gamma((2+ +m)/(1+m))\Gamma((2m+1)/(m+1))$.Here $\Gamma()$ is the gamma function, and $f'(\pi)$ must be taken from the solution (15). Calculating τ_e we have the following equation for $r(\theta)$ at τ_e=const

$$rG\Omega(\tau e)^{m+1}/\tau^2 l = \pi(\sqrt{(sf)^2+(f')^2})^{m+1}/2I(m)/f'(\pi)/^m \qquad (16)$$

From Fig.6 where condition (16) is represented by solid, dashed and interrupted by poihts lines for m = 1, 3, 15 we can see that with the growth of *m* the plastic zone increases and it moves out of the crack, It confirms the solution of the previous chapter for the perfect elastic-plastic body.

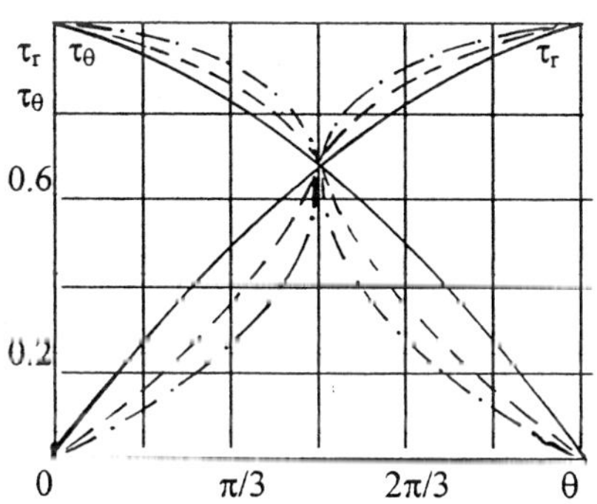

Figure 5. Diagrams $\tau_r=\phi_1(\theta)$, $\tau_\theta=\phi_2(\theta)$

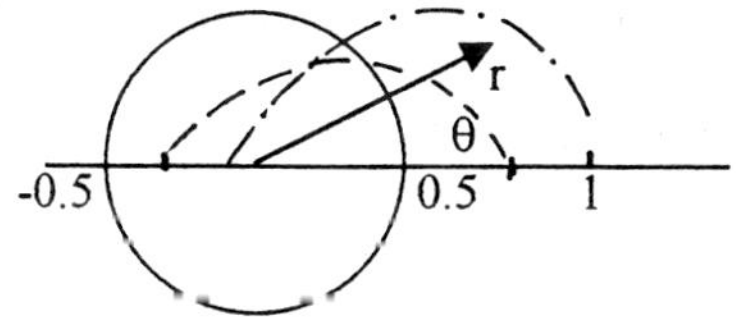

Figure 6. Diagrams $r(\theta)$ at τ_e=const

In the case of the stamp we have boundary conditions $f(0)=0$, $f(\pi)=1$ and from (15) we find $C=\pi$ and the same D-value as before. It means that in the previous sölution angle θ must be replaced by $\theta-\pi$ and diagrams in Figures 5, 6 must be reflected relatively to axis $\theta=\pi/2$. So, with the growth of m-value the plasic zone moves under the stamp and it corrects the position of the inelastic district for the same problem in the previous chapter.

In a similar manner the solution for the crack and the stamp pressure may be found. We take the potential function in the same form $F=Kr^s f(\theta)$ as in the previous problem and as $\sigma_{ij}\varepsilon_{ij}$ is proportional to r^{-1} we have $s = (2m+1)/(1+m)$, and according to the well-known formulae for stresses we find $\sigma_r = Kr^{-1/(m+1)}(sf+f'')$, $\sigma_\theta = Kr^{-1/(m+1)}s(s-1)f$, $\tau = Kr^{-1/(m+1)}(1-s)f'$. Now we have from (4) $\varepsilon_r = -\varepsilon_\theta$, γ and from the fourth law (2) - a very complex 4th order non-linear differential equation which must be solved together with boundary conditions which for the problems of the crack in tension and pressure of the stamp are: $f(0)=1$, $f'(0) = f'(\pi) = f(\pi) = 0$ and $f'(\pi)= -1$, $f'(\pi) = f'(0) = f(0) = 0$. From consequent elastic solutions ($f=\cos^3\theta/2$ and $f=\sin^3\theta/2$) we compute $f'''(0) = 0$. Deriving the equation for $f(\theta)$ at $f'=f'''=0$ we have expression $f^{V} + (3m+7)f''/(1+m)^2 + (2+m)(2m+1)f/(1+m)^4 = 0$ with an obvious solution $f = A_1\cos\beta_1\theta + A_2\cos\beta_1\theta + A_3\cos\beta_2\theta + A_4\cos\beta_2\theta$. Here $\beta_k=\beta_k(m)$, factors A_i should be detailed by the boundary conditions above. So, we know stresses and according to (4), (2) - the strains and the displacements respectively. Using the independence of $J = d\Pi/dl$, where

$$d\Pi = \int_0^{dl}\sigma_\theta(x, 0)u_\theta(dl-x, \pi)dx,$$

on the mechanical properties of the material, we find the K-value. Then we compute τ_e-value and similar to (16) – the expression for $r = r(\theta)$ at $\tau_e = \text{const}$ as

$$rG\Omega(2\tau_e)^{m+1}/\sigma^2(1+k)l = \pi(\sqrt{(f''+f(2m+1)/(m+1)^2)^2+4mf'^2/(m+1)^2})^{m+1}/4/f'(\pi)/^{m-1}f''(\pi)I(m).$$

Here $I(m)= (1+m)(2m+1)\Gamma((2m+1)/(m+1))\Gamma((2+m)/(m+1))$. We can also find K from the condition $\tau_e=\tau_{yi}$. Constructed in this way curves τ_e=const for m=15 are shown in Figures 1, 3 by pointed lines. Calculations show that the curves σ_r/σ_{yi}, $\sigma_\theta/\sigma_{yi}$, τ/σ_{yi} for m=15 are near to the corresponding ones in Figure 4. The same lines in these Figures for the stamp problem are symmetrical with the previous ones relatively to the axis $\theta = \pi/2$,

REFERENCES

Buryshkin, M, Elsoufiev S.(1994). Ultimate bearing capacity of structures at cycling loading in a complex stress state. In Proc.of 4th ICB/MF. St.-Germain en Laye (France), p.2, 263-268.

Elsoufiev, S.(1982). On simplified rheological laws for the description of deformation and fracture of materials.// Sov.Materials (USA).Transl from Phys.-Chem.Mech.of Materials, 1, 72-76.

Elsoufiev, S.(1997). On bearing capacity of densified and reinforced grounds. In Proc.of Conf.on Ground Improvement Geosystems. London, 329-334.

Elsoufiev, S, Nesterova L.(1997). Ultimate state of wedge in soil structures.In Proc.of the Ukranian National Conference on SMFE. Odessa.44-47.

Rice G. (1968) Mathematical Analysis in the Mechanics of Fracture. In “Fracture” V.2.Academic Press, N.-Y, L-n.

Sadowski, M. (1928).Zeitschrift für Angewandte Mathematik und Mechanik. B 8, 107-121.

Stabilization of Landslides using Piles

M. U. ERGUN, Middle East Technical University, Civil Engineering Department, Ankara, Turkey

INTRODUCTION

Although the subject of stabilization of landslides by piles or similar inclusions is very popular a well-established calculation and design procedure is not yet available. There are several procedures proposed for this purpose. Fukuoka (1977), Viggiani (1981), Winter, Schwarz and Gudehus (1983), Ito and Matsui (1975,1977), Reese et.al (1992) may be mentioned among others. There are successful applications reported by De Beer and Wallays (1970), Ito and Matsui (1975), Sommer (1977), Snedker (1985).

The piles used in slope stabilization works are subjected to lateral forces by horizontal movements of the surrounding soil and usually referred as passive piles. Passive piles are drilled through a sliding mass and penetrate into a stable soil, and stabilizing effect of the pile above the slip surface is provided by the passive resistance of the pile below the slip surface. This resistance is introduced into the conventional slope stability calculation of the slope.

FACTORS TO BE CONSIDERED IN STABILIZATION WORKS WITH PILES

Design procedure of slope stabilization with piles is a complex problem and involves many factors. The factors should be considered in stabilization works are:

a. Development of lateral pressures on the pile above and below the slip surface
b. Contribution of piles to slope stability
c. Modes of failure of the soil-pile system and slope
d. Determination of embedded depth of pile below the slip surface
e. Determination of pile diameter and structural capacity
f. Pile head fixity conditions
g. Most suitable location of piles within the slope
h. Pile spacing and group action reduction

The development of lateral stress on a stabilizing pile is trapezoidal-like and the total area under such a distribution may be converted to a rectangular one to simplify the calculations and the total shear force at the slip level will be almost the same. The ultimate pressure developed on a pile due to a moving cohesive soil can be predicted by Eq. 1 which is recommended by several researches (Brinch Hansen, 1961; Broms, 1964; De Beer, 1977; Viggiani, 1981).

$$p = k.c_u.d \qquad (1)$$

Landslides in research, theory and practice, Thomas Telford, London, 2000.

where p is the lateral pressure along pile , c_u is undrained strength, k is a bearing capacity factor and d is pile diameter. Different k values are given by various authors for active and passive piles. The values are also different above (k_1) and below (k_2) the slip surface. k values should be used with caution because they are limiting values and large soil-pile relative displacements are needed for full mobilisation. In many actual cases piles stop land movements at relatively smaller displacements and mobilised k values are smaller than the reported values of 2.0 – 4.0 (k_1) and 5.6 – 8 (k_2). Especially in stiff clays much lower values may be possible before moment capacities of piles are reached and even they are reached limiting k values still may not be mobilised. k is also dependent on penetration depth of pile in the lower soil and ratio of the undrained strengths of the upper and lower soils.

The total lateral force needed to increase the factor of safety of a slope by a desired amount may be evaluated by means of a back analysis technique. For failed slopes factor of safety of unity should be increased to a level to arrest the movements. This level is claimed to be around 1.05 – 1.10. Contribution of resisting forces of piles is added to the resistance of the soil along the slip surface, which is already specified in the analysis. Moment and shear loading of the piles should be below or equal to the allowable values. Load distribution on the pile above sliding surface can be estimated using interslice force (E) at the location of the pile.

Piles should be penetrated into a stable soil by such an amount that reaction needed to stop the slide could be developed. Depth of penetration required can be estimated from the structural solution of pile such that the point of rotation should be within the depth of embedment and passive pressure developed on the pile in the penetrated depth should be within allowable limit (k_2). One way to find the appropriate depth is to the analyze a pile of infinite length and to locate the depth where bending moment and shear are near zero. Embedding the pile deeper than this point will not increase its stability. Depth of penetration in stiffer lower soil is usually equal to or less than half of the sliding upper depth.

When piles are near to each other "arching" phenomenon occurs. Development of arching and total load on piles depend on the geometry of the system i.e. spacing to diameter ratio of piles. Deformation mode of piles, shearing resistance of soil, support of pile head and toe, stiffness of piles and soils also affect the group behaviour. A pile in a closely spaced row takes less load compared to a single pile of the same diameter. Group reduction for passive piles in cohesive soils is studied by Nalçakan (1999). Reduction coefficients down to 0.5 and less are reported. The extent of soil-pile relative displacements affects the results.

After diameter, rigidity and length of the stabilizing pile are specified, load at the displacing upper part is applied to the pile and the conventional laterally loaded pile (p-y) solution is obtained. All calculations are done per meter. The lower (stable) soil reaction is represented by a horizontal modulus of subgrade reaction.

The selected spring coefficient must be in accord with the stress-strain behaviour of the lower soil and the mobilised k_2 value. There is also a need for the use of soft springs on back of the pile in the upper soil due to continuity of the mass. The solutions are not very sensitive to these spring coefficients but they improve the results and they are realistic because the contact behind the upper part of the piles is not fully lost (Sommer, 1977). If the upper and lower soils are similar soft spring values can be taken 1/20 to 1/40 of those of lower soil.

It is clearly known in structural mechanics that end conditions of structural members affect the solution. Use of reinforced concrete beams connecting the pile tops or use of top anchors will affect the top displacements and this is in turn will improve the capacity of the piles (Ito and Matsui, 1977; Popescu, 1991).

Piles are recommended to be placed at locations where interslice forces are relatively greater i.e. far from crest and toe of slips (Yamagami et. al, 1992).

Various modes of failure of slopes and pile-soil systems are possible. This is the factor responsible for the lack of a well-accepted design procedure of piles to stabilize landslides. The relative properties of upper and lower soils, possibilities of passive failure of lower soil or structural failure of pile make this subject a complex one. Viggiani (1981) tried to categorize these possibilities.

PROPOSED DESIGN PROCEDURE

Design process is iterative and a trial and error routine. The steps are summarized below:

1. Make sure that topography of slip and depth of actual or expected slip plane, water conditions are known. Perform conventional slope stability calculations without piles.
2. Select a trial diameter and spacing and a trial depth of penetration below the slip plane. Locate the row(s) at (a) suitable location(s) preferably near the central part of the body.
3. Decide on a simple (uniform) upper rectangular soil reaction on piles. Consider both limiting values (k_1) and interslice forces. Except in case of rock-socketed piles and large penetration depths in stiff lower soils k_1 values will be around 2.0 – 4.0 (4.0 for stiffer soils).
4. Select a (group) reduction coefficient for the row of piles and decrease the distribution concluded in (3).
5. Select spring coefficients for the lower stable soil and the upper soil (soft springs). In most cases a beam connects the pile heads. This may be considered as an end restraint that allows lateral displacements without rotation. Solution of the pile is made by p-y analysis providing rigidity (EI) of the pile.
6. Now moments, shears and displacements are known along the pile. Compare them with the allowable values. Check upper and lower lateral pressure distributions. Compare the soil spring reactions from initial elastic analysis with the ultimate passive pressure. Replace overstressed springs (if any) by ultimate values. Check that k_2 values are less than the reported maximum values in the literature (5.6 – 8.0).
7. If the pile loads are excessive change pile diameter and/or spacing.
8. Make new trials after changes in (6) and (7).
9. Introduce the calculated shear force at the sliding surface or maximum allowable shear capacity of the pile section (they must be close for economical applications) into the slope stability calculation and obtain an increased factor of safety.
10. Make another trial if the calculations about the pile and/or the slope stability are not satisfactory.

CASE STUDY

Stabilizing piles were used to protect the main natural gas line from slope movements due to a large excavation at the lower elevations near Istanbul. It was a flow type of slide consisting of a soft saturated cohesive debris. Brown colored very plastic (PI = 50-63%) clays were encountered down to 6-8m depth from the ground surface. Undrained shear strength values were measured between 23 – 40 kPa. Gray to green plastic stiff clays underlie this soft formation (c_u = 125 kPa). More than ten boreholes were opened (Fig. 1). Three inclinometer pipes were installed. The slip plane was found at 6m depth at the flow area and the displacements were excessive. 3m high selective granular material was placed in front of the pipeline as an emergency measure.

Outside the flow slide area there was also a potential risk because the excavation was very large. The inclinometer pipes at this side showed smaller displacements up to 8m depth.

Reinforced concrete piles with a diameter of 1.2m and a length of 16m including the one-meter thick capping beam were selected. They were tangent to each other at the flow area and were located at a centre to centre distance of 2.4m (s/d=2) (s spacing, d diameter) at the potential slide area. Piles were penetrated 7.0 to 8.0m in the stable soil. Construction of the piles were completed in June 1997. Three new inclinometer pipes were installed in the piles. Readings have been taken for the last 24 months.

Slope stability analyses were performed at the both sides. Loads on piles at the slide (flow) area and the potential slide area were estimated from the interslice equilibrium as 54 kPa/m and 60.6 kPa/m respectively. These are simple rectangular distributions which correspond to k_1 values of 1.80 and 4.04 respectively. Applying group reduction factors of 0.45 and 0.57 for s/d=1 (slide area) and for s/d=2 (potential slide area) reduced loading on the piles are found 24.3 kPa and 34.5 kPa and these values correspond to k_1 values of 0.81 and 2.30 respectively. These are less than the maximum values. The potential loss of material in the flow zone in front of the piles was considered, and moments and shear capacity of the piles were designed against full loss of material above the slide plane. Spring coefficients in the lower soil changed between 50 000 – 80 000 kN/m^3. Structural solutions were obtained and shear and bending moment values were found less than the allowable capacities. The maximum soil spring reaction is calculated in the lower soils as 72.6 kN/m^2, which corresponds to k_2 value of 0.97 which is much less than the recommended values. Calculated shear forces at the slip plane level increased the factor of safety of the slope up to 1.09. The allowable shear capacity of the piles can increase the factor of safety up to 1.28.

The inclinometer profiles show maximum deformation values of 11mm and 6mm at the surface at the slide area (HAD-6) and potential slide area (HAD-4) respectively (Fig. 2). The inclinometer profiles were back-evaluated for 24 months period and the load on the piles (upper) were found to be 18 kPa and 30.8 kPa at the slide and potential slide sections. These values are less than the design values (k_1 values 0.60 and 2.05) and tend to approach them.

CONCLUSIONS

A design routine has been proposed after a review of the subject of stabilization of landslides by piles. The subject is complex and involves several factors. It is believed that the routine recommended considers all crucial aspects of this frequently encountered geotechnical problem. A case treated with the proposed routine is described, and the field data are evaluated and compared with the calculations and predictions.

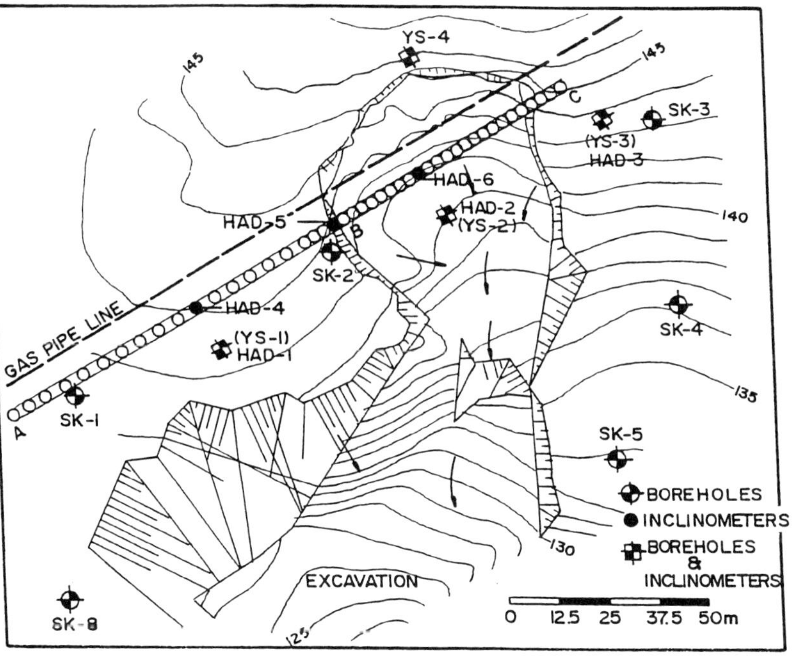

Fig.1 Plan View of the Landslide in İstanbul

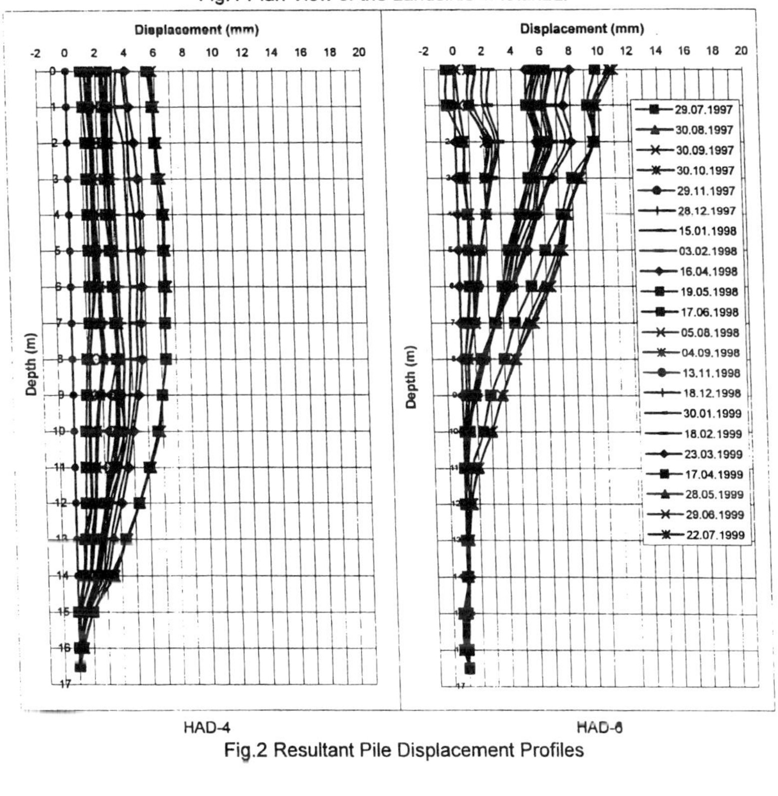

Fig.2 Resultant Pile Displacement Profiles

REFERENCES

Brinch Hansen, J., "The Ultimate Resistance of Rigid Piles against Transversal Forces", Bulletin of the Danish Geotechnical Institute, No: 12, 1961

Broms, B.B., "Lateral Resistance of Piles in Cohesive Soils", ASCE Journal of the SMFE, Vol.90, No: SM2 pp: 27-63, March, 1964

De Beer, E., "Piles Subjected to Static Lateral Loads", Proceedings, 9th I.C.S.M.F.E., Specialty Session 10, Tokyo, 1977

De Beer, E. and Wallays, M., "Stabilization of Slope in Schist by Means of Bored Piles Reinforced with Steel Beams", Proceedings, 2nd International Congress Rock Mechanics, Vol. 3, pp: 361-369, Belgrade, 1970

Fukuoka, M., "The Effects of Horizontal Loads on Piles due to Landslides", Proceedings, 9th I.C.S.M.F.E., Specialty Session 10, Tokyo, 1977

Ito, T. and Matsui, T., "Methods to Estimate Lateral Force Acting on Stabilizing Piles", Soils and Foundation, Vol.15, No.4, pp: 43-59, December 1975

Ito, T. and Matsui, T., "The Effects of Piles in a Row on the Slope Stability", Proceedings, 9th I.C.S.M.F.E., Specialty Session 10, Tokyo, 1977

Ito, T. and Matsui, T., "Discussion: Methods to Estimate Lateral Force Acting on Stabilizing Piles ", Soils and Foundation, Vol.18, No.2, pp: 41-44, June 1978

Nalçakan, M.S., "Stabilization of Landslides by Piles in Cohesive Soils With Special Reference to Group Action Reduction", Middle East Technical University, Ph. D. Thesis, 1999.

Popescu, M.E., "Landslide Control by Means of a Row of Piles", Slope Stability Engineering, Thomas Telford, pp: 389-394, London, 1991

Reese, L. et.al, "Use of Shafts in Stabilizing a Slope", Slopes and Embankments-II A.S.C.E., Speciality Session, 1992

Snedker, E.A., "The Stabilization of a Landslipped Area to Incorporate a Highway by Use of a System of Bored Piles", Failures in Earthworks, Thomas Telford, pp: 436-438, London, 1985

Sommer, H., "Creeping Slope in a Stiff Clay", Proceedings, 9th I.C.S.M.F.E., Specialty Session 10, Tokyo, 1977

Viggiani, C., "Ultimate Lateral Load on Piles Used to Stabilize Landslides", ", Proceedings, 10th I.C.S.M.F.E., Vol.3, pp: 555-560, Stockholm, 1981

Winter, H., Schwarz, W. and Gudehus, G., "Stabilization of Clay Slopes by Piles", Proceedings, 8th European Conference on S.M.F.E., pp: 545-550, Helsinki, 1983

Yamagami, T et.al., "A Simplified Estimation of the Stabilizing Effect of Piles in a Landslide Slope Applying the Janbu Method", Proceedings, 6th International Symposium on Landslides, pp: 613-618, Christchurch, New Zealand, 1992

Slide Activity along the River Göta älv, SW Sweden - An Evaluation of the Effects of Bank Revetment

A. ERIKSSON, Mid-Sweden University, Härnösand, Sweden

INTRODUCTION

The river Göta älv is one of the largest rivers in Sweden with a mean annual discharge of 550 m^3/s. It runs southwards from Lake Vänern (44 m.a.s.l) down through the city of Gothenburg, a distance of some 90 km (Fig. 1). The gradient is very gentle except at rock thresholds such as Vargön (5-m waterfall), Trollhättan (33 m), and Lilla Edet (4 m).

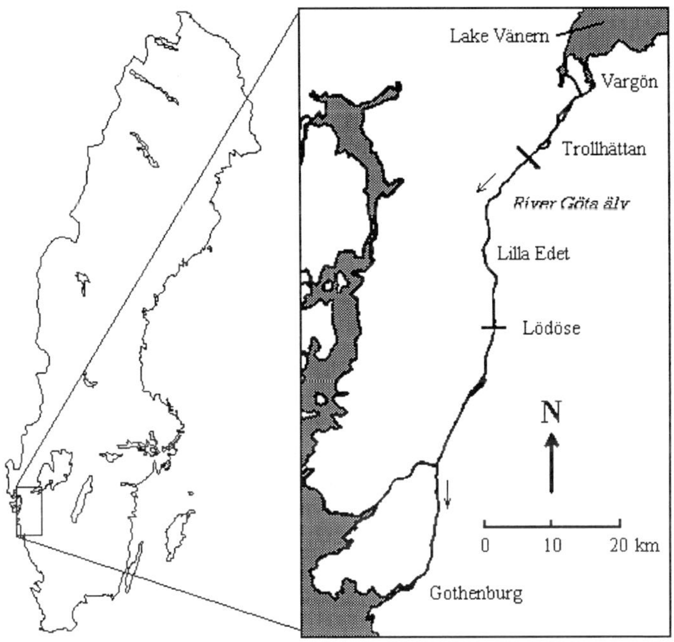

Figure 1. Map showing the area of concern, a 30-km distance between Trollhättan and Lödöse, south-western Sweden.

The river valley is the area in Sweden most sensitive to clay slides. Since the 17th century three slides per century larger than 3 ha have been documented (Sundborg & Norrman, 1963). The stability problem along River Göta älv was given much attention in the 1950s as two slides greater than 25 ha struck a village and a factory site respectively. The Swedish

Landslides in research, theory and practice, Thomas Telford, London, 2000.

government decided to apply some countermeasures in the form of bank revetment. This study evaluates the effects this action has had on slide frequency.

METHODS

The study is based on an old field survey (Järnefors, 1957) along with aerial photographs from 10 occasions in the period 1960-95 and ranges over the most sensitive stretch of the river between Trollhättan and Lödöse (Fig. 1). The aerial photographs have an approximate scale of 1:30,000 which suggests that slides of 100 m^2 or greater can be properly detected. Slide scars were plotted on transparent film and the area was measured with a ruler. As this study does not include any field survey the depths of the slide scars have been set equal to the crest of the slope along a simplified height profile. The 30-km profile has been divided into 8 sections for which the mean depth has been estimated (table 1). Thus the tools for determining the sediment yield are rather blunt, but may serve for a relative assessment.

Table 1. Approximate height above the river surface of the slope crest at various sections (Rasriskerna i Götaälvdalen, 1962). The distance is measured from the Lake Vänern outlet.

Section (km)	20-22	22-27	27-29	29-32	32-36	36-39	39-44	44-50
Height (m)	20	23	20	15	12	8	5	2

All slide scars that were found in the aerial photograph from 1960 but not in the Järnefors map were assumed to have occurred during the period 1957-60. At each following observation year, the new scar map was compared to the preceding one. Two types of slides were distinguished, slumps and "slope falls". Slope fall is here defined as slides having a lesser extension in width (from the shore) than depth, *i.e.* having a slope angle greater than 45°. All other scars were defined as slumps. "Slide" is hereafter used as a general term for both slumps and slope falls. The slides were identified by the bright white colour from the exposed soil since most of the scars were too small to be recognised by morphology. Scars that were revegetated could therefore "disappear" before any air photographs were taken. This problem is most pronounced in later years since the period between flights has increased (Fig. 2).

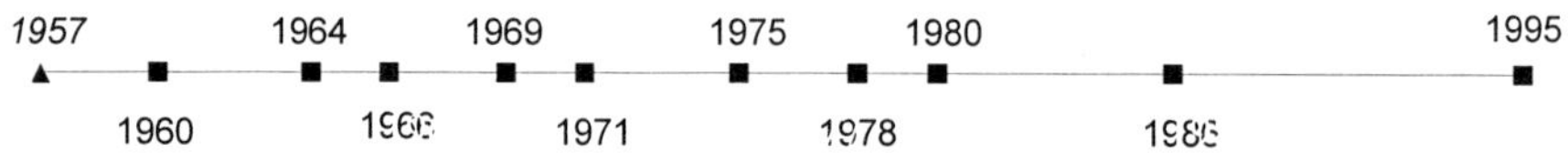

Figure 2. Time line indicating the years when aerial photographs were taken. Only three photo flights have been performed since 1980. The long time between flights may enable some slide scars to be revegetated and thus invisible in the photograph. The Järnefors field survey map is from 1957.

AREA DESCRIPTION

Geology and river morphology

The river Göta älv is incised into thick (up to 100 meters) glaciomarine clays (Rasriskerna i Götaälvdalen, 1962). At some localities, especially in the northern parts of the valley, these clays have been leached to the extent that they are defined as quick clays. It is also in the northern part that the slopes are highest (table 1). The river exhibits a rather straight course and there are no signs of major shifts in the channel's position during the Holocene. Because of the lack of meander bends, natural stream erosion on banks is low (Sundborg & Norrman, 1963).

The channel morphology consists of three easily distinguished entities (Fig. 3). First there is the *deep channel* with depths of 5-25 meters. Moving towards the shore there is a *subaquatic slope*. The angle of these slopes is often 10-20°, at some localities even steeper. The upper part of this slope shifts into the shallow (~1m) *platform*. The *shore notch* marks the high-water limit.

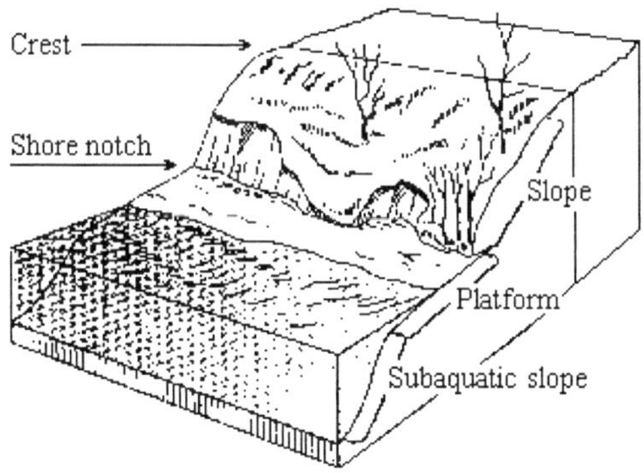

Figure 3. Schematic drawing of the channel and bank morphology along River Göta älv (Sundborg & Norrman, 1963).

Anthropogenic influence on hydrology and erosion

During the 20th century there have been three major events of anthropogenic influence on the river.

1) In 1916 a new lock was built at Lilla Edet which raised the upstream water level by approximately 2 meters. The increase in water depth decreased the flow velocity, which gave less erosion at depth. Bank erosion was however enhanced as a new platform began to develop. The banks were quickly undermined and slides became everyday events (Sundborg & Norrman, 1963).

2) The regulation of Lake Vänern in 1937 changed the hydrological regime of the river. The duration of high discharges was increased, which enhanced the depth erosion on the southern part (Sundborg & Norrman, 1963). The swift changes in water level leaves the platforms to dry and crack up at lowstand. The cracks and flakes makes the platform more sensitive to erosion as it is flooded again.

3) The ships frequenting the river have increased their tonnage rapidly since the 1960s. The larger vessels necessitated dredging on several stretches. More passages of large ships increased the duration of great waves, which led to pronounced bank and platform erosion, especially along the northern part.

The river has thus been divided into two stretches with different characteristics: the northern part with bank erosion as the primary problem, and the southern where depth erosion is prevailing. The platforms that used to be areas of deposition are now eroded and there are extremely few spots where sediment can accumulate before it reaches the Gothenburg harbour (Sundborg & Norrman, 1963).

Bank revetment construction
The bank revetment consists of boulders, 30-100 cm in size, dumped at the shore by truck. The general configuration is shown in Figure 4.

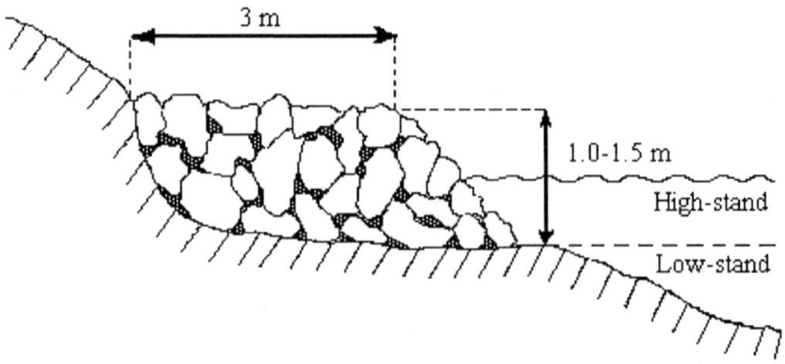

Figure 4. A principal sketch of the bank revetment as it is featured along river Göta älv.

The revetment was constructed during 1962-68 (4 km) and 1972-82 (79 km), respectively. In 1982 even stretches that were not considered as particularly sensitive to sliding were reveted. There were also some other measures taken such as lowering of the water table or ground level, but this was rare.

RESULTS
A total of 261 slides were documented for the period 1957-95. There were no extraordinarily large slides, the greatest being c. 0.9 ha. A summary of the results is given in table 2. It can be seen that the slide activity was high during the 1960s with a mean of 13 slides per year. During the 70s the number of slides were fewer with an annual mean of 8. In this period the reduction of slumps was more pronounced than the effect on slope falls. A possible explanation is that development of slumps is dependent on erosion of base support while slope falls occur where the slope is steep. A slope fall can therefore develop through soil creep and not necessarily by undermining. Thus there can be a lag time before the revetment has any effect on slope falls. As a slope fall has occurred the slope angle is reduced and further falls are less probable since the revetment protects the slide debris from being eroded. The prevention of slumps is however more immediate since they are more dependent on bank erosion. From 1980 to 1995 the number of slope falls was significantly reduced as well and there was only 1 slide per year on average.

Table 2. The number of slides during each period, specified as slope falls and slumps.

	1957-60	1960-64	1964-66	1966-69	1969-71	1971-75	1975-78	1978-80	1980-86	1986-95
Number of slides	34	53	38	25	40	27	19	11	9	5
Slope falls	25	33	31	15	38	24	14	10	4	4
Slumps	9	20	7	10	2	3	5	1	5	1
Slides/year	**11,3**	**13,3**	**19,0**	**8,3**	**20,0**	**6,8**	**6,3**	**5,5**	**1,5**	**0,6**

Figure 5 displays the steady reduction in slide activity during the 1970s. This of course had a great effect on soil loss. It is however not only the *number* of slides that have been reduced but also the *size*. Figure 6 indicates that the average length of slope falls was declining steadily before any reduction in numbers could be seen (*cf.* table 2). This is probably a positive feedback of reduced slide activity. Fewer slides enable vegetation to colonise the slopes, thereby binding the soil. There seems to be a similar trend for the area of slump scars except for the increase in 1995. One should however keep in mind that as the number of data is reduced with time in this investigation the significance of each singular event increases. The 5 slides in the period 1986-95 are not a sound base for any statistical analysis.

There is no documentation of when exactly the bank revetment was constructed. It is therefore not possible to perform any statistical correlation between revetment length and slide activity.

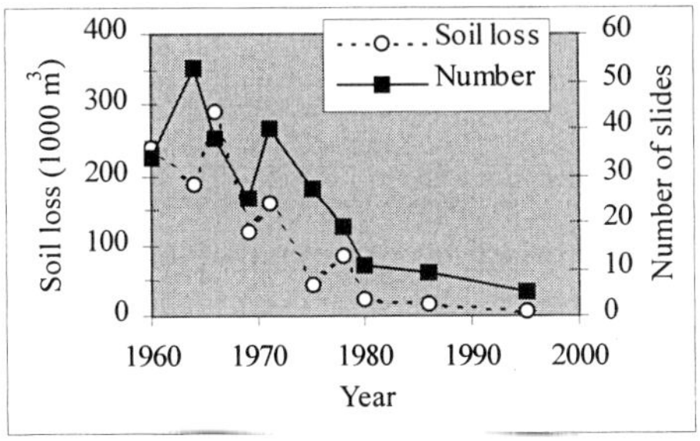

Figure 5. Slide activity presented as "number of slides" and "soil loss" from 1960 to 1995

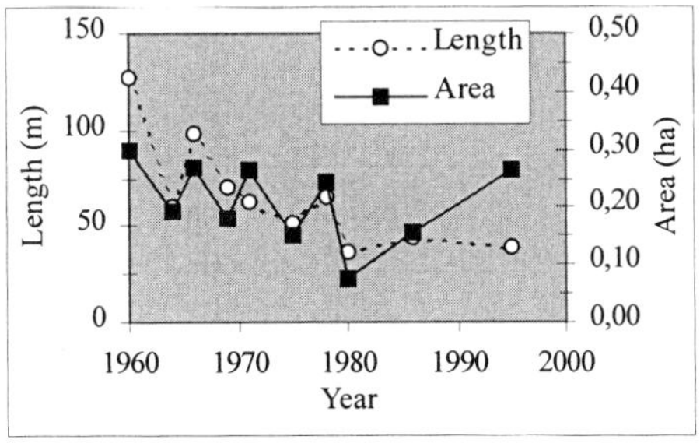

Figure 6. The size of the slides presented as length of slope falls along the river and slump area

DISCUSSION

It is obvious that the number of slides has declined, but how can one be certain that this is an effect of the bank revetment and not a natural fluctuation? A period of intense erosion was initialised after the new lock at Lilla Edet was built in 1916. It is possible that after 80 years of severe erosion the slopes are now finding a new equilibrium. There are however signs indicating something else. The high-water mark is usually seen as a distinct "shore notch" (Fig. 3). As a consequence of the raised water level following the lock construction, erosion began to work out a new shore notch. Such a notch could be seen in the late 50s, but it was situated 30-50 cm *below* the high-water mark (Sundborg & Norrman, 1963). It was therefore argued that the erosion would continue for a long time ahead and probably get worse as the slopes were getting steeper. This has not been the case and the most significant reduction in slide activity was coincident with the most intense construction of revetment.

These results point in the direction of success for preventing slides with the aid of bank revetment. As displayed in Figure 5, the sediment yield from the northern stretch is now only a fraction of the yield 30 years ago. This sediment deficit is probably compensated for downstream. As mentioned above the southern part of the river experiences erosion on depth mainly. If this is enhanced, slides in the subaquatic slope might become a problem, and these are more difficult to handle. Such a slide occurred north of Gothenburg in 1993 (Fig. 7).

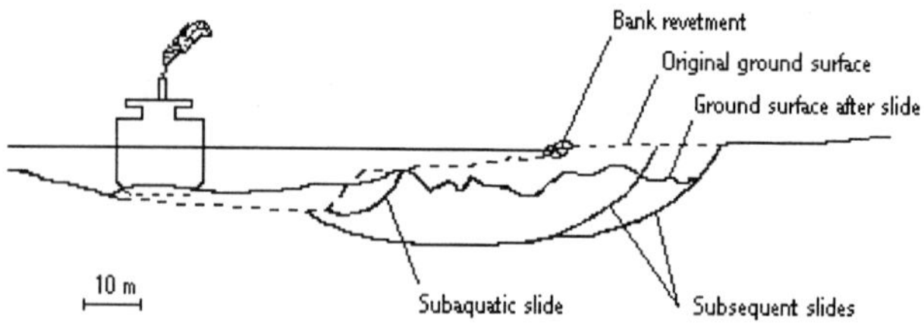

Figure 7. Assumed line of events during the Agnesberg slide in 1993. The sliding began with a slump in the subaquatic slope which was followed by two larger slides that extended into the shore (Sandebring & Ottosson, 1994).

REFERENCES

Järnefors B., 1957: Jordartskarta över Götaälvdalen, SGU ser. Ba nr 20.

Rasriskerna i Götaälvdalen, 1962. SOU 1962:48.

Sandebring H. & Ottosson E., 1994: Agnesbergsskredet - skredförebyggande åtgärder, SGI rapport no 45, Linköping.

Sundborg Å. & Norrman J., 1963: Göta älv – hydrologi och morfologi med särskild hänsyn till erosionsprocesserna (with an English summary), SGU ser. Ca nr 43, Stockholm.

An Anthropogenic Triggering of a Landslide: Ariklar Landslide (NW Turkey)

T. A. ERTEK, Istanbul University, Literature Faculty, Geography section, Physical Geography Department, Istanbul, Turkey

LOCATION

Arıklar Landslide is a rotational landslide that occurred at the east of the Arıklar village. This village bounds to Kandıra district of Kocaeli province in the northwestern part of Turkey (Figs. 1 and 2).

DAMAGES

The landslide has occurred at 20th of October 1997 on the east slope of the Karaağaç Stream, due to the heavy seasonal rainfall. Three animal barns, cultivation areas, groves (small wood), spring waterline and graveyards have been seriously damaged. The small dam body was partly and overflow sluice system was completely destroyed (Figs. 3 and 4). The dam was built at the bottom of the landslide slope in 1992. Karaağaç stream valley has not been covered by landslide mass, but the movement has not stopped completely. The death rate was none, however the estimated damaged cost was approximately 2.2 million $.

FACTORS OF LANDSLIDE

The main reason for the occurrence of Arıklar Landslide is the dam site which placed on a landslide susceptible area. During the dam construction, the slope stability has been disturbed (EROL 1985, ERİNÇ 1996). In this site, the triggering factors are (i) steep slope of the valley; (ii) heavy seasonal rainfall (Figs. 5, 6 and 7); (iii) Eosen flysch formation; (iv) W-NW trending rock slopes.

Subsequent activity of the landslide was caused by (i) tixotrophy of the clays, and (ii) anthropogenic activities during construction works of dam, disturbing the slope stability. These two factors appear to be governing factors on the landslide movement.

Destruction of vegetation (Quercus forest) (DÖNMEZ 1979) and recent agricultural activities are also supporting factors. The fast mass movement was developed in the spoon shaped meander scar. The slide trend follows the NW dominant wind direction. While recently agricultural activity cluster root was planted, instead of taproot. These are the seconder reasons of the Arıklar Landslide.

MECHANISM OF FAILURE AND TYPE OF MOVEMENT

Arıklar Landslide was developed as a rotational landslide type. The small cracks were appeared firstly, landslide was occurred along the rotational or curval surface. The resulting mass has continued to rotational move towards the dam. Slide surface has been bordered by a continuous main crack. This main crack, that forming landslide scarp is placed at 150 m level. Slide surface has developed on a smooth slope in the form of parabolic curve with a length of 900 m. The material of the main landslide toe ends at the 90 m in the stream valley floor. The sliding with a 300 m length, effects 171,250 m^2 of the

Landslides in research, theory and practice, Thomas Telford, London, 2000.

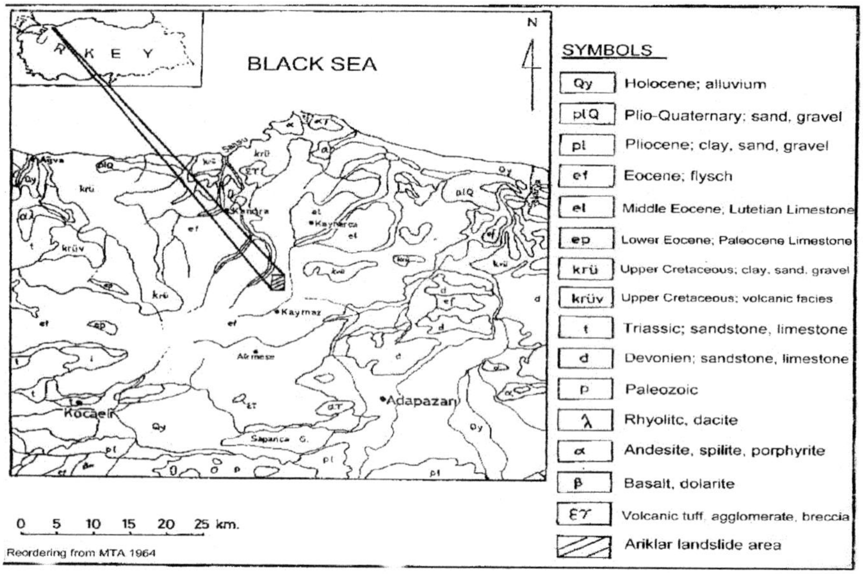

Fig. 1. The geological map of Kandıra and surroundings.

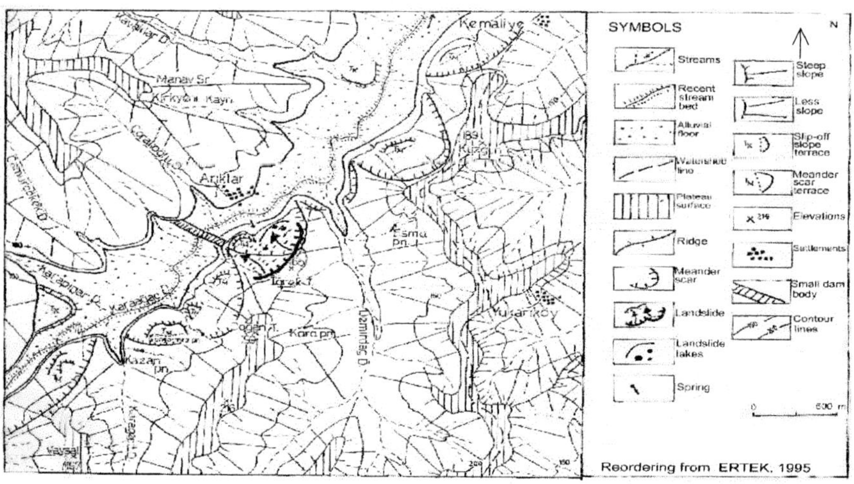

Fig. 2. The geomorphological map of the areas surrounding the Arıklar Landslide

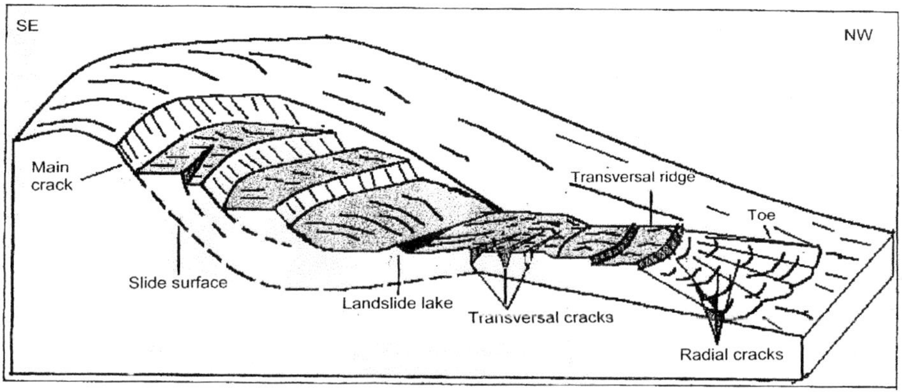

Fig. 3. The blockdiagram of Ariklar landslide.

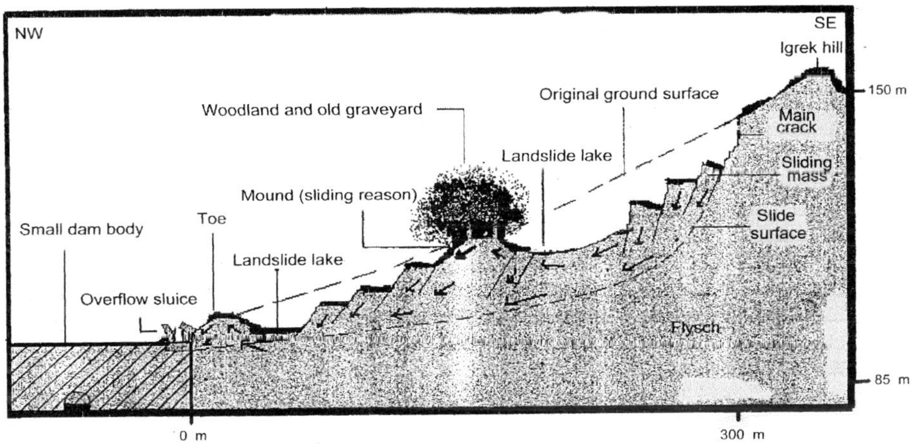

Fig. 4. The cross section of Arıklar Landslide. (20.10.1997)

Fig. 5. The monthly average precipitation (1965-1990,Kandira)

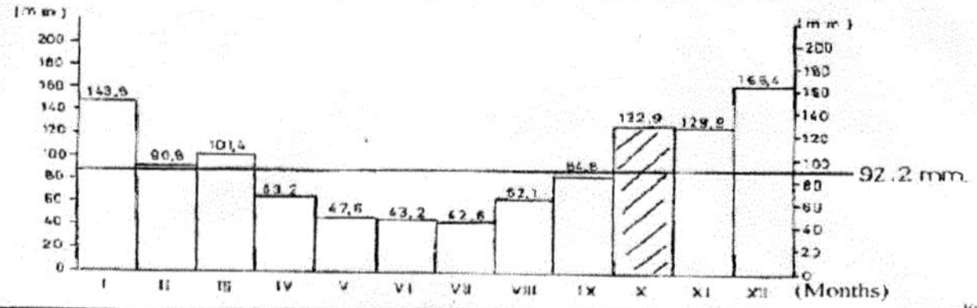

From D.M.I 1997

Fig. 6. The dayly average precipitation (Kocaeli / October 1997)

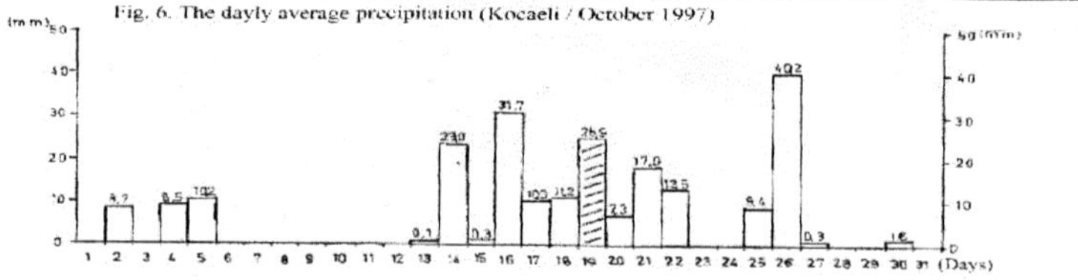

From D.M.I 1997

Fig. 7. The reducing dayly average precipitation(Kandira / October 1997) *

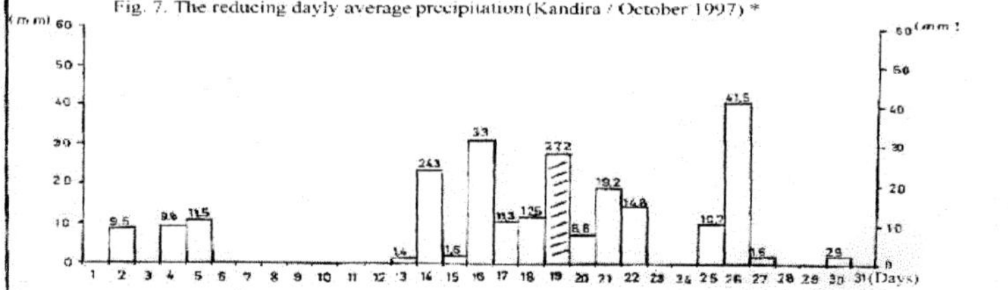

* The precipitation wasn't registered in Kandira by D.M.I. since 1990. In this reason, Kandira's the dayly average precipitation had been calculated by rh=ro + (5,4+h) formula (EROL 1988).

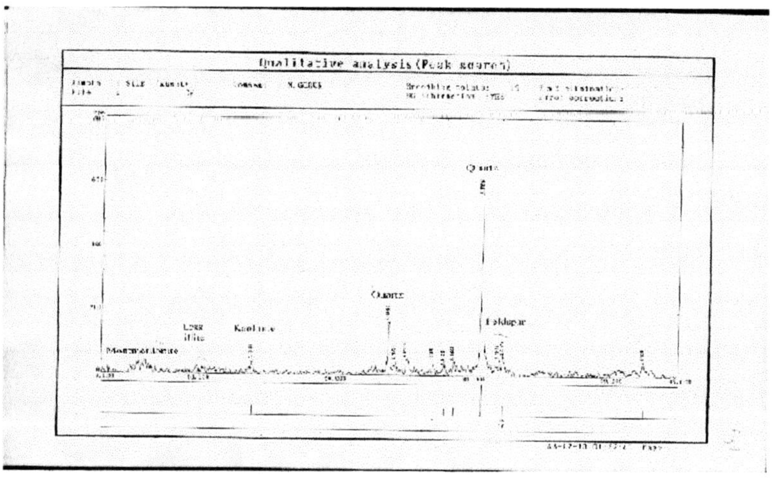

Fig. 8. The XRD results of the Arıklar landslide mass.

area. The "in sight" reason of the slide is heavy rain in Arıklar. However, the main "real" reason seems to be the movement of the clays (kaolinite group) (Fig.8) in the flysch formation (ABDÜSSELAMOĞLU 1963) due to rain, and building of a dam body and its sluice in the meander scar. Consequently, removing of the lateral prop can be the reason for this movement.

MITIGATION MEASURES

There are two main ways for the drainage organisation at landslide area. One of them is to take the ground water which comes in a closed drainage system, to store in a well or draining area and to decrease the ground water table by carrying it away. The other one is to decrease the water flow by laying pipe in a draining area. Such a drainage which will hold the water back, is done by preparing the deep gallery or trenches which are perpendicular digged to the flow direction and their insides are filled with permeable material. In the latter, it is carried towards a distant point, although the surface and ground water collects (ZARUBA and MENCL 1969, COOKE and DOORNKAMP 1974, ERGUVANLI 1982, KURTER 1985). Particularly, the second suggestion is more effective in saving works which will be able to applied.

CONCLUSION

In conclusion, building the dammed lake in Arıklar can not be considered to move to another place. However, the small dam body can be saved from the edge of the meander scar and also founding of the overflow sluice to the opposite side of the landslide is suggested (www.tuat.ac.jp/).

ACKNOWLEDGEMENTS

I thank to Prof.Dr. N. ÇAĞATAY (İ.T.Ü., İst. Tech. Univ.) for reviewing this article and providing XRD analysed in İ.T.Ü. metallugry laboratory for determination of clay minerals. Dr. O. ALGAN (İst. Univ. Inst. Mar. Sci. And Manag.), geol. eng. E. SARI and İ.GÖKÇE and also workers of D.M.İ. (State Meteorological Service) in Kocaeli are greatly acknowledged for their helps.

REFERENCES

ABDÜSSELAMOĞLU, Ş.1963. Kocaeli Yarımadasının Jeolojisi. M.T.A. Rapor No:3249, unpublished, Ankara.

COOKE, R., DOORNKAMP, J.C. 1974. Geomorphology in Environmental Management. Oxford Univ. Pp.128-166, Oxford.

DÖNMEZ,Y. 1979. Kocaeli Yarımadasının Bitki Coğrafyası. İ.Ü.Coğr.Enst.Yay. No:112, İstanbul.

ERGUVANLI, K. 1982. Mühendislik Jeolojisi. İ.T.Ü. Kütüp Sayı:1227, İstanbul.

ERİNÇ, S. 1996. Jeomorfoloji I. Genişletilmiş 4.Baskı. ISBN975-8004-21-2, pp.351-367, İstanbul.

EROL, O. 1985. Baraj Yeri Seçimi ve İnşaatında Gözönünde Bulundurulması Gereken Jeomorfolojik Konular. Geomorphological Aspects of Dam Site Selection and Dam Building. Türkiye Jeom. Dern. 9. Türkiye Bilimsel ve Teknik Kurultayı, Bildiri özleri, pp.27-29, Ankara.

EROL,O. 1988. Genel Klimatoloji. İ.Ü.Dz. Bil. Ve Coğr. Enst. Yay. No:8, pp. 219-221, İstanbul.

ERTEK, T. A. 1995. Kocaeli Yarımadasının Kuzeydoğu Kesiminin Jeomorfolojisi. ISBN 975-7206-08-3, İstanbul.

http://www.tuat.ac.jp/ sabo/lj/ljidx.htm.

KURTER, A. 1985. Uygulamalı Jeomorfoloji. İ.Ü. Dz. Bil. Ve Coğr. Enst. Yüksek Lisans Ders Notları, unpublihed, İstanbul.

M.T.A. (Inst. of Mineral Research and Exploration) 1964. Geological Map of Turkey. (İstanbul and Zonguldak sheets). Ankara.

ZARUBA, Q., MENCL, V. 1969. Landslide and Their Control. Publishing House of Czechoslavak Academy of Sciences, Prag.

Case of a Slope Failure in Lacustrine Deposits

A. J. ESSER, EDP Consultants, Inc., Kirtland, Ohio, USA

INTRODUCTION

This paper examines a slope failure in an area of deep lacustrine deposits overconsolidated at the surface by desiccation. The landslide damaged four homes in a large residential development. Two have been moved to other sites and repaired, a third that was damaged beyond repair was burned by the local fire department as a practice exercise. Ownership of the fourth home was transferred to an associate of experts involved in the case as part of the settlement reached in pretrial negotiations. The foundations of this home have been underpinned and other work has been undertaken to reduce lateral earth pressures on the basement walls.

Meetings were held early in the development of the project during which potential slope problems were discussed. These meetings led to the preparation of general guidelines for building on sloping sites. The guidelines generally advised caution when building on slopes greater than 40% and warned against, but did not forbid, the placement of fill on sloping sites. The guidelines were not specific to any property or area of the project and apparently were not based on geotechnical exploration, stability analysis, or historical research of documented failures. No effort was made to research the literature for engineering data that might be applicable to the soils in the area. The only borings drilled for the project prior to the failures appear to have been a series of shallow, approximately 5 meter deep, borings for roadways. None of the borings was of sufficient depth to reach the plastic clays that later proved to be critical to stability. Such efforts were bypassed despite the fact that the entrance to the project paralleled a highway cut with numerous active surficial landslides.

The project geotechnical engineer was also hired to evaluate a number of lots in the development. Most were evaluated based on surface observations without exploration. Those sites that were evaluated in this way included three of the four properties that are the subject of this paper. The topography of these properties is shown in Figure 1. The map was based on an aerial photograph that was taken prior to the construction of the house on lot 2 which is included in this discussion. Borings and soil tests were completed by the engineer in evaluating lot 2 because problems had already developed on the remaining three lots, and two of the three were in litigation. Prior to construction of the first three houses, the hillside was evaluated by the geotechnical engineer and judged to be stable. The engineer recommended spread footings to support the houses. The engineer's reports recognized that fill would be needed on the slope and provided recommendations for its placement, not limiting the amount of fill that could be placed. The engineer's evaluation apparently relied on surface observations at the site, a review of a topographic map of the property, building plans provided by the homebuilder, and the roadway borings. Pressure had been placed on the engineer to forego exploration specific to the properties to reduce costs to the developer. The engineer had been insistent that a slope stability evaluation was necessary but accepted the assignment and wrote a favorable report without any exploration of the properties.

Landslides in research, theory and practice, Thomas Telford, London, 2000.

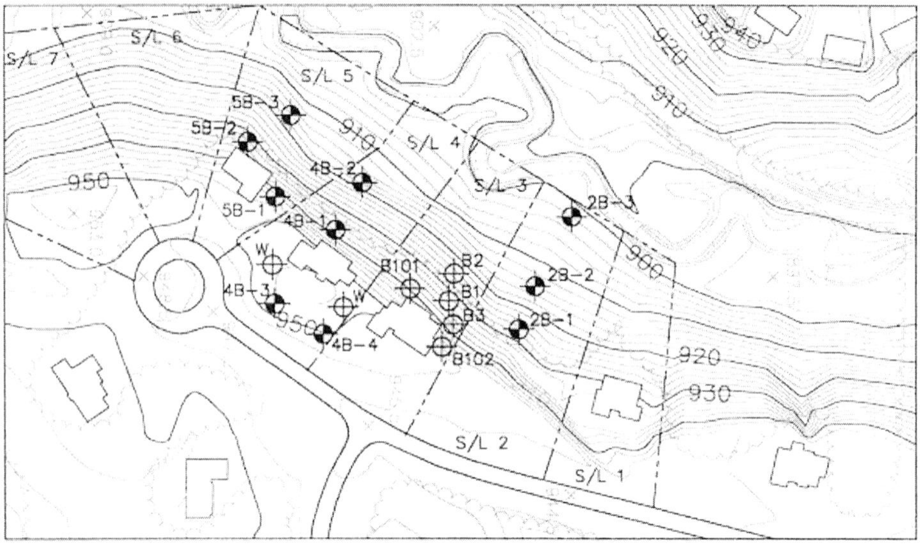

Figure 1. Topographic plan of the area that includes the four properties being discussed

SITE CONDITIONS

The wooded properties were level from the street to approximately 5 meters beyond the minimum building line where the slope began with a relatively steep drop, at about a 2.5:1 angle, to a narrow terrace. An approximately 3 to 4 meter drop in elevation occurred between the flat upper part of the properties and the terrace. The elevation differential and the terrace were obvious topographic features of the properties before construction began. The lower part of the slope was hummocky and averaged about 4:1 to 4.5:1 with some areas as flat as 6:1 and 8:1. Two shallow landslide ponds existed on the lower part of the hillside on Lot 5. The slope is approximately 400 m long with an average height of about 15 m. Figure 2 is a generalized cross section of the slope. The homes on lots 2 through 4 are located on the slope approximately as shown with the main scarp passing through the houses. The home on lot 5 was located on the slide mass with only the front wall of the attached garage supported on the stable ground above the landslide.

EVALUATIONS DURING LITIGATION

The houses on lots 3 and 4 were completed and occupied in October and November 1992. Lot 5 was completed in September 1993, and lot 2, the last of the four to be completed, was occupied in March 1995. Immediately after construction was completed, as the new owners were moving into the homes, cracks began to appear. The condition worsened as the back of the homes moved downslope, walls buckled, and doors and windows began to bind. The most dramatic damage occurred at lot 4 where the rear wall of the garage failed moving downslope with a horizontal displacement component of about 25 centimeters. The wall was excavated. An 8 centimeter void and approximately 1.6 meters of soft soil were found beneath the footings. Several 4 to 6 meter deep test holes were completed and unconfined compression and consolidation tests were undertaken. Stability was checked but the computed factor of safety exceeded 2. Consolidation of the loose soil under the footings and unbalanced lateral pressure attributed to a difference in lateral pressure from loose uncompacted backfill on the outside of the wall and better compacted and a somewhat greater depth of backfill on the

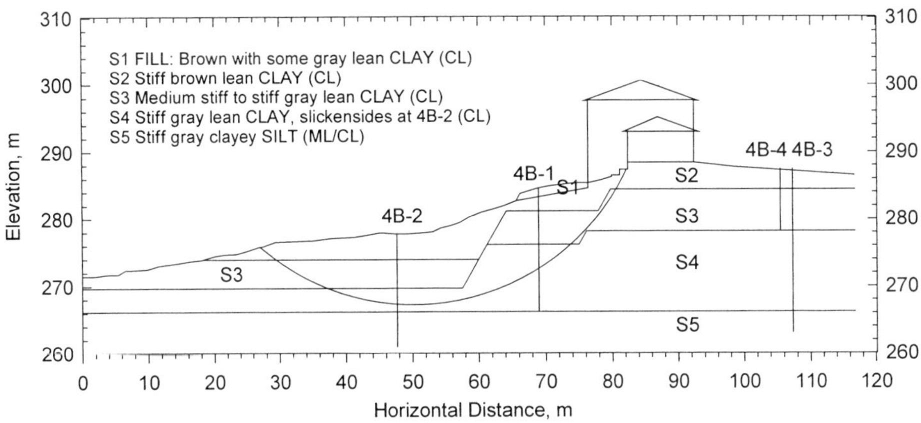

Figure 2. Cross section of the slope showing major soil strata. Layers S2 and S3 have been overconsolidated by desiccation. Layer S4 is normally to very lightly overconsolidated

inside of the wall were blamed for the failure.The wall was rebuilt and the foundations for the house were widened to reduce bearing pressures. Despite these efforts, the cracking and displacement continued. The rear wall of the garage failed a second time and the back of the house continued to move downslope. The front and rear walls of the house eventually lifted off the foundations and the house had to be braced to prevent collapse. The second failure of the garage wall is shown in Figure 3. Damage like this was common in all four homes.

During the litigation for lot 3, the geotechnical engineers hired as experts evaluated the site's soil deposits and installed slope tubes. Shear strength was evaluated through unconfined compression, direct shear, and UU triaxial tests. Ten meter long slope tubes were installed by one firm but the readings were inconclusive. Additional slope tubes were installed to greater depths by a second firm. Readings during 1994 showed that the first tubes were installed with the bottoms in the slide zone. Movement was detected but it was concluded that the movement was not significant enough to account for the observed structural damage.

The authors firm installed slope tubes in lots 2, 4, and 5. Readings were obtained over a longer period and clearly showed that significant slope movement was taking place. The slope tube readings were normalized by calculating annual displacement velocities that are plotted for all lots over a five year period in Figure 4. This shows that the slope tube readings made by the first two firms on lot 3 were taken during a period of extremely slow movement thus leading to the conclusion that slope movements were not significant.

SLOPE STABILITY ANALYSIS

The author first became involved in the project as the technical expert for the owner of Lot 4. The prior damage to the home, the failure of attempted repairs, the geology of the area, and surficial indications of ground movement were sufficient to conclude that the slope was unstable. It was clear, based on the work of Skempton (1977), that the past displacement of the slide mass was sufficient to reduce the shear strength of the soils along the slip surface to the residual value. Shear strengths had been obtained from UU triaxial, unconfined compression, and direct shear tests. Post-peak strengths in the UU triaxial and direct shear tests had been taken as residual values. Some cohesion component of shear strength was also

Figure 3. Failure of garage wall at lot 2.

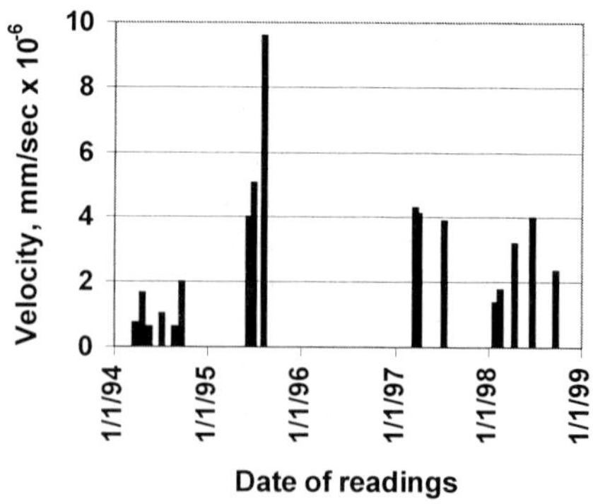

Figure 4. Annualized displacement velocity for 5 years of slope tube readings.

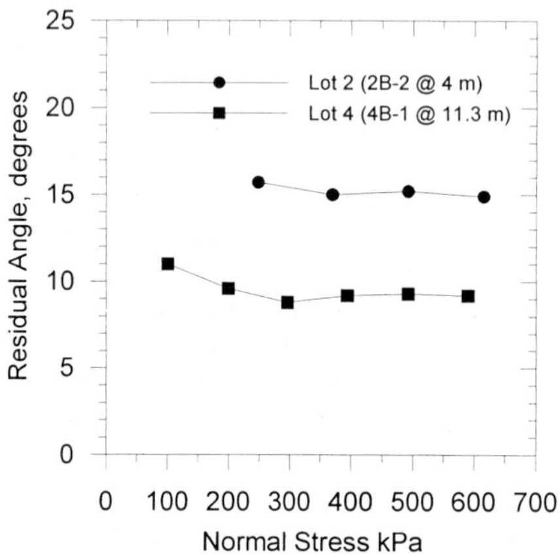

Figure 5. Results of ring shear tests.

Table 1. Soil properties used in analysis.

Soil Layers	γ	ϕ_r	c_r
Fill	125	0-25	0
OC lacustrine clay	127	16-24	0
NC lacustrine clay	122	13-14	0
Silt	135	32-35	0

assumed. However, it was doubtful that the 20% strain displacements in these tests were sufficient to reach residual strengths. It was considered more likely that these tests might have approach softened strengths. Based on Stark and Eid (1997), it appeared likely that for the soils at depths critical to stability, with liquid limits in the range of 40 to 50, and an effective normal stress of about 100 to 150 kPa, that the softened strengths might be as much as 8 degrees greater than residual.

Supplementary exploration and testing was proposed. Two borings were drilled on the slope and instrumented with 20 meter long slope tubes. Additional test holes were drilled and multiple-level piezometers were installed. Laboratory testing included index property and Bromhead ring shear tests. The results of the ring shear tests are given in Figure 5. Residual strengths for thin strata and strata near the surface that would not significantly affect stability were determined from correlations to soil index properties given by Lupini et al. (1981), Skempton (1985), Mesri and Cepeda-Diaz (1986); see also Mitchell (1993).

Similarities in the Atterberg limits, unit weights, water contents, grain size analysis and visual classifications of the soils from all of the adjacent properties demonstrate that the soils are similar across the area and consequently the data from one property can be used in evaluating the others. A substantial database of soils information was developed over a five year period with results from more than 80 tests being available. The properties used in the slope stability analysis are listed in Table 1. The slope was profiled on each lot to develop a surface cross section for the stability analysis. Overlying the plotted surface profiles showed that the cross sections were very similar on each property, and thus, the assumption of plain strain conditions is probably reasonably accurate.

Factors of safety calculated for the pre construction and post construction geometries ranged from 1.04 to 1.07 for the pre construction condition, and 0.97 to 1.01 for the final condition with small amounts of fill added to the slope. A factor of safety of 1.06 was calculated for the intermediate condition with the house completed but with no fill added to the slope. The intermediate condition was considered because fill placed on the upper part of the slope behind the houses was blamed in most of the litigation as the cause or contributing cause of the landslide. It is clear from the calculations that residual strengths are appropriately used. It is also apparent that the fill placed on the properties reduces the factor of safety by about 5% but that the factors of safety were unacceptable prior to construction.

Duncan (1996), and Gilbert et al. (1996), considered the uncertainty in the available limit equilibrium analysis methods and found that an error of about 15% could be expected in the calculations. Additional error arises from a number of sources including strength variations in the soil mass that cannot practically be modeled, variations in strength along the slip surface, assumptions about the shape of the failure surface to make the calculations tractable, and others. Thus, the minimum desired factor of safety for slopes such as this one where the risk of significant property loss is great, should be 1.5. All of the factors of safety calculated by all of the firms that served as technical experts in the litigation of these four properties, and that used non peak shear strengths, were below this minimum value. When residual strengths are used, it is clear that the slope was near equilibrium before construction started and that the construction of the homes and the placement of small amounts of fill on the slope did not have a significant effect on stability.

Many clients resist spending significant sums on exploration. However, slopes in general cannot be properly evaluated without a substantial investment in engineering and testing. Correlations of residual strength to index properties can be used to advantage in those cases where the engineer is asked to evaluate a failed slope. With a reasonable investment in exploration, samples can be obtained, index properties determined and a preliminary analysis can be made assuming an infinite slope. The borings must be drilled to sufficient depths (stopping 2.5 to 10 m below the toe of the slope), and sample intervals must be small enough that all important strata are sampled. Index tests, should be completed on all soil strata likely to control stability. For the slope discussed in this paper, clay fractions for the lacustrine clays ranged between about 45 and 61 percent. Using these soil indices and the correlations discussed earlier, residual friction angles are estimated to be between 11 and 15 degrees. A preliminary analysis assuming an infinite slope where the factor of safety is given by equation (1) yields factors of safety of 0.45 to 0.63 if water is assumed at the surface and 0.87 to 1.21 if groundwater is ignored.

$$FS = \left(1 - \frac{\gamma_w}{\gamma}\right)\frac{\tan\varphi'}{\tan\beta}$$

where: γ_w is the unit weight of water

γ is the total unit weight of the soil

ϕ' should be taken as the effective (residual) strength of the critical soil layer(s)

β is the average angle of the slope

This simple calculation is sufficient to indicate that stability of the slope is a concern that warrants careful analysis. Had the original geotechnical engineer performed such a calculation, significant property loss might have been averted. Slosson and Slosson (1991) indicate that at least 90% of slope failures are recognizable, predictable, and preventable. This is certainly such a case.

CONCLUSIONS

1. The site walkover and visual evaluation in the field is extremely important to the evaluation of slope stability. In evaluating slope stability, the engineer should be able to explain how the slope came to be and then use this information to select appropriate methods for measuring shear strength.
2. Correlations using clay fraction provide a reliable estimate of residual shear strength and are therefore useful in preliminary analysis where pressure exists to keep costs low. Index properties tests can also be used to select the strength parameters of minor strata or strata that have been shown by the preliminary analysis not to have a significant effect on the stability calculations. Preliminary analysis and the use of correlations in the selection of soil parameters are not a substitute for proper testing and slope stability analysis.
3. Soil borings and instrumentation must extend to depths that are sufficient to provide meaningful data. For most and perhaps all slopes, this will mean that the borings must extend well below the bottom of the slope.
4. Slopes do not move at a constant rate and not all parts of the slope move at the same rate. Observing rates of movement for too short a time or at too few locations can lead to erroneous conclusions.
5. The engineer must consider the long-term effects of even very slow rates of slope movement on structures built on or near slopes.

REFERENCES

Duncan, Michael James (1996), "State of the Art: Limit Equilibrium and Finite-Element Analysis of Slopes", Journal of Geotechnical Engineering, ASCE, 122(7), 577-596.

Gilbert, Robert B., Wright, Stephen G., and Liedtke, Eric (1996), "Uncertainty in Back Analysis of Slopes", Proceedings of Uncertainty '96, Shackelford, Nelson and Roth (eds.), v1, ASCE, Madison, Wisconsin, 494-517.

Lupini, J.F., Skinner, A.E., and Vaughan, P.R. (1981), "The drained residual strength of cohesive soils", Geotechnique, 31(2), 181-213.

Mesri, G., and Cepeda-Diaz, A.F. (1986), "Residual shear strength of clays and shales", Geotechnique, 36(2), 269-274.

Mitchell, James K. (1993), Fundamentals of Soil Behavior, 2nd ed., John Wiley & Sons, Inc., New York.

Skempton, A.W. (1977), "Slope stability of cuttings in brown London clay", Proc., 9th ICSMFE, Tokyo, Japan, v3, 261-170.

Skempton, A.W. (1985), "Residual strength of clays in landslides, folded strata and the laboratory", Geotechnique, 35(1), 3-18.

Slosson, J.E. and Slosson, T.L. (1991), "Standard practice equals continuing losses", Landslides, Bell (ed.), Rotterdam: Balkema, v1, 227-229.

Stark, T.D., and Eid, H.T. (1997), "Slope stability analysis in stiff fissured clays", Journal of Geotechnical and Geoenvironmental Engineering, ASCE, 123(4), 335-343.